MODERN ECOLOGY

Basic and Applied Aspects

MODERN ECOLOGY

Basic and Applied Aspects

Edited by

G. Esser
International Institute for Applied Systems Analysis (IIASA),
Schlossplatz 1, A-2361 Laxenburg, Austria

and

D. Overdieck
Institut für Ökologie der Technischen Universität,
Köningin-Luise-Strasse 22, D-1000 Berlin 33, Germany

Elsevier
Amsterdam – London – New York – Tokyo 1991

ELSEVIER SCIENCE PUBLISHERS B.V.
Molenwerf 1
P.O. Box 211, 1000 AE Amsterdam, The Netherlands

Distributors for the United States and Canada:

ELSEVIER SCIENCE PUBLISHING COMPANY INC.
655, Avenue of the Americas
New York, NY 10010, U.S.A.

ISBN 0-444-89183-8

This book is printed on acid-free paper.

Printed in The Netherlands

Preface

This book is dedicated by the editors and authors to the university teacher and outstanding researcher in ecology, **Helmut Lieth**, on the occasion of his 65th birthday and retirement.

Helmut Heinrich Friedrich Lieth was born in 1925 and grew up in Rhenania, Western Germany. He started his ecologically oriented education in the field of agriculture, receiving his apprenticeship at the agricultural school in Unterbrösch (Bergischer Landkreis), and completing his school education there in 1947. From 1947 to 1949 he studied philosophy and natural sciences at the "Philosophical and Theological College" in Bamberg (Bavaria), followed by two years of undergraduate studies at the Botany Department of the University of Cologne.

He continued graduate studies at the same university where he obtained his Ph.D. in 1953. The title of his thesis was "Untersuchungen über die Bodenstruktur und andere vom Tritt abhängender Faktoren in den Rasengesellschaften des Rheinisch-Bergischen Kreises" (Investigations into the effects of footstep frequency on the soil structure and other factors in the turf communities of Rheinisch–Bergisch county).

Early in his university career he served as a scientific assistant in the Botany Departments of the University of Cologne and the Agricultural University of Stuttgart–Hohenheim. In 1960, he was granted the title of lecturer (Privatdozent) by the University of Stuttgart–Hohenheim and was also appointed lecturer of ecology at the Technical University of Stuttgart.

During those years Helmut Lieth broadened not only his scientific interests but also his very practical personal activities enthusiatically to a global scale, starting from 1960 to 1961 with a National Research Fellowship at the Botany Department of the Université de Montréal, Canada. Back in Germany, he was appointed a senior lecturer (Diätendozent) at the Agricultural University of Stuttgart–Hohenheim in 1964, and soon became assistant professor (Außerplanmäßiger Professor) at the same institution. During those years he served as guest lecturer at the Universidad Central de Venezuela, and the Universidad del Tolima at Ibaque, Colombia. After a stay in Hawaii during 1967, he found his second home at the University of North Carolina, Chapel Hill (U.S.A.), where he stayed until 1977.

In the U.S.A., he initiated numerous studies in general ecology, concentrating on plant ecology and vegetation science, always taking into account the interactions and relationships between organisms on the ecosystem level, and looking for formulae and mathematical models in order to find general ecological rules which allow us to quantify the organismic response on varying environmental parameters. Among the authors of this book are several former students of Helmut Lieth from that period of very fruitful work.

As a guest researcher (1973-1974) at the Jülich Nuclear Research Center (Kernforschungsanlage Jülich, today Forschungszentrum Jülich), Federal Republic of Germany, he maintained his connections to the European scientific community. After teaching at the University of Bochum (F.R. Germany) as a guest lecturer in 1977, he was appointed the first professor of ecology at the newly founded University of Osnabrück in Northwest Germany. There, he was not only the first professor of ecology, but one of the first scientists in any branch of biology. He is a pioneer to whom the University of Osnabrück owes a new building for the Department of Biology. His vigorous efforts also led to the planning and foundation of a new Botanical Garden at Osnabrück, which is developing into a unique attraction both for scientists and for visitors searching for recreation.

Simultaneously, he carried out a great number of research projects at his home university, some of which will be continued beyond his retirement. At times, these projects have employed up to 30 students and junior scientists.

During his last years before retirement he also started a new program of study, Applied Systems Research (Angewandte Systemforschung), in the Mathematics Department of the University of Osnabrück in 1989.

In spite of all his duties in Chapel Hill and Osnabrück, he maintained a considerable number of international activities, serving as a guest professor at the Waseda University, Tokyo, and as a participant in scientific research projects in countries all over the world, including Iceland, Canada, Venezuela, Colombia, Brazil, Hawaii, India, Japan, USSR, U.S.A., Mozambique, and the United Arab Emirates. Many of the activities pertained to tropical ecology and phenology. During the last years of his university career Helmut Lieth gave much energy to a "greenification" project with salt tolerant plants, which was started in Abu Dhabi (United Arab Emirates) in 1980, and which after an interruption has been taken up again on a larger scale in 1989.

Helmut Lieth's international engagement earned for him the Biometeorology Research Foundation Award in 1982; the silver medal for his work in Agrometeorology during the 3rd Indian Agrometeorological Congress in Cochin in 1987; and the award for outstanding achievement of the International Society of Biometeorology during its 11th International Congress at Purdue University (U.S.A.) in 1988.

From 1979 to 1984 he was the elected president of the International Society of Biometeorology, from 1985 to 1988 the president of the International Society of Tropical Ecology, and in 1989 the vice-president of the "Gesellschaft für Ökologie". Since 1986 he has served as treasurer of the International Society of Ecology (INTECOL).

More than a hundred scientific publications prove that Helmut Lieth's work covered a wide range of ecological topics. Especially well known are "Die Stoffproduktion der Pflanzendecke" (1962), "Produktivitätskarte der Erde" (1964), "Klimadiagramm–Weltatlas" (1967-1969) which he developed together with Heinrich Walter, "Modeling the Primary Productivity of the World" (1972), "Primary Productivity of the Biosphere" (1974), and "Patterns of Primary Productivity" (1978). He is editor–in–chief of the book series "Tasks for Vegetation Science" (1979), the "Handbook of Vegetation Science" (1980), the "International Journal of Biometeorology" (1987) and "Vegetatio" (1989). He has been on the editorial board of several other scientific journals, including "Radiation and Environmental Biophysics" and "Journal of Biogeography".

The authors and the editors wish that Helmut Lieth should continue to contribute to ecological research for still many years.

We wish to thank all who contributed to the production of this book, which was typeset in the project–group of the first editor at the University Osnabrück using LATEX.

We cordially thank Dipl. Biol. Margot Brosch for her organization of the work and for typesetting major parts of the book. Our thanks are also due to Dipl. Biols. Martina Lohmann, Barbara Meyer, and Alfred Stille for typesetting, Andreas Schürmann for reproducing the graphs, and Teresa Enders for English language editing.

Martina Jöstl–Segalla and Carolyn Fuhrmann helped to make final corrections, typesetting of tables, and Cynthia Ramirez assisted the first editor at IIASA.

We would like to give special thanks to all who have finacially supported the work, namely the University Osnabrück and the Biology/Chemistry Department, the Universitätsgesellschaft Osnabrück e.V., the Institute for Ecology of the Technical University of Berlin (West), and the Zentralverband des Deutschen Steinkohlenbergbaus.

Last, but not least, we wish to thank Elsevier Science Publishers for their kindness in having accepted a manuscript, although sizeable, contains a comprehensive amount of information we feel will help understand how Ecology might be organized in the next decade.

May 15, 1991

Gerd Esser
Laxenburg (Austria)

Dieter Overdieck
Berlin (Germany)

Editorial advisors

List of contributors

BETHKE-BEILFUSS, D.
Fachbereich Biologie/Chemie
der Universität
Barbarastrasse 11
D-4500 Osnabrück
Germany

BEYER, L.
Institut für Pflanzenernährung
und Bodenkunde
der Universität
Olshausenstrasse 40
D-2300 Kiel
Germany

BILBAO, B.
Centro de Ecologia
Instituto Venezolano
de Investigaciones Cientificas
Apartado 21827
Caracas 1020-A
Venezuela

BLUME, H.-P.
Institut für Pflanzenernährung
und Bodenkunde
der Universität
Olshausenstrasse 40
D-2300 Kiel
Germany

BOGENRIEDER, A.
Biologisches Institut II
der Universität
Schänzlestrasse 1
D-7800 Freiburg/Br.
Germany

DE BOK, C.
Department of Plant Ecology
and Evolutionary Biology
Utrecht University
Lange Nieuwstraat 106
NL-3512 PN Utrecht
The Netherlands

BOX, E. O.
Department of Geography
University of Georgia
Room 204, GGS Building
Athens, Georgia 30602
U.S.A.

BREYMEYER, A. I.
Institute of Geography
and Spatial Organisation
Polish Academy of Sciences
Krakowskie Przedmieście 30
PL-00 927 Warszawa
Poland

CABRERA CANO, E.
Smithsonian Environmental
Research Center
P.O. Box 28
Edgewater, MD 21037–0028
U.S.A.

CERNUSCA, A.
Institut für Botanik
der Universität
Sternwartestrasse 15
A–6020 Innsbruck
Austria

CHEN, T.
Abteilung Pflanzenökologie
FB Biologie/Chemie
der Universität
Achterstrasse 2
D–2800 Bremen 33
Germany

DIERSCHKE, H.
Lehrstuhl für Geobotanik
Systematisch–Geobotanisches
Institut
der Universität
Untere Karspüle 2
D–3400 Göttingen
Germany

EBER, W.
Fachbereich Biologie
der Universität
Ammerländer Heerstrasse 114–118
Postfach 2503
D–2900 Oldenburg
Germany

ENGELS, M.
Systematisch–Geobotanisches
Institut
der Universität
Untere Karspüle 2
D–3400 Göttingen
Germany

ERNST, W.
Department of Ecology
and Ecotoxicology
Faculty of Biology
Vrije Universiteit
De Boelelaan 1087
NL–1081 HV Amsterdam
The Netherlands

ESSER, G.
International Institute for
Applied Systems Analysis (IIASA)
Schlossplatz 1
A–2361 Laxenburg
Austria

FANGMEIER, A.
Institut für Pflanzenökologie
Heinrich–Buff–Ring 38
D–6300 Giessen
Germany

FORSTREUTER, M.
Institut für Ökologie
der Technischen Universität
Königin–Luise–Strasse 22
D–1000 Berlin 33
Germany

FRIEDRICH, F.
Fachbereich 11
Technische Fachhochschule
Königin–Luise–Strasse 22
D–1000 Berlin 33
Germany

FUNKE, W.
Abteilung Ökologie und
Morphologie der Tiere
Universität Ulm
Oberer Eselsberg M 25
Postfach 4066
D–7900 Ulm
Germany

GRACE, J.
Institute of Ecology
and Resource Management
University of Edinburgh
Darwin Building
Kings Buildings, Mayfield Road
Edinburgh EH9 3JU
United Kingdom

HALL, D. O.
Division of Biosphere Sciences
King's College London
University of London
Campden Hill Road
London W8 7AH
United Kingdom

HELLMUTH, U.
Systematisch-Geobotanisches
Institut
der Universität
Untere Karspüle 2
D-3400 Göttingen
Germany

HURKA, H.
Fachbereich Biologie/Chemie
der Universität
Barbarastrasse 11
D-4500 Osnabrück
Germany

JAKUCS, P.
Ecological Institute
L. Kossuth University
H-4010 Debrecen
Hungary

JANZEN, D.
Pflanzensoziologie und Ökologie
der Universität GHS
Postfach 103764
D-4300 Essen 1
Germany

JOCHIMSEN, M.
Pflanzensoziologie und Ökologie
der Universität GHS
Postfach 103764
D-4300 Essen 1
Germany

KAPPEN, L.
Lehrstuhl für Ökophysiologie
Botanisches Institut
der Universität
Olshausenstrasse 40
D-2300 Kiel 1
Germany

KLOETZLI, F.
Geobotanisches Institut
ETH
Zürichbergstrasse 38
CH-8044 Zürich
Switzerland

KÖHLER, W.
Biometrie und Populationsgenetik
der Universität
Ludwigstrasse 27
D-6300 Giessen
Germany

KREEB, K. H.
Abteilung Pflanzenökologie
FB Biologie/Chemie
der Universität
Achterstrasse 2
D-2800 Bremen 33
Germany

KUTSCH, W. L.
Botanisches Institut
der Universität
Olshausenstrasse 40
D-2300 Kiel 1
Germany

KUTTLER, W.
FB 9 – Institut für Ökologie
Abt. Landschaftsökologie
Universität-GHS
Postfach 103764
D-4300 Essen 1
Germany

LANDOLT, E.
Geobotanisches Institut
ETH
Zürichbergstrasse 38
CH-8044 Zürich
Switzerland

LARSON, D. W.
Department of Botany
University of Guelph
Guelph
Ontario
Canada N1G 2W1

LESER, H.
Forschungsgruppe Physiogeographie
und Landschaftsökologie
Geographisches Institut
der Universität
Klingelbergstrasse 16
CH-4056 Basel
Switzerland

MARKERT, B.
AG Angewandte Systemwiss.
der Universität
Artilleriestrasse 34
D-4500 Osnabrück
Germany

MEDINA, E.
Centro de Ecologia
Instituto Venezolano
de Investigaciones Cientificas
Apartado 21827
Caracas 1020-A
Venezuela

MEENTEMEYER, V.
Department of Geography
University of Georgia
Room 204, GGS Building
Athens, Georgia 30602
U.S.A.

NAU, W.
Biometrie und Populationsgenetik
der Universität
Ludwigstrasse 27
D-6300 Giessen
Germany

NEUFFER, B.
Fachbereich Biologie/Chemie
der Universität
Barbarastrasse 11
D-4500 Osnabrück
Germany

NUMATA, M.
Laboratory of Ecology
Faculty of Science
Chiba University
74-13 Bentencho
Chiba 260
Japan

ORANJE, B.
Department of Plant Ecology
and Evolutionary Biology
Utrecht University
Lange Nieuwstraat 106
NL-3512 PN Utrecht
The Netherlands

OVERDIECK, D.
Institut für Ökologie
der Technischen Universität
Königin-Luise-Strasse 22
D-1000 Berlin 33
Germany

PONTAILLER, J.-Y.
Laboratoire d'écologie végétale
URA CNRS N° 121
Bâtiment 362, Université Paris Sud
91405 Orsay cedex
France

READER, R. J.
Department of Botany
University of Guelph
Guelph
Ontario
Canada N1G 2W1

RODD, H.
Forschungsgruppe Physiogeographie
und Landschaftsökologie
Geographisches Institut
der Universität
Klingelbergstrasse 16
CH-4056 Basel
Switzerland

ROTH-HOLZAPFEL, M.
Abteilung Ökologie und
Morphologie der Tiere
Universität Ulm
Oberer Eselsberg M 25
Postfach 4066
D-7900 Ulm
Germany

RUTHERFORD, M. C.
National Botanical Institute
c/o University of Cape Town
Cape Town
South Africa

RUTHSATZ, B.
Fachbereich III, Geobotanik
der Universität
Postfach 3825
D-5500 Trier
Germany

RYSZKOWSKI, L.
Department of Agrobiology
and Forestry
Polish Academy of Sciences
Świerczewskiego 19
60-809 Poznań
Poland

SAUGIER, B.
Laboratoire d'écologie végétale
URA CNRS N° 121
Bâtiment 362, Université Paris Sud
91405 Orsay cedex
France

SCHLESER, G. H.
Abteilung Biophysikalische Chemie
Institut für Chemie
Forschungszentrum Jülich GmbH
Postfach 1913
D-5170 Jülich
Germany

SCHMIDT, W.
Lehrstuhl für Geobotanik
Systematisch-Geobotanisches
Institut
der Universität
Untere Karspüle 2
D-3400 Göttingen
Germany

SCURLOCK, J. M. O.
Division of Biosphere Sciences
King's College London
University of London
Campden Hill Road
London W8 7AH
United Kingdom

STEUBING, L.
Institut für Pflanzenökologie
Heinrich-Buff-Ring 38
D-6300 Giessen
Germany

TAPPEINER, U.
Institut für Botanik
der Universität
Sternwartestrasse 15
A-6020 Innsbruck
Austria

TAYLOR, K. C.
Department of Botany
University of Guelph
Guelph
Ontario
Canada N1G 2W1

WERGER, M. J. A.
Department of Plant Ecology
and Evolutionary Biology
Utrecht University
Lange Nieuwstraat 106
NL-3512 PN Utrecht
The Netherlands

WESTHEIDE, W.
Fachbereich Biologie/Chemie
der Universität
Barbarastrasse 11
D-4500 Osnabrück
Germany

WHIGHAM, D. F.
Smithsonian Environmental
Research Center
P.O. Box 28
Edgewater, MD 21037-0028
U.S.A.

WICKLAND, D. E.
Ecosystem Dynamics and
Biogeochemical Processes Branch
Earth Science and
Application Division
NASA
600 Independence Ave., S. W.
Washington, D. C. 20546
U.S.A.

WILMANNS, O.
Biologisches Institut II
der Universität
Schänzlestrasse 1
D-7800 Freiburg/Br.
Germany

Contents

Part I

Morphology, stand structure, and competition

Chapter 1

Morphology in modern ecological research

W. EBER

1.1 Introduction

Both morphology and anatomy of plants belong to the basic subdisciplines of botany with a long tradition. Their subject is the structure of plants on the level of organs (external structure) as well as on the level of tissues (internal structure). Though botanists, especially morphologists (ESAU 1977, FOSTER and GIFFORD 1974, FAHN 1982), have always been aware of the intimate relation between structure and function, one scarcely finds examples of an appropriate treatment of both aspects on the same subject. Plant morphologists have nearly exclusively been concerned with the description and phylogenetic interpretation of form and structure, whereas plant ecology has mainly dealt with the analysis of environmental factors and the physiological response of plants.

During the past decades at least the external structure of plants has found increased interest in two special fields of ecology: on the one hand, in the study of vegetation pattern and on the other hand, in the analysis of plant populations. Nevertheless, the study of the internal structure is still a neglected field of ecological research, though improved histological techniques and the availability of image analysis devices have made anatomical studies more comfortable and attractive.

The contribution presented here is based on experiences in the study of primary production gained during our participation in the German IBP–Project and and subsequent studies in population ecology with increasing emphasis on plant architecture. The advantage or even necessity of a more intensive understanding of morphological aspects in modern plant ecology is emphasized.

1.2 Models of plant construction

The body of vascular plants can be regarded as constructed of only three principal plant organs: root, stem and leaf. The whole variety of plant form is due to a varied development or modification of size, form, proportions and arrangement of these organs. The special performance of each organ is determined by phylogenetical processes, its specific function or requirements of a specific environment. In addition to this classic phylogenetic model of the vascular plant, RITTERBUSCH (1977) has presented an "ecological model" which is based on the specific functions of each external structure such as storage, assimilation, absorption of water and mineral nutrients, elevation of leaves and flowers, regeneration (rejuvenalisation) of shoots and dispersion of propagules. Obviously this model has a greater ecological relevance and is of special advantage where resource allocation is investigated.

Another model has been developed by population biologists and is widely used in plant demographic studies. According to this model (HARPER 1977, WHITE 1979) the plant is regarded as a population of modular units which may remain attached to each other or may be separated in clonal growth. All parts derived from one zygote are termed genetic individual or genet. Genets may extend over a large area and may be fragmented into numerous parts of different size. Modules and phytons may be regarded as the fundamental modular units of construction into which the shoot system can be completely divided. The module in its original meaning (WHITE 1979, HALLE et al. 1978) is defined as the product of activity of one apical meristem terminated by either flowering or parenchymatization. Modules have proved to be especially valuable in the study of plant architecture and life history of clonal plants. Unfortunately, this term has also been used without any need in a more general sense synonymously with modular unit (HARPER and BELL 1979). Phytons are segments of the shoot consisting of a node with its leaf, axillary bud and internode. As the life history of individual phytons cannot be separated from that of the complete shoot, this unit has been used only in studies of plant construction. *Equisetum hyemale*, an evergreen horsetail, the shoots of which die internode by internode, is one of the rare examples where the internode or phyton is the appropriate unit for demographic studies (LEHMANN and EBER, in press). Axillary buds ("bud bank") and leaves, however, are of major importance in demographic studies (BAZZAZ and HARPER 1977, HUISKES and HARPER 1979, KOTANEN and JEFFERIES 1987, EBER and VEENHUIS, in press).

The ramet represents another type of modular unit, as it only includes parts of the shoot system and is more complex. HARPER (1977) defines the ramet as a shoot with its own root system which may follow an independent existence if severed from the parent system. According to this definition those subterranean parts which are not absolutely necessary for the independent existence of aerial shoots are not included. Ramets are regarded as the effective units of clonal growth that are readily counted in the field (HARPER 1977). In ecological research aerial shoots or ramets are predominantly used as "individual plant". As each ramet can represent a whole genet or be a physiologically dependent part of a genet fragment or an integrate genet, a precise characterization of the ramets in question is necessary. Where ramets occur aggregated, they may be supposed to belong to the same genet. The size of the area occupied by individual genets or genet fragments and the density of their ramets, which make up the sociability according to BRAUN-BLANQUET (1964), are often species characteristics within a certain range. Nevertheless, there are some species where no ramets can be recognized as they merely develop prostrate and often subterranean shoots. In species such as *Oxalis acetosella* the leaf is the only counting unit available (EBER 1982).

As in the majority of perennial plants the subterranean component predominates, it is obvious that this component must not be neglected, though its investigation cannot be achieved without destructive sampling.

1.3 Diversity of individuals within populations

Plant populations do not consist of uniform members; on the contrary, the population structure is characterized by individuals which differ in age, size and developmental stage, and all these subdivisions respond to environmental factors in a specific way. Though both size and developmental stage are age dependent to a certain degree, close relations between these three parameters are rarely found under field conditions. In most species age cannot be determined at all or, as in trees or perennial herbs like *Limonium vulgare* (EBER 1987), only after destructive harvesting; nevertheless, it is of lesser ecological significance than size and developmental state. Without any doubt the investigation of each of them, of age and size classes as of developmental states has its own merits depending on the aim of the studies, and detailed population analysis can be regarded as a highly sensitive method to detect environmental changes. Population biologists have developed more or less detailed systems of ontogenetic states occuring during the life cycle of plants. PELTON (1953) distinguished the following stages which have been widely used by other authors: seed, seedling, juvenile, reproductive and senescent stage. Some authors (GATSUK et al. 1980, BALLEGAARD and WARNCKE 1985) have developed finer subdivisions of the juvenile (also termed vegetative or prereproductive) and senescent (postreproductive) stage. Enormous differences in size can even be found in mature stands of homogeneous age composition due to density dependent competition. This heterogeneity can be characterized by a scale of vitality classes ranging from dominant to suppressed individuals (RABOTNOV 1978). With increasing plant density the transition of the bell-shaped frequency curve of low–density stands to the L–shaped curve of high–density stands can be observed (KOYAMA and KIRA 1964). Due to density–dependent mortality a close relationship can be found between plant density and mean plant biomass (WHITE 1980).

The diversity within a population has to be taken into account where mean values for individual plants are calculated and is absolutely necessary in the study of population dynamics or succession since life cycle events such as birth and death rates are state and class specific. Populations of clonal plants may be rather homogeneous in size compared with the majority of annual plants and in extreme environments one ontogenetic state, the mature vegetative plant, may be represented nearly exclusively. This was the case in our first investigations on primary production of the herb layer of a forest (EBER 1971), where juvenile as well as flowering plants were rare during the period of intense studies. But in a subsequent year with high irradiation the proportion of flowering shoots increased from 5 up to 15 % (EBER 1986) and it is evident that the transition rates from one ontogenetic state to another have to be taken into account. On the contrary in populations of *Calla palustris* size differences are extreme and mean values are without any information, because there is a high annual recruitment of new members originating from branches of different size, brood shoots and seeds due to rapid fragmentation of the rhizome system (EBER 1983a). Furthermore, population structure changes from pioneer to mature stands and along environmental gradients (EBER, in preparation).

1.4 Application of morphological aspects in the study of primary production and related processes

Consideration of morphological aspects has proved to be of great advantage in many fields of ecology and has already been applied in most of them for some time. From our own experience, however, we think that especially in the study of primary production some progress will be possible by the application of methods based on a thorough morphological analysis of plant organization. These methods have been developed and tested by population biologists and may be adopted for the study of primary production as it is obvious that production biology and population biology are closely related fields of science. Both deal with the same process as is evident from their fundamental equations:

1. Production equation
   Net primary production = annual change in dry matter + losses to detritus + grazed material (LARCHER 1983)
2. Population equation
   $N_{t+1} = N_0 + \text{births} - \text{deaths}$ (HARPER 1977)

Immigrants and emigrants as well as import and export of biomass may be neglected for the majority of plants and ecosystems.

While production ecologists investigate changes in weight, population biologists focus on changes in number. But differences have diminished during the past decades, and in the study of resource allocation, for example, population biologists, too, are occupied with estimating biomass and production.

The most serious drawback in the study of primary production of vascular plants is that, in contrast to population biology, the majority of methods are indirect and mostly based on destructive harvesting. In the harvested quadrat method (NEWBOULD 1967) the vegetation is harvested at intervals from quadrats which are thought to be comparable, whereas in the individual plant method (NEWBOULD 1967) biomass contribution of each species is estimated separately as the product of individual plant weight and density data from countings on permanent plots. Reliability of the quadrat method depends very much on the homogeneity of the vegetation. The individual plant method, however, has no limitations of this kind. As it is more time–consuming and numerous individuals have to be sampled it is restricted to species–poor vegetation types and common species. In both methods primary production is usually calculated as the difference between maximum and minimum biomass. Two major sources of underestimation must be kept in mind. First, if species are not treated separately there may arise serious errors if the periods of their individual maxima differ significantly, and secondly, there may be considerable losses due to mortality between sampling dates, which are extremely high in species with a high turnover of shoots or leaves (MATTHEWS and WESTLAKE 1969). Many attempts have been made to develop adequate methods for the estimation of mortality (e.g. SMALLEY 1959, WIEGERT and EVANS 1964, MILNER and HUGHES 1968). These methods have been compared and modified by several authors (LOMNICKI et al. 1968, LINTHURST and REIMOLD 1978, WALLENTINUS 1973, GROENENDIJK 1984). One of those factors supposed to influence the results of any single method is the morphology of species (LOMNICKI et al. 1968), and this corresponds to the intentions of this paper.

The fundamental problem in this context is the heterogeneity of the samples. Considerable differences in age, activity and fate of the material make it impossible to get reliable results on increment and losses which are the basic factors of primary production. Therefore it is obvious that the accuracy of estimations and the extent of information increases with the degree of appropriate subdivisions. Until now the harvested material has predominantly been separated into above–ground and below–ground material which might be adequate for practical reasons to facilitate harvesting and to distinguish between autotrophic and heterotrophic parts. Unfortunately, many investigations are confined to the aerial parts alone which must be regarded as problematic considering the manifold translocation processes between subterranean and aerial organs. When the contributions of individual species are separated, quadrat samples often do not contain enough material to allow exact estimations of peak biomass of each species. Rare examples of a more detailed separation into shoots, leaves, rhizomes and roots are found. These conventional modes of subdivision are unspecific and do not provide subsamples suitable for the estimation of increments and losses. Therefore a more promising approach is suggested here which is based on the modular construction of plants and the life history of modular units. A characteristic feature of this approach is the combination of the non-destructive methods of population biology with the destructive harvesting of production biology. These ideas are not completely new; substantial elements can already be found in the studies of CALLAGHAN (1976), GRABHERR et al. (1978) and JACKSON et al. (1986).

1.5 The analysis of plant architecture

At the beginning of each investigation a careful investigation of the architecture of each individual species should be carried out as the basis for the design of an appropriate programme for the study of the production process.

The study of the organization of whole genets or genet fragments of herbaceous plants demands a procedure very similar to that one used in the study of primary production. The biomass is harvested as completely as possible, but in addition attention must be paid to keep the connecting rhizomes intact. The period between the end of the season and early spring, when the annual growth is finished and the tender annual organs as leaves and fine roots are no longer present, has proved to be the most suitable time. Ramets should be marked and their original position in the field be fixed in a scetch before excavation. In the laboratory plants have to be cleaned thoroughly and attached dead roots and leaves have to be removed.

After this procedure a great variety of characters become visible which were obtected before: scars of leaves, branches, inflorescences and roots, regular changes in length and width of internodes and a specific pattern of branching and flowering indicating rhythmic growth. Many species with a high proportion of subterranean organs have a highly jointed rhizome system the analysis of which yields information on

1. annual increments
2. age of the genet or number of generations still present
3. life span of individual organs or structures
4. birth rates and mortality of modular units
5. spatial pattern of modular units
6. resource allocation

In species with long-lived organs the history of genets or genet fragments can be reconstructed with some accuracy as demonstrated for entire genets of *Limonium*

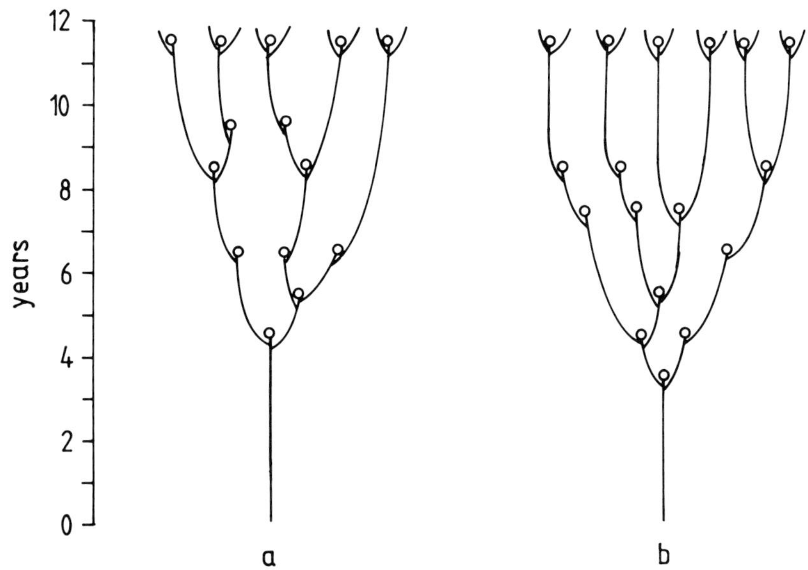

Figure 1.1: Schematic diagram of the rhizome organization of 12 years old individuals of *Limonium vulgare*. Circles indicate inflorescences or their remainders (from EBER and VEENHUIS, in print)

vulgare (Fig. 1.1) and a genet fragment of *Luzula luzuloides* (Fig. 1.2), which are subjects of detailed studies.

The principles of construction are species–specific and genetically fixed, but the specific form of the shoot system of each individual genet is subject to modifications by the influence of environmental factors. Size, birth rates, mortality of structures, number of inflorescences and flowers and the direction of growth are influenced by the climatic conditions of the season and soil environment.

In studies on primary production it is important to distinguish between the current year's increments and older parts which are subject to gradual destruction. If this separation has been carried out carefully, it should be possible to get reliable results on the subterranean biomass as well. Methods based on this principle have rarely been applied so far (IWAKI and MIDORIKAWA 1968, PERSSON 1975).

The estimation of the aerial primary production can also be done by separate weight estimations of current year's and previous years' increments (Fig. 1.3) and the application of life–table and other demographic analyses to shoot and leaf birth and death. This has been shown to be feasible and to permit ecological interpretations at a more sophisticated level than is possible from classical growth analysis (BAZZAZ and HARPER 1977). At intervals the lengths of stems and leaves

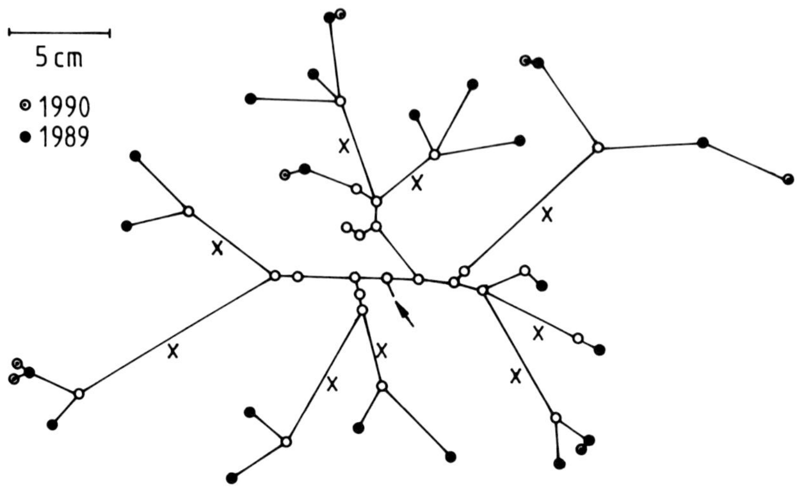

Figure 1.2: Schematic diagram of the rhizome system of a clone fragment of *Luzula luzuloides* with the growth of seven years. Circles indicate ramets, the arrow the oldest parts of the system. Note the enormous increments of the year 1988 (×).

and, if possible, leaf area are to be measured for each developmental state and leaf cohort (leaves of the same age) separately. For the conversion to weight units length: weight and area: weight regressions have to be established. Thus it is possible to estimate increments between sampling dates from weight differences of organs already present before and the difference between parts produced and lost within the sampling interval.

It is obvious that this method has some limitations as it is very time–consuming and some morphological provisions must be made. It is of special value in species with clonal growth, moderate mortality, density and sufficient size of modules and leaves. In our intended long–term studies on the population biology and primary production of salt marsh angiosperms and plant communities (EBER in preparation) it has already proved to be applicable to *Halimione portulacoides* (FISCHER and EBER in press), *Aster tripolium* (STRUTZ–FISCHER and EBER, in preparation), *Limonium vulgare* (EBER and VEENHUIS in press, EBER 1987), *Plantago maritima* and *Spartina anglica* (EBER, unpublished), but we suppose there will be some problems with *Puccinellia maritima* and *Festuca rubra*. Nevertheless, these restrictions in mind this method will be of advantage where a high accuracy of estimations is wanted, and as a variety of rough estimates is already available for many ecosystems and vegetation types, there seems to be no need of further ones. On the contrary, there is need of more precise information which allows broader ecological interpretations and applications for various purposes.

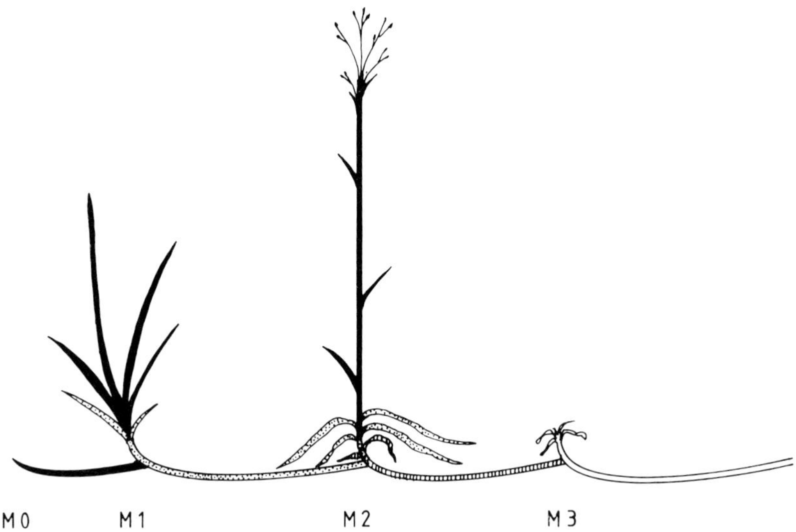

Figure 1.3: Schematic diagram of the module system of *Luzula luzuloides*. M0–M3 represent modules of subsequent years. Black = increment of the current year, dotted = increment of the previous year.

1.6 Resource allocation

The detailed compartmentation of plant material recommended above will be of special value in the study of resource allocation and, based on it, mineral cycling and energy flow. The analysis of resource allocation has found increasing interest among population biologists (e.g. HARPER and OGDEN 1970, OGDEN 1974, ANDEL and VERA 1977), and particularly reproductive allocation ("reproductive effort") has become one of the fundamental subjects of population biology and has been intensively discussed under the aspect of evolutionary strategies (GADGIL and SOLBRIG 1972). It is difficult to make a clear delimitation between reproductive and non–reproductive structures and results differ depending on whether seeds, fruits or entire inflorescences are regarded as sexual reproductive organs. Comparableness is only guaranteed if comparable structures are treated. As reproductive effort is usually expressed as percentage of net production, the accuracy of production estimates is a crucial point. The majority of investigations exclusively deal with dry matter allocation though, as stressed by HARPER and OGDEN (1970) and ABRAHAMSON and GADGIL (1973) a particular mineral nutrient may be likewise the limiting resource and important in the evolution of strategies. There are only few examples for a more comprehensive treatment of resource types (e.g. EBER 1983b, OHLSON and MALMER 1990, FITTER and SETTERS 1988, BENNER and

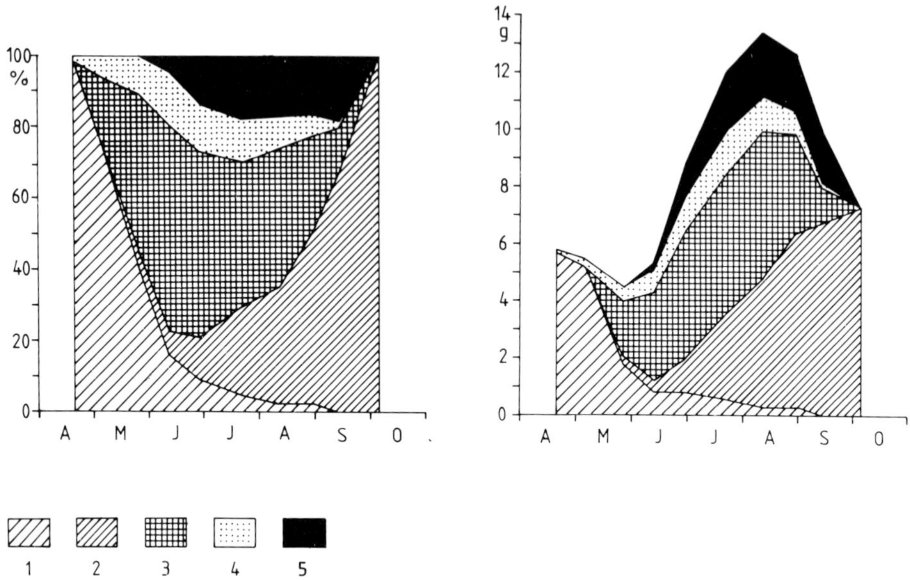

Figure 1.4: Dry matter allocation patterns in *Calla palustris*:
a) proportional representation (EBER 1983a, modified)
b) absolute values (EBER, in preparation).
1 rhizome of the previous year, 2 rhizome of the current year, 3 leaves, 4 roots, 5 inflorescences.

BAZZAZ 1988).

The annual course of the allocation process may be presented by allocation pattern diagrams. Proportionate representations (Fig. 1.4a), which are much easier to establish, serve best the purposes of population biology. Allocation pattern diagrams with absolute values (Fig. 1.4b), however, provide additional information on translocation processes and resource budgets, but are influenced by size differences within populations (OHLSON and MALMER 1990, EBER, in preparation). Allocation patterns can be estimated for each developmental state separately as well as for the whole population. The allocation to sexual reproduction and clonal growth have proved to be competitive activities. Individuals of *Geum reptans*, the flowers of which had been removed, developed longer stolons (BRZOSKA 1983) than the control, and in *Calla palustris* populations the production of extremely long rhizomes was accompanied by poorer flowering (EBER 1983a). As has been shown in field investigations (EBER 1983b) and experiments (HARPER and OGDEN 1970, OGDEN 1974, ANDEL and VERA 1977) resource allocation patterns are species-specific, but vary in a high degree with nutrient supply and according to the resource form (dry matter, energy, mineral nutrients) under consideration.

1.7 The growth form, clonal growth and lateral spread

Many herbaceous perennials are able to develop aerial or subterranean plagiotropic shoots (stolons or rhizomes) which serve lateral spread as well as the multiplication of ramets. The term "clonal growth" is now widely used instead of "vegetative reproduction", which has been rejected by HARPER (1977). According to his opinion reproduction should be confined to sexual reproduction only. Among species with clonal growth two categories can be distinguished according to the length of connections between individual ramets: one with modular units closely packed together and another one with widely–spaced modular units. Both categories can be found between species as well as within species, in the latter case as a plastic response to differences in the environment of the population. These different growth forms have been described as strategies by LOVETT DOUST (1981): the "guerilla" strategy of plants which infiltrate the surrounding vegetation with long stolons, and the "phalanx" strategy of plants which maximize their interspecific contacts. The interpretation of the function implied in these terms, which seem to have been coined by CLEGG (HUTCHINGS and SLADE 1988) is doubtful. In general the conservative "phalanx" type as a consequence of its low mobility must be able to exploit its area intensively and to withstand competitors, whereas the more opportunistic "guerilla" type with its ability of rapid and extensive stolon elongation is capable of colonizing adjacent areas.

The analogy between the pattern of growth of a stoloniferous or rhizomatous plant and the search path of a foraging animal has recently been emphasized, and, in consequence, the term "foraging" has been applied for plants to describe the activities of acquiring essential resources (SLADE and HUTCHINGS 1987b). It has been assumed that plants growing in a patchy environment branching probability should be higher in better patches and internode length should decrease with patch quality, but only the first assumption was consistently supported by a literature review (SUTHERLAND and STILLMAN 1988).

As both types may be developed within the same population, it is necessary to examine the conditions that favour the one or the other. The experiments of SLADE and HUTCHINGS et al. (1987, 1987a, 1987b) demonstrate that in *Glechoma hederacea* low light intensity and low soil nutrient availability produced extensive foraging, whereas a greater supply of nutrients resulted in a more intensive foraging and consolidation of site occupation. Changes in the water–table also influence growth form. In an experiment (EBER, unpublished) stolon elongation in *Ranunculus repens* increased rapidly when the water level was raised. As CALLAGHAN (1988) states in *Lycopodium annotinum*, the "guerilla" form may have been selected to avoid competition rather than invade and efficiently exploit nutrient–rich pockets where intense competition is likely to occur or develop. The mobility of *Potentilla anserina* has also been interpreted as a mechanism of escape from superior competitors (ERIKSON 1986). Effective vegetative spread is also an important factor in colonization processes, as has been shown for *Prunella vulgaris* in the colonization of bare ground (SCHMID 1985). Correspondingly, in *Calla palustris* extremely long rhizomes were found at the border between the population territority and the open water surface, whereas the internal area and parts adjacent to *Phragmites* are occupied by dense stands of very compact short rhizomes (EBER 1983a). These parts of the population have been termed pioneer and mature stand.

Summing up,we can state that the ability of intense and rapid stolon elongation is advantageous for three specific purposes:

1. to escape adverse environmental conditions or to search for an optimal territory in gradient environments as canopy gaps and borders of woodlands (*Fragaria vesca, Glechoma hederacea*) or embankments of waters (flood meadow species like *Ranunculus repens, Potentilla anserina, Agrostis stolonifera*)

2. to escape interspecific and intraspecific competition

3. to colonize bare areas

Item 2 and 3 seem to be different aspects of the same process and, if working on a fine scale, contribute to maintain the species equilibrium in plant communities. The extent of clonal growth is limited by the number of available meristems (WATSON 1984, BISHOP and DAVY 1985, ERIKSSON 1985, CARLSSON and CALLAGHAN 1990).

Acknowledgments

I thank G. Wiegleb (Oldenburg) for critical comments on the manuscript.

Bibliography

ABRAHAMSON, W. G.; GADGIL, M., 1973, Growth form and reproductive effort in goldenrots (*Solidago, Compositae*).
Amer. Natur. **107**, 651–661.

ANDEL, J. VAN; VERA, F., 1977, Reproductive allocation in *Senecio sylvaticus* and *Chamaenerion angustifolium* in relation to mineral nutrition.
J. Ecol. **65**, 747–758.

BALLEGAARD, T. K.; WARNCKE, E., 1985, The age distribution of a *Cirsium palustre* population in a spring area in Jutland, Denmark.
Holarct. Ecol. **8**, 59–62.

BAZZAZ, F. A.; HARPER, J. L., 1977, Demographic analysis of the growth of *Linum usitatissimum*.
New Phytol. **78** 193–208.

BENNER, B. L.; BAZZAZ, F. A., 1988, Carbon and mineral element accumulation and allocation in two annual plant species in response to timing of nutrient addition. J. Ecol. **76**, 19–40.

BISHOP, G. F.; DAVY, A. J., 1985, Density and commitment of apical meristems to clonal growth and reproduction in *Hieracium pilosella*. Oecologia **66**, 417–422.

BRAUN–BLANQUET, J., 1964, Pflanzensoziologie. 3rd edn., Springer, Wien New York.

BRZOSKA, W., 1983, Untersuchungen zur Wachstums- und Reproduktionsstrategie von *Geum reptans*. Verh. Ges. Ökol. **10**, 499–503.

CALLAGHAN, T. V., 1976, Growth and population dynamics of *Carex bigelowii* in an alpine environment. Oikos **27**, 402–413.

CALLAGHAN, T. V., 1988, Physiological and demographic implications of modular construction in cold environments. In: Davy, A. J.; Hutchings, M. J.; Watkinson, A. R. (Eds.): Plant Population Ecology. Blackwell Sci. Publ., Oxford.

CARLSSON, B. A.; CALLAGHAN, T. V., 1990, Effects of flowering on the shoot dynamics of *Carex bigelowii* along an altitudinal gradient in Swedish Lapland. J. Ecol. **78**, 152–165.

EBER, W., 1971, The primary production of the ground vegetation of the Luzulo–Fagetum. Ecol. Stud. **2**, 53–56.

EBER, W., 1982, Struktur und Dynamik der Bodenvegetation im Luzulo–Fagetum. In: Dierschke, H. (Ed.): Struktur und Dynamik von Wäldern. Ber. Int. Symp. Int. Ver. Vegetationsk., p. 495–511.

EBER, W., 1983a, Untersuchungen zur Populationsbiologie von *Calla palustris*. Tüxenia **3**, 417–421.

EBER, W., 1983b, Wachstum und Nährstoffhaushalt von *Calla palustris* in Abhängigkeit vom Standort. Verh. Ges. Ökol. **10**, 505–512.

EBER, W., 1986, Jahresrhythmus und Produktion der Kraut- und Moosschicht. In: Ellenberg, H,; Mayer, R.; Schauermann, J. (eds): Ökosystemforschung — Ergebnisse des Solling–Projektes. Eugen Ulmer, Stuttgart, p. 127–136.

EBER, W., 1987, The growth form of *Limonium vulgare* MILL: Structural demography and ecological significance.
XIV Int. Bot. Congr. Abstr. p. 436.

EBER, W., (in preparation), Population differentiation in *Calla palustris* along a bog – fen gradient.

EBER, W., (in preparation), Populations– und produktionsbiologische Untersuchungen in Salzwiesen der deutschen Nordseeküste.

EBER, W., (in preparation), Resource allocation patterns in *Calla palustris.*

EBER, W.; VEENHUIS, B., 1991, Natality and mortality in *Limonium vulgare.*
In: Schmid, B.; Stöcklin, J.: Populationsbiologie der Pflanzen. Birkhäuser–Verlag, Basel, 62–73.

ERIKSSON, O., 1985, Reproduction and clonal growth in *Potentilla anserina* L. (*Rosaceae*): the relation between growth form and dry weight allocation.
Oecologia **66**, 378–380.

ERIKSSON, O., 1986, Mobility and space capture in the stoloniferous plant *Potentilla anserina.*
Oikos **46**, 82–87.

ESAU, K., 1977, Anatomy of seed plants.
John Wiley & Sons, New York.

FAHN, A., 1982, Plant Anatomy.
3rd edn. Pergamon Press, Oxford.

FITTER, A. H.; SETTERS, N. L., 1988, Vegetative and reproductive allocation of phosphorus and potassium in relation to biomass in six species of *Viola.*
J. Ecol. **76**, 617–636.

FOSTER, A. S.; GIFFORD, E. M., 1974, Comparative morphology of vascular plants.
W. H. Freeman and Co., San Francisco.

GADGIL, M.; SOLBRIG, O. T., 1972, The concept of r– and K– selection: evidence from wild flowers and some theoretical considerations.
Amer. Natur. **106**, 14–31.

GATSUK, L. E.; SMIRNOVA, O. V.; VORONTZOWA, L. I.; ZAUGOLNOVA, L. B.; ZHUKOVA, L. A., 1980, Age states of plants of various growth forms: a review.
J. Ecol. **68**, 675–696.

GRABHERR, G.; MÄHR, E.; REISIGL, H., 1978, Nettoprimärproduktion und Reproduktion in einem Krummseggenrasen (Caricetum curvulae) der Ötztaler Alpen, Tirol.
Oecol. Plant. **13**, 227–251.

GROENENDIJK, A. M., 1984, Primary production of four dominant salt–marsh angiosperms in the SW Netherlands.
Vegetatio **57**, 143–152.

HALLE, F.; OLDEMAN, R. A. A.; TOMLINSON, P. B., 1978, Tropical Trees and Forests: an Architectural Analysis.
Springer–Verlag, Heidelberg.

HARPER, J. L., 1977, Population Biology of Plants.
Academic Press, London.

HARPER, J. L.; BELL, A. D., 1979, The population dynamics of growth form in organisms with modular construction.
In: Anderson, R. M.; Turner, B. D.; Taylor, L. R. (Eds): Population Dynamics. Symp. Brit. Ecol. Soc. **20**, 29–52.

HARPER, J. L.; OGDEN, J., 1970, The reproductive strategy of higher plants. I. The concept of strategy with special reference to *Senecio vulgaris* L.
J. Ecol. **58**, 681–698.

HUISKES, H. L.; HARPER, J. L., 1979, The demography of leaves and tillers of *Ammophila arenaria* in a dune sere.
Oecol. Plant. **14**, 435–446.

HUTCHINGS, M. J.; SLADE, A. J., 1988, Morphological plasticity, foraging and integration in clonal perennial herbs.
In: Davy, A. J.; Hutchings, M. J.; Watkinson, A. R. (eds.): Plant Population Econology.
28th Symp. Brit. Ecol. Soc. 1987. Blackwell Scient. Publ., Oxford, 83–109.

IWAKI, H.; MIDORIKAWA, B., 1968, Principles for estimating root production in herbaceous perennials.
In: Methods of productivity studies in root systems and rhizospere organisms. Int. Symp. USSR Acad. Sci., Nauka, Leningrad.

JACKSON, D.; LONG, S. P.; MASON, C. F., 1986, Net primary production, decomposition and export of *Spartina anglica* on a Suffolk salt–marsh.
J. Ecol. **74**, 647–662.

KOTANEN, P.; JEFFERIES, R. L., 1987, The leaf and shoot demography in grazed and ungrazed plants of *Carex subspathacea*.
J. Ecol. **75**, 961–975.

KOYAMA, H.; KIRA, T., 1964, Intraspecific competition among higher plants. VIII. Frequency distribution of individual plant weight as affected by the interaction between plants.
J. Inst. Polytechn. Osaka City Univ. Ser. D **7**, 73–94.

LARCHER, W., 1983, Physiological Plant Ecology.
2nd edn., Springer, Berlin Heidelberg New York Tokyo.

LEHMANN, A.; EBER, W. (in press), Zur Populationsbiologie von *Equisetum hyemale*.
Verh. Ges. Ökol. XIX.

LINTHURST, R. A.; REIMOLD, R. J., 1978, An evaluation of methods for estimating the net aerial primary productivity of estuarine angiosperms.
J. appl. Ecol. **15**, 919–931.

LOMNICKI, A.; BANDOLA, E.; JANKOWSKA, K., 1968, Modification of the Wiegert–Evans method for estimation of net primary production.
Ecology **49**, 147–149.

LOVETT DOUST, L., 1981, Population dynamics and specialisation in a clonal perennial (*Ranunculus repens*). I. The dynamics of ramets in contrasting habitats.
J. Ecol. **69**, 743–755.

MATTHEWS, C. P.; WESTLAKE, D. F., 1969, Estimation of production by populations of higher plants subject to high mortality.
Oikos **20**, 156–160.

MILNER, C.; HUGHES, R. E., 1968, Methods for the measurement of primary production of grassland.
IBP Handbook No. **6**. Oxford.

NEWBOULD, P. J., 1967, Methods for estimating the primary production of forests.
IBP Handbook No. **2**. Blackwell Sci. Publ., Oxford, Edinburgh.

OGDEN, J., 1974, The reproductive strategy of higher plants. II. The reproductive strategy of *Tussilago farfara* L.
J. Ecol. **62**, 291–324.

OHLSON, M.; MALMER, N., 1990, Total nutrient accumulation and seasonal variation in resource allocation in the bog plant *Rhynchospora alba*.
Oikos **58**, 100–108.

PELTON, J., 1953, Ecological life cycle of seed plants.
Ecology **34**, 619–626.

PERSSON, H., 1975, Deciduous woodland at Andersby, Eastern Sweden: Field-layer and below-ground production.
Acta Phytogeograph. Suec. **62**, 1–71.

RABOTNOV, T. A., 1978, On coenopopulations of plants reproducing by seeds. In: Freysen, A. H. J.; Woldendorp, J. W. (eds.): Structure and function of plant populations, p. 1–26, North Holl. Publ. Comp., Amsterdam.

RITTERBUSCH, A., 1977, Homolog- und Analog-Modell einer spermatophyten und einer terrestren Pflanze.
Ber. Deutsch. Bot. Ges. **90**, 363–368.

SCHMID, B., 1985, Clonal growth in grassland perennials. II. Growth form and fine-scale colonizing ability.
J. Ecol. **73**, 809–818.

SLADE, A. J.; HUTCHINGS, M. J., 1987, The effects of nutrient availability on foraging in the clonal herb *Glechoma hederacea*.
J. Ecol. **75**, 95–112.

SLADE, A. J.; HUTCHINGS, M. J., 1987a, Clonal integration and plasticity in foraging behaviour in *Glechoma hederacea*. J. Ecol. **75**, 1023–1036.

SLADE, A. J.; HUTCHINGS, M. J., 1987b, The effects of light intensity on foraging in the clonal herb *Glechoma hederacea*.
J. Ecol. **75**, 639–650.

SMALLEY, A. E., 1959, The growth cycle of *Spartina* and its relation to insect populations in the marsh.
Proc. Salt Marsh Conf. March 1958, Sapelo Island, Georgia.

STRUTZ-FISCHER, H.; EBER, W., (in press), Zur Populationsbiologie von *Halimione portulacoides*.
Verh. Ges. Ökol. XIX.

STRUTZ-FISCHER, H.; EBER, W., (in press), Lebenszyklus, Reproduktion und klonales Wachstum von *Aster tripolium*.
Verh. Ges. Ökol. XX.

SUTHERLAND, W. J.; STILLMAN, R. A., 1988, The foraging tactics of plants.
Oikos **52**, 239–244.

WALLENTINUS, H.-J., 1973, Above-ground primary production of a Juncetum gerardi on a Baltic sea-shore meadow.
Oikos **24**, 200–219.

WATSON, M. A., 1984, Developmental constraints: effect on population growth and patterns of resource allocation in a clonal plant.
Amer. Natur. **123**, 411–426.

WIEGERT, R.; EVANS, F., 1964, Primary production and the disappearance of dead vegetation on an old field.
Ecology **45**, 49–63.

WHITE, J., 1979, The plant as a metapopulation.
Ann. Rev. Ecol. Syst. **10**, 109–145.

WHITE, J., 1980, Demographic factors in populations of plants.
In: Solbrig, O. T. (ed.) Demography and evolution in plant populations. Blackwell Sci. Publ., Oxford London Edinburgh Melbourne, p. 21–48.

Chapter 2

Vegetation structure, phytomass and phenology of the dry thorn scrub of Curaçao, West Indies

M. J. A. WERGER, C. DE BOK and B. ORANJE

2.1 Introduction

Curaçao is the largest island (472 km^2) in the Netherlands Antilles, Leeward Islands, West Indies. The island lies 60 km away from the coast of Venezuela and measures maximally 60 by 14 km. It is situated in the North Eastern Trade Wind System. These winds bring in clouds during most of the year but relatively little precipitation. The island's climate is semi-arid. Only one small area is sufficiently elevated (Mnt. Christoffel, 372 m) to yield the amount of precipitation allowing forest growth. Most of the island is below 120 m and the average annual precipitation in this zone is 571 mm or less (STOFFERS 1956). November and December are the wettest months with 120, respectively 100 mm, and March to May are driest, with about 18 mm per month. As in other dry areas, year–to–year differences in rainfall are rather large, and at any one time local differences can be large too.

The island consists mainly of mesozoic diabase as well as some conglomerates, shales and sandstones which, on the higher parts, are crowned with fairly substantial crusts of pleistocene coral limestones. These limestones show clear terraces that correlate with previous sea levels (BEETS 1972). Presently soils are very poorly developed on most of the island, possibly because of severe erosion during the past centuries.

The vegetation of Curaçao is mainly a dry, thorny, scrubby woodland, up to about 4 m tall, with few emergent, low trees. BEARD (1953, 1955) classified it as thorny woodland and STOFFERS (1956) distinguished between evergreen thorny woodland on the limestone plateaus and deciduous thorny woodland on the diabase. He pointed out that the limestone and diabase have very few woody species in common and that all evergreen woody species are restricted to the limestone except for *Balanites aegyptica*, a tall, thorny evergreen shrub introduced from arid northern Africa in the 19th century (ARNOLDO 1971).

The vegetation on both substrates consists nearly entirely of shrubs with a few giant cacti (*Cereus repardus*) on the diabase and the small *Melocactus* sp. in fissures on the limestone; some *Opuntia* spp. invade both habitats. Subshrubs and herbs form an open ground layer that is locally well developed, particularly on diabase.

The differences in seasonal phenologies between the vegetation types of the two substrates are striking. Since both types meet along a narrow and long fringe line that coincides with the geological boundary, the differences in vegetation are generally explained as resulting from the different substrates perhaps affecting the water supplies (e. g. STOFFERS and MANSOUR–ELASSAISS 1967). However, no work has been carried out to substantiate this claim. As a preliminary to investigations on the water economy of the vegetation types on both substrates we studied the species composition, structure, phytomass and phenology of both types in two adjacent areas on different substrates. The areas are adjacent to each other, and between them there is no difference in rainfall, radiation regime, temperature, and other major climatic factors. The only climatic difference between the sites might be a somewhat stronger wind exposure of the vegetation on the limestone plateau.

2.2 Area and methods

2.2.1 Area and floristic composition

The study has been carried out at Malpais (ca. 35 m a. s. l.) on the diabase and at adjacent Ceru Jamanica (ca. 90 m a. s. l.) on the limestone plateau. Both areas are situated at 12° 12' N and 68° 56' W, about 10 km NW of Willemstad. At both localities apparently homogeneous study sites of about 20 000 m^2 were delineated. In these sites plots of 10 x 10 m, systematically distributed over the sites, were sampled to inventory the floristic composition. Systematic distribution of sample plots was chosen because of the apparent internal homogeneity of the study sites. Cover–abundance of all species of phanerogams were rated on the usual

Braun–Blanquet scale (MUELLER–DOMBOIS and ELLENBERG 1974), and values were transformed (VAN DER MAAREL 1979) for analysis using the TWINSPAN and DECORANA programs (HILL 1979, GAUCH 1982). Species nomenclature follows ARNOLDO (1964).

2.2.2 Biomass and Leaf Area

At both sites a plot of 5 × 5 m was laid out in a patch that appeared representative of the site in species composition, overall structure and cover. In each plot all individuals of woody and semi–woody species were mapped and their total height and height at widest canopy diameter recorded. Subsequently in each plot the vegetation was arbitrarily divided into four layers: 0–0.5 m, 0.5–1.0 m, 1.0–1.5 m, and > 1.5 m. Very few individuals were taller than 2.0 m. Per layer the above-ground biomass of all species was harvested and pooled per species. Since herbs contributed rather little biomass they were pooled in one group. All individuals taller than 1.5 m were measured individually.

At each site another homogeneous plot of 100 × 100 m was delineated. Frequency countings and visual assessment revealed that only 17 woody and semi-woody species contributed importantly to vegetation and biomass in these plots; other woody and semi–woody species contributed little to the total phytomass and so did the herbs, though their species number was considerable (see Tables 2.1–2.2). Above-ground biomass of these 17 most important species (see Table 2.3) was determined as follows. All individuals of species with a low to medium high frequency were harvested. From species with high frequency only 15 to 30 individuals were harvested after the volume of all individuals had been determined. Volume was calculated by considering each individual a truncated cone with the narrow end at soil surface and topped by half a globe; measurement of total height, widest crown diameter, height at widest crown diameter and width of the 'cone' at soil surface allowed calculation of volume. Measured biomass values were fitted to volume data by first, second and third order polynomials, but simple linear regressions proved to give the best fit. The regressions were used to calculate biomass values of non-harvested individuals.

Biomass of trees was measured using the method of VALENTINE et al. (1984) which resulted in volume estimates per individual. Based on measured fresh weights and specific weights of dry branch discs per species, total dry weights could be calculated. One individual per species was completely harvested to calibrate the method. All biomass values have been calculated on a dry weight base per hectare of ground area.

In all harvested individuals of all species leaf mass has been determined by harvesting all or part of the foliage and extrapolate from the measured values. Precise measurements of leaf size and weight were taken from 25 randomly sampled leaves in each individual. These values were used to calculate total leaf area per individual. For each species leaf area of fully foliated plants have been calculated using the data collected in the phenological study. All leaf area values were calculated per hectare of ground area.

2.2.3 Phenology

In the phenological study the 17 most important species have been studied during 29 weeks from week 12 till week 41 in 1985. Of each species at each site 10 individuals distributed over all size classes were studied simultaneously. At Malpais this concerned 10 species (13.5% of the total number of species at Malpais) which together contributed just over half of the total vegetation cover of 55% of the ground area there. At Ceru Jamanica it concerned 12 species (15% of the total number) which together made up three quarters of the total vegetation cover of 35% there. The species are listed in Table 2.3.

Phenological observations were made fortnightly from week 12 till week 21 and subsequently weekly. Plants were scored on the categories v (vegetative), fl (flowering), and fr (fruit bearing), and per species the percentage of individuals in each category at each date was calculated. Category v was subdivided into v1 (young leaf), v2 (full grown leaf), v3 (withering and browning leaf), v4 (dead leaf), and on each sampling day the percentages of leaves in these subcategories were estimated per individual.

Values for subcategories v1 to v4 were used to calculate a leaf phenological index (LPI) per individual and per species: to this end the proportion of leaves in a certain subcategory was multiplied by a value given to that subcategory and expressing the physiological functionality of the leaf in that phenological stage. Values given were v1 = 3, v2 = 6, v3 = 2, and v4 = 0. Consequently LPI ranges between 0 and 6. During the full period of the phenological study rainfall was measured daily at both sites and summarized per week. Per site LPI values and weekly rainfall data have been correlated.

2.3 Results

2.3.1 Floristic composition

The vegetation at Malpais and Ceru Jamanica differed rather strongly in floristic composition. Separate analysis of the two data sets by TWINSPAN and DECORANA showed no interpretable pattern of variation within the set of plots from the diabase at Malpais and from the limestone plots at Ceru Jamanica (Tables 2.1 and 2.2).

At Malpais 64 species were recorded and at Ceru Jamanica 54 species, omitting the 24 species present only in the stands at the plateau's edge. Of the 29 species common to both sites about 10 were equally important at both sites, while the others showed a preference for either the limestone or the diabase site. At Malpais the vegetation covered on average about 55% of the ground area and at Ceru Jamanica about 35%, these cover values being based on crown circumferences.

2.3.2 Biomass, structure and leaf area

In the small plots total above–ground biomass was determined at 16,300 $kg{\cdot}ha^{-1}$ at Malpais and 12,700 $kg{\cdot}ha^{-1}$ at Ceru Jamanica. In the large plots the measurements on the 17 most important species gave above–ground biomass values of 12,100 $kg{\cdot}ha^{-1}$ for Malpais and 13,300 $kg{\cdot}ha^{-1}$ for Ceru Jamanica (Table 2.3). The discrepancy between the values for Malpais indicates the contribution of the less important species to the total above–ground biomass in the vegetation on diabase. These were some species of small subshrubs and shrubs and quite a number of herbs. At Ceru Jamanica the unimportant species hardly contributed to the total above–ground biomass and there were very few herbs there.

Calculated for maximum leaf development per species, total leaf weight, and particularly total leaf area, is different between the two areas: Malpais has 532 $kg{\cdot}ha^{-1}$ or 7,598 $m^2{\cdot}ha^{-1}$, while Ceru Jamanica has 479 $kg{\cdot}ha^{-1}$ or 3,281 $m^2{\cdot}ha^{-1}$.

Species which occur both at Malpais and at Ceru Jamanica have higher biomass values per individual at Malpais. Individuals of *Caesalpinia coriaria, Cordia curassavica* and *Croton flavens* are three to four times bulkier at Malpais than at Ceru Jamanica, while those of *Casearea tremula* are nearly eight times and those of *Phyllanthus botryanthus* are only little bulkier at Malpais. In total leaf weights differences are similarly large, though the percentage of the plant's dry weight invested in leaves is generally less than two times higher in Malpais than in Ceru Jamanica in the species common to both sites. Only in *Croton flavens* differences are somewhat larger (Table 2.4). Specific Leaf Weights (SLW) of the species common to both sites (data not given) appear to be 10% higher at Ceru Jamanica, while the average SLW over all species at Ceru Jamanica is double that at Malpais.

At both sites more than one third of the total above–ground biomass is contained in the lowermost half meter and more than two thirds in the lowermost meter of vegetation. Only about 12% of the above-ground biomass is contained in the layer above 1.5 m (Figure 2.1, Table 2.4).

Fig. 2.1 shows the height distribution of the widest canopy diameters. It is clear that despite a comparable total amount of above–ground biomass the structure of the dry scrubby woodlands at Malpais and Ceru Jamanica is different: at Malpais crown widths seem more evenly distributed and there is no interrupted stratification. At Ceru Jamanica the vegetation is made up of a layer of rather small shrubs up to 1 m tall and a sparser layer of taller shrubs with relatively wide canopies.

Table 2.1: TWINSPAN sequence of species and plots of the vegetation at Malpais.

Species	1	2	3	4	5	6	7	8	9	10	11
Mentzelia aspera	1				2	2					
Elytraria imbricata	1	1	2	1		2		1			
Piriqueta ovata	2	2	2	2	2			2			
Ayenia magna	1	2	2								
Passiflora foetida			1			2					
Cassia absus		1									
Krameria ixinia					2						
Matelea rubra			2								
Solanum americanum				1							
Acacia glauca	2			1	2						
Heliotropium fruticosum	2			1							1
Ipomoea incarnata	3					2		1			
Convolvulus nodiflorus	3	2									
Spigelia anthelmia	1	2	2	2	2	2		2	2	2	
Lantana camara	2	2					2				
Polygala minticola	1				1		1				
Turnera pumilea			1	1				1			
Borreria ocimoides	1	2	2	2		2	2	2		2	2
Physalis cordata		2	2	2	2	2	1	2	2	2	2
Hyptis suaveolens	1	2	2		2	2	2	2	2	2	1
Euphorbia hypericifolia	1					1		2			
Indigofera suffruticosa						1	1				
Crescentia cujete	2	3		2	2		1	2	3		
Adelia ricinella							1				
Calotropis procera										1	
Porophyllus ruderale										2	
Balanites aegyptica					2		2	2	2		3
Pectis prostrata											1
Acalypha poiretii								2	2	2	
Melochia tomentosa	2	1			2	2	2	1			2
Solanum agrarium	1		2	2	1	2	2	1		2	2
Croton flavens	2	3	3	2	3	3	1				
Lemaireocereus griseus						1					
Randia aculeata	2	3	3	3	2	2	2	2	2	2	2
Cordia curassavica	3	2	3	2	3	3	3	3	2	3	3
Tournefortia volubilis	2	3	2	2	2	3		1	2	2	2
Rhynchosia minima		2	2		2	2		2	2	3	1
Acacia tortuosa	2	2	2	3	2	3	3	2	3	3	2
Cereus repandus	1				1			1	2		
Commicarpus scandens					1					3	2
Casearea tremula	1		2	2			1	2	2	2	
Bursera tomentosa				1		1	1	1	1	2	
Caesalpinia coriaria					2		2	2			
Euphorbia thymifolia											2
Lantana involucrata	2										1
Capparis odoratissima				2							
Cucumis dipsaceus	1										
Croton ovalifolius			2	1	2		3	2			
Egletes prostrata											1
Ipomoea nil		2	2	2	2	2	1	2	2	3	
Lactuca intybacea				1							
Opuntia wentiana	2	2	2		2	2	2	1	2	2	2
Phyllanthus botryanthus	2	2					2	3		2	
Cuscuta americana	1		1							2	
Bursera bonairensis							1	3			
Cordia globosa									2	2	
Tillandsia recurvata						1					
Haematoxylon brasiletto	1				1						
Malpighia punicifolia	1										
Trichilia trifolia		1									
Cordia alba		2									
Melampodium bonairensis									1		
Cassia emarginata											2

Table 2.2: TWINSPAN sequence of species and plots of the vegetation at Ceru Jamanica.

Corchorus hirsutus	2	2	1	2				1								
Cephalocereus lanuginosus	1															
Metopium brownei			1	1	1	2	1	2				1				
Morinda royoc				2		2	1									
Passiflora foetida				1			1									
Ipomoea incarnata								1								
Usnea spec.		1	1	1	1	1	1	1	2			1				
Acacia glauca												1				
Calymperes richardii												2				
Galactia striata											1					
Pithecellobium unguis-cati									1					2	2	
Capparis flexuosa													1			
Ipomoea nil													1			
Passiflora suberosa														1		
Opuntia curassavica	1	1								1	1	1	1	1	1	
Serjania curassavica							1					1		1		
Heliotropium fruticosum	1	2	2	2	2			1								
Haematoxylon brasiletto	2	3	2	2	3	3	3	3	2	2	2	2	2	1	3	3
Crossopetalum rhacoma	2	3	3	2	1	2	2	2	3		2	2	1	1		1
Coccoloba swartzii	2	3	3	3	1	2	3	2	2	2		2	1		2	
Anthirroea acutata	1	2	2	2	1	2	2	2			1	2	3		3	
Euphorbia thymifolia	2	2	2	2	2	2	2	2	2	2	2	2	2	2		2
Jacquinia barbasco		2	2	2	2	2	1	1	2		2	2		1	2	2
Guaiacum sanctum	1	1			1			1	1	2	3	2	2	2	2	2
Bursera tomentosa		1			1				1	1		1				
Bursera bonairensis	1	1			1				1	2	1	1	1		1	1
Samyda dodecandra						1		1	1		1		1		1	
Lantana involucrata			2													2
Croton flavens	2	2	2	2	2	2	2	2	3	2	2	3	2	3	2	3
Lantana camara	2	2	2	2	2	1	1	2	3	3	3	2	2	1	2	2
Cordia curassavica	2	2	2		2	2	2	2	3	2	2	2	3	3	2	3
Erithalis fruticosa	2	2	3	2	2	2	2	2	3		3	2	2	1	3	3
Phyllanthus botryanthus	2	1	2		1	2	2	2	2	2	2	2	2	1	2	2
Opuntia wentiana	1				1	1	2	1	1	1	1	1	2	1	1	2
Spigelia anthelmia	1		2	2	1	1			2	1						
Casearea tremula		1			1	1		1	1	1		1	2	2	2	2
Jatropha gossypifolia							1				1					2
Caesalpinia coriaria	1				1				1	2	2	1	1	1	2	2
Trichilia trifolia	1					1				2	2	1	1	2	2	2
Tillandsia recurvata									2				1	1	1	2
Capparis odoratissima										1						1
Cereus repandus				1												
Melocactus spec.								1								
Bastardia viscosa														1	1	
Acalypha poiretii	1															
Cordia globosa													1			
Spilanthes urens														2		2
Condalia henriquezii															1	
Bourreria succulenta						1			2	1		2			2	1
Randia aculeata					1					1	1	1		1	1	
Bumelia obovata										1		1				
Melochia tomentosa	2					2				2	2	1		2		
Rhynchosia minima	2										1	1		1		
Acacia tortuosa	1					1					1				1	
Commelina elegans	1															

Table 2.3: Total above–ground biomass (kg·ha^{-1}) of the 17 most important species at Malpais and at Ceru Jamanica.

	Malpais	Ceru Jamanica
Evergreen		
Acacia tortuosa	3,319.7	—
Balanites aegyptica	685.0	—
Crossopetalum rhacoma	—	2,287.0
Erithalis fruticosa	—	248.5
Guaiacum sanctum	—	1,942.8
Jaquinia barbasco	—	505.5
subtotal	4,004.7	4,983.8
Semi-deciduous		
Anthirroea acutata	—	51.1
Caesalpinia coriaria	185.4	68.0
Coccoloba swartzii	—	569.2
Cordia curassavica	214.4	474.1
Crescentia cujete	1,339.3	—
Croton flavens	259.5	177.0
Haematoxylon brasiletto	—	6,806.4
Tournefortia volubilis	183.5	—
subtotal	2,182.1	8,145.8
Deciduous		
Casearea tremula	110.7	69.9
Phyllanthus botryanthus	117.4	83.9
Randia aculeata	5,696.5	—
subtotal	5,924.6	153.8
Total	12,111.4	13,283.4

2.3.3 Phenology

Rainfall distribution and totals at Malpais and Ceru Jamanica were very similar in the period of measurement. Fig. 2.2 illustrates the LPI for all species studied at the two sites as well as the rainfall over the total period of study. The average LPI for Malpais seems slightly lower (4.4 ± 1.2) than for Ceru Jamanica (4.7 ± 0.9), but the maximal values (5.8) are the same for both sites. The beginning of the period of observation coincides with the start of the dry season. As the season

Table 2.4: Distribution of above-ground biomass in g·m^{-2} dry weight (in parentheses as percentage of total per layer) for two species common at both sites, *Cordia curassavica* and *Croton flavens*, on diabase (Malpais) and limestone (Ceru Jamanica).

	Cordia curassavica					
	Malpais			Ceru Jamanica		
layer (m)	leaf	stem	total	leaf	stem	total
0.–0.5	1.7 (6)	25.9 (94)	27.6	2.7 (8)	31.1 (92)	33.8
0.5–1.0	18.3 (47)	20.7 (53)	39.0	20.0 (48)	21.8 (52)	41.8
1.0–1.5	11.6 (73)	4.2 (27)	15.8	14.1 (57)	10.4 (43)	24.5
Total	31.6 (38)	50.8 (62)	82.4	36.8 (37)	63.3 (63)	100.1

	Croton flavens					
	Malpais			Ceru Jamanica		
layer (m)	leaf	stem	total	leaf	stem	total
0.–0.5	1.8 (18)	8.2 (82)	10.0	0.7 (8)	8.4 (92)	9.1
0.5–1.0	4.6 (61)	2.9 (39)	7.5	0.9 (38)	1.4 (62)	2.3
Total	6.4 (37)	11.1 (63)	17.5	1.6 (14)	9.8 (86)	11.4

progresses LPI decreases as leaves wither and fall. Leaf flush appears again when substantial rains come from week 28 onwards. However, the withering and fall of leaves at Malpais is more severe than at Ceru Jamanica, while the recovery at Malpais following the major rains is also stronger. By the end of the dry period LPI at Malpais has diminished by 35% while at Ceru Jamanica this is only 20%. The light rains following week 20 show a clear leaf flush at both sites, but the short dry period from week 33 till 39 shows a decreasing LPI only at Malpais.

LPI data and rainfall data were analyzed with Pearson's correlation analysis. Simple rainfall data (in mm) and log transformed (mm +1) rainfall data were used. No correlation was found between LPI data and rainfall data obtained in the same week the LPI data referred to, nor with the rainfall data of the previous week or two weeks earlier. Strong correlations were found between LPI and log transformed (mm +1) cumulative rainfall data of the two weeks prior to the week the LPI was measured. For Malpais this gave the following regression line:

$$LPI = 1.56(\log(mm + 1)) + 2.86 \qquad (r^2 = 0.36, n = 19, p < 0.01)$$

and for Ceru Jamanica:

$$LPI = 1.46(\log(mm + 1)) + 3.28 \qquad (r^2 = 0.58, n = 18, p < 0.001).$$

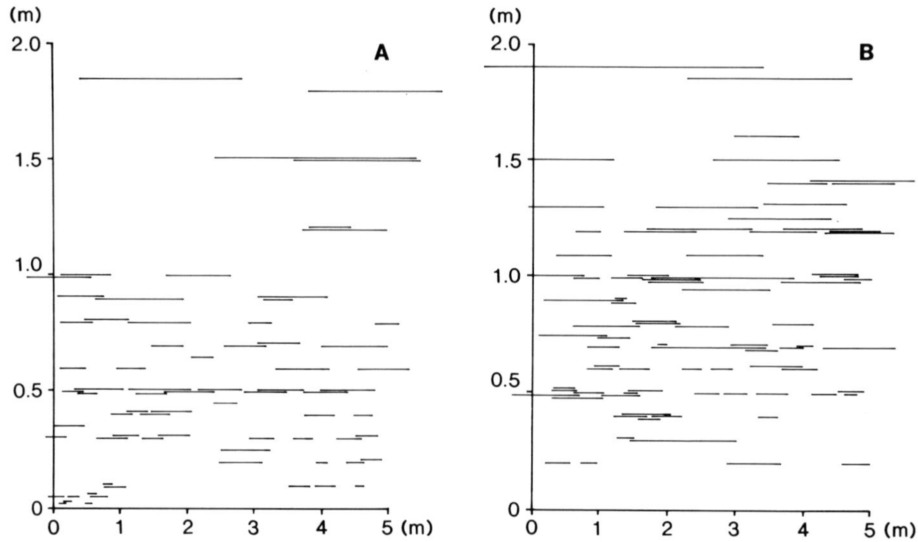

Figure 2.1: Height distribution (ordinate) of maximum crown widths (abscissa) of woody and semi-woody plants in a plot of 5 x 5 m at Ceru Jamanica (A) and at Malpais (B).

The two lines were not significantly different from each other.

Fig. 2.3 shows the LPIs of various species at both sites. Clearly three groups of species can be distinguished: a number of species maintain a LPI above 5 throughout the period of observation; these we call the evergreens. Another group shows an eventual sharp drop in LPI as the dry spell continues but they keep their LPI above 1; these we call the semi-deciduous species. A third group shows a drop in LPI immediately after the start of the dry spell and continue to decrease to eventually very low values; these are the truly deciduous species. Species differ in the recovering potential of their foliage. In many species drought causes leaves to yellow, and as the rains start some species re-green these leaves (e. g. *Croton*) while others drop them and produce new leaves (e. g. *Randia, Crescentia*).

The species studied phenologically at Malpais had a total ground surface cover of 29% which is just over half of the total cover of the stand. At Ceru Jamanica the species studied phenologically had a ground cover of 26% which amounted to three quarters of the total ground cover of the stand. In Table 2.5 it is shown that the common assessment of the vegetation at Malpais as deciduous and at Ceru Jamanica as evergreen is not based on a difference in real cover of the evergreen species at the sites. Only their contribution to the stand's total cover is higher at Ceru Jamanica as compared to Malpais. The ground cover of the semi-deciduous

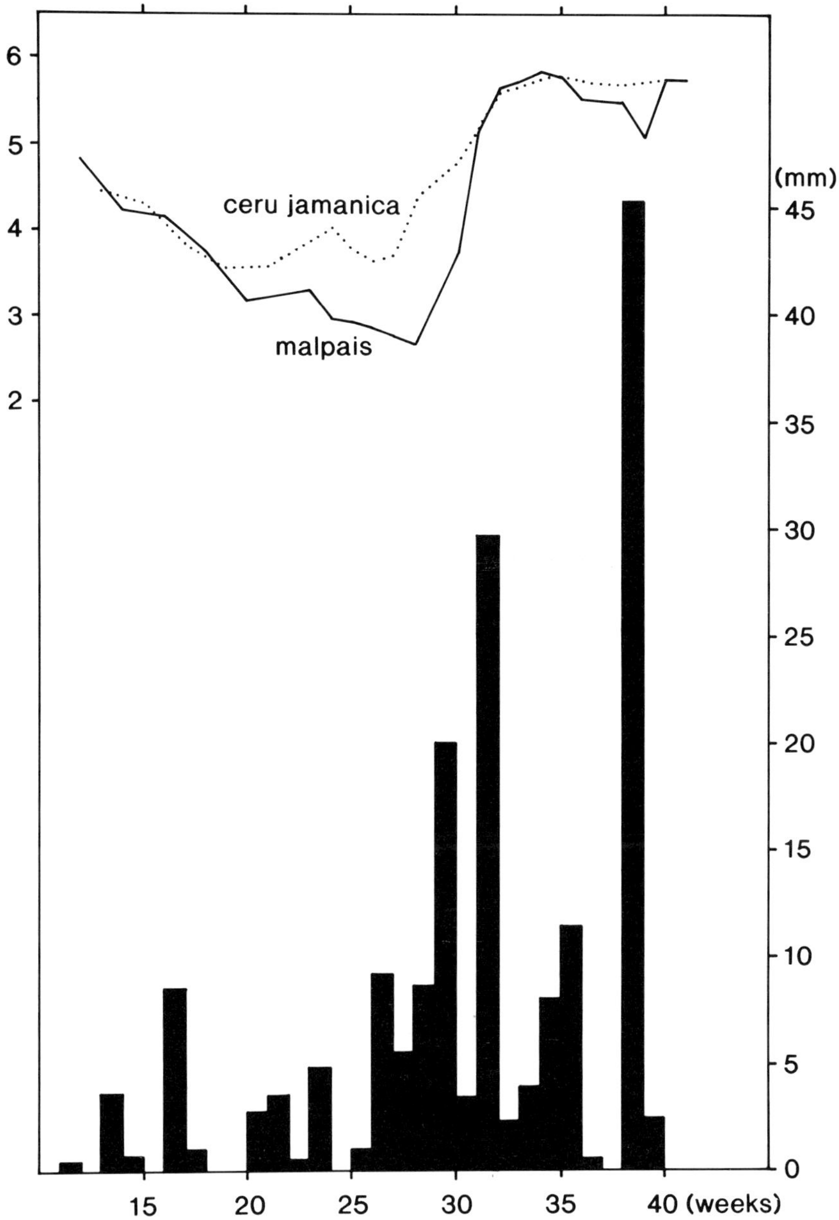

Figure 2.2: Weekly rainfall (scale right) and Leaf Phenological Index (scale left) at Malpais and Ceru Jamanica.

species is much higher at Malpais than at Ceru Jamanica, but proportionally these species are equally important at both sites.

As regards flowering species show two patterns: Some species flower permanently throughout the period of observation or nearly so, or species flower seasonally. Permanent flowerers at Malpais are *Acacia tortuosa, Balanites aegyptica, Casearea tremula, Cordia curassavica, Croton flavens* and *Crescentia cujete*, and at Ceru Jamanica, *Caesalpinia coriaria, Casearea tremula, Guaiacum sanctum* and *Haematoxylon brasiletto*. Thus, of the five species studied that are common two both sites (Table 2.3) three show a different flowering behavior between the sites: while *Cordia* and *Croton* flower permanently at Malpais, they flower seasonally at Ceru Jamanica; for *Caesalpinia* it is exactly opposite. Most species, but not *Randia aculeata*, show relatively long periods of flowering. Nearly all species show a clear increase in flowering shortly after the period of rainfall in the middle of the period of observation. This effect is somewhat stronger at Ceru Jamanica than at Malpais. Only *Casearea tremula* and *Tournefortia volubilis* show increased flowering following dry spells. Most species that became completely bare in the dry spell first grew some leaves when rains fell again and then started flowering, but there was no clear evidence of synchronization in flowering.

In many species fruits were present throughout the period of observation, although some species only temporarily bore fruits following their peaks in flowering, e. g. *Croton flavens, Randia aculeata, Phyllanthus botryanthus* and *Erithalis fruticosa*.

2.4 Discussion

Dry thorn scrub is a widespread formation in the area of the Antilles. It occurs in the lower altitudinal belt on most of the islands and in parts of the northern edge of South America. This formation shows considerable local variation in species composition, but also interesting, though complicated patterns of floristic similarity (BEARD 1949, 1953, 1955; ASPREY and ROBBINS 1953; STOFFERS 1956, HOWARD 1973).

Floristic composition on the diabase at Malpais and the limestone at Ceru Jamanica were quite different. Though the two substrates had about 50% of their species in common, only few were presented in about equal numbers at both sites. Vegetation cover on the diabase was nearly double that on the limestone.

Despite this strong difference in cover, amounts of above-ground biomass were rather similar. This amount is well within the normal range given for semi-desert scrub (WHITTAKER 1975). It is similar to the above-ground dry mass of Australian dry shrub- and heathlands (SPECHT 1981), that of dry fynbos shrub communities under a similar total rainfall in the Cape of South Africa (RUTHERFORD 1978) and the dry matorral of central Chile (KUMMEROW et al. 1981), even though rainfall seasonality and plant phenological patterns are totally different in those areas (e.g. BELL and STEPHENS 1984). The Curaçao biomass values are at the lower edge of

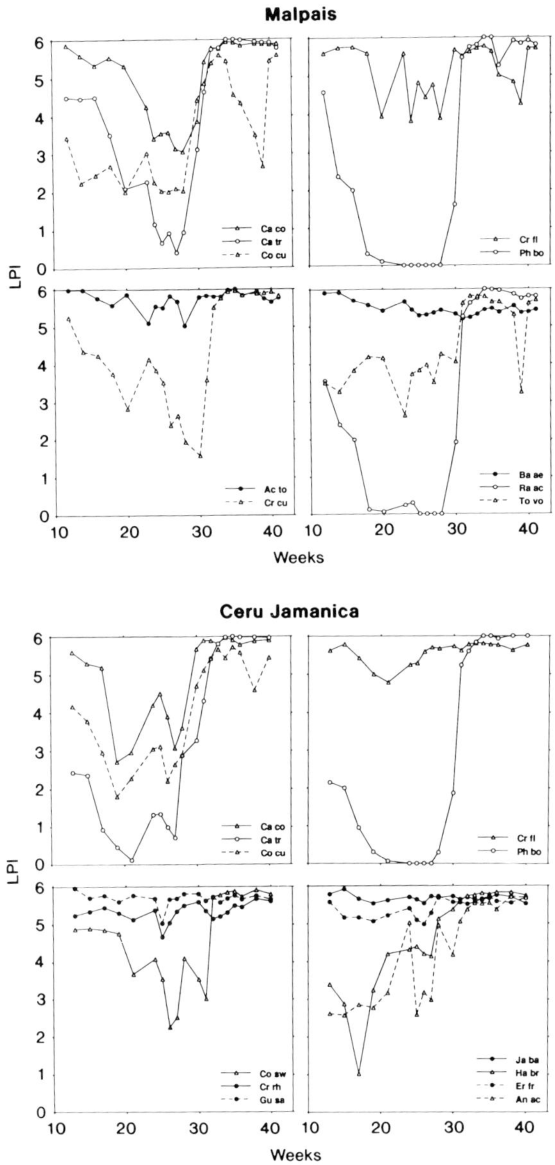

Figure 2.3: Weekly pattern of Leaf Phenological Indices (LPI) of ten species at Malpais and twelve species at Ceru Jamanica. Throughout the period of observation the same individuals have been sampled (see text). The symbols mean: • evergreen species, △ semi-deciduous species, ∘ deciduous species. Different line signatures distinguish different species. Species names are indicated by the first two letters of genus and species names. For full names see Table 2.3

Table 2.5: Ground cover (%) and proportion of total cover of the stand (in parentheses) of three phenological species groups on diabase (Malpais) and limestone (Ceru Jamanica)

species group	Malpais	Ceru Jamanica
evergreen	8.4 (15)	8.5 (25)
semi-deciduous	41.2 (73)	24.5 (71)
deciduous	6.5 (12)	1.4 (4)
total	56.1 (100)	34.4 (100)

the range for dry mopane scrub communities in the dry savanna area of Zimbabwe (RUTHERFORD 1978) and the dry *Burkea* savanna in South Africa though the proportion of leaves in the above-ground biomass is similar here (RUTHERFORD 1982) and considerably lower than in the mediterranean-type plant communities of the winter rainfall areas mentioned above (KUMMEROW et al. 1981).

The methods used to determine the total above-ground biomass were rather different, but produced closely similar results in the area where few small, mainly herbaceous undergrowth plants occurred (Ceru Jamanica). The second method, using the most important species only, is less laborious than total harvesting and seems to yield precise results in scrub vegetation with little undergrowth.

The spatial arrangement of the biomass is quite different between the sites. On the diabase at Malpais crown-sizes and heights are regularly distributed and biomass arrangement shows no evidence of layering. On the limestone at Ceru Jamanica, however, much of the biomass is concentrated in a low layer of small shrubs with small crowns while a few trees or shrubs with bigger crowns occur isolated or in small clumps, and clearly emerge from that layer.

Three phenological species groups could clearly be distinguished on the bases of leaf phenology (Table 2.3). At Malpais the deciduous species contribute most to the above-ground biomass, particularly *Randia aculeata*, with a substantial contribution by the evergreens, particularly *Acacia tortuosa*. At Ceru Jamanica, on the contrary, semi-deciduous species contribute overwhelmingly, again with a substantial additional contribution by the evergreens. The difference in phenological appearance between the vegetation types at the two sites is thus brought about by the contributions of the semi-deciduous and deciduous species (Table 2.3). Additionally, the evergreen species are more apparent at Ceru Jamanica because their relative cover values, not their real cover values, are higher, while the area is more bare than Malpais (Table 2.5). Thus, the assessment of the thorn scrub at Ceru Jamanica and at Malpais as evergreen, respectively deciduous by STOFFERS (1956) is probably made on the basis of the appearance of the evergreens in these vegetation types, while biologically the most important differences are the contributions of the semi-deciduous and deciduous species to the phytomass of the two types of stands.

The close correlation between LPI and the total amount of rainfall in the two weeks prior to the week of the LPI measurement indicates an opportunistic leaf flush pattern that rapidly responds to available moisture. The correlation is strong considered that not only total amount of rainfall influences the available amount of moisture, but also length of the dry intervals between rainy days.

Opportunistic leaf flush following rains is particularly common in desert dwarf shrubs (e. g. CUNNINGHAM and STRAIN 1969, HOFFMAN 1989), but often deciduous tropical trees and shrubs of semi-dry and dry areas also show a strong correlation of leaf flush with rainfall and leaf fall when soil moisture drops low again. However, in the tropics species with pre-rainfall leaf flushes are also common, particularly in areas with a strictly seasonal, well-defined rainfall pattern (e. g. VOLKENS 1912, DAUBENMIRE 1972, MALAISSE 1974, SARMIENTO and MONASTERIO 1983, CHILDES 1989). In perhumid subtropical areas even temperature-controlled leaf fall and flush seem to occur (ESKUCHE 1989).

In our study, there was no clear evidence of synchronization in flowering and in most species flowering occurred over a rather long period of time. This fits AUGSPURGER's (1983) findings who reported from Panama a decrease in flowering synchronization as flowering duration became longer and it fits the reported relaxation of seasonality in tropical thorn scrub in South Africa as aridity decreases (HOFFMAN 1989). Semi-dry and dry tropical areas with synchronous flowering often seem to show their peak flowering in the dry seasons (e.g. DAUBENMIRE 1972) but peaks at the onset of the rainy season also have been reported (FRANKIE et al. 1974, OPLER et al. 1980).

Our results suggest that plants at Malpais react to strongly and rapidly varying water resources in the substrate, while at Ceru Jamanica they react to a continuously limited water supply that is more equally spread in time. At Malpais the plants tend to deciduousness. They opportunistically grow an abundant crop of leaves with relatively low SLWs when, after a dry spell, the diabase-derived soil is recharged with water by rain. Soil and plants rapidly loose water again during dry periods. The LPI drops rapidly and severely, most leaves roll within four days after the start of the dry spell, and soon after they wither and are dropped. Many plants remain bare till rains fall again. Soil depth seems to be rather equal throughout the area, since the vegetation is homogeneously structured.

At Ceru Jamanica the porous limestone buffers input and loss of water. It seems to cause an almost permanent water stress to the plants but, at the same time, releases some water during long periods of time. The stress is indicated by leaf crops that are smaller than at Malpais and SLWs that are higher. When dry spells occur plants do not roll their leaves within 2 weeks and plants tend to be more permanently foliated than at Malpais, indicating the longer time span of water uptake from the limestone. Also the smaller fluctuations in LPI at Ceru Jamanica indicate this. Competition for water seems indicated by the plants alongside the plots harvested for biomass measurements: most of them retained their leaves at a time when individuals further away from the plots' edges already had shed theirs. REICH and BORCHERT (1984) earlier demonstrated such a link between tree water status and leaf phenology in the dry parts of Costa Rica. Within the limestone substrate water availability seems spatially heterogeneous: in places plants seem to succeed in tapping richer water resources and grow out while elsewhere continuous limited water supply leads to dwarf growth of the woody vegetation.

Bibliography

ARNOLDO, M., 1964, Zakflora.
Natuurwetenschappelijke Werkgroep Nederlandse Antillen **16**, Curaçao.

ARNOLDO, M., 1971, Gekweekte en nuttige planten van de Nederlandse Antillen.
Natuurwetenschappelijke Werkgroep Nederlandse Antillen **20**, Curaçao.

ASPREY, G. F.; ROBBINS, R. G., 1953, The vegetation of Jamaica.
Ecological Monographs **23**, 359–412.

AUGSPURGER, C. K., 1983, Phenology, flowering synchrony and fruit set of six Neotropical shrubs.
Biotropica **15**, 257–267.

BEARD, J. S., 1949, The natural vegetation of the Windward and Leeward Islands.
Oxford Forestry Memoirs **21**.

BEARD, J. S., 1953, The savanna vegetation of northern tropical America.
Ecological Monographs **23**, 149–214.

BEARD, J. S., 1955, Classification of tropical American vegetation types.
Ecology **36**, 89–100.

BEETS, D. J., 1972, Lithology and stratigraphy of the Cretaceous and Danian succession of Curaçao.
Natuurwetenschappelijke Studiekring voor Suriname en de Nederlandse Antillen **70**, Utrecht.

BELL, D. T.; STEPHENS, L. J., 1984, Seasonality and phenology of Kwongan species.
In: Pate, J. S.; Beard, J. S. (eds.), Kwongan, Plant life of the Sandplain, University Western Australia Press, Nedlands, 205–226.

CHILDES, S. L., 1989, Phenology of nine common woody species in semi-arid, deciduous Kalahari Sand vegetation.
Vegetatio **79**, 151–163.

CUNNINGHAM, G. L.; STRAIN, B. R., 1969, An ecological significance of seasonal leaf variability in a desert shrub.
Ecology **50**, 400–408.

DAUBENMIRE, R., 1972, Phenology and other characteristics of tropical semi–deciduous forest in North–western Costa Rica.
Journal of Ecology **60**, 147–170.

ESKUCHE, U., 1989, Fenologia de follaje de los bosques en el nordeste Argentino.
Folia Botanica et Geobotanica Correntesiana **1**, 2–22.

FRANKIE, G. W.; BAKER, H. G.; OPLER, P. A., 1974, Comparative phenological studies of trees in tropical wet and dry forests in the lowlands of Costa Rica.
Journal of Ecology **62**, 881–920.

GAUCH, H. G., 1982, Multivariate analysis in community ecology.
Cambridge University Press, Cambridge.

HILL, M. O., 1979, TWINSPAN. Ecology and systematics.
Cornell University, Ithaca New York.

HOFFMAN, M. T., 1989, Vegetation studies and the impact of grazing in the semi–arid Eastern Cape.
PhD thesis, University of Cape Town, unpubl.

HOWARD, R. A., 1973, The vegetation of the Antilles.
In: Graham, A. (ed.), Vegetation and vegetational history of northern Latin America, Elsevier, Amsterdam, 1–38.

KUMMEROW, J.; MONTENEGRO, G.; KRAUSE, D., 1981, Biomass, phenology, and growth.
In: Miller, P. C. (ed.), Resource use by chaparral and matorral, Springer, New York, 69–96.

MALAISSE, F. P., 1974, Phenology of the Zambezian woodland area.
In: Lieth, H. (ed.), Phenology and seasonality modelling, Springer, New York, 269–286.

MUELLER–DOMBOIS, D.; ELLENBERG, H., 1974, Aims and methods of vegetation ecology.
Wiley and Sons, New York.

OPLER, P. A.; FRANKIE, G. W.; BAKER, H. G., 1980, Comparative phenological studies of treelet and shrub species in tropical wet and dry forests in the lowlands of Costa Rica.
Journal of Ecology **68**, 167–188.

REICH, P. B.; BORCHERT, R., 1984, Water stress and tree phenology in a tropical dry forest in the lowlands of Costa Rica.
Journal of Ecology **72**, 61–74.

RUTHERFORD, M. C., 1978, Primary production ecology in southern Africa.
In: Werger, M. J. A. (ed.), Biogeography and ecology of southern Africa, Junk, The Hague, 621–659.

RUTHERFORD, M. C., 1982, Woody plant biomass distribution in *Burkea africana* savannas.
In: Huntley, B. J.; Walker, B. H. (eds.), Ecology of tropical savannas, Springer, New York, 120–141.

SARMIENTO, G.; MONASTERIO, M., 1983, Life forms and phenology.
In: Bourliére, F. (ed.), Tropical savannas, Elsevier, Amsterdam, 79–108.

SPECHT, R. L., 1981, Responses to fires in heathlands and related shrublands.
In: Gill, A. M. et al. (eds), Fire and the Australia biota, Australian Academy of Science, Canberra, 395–415.

STOFFERS, A. L., 1956, The vegetation of the Netherlands Antilles.
Mededelingen Botanisch museum en Herbarium, Universiteit Utrecht **135**, and Natuurwetenschappelijke Studiekring voor Suriname en de Nederlandse Antillen **15**, Utrecht.

STOFFERS, A. L.; MANSOUR–ELASSAISS, C. J. A., 1967, On the water relation in limestone and diabase vegetation in the Leeward islands of the Netherlands Antilles.
Acta Botanica Neerlandica **15**, 539–556.

VALENTINE, H. T.; TRITTON, L. M.; FURNIVAL, G. M., 1984, Subsampling trees for biomass, volume or mineral content.
Forest Science **30**, 673–681.

VAN DER MAAREL, E., 1979, Transformation of cover–abundance values in phytosociology and its effects on community similarity.
Vegetatio **39**, 97–114.

VOLKENS, G., 1912, Laubfall und Lauberneuerung in den Tropen.
Borntraeger, Berlin.

WHITTAKER, R. H., 1975, Communities and ecosystems.
2nd edn. MacMillan, New York.

Chapter 3

Structure and phytomass production of a pioneer community

M. JOCHIMSEN and D. JANZEN

3.1 Introduction

Realizing that the condition of nature has deteriorated, plant ecologists have to face a further problem: landscapes and soils created by man, which have to be restored. The establishment of a new plant cover is required by law, and hitherto reclamation has been the business of foresters and landscape gardeners only. But in a way they have neglected nature's own principles. We are now trying to solve the problems with the support of vegetation science; that means that we should mimic natural succession when dealing with recultivating raw soils. The most promising way to bring plant communities in balance with the environment is the development of a natural vegetation. With respect to contaminated but technically cleaned soils and mine spoil, it is primary succession which happens, implying very harsh site conditions. Therefore, technicians make every effort to reforest these soils, but so far with little success. Regarding vegetation, naturally developed trees are the latest to come. Nature starts with pioneers, which prepare the site for the following settlers. On that account it seems reasonable to do reclamation by mimicking and accelerating natural succession. Of course, this intention requires much more information about the early stages of plant settlement than yet available. On that score those spoil heaps produced by coal mining offer an ideal opportunity to study the mechanisms ruling the colonization of an undeveloped bare ground.

Research on vegetation development is usually done by phytosociologists, who estimate cover and abundance of species. Our respective studies on mine spoil have revealed that there are further characteristics, which have to be considered, and which do not emerge from floristic relevés (BORNKAMM and HENNING 1982). No doubt, plant cover degree refers to site conditions, but above ground biomass may differ despite the same value for coverage, because it is not only a result of the horizontal development but also of the vertical structure of vegetation. "Productivity values from primary successional stages are rarely measured". "The productive capacity of successional stages", however, "is of primary importance for the early establishment of climax community" (LIETH 1974).

3.2 Research area and methods

In cooperation with the Kommunalverband Ruhrgebiet and Ruhrkohle AG mine spoil from different collieries (MONOPOL and ACHENBACH) was heaped up at Waltrop (Ruhr Area) for research on reclamation. This waste is substantially formed by larger pieces of clay shale and sandstone, which disintegrate within a year, leaving coarse soil. Water capacity and nutrient supply, however, leave much to be desired. At the beginning of the research period substrate characteristics were studied by geologists (SCHNEIDER 1989) and pedologists (BURGHARDT 1989), and the values are reported by them.

Preceding investigations proved (JOCHIMSEN 1986, JOCHIMSEN 1987) that in order to mimic vegetation development on waste produced by coal mining it is necessary to supply it with seeds, because qualified propagules for a spontaneous colonization are no longer available due to the present state of our environment. Moreover, they revealed that species belonging to the phytosociological alliance Dauco–Melilotion are highly adapted to colonize this ruderal site. The mixture comprised 49 species composed quantitatively according to their 1000 grain weight and to their frequency in floristic relevés of similar sites. Sowing was done in autumn 1986 (10 $g{\cdot}m^{-2}$). In order to study the vegetation development dependent on exposition and substrate, the research area (about 1 ha) was arranged correspondingly. South (SS) and north facing slopes (NS) had an inclination between 20° and 40° and compartments of 200 m^2, whereas the plateau (PL) was divided into areas 116.6 m^2 in size, each treatment repeated 3 times. Substrate variation resulted both from admixtures to mine spoil and from fertilization (D = 25 $g{\cdot}m^{-2}$ NPKMg). In case of the SD–treatment Agrosil, a phosphate containing gel, with storage capacity was given additionally (120 $g{\cdot}m^{-2}$). Sand (S) was used in order to improve the water capacity (within the scope of this stony material) and to prevent the seeds from being blown away, while sandy loam covered the waste as a 15 cm thick top layer (E) or was mixed with it in proportion 1:3 (U). The abbreviation "Z" means that the area did not get any artificial seed.

S – mine spoil, sand added
SD – mine spoil, sand, Agrosil and fertilizer added
ED – mine spoil with top layer of sandy loam and fertilized
UD – mine spoil mixed with sandy loam and fertilized
U – mine spoil mixed with sandy loam not fertilized
UZ – mine spoil mixed with sandy loam; no seed!

Besides phytosociological analyses according to the method of Braun–Blanquet, vertical structure and above ground phytomass production were studied too. (As digging disturbs future vegetation development, below ground phytomass was not determined.) During the first growing season it became already obvious that the differences between the various treatments did not only consist in plant cover and species abundance but most conspicuously in physiognomic qualities. Therefore stratification, the spatial arrangement of plant individuals, was recorded by precise hand drawings, and above ground phytomass was harvested from small plots (see Table 3.2 MJ) representing the full range of shape within 1 m^2 typical for the respective vegetation. Sometimes the growth form of particular plants, especially *Melilotus alba*, compelled to change the plot size. For the purpose of comparison total production was calculated for 1 m^2. In order not to affect the plant cover too much we refrained from taking replicates. (In consequence no reliable statistical analysis could be realized.) At the time of harvest (end of July) each individual was promptly counted, and each species treated separately. That is why the former drying procedure, 24 hours at 105 °C, had to be maintained. All samples originated from MONOPOL waste only.

3.3 Phytomass

Productivity is a most valuable standard for assessing efficiency, and therefore it is of great interest to know how phytomass production develops in the course of succession. Weather conditions which may have an influence on productivity were similar; the early summer of 1988 and 1989 being marked by a dry period.

Figure 3.1 MJ displays the results obtained for each treatment and exposition during 1987 to 1989, and the differences in absolute yield are obvious. But because of the special characteristics of the population, the increase was not linear. Already during the first investigation period the developing vegetation produced a remarkable yield. The highest values were recorded for those plots which had been fertilized (SD, ED, UD), but differences emerged from exposition. In early spring, vegetation on the south facing slope turned higher insolation to profit, thus getting a margin in growth (SD, ED). At the moment it does not stand to reason why that did not happen to the respective UD–plot. The second factor influencing phytomass production was provided by water supply, which is likewise correlated to exposition but disguised in the case of the southern slope by the process mentioned before. As lower temperatures not only mean a later start of growing but also less evapotranspiration, the northern slope produced a higher yield than the plateau. In a way this result was also true for the S– and U–plots, which did not get any fertilizer. Another striking relation was met with all U–treatments, the substrate of which consists of mine spoil mixed with sandy loam. The data for the plateau and northern slope plots (UD, U, UZ) did not differ much, though the values were lower, of course, without fertilization. That applied to the UZ–plots particularly, which did not get any seed but by wind, and therefore the southern slope produced so little phytomass. Moreover, for restoration purposes we found that the values for the S– and U–treatment did not beat each other.

With respect to succession, and as the rate of productivity depends on the state to which vegetation has already come up, it seems reasonable to discuss the further development of phytomass on the basis of the first yield, that means proportionally. As shown by Figure 3.1 MJ in 1988, the standing crop increased obviously, in some instances spectacularly. The most striking values were obtained from the U–plots of southern and northern exposure, which produced a rate of phytomass more than twelvefold the first yield. It is to be supposed that because of the poorly developed plant cover in 1987, in the following growing season there was much more space for particular species to develop compared to the other areas. The values for the fertilized plots seem to be connected in a similar way. If the results of the first and second vegetation period are ranked, a nearly inverse sequence is obtained. That means that in the second season the yield was controlled among others by the extent of the vegetation development already achieved during the first period, though the enormous increase varying between +881.0% (PL/SD) and +166.2% (SH/SD) mainly resulted from the dominant *Melilotus alba*, a biennial plant concluding its life cycle. Sometimes sampling was affected by the growth form of this species, which called for changing the plot size (see Table 3.2 MJ).

Considering this fact, in 1989 the reduction of phytomass was not surprising, since with three exceptions (NS/S, NS/UD, SS/SD) it did not go down to the original amount. Hence looking at the results from the starting point, the phytomass production still increased. The highest rates were observed with the UZ–plots (200–300%) followed by those of the U–treatment (100–200%), which was probably due to some space left for colonization. The values obtained for the fertilized plots exhibited more or less a similar trend with an increase in ratio of about 40%. As to the rank order according to the harvested phytomass it was surprising that there were only small variations compared with the preceding year, especially with regard to the fertilized areas. At the top the first and the second (PL/SD × SS/UD) changed their positions, whereas at the end NS/UD and SS/SD stayed in their places. In between shifting only meant two ranks in the course of which formerly lower positions were improved and vice versa.

In this context, the results for phytomass have to be compared with those for

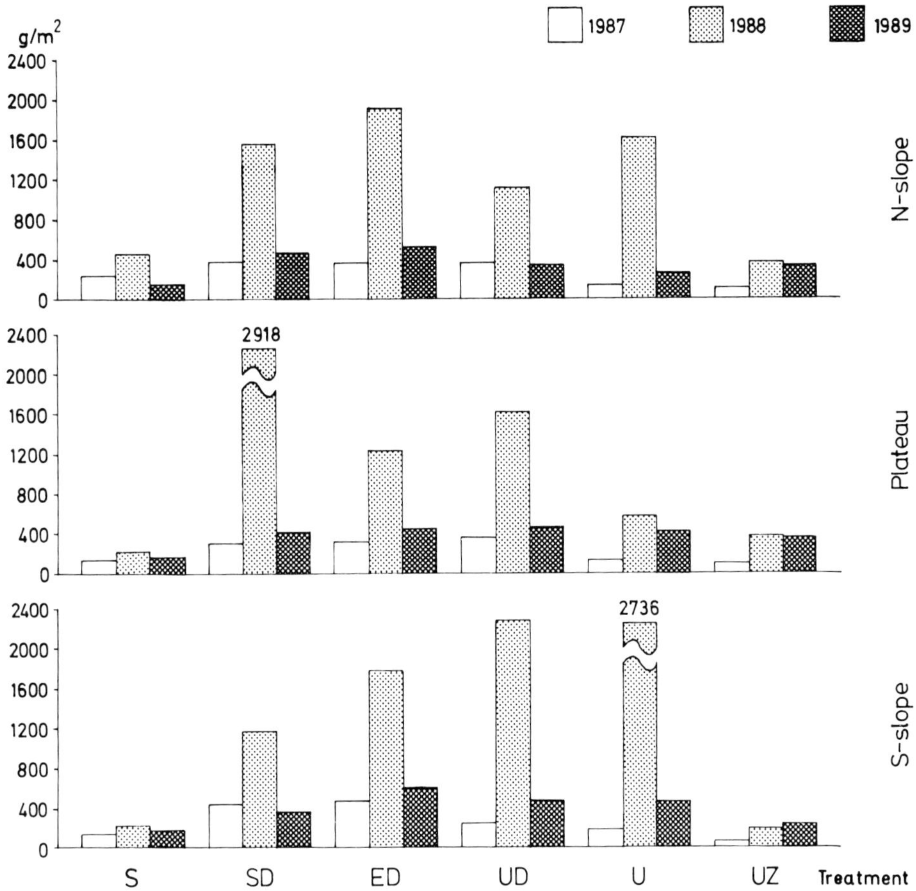

Figure 3.1: MJ Total yield of above ground phytomass dependent on treatment, exposition and year (dry matter calculated g·m^{-2}).

coverage. Table 3.1 MJ represents the mean values estimated for the total area of the particular treatment, while those data which are the base of Figure 3.1 MJ only refer to the phytomass harvested from the respective plots out of that area.

Though both phytomass and coverage were controlled by fertilization, there was no direct correspondence between these two parameters when comparing the rank order of their results. As already mentioned, in 1987 the highest yield occurred with the ED–plot from the southern slope, but with respect to coverage it only held the 6th position. We observed similar conditions in the second vegetation period,

Table 3.1: MJ Mean values for plant cover degree (%) dependent on treatment, exposition and year.

treatment		S	SD	ED	UD	U	UZ
expo-sition	year						
	1987	40	90	70	75	45	30
NS	1988	45	95	85	95	85	40
	1989	35	90	85	90	90	50
	1987	50	80	65	80	55	45
PL	1988	45	90	80	80	75	45
	1989	50	75	80	90	80	35
	1987	50	75	75	75	55	35
SS	1988	40	75	80	95	80	30
	1989	40	60	65	90	75	45

when a coverage of 95% was connected with the 2nd, 6th and 9th rank for biomass, and the same was found to be true in 1989. To answer this problem it is helpful to look at Figures 3.2 MJ – 3.4 MJ, which show the typical vertical structure of the UD–vegetation on the northern slope by a transect. The respective results for coverage and yield are

1987	75%	354.25	$g{\cdot}m^{-2}$
1988	95%	1104.75	$g{\cdot}m^{-2}$
1989	90%	347.34	$g{\cdot}m^{-2}$

and from the drawings, though they do not represent permanent plots, it becomes obvious that species composition and structure of plant cover differed each year, thus creating a changing degree of coverage and variable yields.

Species list (valid for all figures and tables).

A *Artemisia vulgaris* L.
Aa *Artemisia absinthium* L.
Ae *Arrhenatherum elatius* (L.) J. & C. PRESL
Am *Achillea millefolium* L.
B *Bromus erectus* HUDS.
C *Chenopodium album* L.
Ci *Cichorium intybus* L.
Cs *Centaurea scabiosa* L.
D *Daucus carota* L.
E *Echium vulgare* L.
E/A *Echium vulgare* or *Anchusa officinalis* L.[1]

H *Holcus lanatus* L.
Hp *Hypericum perforatum* L.
I *Isatis tinctoria* L.
L *Linaria vulgaris* MILL.
Lc *Lotus corniculatus* L.
M *Melilotus alba* MED. or
Melilotus officinalis (L.) PALL.[1]
Ma *Melilotus alba* MED.
Mo *Melilotus officinalis* (L.) PALL.
O *Oenothera biennis* L.
Pa *Polygonum aviculare* L.
Pc *Poa compressa* L.
Pl *Plantago lanceolata* L.
Pm *Plantago major* L.
Ps *Pastinaca sativa* L.
R *Reseda lutea* L.
Ro *Reseda luteola* L.
S *Sisymbrium officinale* (L.) SCOP.
Se *Senecio vulgaris* L.
Si *Silene vulgaris* (MOENCH) GARCKE
T *Tanacetum vulgare* L.
(= *Chrysanthemum vulgare* (L.) BERNH.)
V *Verbascum densiflorum* BERT.
Vo *Verbena officinalis* L.
(v) wilted from last year
[1] seedlings, not distinguishable
(nomenclature according to OBERDORFER (1983))

3.4 Number of individuals

Biomass values do not allow any visualization of structure and composition of a plant cover. From an ecological point of view, however, it is important how many species and individuals contribute to phytomass. Therefore harvesting the respective plot, each individual, independent of its growth, was counted; Figure 3.5 MJ gives an impression of the amounts and alterations which took place during a 3 year vegetation development. As the growth form of *Poa compressa* propagating vegetatively by rhizomes does not allow a shoot to be assigned to a particular individual originating from the seeding, we excluded the grasses (*Holcus lanatus* only present in the plot SS/UZ in 1988) of the individual numbers presented in Figure 3.5 MJ. Their rank order with respect to above ground phytomass production is quoted in Table 3.3 MJ.

The seed mixture used in our trials included ca. 29,000 seeds$\cdot$m^{-2}, by weight

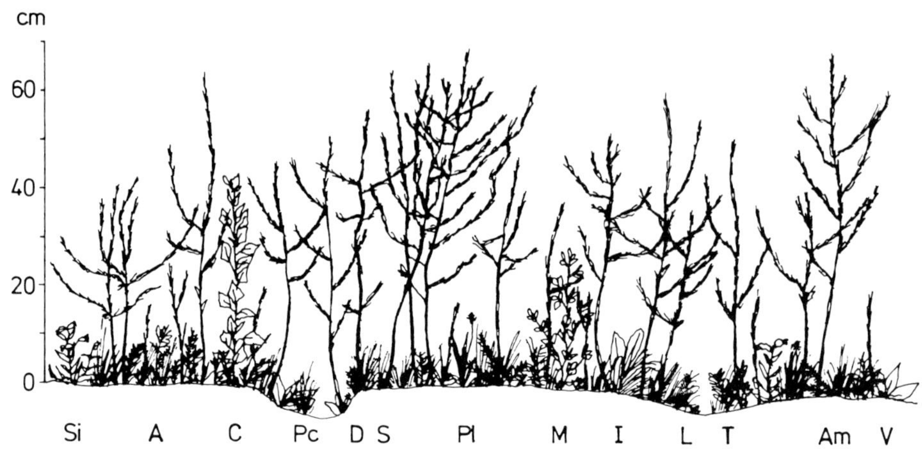

Figure 3.2: MJ Structure of vegetation developed on mine spoil mixed with sandy loam and fertilized on the northern slope (UD/NS) in 1987 (abbreviations of names see species list).

10 g·m^{-2}. We did not expect each seed to germinate and grow, because irrespective of site conditions there is always a loss largely due to wind and animals and contingent on the germination rate. That is why landscape gardeners advise you to sow 25 g·m^{-2}. But it is interesting to compare the number of possible seedlings with that of plant individuals counted on the respective plot at the time of harvest (Fig. 3.5 MJ). The great differences between these records as well as the little space left between the several plants, which make it difficult for new settlers to become integrated within this developing vegetation, made us decide to diminish the amount of seeds·m^{-2} in a fresh trial started in autumn 1989.

With respect to the present study a large number of individuals means that this site, called hostile to plants, does not hinder germination, and hence it follows that a potentiality for vegetation exists. As Figure 3.5 MJ demonstrates, exposition played an important role with respect to the number of individuals, and so did the time after the initiation of vegetation development by sowing. At the south facing slope with the lowest overall values the situation was very simple. The first growing season produced the highest numbers of individuals, but there were differences due to the kind of treatment. The maximum occurred with the UZ–plot, even though this area did not get any artificial seed. Most of the seedlings either belonged to *Betula pendula* — a tree standing close to the area — or were too small to be determined. That was in part also true for the U– and ED–plot, making the speculation more certain that these individuals emerged from the seedbank

Figure 3.3: MJ Structure of vegetation developed on mine spoil mixed with sandy loam and fertilized on the northern slope (UD/NS) in 1988 (Abbreviations of names see species list).

provided by the sandy loam, which had been mixed with mine spoil. Besides that, fertilization controlled the number of seedlings, too, as becomes apparent by comparing the respective data with those from the S–plot. During the second vegetation period the amount of individuals decreased more or less largely, but by combining Figure 3.1 MJ and 3.5 MJ it becomes obvious that this process did not mean a loss of vegetation nor phytomass, though, of course, lots of individuals had perished for the benefit of stouter plants producing a better yield. The third growing season was characterized by a rise in the number of individuals, save the ED– and UZ–plots, reflecting an improvement of site conditions and population development.

In this study large numbers of individuals are not only caused by a lot of small "indifferent" seedlings but also by numerous individuals of a certain species, particularly *Achillea millefolium* and/or a high number of plant species themselves (Tab. 3.2 MJ). This case is represented by the SD– and UD–plots from the north

Figure 3.4: MJ Structure of vegetation developed on mine spoil mixed with sandy loam and fertilized on the northern slope (UD/NS) in 1989 (Abbreviations of names see species list).

facing slope where the highest counts were also received during the first vegetation period. According to the findings on the southern slope the smallest amount of individuals was achieved by the S–treatment, and the results concerning the UZ–plots are comparable, too. Further conformity with respect to the south facing slope is due to the results referring to the second vegetation period, except those of the ED– and U–plots, the numbers of which did not change. The third growing season brought an increase only to the non–fertilized plots, while the values for the SD– and ED–substrate dropped beneath those of the respective south facing areas.

In general the development of individual numbers on the plateau resembled that of the north facing slope. But it was not only the first season that produced the highest results. The SD–plot, for instance, had its maximum in the second year, while that of the S–plot did not happen until the third vegetation period brought

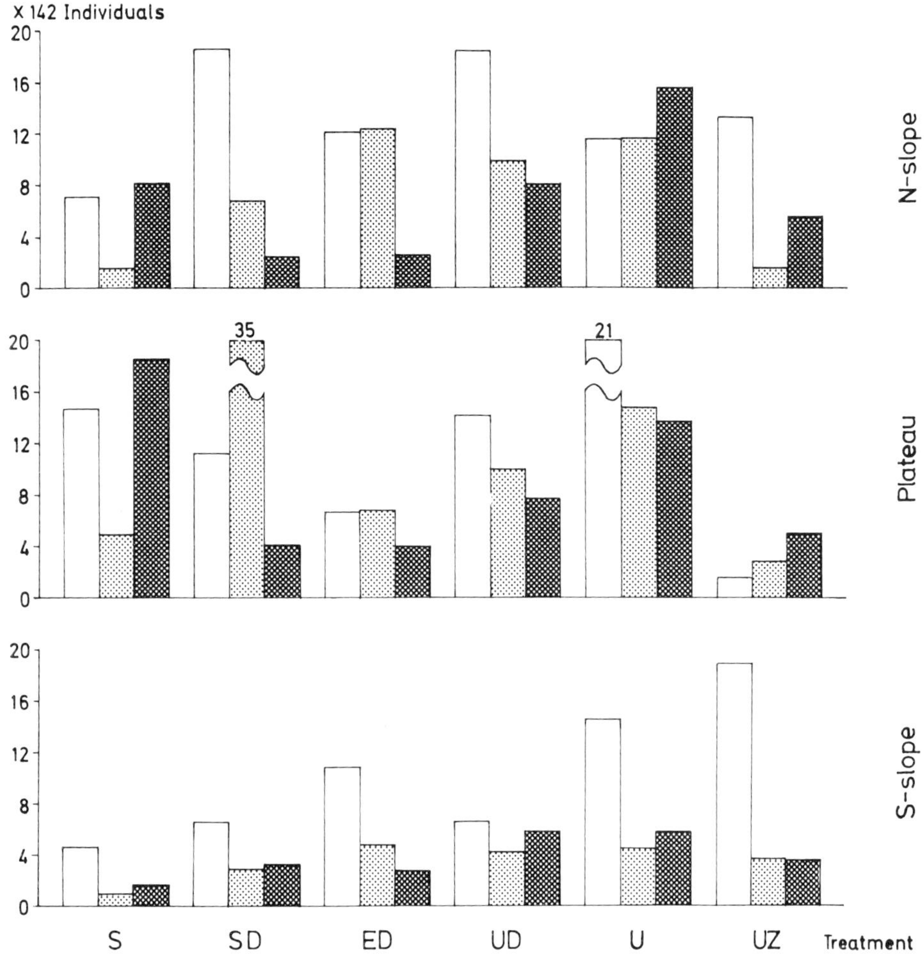

Figure 3.5: MJ Total numbers of individuals calculated per 1 m^2 (disregarding the shoots of *Poa compressa*).

about by an enormous number of *Melilotus alba* and *M. officinalis* seedlings.

3.5 Species numbers

Presence or absence of species does not allow any calculation for a certain plot size (m^2). That is why Table 3.2 MJ refers to both, the harvested plot and the

Table 3.2: MJ Species numbers resulting from the analysis of harvested plots and floristic composition of the respective area.

treatment		S		SD		ED	
expo-sition	year	plot	area	plot	area	plot	area
	1987	17 (1)	24	31 (1)	42	27 (1)	32
NS	1988	19 (2)	33	22 (1)	35	23 (1)	35
	1989	16 (3)	31	21 (2)	32	19 (2)	35
	1987	27 (1)	33	24 (1)	36	21 (1)	30
PL	1988	26 (4)	29	25 (1)	31	20 (1)	33
	1989	20 (3)	27	22 (3)	32	24 (3)	30
	1987	13 (1)	23	21 (1)	38	20 (1)	40
SS	1988	6 (1)	34	16 (1)	34	15 (1)	30
	1989	19 (3)	42	16 (3)	36	17 (3)	32
treatment		UD		U		UZ	
expo-sition	year	plot	area	plot	area	plot	area
	1987	32 (1)	41	26 (1)	35	24 (4)	33
NS	1988	21 (1)	35	19 (1)	30	17 (2)	46
	1989	18 (2)	34	27 (2)	34	29 (3)	35
	1987	21 (1)	30	26 (1)	34	27 (4)	34
PL	1988	21 (1)	35	24 (1)	27	19 (4)	39
	1989	20 (2)	35	25 (2)	32	21 (3)	32
	1987	26 (1)	35	26 (1)	39	26 (4)	37
SS	1988	19 (1)	31	18 (1)	28	21 (3)	45
	1989	19 (2)	42	24 (3)	42	26 (3)	52

The numbers in brackets mean plot category: (1) 3×0.04 m^2; (2) 2×0.25 m^2; (3) 3×0.25 m^2; (4) 1×1.00 m^2.

whole area. The seed mixture sown in the beginning of the trial comprised 49 species, and besides them an additional input from the air and from the seedbank, concerning the E– and U–treatments, had to be expected. Despite the small areas that were harvested, the number of species included in the yield was respectable, thus probably proving a well–organized plant community.

3.6 The five most productive species

To some extent the structure of a plant community is formed by its dominant species. The mosaic which they exhibit is not random but caused by competition, at least in our case study. Therefore the question arose as to whether any alteration of phytomass production was connected with the behavior of single species dependent on site and time. In order to throw light upon this, the yield of each species was treated separately, so that a rank order could be established. Though the list comprises 49, only the 5 most productive species will be discussed here.

With respect to phytomass it is of interest to what extent these 5 species contributed to the total yield. Figure 3.6 MJ represents the percentages dependent on site conditions and year of harvest, and Figures 3.7 MJa–c give an example of their changing portion and the variability of the species participating in it. Increasing amounts mean an intensification of dominance and a decrease in species diversity, and vice versa, of the respective vegetation. Regarding the graph, different patterns of behaviour can be recognized.

In general the 5 most productive species augmented their phytomass in conformity with the results found for the total yield (Fig. 3.1 MJ). In other words, in the second vegetation period the values rose mostly due to the biennial plant *Melilotus alba*, and then decreased – apart from 2 exceptions – without going back to the basic yield. In contrast to the findings represented in Figure 3.1 MJ, a continually increasing yield was only observed on the northern slope (ED), and was most obvious from the southward exposed UZ–area. However, we received diminishing values from the S–treatment only, excepting the southern exposure, and in a very moderate way from the southern SD–area. An intermediate, and compared with the results for the total yield inverse position – the decrease followed by a rise – was held by two plots (U and UZ) from the plateau.

Further information is provided by Table 3.3 MJ. Here the 5 most productive species are listed according to their rank with respect to phytomass production, and moreover, the number of individuals incorporated is mentioned. This inventory provides a better perspective concerning matters in process with this developing vegetation. During the first growing season, irrespective of the exposition, the treatments presented their particular dominant or twice subdominant species. The mine spoil areas (S, SD) were governed by *Reseda lutea*, whereas *Sisymbrium officinale* dominated those treated with sandy loam (ED, UD, U). In this context the UZ–treatment has to be left out of discussion, because it was not seeded artificially like the others. While *Reseda lutea* stayed in the group of the 5 most productive species for three or two years respectively, fertilization shortening its time of presence, *Sisymbrium officinale*, an annual, however, had gone immediately after the first vegetation period. Comparing the species replacement within this group of most productive species from one year to the other revealed 3 categories of behaviour. The first refers to species staying all the years, called lasting species, the second to those only present in 1987 and 1988 (true pioneers),and the last to

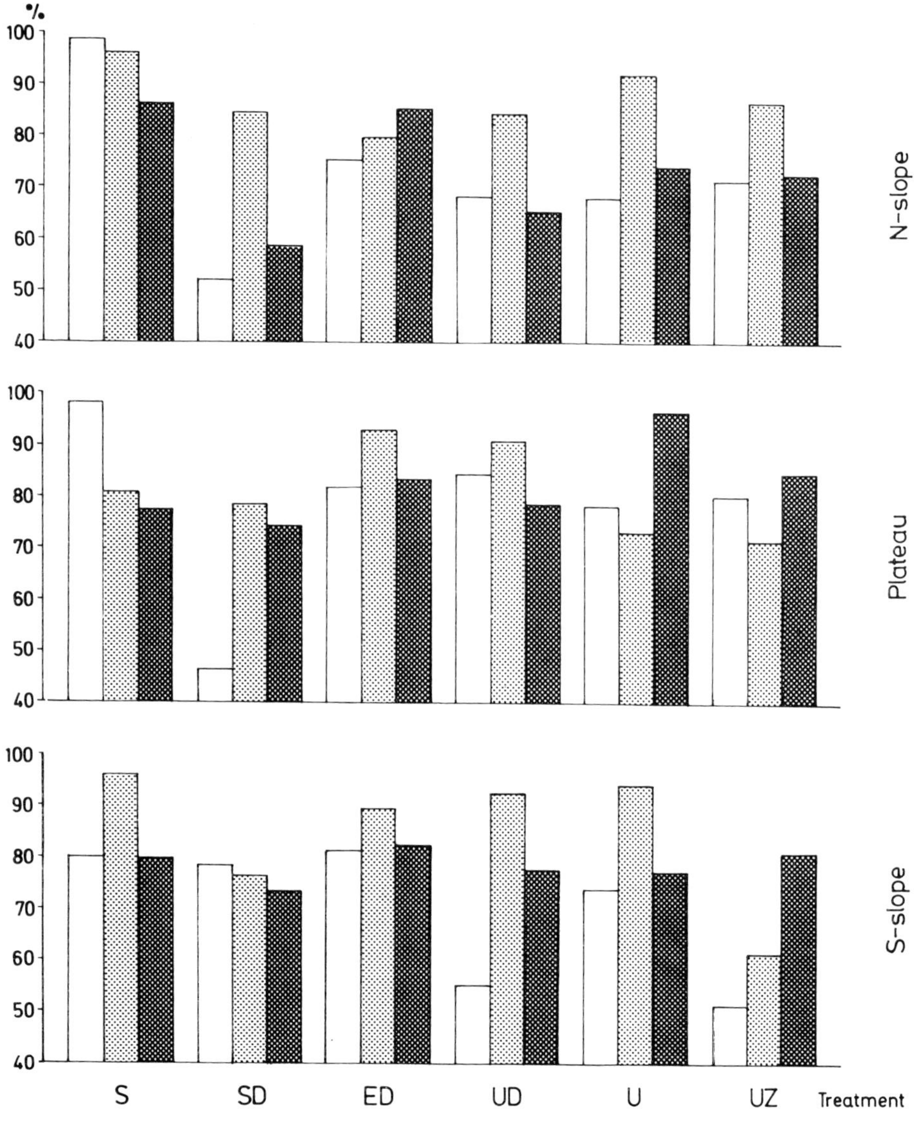

Figure 3.6: MJ Percentages of total phytomass of the 5 most productive species dependent on treatment, exposition and year.

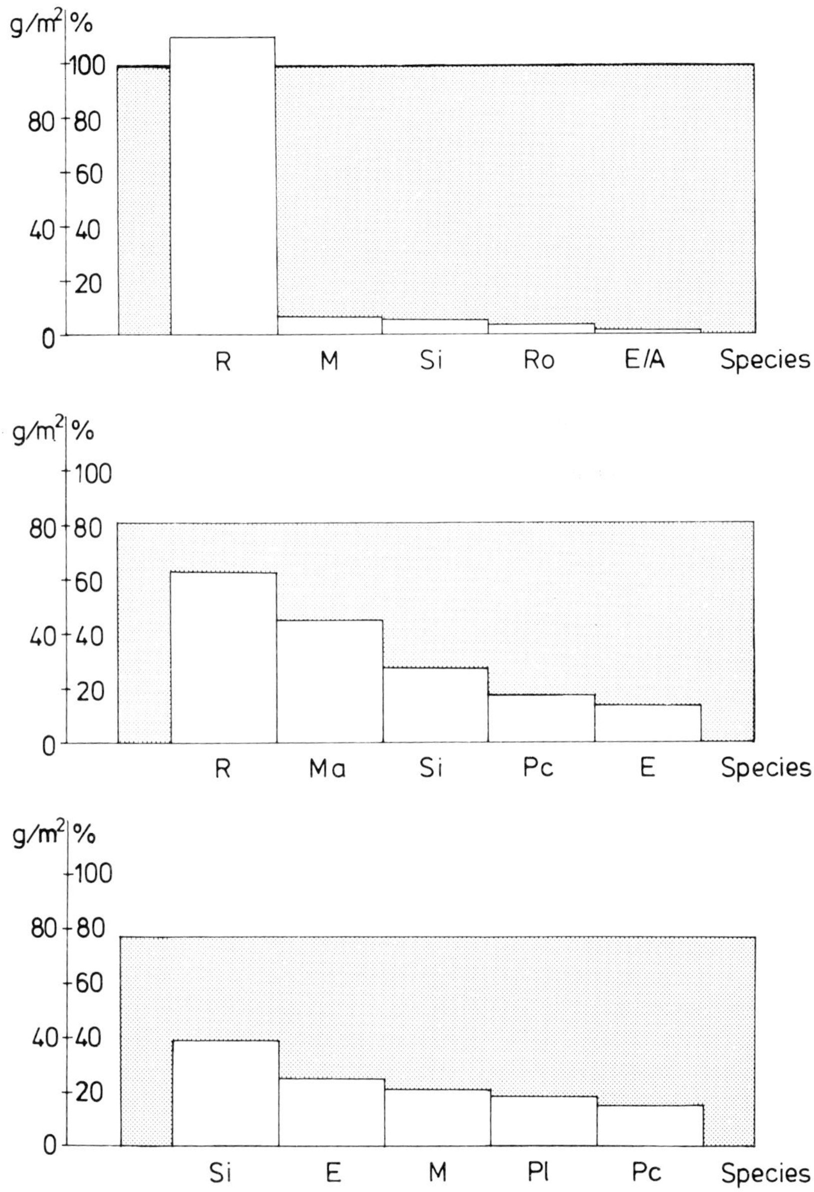

Figure 3.7: MJ a–c. The 5 most productive species and their contribution to the absolute yield of the whole plot (S–treatment/plateau) dependent on the growing season. The columns figure the absolute yield (g dry matter) of the 5 most productive species, and the grey area indicates to what proportion they participate in the phytomass production of the whole plot.

a group finally coming up in the second growing season (newcomers). Apart from these, the group of most productive plants also comprises species with a so-called indifferent behaviour, and it is striking that the most dominant and productive *Melilotus alba* played its role only once, in the second vegetation period, because of its life cycle and the date of seeding. However, on condition that fertilization did not happen (PL/S), which implies a low coverage, it became clear that this species still remained in the background.

It was the second vegetation period when the great change occurred. *Melilotus alba* emerging from the void ranked first, and in the northern SD-area *Reseda lutea* and *Artemisia vulgaris* prepared themselves to leave this productive species group, whereas *Linaria vulgaris* and *Verbascum densiflorum* started to join it. This pattern was also true for the southern slope with the only exception that *Echium vulgare* worked as true pioneer, and for the plateau, where *Reseda lutea* and *Achillea millefolium* marked the first period passed by *Poa compressa*. We consider it important that *Echium vulgare* held its position from the very first beginning to the end, even slightly improving its values. This mode of lasting characterized the group of the most productive species on the unfertilized mine spoil (S) as well. Once more both slopes bore resemblance to each other more than to the plateau. While *Reseda lutea* persisted on both sites, it was *Silene vulgaris* which accompanied it on the northern, but *Oenothera biennis* on the southern slope. As the newcomers differed also, it looked as if exposition exerted its influence by microclimatic tools. No doubt, the northern site turned out to favour vegetation, likewise proven by *Melilotus alba* only ranking first in 1988, thus proving a relationship to the conditions caused by the SD-treatment. The other newcomer was presented by *Daucus carota*, while on the southern slope only *Echium vulgare* played this part. At the plateau three lasting species *Melilotus alba*, 1987 already present in the respective group, *Echium vulgare* and *Silene vulgaris* faced one true pioneer, *Reseda lutea*, and one newcomer, *Poa compressa*, already known from the respective SD-plot. By disclosing these relations, it becomes evident that this kind of species behaviour, lasting or changing, means something to vegetation development. If the first group prevails, site conditions do not favour succession. If, however, newcomers relieve species working as true pioneers, conditions have changed for the better, thus accelerating vegetation development. It seems that the decision as to whether a species acts as a true pioneer or a newcomer is taken by particular environmental factors, such as for instance the microclimate by exposition or nutrient supply, utilizing the full ecological range of each species.

With these findings in mind, it is easy to realize what was going on with that waste which got a top layer (ED) or was mixed with sandy loam (UD, U). The prevailing impression is that of a stable vegetation marked by lasting species particularly *Achillea millefolium* and *Poa compressa*. In case of the northern UD-plot they were accompanied by *Verbascum densiflorum*, another representative of this category. In the same way those samples arising after the ED-treatment at the plateau did not show any alteration with respect to the behavioural groups. *Reseda*

Table 3.3: MJ Rank order and individual numbers of the 5 most productive species dependent on treatment (Tr:), exposition (Ex:), and year.

year	1987		1988		1989		1987		1988		1989	
Tr:				S						SD		
Ex:	R	192	Ma	2	Si	57	R	175	Ma	8	Aa	30
	I	25	Si	76	R	65	S	208	R	108	O	28
NS	Ro	8	Si	16	D	11	A	75	V	117	V	36
	Si	17	Am	58	M	933	PL	83	A	117	T	32
	Se	133	D	10	O	7	Ro	275	L	108	L	58
	R	775	R	85	Si	85	S	33	Ma	17	V	11
	M	17	Ma	3	E	29	R	83	Pc	–	Pc	–
PL	Si	50	Si	37	M	2079	T	175	Am	2017	E	4
	Ro	67	Pc	–	Pl	28	Am	142	R	242	Aa	40
	E/A	92	E	21	Pc	–	E/A	17	E	117	O	12
	R	67	R	75	O	9	R	17	Ma	8	V	52
	Ro	58	E	17	D	29	S	33	V	8	Am	216
SS	I	8	O	17	R	57	C	58	E	25	L	49
	O	8	Si	17	Pl	8	A	75	L	125	T	13
	Se	392	L/Hp	16	E	48	E/A	108	Ae	25	A	32
Tr:				ED						UD		
Ex:	S	125	Ma	17	T	92	Pc	–	Ma	8	Am	326
	Ro	317	Am	400	O	72	S	267	D	33	Aa	76
NS	Am	367	Pc	–	A	16	Si	75	Pc	–	T	130
	T	300	Ro	50	Aa	34	Am	433	Am	383	V	44
	Pc	–	Pl	83	R	20	V	200	V	83	Pc	–
	S	75	Ma	8	Pc	–	S	183	Ma	8	Pc	–
	Pc	–	Mo	8	Am	325	Pc	–	Ro	42	Am	370
PL	Ro	150	Pc	–	Aa	16	Am	258	E	8	A	74
	Am	192	Am	383	T	40	Ro	208	Am	450	T	114
	V	42	I	25	V	4	V	117	Pc	–	Aa	64
	S	108	Ma	8	V	5	S	17	Ma	8	Aa	26
	Ro	117	Ro	108	T	51	Ro	150	Aa	75	V	16
SS	Pc	–	Am	183	Pc	–	Am	150	T	75	Am	250
	Am	217	Pc	–	Aa	15	T	125	L	133	Pl	48
	R	75	T	100	Am	187	Pc	–	E	8	L	172

continued next page

Table 3.3, continued

year	1987		1988		1989		1987		1988		1989	
Tr:			S						SD			
Ex:	S	58	Ma	8	Am	232	L	308	O	8	T	13
	Pc	–	D	25	T	88	V	50	Aa	2	Si	15
NS	Am	150	Lc	17	O	42	Ro	42	L	118	Aa	4
	L	267	Am	658	Aa	32	Aa	50	Ro	8	Am	88
	T	142	Pl	33	M	1280	O	25	A	4	Lc	4
	S	325	Am	583	Lc	112	Ro	50	V	9	T	36
	Pc	–	Ro	58	Am	512	V	75	Ro	4	V	99
PL	Am	867	Lc	92	D	106	S	25	Am	171	Pc	7
	T	658	Ma	8	Pc	–	R	50	A	7	Aa	32
	Ro	225	Pc	–	E	8	L	533	L	177	Am	96
	M	8	Ma	8	Aa	36	O	42	Aa	4	Aa	3
	Si	42	Pl	8	T	40	C	17	O	3	T	17
SS	S	67	D	8	Ae	3	Pm	8	Am	45	Am	246
	Pc	–	Ps	8	Pl	15	V	8	H	–	V	10
	T	292	E	17	O	11	Pa	25	T	5	D	93

luteola, almost solely acting as true pioneer, however, brought some movement into the group of the 5 most productive species. But the level of noise was low at the plateau in case of the UD–treatment, and it is interesting to trace the line forward to the respective U–plot, the species group whose behaviour took after that of the vegetation growing on the SD–treated mine spoil, though the species were different. As the southern ED–plot owned the same pattern of replacement, it is tempting to conclude that a principle underlies this process, the more so, since correspondence occurred between the southern UD– and the SD–groups of both slopes, just as between the U– and S–treatment from the northern slope.

3.7 Conclusion

Pioneer communities said to be randomly gathered assemblages of species are not in the limelight of botanical interest, though being of great ecological importance with regard to natural restoration of degraded areas. Plant cover degree and humus accumulation, the most essential factors for the improvement of site conditions under these circumstances, depend on structure and biomass production of the respective community. Hence one of the questions to be answered by this trial was: Is there any chance of stimulating the development and productivity of natural vegetation? The idea of imitating it artificially by sowing seeds according to phytosociological principles turned out to be profitable. As sowing constituted plant settlement, fertilization was the second step improving phytomass production. While BORNKAMM (1985), who studied a spontaneous primary succession on sand and gravel within a lignite mining area, recorded about 30 g dry matter$\cdot$m^{-2} above ground, we obtained mean yields of 245.5 – 1238.0 g$\cdot$m^{-2} from the waste produced by coal mining owing to site conditions and the year of harvest (JOCHIMSEN 1990). These values resemble those reported by LIETH (1972) for natural grasslands within temperate zones and by BOEKER (1966) for a late herbaceous successional stage (3–10 years) within a German deciduous forest. Average values, however, do not supply any detailed information. This became already evident during the first vegetation period. Exposition varied productivity, as in the south facing areas growing started earlier than on the northern slope, but later on vegetation development here profited by lower evapotranspiration, almost neutralizing the former margin. The second year brought an enormous increase in phytomass, mostly due to *Melilotus alba*, which concluded its life cycle. But the rank order, a proportionate measure taken as a basis, revealed that those areas the vegetation of which was poorer developed in the beginning, now made up leeway. That means that the yield was controlled by the state of development already achieved by growth and the space left, neither of which was necessarily connected with the other. The third year after seeding, phytomass production decreased obviously, but compared to the first harvest there was still a slight increase. The rank order showed little alteration, and improvement only occurred on a small scale concerning those areas with some space left. At the moment the state of vegetation seems to be in balance, but time is going on, and things will change again, because plant matter production is not only a result of the growth rate of a certain community, but also dependent on the life cycle of its members, which may alter owing to site conditions (WERNER 1977) as proven by this trial, and on the horizontally and vertically existing space (illustrated by respective drawings) being reduced to niche in the course of succession (PARRISH and BAZZAZ 1982). If, however, phytomass production is controlled by growth, which in its part is dependent on the resources and microclimate, population dynamics and stratification, it becomes evident that the values for dry matter and degree of coverage are not necessarily correlated with each other (BORNKAMM 1981, BORNKAMM 1985).

With respect to a recently established vegetation the number of individuals colonizing a particular area is of great interest, though they may differ in size. As this trial proves, the most important factor is represented by the availability of seeds, but exposition, that is to say microclimate, plays a role too, where regeneration is concerned. That is why the southern slope presented the smallest amount of individuals during the first vegetation period, whereas the maximum occurred on the north facing slope. The second growing season was characterized by a decreasing number of individuals. This process might be caused by reduced resources and unfavourable site conditions, as the next year brought rising values again, perhaps due to the improvement of microclimate by the vegetation. Growth, however, provokes competition, and in order to get a better understanding of it, whether it concerns exploitation and/or interference, it will be necessary to examine the individual numbers in particular with respect to species and life cycle. In addition to the factors already mentioned it was population dynamic which contributed to this increase to a large extent.

The structure of a plant community is formed quite visually by its governing species (BORNKAMM 1981). In the context of this trial the questions arose, (i) whether the species which dominate with respect to their phytomass production are common to each treatment, (ii) which species are incorporated in the group of the five most productive plants, (iii) to what extent do they contribute to the total yield, and (iv) is there any alteration with time? Though this study made it probable that a lot of interactions exist, which do not allow any easy answer, it was obvious that site conditions, as given by pure mine spoil (S) and mixed substrate (E,U), had a group of dominants of their own. That means already at this early stage of development it was the substrate that took the decision about species to be favoured out of the given set, or the other way round, those species more adapted to it got a better chance, but we must still examine whether the species group forming the vegetation on the "mixed areas" already represents a higher level as defined by phytosociological progression.

As different patterns occurred with the replacement within the group of the five most productive species, three types of behaviour — for working reasons only — were established. Looking at the three years of this development, a high proportion of "lasting species", mainly characteristic in connection with added soil and fertilization, means site conditions seem to be in balance. A preponderance of "true pioneers", however, may signalize that vegetation development is found at the start of a particular stage, whereas the appearance of "newcomers" could announce that the development is going on. The question is, whether the combined presence of these three types means a speedy development or chaos. Provided the first is true, using a top layer (ED) would accelerate vegetation development in areas with a southward exposition, while under condition of this trial pure and fertilized mine spoil (S, SD) as well as mine spoil mixed with sandy loam (U) would take advantage of a horizontal exposition. But there is still some precariousness, because the type of behaviour is not fixed. Species may act differently using their whole ecological range, and the regularity which seems to exist between type or behavior and site conditions has to be examined in the future.

As already mentioned, plant cover of mine spoil areas (S/SD), representing a true primary succession, differed qualitatively from those whose waste was combined with soil (U,UD,ED). Whereas vegetation development on mine spoil covered with top soil (ED) tends to be a secondary succession, that of the "mixed" areas takes an intermediate position. This reflects MILES's (1982) view, which rejects the conventional classification, as our understanding of vegetation dynamics is still rudimentary. That applies to this study, too, because a period of three years is an extraordinarily short time with respect to succession, which does not yet allow any final assessment. That is why this paper can only be valued as a preliminary report. There is little possibility for comparison, most studies are concerned with secondary succession (SCHMIDT 1981, BORNKAMM 1984, BORNKAMM 1986), and in addition to that our vegetation development was initiated artificially. The most important result as to reclamation purposes is the fact that the long time normally taken by a spontaneous colonization was shortened. The phase of therophytes observed and postulated by GRIME (1986) and others (BORNKAMM and HENNING 1982, BORNKAMM 1984) was skipped, provided it really exists in the case of primary succession.

Bibliography

BOEKER, P., 1966, Leistungsvergleich zwischen Wiese und Weide, dargestellt am Beispiel eines Grundwasserstandsversuchs.
Bayer. Landwirtsch. Jhb. **43**, 223–230.

BORNKAMM, R., 1981, Rates of change in vegetation during secondary succession.
Vegetatio **47**, 213–220.

BORNKAMM, R.; HENNING, V., 1982, Experimentell–ökologische Untersuchungen zur Sukzession von ruderalen Pflanzengesellschaften auf unterschiedlichen Böden. I. Zusammensetzung der Vegetation.
Flora **172**, 267–316.

BORNKAMM, R., 1984, Experimental–ecological investigations on the succession of ruderal plant communities. II. Quantity and quality of the phytomass.
Flora **175**, 45–74.

BORNKAMM, R., 1985, Veränderungen der Phytomasse und Vegetationsentwicklung.
In: Wolf, G. (ed.) Schr. Reihe Veg.kde **16**, Bonn–Bad Godesberg, 111–151.

BORNKAMM, R., 1987, Changes of phytomass during the first two years of succession on differing soils.
Flora **179**, 179–192.

BURGHARDT, W., 1989, Bodenbeschaffenheit und Pflanzennährstoffgehalte der Bodenvarianten des Begrünungsversuches Bergehalde Waltrop.
Mitt. Dtsch. Bodenkundl. Ges., Bd. **58**, 219–225.

GRIME, J. P., 1986, Plant Strategies and Vegetation Processes.
John Wiley & Sons, Chichester.

JOCHIMSEN, M., 1986, Begrünungsversuche auf Bergematerial der Halde Ewald/Herten.
Ver. Ges. Ökologie **XIV**, 223–228.

JOCHIMSEN, M., 1987, Vegetation development on mine spoil heaps — a contribution to the improvement of derelict land based on natural succession.
In: Miyawaki, A.; Bogenrieder, A.; Okuda, S.; White, J. (eds.) Vegetation ecology and creations of new environments, Proceed. Int. Symp. Tokyo, 245–252.

JOCHIMSEN, M., 1990, Recultivation of raw soils according to natural succession.
In: Ravera, O. (ed.), Perturbation and recovery of terrestrial and aquatic ecosystems, Ellis Horwood Ltd., Chichester, in press.

JOCHIMSEN, M., 1990, Advantages and possibilities of recultivating fallow land in accordance with natural succession.
Options Méditerranéennes, (in press).

LIETH, H., 1972, Über die Primärproduktion der Pflanzendecke der Erde.
Angew. Bot. **46**, 1–37.

LIETH, H., 1974, Primary productivity of successional stages.
In: Knapp, R. (ed.), Vegetation Dynamics, Dr. W. Junk b. v., The Hague, 185–193.

MILES, J., 1982, Vegetation Dynamics.
Chapman and Hall, London.

OBERDORFER, E., 1983, Pflanzensoziologische Exkursionsflora.
Eugen Ulmer, Stuttgart.

PARRISH, J. A. D.; BAZZAZ, F. A., 1982, Competitive interactions in plant communities of different successional ages.
Ecology **63**, 314–320.

SCHMIDT, W., 1981, Ungestörte und gelenkte Sukzession auf Brachäckern.
Scripta Geobotanica **XV**, 5–199.

SCHNEIDER, S., 1989, Bodenkundliche Untersuchungen auf der Bergehalde Waltrop im Rahmen eines Großversuchs zur Haldenrekultivierung.
Mitt. Dtsch. Bodenkundl. Ges., Bd. **58**, 211–218.

WERNER, P. A., 1977, Colonization success of a "biennial" plant species: experimental field studies of species cohabitation and replacement.
Ecology **58**, 840–849.

Chapter 4

Survival and growth beneath and near parents: The case of *Myrcianthes fragrans* (Myrtaceae)

D. F. WHIGHAM and E. CABRERA CANO

4.1 Introduction

Discussions of the fate of seedlings in tropical forests have mostly focussed on whether or not they become established beneath the canopy of their parents and the importance of gap versus non–gap environments to their growth and survival (CLARK and CLARK 1984). In the early 1970's, CONNELL (1971) and JANZEN (1971) independently proposed that seedling establishment would be lower near parents because of disproportionately high mortality associated with seed predation, herbivory, and pathogens. CLARK and CLARK (1984) reviewed 24 studies on this topic published between 1970 and 1984 and found general support for either density–dependence or distance–dependence in progeny mortality. Additional support for the CONNELL–JANZEN hypothesis was provided by AUGSPURGER (1984a) who found that seedlings of eight out of nine wind–dispersed species in Panama had higher survival rates away from parents. Others have found different results. KITAJIMA and AUGSPURGER (1989) found that, seedlings of *Tachigalia versicolor* had a higher survival rate with increasing distance from the parent, but that few seedlings persisted beyond 40 m from parents. SCHUPP (1988) found that seed survival of *Faramea occidentalis* was greatest beyond the canopy of the parent but that mortality of established seedlings was similar beneath and away from parents. HUBBELL (1980) also noted that juveniles can reach maturity immediately adjacent to an adult of the same species.

A number of investigators have examined the role of tree gaps as sites for successful establishment and growth of seedlings (SCHUPP 1988, AUGSPURGER 1983, AUGSPURGER 1984b, AUGSPURGER and KELLY 1984, BROKAW 1985, BROKAW 1987, DENDLOW et al. 1990, DE STEVEN and PUTZ 1984, UHL et al. 1988). In most of these studies, seedling survival and growth rates were greater in tree gaps. SCHUPP (1988), however, found that predation was higher in gaps and concluded that long–term seedling survival might be best in the forest understory.

Given the high diversity of trees in most tropical forests compared to the few species that have been studied, it is not surprising that a variety of relationships between adults and their offspring have been described. More studies are needed before generalizations can be made either about the JANZEN–CONNELL hypothesis or about the importance of tree gaps to the dynamics of humid and dry tropical forests.

In this paper we address the issues of parent–offspring growth and survival by using results of a six year study of *Myrcianthes fragrans* (SW.) MCVAUGH (Myrtaceae) in a dry tropical forest in the Yucatan Peninsula of Mexico. *Myrcianthes fragrans* occurs from northern South America through the West Indies and into southern Florida (MCVAUGH 1963). We have three objectives: 1) to evaluate factors affecting seedling survival by: a. characterizing distance relationships between adults and between adults and smaller individuals, and by b. evaluating the growth and mortality of seedlings in long-term plots located at three distances from parent trees, 2) to determine the relationship between seedling growth and soil depth. Soils at our study site are shallow, usually less than 20 cm, and we hypothesized that seedling growth would be negatively related to soil depth, and 3) we address the hypothesis that seedlings respond positively to canopy opening using measurements of seedling growth and mortality prior to and following complete canopy defoliation caused by Hurricane Gilbert in September, 1988.

4.2 Study site

The studies were conducted in a dry tropical forest ("selva mediana subperennifolia" (MIRANDA 1958) on the Yucatan Peninsula. The site, hereafter called Rancho San Felipe, is located approximately 10 km south of the village of Puerto Morelos (20° 49' N, 87° 7' W) in the Mexican state of Quintana Roo. Canopy height at the site varies between 15 and 25 meters. Twelve long–term study plots (40 × 40 m) were established in February 1984. All trees with a diameter (DBH) greater than 10 cm were tagged and identified in each plot. The twelve plots contained 1508 individual trees that belonged to 79 species. Nine species (*Manilkara zapota* (L.) VAN ROYER, *Talisia olivaeformia* (HBK.) RADLK., *Gymnanthes lucida* (SWARTZ), *Brosimum alicastrum* SWARTZ, *Drypetes lateriflora* (SWARTZ) DRUG and URBAN, *Sapindus saponaria* L., *Coccoloba diversifolia* JACQ, *Beaucarnea pliabilis* (BAKER) LUNDELL), and *Myrcianthes fragrans* accounted for approximately 60% of all individuals. Thirty-eight *Myrcianthes* trees occurred in eleven of the plots and its density ranged from 1 to 6 per 1600 m^2. DBH of *Myrcianthes* ranged from 14.7 to 69.2 cm (mean 36.4 cm; SD±14.0).

There are no long-term climatic data from Rancho San Felipe but relevant data are available from the Cancun airport (25 km), Felipe Carillo Puerto (185 km) (LOPEZ ORNAT 1983), and Valladolid (135 km) (WALTER and LIETH 1967). The climate is strongly seasonal with a distinct period of low precipitation that usually lasts from November through May (Fig. 4.1). Annual precipitation is approximately 1100 mm, but the amount and seasonal distribution are highly variable (WHIGHAM et al. (in press)). The average annual temperature at Valladolid and Felipe Carillo Puerto is 25°C and the monthly mean varies from approximately 26 to 22°C. Maximum temperatures occur between May and August (ca. 40–33°C) and minimum temperatures between January and March (ca. 18–8°C).

The shallow azonal soil, classified as a litosol-rendzina and derived from weathered Miocene and Pliocene parent material (WEST 1964, BACK and HANSHAW 1970), occurs in pockets between exposed limestone rocks. The soil has an almost neutral pH (7.1 ± 0.1) and the organic matter content is high (51.5 ± 5.1%).

Average light conditions in the forest understory were determined prior to Hurricane Gilbert. Shaded areas in long-term seedling plots (see page 65) received an average of 1.0 ± .5% (1 standard error) of full sunlight (N= 34; PAR measurements made in a nearby field and at the soil surface in seedling plots during cloudless conditions on 16 February, 1985 between 11AM and 12PM using a LI-COR Model 185B light meter). During the same sampling period, sunflecks (N= 14) in the seedling plots received an average of 12.3 ± 11.4% (Range: 40.0–1.7%) of full sunlight. No light measurements were made immediately after Hurricane Gilbert but the canopy was completely defoliated and light intensities were accordingly quite high (Ingrid Olmsted, personal communication).

4.3 Methods

Distances between juveniles, saplings, and trees. We measured interplant distances between all *Myrcianthes* individuals > 10 cm DBH in the 1600 m^2 plots. We also measured the distances between each individual tree and all juveniles (plants less than 1 m in height), saplings (plants greater than 1 m in height and less than 10 cm DBH), and trees within a ten meter radius. A ten meter sampling radius was chosen to represent areas beneath (0–5 m) and beyond (5–10 m) the

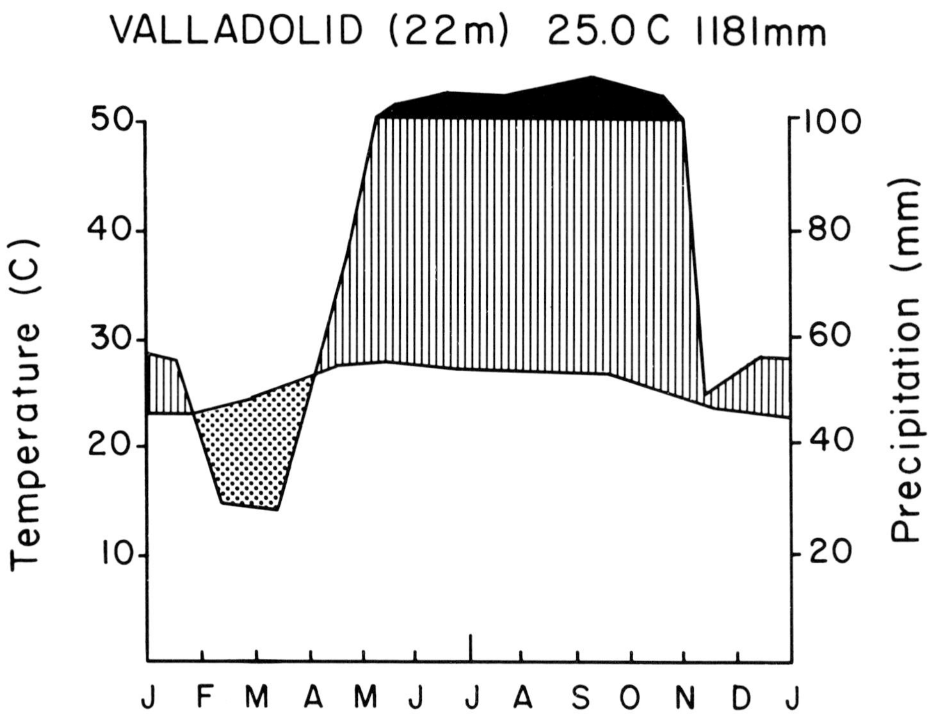

Figure 4.1: Climate diagram (WALTER and LIETH 1967) for Valladolid (elevation 22 meters). The long–term mean for temperature and precipitation are given in the upper right corner of the diagram. The relationship between temperature and precipitation suggests periods of excess or deficit water conditions. The dotted area is a period of relative drought when the precipitation curve falls below the temperature curve. The vertically shaded area is the humid season when the precipitation curve is above the temperature curve. The black area represents the part of the year when precipitation exceeds 100 mm per month and the scale is reduced to 1/10.

canopy of each tree, and to include the distribution of virtually all seedlings. Differences between the density of *Myrcianthes* juveniles, shrubs, and trees within the 10 m sampling radius were compared with ANOVA and the Tukey–Kramer test (WILKINSON 1989).

Seedlings in long–term plots. In February 1984, we established nine plots (1 × 1 m) around each of five trees. Plots, three per transect, were positioned between 1–2, 5–6, and 10–11 meters from the boles of the putative source trees. These plots will hereafter be referred to as 1, 5, and 10 meter distances. A cohort of *Myrcianthes* seedlings had germinated during the pervious 2–3 months. Cotyledons were still attached and most seedlings had produced the first true leaves. We counted the number of seedlings in each plot and fitted their stems with small numbered aluminum tags. In most plots we tagged all seedlings but in plots with more than 100 individuals, only a subset (approximately 50%) of the seedlings were tagged. Since 1984, the plots have been censused yearly at approximately the same time to determine the number and height of live seedlings. Height and density data were analyzed with a nested ANOVA model (quadrat nested within tree and distance) and means compared with the Tukey test (JOYNER (ed.) 1985). Seedling survival at 1, 5, and 10 meters was compared with the Peto and Peto's Logrank Test (PYKE and THOMPSON 1986).

Depth of the soil–litter layer. The potential rooting depth available to seedlings was evaluated by measuring the depth (cm) of the litter–soil layer at 10 cm from the trees and then at 50 cm intervals along three randomly located 10 m transects per tree (Total = 15 transects). The litter–soil layer was defined as the distance between the top of the litter layer and solid rock. Depth was measured by inserting a wooden stick into the soil and marking it at the top of the soil–litter layer after the stick had reached solid rock. The distance from the bottom of the stick to the mark was then measured. Statistical comparisons of the depth of the litter–soil layer at different distances from parent trees were made using ANOVA (quadrats nested within trees) and means were compared with the Tukey–Kramer test (WILKINSON 1989).

Seedling growth and microhabitat characteristics. In February 1990, we harvested 300 seedlings from the 1984 cohort. Twenty seedlings were collected along each of three randomly located 10 m transects at each of the five trees described on page 65. The following data were collected in the field: 1. distance (m) between parent and seedling, 2. qualitative assessments of light conditions were made by evaluating the percent canopy above each seedling. Three categories were used: A. shade – seedling judged to be completely shaded by leaves of overhead vegetation, B. partial shade – seedlings mostly shaded by overhead vegetation but judged to receive some direct light during the day, C. sun – seedlings covered by little overhead vegetation and judged to receive large amounts of direct light during the day, 3. depth of soil (cm) immediately beneath the seedling, 4. number of leaves and branches, 5. shoot height (cm), 6. length (cm) of the root divided into vertical and/or horizontal components. The latter division was necessary because *Myrcianthes* roots begin to grow horizontally after they have reached the bottom of the soil profile. Plants were dried to constant weight in the sun, and biomass (g)

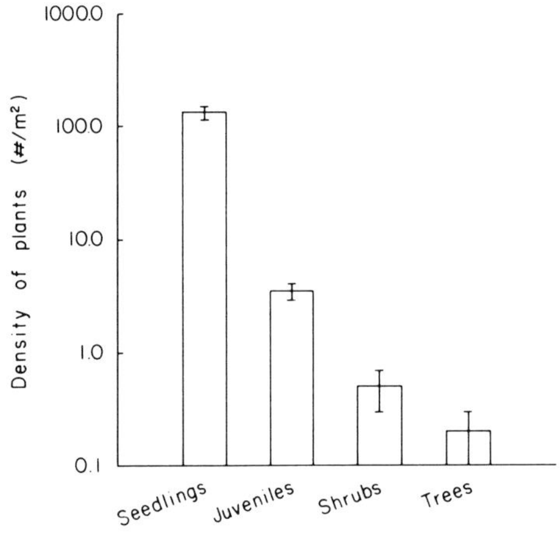

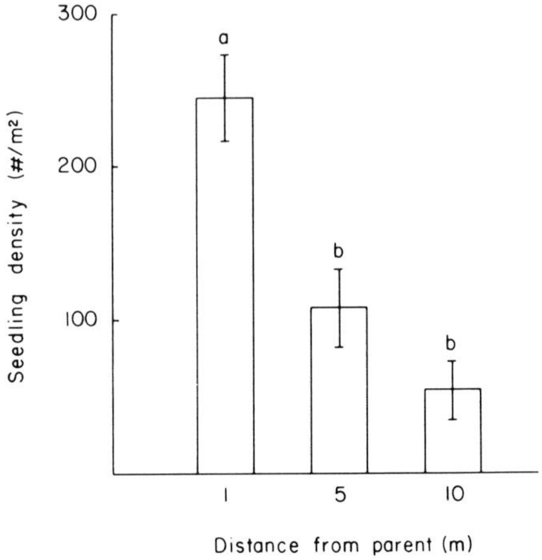

Figure 4.2: Density (log scale) of *Myrcianthes fragrans* seedlings, juveniles, saplings, and trees. Values are means ± 1 standard error.
(a) Seedling data are from the long-term plots (see page 65). Juvenile, sapling, and tree data represent densities within 10 meters of individual trees (see page 63).
(b) Mean density (± 1 standard error) of seedlings in long-term study plots. Means that are not different at P < .05 share the same letter.

determined for leaves, shoots, branches, vertical roots, and horizontal roots. Using a nested (transect nested within tree) ANOVA model, the data were analyzed using the GLM procedure in SAS (JOYNER (ed.) 1985).

4.4 Results

Density and distance relationships The mean and standard deviation of the distance between trees in the 11 plots was 22.8±11.2 meters (N= 62; max= 51.3 m, min= 1.9 m). Within 10 m of each tree, the average distances to juveniles (7.0±1.9), shrub-sized individuals (6.4 ± 2.0), and tree-sized individuals (6.2 ± 1.5) were not statistically different. Densities were, however, significantly different (P< .0001) with juveniles > shrubs = trees (Fig. 4.2a). Within the long-term plots, seedling densities in 1984 differed significantly (P < 0.01) as a function of distance to the source tree, with 1 > 5 = 10 m (Fig. 4.2b).

Growth and mortality in long-term plots. The shape of the mortality curves were similar for all three distances but the slopes were significantly different (P < .001) and survivorship was greater at 5 and 10 m (Fig. 4.3). After almost 60

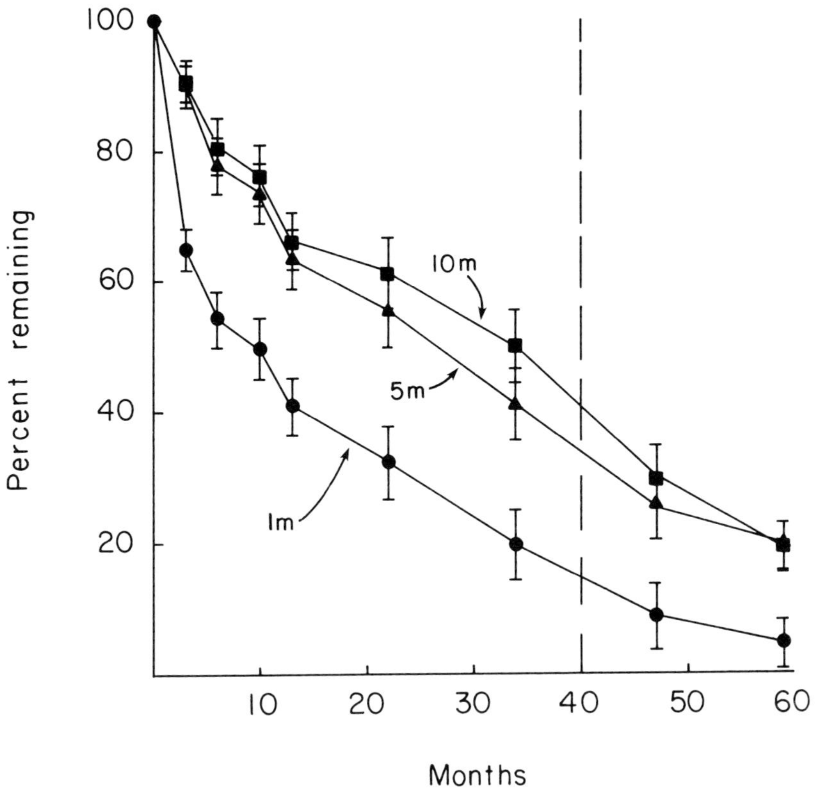

Figure 4.3: Survivorship curves for seedlings in long–term study plots at 1, 5, and 10 meters from parent trees. The vertical dashed line indicates when Hurricane Gilbert struck the study area in September, 1988.

months, only $4.5 \pm 1.2\%$ of the original cohort of seedlings were alive at 1 m, compared with 19.5 ± 3.8 and $19.2 \pm 3.8\%$ at 5 and 10 m respectively. At each census, we were able to locate almost all seedlings and it was apparent that few had suffered from herbivore damage. Most dead seedlings were upright and had dried intact leaves attached. Most seedlings appeared to have died from drought stress although a few had been physically damaged from falling debris and a few others appeared to have died from shading as a result of being covered by fallen bark.

From April, 1985 to February, 1988, mean seedling height increased less than 1 cm at 1 and 5 m and only about 1.5 cm at 10 m (Fig. 4.4). Average annual height growth prior to the hurricane did not differ significantly between the three distances (Fig. 4.5). Annual growth rates increased significantly following the hurricane but

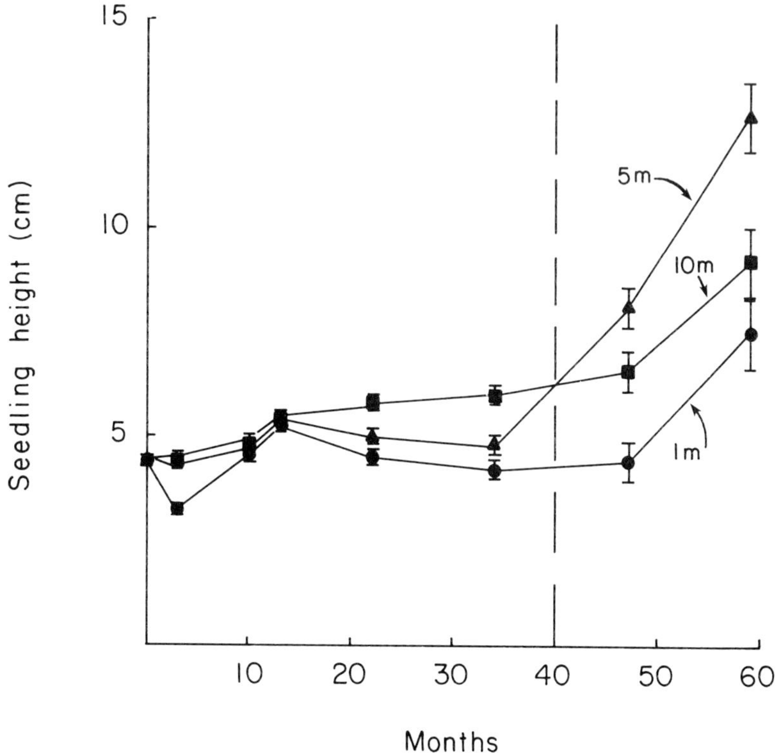

Figure 4.4: Height (cm ± 1 standard error) of seedlings at 1, 5 and 10 meters from parent plots. The vertical dashed line indicates when Hurricane Gilbert struck the study area in September, 1988.

the three distances were not significantly different even though the growth rate was less at 1 m (Fig. 4.5).

Soil depth and distance from parent trees. Large amounts of bark exfoliates from the trunk and large branches of *Myrcianthes* and a mound of litter develops that extends outward from the base of the tree to about 2.5 m (Fig. 4.6). Depth of the litter–soil layer at 0.1 m was significantly more ($P < 0.01$) than greater distances. Depth of the litter–soil layer was also significantly greater at 0.5, and 1 m but there were no significant differences beyond 1.5 m where the soil depth was approximately 10 cm.

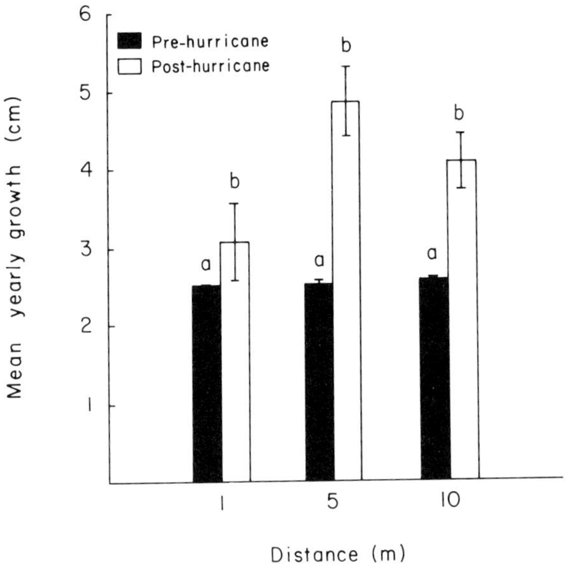

Figure 4.5: Average annual growth in height of *Myrcianthes* seedlings before (black bars: March 1984–March 1987) and after (Open bars: September 1988–March 1990) Hurricane Gilbert. Values are means ± 1 standard error.

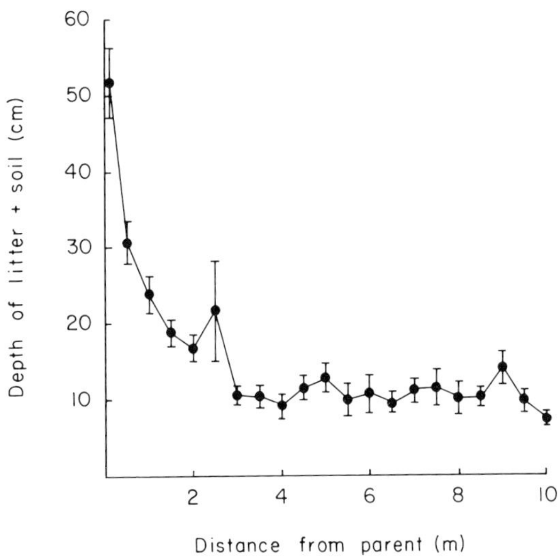

Figure 4.6: Depth of the litter–soil layer as a function of distance from trees. Values are means ± 1 standard error. Methods for sampling see page 65.

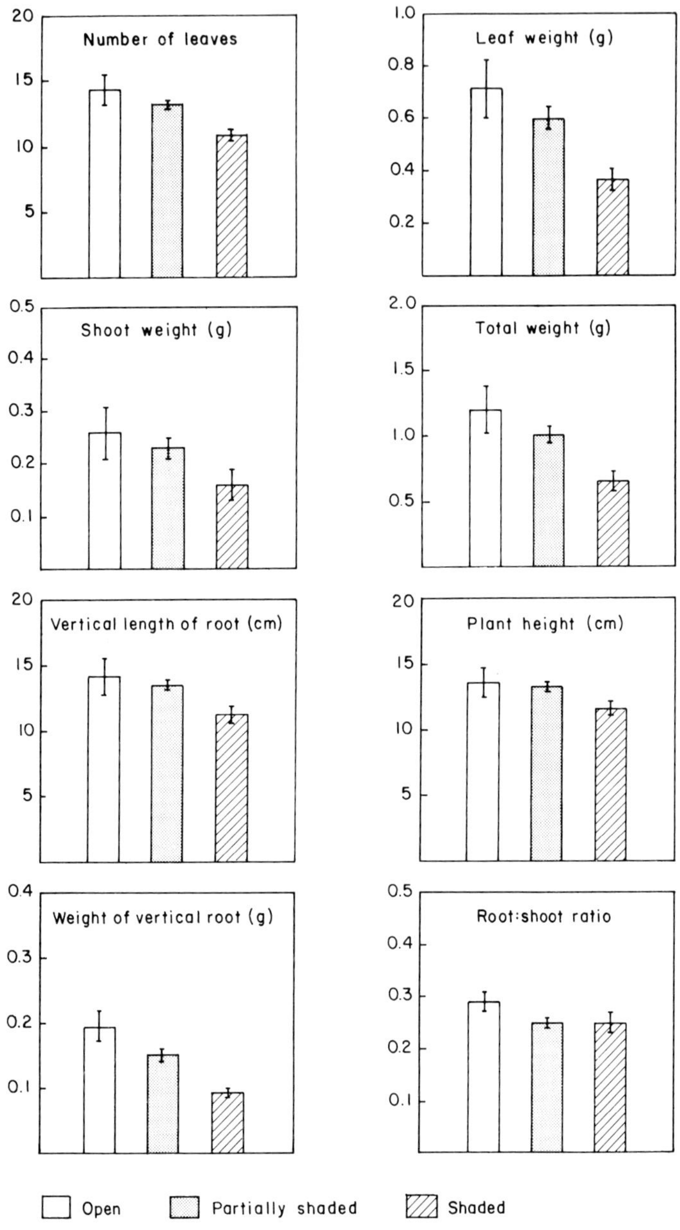

Figure 4.7: Growth parameters for *Myrcianthes* seedlings in shaded, partially shaded, and open microhabitats. Values are means ± 1 standard error. Microhabitat designations are as shown at the bottom of the figure.

Table 4.1: Results of linear regression analysis of growth variables against soil depth. N= 300.

VARIABLE	R^2	INTERCEPT	SLOPE
Length of vertical root (cm)	.358	5.678	.560
Weight of vertical root (g)	.090	0.074	.005
Number of leaves	.058	10.475	.188
Length of horizontal root (cm)	.053	4.381	−.162
Plant height (cm)	.036	10.990	.158
Number of branches	.035	.144	.024
Weight of horizontal root (cm)	.035	.050	−.002
Weight of leaves (g)	.030	.395	.013
Weight of branches (g)	.023	.157	.005
Root:shoot ratio	.004	.266	−.001

Seedling growth and microhabitat characteristics. After 6 years of growth, soil depth showed almost no statistically significant relationship to any of the variables measured on harvested seedlings (Table 4.1). Length of the vertical root was the only variable for which soil depth accounted for more than 10 percent of the variance.

Although there were no significant differences in seedling growth rates with distance, there were significant microhabitat effects on growth following Hurricane Gilbert. In comparison with seedlings in open and partially shaded microhabitats, seedlings in shaded microhabitats were significantly shorter ($P < .001$), had fewer leaves ($P < .008$), less leaf biomass ($P < .009$), shorter and smaller vertical roots ($P < .046$), and less shoot and total biomass ($P < .009$) (Fig. 4.7).

4.5 Discussion

Initially, seedling densities were highest beneath parent trees. Seedling mortality was greatest near the parent tree as predicted by the JANZEN–CONNELL hypothesis. However, herbivory or pathogens probably were not responsible for the higher levels of mortality observed near parent trees. As indicated, almost all marked seedlings appeared to have died from physical damage (crushing or breaking of the stems), shading, drought stress or some combination of those factors.

AIDE (1987) and CLARK and CLARK (1989) found that seedling mortality due to physical damage was common in moist tropical forests. Mortality due to direct physical damage was of minimal importance in this study, even during the most intense hurricane ever measured in the Western Hemisphere. We found very few seedlings that had died as a result of direct impact (i. e., broken stems) from fallen branches. A few seedlings were killed by direct impact from fallen objects before and after the hurricane but most appeared to die from shading that resulted from being covered by exfoliated bark (before the hurricane) or by woody debris (after the hurricane).

Although we have no direct measurements of seasonal variation in available soil water, we believe that drought was the primary cause of mortality. The soils are very shallow (Fig. 4.6) and contain very little water, especially during the dry season when many plants appear to be water stressed. These results contrast with AIDE and ZIMMERMAN (1990) who suggested that seedling growth in the forest understory may be enhanced during dry periods because of increased light and soil moisture due to less evapotranspiration from leafless canopy trees. Other factors such as nutrient availability may also be important but soil depth is not as it was related to only one seedling growth parameter, length of the vertical root.

Light also appeared to be an important factor that limited seedling growth and perhaps survivorship (AUGSPURGER 1984a, AUGSPURGER 1984b, BROKAW 1985, DENSLOW et al. 1990, UHL et al. 1988, CLARK and CLARK 1989, DENSLOW 1987). Prior to the hurricane, light levels were very low (mostly $< 1\%$) in the forest understory and the average seedling height increased very little during the 40 months before the hurricane. Seedling growth measured after the hurricane indicated that seedling in shaded microhabitats grew significantly less than seedlings growing in open and partially shaded areas. These findings agree with those of others (e. g., UHL et al. 1988, AIDE and ZIMMERMAN 1990) who have found that variations in light conditions played an important role in controlling the growth of woody seedlings in moist tropical forests.

Following the seedlings stage, patterns of mortality must remain spatially constant for many years. We found surprisingly constant mean distances (ca. 7 m) from parent trees to juveniles and saplings. Survival to tree size near parent trees can occur (HUBBELL 1980) in tropical forests but is probably uncommon. The average distance between adults exceeded 20 meters and, within the 1600 m^2 plots, only three of the sixty-two nearest neighbor interplant distances were less than 5 meters.

We do not know the causes of mortality for individuals larger than seedlings but hurricanes may be more important than all other factors combined. Of 38 *Myrcianthes* trees in the eleven 1600 m^2 plots, one (2.6%) died (cause of mortality unknown) between 1984–1988, compared with 12 (31.6%) as a result of Hurricane Gilbert. Trees of all species in the forest were defoliated and suffered varying degrees of canopy damage during the hurricane. For *Myrcianthes*, two large (DBH > 50 cm) individuals were uprooted during the hurricane, two completely defoliated trees never sprouted, and the other eight individuals died even though they had initially sprouted.

Perhaps more damaging to trees, however, were widespread fires which occurred during the dry season after the hurricane. Although Rancho San Felipe itself was mostly protected, two small areas burned when fire jumped the fire lane that had been cut around the ranch. In one of these areas, we established four 1600 m^2 plots and sampled all trees greater than 10 cm DBH. The number of trees in the four plots averaged 93 ± 15 (1 standard error) and mortality averaged 89.3 $\pm 3.9\%$, a mortality level typical of other areas in northern Yucatan that burned during the same period (Ingrid Olmsted, personal communication). All 18 *Myrcianthes* trees in the plots were killed and we found only two surviving saplings in one of seventy-two plots (each plot was 4 x 4 m) sampled within the four larger plots. Clearly, the distribution and abundance of *Myrcianthes* can be influenced by both hurricanes and subsequent fires.

In summary, we found evidence to support almost every hypothesis about the distribution, abundance, and growth of seedlings and trees. Data from long-term plots support the hypothesis that mortality is greatest near parents. Mortality of *Myrcianthes* seedlings was greatest within one meter of parent trees but the primary cause did not appear to be herbivory and disease but physical damage and shading caused by exfoliated bark. These findings support the hypothesis that direct and indirect damage caused by fallen objects may be one of the most important causes of seedling mortality (AIDE 1987, CLARK and CLARK 1989). Between-tree distance data support the hypothesis that few plants survive beneath their parent but, like HUBBELL (1980), we found a few individuals growing close to each other.

Our data suggest that *Myrcianthes* has adopted a "sit-and-wait" approach to reproduction and growth. Individuals do not produce seeds regularly (ca. only once in the six years of this study) and seedlings growth in the understory is very slow and appears to be primarily light limited. If seedlings can survive long enough, however, a canopy disturbance (i. e., single-tree death or large disturbances such as those caused by the hurricane) may occur and subsequent growth may no longer be as strongly light-limited. In the case of single-tree disturbances, we agree with UHL et al. (1988) who suggested that individuals that are recruited into gaps or are present as seedlings would only reach the canopy after multiple disturbances. Seedlings, juveniles, and saplings would have a greater probability of reaching the sub-canopy or canopy following a single large-scale disturbance that defoliates and causes extensive damage to trees. Even in this instance, however, it seems likely that only larger (i. e. small tree-sized individuals) would reach the canopy after a single disturbance event because the canopy reforms rather quickly (unpublished data). Perhaps most important, this study demonstrates that endogenous (e. g., seed dispersal, seedling predation and herbivory etc.), factors that control the distribution patterns of trees in dry tropical forests may be less important than exogenous factors (e. g., hurricanes, fires). Because most seedling studies have been conducted in moist tropical forests, many more species-specific comparisons of this sort are needed before we can begin to understand the dynamics of tree populations in dry tropical forests.

Acknowledgments

Jay O'Neill and Elizabeth Ley helped count and measure seedlings in the permanent plots. Volunteers with the Smithsonian Expedition Program (Owen Lavin, Sol Courtman, Sherley Higuera, Margery Plymire, Shirley and James Virosco, and William McNiven) helped harvest seedlings and measure interplant distances. Jess Zimmerman, Jay O'Neill, Julie Denslow, and Jim Lynch provided comments on drafts of the manuscript. Financial support was provided by the World Wildlife Fund, Smithsonian Environmental Sciences Program, and Smithsonian International Environmental Sciences program. The logistical support of Patricia Zugasty Towle, Felipe Sanchez Roman, and Ingrid Olmsted are appreciated.

Bibliography

AIDE, T. M., 1987, Limbfalls: a major cause of sapling mortality for tropical forest plants.
Biotropica **19**, 284–285.

AIDE, T. M.; ZIMMERMAN, J. K., 1990 Patterns of insect herbivory, growth, and survivorship in juveniles of a neotropical liana.
Ecology **71**, 1412–1421.

AUGSPURGER, C. K., 1983, Offspring recruitment around tropical trees: changes in cohort distance with time.
Oikos **40**, 189–196.

AUGSPURGER, C. K., 1984a, Light requirements of neotropical tree seedlings: a comparative study of growth and survival.
J. Ecol. **72**, 777–796.

AUGSPURGER, C. K., 1984b, Seedling survival of tropical tree species: interactions of dispersal distance, light-gaps, and pathogens.
Ecology **65**, 1705–1712.

AUGSPURGER, C. K.; KELLY, C. K., 1984, Pathogen mortality of tropical tree seedlings: experimental studies of the effects of dispersal distance, seedling density, and light conditions.
Oecologia **61**, 211–217.

BACK, W.; HANSHAW, B. B., 1970, Comparison of chemical hydrogeology of the carbonate peninsulas of Florida and Yucatan.
J. Hydrol. **10**, 330–368.

BROKAW, N., 1985, Gap-phase regeneration in a tropical forest.
Ecology **66**, 682–687.

BROKAW, N., 1987, Gap-phase regeneration of three pioneer tree species in a tropical forest.
J. Ecol. **75**, 9–19.

CLARK, D. A.; CLARK, D. B., 1984, Spacing dynamics of a tropical forest tree: evaluation of the Janzen-Connell model.
Am. Nat. **124**, 769–788.

CLARK, D. A.; CLARK, D. B., 1989, The role of physical damage in the seedling mortality regime of a neotropical forest.
Oikos **55**, 225–230.

CONNELL, J. H., 1971, On the role of natural enemies in preventing competitive exclusion in some marine animals and in rain forest trees.
In: DEN BOER, P. J.; GRADWELL, G. R. (eds.), Dynamics of populations. Proceedings of the Advanced Study Institute on dynamics of numbers in populations, Oosterbeek, 1970, Centre for Agricultural Publishing and Documentation, Wageningen, pp. 298–310.

DENSLOW, J. S., 1987, Tropical rainforest gaps and tree species diversity.
Ann. Rev. Ecol. Syst. **18**, 431–451.

DENSLOW, J. S.; SCHULTZ, J. C.; VITOUSEK, P. M., STRAIN, B. R., 1990, Growth responses of tropical shrubs to treefall gap environments.
Ecology **71**, 165–179.

DE STEVEN, D.; PUTZ, F. E., 1984, Impact of mammals on early recruitment of a tropical tree, *Dipteryx panamensis*, in Panama.
Oikos **43**, 207–216.

HUBBELL, S. P., 1980, Tree dispersion, abundance, and diversity in a tropical dry forest.
Science **203**, 1299–1309.

JANZEN, D. H., 1971, Herbivores and the number of tree species in tropical forests.
Am. Nat. **104**, 501–528.

JOYNER, S. P. (ed.), 1985, SAS/STAT guide for personal computers. Version 6 edition, SAS Institute Inc., Cary, NC.

KITAJIMA, K.; AUGSPURGER, C. K., 1989, Seed and seedling ecology of a monocarpic tropical tree, *Tachigalia versicolor*.
Ecology **70**, 1102–1114.

LÓPEZ ORNAT, A., 1983, Localisación y medio físico,
In: SIAN KA'AN, Estudios preliminares de una zona en Quintana Roo propuesta como Reserva de la Biósfera. Centro de Investigaciones de quintana Roo, A. C. Puerto Morelos, Quintana Roo, Mexico, pp. 20–49.

MCVAUGH, R., 1963, Flora of Guatemala.
Fieldiana Botany **24**, 377–379.

MIRANDA, F., 1958, Vegetación de la peninsula Yucateco,
In: BELTRAN, E. (ed.), Los recursos naturales del sureste y su aprovechamiento. Volume II, Instituto Méxicana Recursos Nautrales Ren. México, D. F. , México, pp. 215–227.

PYKE, D. A.; THOMPSON, J. N., 1986, Statistical analysis of survival and removal rate experiments.
Ecology **67**, 240–245.

SCHUPP, E. W., 1988, Seed and early seedling predation in the forest understory and in treefall gaps.
Oikos **51**, 71–78.

UHL, C.; CLARK, K.; DEZZEO, N.; MAQUIRINO, P., 1988, Vegetation dynamics in Amazonian treefall gaps.
Ecology **69**, 751–763.

WALTER, H.; LIETH, H., 1967, Klimadiagramm-Weltatlas, VEB. Gustav Fischer Verlag, Jena.

WEST, R. C., 1964, Surface configuration and associated geology of Middle America,
In: WEST, R. C. (ed.), Handbook of Middle American Indians Vol. 1, University of Texas Press, Austin, pp. 33–83.

WHIGHAM, D. F.; ZUGASTY TOWLE, P.; CABRERA CANO, E.; O'NEILL, J.; LEY, E., The effect of annual variation in precipitation on growth and litter production in a tropical dry forest in the Yucatan of Mexico.
Tropical Ecology (in press).

WILKINSON, L., 1989, SYSTAT: The system for statistics.
Evanston, IL.

Chapter 5

Colonizing success in plants: Genetic variation and phenotypic plasticity in life history traits in *Capsella bursa–pastoris*

H. HURKA and B. NEUFFER

5.1 Introduction

Weeds can be defined as colonizing species that grow in habitats markedly disturbed by man (BAKER 1965). They share or are characterized by a number of genetic features which contribute to their success and spread in disturbed habitats. These include polyploidy, self–fertilization or clonal reproduction with a high reproductive capacity, high levels of enzyme multiplicity at duplicated gene loci and genetically depauperate populations, substantial interpopulation differentiation and high levels of phenotypic plasticity (BROWN and MARSHALL 1981; BARRETT and RICHARDSON 1986). However, it seems difficult to predict the colonizing success of a population introduced into a given habitat. "The relative impact of evolutionary forces acting on colonizing species seems to vary with species and event such that each case is unique. Colonizing species are of interest not so much because they are a homogeneous group, but because they display a wide range of possible evolutionary outcomes" (BROWN and MARSHALL 1981).

Adaptation to a wide range of environments — a common feature of weedy plants — can be achieved by considerable phenotypic plasticity as well as genetic differentiation. It has been postulated that genetic variation and phenotypic plasticity may be alternative strategies. Phenotypic plasticity appeared to replace genetic variation in three pairs of congeneric *Avena*-, *Bromus*- and *Limnanthes*-species (see JAIN 1979). However, reviewing the more recent literature, QUINN (1987) argues that there is "no reason to suspect that selection for genetic variation and phenotypic plasticity would always be naturally exclusive".

Extensive studies on *Capsella bursa-pastoris* (L.) MED. (*Brassicaceae*) revealed a high degree of genetic and phenotypic variation in life history traits. This paper concentrates on traits involving survival, reproduction and dispersal. The role of genetic polymorphism and phenotypic plasticity for colonization success is discussed. Evidence is provided that *Capsella bursa-pastoris* followed different adaptive strategies associated with different life history traits.

5.2 History of *Capsella bursa-pastoris*

Capsella bursa-pastoris is a very successful colonizer on arable soils and in ruderal habitats. Its distribution is world-wide. It is a tetraploid ($4x = 2n = 32$ chromosomes). Question arises as to whether *Capsella bursa-pastoris* is of auto- or alloploid origin. Studies at the isozyme level helped to clarify this question. The inheritance of Aspartate Aminotransferase (AAT) allozymes was disomic. This disomic inheritance and the true-breeding multiple-banded patterns ("fixed heterozygotes") would suggest an allotetraploid origin for *Capsella bursa-pastoris* (HURKA et al. 1989). This conclusion is further corroborated by studies of the enzyme systems PGI, PGM, LAP, GDH, MDH (HURKA et al., in preparation). Support also comes from isoelectric focusing (IEF) of Ribulose-1,5-bisphosphate carboxylase/oxygenase (= Rubisco). Rubisco is composed of large subunits (LSU) coded by chloroplast DNA, and of small subunits (SSU) coded by nuclear DNA. IEF patterns of Rubisco and its subunits were studied in the genus *Capsella* (MUMMENHOFF and HURKA 1990). The two diploid species *C. rubella* and *C. grandiflora* and the tetraploid *Capsella bursa-pastoris* share the same LSU banding pattern, indicating a close relationship. However, the species differ in their SSU patterns. The two diploid species have identical SSU patterns, whereas that of *Capsella bursa-pastoris* is different. The SSU pattern of the tetraploid may indicate an ancient alloploid from which the original SSU multigene family has diverged with time.

The genus *Capsella* almost certainly originated in the Old World, presumably in the East Mediterranean area including Asia Minor. Early Neolithic records of *Capsella bursa-pastoris* seeds are reported in Turkey between 5,850-5,600 BC, whereas in Central Europe the earliest records date from the Middle and Late Neolithicum, approximately from 4000 BC (WILLERDING 1986). It seems that *Capsella* was brought to Central Europe with the spread of agriculture from the Middle East. However, there are two records of *Capsella* seeds from the Pleistocene in Great Britain (GODWIN 1975). This would rather suggest that *Capsella* may constitute part of the late glacial or preboreal flora. Its original habitats might have been river banks which were periodically flooded (LOHMEYER 1954). From Eurasia, *Capsella* extended its range into the New World and Australasia by following European colonists.

5.3 Breeding system of *Capsella bursa-pastoris*

The mating system is crucial in determining population genetic structure. It has long been known that among highly successful colonizers predominant self fertilization or apomixis is common. Reasons suggested for this are to ensure reproduction and greater genetic similarity between progeny. It can be argued that evolution under colonization would promote reduced levels of outcrossing. However, the opposite trend of an increase in outcrossing, has also been documented (BROWN and BURDON 1987). It appears "that despite the initial advantages of a self sufficient breeding system colonizing populations may have to increase outcrossing to generate heterotic combinations needed for survival in harsh environments," (BROWN and MARSHALL 1981).

SHULL (1929) reported for *Capsella bursa-pastoris* that self fertilization takes place automatically and almost exclusively, the percentage of outcrossing in greenhouse plants being practically zero. He assumes outcrossing rates of 1–2% under field conditions. Using isozyme markers HURKA et al. (1989) estimated outcrossing rates for different populations between 2 and 10%, still low but higher than SHULL would have predicted. Previous studies using morphological markers have also indicated that outcrossing rates in *Capsella bursa-pastoris* are variable (HURKA and WÖHRMANN 1977). In general, the heterogeneity of outcrossing rates may be related to spatial and temporal variation, population density and maternal genotype. We have strong evidence for all of these variation sources in *Capsella bursa-pastoris*. Temporal variation, for instance, is related to climatic factors. Such aspects of flower morphology as the opening of the flower, exposure of the style, position of the anthers and length of anthesis depend on temperatures and air humidity (HURKA et al. 1976). Plant to plant variation caused by genetic differences between individuals within a population can at least be inferred by the occasional occurrence of pollen–sterile individuals, which is reported in the literature and confirmed by our own observations.

5.4 Reproductive capacity

Characteristic of the ruderal strategy (C–S–R model of primary plant strategies, GRIME 1979) is the tendency for early and heavy resource allocation to reproduction, evident in the early onset of flowering and high seed production. The total seed output depends on the number of seeds per fruit and on the number of fertile fruits per plant. One would expect a priori that growing conditions may highly influence total seed production. Consequently, in determining the reproductive capacity of *Capsella bursa-pastoris* we compared only plants grown under similar conditions, either in open-field experiments, in glasshouses or in growth chambers.

5.4.1 Number of seeds per fruit

In general, the number of seeds per fruit varied between < 10 and > 35, sometimes approaching 40. The seed content depends on several factors, namely (a) pollination and fertilization success, (b) fruit and seed dimensions, (c) the environment and (d) the genotype.

Capsella bursa-pastoris is a predominantly self-pollinated plant. However, fertilization of the ovules depends on temperature, so that in spite of sufficient pollination, seed production may be hampered or even fail. The critical temperatures are often near 5 °C, but vary apparently with genotype (HURKA et al. 1976). High temperatures may also prevent seed-setting. Plants kept at temperature regimes 30:10 °C produced a higher number of sterile fruits than those kept at 10:5 °C (NEUFFER and HURKA 1986b). One would a priori anticipate that the dimensions of the fruit and seeds would determine the maximum number of seeds per fruit. This has been shown for several populations. However, fruit measurements may depend on growth form parameters like the plant height (HURKA and WÖHRMANN 1977). Fruit measurements are also significantly influenced by the environment, especially by temperature. Growth chamber experiments were in accordance with field observations, in that fruits are long and small in midsummer and tend to form various shapes in autumn. Nevertheless, genotypic differences are obvious and, moreover, genotype–environment interactions were found to be different for individual genotypes (NEUFFER and HURKA 1986b; see also ALMQUIST 1923, 1929).

The number of seeds per fruit depends on the environment in which the plants were grown. Parent–offspring comparisons revealed great differences between generations, if parent and daughter plants were kept under different conditions. However, only slight differences were found when the plants were raised under similar conditions (Tab. 5.1).

Genetic variation of the number of seeds per fruit is also detectable (Tab. 5.2). However, it seems that the phenotypic response to environmental conditions can easily mask and exceed genetic differences (cp. Tab. 5.1). The situation is further obscured by correlations between seed set per fruit and the onset of flowering.

Table 5.1: Number of seeds per fruit (average of 10 countings per plant). Parent generations ($n = 10$ plants) raised from single wild plants. Offspring: average of 5 F_1–families ($n = 10$ each) raised from randomly selected plants of the parent generation. In Pop. "Liggeringen" parent generations were grown in narrow flower pots in the greenhouse, F_1–families in an experimental garden. In Pop. "Nabburg" both generations were raised under open–field conditions.

Population	Parent generation ($n = 10$) raised from wild plants		5 offspring families ($n = 10$ each)	
	$\bar{x}$	CV	$\bar{x}$	CV
Liggeringen	9	0.495	23	0.188
	10	0.322	26	0.153
	14	0.370	23	0.256
	15	0.315	23	0.222
	15	0.412	26	0.316
	18	0.349	33	0.120
Nabburg	31	0.211	28	0.212
	29	0.104	25	0.162
	25	0.349	26	0.162
	31	0.178	32	0.112
	28	0.074	28	0.147

Table 5.2: Number of seeds per fruit. Mean values $\bar{x}_i$ of progeny ($n_i = 10$ with 10 counts per plant) from 25 plants taken randomly from a natural population. CV_i = coefficient of variability within progenies, CV_p between progenies within the population. $\bar{\bar{x}}$ = population mean.

$\bar{x}_i$	38	30	30	31	32	31	31	31	32
CV_i	0.095	0.203	0.242	0.224	0.298	0.265	0.270	0.198	0.140
$\bar{x}_i$	33	35	34	33	32	30	36	36	33
CV_i	0.149	0.068	0.158	0.213	0.136	0.211	0.079	0.113	0.129
$\bar{x}_i$	32	31	33	20	28	29			
CV_i	0.138	0.159	0.166	0.286	0.121	0.337			

$\bar{\bar{x}} = 32$

$CV_p = 2.02$

Table 5.3: Estimated seed production per plant (rounded off to the nearest thousand) of individual plants from different populations. All plants were grown in the same experiment (open–field). Plants were single offspring of individual seed collections sampled at random from natural populations. Origin of all populations is Switzerland. $\overline{x}$ = total mean.

Population	279	280	281	282	260	264	267
	7,000	4,000	9,000	6,000	46,000	2,000	16,000
	7,000	12,000		8,000	38,000	2,000	11,000
	4,000	100		1,000	24,000	1,000	2,000
	7,000	1,000		22,000	24,000	2,000	61,000
	17,000			91,000	31,000	4,000	24,000
	65,000			3,000	29,000	3,000	40,000
				11,000	22,000	6,000	69,000
				33,000	19,000	5,000	17,000
				33,000	9,000	2,000	51,000
					25,000	3,000	
$\overline{x}$	17,800	4,300		23,100	26,700	3,000	32,300
CV	1.319	1.266		1.220	0.381	0.521	0.735

Later flowering correlates with less seeds per fruit (HURKA and WÖHRMANN 1977). However, populations react differently in different environments.

5.4.2 Total seed output

The total range of seed production is tremendous: In one common garden experiment, the lowest number of seeds produced by a single plant was 100, the highest 91,000 (Tab. 5.3). There was extremely high variation between the plants within populations (CV > 1 in Pop. 279, 280, 282!). Nevertheless, significant differences in total means between the populations are also obvious (cp. Pop. 260 with 264). As the main inflorescence and all side branches of *Capsella* produce flowers, the total seed output per plant highly depends on the growth form, and the differences in reproductive capacity are mainly due to different growth forms. Although the phenotypic plasticity of growth form parameters is high, genetic differences are nevertheless pronounced. Plant height and number of branches in *Capsella* may depend on the place of origin, either displaying clinal or patchy variation patterns. The growth form is also strongly influenced by the time of flower begin (STEINMEYER et al. 1985; NEUFFER and HURKA 1986b; NEUFFER and BARTELHEIM 1989). The web of interactions determining reproductive capacity is too sophisticated to justify simple generalizations.

5.5 Founding of populations in new habitats

5.5.1 Seed sizes, seed polymorphism

The establishment of new populations in *Capsella* is solely via seeds. Seed characteristics and seed dispersal mechanisms are therefore key characters in understanding the colonizing ability of *Capsella*. Colonizing plants as a group have small seeds. *Capsella* fit well into this general picture. An extensive survey of seed sizes from 72 natural populations of *Capsella bursa-pastoris* from North and Central Europe revealed individual seed weights ranging from about 0.075 mg to 0.16 mg, with an average of 0.12 mg per seed (HURKA and BENNEWEG 1979). The range of seed weights within single plants was about 0.03 mg and less than half of the observed interindividual ranges (0.08 mg). There is a clear correlation between the seed size and the position of the flower in the inflorescence. The heaviest seeds were generally amongst those flowers which were first to open. There was a decrease in seed sizes during the end of the flowering span. This suggests that competition for nutrients is involved.

Variation of seed sizes between as well as within populations was considerable. Cluster analyses were carried out. The Kolmogoroff–Smirnoff–Test was used to test for homogeneity of the clusters. These tests show that a **discontinuous variation** pattern underlies the observed cluster-pattern, thus demonstrating the existence of different morphs and sizes of seed. It is evident from progeny tests that the clustered samples of progenies are preferably located in the parents' clusters and that the progenies are almost completely homogeneous, thus demonstrating the inheritance of these different seed size morphs in general. To get a more profound view into possible genotype–environment interactions, heritability was estimated from parent–offspring regressions for six populations from different climates. It turned out that h^2–values varied considerably, indicating different degrees of genotype environment interactions and/or different genetic structure within single populations. On the whole, evidence was provided for the existence of different genotypes coding for different seed sizes. Secondly, populations can be differently structured with regard to the seed size controlling the genotypes. Thirdly, genotype–environment interactions can be high, providing the plants with an additonal high degree of phenotypic plasticity. — *Capsella bursa-pastoris* displays seed polymorphism, not easy to detect, but nevertheless very pronounced.

It was surprising that no correlation between resource availability in the soils where the plants had grown and seed sizes could be detected; (soils were analyzed for pH, phosphorus, potassium, magnesium, total nitrogen, manganese, iron and sodium). However, by plotting seed sizes against climate types characterized by the symbols of WALTER and LIETH (1967), it was demonstrated that *Capsella* from rough climates are characterized by larger seeds. This clinal pattern may be superimposed by a discontinuous variation pattern. Whenever the soils were highly disturbed, the mean population values for seed sizes were above the average.

We interpret the climate-conditioned variation pattern as ecotypic variation controlled by the genotype, and of adaptive significance. The discontinuous variation pattern seems to be more the outcome of phenotypic plasticity. In highly disturbed habitats where *Capsella* is one of the first plants to emerge, *Capsella* produces much more photosynthetic tissue than when grown under density and successional stress. It seems reasonable to assume that in these environments primary production is enhanced and seeds can be supplied with more reserve products.

Seed sizes affected the germination rate and hypocotyl length of the seedlings of *Capsella bursa-pastoris* (HURKA und BENNEWEG 1979). 80% of the larger seeds from the basal fruits of the main inflorescences of single plants had germinated after 15 days, compared with 50% of the smaller seeds of the apical fruits. Hypocotyls were remarkably longer in seedlings from larger seeds ($\overline{x} = 3.71 \pm 0.16$ mm), compared with those from smaller seeds of the same plants ($\overline{x} = 2.78 \pm 0.15$ mm).

Aspects of different seed sizes produced within and between strains of the same species have long been of concern. The seed size may affect germination, the emergence of the seedlings and seedling growth rate within cultivars and species. The various properties of seed size involve problems of water uptake, the response to the depth of sowing, the extent of light sensitivy in germination, interference with soil micro-topography, seed predation by animals and attack by diseases (for review and discussion see SALISBURY 1942, 1961; HARPER 1977; KHAN 1977; FENNER 1985). There is convincing evidence that the "germination polymorphism" frequently present in weedy species, is associated with seed polymorphism (cp. HARPER 1977). As differences in germination time and seedling performance most critically determine the success or failure of plant populations in uncertain environments, it may be assumed that variations in the expression of seed polymorphism are extremely sensitive indicators of evolutionary changes within weeds. It does not seem surprising then, that seed heteromorphy is common only in the families of *Compositae, Chenopodiaceae, Fabaceae, Brassicaceae* and *Poaceae* the so-called "weedy" families.

A question of substantial interest is whether intraspecific seedsize variation may also affect the final plant growth and related fitness parameters of the mature plant. Most experiments which address this question have been carried out in greenhouse environments or in stands that are homogeneous with respect to seed size. No significant effect on the plant size could be detected. However, when working within environments that closely resembled natural growth conditions, an effect of the seed size on fitness could be demonstrated. Experiments with wild radish (*Raphanus raphanistrum*) by STANTON (1984), for instance, revealed that seedlings from large seeds grew more rapidly and produced more flowers than those from related smaller seeds. These results from field experiments contrasted with those obtained in greenhouse studies, where the seed size had no significant effect on the final plant size. It has been shown in other studies that seedling performance which depends on the seed size has an impact on stress tolerance. Larger seedlings were not as susceptible to water stress or shading by competitors as the related smaller seedlings (HARPER 1977; COOK 1979). It would appear that in natural habitats the seed size can significantly influence the probability of emerging from the soil and the final plant size and, therefore, the reproductive output. The seed size may also influence seedling mortality under certain stress situations.

5.5.2 Establishment near to the previously seeded range

The ripe fruits of *Capsella* split along their septa. The opening of the fruit, however, does not cause the seeds to fall off immediately. The seeds may remain attached to the replum for a while. The actual seed scattering, as well as the opening of the fruit, is enhanced mainly by vibration and contact by men, animals, wind, rain drops, etc. and by fluctuating humidity (HURKA and HAASE 1982). Seeds of test plants kept in greenhouses scattered their seeds no further than 20 cm. In the field, the distances travelled by seeds are often in the same order of magnitude, which can be deduced from the following data. An experimental field of 5 × 6 m was divided into plots of 0.5 × 0.5 m and distribution patterns of *Capsella bursa-pastoris* plants were accurately mapped over a period of four years. The initial distribution pattern was still recognizable at the end of the experiment, despite removing competing species and breaking the ground several times. In the light of the heavy seed output of *Capsella* and recalling the narrow scale of the experimental field, it is reasonable to assume only short seed dispersal distances.

5.5.3 Soil seed bank

By far the best way for a colonizing species to ensure that it is there first when an opening occurs, is to be there already in anticipation of the disturbance. Not surprisingly, then, many colonizing species possess a reserve of dormant seeds in the soil waiting to germinate as soon as the opportunity arises. It has long been known that *Capsella* seeds may constitute a prominent part of the persistent seed banks in arable soils (cp. SALISBURY 1961; ROBERTS 1981). We repeatedly analyzed the buried seed populations of *Capsella* in different localities and found reserves of buried viable seeds in many habitats. The methods used for assessment were direct counting of seeds under a microscope after washing and seiving procedures and counting seedlings as they emerged from the collected soil. In general, our estimates of numbers per m^2 are in accordance with the numbers given in the literature (Tab. 5.4). However, enormous differences in the size of the soil seed banks are obvious from that table. The differences in quantities do not only refer to different habitats, as one would anticipate (compare site B and D, for instance), but there are also highly localized differences within the habitats themselves (see site F) of such orders of magnitudes which cannot be explained by the inherent errors of the estimation procedure. This is of considerable consequence for the genetic structure of a given population. The founding and recruitment of *Capsella*-populations can only be described in general terms. The exploitation of open habitats in a given locality may completely, partly or not at all start from the buried seed population.

Table 5.4: Buried seed populations of *Capsella bursa–pastoris* extracted from soil samples of different locations in Northwest Germany: Münster (A) and Osnabrück (B–G). A dash indicates that no seeds were found. The soil surface area sampled was always 1 m^2 (data from HURKA and HAASE (1982) and hitherto unpublished results).

Site	Soil volume [dm^3]	Estimated seed number	Remarks on habitat and sampling
A	21	50,000	*Capsella* field station, arable soil, disturbed; average of 100 samples randomly taken from a 5 × 6 m plot; Botanical Garden Münster.
B	20	280,000	Experimental field; Botanical Garden Osnabrück.
C 1	16	6,000	Freshly filled up with soil, origin unknown; subplots regularly placed over an area of 18 × 19 m; Botanical Garden Osnabrück.
2	16	–	
3	16	–	
4	16	–	
5	16	500	
6	16	500	
7	16	–	
8	16	–	
9	16	–	
D 1	16	–	Long–established cattle pasture, distance between each subplot 4 m; near Osnabrück.
2	16	–	
3	16	–	
4	16	–	
5	16	–	
6	16	500	
E 1	16	2,000	Reimproved pasture, freshly sown a year before. Subplots 1, 2, 4 and 5 from the corners of the 13 x 20 m area, 3 from the centre; near Osnabrück.
2	16	500	
3	16	500	
4	16	500	
5	16	500	
F 1	16	300,000	Poultry yard for 8 years, abandoned for one year. Distance between F1 and F2 is 4m; near Osnabrück.
2	16	18,000	
G 1	16	14,000	Sheep grazing and mowing area; distance between G1 and G2 is 4 m; near Osnabrück.
2	16	15,000	

Even the subpopulation structure may be highly influenced by localized differences in the soil seed bank.

How can *Capsella* seeds enter the soil? Depending on the texture of the soil, the seeds of *Capsella* can undoubtedly be passed down the soil profile with rain water. In man-operated soils, ploughing and similar activities are important factors. The soil may also be turned over by animals. Especially interesting is the activity of earthworms. Earthworms collected from soils known to contain *Capsella* seeds in high numbers, were kept on wet filter paper and their casts were searched for *Capsella* seeds. A number of seeds could be detected, some of which germinated on wet filter paper. This preliminary test was followed by a number of laboratory experiments, in which earthworms were fed with *Capsella* seeds (HURKA and HAASE 1982). On average, one worm ingested 17 seeds per day, 8 of which were recovered from the casts, the loss amounting to 50%. Consumption by earthworms may be one cause for the loss of *Capsella* seeds from the soil. Other causes are deterioration and germination. Germination experiments proved that the seeds passing through the earthworms did not loose their viability (germination was 88% as compared with 99% of the original seed material). — In another set of experiments it was demonstrated that earthworms do indeed transport *Capsella* seeds vertically through the soil, from top to bottom and from the bottom to the top layers of an experimental container. All findings indicate that under natural conditions earthworms can strongly influence the buried seed population of *Capsella bursa-pastoris*, and that they may constitute an important part in the formation and dynamics of the soil seed bank.

5.5.4 Long distance seed dispersal

According to COQUILLAT (1951), the five most common species of flowering plants in the world are *Polygonum aviculare, Capsella bursa-pastoris, Chenopodium album, Stellaria media* and *Poa annua*. All these species are associated with disturbed habitats. Surprisingly, these plants are not wind dispersed and appear to have no special means of seed dispersal except their small seeds. Although wind is a very widespread and most effective dispersal agent in flowering plants, it seems that internal and external transport by animals and humans is also an extremely effective dispersal mechanism, at least in the case of the above-mentioned species.

In *Capsella bursa-pastoris*, a passive transport of seeds is facilitated by the mucilaginous seed epidermis. In a number of experiments, it has been shown by HURKA and HAASE (1982), that repeatedly wetting and drying the seeds did not result in a measureable decline of mucilage production and that repeated soaking did not influence the germinating capacity. Imbibing the seed mucilage with water vapour does not cause the beginning of germination. These properties can promote long distance transport via myxospermy. Reports in the literature confirm external and internal transport of *Capsella* seeds by animals: (1) The seeds may directly stick to the feet of birds (BERNHARDT 1989) or alternatively, mud containing seeds may be carried on their feet. In general, the carriage of seeds in mud and soil, on transport vehicles, on human footwear, on the hooves of cattle and sheep, on the birds' feet, etc. seems to be an important means of seed dispersal (cp. SALISBURY 1961). (2) The internal transport of *Capsella* seeds is also significant: seedlings could be raised from the dung of cattle, horses, pigs and goats and from the excreta of various birds, including gulls (cp. SALISBURY 1961).

5.6 Germination behavior

While a considerable amount of literature on the germination behavior of *Capsella bursa-pastoris* has been accumulated, the results are far from consistent. Temperature, seed age and geographical origin may strongly influence germination behavior (NEUFFER and HURKA 1988). High temperatures seem to increase germination capacity on the whole. Temperature fluctuations around the daily mean seem to be more important than the mean itself. Relationships between germination capacity and temperature treatments, however, were different, depending on the provenances of the populations. Germination rates, in contrast to germination capacity, were nearly always linked to the same sequence of temperature regimes. This results in a low correlation between germination capacity and germination rate. Seed age has a pronounced influence on both germination capacity and rate. Within all temperature regimes tested, a general decrease in germination capacity with seed age was observed. The rate and uniformity of germination were also affected. However, data indicated that within some populations older seeds may show greater germination than younger ones; germination behavior of equally aged seeds may significantly vary between temperatures; seeds of different ages may germinate to the same extent. It appears that seed age is more likely to influence quantitative aspects of germination than to alter temperature requirements qualitatively.

The effect of temperature on germination alone, however, and the role of temperature in terminating dormancy must be considered separately. When testing seed material which can be regarded as nondormant, it turned out that all populations were able to germinate over the whole range of temperature regimes, although with different germination peaks. The differential temperature preferences among populations would suggest genetic heterogeneity. However, we could not establish convincing evidence for a correlation between this genetic variation and any environmental pattern. There seems to be no ecotypic variation with regard to germination. Our data rather suggest that germination in *Capsella bursa-pastoris* in the field is mainly regulated by induction and emergence from dormancy.

Capsella displays innate or primary dormancy, secondary or imposed dormancy and relative dormancy (terminology following BEWLEY and BLACK 1985). Primary dormancy in *Capsella* may be terminated by chilling imbibed seeds or by after-ripening of dry seeds. However, the extent of primary seed dormancy in *Capsella* varies greatly. Whereas some authors found that freshly matured seeds of *Capsella bursa-pastoris* did not germinate in light or darkness before an after-ripening period of several months, others report a much shorter after-ripening period or none at all. It has long been known that the degree of primary dormancy may be

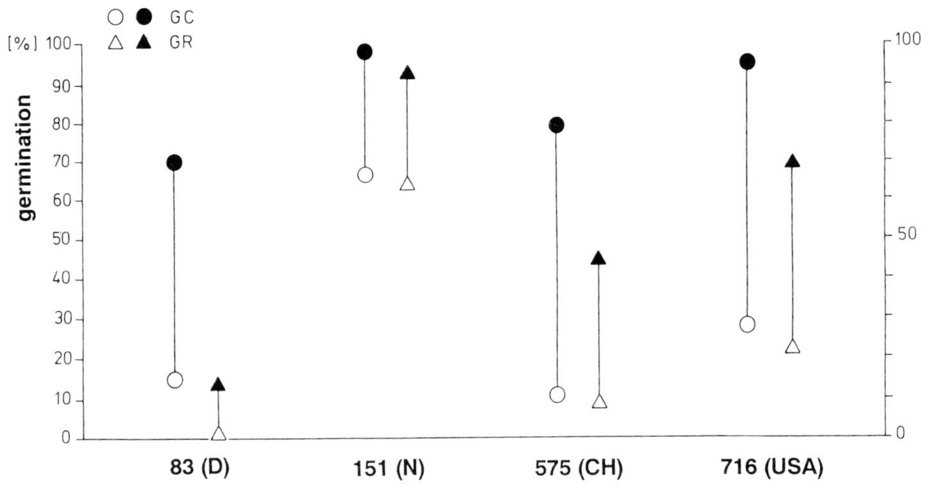

Figure 5.1: Maternal effects on germination capacity GC and germination rate GR. Sister plants raised from individual seed material collected in Germany (D), Norway (N), Switzerland (CH), and North–America (USA) were kept in either artificial light or dark at the end of seed maturation period. Seeds ripened in the dark (black symbols) had higher germination capacities and rates as compared to those ripened in light (open symbols). Seeds were sown immediately after harvesting. Numbers at the X–axis represent population numbers (Data from NEUFFER and SCHULTES).

strongly influenced by environmental and physiological factors operating during the ripening of the seeds on the mother plants ("maternal effects"). These preharvest factors include temperature, photoperiod, light quality, nutrition, maturity at harvest, seed size and seed morphology. The degree of primary dormancy may also be of a genetic nature (for review cp. BEWLEY and BLACK 1982, 1985, for instance). There is no doubt that all or any of these factors might contribute to the variation in germination behavior of *Capsella*. NEUFFER and SCHULTES (in press) showed that seeds ripened on the mother plants in darkness displayed higher germination capacities and rates, compared with the seeds of sister plants ripened in light (Fig. 5.1). These authors also presented strong evidence for endogeneous yearly activity rhythms for drystored seeds. Endogeneous rhythms of germinability must not be confused with seasonality and flushes of germination. Most reports on seasonality in *Capsella* are related to inception and the release of dormancy (FROUD–WILLIAMS et al. 1984, BASKIN and BASKIN 1989) or to light–flashes in connection with soil disturbance (BOSBACH et al. 1982).

Secondary dormancy in *Capsella* develops in dispersed seeds in response to unfavourable germination conditions. Although secondary dormancy is of great importance in germination ecophysiology, reports given in the literature are contradictory for *Capsella*. It seems that low and high temperatures can induce secondary dormancy. A further complication arises by the fact that *Capsella* seeds do not suddenly become non–dormant, but pass through a period of slow germination rate and/or low capacity. This period is also characterized by an enlargement of its temperature limits until the full potential is gradually reached (relative dormancy, conditional dormancy).

Dormancy and germination in relation to seed bank ecology is an aspect of special interest. As already discussed, *Capsella* seeds may constitute a prominent component of the persistent soil seed banks and are reported to be viable for more than 30 years buried in the soil (SALISBURY 1961; 35 years in Beal's long–termed burial studies, KIVILAAN and BANDURSKI 1973). Patterns of change in dormancy during the burial of seeds in soil are discussed by KARSSEN (1980/81) in general, and by BASKIN and BASKIN (1989) for *Capsella* in particular. In their experiments buried seeds of *Capsella bursa–pastoris* exhibited an annual conditional dormancy/non–dormancy cycle. Seeds after–ripened during summer became conditionally dormant, and remained non–dormant during autumn and winter. Seeds entered conditional dormancy in early spring and became non-dormant during the second summer of burial, and the cycle was then repeated. Although seeds entered conditional dormancy in early spring, they did not lose the ability to germinate at March and April temperatures, which means that seeds brought to the soil surface should germinate at any time during the growing season in regions where summers are relatively cold and moist (see for instance BOSBACH et al. 1982).

Germination strategy of *Capsella* appears to be different from variation patterns of other life history traits such as growth form and flowering. There is considerable evidence for genetic heterogeneity for germination, but no convincing evidence for ecotypic variation. Our data suggest that the germination of *Capsella bursa–pastoris* in the field is mainly regulated by dormancy and by the influence of temperatures on the rate of germination. Since dormancy inception and release is highly influenced by environmental factors, including pre– and postharvest conditions, this appears to be a satisfactory way to regulate seedling emergence in time and space for *Capsella*, which occupies a wide range of soil types, climatic conditions and which has to cope with the unpredictability of its habitats. Such a germination strategy comes close to an "all purpose genotype" (BAKER 1974).

Conclusion

Capsella bursa–pastoris shares some features common to many weedy plants. It is a tetraploid plant with an increased number of isozymes and "internal hybridity" (HURKA et al. 1989). As argued by several authors, this "fixed heterozygosity" might be a key element in the colonizing ability of polyploids. The mating system is predominantly autogamous. Outcrossing rates, however, vary spatially and temporally. The predominantly self–fertilizing, yet flexible breeding system combined with polyploidy would contribute to the colonizing ability of *Capsella bursa-pastoris*.

Life history traits of *Capsella bursa-pastoris* exhibit a very complex variation pattern characterized by a multitude of interacting factors. Life history is influenced by the effective breeding system, seed dispersal mechanisms, and properties of the soil seed bank. Adaptive strategies as reflected in variation patterns depend on the trait under study and on the amount and pattern of phenotypic plasticity.

Most seeds fall to the ground near the mother plant and are incorporated into the soil seed bank. Therefore, populations in occupied sites are primarily recruited from local seed material. Selection can operate over many years on the same gene pool and adaptation to local environments may evolve. Chance long distance transport of a small number of seeds to hitherto unoccupied sites can create new populations. Such populations are often genetically rather uniform. If different populations were established by small groups of founders drawn from a polymorphic parental populations, variations between the founder populations could be expected. It can be assumed that the existence of genetic divergence between *Capsella* populations is partly due to such a mechanism. The soil seed bank can serve as a permanent source for the introduction of "new" genotypes into the actual plant population, depending on how well it is shaken up and on the degree of polymorphism stored in the soil. This introduction of genetic variation will be counteracted, however, by the competitive elimination of seedlings and plants under conditions of high successional displacement stress. Therefore, one would expect less variable populations in places where succession has already begun. It is also to be expected that in freshly filled or newly broken soils more seeds are brought up to the surface than in bare, but otherwise nondisturbed soils. This disturbance increases the probability of different genotypes in highly disturbed soils. Differences in polymorphism between populations on highly and less disturbed soils have been demonstrated (BOSBACH et al. 1982).

Patterns of variation may thus be patchy and random. Nevertheless, overall variation patterns might reflect genotypic response to the environment and hence ecotypic variation. However, phenotypic plasticity may obscure genotypic differentiation. The recognition of ecotypic variation may also depend on the characters which are measured. Overemphasis on single characters may confuse rather than clarify the understanding of variation patterns. Adaptive strategies as reflected in variation patterns depend on the trait under study and on the amount and pattern of phenotypic plasticity. Germination, for instance, although displaying genetic polymorphism, seems to be regulated mainly by the factors contributing to the inception and breaking of seed dormancy, which depends on the pre- and postharvest conditions. As yet, no correlation between the genotype and environmental parameters has been detected so far. Other life history traits, such as the beginning of flowering, and growth form parameters may respond with genotypic variation to the environment (NEUFFER and HURKA 1986a and b), both on a macrogeographic and a microgeographic scale (NEUFFER and BARTELHEIM 1989; NEUFFER 1990).

Nevertheless, the reaction norm of these different genotypes is high. In addition, NEUFFER and HURKA (1986a) have provided evidence for genetic polymorphism for phenotypic plasticity. Obviously, *Capsella bursa-pastoris* has adopted different adaptive strategies associated with different life history traits.

Germination behavior in *Capsella* comes close to that of a "general purpose genotype", as do some traits involved in reproduction. Other life history components rely more on adaptive genetic polymorphism. However, it would appear from our results that selection for adaptive genetic variation and phenotypic plasticity are not mutually exclusive. Rather, response in genetic polymorphism and a concomitant increase in phenotypic plasticity seems to account for the colonization success of *Capsella bursa-pastoris*.

Genetic variation at the molecular level is different in quality from that at the phenotypic (polygenic) level (HURKA et al. 1989). Allozyme variation does not seem to reflect adaptations to local environments, nor does it provide any image of the adaptive diversity of the species. Instead, allozyme variation is more likely to reflect genetic drift effects, gene flow events, and the effective mating system. Variation at the allozyme level seems to be more the result of the origin and evolutionary history of the various populations than a result of their local ecotypic adaptation (HURKA 1990). Studies at the phenotypic and molecular level provide different information. Both are necessary for understanding the evolutionary biology of a species.

Acknowledgments

We thank K. Bachmann, Amsterdam, and K. Wöhrmann, Tübingen, for valuable comments on an early draft of the manuscript. Supported by the German Research Foundation (DFG).

Bibliography

ALMQUIST, E., 1923, Studien über *Capsella bursa-pastoris* (L.) II.
Acta Horti Bergiani **7**, 41–95.

ALMQUIST, E., 1929, Zur Artbildung in der freien Natur.
Acta Horti Bergiani **9**, 37–75.

BAKER, H. G., 1965, Characteristics and modes of origin of weeds.
In: Baker, H. G.; Stebbins, G. L. (eds.), The Genetics of Colonizing Species, Acad. Press, New York and London, 147–172.

BAKER, H. G., 1974, The evolution of weeds.
Ann. Rev. Ecol. Syst. **5**, 1–24.

BARRETT, S. C. H.; RICHARDSON, B. J., 1986, Genetic attributes of invading species.
In: Groves, R. H.; Burdon, J. J. (eds.), Ecology of Biological Invasions, Cambridge Univ. Press, Cambridge, London, 21–33.

BASKIN, J. M.; BASKIN, C. C., 1989, Germination responses to seeds of *Capsella bursa-pastoris* exposed to seasonal temperature changes.
Weed Research **29**, 205–212.

BERNHARDT, K.-G., 1989, Pflanzliche Strategien der Pionierbesiedlung terrestrischer und limnischer Sandstandorte in Nordwestdeutschland.
Drosera 1989, 113–124.

BEWLEY, J. D.; BLACK, M., 1982, Physiology and Biochemistry of Seeds II: Viability, Dormancy and Environmental Control.
Springer, Berlin, Heidelberg, New York.

BEWLEY, J. D.; BLACK, M., 1985, Seeds. Physiology, Development and Germination.
Plenum Press, New York, London.

BOSBACH, K.; HURKA, H.; HAASE, R., 1982, The soil seed bank of *Capsella bursa-pastoris* (Cruciferae): its influence on population variability.
Flora **172**, 47–56.

BROWN, A. H. D.; BURDON, J. J., 1987, Mating systems and colonizing success in plants.
In: Gray, A. J.; Crawley, M. J.; Edwards, P. J. (eds.), Colonization, Succession and Stability, Blackwell, Oxford, London, 115–131.

BROWN, A. H. D.; MARSHALL, D. R., 1981, Evolutionary changes accompanying colonization in plants.
In: Scudder, G. G. E.; Reveal, J. L. (eds.), Evolution Today, Proc. Sec. Intern. Congr. Syst. Evol. Biol. Hunt Instit. Bot. Document. Pitsburgh, Pensylv. USA, 351–363.

COOK, R. E., 1979, Patterns of juvenile mortality and recruitment in plants.
In: Solbrig, O. T.; Jain, S. K.; Johnson, G.; Raven, P. H. (eds.), Topics in Plant Population Biology, Columbia Univ. Press, New York, 207–231.

COQUILLAT, M., 1951, Sur les plant les plus communes de la surface du globe.
Bull. Mens. Soc. Linn. Lyon **20**, 165–170.

FENNER, M., 1985, Seed Ecology.
Chapman and Hall, London.

FROUD-WILLIAMS, R. J.; CHANCELLOR, R. J.; DRENNAN, D. S. H., 1984, The effect of seed burial and soil disturbance on emergence and survival of arable weeds in relation to minimal cultivation.
J. Appl. Ecol. **21**, 629–641.

GODWIN, H., 1975, The History of the British Flora.
Cambridge Univ. Press, Cambridge.

GRIME, J. P., 1979, Plant Strategies and Vegetation Processes.
Wiley and Sons, Chichester, New York.

HARPER, J. L., 1977, Population Biology of Plants.
Acad. Press, New York.

HURKA, H., 1990, Differentiation and adaptation in the genus *Capsella* (*Brassicaceae*).
In: Kawano, S. (ed.), Biological Approaches and Evolutionary Trends in Plants, Acad. Press, London, 19–32.

HURKA, H.; BENNEWEG, M., 1979, Patterns of seed size variation in populations of the common weed *Capsella bursa-pastoris* (*Brassicaceae*).
Biol. Zentralbl. **98**, 699–709.

HURKA, H.; HAASE, R., 1982, Seed ecology of *Capsella bursa-pastoris* (*Cruciferae*): dispersal mechanism and the soil seed bank.
Flora **172**, 35–46.

HURKA, H.; WÖHRMANN, K., 1977, Analyse der genetischen Variabilität natürlicher Populationen von *Capsella bursa-pastoris* (*Brassicaceae*).
Bot. Jahrb. Syst. **98**, 120–132.

HURKA, H.; FREUNDNER, S.; BROWN, A. H. D., PLANTHOLT, U., 1989, Aspartate aminotransferase isozymes in the genus *Capsella* (*Brassicaceae*): subcellular location, gene duplication, and polymorphism.
Biochem. Genetics **27**, 72–90.

HURKA, H.; KRAUSS, R.; REINER, TH.; WÖHRMANN, K., 1976, Das Blühverhalten von *Capsella bursa-pastoris* (*Brassicaceae*).
Plant Syst. Evol. **125**, 87–95.

JAIN, S. K., 1979, Adaptive strategies: polymorphism, plasticity and homeostasis.
In: Solbrig, O. T.; Jain, S. K.; Johnson, G.; Raven, P. H. (eds.), Topics in Plant Population Biology, Columbia Univ. Press, New York, 207–231.

KARSSEN, C. M., 1980/81, Patterns of change in dormancy during burial of seeds in soil.
Israel J. Bot. **29**, 65–73.

KHAN, A. A., (ed.), 1977, The Physiology and Biochemistry of Seed Dormancy and Germination.
North Holland Publ. Co, Amsterdam.

KIVILAAN, A.; BANDURSKI, R. S., 1973, The ninety-year period of Dr. Beal's seed viability experiment.
Amer. J. Bot. **60**, 140–145.

LOHMEYER, W., 1954, Über die Herkunft einiger nitrophiler Unkräuter Mitteleuropas.
Vegetatio **5/6**, 63–65.

MUMMENHOFF, K.; HURKA, H., 1990, Evolution of the tetraploid *Capsella bursa-pastoris* (*Brassicaceae*): isoelectric focussing analysis of Rubisco.
Plant Syst. Evol. **172**, 205–213.

NEUFFER, B., 1990, Ecotype differentiation in *Capsella.*
Vegetatio **89**, 165–1719.

NEUFFER, B.; BARTELHEIM, S., 1989, Genecology of *Capsella bursa-pastoris* from an altitudinal transsect in the Alps.
Oecologia **81**, 521–527.

NEUFFER, B.; HURKA, H., 1986a, Variation of development time until flowering in natural populations of *Capsella bursa-pastoris* (*Cruciferae*).
Plant Syst. Evol. **152**, 277–296.

NEUFFER, B.; HURKA, H., 1986b, Variation of growth form parameters in *Capsella* (*Cruciferae*).
Plant Syst. Evol. **153**, 265–279.

NEUFFER, B.; HURKA, H., 1988, Germination behavior in populations of *Capsella bursa-pastoris* (*Cruciferae*).
Plant Syst. Evol. **161**, 35–47.

NEUFFER, B.; SCHULTES, J., in press, Das Keimverhalten der annuellen Unkraut–Pionierpflanze *Capsella bursa-pastoris* (L.) MED. Maternale Effekte und endogene Rhythmik.
Verh. Ges. f. Ökologie.

QUINN, J. A., 1987, Complex patterns of genetic differentiation and phenotypic plasticity versus an outmoded ecotype terminology.
In: Urbanska, K. M. (ed.), Differentiation Patterns in Higher Plants, Acad. Press, London, 95–113.

ROBERTS, H. A., 1981, Seed banks in soil.
Advanc. Appl. Biol. **6**, 1–55.

SALISBURY, E. J., 1942, The Reproductive Capacity of Plants.
Bell and Sons, London.

SALISBURY, E. J., 1961, Weeds and Aliens.
Collins, London.

SHULL, G. H., 1929, Species hybridizations among old and new species of shepherd's purse.
Proc. Internat. Congr. Plant Sci. **1**, Collegiate Press, George Banta Publ. Co., 837–888.

STANTON, M. L., 1984, Seed variation in wild radish: effect of seed size on components of seedling and adult fitness.
Ecology **65**, 1105–1112.

STEINMEYER, B.; WÖHRMANN, K.; HURKA, H., 1985, Phänotypenvariabilität und Umwelt bei *Capsella bursa-pastoris* (*Cruciferae*).
Flora **177**, 323–334.

WALTER, H.; LIETH, H., 1960–1967, Klimadiagramm Weltatlas.
VEB G. Fischer, Jena.

WILLERDING, U., 1986, Zur Geschichte der Unkräuter Mitteleuropas.
K. Wachholtz Verl., Neumünster.

Chapter 6

Niches of longevity and stress

F. KLÖTZLI

6.1 Introduction

Spectacular old trees in our daily environment are probably a few hundreds of years old, and, very rarely, an old single tree can be found passing the age of 500 years. Such monuments of nature often have a historical background, as e.g. some oaks growing since almost pre–Saxon times (1500 a) or some linden trees (> 800 yrs), or when by some chance they have escaped natural disasters as for e.g. chestnut (~ 1000 a) or some conifers: Norway spruce (400), Scotch pine (up to 600), juniper (probably more than 600), yew or fir (see e.g. KANNGIESSER 1907, 1909). We may add some famous tourist attractions to that bunch of old world trees, often overestimated in age, as e.g. the dragon tree (*Dracaena draco*) on Teneriffa/Canary Islands (around 200 yrs old) or the Montezuma cypress of Sta. Maria del Tule, near Oaxaca/Mex., a triple fused stem (only around 1500 yrs), one of the largest stems existing.

More often such old individuals are met with under more stressed conditions, such as those determining timberlines, especially in drier areas (stone pine > 750, European larch > 400, Bristlecone pine > 4600, Centralasian junipers > 1000, *Cupressus dupreziana* of the Saharan Tassili and *Welwitschia* of Namibia, both for

Figure 6.1: *Cupressus dupreziana* in the shrub–semidesert at Tamrit–Tassili range, Algerian Sahara, about 1,500 m a.s.l. The shrubs in the foreground are *Zilla spinosa*. Photo taken by J. Üblagger on February 25, 1990.

Figure 6.2: *Pinus longaeva* stand in the White Mountains, about 3200 m a.s.l., August 1988.

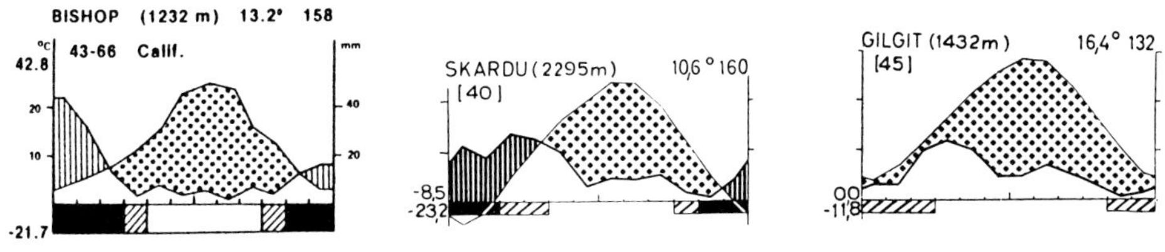

Figure 6.3: Climate diagrams for representative stations in the Great Basin–Colorado Plateau sagebrush semi-desert (Data from National Weather Service network, in WEST et al. 1983, Bishop) and in the Northern Areas of Pakistan (WALTER and LIETH 1960 ff.).

certain > 1500). But also the legendary giant sequoias (Sierra redwood *Sequoiadendron giganteum* > 2000 yrs) grow under very specific site conditions. Certainly, human influence is often to be blamed for the scarcity of such trees in the more accessible areas. However, the existence of particularly old individuals (or even stands) is more readily given under specific climatic conditions, definite site conditions with lack of water (and therefore often also lack of nutrients) or then — on the other hand — bog complexes. Thus marked longevity is mostly in liaison with marked stress of water under limiting site conditions, temperature and/or nutrients and therefore slow growth (WARD 1982, compare *Juniperus communis*, N–Kola Peninsula/USSR, examples of 8.3 cm radius, 544 yrs old, another with 275 cm circumference).

In this context we will not discuss problems around genetic fixation of life span nor signs of senescence or the ageing process as such or even definitions of age (compare e. g. WOOLHOUSE 1972, BEHNKE et al. 1978, COMFORT 1978). The only goal of this paper is to review some aspects of living conditions of old groves of some particular trees, especially comparing growth conditions of Bristlecone pine with those of Centralasian *Juniperus macropoda* discovered recently and existing in similar ecological niches (KLÖTZLI et al. 1989, annex; see also WEST 1983).

6.2 Circumstances of longevity

6.2.1 Site conditions and ecophysiological reactions of Bristlecone pine

Longevity of stands and their eco–physiological reactions have been particularly well analysed in populations of Bristlecone pines (especially in *Pinus longaeva* but also in the more eastern *P. aristata*, compare MOONEY et al. 1966, SCHULZE et

Table 6.1: Comparison of Bristlecone pine and Great juniper sites.

	White Mountains (nr. and above Bishop Calif.)	Hunza Valley, Karakorum (nr. Boibar above Morkhoon)
Climate Diagram (WALTER and LIETH, 1960ff.)	compare Bishop	compare Gilgit and Skardu of the Northern Areas (Pakistan) giving an approx. view on the prevailing conditions
altitude [m]	2900–3500	3200–3900
precip. [mm]	> 400	estim. at about 300–400
peak(s) of prec.		IV/V (VIII)
lower belt	Juniper Piñon woodland, mostly above *Artemisia* semi-desert communities	*Artemisia*–semi-desert
higher belt	*Trifolium/Koeleria*	*Kobresia/Potentilla*
typical herb genera in woodland belt	*Artemisia* *Arenaria* *Astragalus* *Erigeron* *Festuca* *Koeleria* *Oxytropis* *Poa* *Ribes*	*Artemisia* *Arenaria* *Astragalus* *Erigeron* *Festuca* *Koeleria* *Oxytropis* *Poa* (*Ribes*)
	Chrysothamnus *Haplopappus* *Phlox*	*Acantholimon* *Agropyron* *Androsace* *Boraginaceae* *Cruciferae* div. spec. *Ephedra* *Kobresia* *Myosotis* *Polygonum* *Potentilla* *Stipa*

Genus & species names acc. to POLININ and STAINTON, 1984.

al. 1967, compiled in TRANQUILLINI 1979, BARBOUR and MAJOR 1988). The oldest individuals are near the limit of any tree growth and the below–mentioned measurements show that their carbohydrate reserves are exhausted in years of severe summer drought. But under special conditions *P. aristata* (3150 m, Colorado) is still forming logs above krummholz of *Picea engelmannii* and *Abies lasiocarpa*, proving the genetic fixation of definite stress endurance.

Photosynthetic activity and transpiration can be very low in summer but working at a steady level, even in periods of extreme drought at a suction force in the upper soil horizon of up to 15 bar (photosynthetic activity under optimal light and temperature conditions 1.0–1.3 mg $CO_2 \cdot g^{-1} DW \cdot h^{-1}$, transpiration rate only 140 $mg \cdot h^{-1}$ or 1.18 $g \cdot d^{-1}$, compared to *P. flexilis* growing under the same conditions with 400 to 3.22 (transpiration rate) respectively (stations being at 3100–3300 m, with cover values of around 30 %). Although winter can be very severe (T $< 0°C$ from January to April) these trees behave unlike other conifers and depressed summer rates are maintained until late winter. Very low activities are carried on later on and provoke losses of around 140 mg $CO_2 \cdot g^{-1}$ needle DW. These losses due to high respiration rate and low photosynthetic activity give a very low C–balance and may bring individuals to the brink of their existence, and for about half the growing season they have to make up for these respiration losses (117 days of photosynthesis at peak summer rate of 1.2 mg $CO_2 \cdot g^{-1} DW \cdot h^{-1}$ at 15 °C). Warmer days in winter do apparently not break a state of near–dormancy to equalize the losses ("be–prepared"–state as with a number of European conifers; KLÖTZLI 1975). Needles must be well adapted to such harsh conditions: although cuticular transpiration is not stopped during winter, needle replacement only amounts to 1/20 (up to 1/3 in abnormal years). (In the case of *P. flexilis* it is 1/4.) A special cell structure in the xylem also makes it resistant to moisture stress (more details in TRANQUILLINI 1979, HOLTMEIER 1989).

Above all, also nutrient conditions can be restrictive on dolomite or limestone (or even granite). Under these severe conditions, however, special resistant dense resinuous wood is formed, saving it from heart rot or any other parasites (insects, like e.g. bark beetles, mites, fungi) and reduces water loss. Even dead wood is keeping up its resistance to decay for periods of over 1000 years (On reactions of xylem and phloem anatomy see CONNER and LANNER, 1990).

6.2.2 Comparable niches in Eurasia

Continuous stress under such site conditions definitely creates organic tissues with only little attraction for parasites. Under similar climatic and edaphic conditions — predominantly winter precipitations (300–500 mm in total), summer drought, very cold winter months, high temperature intervals in summer, high irradiation, timberline conditions particularly on dolomite rock or on rubble slopes — not only the *Pinus aristata* group shows pronounced longevity of such exposed populations. Other members of the genus, particularly in mediterranean high mountains, or even cupressoid conifers show similar capacities: viz. *Pinus montana* in the eastern Pyrenees/SP, *P. leucodermis* on Olymp/GR, *P. hartwegii* on Popocatepetl/Mex.

(to a lesser extent *P. cembra* in dry alpine valleys), and then, in extreme continental Asia, the junipers (compare some details in WEST 1983, contributions of WALTER and BOX, BRECKLE resp., from northern Tien Shan [*J. turkestanica, J. sibirica*], Pamir Alay [*J. turkestanica, J. seravschanica*], Turkestan Range [*J. turkestanica, J. semiglobosa*], West Pamir [*J. semiglobosa, J. seravschanica*] and also Tibet [with *J. squamata, J. tibetica, J. semiglobosa, J. przewalskii, J. convallium*] or even Elburz, open stands, small trees, krummholz– or cushion–like) which mostly grow under timberline conditions and above arid valleys with *Artemisia* semi–desert.

6.2.3 Junipers of the Karakorum ranges

A nearly identical niche to that of the White Mountains (Inyo Co. Cal., USA) is given in some dry valleys of the inner Karakorum (compare also Pamir and Hindukush in WEST 1983, or WALTER and BRECKLE 1986, report of KLÖTZLI et al. 1989; see annex and Fig. 6.6). While quite stately stands of *Juniperus excelsa* (= *macropoda*) are known from Balouchistan (AHMED et al. 1990, 20–30 cm, 160±38yrs, 10±3 yrs·cm^{-1}) and mixed with *Picea smithiana, Pinus wallichiana, Betula utilis* also from the southern part of the Hunza Valley, northernmost Pakistan, the physiognomy of this same tree species is somewhat similar to that of *Pinus longaeva* on the more extreme crests of the White Mountains. Scattered individuals, gnarled, struck by lightning and torn by winds, often with single living branches and multistemmed, are interwoven with sage (e. g. *Artemisia maritima* coll., *A. gmelinii*) and its companions from the montane semi–deserts, on windswept base rich slopes also with representatives of the thorny cushion formation with *Acantholimon lycopodioides* (here in contrast to the White Mountains; for further details on the vegetation see HARTMANN 1968, 1977; Table 6.1).

During our work in the upper vegetation belts of the Karakorum Mountains these stands have only been analysed floristically, partly edaphically, only a few stem discs have been cut from fallen logs and the annual rings were counted. While the southern populations had trunks of around 300 yrs of age (40 cm dbh), the northern stands were often multistemmed (compare fused stems of Montezuma cypress, yew etc.), but some true individuals in lower altitudes (3500 m) showed an age of between 800 and 1000 a (25–30 cm). Unfortunately, details on the tree line specimens (3900 m) could, so far, not be assessed, except for some pieces of dead wood lying there on slate rubble. The lowest ring widths were approximately 100–120 rings per cm (~ 0.01 mm being the lowest; compare WARD 1982, *J. communis*, 0.03 mm·yr^{-1} for stems of an age of about 200 yrs). Therefore, I assume that the uppermost single stems at timberline (~ 70 cm dbh) must have an age of more than 2000 yrs (theoretical max. around 3500 yrs; compare also MUKHAMEDSHIN 1977, > 1000 yrs for Russian specimens.

Further work on these populations will be done by MIEHE (G. and Sabine, Göttingen) BOSSHARD and myself of the Karakorum working groups.

A special case is given, by the way, by the Sierra redwood groves of the central

Figure 6.4: *Juniperus excelsa* syn. *J. macropoda*, uppermost specimen on crest, Boibar, about 3900 m a.s.l., August 1989.

Figure 6.5: *Juniperus excelsa* syn. *J. macropoda*, open stand at the timberline, Boibar, about 3800 m a.s.l., August 1989.

and southern Sierra Nevada at 1400–2000 (2450) m. (The case of the coastal redwood will not be treated here.) Although under mediterranean type montane climate they reach one of the largest biomasses for single trees, and the groves are of a remarkable centennial stability. In this case age and biomass are specially supported by additional underground water minimizing some water stress during the summer drought period. These flushed places with summergreen spring fen vegetation are quite frequent between the redwood stems. Very large juniper stems are also found under similar flushed conditions.

6.3 Conclusions

Combining all growing conditions in summer drought areas where extremely old stands or individuals can be seen, one main conclusion is clear: Permanently stressed individuals, growing and living under stress of water, nutrients, or generally, climatic conditions, build up tissues which have a very low attraction for parasites of most kinds, fungi, and insects.

More attraction is given to their seed. Thus parasites concentrate preponderantly on the more easily available parts of these remarkable tree individuals, during the very few years of good yield. By this process, normal regeneration is mostly prevented, adding to the picture we have of such populations, being void of a well thriving younger generation (as an exception, WALTER and BOX in WEST 1983, are aware of good rejuvenation in the Turkestan Range, 2800–3100 m, in mixed juniper stands). Probably, the heavier attack, therefore, on female trees may accentuate the altitudinal shift in the sex ratio, sponsoring only male juniper trees in higher elevations living longer, especially in drier nutrient–poor situations (WARD 1982).

Thus under the prevailing conditions of reduced regeneration, juniper (including *J. macropoda*) grows in rather even–aged stands, produced by an initial flush of colonisation. Quite often such stands are single senescent populations, as investigated for a number of juniper species.

In such populations younger trees are either fully absent or they rarely occur, in special niches, rocky crevasses, on the border of flushed sites, etc. During the survey no saplings were found in the whole investigation area, except in such special niches.

Considering the special slope conditions and glacial dynamics in the Karakorum (RÖTHLISBERGER 1986, GOUDIE 1989), the invasion of *Juniperus* into these valleys was probably a most peculiar event, happening only in periods of slightly milder climatic conditions. According to MOONEY et al. (1966), there are metabolic shifts in the behaviour of lowland species when temperatures and moisture are higher.

In other words: energy costs to set up a new generation is generally preventing regeneration, all the more under severe timberline conditions, bringing late and early frost, winter desiccation and pronounced mechanical effects of snow and wind. All these negative influences are all the more more dangerous due to incomplete development of cuticular layers, incomplete tissue maturation, bud formation etc. (TRANQUILLINI 1979, WARDLE 1974, HOLTMEIER 1989). Seed development and young trees must get a prolonged chance to subsist under more favourable conditions, i.e., less drought and less severe winters, such chances growing with the life span of mature trees (females in the case of junipers).

Therefore, some of the more extreme stands are most likely of relict character and not replaceable. Any reproductive strategy is prohibited by the combination of stressing factors, and viable seeds and an ensuing regeneration is only produced by a succession of extraordinary years.

Acknowledgment

On behalf of my colleagues and myself I am grateful to the IUCN (International Union for Conservation of Nature and Natural Resources in Gland/CH) and the Aga Khan Foundation, especially their coworkers in Gilgit/Pakistan (Gen. Man. Shoahib Sultan Khan) for their valuable cooperation and support and for providing additional stem-discs.

Bibliography

AHMED, M.; SHAUKAT, S. SH.; BUZDAR, A. H., 1990, Population strategies and dynamics of *Juniperus excelsa* in Balouchistan, Pakistan.
J. Veg. Sci. **1**, 271–276.

BARBOUR, M. G.; MAJOR, J., 1988, Terrestrial vegetation of California.
New exp. ed. (1st ed. 1977). Cal. Nat. Plant Soc. Spec. Publ. No. **9**, 1030 pp.

BEHNKE, J. A.; FINCH, C. B.; MOMENT, G. B.(eds), 1978, The biology of ageing. Plenum Press, New York and London, 388 pp.

COMFORT, A. 1978, The biology of senescence.
3rd ed. (1st ed. 1956). Churchill Livingstone, Edinburgh and London, 414 pp.

CONNOR, K. F.; LANNER, R. M., 1990, Effects of tree age on secondary xylem and phloem anatomy in stems of Great Basin bristlecone pine (*Pinus longaeva*).
Am. J. Bot. **77**, 1070–1077.

GOUDIE, A., 1989, The nature of the environment.
2nd ed. (1st ed. 1984). Basil Blackwell, Oxford, 370 pp.

HARTMANN, H., 1968, Über die Vegetation des Karakorum. I. Gesteinsfluren, subalpine Strauchbestände und Steppen-Gesellschaften im Zentral-Karakorum.
Vegetatio **15**, 297–388.

HARTMANN, H., 1977, Über die Vegetation des Karakorum. II. Rasen- und Steppen-Gesellschaften im Bereich der alpinen und der höheren subalpinen Stufe des Zentral-Karakorum.
Vegetatio **24**, 91-157.

HOLTMEIER, K. H., 1989, Ökologie und Geographie der oberen Waldgrenze.
Ber. R. Tüxen-Ges. **1**, 15-45 (References).

KANNGIESSER, F., 1907, Über die Lebensdauer der Sträucher.
Flora **97**, 401-420.

KANNGIESSER, F., 1909, Zur Lebensdauer der Holzpflanzen.
Flora **99**, 414-435.

KLÖTZLI, F., 1975, Ökologische Besonderheiten *Pinus*-reicher Waldgesellschaften.
Schweiz. Z. Forstwes. **126**, 672-710.

KLÖTZLI, F.; SCHAFFNER, R.; BOSSHARD, A., 1989, Pasture development and its implications in the Hunza Valley.
- High Pasture Mission 1989. - Aga Khan Rural Support Programme, N-Areas, Pakistan. AKF and IUCN. - Mscr. -

MOONEY, H. A.; WEST, M.; BRAYTON, R., 1966, Field measurements of the metabolic responses of bristle cone pine and big sagebrush in the White Mountains of California.
Bot. Gaz. **127**, 105-113.

MUKHAMEDSHIN, H., 1977, The oldest trees of C-Asia studied by dendrochronology.
Russian papers on dendrochronology and dendroclimatology. **52**, **68**, **72** (transl.). In: Fletcher, J. M.; Kinnard, W., Res. Labor. Archaeol. and History of Art. Univ. Press, Oxford. pp. 36-42.

POLUNIN, O.; STAINTON, A., 1984, Flowers of the Himalayas.
Oxford Univ. Press., Delhi, India, 580pp.

RÖTHLISBERGER, F., 1986, 10 000 Jahre Gletschergeschichte der Erde.
Sauerländer, Aarau, Frankfurt a. M., Salzburg, 416 pp. and annexes.

SCHULZE, E. D.; MOONEY, H. A.; DUNN, E. L., 1967, Wintertime photosynthesis of bristlecone pine (*Pinus aristata*) in the White Mountains of California.
Ecol. **48**, 1044-1047.

TRANQUILLINI, W., 1979, Physiological ecology of the alpine timberline.
Ecol. Stud. **31**. Springer, Berlin, Heidelberg, New York. 137 pp.

WALTER, H.; BRECKLE, S.–W., 1986, Ökologie der Erde. Spezielle Ökologie der gemäßigten und arktischen Zonen Euro–Nordasiens. Zonobiom VI–IX.
G. Fischer, Stuttgart, 587 pp.

WALTER, H.; LIETH, H., 1960 ff., Klimadiagramm–Weltatlas.
VEB G. Fischer Verlag, Jena.

WARD, L. U., 1982, The conservation of *Juniperus*: longevity and old age.
J. appl. Ecol. **19**, 917–928.

WARDLE, P., 1974, Alpine Timberlines.
In: Ives, J. D.; Barry, R. G. (eds), Arctic and alpine environments. Methuen, London, 371–402.

WEST, N. E. (ED.), 1983, Ecosystems of the world 5. Temperate deserts and semi–deserts.
Elsevier Sci. Publ. Co., Amsterdam, Oxford, New York, 522 pp. (Contributions of Walter, Box and Breckle, resp.)

WHITEMAN, P.T.S., 1985, Mountain oases. A technical report of agricultural studies in Hunza, Ishkoman & Yasin valleys of Gilgit district (1982–1984).
FAO/UNDP. PAK/80/009.

WOOLHOUSE, H. W., 1972, Ageing processes in higher plants.
Oxford. Biol. Readers **30**, Oxford Univ. Press, London.

6.4 Annex: Hunza Valley area

One of the main rivers in Gilgit district in the Northern Areas of Pakistan is the Hunza river. Its watershed covers an area of nearly 20,000 square kilometers, as a terrific and wild scenery in the midst of the great mountain ranges of the west Karakorum. Their slopes extend over more than 6000 meters of altitude from the valley bottom at about 1400 m asl near Gilgit up to uncounted peaks of 6—7000 m, which are dominated by Rakaposhi with 7788 m. Steep valleys between the unnamed mountains release wild rivers.

At least 50 % of the area of this watershed lie under glaciers or permanent snow, and another 20–30 % consist of bare rock. Only about 25 % are covered by a more or less dense vegetation. Along the river terraces and on the lower slopes, man created green, fertile oases.

6.4.1 Physiography

The Hunza river drains the most northwestern part of the Himalaya range, to which the Karakorum has to be counted. Although only the higher elevations in the southern and eastern parts of the Himalayas surpass 8,500 m, the topographic energy of the Hunza region is worldwide among the highest. Along the main river, the extraordinarily steep and high mountain slopes with their rare natural terraces are pierced by "nullahs" (steep natured gullies with mostly seasonal streams). These create flatter valley bottoms and, above all, hill ranges in between, which represent the most favourable areas for alpine pastures.

Topography strongly influences soil factors and local climate conditions. The geological instability of one of the most active mountain parts in the world, in connection with the crumbly rock (Karakorum means "black rubble"), leads to a fast degradation of the mountain. The problem of erosion is acute on the steep and often bare slopes, especially during snowmelt or thunderstorms. On the other hand, eroded material accumulates on terraces and new lands, building up a new soil layer. Besides this, the highly siltated water of the irrigation channels has a fertilizing effect.

6.4.2 Climate

Usually the monsoon rains do not surpass the Indus valley. The region gets its rain precipitation through the westwinds. Therefore, we have in this area a climate of the continental–mediterranean type. The multivarious topography and the extreme differences in altitude lead to a great diversity of regional macro–climates. Except for the information of WHITEMAN (1985) and a few other authors, climatic data are missing, especially for the alpine and subalpine regions. In order to fill these gaps, a German research group (DFG–Schwerpunktprogramm "Kulturraum Karakorum", project leader Prof. Dr. M. Winiger) intends to establish transects of climate stations from Chitral over Yasin to Gilgit and Skardu. (Climate diagrams from WALTER and LIETH, 1960 ff. and from WEST, 1983, p. 333. See Fig. 6.6).

Precipitation. Rainfall data from Chilas (165 $mm \cdot yr^{-1}$), Gilgit (132 $mm \cdot yr^{-1}$), Karimabad (145 $mm \cdot yr^{-1}$) and Mishgar (129 $mm \cdot yr^{-1}$) show a maximum for April and May (18–40 mm) with a second peak in August (10–26 mm), while June, October and November are the driest months with generally less than 10 mm of rainfall. The winter and spring precipitation, brought in by cyclones is generally widespread. The summer rain occurs much more locally and its amount varies much due to convectional disturbances and irregular influence of the monsoon. Again, aspect, topography and particularly elevation determine then the actual amount of rainfall. Generally, precipitation decreases towards northwest, due to the increasing rain shadow effect towards the western depressions and to the declining influence of the monsoon. With a few exceptions, the intensity of the rainfall is low with little effect on erosion. But there can be heavy rains, especially in summer when occasionally the monsoon breaks in, with considerable erosion damages. Accordingly, the Gilgit rainfall data show a higher reliability for winter and spring than for summer months.

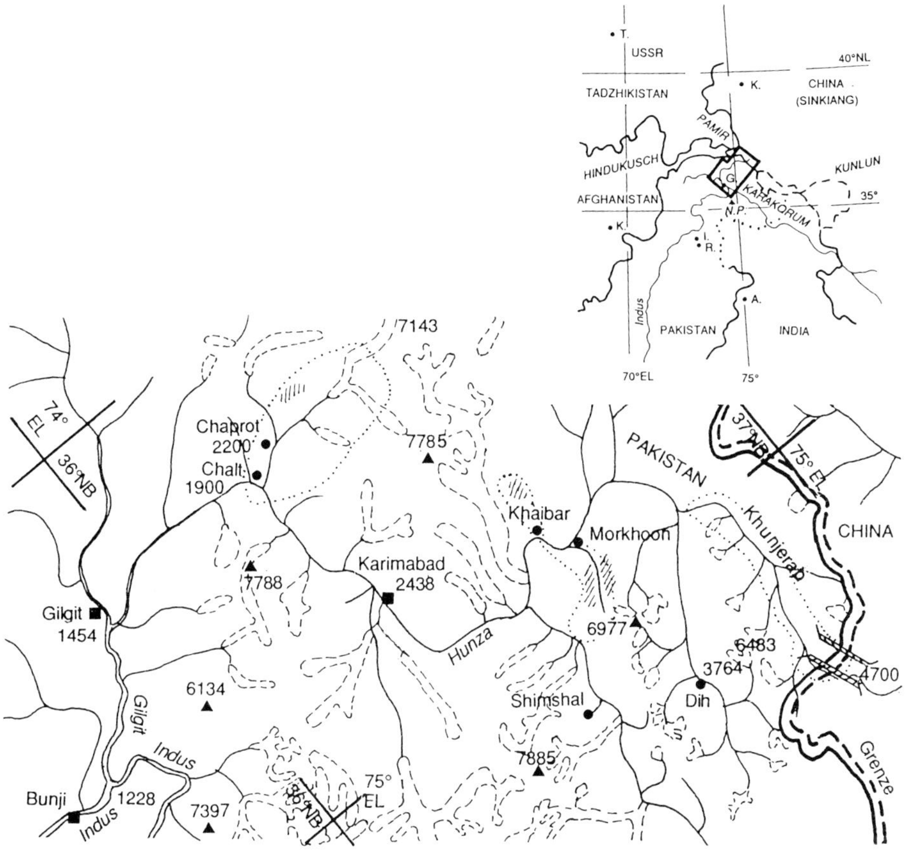

Figure 6.6: Topographic sketch map showing the most important localities in the Hunza valley area.

Temperature. The Northern Areas have temperate conditions with hot summers, cold winters and great seasonal and, in higher altitudes, big daily differences. Again, the local data vary much according to topography (e. g. "stove conditions" in steep valleys), altitude (increasing altitude will decrease average temperatures and increase the differences between day and night), aspect and reflection (which is especially strong e. g. on bright limestone or glaciers).

Evaporation. The yearly variation of evaporation is lower than that of the rainfall. The seasonal variability corresponds with the general weather conditions.

Radiation. Radiation is high in the Hunza region due to the great number of bright or only slightly cloudy days (continentality).

6.4.3 Soil

Artemisia **belt.** The prevailing physical weathering forms primarily coarse soils, mainly silty, gravelly or rocky **regosols** and **xerosols** with a high carbonate content. Due to the aridity of this belt, only a small amount of organic matter is produced, and the high summer temperatures lead to a quick decomposition. Therefore, the humus content is always very low and can be found in the uppermost cm of the soil profile. The soils are relatively inactive, that means that especially nitrogen is a limiting factor for plant growth. Most of the other mineral components are there in a sufficient amount due to the little washout by rainfall. Carbonate content is usually high, and the readily soluble salts, especially gypsum, frequently accumulate on the surface.

Subalpine and alpine zone. Higher precipitation and with this connected an increasing plant growth leads first to a higher humus accumulation, but at the same time, nutritive salts are leached out. The results are rather poor **acrisols** or **cambisols**. With increasing altitude, coarse scree slopes with poor **lithosols** prevail more and more, with a humus deposition not worth mentioning.

Under special conditions, richer humic cambisols or smalls plots of histosols can develop, as it is found for example on Khunjerab Pass.

Part II

Mass and water balances at stand level

Chapter 7

Plasticity of the photosynthetic production of Galium aparine L.

W. L. KUTSCH and L. KAPPEN

7.1 Introduction

Photosynthesis is the essential process of energy and carbon import into ecosystems. Consequently light is the primary environmental factor for all processes. The adaptation of plants to light controls species composition and competition between species. Plant adaptation and response to light has been intensively investigated since the beginning of this century (PFEFFER 1897, WIESNER 1909). Many morphological, physiological and biochemical mechanisms of light adaptation have been described (RABINOWITCH 1951, GABRIELSEN 1960, BOARDMAN 1977, BJÖRKMAN 1981), and the effect of light in plant competition has been investigated by various authors (e.g. BORNKAMM 1961, GLAUNINGER and HOLZNER 1982).

In deciduous forest and agricultural ecosystems the availability of light in the canopy varies drastically with the season. Niches controlled by strong seasonal variations of resource availability are used in different ways: by a single species during the phase of better resource availability, by two species that utilize the resource differently and that change temporally, or by one persistent species that is able to adjust to these variations (PIANKA 1980, SCHULZE and CHAPIN 1987). In the present study a species is considered that stays by plastic response in a niche with seasonally contrasting light and temperature. *Galium aparine* is a nitrophilous, annual plant that grows in open forest and hedges. The species is distributed synanthropically within moderate climate zones of the world. The most important parameters of environmental plasticity are productivity, growth–form, frost resistance and timing of the seed germination. Studies on seed ecology and population dynamics of *Galium aparine* were carried out by KOCH (1969) and RÖTTELE (1980). Investigations about ecotype formation in *G. aparine* were carried out by BERKEFELD (1988) and by GROLL and MAHN (1986). Plasticity of the growth form and aspects of the biomass production were investigated by GROLL and MAHN (1986) and AUGE and MAHN (1988). These investigations suggest that *G. aparine* has different responses to the environment corresponding to different environmentally conditioned life cycles.

Most conspicuous are the different ontogenetic cycles that depend on the date of seed germination. Seeds of *G. aparine* are able to germinate between September and May. If one assumes that germination is inhibited during the cold season, then a population can be devided in "winter–annual" individuals that germinate in fall and "summer–annual" individuals that germinate in spring. No data on the plasticity of the photosynthetic apparatus of *G. aparine* exist. Considering the changing irradiance during the season, the question should be raised as to what extent *G. aparine* reacts as a sun species or as a shade species. Sun species are adjusted to a wide range of light intensities. Shade species with a smaller margin of adaptation have a competitive advantage in extreme dark habitats (BJÖRKMAN 1981). Because *G. aparine* is a pioneering species, it is expected, that its photosynthetic capacity shows a wider range of adaptability to the irradiance than that of later successional species (BAZZAZ and CARLSON 1982).

Mathematical modelling was used to quantitatively describe adaptability of CO_2 exchange to various light or temperature conditions. CO_2 exchange was modelled for different developmental stages of the plant, and model parameters and constants were calibrated with laboratory and field data. Model outputs for different developmental stages were compared. These models may also form the basis for a comparison between potential light competitors.

7.2 Material

All plants were part of a population growing at the edge of an open forest east of the Ahrensee near Kiel (UTM: 32UNF643178). The soil of this forest is very eutrophicated as a result of drainage and import of nutrients and fertilizer from the surrounding crop fields. Seeds and emergent plants were sampled at this site in October and November 1986. Emergent plants were potted and planted into soil in the Botanical Garden of the University of Kiel at sites with different light conditions. Plants grown in an openly exposed bed were called "sun plants". "Semi–shade plants" grew at the edge of a hedge and were shaded half-day when the woody plants were in leaf. "Shade plants" grew in the centre of the hedge with a relative light yield (WIESNER 1909) of 4% when in leaf. In October and in spring, seeds were sown in pots and in the ground at the same sites, and the seedlings were accordingly marked as winter-annuals or as summer-annuals depending on the time of germination. CO_2 exchange was measured on potted plants in temperature-controlled cabinets and in the field. *Galium aparine* passes a characteristic sequence of developmental stages (see GROLL and MAHN 1986):

1. Germination and emergence: G/E.
2. Vegetative growth; first vegetative developmental stage: V1.
   In this stage plants have short internodes (mostly shorter than 1 cm) and relatively small spheric leaves. In general the plants reach a height of only 10–20 cm. The winter-annual plants hibernate in this stage. For summer-annuals, this stage is strongly reduced. Only the leaves of the first two nodes can be related to this stage.
3. Extension growth; second vegetative developmental stage: V2.
   This stage is characterized by long internodes (up to more than 10 cm) and large, lanceolate leaves with hairs. While the lanceolate leaves are formed the small spheric leaves die.
4. Flowering, fruit development; first generative-reproductive stage: R1.
   The beginning of this stage is defined by the appearance of the first flowers. The flowers are found mostly at the ends of side axes while the main shoot still continues to grow.
5. Fruit ripening and death of the mother plant; second generative-reproductive stage: R2.
   The shoots end in terminal inflorescences. The transition to this stage is marked by the appearance of these inflorescences and the cessation of elongation.

Light and temperature dependence of net photosynthesis were measured and modelled for these developmental stages (except the largely suppressed V1 of the summer-annuals). The "semi-shade plants" were used as the baseline group for the comparison of the photosynthetic performances. Comparisons were started with the beginning of stage R1 in summer-annual plants.

7.3 Methods

CO_2 exchange was measured in an open gas stream by using an infrared gas analyser as the difference between the CO_2 concentration of the ambient air and that of the air which passed through the plant cuvette (WALZ, Effeltrich, F.R.G). The temperature-controlled cuvettes had a volume of about 3000 ml and an air circulation (KAPPEN et al. 1986). Plants were illuminated by two incandescent lamps (HQIL, Norka, 400 W) in a temperature-controlled growth cabinet. Quantum flux density was varied by colorless grey filters and was measured by a LI190Sb quantum sensor (LICOR, Nebraska, U.S.A.) in the CO_2 exchange cuvettes. The same system was also used for measurements of net photosynthesis under natural light and temperature conditions in the Botanical Garden. During these field measurements the temperature in the cuvette followed the ambient temperature by a feedback-controlled system. Leaf area was measured by using a digiplan area meter (KONTRON). Leaf chlorophyll was extracted using an acetone/TRIS buffer solution (80:20) and was quantified in a HITACHI-photometer according to ZIEGLER and EGLE (1965).

7.4 Mathematical Modelling

7.4.1 Photosynthetic light response and light acclimation

A quantitative measure of the adaptability of plant photosynthesis to different light regimes is described by the following model:

$$NP = GP \cdot (1 - \exp(-K \cdot I)) - R \tag{7.1}$$

where

NP	net photosynthesis at given temperature and quantum flux density I.
GP	maximum gross photosynthesis at that temperature. Simplified: $GP =$ light saturated net photosynthesis + dark respiration.
K	empirical coefficient, describes adaption to the light climate.
I	quantum flux density.
R	dark respiration.

This equation was suggested by PEAT (1970) in a slightly different form. Compared with other equations (see BLACKMAN 1905, SMITH 1938, CHARTIER 1970, THORNLEY 1976, MARSHALL and BISCOE 1980) it is simple and has a high accuracy (Fig. 7.1, 7.2). The light reponse curve was fitted to the data using a least sum of squares procedure. The equation is suitable for estimating GP if light saturation is not reached during the gas exchange measurements. Maximum quantum yield of the incident light can be estimated by the equation:

$$\Phi_i = K \cdot GP \tag{7.2}$$

where K is the same empirical coefficient as in Eq. (7.1).

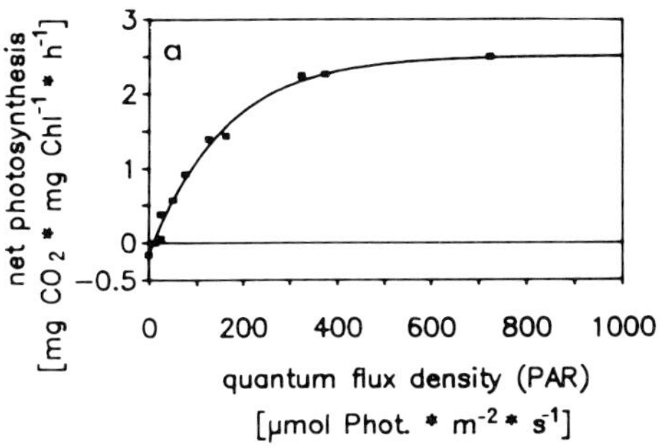

Figure 7.1: Light response curve of the net photosynthesis of *Galium aparine* fitted to data measured in December 1986 at 16 °C in the temperature–controlled growth cabinet.

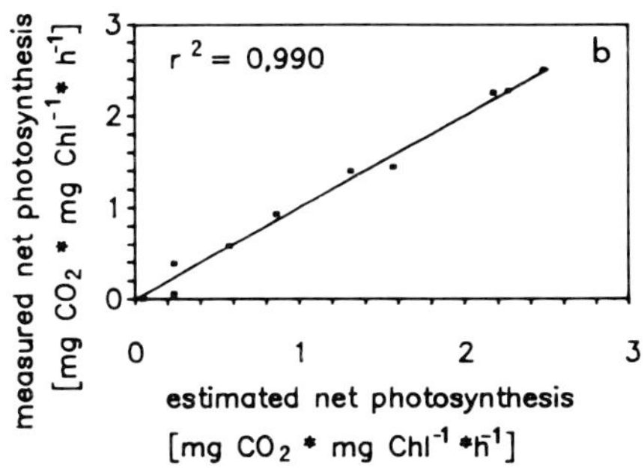

Figure 7.2: Correlation between measured photosynthesis and values of photosynthesis estimated with Eq. (7.1).

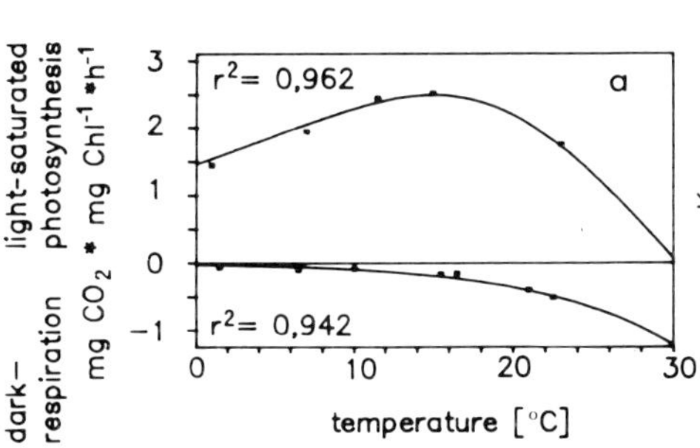

Figure 7.3: Temperature response curve of light–saturated net photosynthesis and dark respiration of *Galium aparine* fitted to data measured in December 1986.

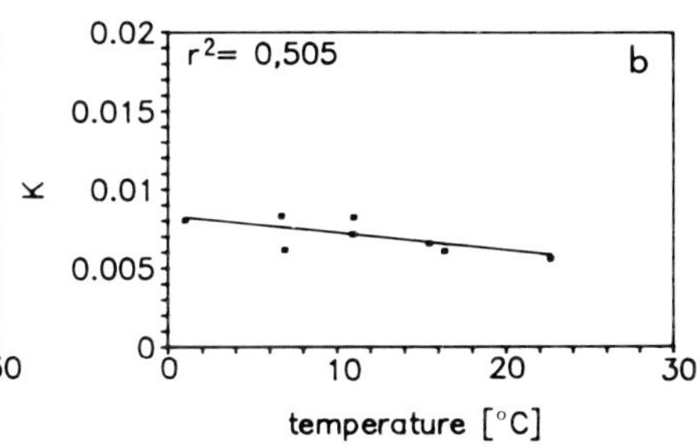

Figure 7.4: Temperature dependence of the empirical coefficient K used in Equation (7.1) for different light response curves fitted to data measured in December 1986.

7.4.2 Temperature dependence of dark respiration and photosynthesis

Temperature dependence of dark respiration was modelled as an Arrhenius plot (ARRHENIUS 1915) using the equation:

$$W = \exp(-\frac{E}{R \cdot T_K} + A) \qquad (7.3)$$

where

W rate of respiration at temperature T_K.
E apparent activation energy of the process.
R universal gas constant.
T_K absolute temperature.
A empirical coefficient.

The apparent activation energy (E) of the process is not measurable by gas exchange, so E was estimated like A as an empirical coefficient by a least sum of squares procedure. The temperature dependence of photosynthesis was modelled by the following equation (TENHUNEN, YOCUM, GATES 1976; TENHUNEN, WEBER, YOCUM, GATES 1976):

$$GP = \frac{C \cdot T_K \cdot \exp(-\frac{\Delta H^*}{R \cdot T_K})}{1 + \exp(-\frac{\Delta H_1}{R \cdot T_K}) \cdot \exp(\frac{\Delta S}{R})} \qquad (7.4)$$

where

GP calculated gross photosynthesis at light saturation and temperature T_K, GP in percent of the rate at optimum temperature.
C is an empirical coefficient.
ΔH^* energy of activation for the enzyme–catalyzed reaction.
ΔH_1 energy of activation for denaturation equilibrium.
ΔS entropy of the denaturation equilibrium.
T_K absolute temperature.
C, ΔH^*, ΔH_1, ΔS were estimated by a least sum of squares procedure.

As an example Fig. 7.3 shows data–fitted curves of light saturated net photosynthesis and dark respiration as response to temperature. Measurements were carried out with *G. aparine* in December 1986. Measured data and model equation values were correlated in the same way as in the description of the light dependence of net photosynthesis (Fig. 7.2). Fig. 7.4 shows the temperature dependence of the empirical coefficient K used in Eq. (7.1).

7.5 Results

7.5.1 Seasonal course of the developmental stages

Fig. 7.5 shows the course of developmental stages according to our phenological observations during a period between September 1986 and August 1987. When

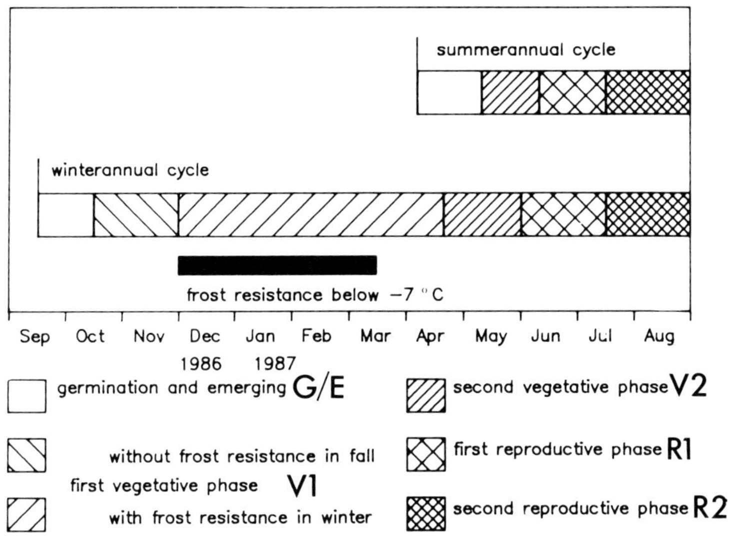

Figure 7.5: Seasonal course of the developmental stages of *Galium aparine* according to phenological observations during a period between September 1986 and August 1987.

the winter–annual plants gained freezing tolerance, the photosynthetic capacity decreased but did not cease. For this reason the stage V1 is devided in a fall period (no frost, September – November) and a the winter period (December – March). Details about the freezing tolerance and the influence of frost on photosynthesis of *G. aparine* are reported elsewhere (Kutsch and Kappen 1989). The winter was comparatively cold for the area. Consequently the winter–annual plants did not change from V1 to V2 until the last third of April. The cold winter also led to a late germination of the summer–annual plants (early April). The summer–annual plants passed from V1 to V2 in early May. The periods V2 and R1 were shorter than those of the winter–annual plants.

7.5.2 Seasonal courses of the CO_2 exchange parameters

The seasonal courses of the CO_2 exchange and of the controlling parameters are summarized in Fig. 7.6, a–e, for the winter–annual and the summer–annual plants. They show net photosynthetic capacity and dark respiration rates at the optimum temperature of net photosynthesis. Due to changes of this optimum temperature the dark respiration rates of different developmental stages are not directly comparable. Therefore the dark respiration rate at 20 °C is additionally plotted by (dotted line). Also the empirical coefficient K (Eq. (7.1)) and the light compen-

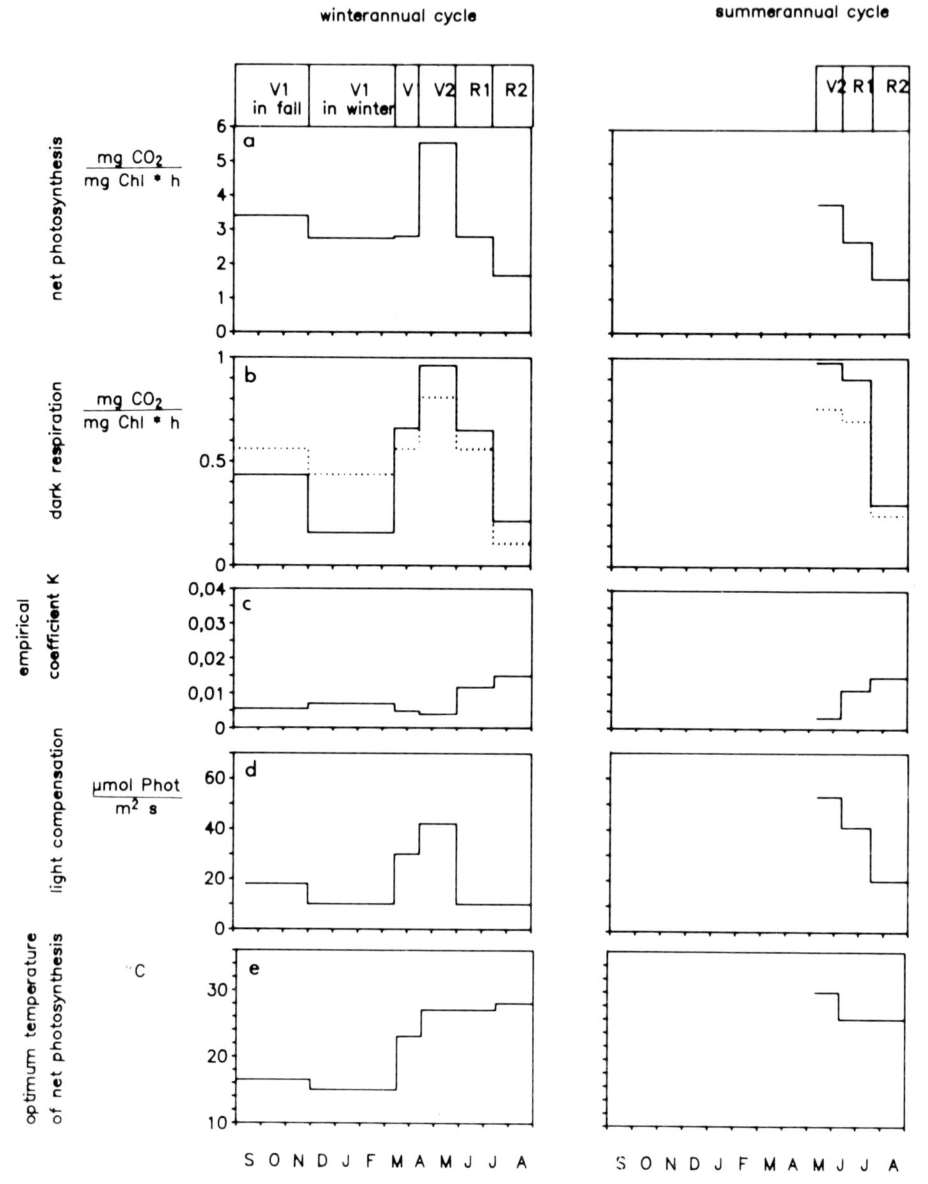

Figure 7.6: Seasonal course of net photosynthetic capacity, dark respiration rate, empirical coefficient K, light compensation point, and optimum temperature of net photosynthesis for the winter–annual and the summer–annual cycle of *Galium aparine* during a period between September 1986 and August 1987.

sation point of the CO_2 exchange are related to the optimum temperature of net photosynthesis. The winter–annual population of *G. aparine* reveals the highest net photosynthetic capacity (P_{MAX} = 5.5 mg $CO_2 \cdot mg^{-1}$ chlorophyll $\cdot h^{-1}$, measured in the stage V2, Fig. 7.6 a). The photosynthetic capacity in V1 is 3.4 mg $CO_2 \cdot mg^{-1}$ chlorophyll $\cdot h^{-1}$ (fall) and decreases to 2.9 mg $CO_2 \cdot mg^{-1}$ chlorophyll $\cdot h^{-1}$ during the period of frost hardening (the vital freezing tolerance limit was at −17 °C in December). It was slightly increased in the overwintered leaves in spring. After the maximum during V2, P_{MAX} decreases continuously during the reproductive stages.

Within their shorter life cycle, the summer–annual plants show a similar course of P_{MAX} with a maximum of 3.9 mg $CO_2 \cdot mg^{-1}$ chlorophyll $\cdot h^{-1}$ in V2 and a decrease during the reproductive stages.

Dark respiration at the optimum temperature of net photosynthesis (Fig. 7.6 b), like P_{MAX}, has two maxima during the winter–annual life cycle. Respiration is high in stages V1 in fall and in stage V2. During stage V1 in winter and in the reproductive stages the respiratory rate was comparatively low. The dark respiratory rates of the summer–annual plants run parallel to that of the winter–annual plants. The values of the empirical coefficient K are given in Fig. 7.6 c. When the woody plants of the hedge had no leaves (October–May), K of the winter–annual form varied between 0.005 and 0.008. K was at its minimum in the stage V2 in both populations when the trees were still leafless. The increase of K to 0.015 in both populations corresponds with the reduced irradiance of the woody plants in summer.

The light compensation point of net photosynthesis in the winter–annual form was low during V1 in winter and in both populations during the reproductive stages (Fig. 7.6 d), both forms needed more light to pass the light compensation point. In most cases the course of the light compensation point parallels that of the dark respiration. The optimum temperature of net photosynthesis (Fig. 7.6 e) shows an adaptation to the ambient temperature in the winter–annual population: it varies between 15 °C in fall and winter and 28 °C in R2. In the summer–annual population it decreases slightly from 30 °C in V2 to 26 °C in the reproductive stages.

7.5.3 Adaptation of the plants to the light conditions

Diurnal courses of the CO_2 exchange, and the parameters quantum flux density and temperature were measured in summer–annual plants in stage R1 in the three habitats during the period June and July, 1987. Light response curves of net photosynthesis were fitted to the data by means of Eq. (7.1) (Fig. 7.7). Changes due to temperature effects were minimized by using data measured in the temperature range 13–17 °C. Specimens of the sun plants showed the highest light compensation point and the highest rates of light–saturated photosynthesis and dark respiration. The empirical coefficient K was 0.008 for the sun plant, 0.012 for the semi–shade

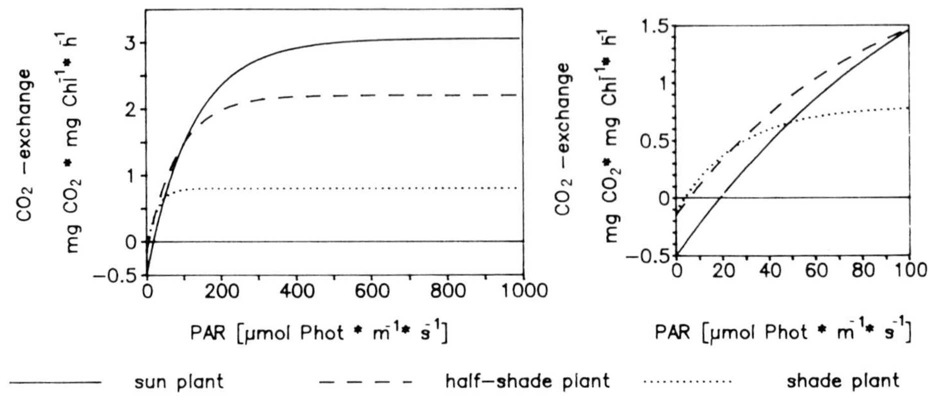

Figure 7.7: Light response curves of net photosynthesis from the different populations of *Galium aparine* developed at different light climates. Curves were fitted to field data by means of Eq. (7.1).

plant and 0.038 for the shade plant. The maximum quantum yield (Φ_i) was nearly the same for all plants.

7.5.4 Parameters of the mathematical modelling and leaf parameters

An attempt was made to correlate specific leaf weight, chlorophyll content per leaf area and per dry mass to the different modelling parameters of the light response of CO_2 exchange. There was no correlation between the photosynthetic capacity and any of these leaf parameters. However a correlation was found between the empirical coefficient K and the natural logarithm of the specific leaf weight (Fig. 7.8).

7.6 Conclusions

The CO_2 exchange of summer–annual populations of *G. aparine* differs in two respects from that of winter–annual populations: 1. In stage V2 the photosynthetic capacity of the summer–annual population reaches only 75% of that of the winter–annual population. 2. The photosynthetic capacity of the summer–annual population declines with a delay after the onset of the reproductive stages. This decline in photosynthetic capacity parallels the decline of dark respiration rate in both populations. This confirms the assumption that photosynthesis and dark respiration are coupled through the maintenance metabolism (BJÖRKMAN 1981).

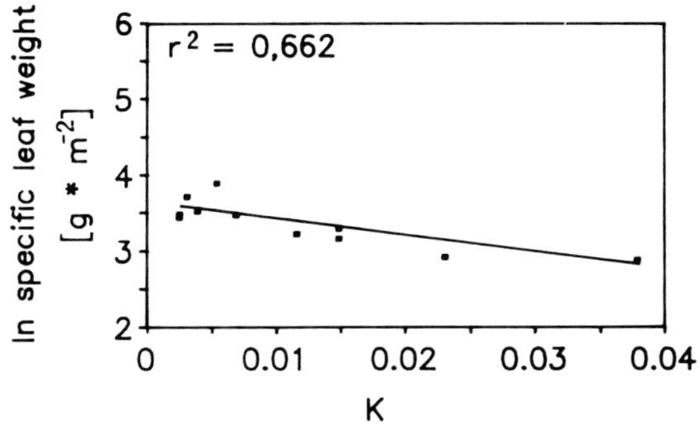

Figure 7.8: Correlation between the empirical coefficient K (Eq. (7.1)) an the natural logarithm of the specific leaf weight.

These differences do not sufficiently justify the assumption of genetic differences between winter–annual and summer–annual populations. All the different performances of the plant discussed above are of modificative nature (GROLL and MAHN 1986). However BERKEFELD's (1988) and GROLL's and MAHN's (1986) hypothesis about different ecotypes of *G. aparine* is not contradicted, because they found an ecotype differentation due to the environmental conditions of agricultural ecosystems. A great plasticity of *G. aparine* is obvious in both a capacity for adjusting the onset of different developmental stages to the seasonal conditions and in the adaptability of the photosynthetic apparatus to different environmental conditions. Plasticity determines the reproductive success of the individuals and consequently the niche competition of this species. Mathematical modelling should provide a reliable measure of light adaptation. For this purpose a physiological discussion of the adaptation phenomena is necessary. Four characteristics of the light–dependence curves are important.

1. The photosynthetic capacity is much higher in sun–adapted plants than in shade–adapted plants. Photosythetic capacity is the net photosynthetic rate at light saturation, normal CO_2 concentration of the external air, and all other factors in optimum for photosynthesis (temperature, humidity, leaf water potential). Adaptation processes can cause intra–specific differences of up to a factor of seven (BJÖRKMAN et al. 1972, PATTERSON et al. 1978). Aftereffects of certain events such as freezing may reduce the indicator value for the adaptation of plants to the light regime (KUTSCH and KAPPEN 1989).

2. The dark respiration rates of the shoots — generally measured at 20 °C — are negatively correlated with photosynthetic capacity. The dark respiration rate is extremely low in shade plants. This allows a positive carbon balance even for plants from shaded habitats. As BJÖRKMAN (1981) has hypothesized, the

limiting factor for the adaptability of plants to shade habitats is the ability to reduce dark respiration.

3. The light compensation point of shade plants is normally reached at extremely low values due to minimal dark respiration. In extreme shade specialists, this point is already reached at a quantum flux density of 0.5 to 2 μmol·m^{-2}·s^{-1} (BOARDMAN 1977).

4. At low light, the photosynthesis rate increases proportionally with quantum flux density. At this point the quantum yield is at its maximum and shows the maximum efficiency of light energy conversion (CHARTIER 1970).

If maximum quantum yield of photosynthesis is used as a measure of plant light adaptation, a clear distinction has to be made between the quantum yield on the basis of incident quanta (Φ_i) and the quantum yield on the basis of absorbed quanta (Φ_a). Significant changes of Φ_a have not been shown in comparisons between sun and shade forms of C_3 plants (BJÖRKMAN 1968, GAUHL 1976, ÖQUIST et al. 1982) nor in comparisons of C_3 and C_4 plants (EHLERINGER and BJÖRKMAN 1978). BOARDMAN (1977) records studies that suggest an increase of Φ_i due to shade adaptation. However, BAZZAZ and CARLSON (1982) found, in 8 out of 14 species, that shade grown plants had a decreased Φ_i. GABRIELSEN (1948) showed that the light absorption capacity of leaves, and therefore Φ_i, depends primarily on pigment composition and its relation to leaf surface area. Theoretically, the fraction of absorbed light should increase with chlorophyll concentration. Hence an increase of chlorophyll content per leaf area can be interpreted as an adaptation to low light conditions. However, this expected effect is superimposed by another mechanism of light adaptation. Sun-adapted plants normally have much thicker leaves than shade-adapted plants. In many species, a considerably thicker palisade parenchyma layer, and accordingly more chloroplasts per area are developed. This can be discussed by a "cost-benefit analysis": Chloroplasts at a deeper layer in the leaf receive less quantum flux density. Hence, they produce a lower amount of assimilates and use relatively more for their maintenance metabolism. Optimum light adaptation is dependent on leaf thickness, chlorophyll content per surface area and the absorption properties. Therefore it can be concluded that Φ_i does not vary between sun and shade adapted plants of one species.

All those phenomena should be reflected by a mathematical model. The predictive capacity of a model can be additionally estimated from the extent to which a measure of light adaptation can be determined from modelled light-dependence curves and whether this measure can be related to leaf parameters like chlorophyll content or specific leaf weight. The predictive capacity of our equation for the light dependence of net photosynthesis (Eq. (7.1)) can be demonstrated by some light response curves of theoretical character (Fig. 7.9).

The curves in Fig. 7.9 a were calculated by varying the gross photosynthesis at light saturation (GP) while all other variables in the equation were left constant. Φ_i

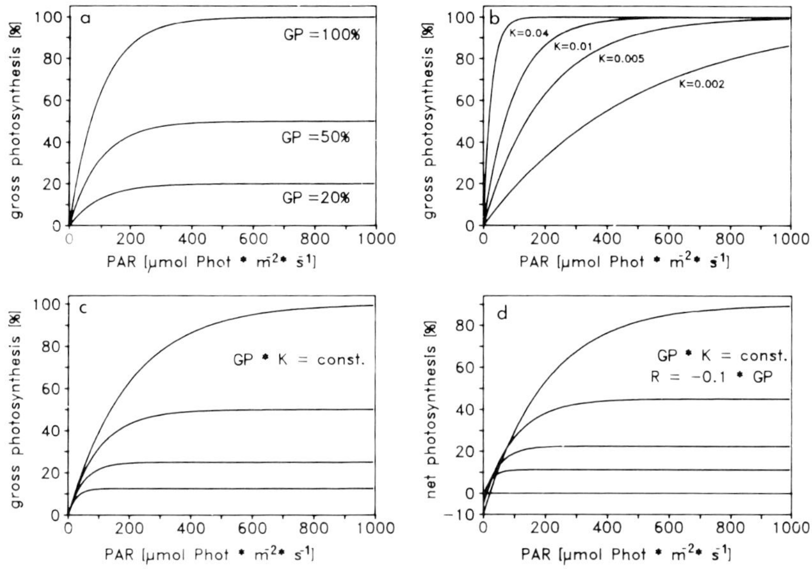

Figure 7.9: Light response curves of theoretical character constructed by using Eq. (7.1).

is correlated with the maximum photosynthetic rate. If GP is always set at 100% (Fig. 7.9 b), the light response curves change between sun type and shade type when the value of the empirical coefficient K increases. However Φ_i also changes. In this case, the variability of R and GP is still not accounted for. If one assumes that light adaptation changes R as well as GP, then the light response curves of gross photosynthesis (Fig. 7.9 c) and of net photosynthesis (Fig. 7.9 d) must be calculated by keeping the product of K and GP constant. Now their shape is typical for that of light– and shade–adapted leaves: Dark respiration, compensation point and the maximal photosynthetic rate decrease as shade adaptation increases. The maximum quantum yield of net photosynthesis remains constant while K increases. It is therefore a reasonable assumption that the empirical coefficient K is a measure of light adaptation, though it can not be related easily to a physiological meaning. Correlation between K and the specific leaf weight is however apparent and correlations between leaf parameters and CO_2 exchange were shown by several authors (e.g. NOBEL et al. 1975, TSEL'NIKER 1979, VON CAEMERER and FAQUHAR 1981).

OREN et al. (1986) proposed the use of such relations for estimating the photosynthetic rates and annual carbon gain of the whole canopy. They found the maximum photosynthetic rate was highly correlated to specific leaf weight in different *Larix* species. On the other hand, specific leaf weight was correlated to the natural logarithm of daily quantum input at various positions in the crown. This shows the significance of modelling and correlations with leaf parameters beyond intraspecific comparisons of different developmental stages. It offers the opportunity to account for the adaptation of plants and plant parts to the light climate in very heterogeneous stands if stand production is calculated. This could be done by combining CO_2 exchange measurements with the determination of leaf parameters. The comparative measurements of specimens of *G. aparine* grown at habitats of different light climate show the expected phenomena of photosynthetic light adaptation: Photosynthetic capacity and dark respiration rate increase with increasing quantum flux density at the natural stand. Also the empirical coefficient K allows a good classification of the individuals to the different classes of light conditions.

Because of the higher photosynthetic capacity and the high quantum flux densities *G. aparine* attains the highest daily carbon gains in sunny habitats. This agrees with GROLL and MAHN (1986) who found a production up to 1000 seeds per individual. The plasticity of *G. aparine* allows an adaptability of the photosynthetic apparatus in the range of 4% to 100% of relative light yield. All these results agree with the findings of BAZZAZ and CARLSON (1982), who found a wide range of adaptability to the light conditions in pioneering plants. Seasonal changes of light and temperature dependence of CO_2 exchange in the winter–annual semishade plants are illustrated in Fig. 7.10. The two reproductive stages were combined because of their similarity. *G. aparine* exhibits a plastic reaction to altered light and temperature conditions. This is true for emerging as well as fully developed plants responding to environmental conditions changing during their life cycle.

The plasticity concerning the starting of the developmental stages, the freezing tolerance, and the light and temperature dependence of photosynthesis enables the species to establish itself in a great variety of habitats: In deciduous forests and hedges the time period is utilized by the winter–annual population whilst other light competitors are leafless. While woody plants have set leaves, *G. aparine* is able to adjust to shade at least to a limited extend. A comparison of daily carbon gains show that *G. aparine* is not able to cope with deep shade like *Athyrium filix-femina* and *Oxalis acetosella* (SCHULZE 1972). However it reaches a shade characteristic which is comparable with other forest–floor species such as *Deschampsia flexuosa* (SCHULZE 1972) and *Lamiastrum galeobdolon* (MIETH 1985). In extreme cases, *G. aparine* suffers from light competition by trees, as has been shown by GROLL and MAHN (1986). Plants in a wetland forest revealed shorter periods in their developmental stages and, as a consequence, had very low seed production. In weed communities and crop fields with less light competition, *G. aparine* is more productive. Especially during the stage V2 the plants have a distinct sun plant chararacteristic with a high photosynthetic capacity. Its capacity of supressing stage V1 if germinated in spring enables *G. aparine* to exist permanently in spring crop agroecosystems.

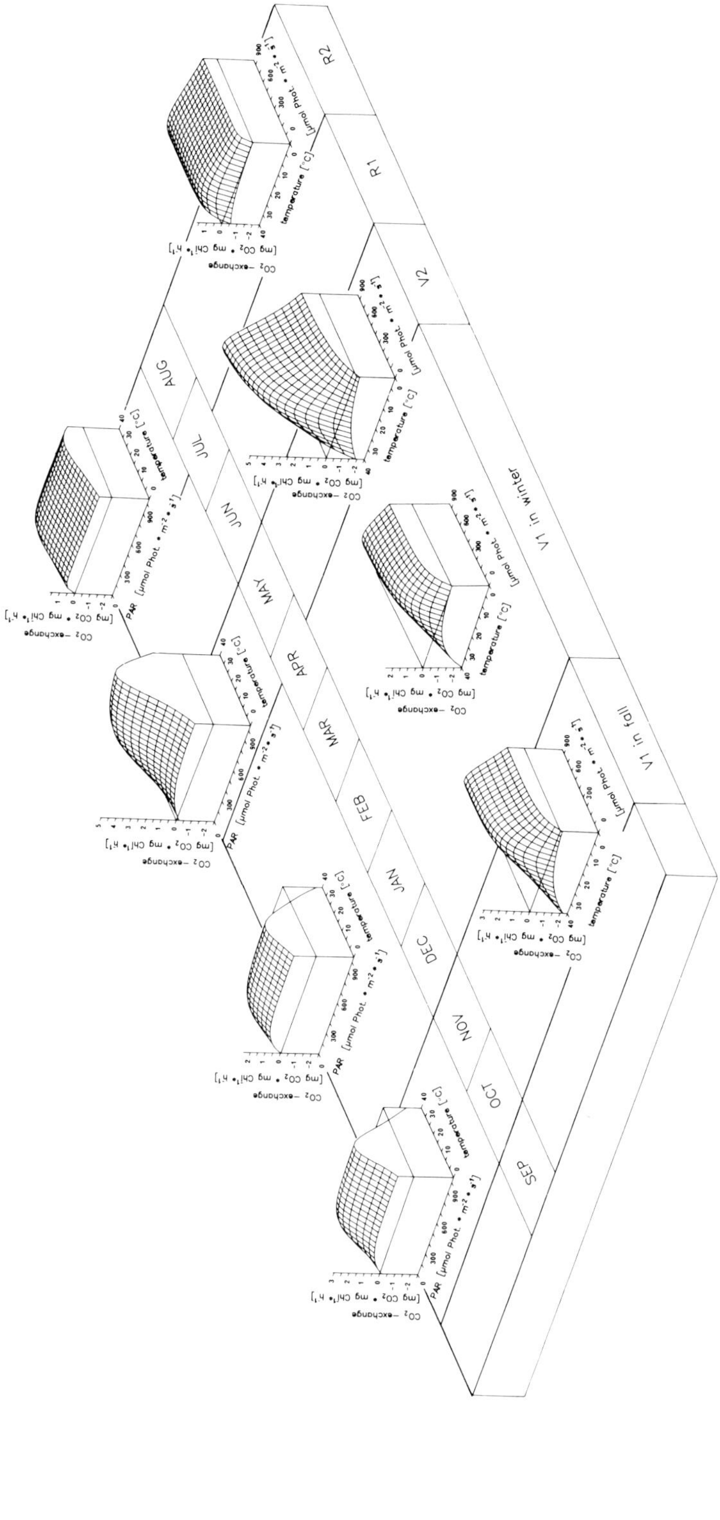

Figure 7.10: Seasonal changes of light and temperature dependence of net photosynthesis in the winter–annual cycle of *Galium aparine*. A data response surface of the model is shown in two perspectives for each developmental stage.

Acknowledgements

The authors thank Prof. Dr. Rainer Lösch for his constant support during this work and Dr. Robert J. Palmer and Till Spranger for their critical reading of the manuscript and many helpful suggestions.

Bibliography

ARRHENIUS, S., 1915, Quantitative laws in biological chemistry.
Bell, London. 164 pp.

AUGE, H.; MAHN, E. G., 1988, Zur Plastizität von *Galium aparine* L. Populationen — Wachstum und Biomasseverteilung bei sommerannuellem Zyklus.
Flora **180**, 429–443.

BAZZAZ, F. A.; CARLSON, W. C., 1982, Photosynthetic acclimation to variability in the light environment of early and late successional plants.
Oecologia **54**, 313–316.

BERKEFELD, K., 1988, Investigations about ecotype formation in *Galium aparine* L. (Rubiaceae) and *Lapsana communis* L. (*Compositae*).
Flora **181**, 111–180.

BJÖRKMAN, O., 1968, Further studies on differentation of photosynthetic properties in sun and shade ecotypes of *Solidago virgaurea* L.
Physiol. Plant **21**, 84–89.

BJÖRKMAN, O.; BOARDMAN, N. K.; ANDERSON, J. M.; THORNE, S. W.; GOODCHILD, D. J.; PYLIOTIS, N. A., 1972, Effects of light intensity during growth of *Atriplex patula* on the capacity of photosynthetic reactions, chloroplast components and structure.
Carnegie Inst. Washington Yearb. **71**, 115–135.

BJÖRKMAN, O., 1981, Responses to different quantum flux densities.
In: Lange, O. L.; Nobel, P. S.; Osmond, C. B.; Ziegler, H. (eds.): Encyclopedia of Plant Physiologie, New series, **12A**, Springer, Berlin, 57–107.

BLACKMAN, F. F., 1905, Optima and limiting factors.
Ann. Bot. **19**, 281–295.

BOARDMAN, N. K., 1977, Comparative photosynthesis of sun and shade plants.
Ann. Rev. Plant. Physiol. **28**, 355–377.

BORNKAMM, R., 1961, Zur Lichtkonkurrenz von Ackerunkräutern.
Flora **159**, 126–143.

CHARTIER, P., 1970, A model of CO_2 assimilation in the leaf.
In: Prediction and measurement of photosynthetic productivity, Pudoc, Wageningen, The Netherlands, 307–315.

EHLERINGER, J.; BJÖRKMAN, O., 1978, Quantum yields for CO_2 uptake in C_3 and C_4 plants.
Plant Physiol. **59**, 86–90.

GABRIELSEN, E. K., 1948, Effects of different chlorophyll concentrations on photosynthesis in foliage leaves.
Physiol. Plant **1**, 5–37.

GABRIELSEN, E. K., 1960, Beleuchtungsstärke und Photosynthese.
In: Ruhland, W. (ed.): Encyclopedia of plant physiology, V/2, Springer, Berlin, 27–47.

GAUHL, E., 1976, Photosynthetic response to varying light intensity in ecotypes of *Solanum dulcamara* L. from shaded and exposed environments.
Oecologia **22**, 275–286.

GLAUNINGER, J.; HOLZNER, W., Interference between weeds and crops: A review of literature.
In: Holzner, W.; Numata, M.: Biology and ecology of weeds, Geobotany **2**, 149–159.

GROLL, U.; MAHN, E. G., 1986, Zur Entwicklung ausgewählter Populationen des Kletten-Labkrauts (*Galium aparine* L.).
Flora **178**, 93–110.

KAPPEN, L.; BÖLTER, M.; KÜHN, A., 1986, Field measurements of net photosynthesis of lichens in the Antarctic.
Polar Biol. **5**, 255–258.

KOCH, W., 1969, Einfluß von Umweltfaktoren auf die Samenphase annueller Unkräuter — insbesondere unter dem Gesichtspunkt der Unkrautbekämpfung.
Ulmer, Stuttgart. 204 pp.

KUTSCH, W. L.; KAPPEN, L., (1989), Ein Modell zur Produktivität wintergrüner Kräuter am Beispiel von *Galium aparine* L.
Verhandlungen Gesellschaft für Ökologie **XIX**, Osnabrück, in press.

MARSHALL, B.; BISCOE, P. V., 1980, A model for C_3 leaves describing the dependence of net photosynthesis on irradiance.
J. Exp. Bot. **31**, 29–39.

MIETH, B., 1985, Jahresperiodischer Verlauf des autökologischen Verhaltens von *Lamiastrum galeobdolon* (L.) EHREND. und POLATSCHEK.

Diplomarbeit am Botanischen Institut der Christian–Albrechts–Universität, Kiel. 93 pp.

NOBEL, P. S.; ZARAGOZA, L. J.; SMITH, W. K., 1975, Relations between mesophyll surface area, photosynthetic rate, and illumination level during development for leaves of *Plectrantus parviflorus* HENCKEL.
Plant Physiol. **55**, 1067–1070.

ÖQUIST, G.; BRUNES, L.; HÄLLGREN, J. E., 1982, Photosynthetic efficiency of *Betula verrucosa* acclimated to different light intensities.
Plant, Cell Environment **5**, 9–15.

OREN, R.; SCHULZE, E.–D.; MATYSSEK, R.; ZIMMERMANN, R., 1986, Estimating photosynthetic rate and annual carbon gain in conifers from specific leaf weight and leaf biomass.
Oecologia **70**, 187–193.

PATTERSON, D. T.; DUKE, S. O.; HOAGLAND, R. E., 1978, Effect of irradiance during growth on the adaptive photosynthetic characteristics of velvet leaf and cotton.
Plant Physiol. **61**, 402–405.

PEAT, W. E., 1970, Relationships between photosynthesis and light intensity in the tomato.
Ann. Bot. **34**, 319–328.

PFEFFER, W., 1897, Pflanzenphysiologie.
Vol. 1, 323–338, 2. Auflage, Verlag Wilhelm Engelmann, Leipzig.

PIANKA, E. R., 1980, Konkurrenz und Theorie der ökologischen Nische.
In: May, R. M. (ed.): Theoretische Ökologie, Verlag Chemie, Weinheim, 105–128.

RABINOWITCH, E. I., 1951, The light factor.
In: Photosynthesis II, Wiley–Interscience, New York, 964–1191.

RÖTTELE, M. A., 1980, Populationsdynamik des Klettenlabkrauts (*Galium aparine* L.).
Diss. Univ. Hohenheim. 133 pp.

SCHULZE, E.–D., 1972, Die Wirkung von Licht und Temperatur auf den CO_2–Gaswechsel verschiedener Lebensformen aus der Krautschicht eines montanen Buchenwaldes.
Oecologia **9**, 235–258.

SCHULZE, E.–D.; CHAPIN, T. S., 1987, Plant specialisation to environments of different ressource availability.
In: Schulze, E.–D.; Zwölfer, H.: Potentials and limitation of ecosystem ecology, Ecological Studies **61**, Springer, 120–148.

SMITH, E., 1938, Limiting factors in photosynthesis: Light and carbon dioxide. Gen. Physiol. **22**, 807–830.

TENHUNEN, J. D.; YOCUM, S. C.; GATES, D. M., 1976, Development of a photosynthesis model with an emphasis on ecological applications, I. Theory. Oecologia, **26**, 89–100.

TENHUNEN, J. D.; WEBER, J. A.; YOCUM, C. S.; GATES, D. M., 1976, Development of a photosynthesis model with an emphasis on ecological applications, II. Analysis of a data set describing the PM–surface.
Oecologia **26**, 101–119.

THORNLEY, J. H. M., 1976, Mathematical models in plant physiology.
Academic Press, London. 318 pp.

TSEL'NIKER, J. L., 1979, Resistances to CO_2 uptake at light saturation in forest tree seedlings of different adaptation to shade.
Photosynthetica **13**, 124–129.

VON CAEMERER, S.; FAQUHAR, G. D., 1981, Some relationships between the biochemistry of photosynthesis and the gas exchange of leaves.
Planta **153**, 376–387.

WIESNER, J., 1909, Der Lichtgenuß der Pflanzen.
Leipzig. 322 pp.

ZIEGLER, R.; EGLE, K., 1965, Zur quantitativen Analyse der Chloroplastenpigmente 1: Kritische Überprüfung der spektralphotometrischen Chlorophyllbestimmung.
Beiträge zur Biologie der Pflanzen **41**, 11–37.

Chapter 8

Diversity of photosynthetic responses in the mesic and arid Mediterranean–type climate regions of southern Africa

M. C. RUTHERFORD

8.1 Introduction

The failure of evolution to produce an effective differentially permeable material to gaseous CO_2 and H_2O may be considered to be one of the main limitations in governing options for adaptive strategies for success of land plants. It is ironic that the resource for creating the bulk plant structure should reside in the atmosphere so that plants are forced to adopt an atmospherically open system for uptake of CO_2. Where the planetary surface thermal energy levels promote a dynamic equilibrium between liquid and vapour phases of water, the potential of liquid water–based metabolism and physiology of plants is continuously subject to limitations caused by unrestored loss of water from the open system in the vapour phase. This problem in plant design is exacerbated dramatically in those seasonal environments where periods of adequate energy levels for metabolic reactions are out of phase with periods of water availability. These globally rare, disjunct and relatively short-lived asynchronous Mediterranean-type climate environments support a loosely convergent and unique structural plant vegetation type (CODY and MOONEY 1978).

In southern Africa, Mediterranean-type climate areas with higher and lower winter rainfall have given rise to particularly unique plant assemblages in the Fynbos and Succulent Karoo Biomes respectively (RUTHERFORD and WESTFALL 1986) (Fig. 8.1). The greater Fynbos region is regarded as sufficiently unique to constitute one of the world's six floristic kingdoms (GOOD 1964) while the Succulent Karoo Biome supports the greatest diversity of the world's succulent flora with a wide range of distinctly unusual growth forms (COWLING and CAMPBELL 1980). To what extent do these plant forms represent special environmental adaptive responses and do their carbon use efficiencies carry any competitive or survival advantage? These are open questions that require attention.

The area with predominantly winter half-year rainfall occurs in the west of the region and grades eastwards very gradually around the 21 longitude east into an essentially non-Mediterranean climate type zone with a fairly even distribution of rain throughout the year. Patterns and processes in these ecosystems have attracted much scientific interest (CODY and MOONEY 1978). Annual precipitation of the Mediterranean-type climate zone ranges from about 20 mm in the far north to greater than 3000 mm in the main mountain complex in the south. Thus the concept of Mediterranean-type climate extends further into the arid and mesic domains than does that of ASCHMANN (1973).

Fynbos vegetation is heterogeneous but tends to contain the three recurring sclerophyllous elements of broad leafed evergreen plants of, for example, the Proteaceae, the narrow leafed evergreen ericoid leafed plants, for example, the Ericaceae, and the generally aphyllous evergreen graminoid-type plants of the Restionaceae. Soils of the area are predominantly shallow, nutrient impoverished acid sandy soils with the major exception of the neutral to alkaline sands of the coastal belt.

Vegetation of the Succulent Karoo Biome is not limited to dominance by succulent plants but various sclerophyllous evergreen subshrubs, summer-deciduous shrubs and in places, geophytes and many seasonal annuals are also important. Soils tend to be lime-rich, weakly developed, on rock with generally higher nutrient status than that of most soils of the Fynbos Biome.

The study of environmental adaptation of plants at the gas exchange level in these systems is relatively recent. The two main pioneers in this field from the 1970's have been D. J. VON WILLERT and H. A. MOONEY. Their work has given impetus to further research which has in turn made it opportune to highlight current questions on the diversity and efficacy of photosynthetically related responses of the exceptionally wide array of plant species to the particular environmental stresses of this climatic zone. High species diversity and complex ecosystems rel-

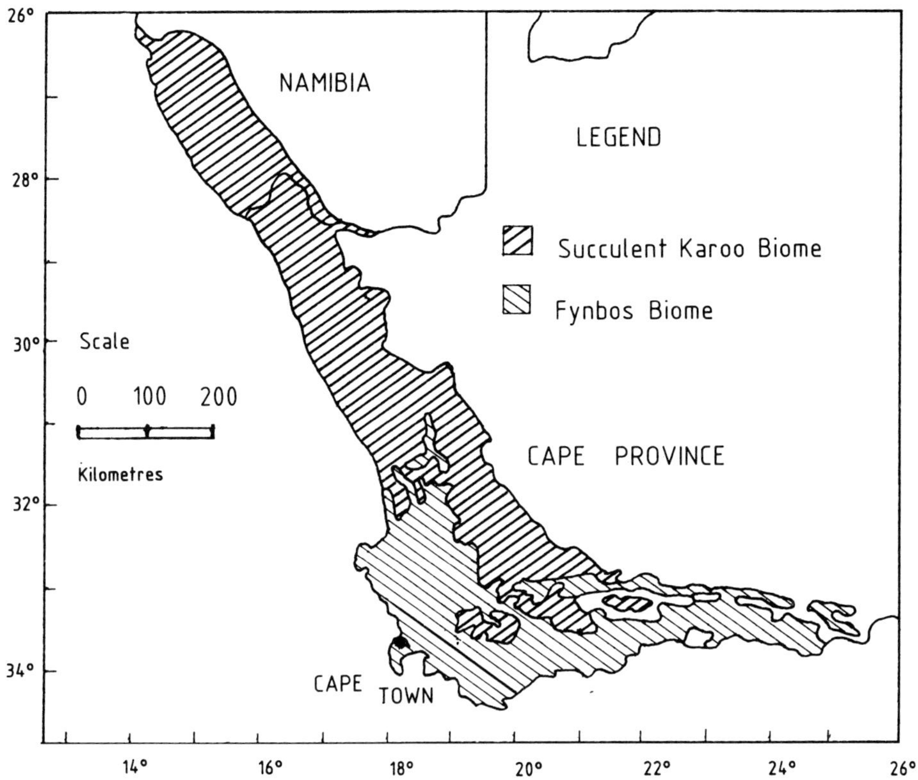

Figure 8.1: Delimitation of the Fynbos and Succulent Karoo Biomes of southern Africa (RUTHERFORD and WESTFALL 1986).

ative to the limited regional research resources, however, do not promote an early definitive synthesis of ecological consequences of the physiological level responses. It is in recognition of this limitation that the diversity of photosynthetic response patterns is emphasized as descriptive and explanatory propositions for concept development.

8.2 Photosynthetic pathways

Many anatomical, $\delta^{13}C$, photosynthetic measurements and biochemical determinations have now clearly established that the Fynbos Biome has the highest proportion of plants with C_3 mode of carbon dioxide fixation in southern Africa (VOGEL et al. 1978; VON WILLERT et al. 1989). On a plant abundance basis, the

Succulent Karoo Biome appears to have the lowest proportion of C_4–type plants and contains about 70% of the plants with Crassulacean Acid Metabolism (CAM) (WERGER and ELLIS 1981). The majority of succulent plant species of this arid biome are constitutive or facultative CAM or CAM–C_3 intermediates (MOONEY et al. 1977). Among grass species, C_3–types tend to predominate in both the Fynbos and Succulent Karoo Biomes (VOGEL et al. 1978; ELLIS et al. 1980).

The counter productive consequences of oxygen competing with carbon dioxide at the active site of the ribulose–1,5–bisphosphate carboxylase/oxygenase (rubisco) enzyme, which catalyses carbon fixation, can only be reduced or avoided if rubisco functions in an environment with high carbon dioxide and low oxygen partial pressure. This is effected in C_4, CAM and C_3–C_4 intermediate plants but at the expense of lower photochemical efficiency (AUSTIN 1989). The lower mean light intensities associated with the more mesic Fynbos Biome favour C_3 CO_2 fixation. With little competition for light in the Succulent Karoo Biome, the short term and long term implications of CAM need to be considered.

8.3 Responses to photosynthetic photon flux density

A range of fynbos evergreen shrubs have been found to respond in the conventional manner to photosynthetic photon flux density (PPFD). As light is increased beyond light compensation point, net CO_2 uptake increases approximately linearly and is presumed subjected to photochemically limited regeneration of ribulose biphosphate (RuBP). At high light intensities, the carboxylation capacity and rate of diffusion of CO_2 to the sites of carboxylation are limiting (AUSTIN 1989) and light saturates at a given rate of photosynthesis. Intermediate between the initial slope and light saturation point both processes are co–limiting. Light saturation point varied from about 1400 $\mu mol \cdot m^{-2} \cdot s^{-1}$ in the evergreen proteoid shrub *Leucospermum rodolentum* under very arid conditions (RUTHERFORD et al. (submitted)) (Fig. 8.2a) to between 800 and 900 $\mu mol \cdot m^{-2} \cdot s^{-1}$ in several of the larger evergreen proteoid shrubs of mesic fynbos (Fig. 8.2b) and as low as about 500 $\mu mol \cdot m^{-2} \cdot s^{-1}$ for the ground–level evergreen *Protea acaulos* under similar conditions (MOONEY et al. 1983).

Numerous published light response curves of fynbos species do not indicate light saturation at full sunlight (Fig. 8.2). Some of these non–saturating responses are produced under uncontrolled field conditions and are possibly artefacts of changing gas diffusion constraints. To avoid periods of low leaf water potential in the field, the stepping down of PPFD is usually performed in the morning over several hours and normally coincides with a steady increase in temperature and leaf to air vapour pressure difference (ΔW) at this time of day. By illustration, a light response curve produced in this way for field populations of *L. rodolentum* shows virtually no light saturation at full sunlight whereas a curve produced using the same population but keeping ΔW and leaf temperatures constant, achieved light saturation at well below full sunlight level (Fig. 8.2a). Discussion of the effects of ΔW is given in a later section. Non-controlled field measurements of riverine tree species and well watered large Proteaceae species (RICHARDSON and KRUGER 1990) show near-saturating responses that are probably linked to the relatively free water supply (Fig. 8.2b). Thus the response under these conditions approaches that for *Protea repens* under constant ΔW and temperature conditions (MOONEY et al. 1983) (Fig. 8.2b). VON WILLERT and co-workers (VON WILLERT et al. 1989) report finding no light saturation for all sun plant species investigated on a dry mountain fynbos site (Fig. 8.2b) and for species studied under controlled conditions in coastal fynbos near Pella (VON WILLERT 1988). In arid Karoo peripheral to the Succulent Karoo Biome near Prince Albert, net CO_2 uptake of various growth forms, including the evergreen shrub *Pteronia pallens* and the drought deciduous succulent *Drosanthemum montaguense*, were found not to saturate at full light intensity under non-constant water vapour pressure conditions (PHILLIPS 1990). If net CO_2 uptake rates do not saturate at full sunlight, at what light level may saturation be expected to occur? In the coastal shrub *Rhus glauca* it was found that photo-inhibition of photosynthesis occurred at light levels above full sunlight. This species showed clear light saturation at about 1000 $\mu mol \cdot m^{-2} \cdot s^{-1}$ and also exhibited typical shade leaf adaptations with reduced light saturation point and lowered compensation point (MIDGLEY et al. (in preparation)) (Fig. 8.2c). Interestingly, VON WILLERT and co-workers (VON WILLERT et al. 1989) found that the shrublet *Rhus dissecta* did not light saturate in the open, but showed both light saturation and typical shade leaf adaptations as an understorey plant on the dry mountain fynbos site.

Given the relatively low carboxylation capacities of typical fynbos species (see below), the apparent photosynthetic sensitivity to changes in bright light in the above set of C_3 species appears anomalous and generally different to responses of plants of Mediterranean-type climate areas elsewhere, and needs elucidation. Potentially interesting differences in apparent quantum yield indicated for diverse plant species and growth forms (Fig. 8.2d) are also deserving of further investigation.

Response time lags for net CO_2 uptake to recover to near saturation levels following a sudden step change from dark to full PPFD are indicated by one early study (CRESSWELL 1978) to be particularly protracted in certain mountain fynbos Proteaceae and even longer in certain Restionaceae. The latter took 1,5 to 2 hours to recover from a half hour's dark treatment. *Rhus glauca* and *R. lucida* recovered well within half an hour of dark treatment (MIDGLEY et al. (in preparation)) while *Leucospermum rodolentum* of the arid west attained about 80% of steady state carbon uptake after six minutes and about 95% after 12 minutes (RUTHERFORD et al. (submitted)). These response times are, as expected, much longer than those corresponding to the steps down from full PPFD to dark but are also much longer than the step up responses for shade species elsewhere (GROSS and CHABOT 1979).

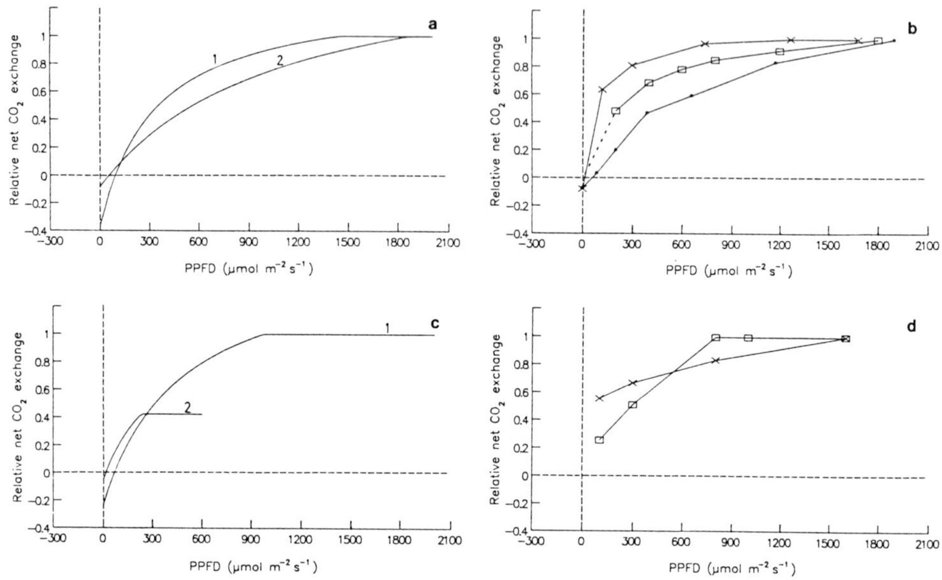

Figure 8.2: Light response curves in a) *Leucospermum rodolentum* under field conditions but with ambient T of 25±1 °C, ΔW of 20±2 mPa·Pa^{-1} (curve 1) (RUTHERFORD et al. (unpublished data)) and under uncontrolled field conditions (curve 2) (RUTHERFORD et al. (submitted)); b) *Protea repens* with leaf T of 21±1,5 °C and ΔW of 15±3 mmol·mol^{-1} (crosses) (MOONEY et al. 1983), *P. nitida* (rectangles) (RICHARDSON and KRUGER 1990) and *P. laurifolia* (points) (VON WILLERT et al. 1989); c) *Rhus glauca* as sun plant (curve 1) and as shade plant (curve 2 relativised to maximum of curve 1) (MIDGLEY et al. (in preparation)); d) *Erica plukenetii* in February with ambient T of 20 °C (rectangles) and *Thamnochortus lucens* in February with ambient T of 20 °C (crosses) (VAN DER HEYDEN and LEWIS (submitted)).

The work on *L. rodolentum* suggests that the delay in photosynthetic response to increased light is governed primarily by intrinsic lags in guard cell movement, particularly in the lower light range.

Plants with significantly delayed responses to light increases would be expected to be ill–adapted to overtopping by other vegetation with radiant energy alternation periods often shorter than the above response times (MIDGLEY et al. (in preparation)). Many of the Proteaceae and Restionaceae do tend to form stands exposed to full radiant energy, with important exceptions.

Although there seems to be a general expectation that light should be an important determinant of stomatal conductance (JARVIS and MORISON 1981; MILLER

1985), there is much evidence to the contrary in plants of both the Fynbos and Succulent Karoo Biomes. For example, in *Rhus dissecta* growing at the arid edge of fynbos, data indicate no relationship between PPFD and leaf conductance (MILLER 1985). No significant response of stomatal conductance to PPFD in the range of 800 to zero μmol·m^{-2}·s^{-1} was obtained for well watered and water stressed potted plants of the proteoid shrub *Leucodendron xanthoconus* (DAVIS et al. (submitted)). Stomatal conductance has been found to remain at day time levels after sunset and after light compensation point has been reached in populations of *L. rodolentum* and the Australian invasive shrub *Acacia cyclops* that co-occur along the arid western coastal belt of southern Namaqualand (RUTHERFORD et al. (submitted)). Conductance in these species was also consistently higher on cloudy low PPFD days than on sunny days. Under constant conditions of ΔW and temperature, stomatal conductance remained unaffected for light levels from full sun to around 150 μmol·m^{-2}·s^{-1} whereafter there was a rapid decrease in conductance to 50% of maximum level at approximately 20 μmol·m^{-2}·s^{-1} (RUTHERFORD et al. (submitted)). Potted *Protea repens* plants have been found to often transpire throughout the night at a level suggestive of not inconsequential conductance levels (SCHÜTTE and ACHLEITNER 1986). In CAM plants of the Richtersveld, relationships between light and stomatal conductance, as one may expect, appear to be non-existent (VON WILLERT et al. 1983 and 1984). Co-occurring sclerophyll shrubs, *Ozoroa dispar* and *Rhus populifolia* exhibit no transpiration in the dark (VON WILLERT et al. 1984). It is interesting to see that apparently extreme sensitivity of guard cells to, for example, first blue light at dawn in *P. repens* (SCHÜTTE and ACHLEITNER 1986), seems not to be reflected in many stomatal responses over the larger range of naturally varying PPFD levels in fynbos and Karoo sclerophyllous plant species.

Although the radiation absorptance values in the photosynthetically active spectral band of some of the Richtersveld succulent plant species, for example, *Cotyledon orbiculata* and *Prenia sladeniana*, tend to be lower than that of mesophyte leaves, light is unlikely to be limiting to plants in this area (ELLER et al. 1983).

8.4 Response to temperature

Temperature optima for net CO_2 uptake by various fynbos perennial plant species have been found to be between 25 and 27 °C (CRESSWELL 1978), between 20 and 25 °C (MOONEY et al. 1983), and generally about 20 °C (VAN DER HEYDEN and LEWIS 1988 and submitted). These optima are broadly in keeping with those of most evergreen plants of Mediterranean climate regions which are reportedly at about 20 °C (CODY and MOONEY 1978).

A much higher temperature optimum of between 30 and 35 °C (Fig. 8.3a) was found for the sclerophyll shrub *L. rodolentum* in a very arid habitat in spring (RUTHERFORD et al. (unpublished data)). A temperature optimum of between 30 and 35 °C has also been obtained for the drought deciduous shrub *Osteospermum sinuatum* of the arid Karoo in sharp contrast to the relatively temperature in-

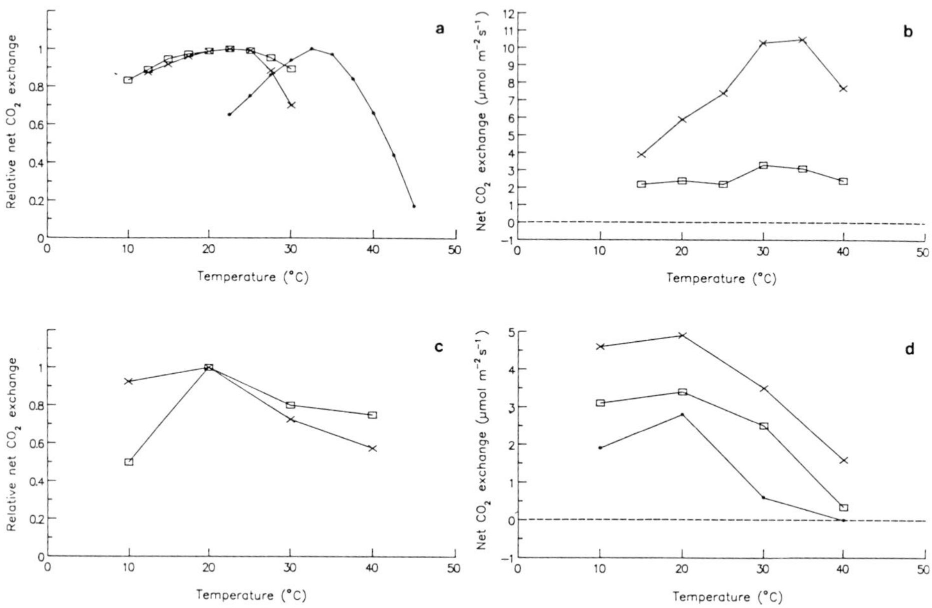

Figure 8.3: Temperature response curves in a) *Protea repens* (rectangles) and *P. acaulos* (crosses) with ΔW of 15±3 mmol·mol^{-1} (MOONEY et al. 1983) and *Leucospermum rodolentum* (points) with ΔW of 20±2 mPa·Pa^{-1} (RUTHERFORD et al. (unpublished data)); b) *Osteospermum sinuatum* (crosses) and *Drosanthemum montaguense* (rectangles) (PHILLIPS 1990); c) *P. laurifolia* (crosses) and *Erica plukenetii* (rectangles) with PPFD of 1600 μmol·m^{-2}·s^{-1} (VAN DER HEYDEN and LEWIS (submitted)); d) *Thamnochortus lucens* in February with PPFD of 1600 (crosses), 300 (rectangles) and 100 μmol·m^{-2}·s^{-1} (points) (VAN DER HEYDEN and LEWIS (submitted)).

sensitive drought deciduous C_3 succulent *Drosanthemum montaguense* (Fig. 8.3b). Relative sensitivity to higher and lower temperature appears to be reversed in *Erica plukenetii* and *Protea laurifolia* (Fig. 8.3c). Temperature effects are also light dependent. At low light intensities, high temperatures can cause plants to reach compensation point (Fig. 8.3d).

In the facultative CAM plants, temperature effects on nocturnal CO_2 fixation appear much less important relative to the dominant effects of water availability (VON WILLERT et al. 1983). Temperatures of 15 and 25 °C during the dark period were found to have no differential effect on malate accumulation in leaf slices of the annual succulent *Mesembryanthemum crystallinum* (WINTER and LÜTTGE 1976). By contrast, the reversal of day temperature of 25 °C and night temperature of 14 °C on highly stressed CAM performing shoots (with CO_2 uptake only at night) of

Prenia sladeniana resulted in conversion to C_3 mode within a day (VON WILLERT 1979). Research on the range of different temperature effects on CAM plants of the zone under consideration appears still to be in its infancy.

Plants have evolved various adaptations to control their heat load. For example, many of the plants in the Richtersveld with high absorptances have vertically inclined leaves (ELLER et al. 1983). In the succulent *Argyroderma pearsonii* of the Knersvlakte there is a significant seasonal reduction of radiation absorptance from spring into the dry summer which may be an adaptation to avoid lethal heat stress during the inactive period with high solar radiation influx (ELLER 1982). A few fynbos sclerophyll shrubs, for example, *Leucospermum salignum* and *Protea nitida*, have exceptionally low absorptance values (KRUGER et al. 1988) which may assist in heat load avoidance during the dry summer periods.

8.5 Responses to water

Leaf water potential has often been assumed to play an important role in limiting net CO_2 uptake under stress conditions. However, summer leaf pressure potentials of the evergreen karroid shrub *Pteronia pallens* can remain below −7. MPa before dawn yet leaves can maintain a low level of photosynthetic activity at midday (PHILLIPS 1990). On the same site, other plants such as the evergreen succulent c. f. *Eberlanzia ferox*, at higher summer predawn pressure potentials than *P. pallens*, remain at or below compensation point throughout the day and night. However, within a species, such as *Osteospermum sinuatum*, there is a tendency for a direct relationship between leaf pressure potential and mid–morning net CO_2 exchange through the year.

In a wide range of karroid dry fynbos shrubs, it appears that there is a threshold minimum value of leaf water potential above which stomatal conductance can vary almost independently of leaf water potential when viewed over the year (MILLER 1985). It has also been reported that following first rains after summer on mountain fynbos sites, pressure potentials rose in many species but leaf conductances did not (MILLER et al. (unpublished, 1984)). Leaf pressure potentials can vary widely according to plant species ability to control water loss. The arid zone herb *Codon royenii*, for example, exercises poor stomatal control which possibly explains its confinement to river beds (VON WILLERT et al. 1984). Most of the proteoid fynbos shrubs investigated typically have high predawn leaf pressure potentials with rapid recovery toward evening (MOLL and SOMMERVILLE 1985). Relatively high stomatal conductances are maintained through the year (MILLER et al. 1983 and 1984), although summer depressions in net CO_2 exchange can occur (RICHARDSON and KRUGER 1988). The smaller ericoid shrubs and many restioid plants, by contrast, tend to have lower pressure potentials especially in summer. This difference has been interpreted as relating to the rooting depth of these forms with proteoid shrubs extending more effectively to permanent water supplies (HIGGINS et al. 1987; VAN DER HEYDEN and LEWIS 1988). The larger proteoid species appear to differ little with respect to their water relations and photosynthetic patterns from riverine trees in the area (RICHARDSON and KRUGER 1990).

It has been found that after watering field populations, the restioid *Thamnochortus lucens* showed a rapid increase in photosynthetic rate whereas *Protea laurifolia* remained unaffected (VAN DER HEYDEN and LEWIS (submitted)). Not all larger shrubs in the Fynbos Biome show this favourable water status. Many shrubs, typical of the neutral and alkaline coastal sands, have very low summer pressure potentials, for example, predawn values of less than −5. MPa in the evergreen *Cassine peragua* (RUTHERFORD and BÖSENBERG 1988). In terms of osmotic potentials, some other species in this guild register summer values of below −7. MPa (GOUWS and AALBERS 1969), compared to only −1.5 to −2.5 MPa (WALTER and VAN STADEN 1965) in proteoid shrub examples. The small ground shrublet *Protea acaulos* had the lowest osmotic potential (−2.7 MPa) and the largest proteoid shrub *P. nitida* the highest (−1.0 MPa) of Proteaceae investigated by VON WILLERT and co-workers (VON WILLERT et al. 1989). They also have found that within the Proteaceae, the contribution by inorganic ions to osmotic potential is much lower than in other fynbos growth forms. Only 20 to 40% of the osmotic potential is accounted for by inorganic ions which indicates that other unidentified osmotically active compounds must occur in the Proteaceae.

On zoogenic earth mounds, which occur throughout much of the Succulent Karoo Biome (LOVEGROVE and SIEGFRIED 1986), it has been found that predawn pressure potentials of *Pteronia paniculata* in summer are far higher off the mounds than those of the allied *Pteronia incana* on the mounds on pole–facing slopes (MIDGLEY and BÖSENBERG (in press)).

Water potentials can also be affected by competitive interaction, where, for example, indigenous shrubs on neutral coastal sands have significantly depressed pressure potentials when overtopped by invasive *Acacia cyclops* individuals under dry soil conditions (RUTHERFORD and BÖSENBERG 1988).

The generally aphyllous Restionaceae seem to have a number of adaptations for reducing water loss and even water gain. These include: sealing off the central stem of the roots of a *Cannamois* species during the summer months (HARDCASTLE and SCHUTTE 1983); development of a thick cuticle on the culms of this species and deeply sunken stomata on the culms of *Thamnochortus punctatus* (MOLL and SOMMERVILLE 1985), development of a thin cuticular layer on the protective cells lining most of the substomatal cavity in *Elegia vaginulata* (BOTHA et al. 1972) and increase of absorptive area by development of capillaroid roots – a functional analogue of the proteoid roots of the Proteaceae (LAMONT 1982).

In certain plant groups of the arid zone, unusual water conserving mechanisms include the imbedding of much of the leaf into the soil in *Lithops* species (ELLER and RUESS 1982) and the presence of a layer of bladder cells on the leaf surface in *Mesembryanthemum crystallinum* and a *Drosanthemum* species from the Richtersveld (BRINCKMANN et al. 1985; RYGOL et al. 1986). The bladder cells in the former species can constitute about 25% of the volume of the aerial plant

(STEUDLE et al. 1977; LÜTTGE et al. 1978), are metabolically not very active and, during osmotic stress, act as a buffering system to protect the photosynthetic tissue from osmotic injury (STEUDLE et al. 1975). Transpiration into the ambient atmosphere is also minimized by the water impermeable cuticle of the bladders (LÜTTGE et al. 1978). In the *Drosanthemum* species, water stress results in a flattening of the bladders to reduce the area of the transpiration passages between the bladder cells by two thirds (BRINCKMANN ET AL. 1985).

Water addition through mist or fog in both the Fynbos and Succulent Karoo Biomes may be highly significant to plant survival at times. It has been found that morning lowland fog can raise midday pressure potentials above that of predawn potentials for larger but not small tussocks of the restioid *Thamnochortus punctatus* (MOLL and ROMOFF 1983). In mountain fynbos, it has been established that many of the Restionaceae that have dense tussocks of wiry culms effectively intercept droplets of moving mist (SNOW 1985).

In arid areas, the graminoid perennial, *Stipagrostis sabulicola* has been shown, using tritiated water, to take up fog precipitation on sandy surfaces (LOUW and SEELY 1980). The succulent shrub, *Trianthema hereroensis*, which fringes the northern extremity of the Succulent Karoo Biome, has been shown to take up fog precipitated on their leaf surfaces (SEELY 1981). Fog during the first few hours of light have been found to increase diffuse light intensity causing a prominent burst of CO_2 uptake that does not occur on a clear day at this time in *Psilocaulon* cf. *subnodosum* on a site in the Richtersveld (VON WILLERT et al. 1983). It has been shown elsewhere that leaf conductance on a foggy morning is twice as high as on days without fog.

Water stress in many succulent plants of the arid zone has particularly interesting facets relating to the CAM photosynthetic pathway. One of the identifying characteristics of CAM is a diurnal fluctuation in organic activity, mainly malate production. Malate is produced in the dark as a result of dark stomatal opening leading to a CO_2 influx that is fixed by PEP–carboxylase to form malate (VON WILLERT 1975). During the day the malate is decarboxylated and the internally generated endogenous CO_2 is used for conventional C_3 photosynthesis. It has been shown that many succulents of the Richtersveld can accumulate malate overnight (VON WILLERT et al. 1977). Treatment of *M. crystallinum* with NaCl results after a few days in significant accumulation of malate accompanying a many-fold increase in PEP–carboxylase activity (VON WILLERT et al. 1976). The increase is independent of whether the osmotic stress is applied by salts, by sugars, or by mixtures of salts and sugars. It, therefore, seems that malate is accumulated in response to decreased water potential rather than in response to increases in concentration of specific ions (VON WILLERT 1975). There is much evidence in support of increasing water stress promoting CAM or inducing CAM in certain C_3 succulents, for example in *Portulacaria afra* (TING and HANSCOM 1977; TING 1981; GURALNICK et al. 1984; GURALNICK and TING 1986 and 1987), *Carpobrotus edulis* (WINTER 1973; EARNSHAW et al. 1987), *Mesembryanthemum nodiflorum* (WIN-

TER and TROUGHTON 1978), and in *M. crystallinum* (WINTER and VON WILLERT 1972; WINTER 1974; WINTER et al. 1978; WINTER and LÜTTGE 1979). With CAM induction in the last mentioned species, it has been shown that there are no changes in the photochemical apparatus of its thylakoids (KÖSTER and ANDERSON 1988).

In CAM exhibiting leaf succulents, such as *Cotyledon orbiculata*, PEP–carboxylase is inactivated by light in the early morning (VON WILLERT 1979) and the malate content starts to decrease later in the morning (VON WILLERT et al. 1984). As soon as the malate pool is exhausted, typically by mid–afternoon, leaf conductance increases sharply and the plant employs photosynthetic uptake by exogenous CO_2 without malate formation until dark (VON WILLERT et al. 1984). In well watered *Psilocaulon* c. f. *subnodosum*, photosynthetic CO_2 uptake starts soon after dawn (VON WILLERT et al. 1983). In *Portulacaria afra* in CAM mode, it is assumed that there is preferential use of exogenous CO_2 during photosynthesis when stomata are open during the day and organic acid levels remain high. It appears that the stomatal closure under temporary water stress directly brings about organic acid decarboxylation during the day (TING 1981).

For most CAM succulents, malate is accumulated only at leaf water contents above 80% (VON WILLERT et al. 1985). This, together with the foregoing, suggests an apparent paradox in CAM as an adaptive response to water stress. VON WILLERT (1979) questioned CAM as an adaptive mechanism to water stress since he found that an increase in water stress in the natural habitat reduced existing CAM. Further, after a 3–year drought, CAM succulents were observed to be far more damaged than most non–CAM succulents. Levels of ion concentration during desiccation in Mesembryanthemaceae were many times higher than that of many other succulents which may explain the disproportional demise of many of the Mesembryanthemaceae in the drought (VON WILLERT et al. 1985). It appears that CAM could be regarded as an adaptive mechanism to drought only in the short term (months) and is not specifically adapted to long term water stress (years).

The Mesembryanthemaceae have a great ability to concentrate ion content to high levels in the leaves irrespective of ionic conditions of most soils (VON WILLERT et al. 1977; 1979). The resulting low osmotic potentials of -1.5 to -4.5 MPa lowers water–vapour saturation in the substomatal cavity by between 1 and 3%. Under natural conditions in the Succulent Karoo Biome on clear, calm nights which allow leaves to cool well below ambient air and with high air humidity, a negative leaf to air vapour pressure gradient (ΔW) can result which must lead to water vapour diffusion into the leaf (reversed transpiration). VON WILLERT et al. (1985) estimate that 10% (or up to 49% (VON WILLERT and BRINCKMANN 1985)) of the average water loss during the light period could be compensated overnight. In addition, it is argued that nocturnal malate formation facilitates not only reverse transpiration as above, but also water uptake through the roots (ELLER and RUESS 1984). Osmotic potential in *Psilocaulon* c. f. *subnodosum* is inversely related to the stress indicator proline (VON WILLERT et al. 1985) which has been found to be present in 94 species in 25 families in a study in the Richtersveld (TREICHEL et al. 1984).

Another question regarding the effectiveness of CAM in terms of carbon capture relates to the relative CO_2 uptake during the night versus that in the day. For example, in CAM induced *Carpobrotus edulis* the dark uptake of CO_2 is only a small fraction of that taken up during the day (WINTER 1973). Even "idealistic daily courses" of gas exchange for a CAM plant shows early morning and late afternoon CO_2 uptake fully equivalent to the sum of the night time uptake (VON WILLERT and BRINCKMANN 1985). OSMOND et al. (1982) have indicated that it is unlikely that CO_2 fixation in the dark can confer any competitive advantage in terms of productivity.

A last imponderable regarding CAM is the question of CAM idling. This is the refixation of respiratory CO_2 or recycling of organic acids, where stomata remain closed day and night but a dampened organic acid fluctuation is still observed (GURALNICK et al. 1984). CAM idling has been reported for a number of CAM plants including CAM-induced *Portulacaria afra* after two months of water stress (GURALNICK and TING 1986). However, in CAM plants investigated in the Richtersveld toward the end of a drought, virtually no acid fluctuation was detected with respiratory CO_2 release throughout day and night (VON WILLERT et al. 1983) and a reduction of net CO_2 uptake was always accompanied by a reduction in malate formation (VON WILLERT et al. 1985). One concludes that there is no evidence for CAM idling in these plants. GURALNICK and TING (1987) state that even in *P. afra*, it is becoming increasingly clear that CAM/CAM idling metabolism may only partially explain the survival of the plant during long-term drought.

Water use efficiency varies with environmental conditions, stomatal conductance and photosynthetic capacity which restricts many comparisons of field data. Available data on instantaneous water use efficiencies (WUE) for a range of plant species under standard conditions in the winter rainfall zone tend to be similar (Table 8.1). However, the successful alien invasive *A. cyclops* has a noticeably higher WUE than most of the other plants especially on a cloudy day with reduced ΔW. Lowered WUE in shaded understorey *Rhus glauca* corresponds with a much smaller reduction in ΔW compared to that of the open sun populations of this species. *Thamnochortus lucens* in summer has an extremely low WUE which is associated with very low levels of net CO_2 uptake at that time (VAN DER HEYDEN and LEWIS 1989). Lower water potential appears to raise WUE in the Karoo evergreen shrub *Pteronia paniculata* (MIDGLEY 1990).

8.6 Responses to leaf to air water vapour pressure deficits

Despite earlier published work on the subject, plant responses to leaf to air water vapour pressure deficit or difference (ΔW) (akin to vapour pressure deficit, VPD)

Table 8.1: (next page) Typical instantaneous water use efficiencies (WUE, CO_2/H_2O [mmol·mol^{-1}]) of species under field conditions in the Fynbos and Succulent Karoo Biomes. Sources: [1] MOONEY et al. 1983; [2] VON WILLERT et al. 1989; [3] VAN DER HEYDEN and LEWIS 1989; [4] RICHARDSON and KRUGER 1990; [5] RUTHERFORD et al. (submitted); [6] MIDGLEY et al. (in preparation); [7] MIDGLEY 1990.

---

have been under-emphasized in much of the ecological work of the area under consideration.

It has been found that leaf conductance decreased with increasing ΔW for most plants investigated in dry mountain fynbos (VON WILLERT et al. 1989), for most plants along a karroid to dry fynbos gradient (MILLER 1985), for plants studied in dry and mesic mountain fynbos (MILLER et al., unpublished, 1984), and for *L. rodolentum* and *A. cyclops* shrubs in very arid conditions (RUTHERFORD et al. (submitted)). Increasing ΔW with the normally concomitant increasing temperature appears to be more important in controlling diurnal photosynthesis than is plant water status (VON WILLERT et al. 1989; RUTHERFORD et al. (submitted)). In these cases, therefore, and under adequate light and nutrient conditions, ΔW may largely control net CO_2 exchange as mediated through stomatal conductance.

From variation in diurnal data of succulents of the arid areas, it is clear how ΔW effects net CO_2 uptake inversely (VON WILLERT et al. 1985). An inverse relationship between ΔW and malate accumulated overnight has also been shown in species such as *Psilocaulon* c. f. *subnodosum* and *Stoeberia beetzii* (VON WILLERT et al. 1985).

Others have successfully tied the development of CAM in *M. crystallinum* to conditions of low ΔW (WINTER 1974; WINTER and LÜTTGE 1979). It is clear from the earlier discussion that ΔW is presumably capable of controlling the balance of use of endogenous and exogenous CO_2 during the day in CAM plants.

8.7 Nutrient considerations

Many fynbos soils have a very low nutrient status with very low nitrogen levels. A number of typical plant forms on these soils have evolved special nutrient conservation measures. For example, 70% of total N present in mature culms of *Thamnochortus punctatus* may be withdrawn into the remainder of the plant before the culm dies back (STOCK et al. 1987). The low soil N levels do not mean that N addition above normal levels has any effect on growth. *Protea repens* has been shown to exhibit minimal growth (STOCK and LEWIS 1984) or even decreased growth (STOCK and MANUEL (submitted)) in response to such N fertilization. Leaf N of six fynbos species has nevertheless been found to relate directly to net CO_2–

Species/ plant type	Condition	Place	Source	WUE
Sclerophyll shrubs	Spring, mesic fynbos	Jonkershoek	[1]	2.2–4.6
Proteoid shrubs	Spring, mesic fynbos	Swartboskloof	[2]	1.3–5.6
Protea laurifolia	Spring, dry fynbos	Cedarberg	[2]	1.5
Paranomus bracteolaris	Spring, dry fynbos	Cedarberg	[2]	1.5
Protea laurifolia	Summer(12:00) mesic fynbos	Bainskloof	[3]	4.1
Protea laurifolia	Winter(12:00)	Bainskloof	[3]	7.1
Thamnochortus lucens	Summer(12:00)	Bainskloof	[3]	0.1
Thamnochortus lucens	Winter(12:00)	Bainskloof	[3]	2.1
Brabejum stellatifolium	Summer(12:00)	Swartboskloof	[4]	1.7–2.0
Protea nitida	Winter(12:00)	Swartboskloof	[4]	3.8
Protea nitida	Summer(12:00)	Swartboskloof	[4]	2.0
Leucospermum rodolentum	Spring(12:00) cloudy day	Southern Namaqualand	[5]	5.0
Leucospermum rodolentum	Spring(12:00) sunny day	Southern Namaqualand	[5]	3.0
Acacia cyclops	Spring(12:00) cloudy day	Southern Namaqualand	[5]	8.0
Acacia cyclops	Spring(12:00) sunny day	Southern Namaqualand	[5]	4.0
Rhus glauca	Spring(12:00) open (sun)	Strandfontein	[6]	3.0
Rhus glauca	Spring(12:00) understorey(shade)	Strandfontein	[6]	1.8
Pteronia paniculata	Ψ_p: −1.0 MPa PPFD: 300–700 [μmol·m^{-2}·s^{-1}]	Worcester	[7]	4.2
Pteronia paniculata	Ψ_p: −3.9 MPa PPFD: 300–700 [μmol·m^{-2}·s^{-1}]	Worcester	[7]	6.5

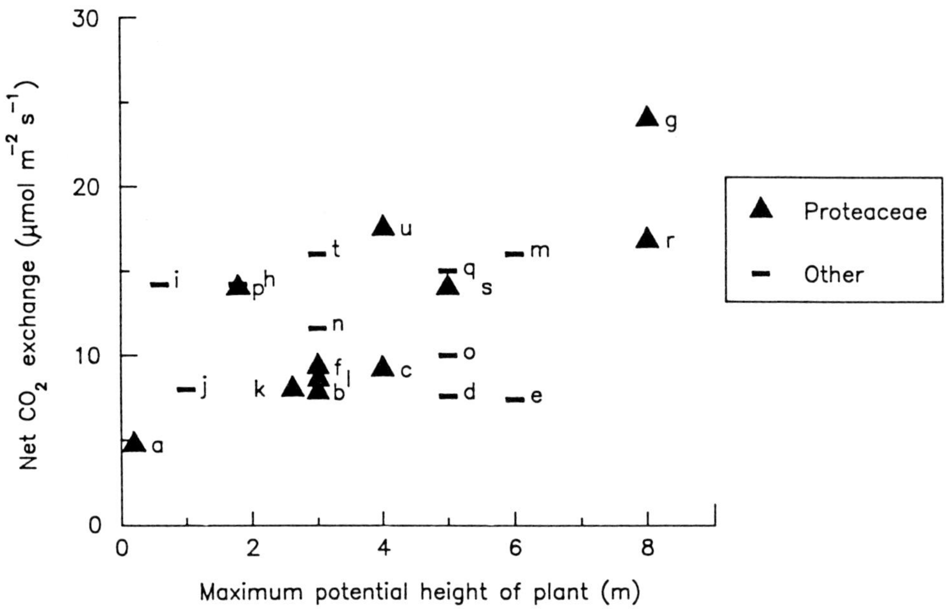

Figure 8.4: Photosynthetic capacity of adult shrubs under non–extreme field conditions according to maximum potential plant height (BOND and GOLDBLATT 1984) in the fynbos biome. Fluxes are standardized per one–sided leaf area. Species are: a, *Protea acaulos*; b, *Protea neriifolia*; c, *Protea repens*; d, *Rhus tomentosa*; e, *Maytenus oleoides*; f, *Paranomus bracteolaris*; g, *Protea laurifolia*; h, *Rhus dissecta*; i, *Elytropappus gnaphaloides*; j, *Diosma oppositifolia*; k, *Leucadendron pubescens*; l, *Leucospermum rodolentum*; m, *Acacia cyclops*; n, *Rhus glauca*; o, *Rhus lucida*; p, *Leucadendron salignum*; q, *Metrosideros angustifolia*; r, *Protea laurifolia*; s, *Brabejum stellatifolium*; t, *Brachylaena neriifolia*; u, *Protea repens*. Sources are: a–e MOONEY et al. 1983; f–k VON WILLERT et al. 1989; l–m RUTHERFORD et al. (submitted); n–o MIDGLEY et al. (in preparation); p–r VAN DER HEYDEN and LEWIS 1989; s–u RICHARDSON and KRUGER 1990.

uptake expressed on a unit dry mass basis (MOONEY et al. 1983). In a study of potted *Protea lepidocarpodendron* grown at different nutrient levels at relatively low light intensity, there was not only greater net CO_2–uptake with nutrient additions but also significantly increased stomatal conductance (STOCK and MANUEL (submitted)). This latter phenomenon is presumably mediated by a form of C_i feedback control effective under the conditions of the experiment. In water stressed populations, the effect of the nutrient amendments disappears possibly through immobilization of the additional nutrients.

The arid karoo soils are poorly developed but are not nutrient poor. The strong representation of succulent plant species on the zoogenic mounds of the Succulent Karoo Biome (KNIGHT et al. 1989) together with the elevated N status of their soils (MIDGLEY and MUSIL 1990) provides fertile opportunity for further testing relationships between net CO_2-uptake and N of alternative magnitudes within different constraints of water availability.

8.8 Photosynthetic capacity

When one considers the large number of factors potentially limiting photosynthetic rates, it is not surprising that photosynthetic capacity is assessed with difficulty, especially from field data. There are the diffusion constraints, the biochemical limitations and factors of plant and plant organ organization and maturation (see next section). There are also treatments that may simulate conditions under which the plant species may have evolved. For example, removal of 40% of leaf material to simulate grazing in the winter rainfall grass perennial *Ehrharta calycina* was found to result in a highly significant increase in net CO_2 uptake of the remaining leaves within 24 hours of the treatment (VAN DER WESTHUIZEN et al. 1981). This increase, which may be related to increased stomatal conductance or creation of new metabolic sinks, exacerbates the problem of what is regarded as the norm for determining photosynthetic capacity of the species concerned. Reduced stomatal densities in overtopped plants, for example in *Rhus glauca* (MIDGLEY et al. (in preparation)), become an intrinsic limitation to CO_2 diffusion when the overtopping vegetation is removed.

Adult plant photosynthetic capacities for a selection of fynbos species under normal field conditions are given in Fig. 8.4. For especially *Proteaceae* shrubs with a potential height of less than 9 m, an empirical tendency may exist for increasing maximum net CO_2-uptake with increasing plant height which may reflect long-evolved relative access to light and deeper soil water. Any such suggested relationship breaks down for species of tree above 9 m height which tend to be associated with more moist habitats or riverine environments in the Fynbos Biome and which vary greatly in apparent photosynthetic capacity. No direct mechanistic relationship is implied between photosynthetic capacity and plant height. The range of photosynthetic capacities documented for species in the Fynbos Biome is generally within the range of capacities reported for Mediterranean climate areas elsewhere (MOONEY 1983).

Reported photosynthetic capacities of fieldgrown evergreen plants of the arid Karoo are lower than that of many fynbos plants and usually show rates of less than 10 $\mu mol \cdot m^{-2} \cdot s^{-1}$. Several succulent plant species have lowest reported photosynthetic capacities. This sequence of groups with decreasing photosynthetic capacities mirrors that provided for other desert and Mediterranean-type climate plants on other continents (EHLERINGER and MOONEY 1983).

8.9 Leaf and plant age effects

It is widely known that degrees of internal organization and maturation of organs and whole plants can have profound effects on carbon acquisition rates of plants. This last section seeks to address briefly effects of leaf age and plant age on net CO_2 exchange and its determining parameters in plants of the region under review.

Maximum photosynthesis in old (about 1 year) leaves of *Paranomus bracteolaris* was found to be about twice that of the young leaves (VON WILLERT et al. 1989). An opposite relationship was found for mature and juvenile (about one year old) plants of *Leucospermum rodolentum* with approximately equal aged leaves (RUTHERFORD et al. (submitted)). Stomatal conductance was also higher in the younger plants although WUE and Ψ_p were essentially no different. In a study of a wide range of fynbos species at several different sites, age of leaves in the one to two year range did not affect either conductances or water potentials (MILLER et al. 1984).

Predawn pressure potential of *Protea nitida* seedlings in the dry season was determined as −2.0 MPa which was similar to that of the mature plants (RICHARDSON and KRUGER 1990). However, resprouting adults of *P. nitida* had much higher potentials of −0.5 MPa which is possibly related to a large root system supplying water to a small total leaf area. This would accord with the improved water status and higher photosynthetic rate found in resprouts of various chaparral shrubs after fire or hand clearing compared to mature plants (HASTINGS et al. 1989). The contrary results in comparing water potentials of mature plants of *Podalyria calyptrata* with resprouting individuals after fire (JEFFERY et al. 1987) seem to be at variance with these findings. Differences in water potential between seedlings and mature plants must largely depend on the current relative extent of the seedling root system towards the sites of available water. Hence, predawn pressure potentials of *Protea repens* seedlings in summer were lowered to −3. MPa relative to the −1.5 to −2.0 MPa of the mature plants (RICHARDSON and KRUGER 1990) whereas pressure potentials of *Leucadendron xanthoconus* seedlings were, as in *P. nitida*, very similar to that of the adult plants (DAVIS and MIDGLEY 1990).

In the Succulent Karoo Biome, there is ample evidence that PEP–carboxylase activity for malate formation is the highest in the older leaves of *Prenia sladeniana*, *Sceletium joubertii*, (VON WILLERT and VON WILLERT 1979) and *Mesembryanthemum crystallinum* (VON WILLERT et al. 1976b). In fact, in all the Mesembryanthemaceae, PEP–carboxylase activity is low in young leaves but increases rapidly with leaf age (VON WILLERT 1979). In *M. crystallinum*, the youngest leaves not only always showed the lowest PEP–carboxylase activity, but never accumulated malate at night (VON WILLERT et al. 1976). Mature leaves of *Psilocaulon* c. f. *subnodosum* have much higher morning contents of malate than young leaves. However, this difference between mature and young leaves disappears under imposition of water stress (VON WILLERT et al. 1985). CAM induction in *Portulacaria afra* seedlings is age dependent. In contrast to six week old seedlings, seedlings under two weeks old were not able to change from C_3 metabolism to full CAM under drought conditions (GERBER 1987). The younger seedlings did take up CO_2 nocturnally but respiratory loss exceeded uptake.

An interesting consequence of leaf age effects in facultative CAM plants such as *M. crystallinum* is that, under conditions of CAM-inducing water stress, the actively growing plant individual may be expected to be utilizing both CAM and C_3 metabolism on different parts of the same plant.

8.10 Conclusions

There is much scope for further research on intriguing processes and patterns in plants of the Mediterranean-type climate zone of southern Africa. Some isssues of particular interest are adaptations in scheduling endo- and exogenous CO_2 utilization during the light period in CAM plants under dynamic environmental conditions, understanding mechanisms underlying the apparent relationship between unusual combinations of environmental factors and depressed carbon assimilation, and investigation of the possible integral effects of co-occuring photosynthetic pathways within the same organism with implications for source sink relationships. The carbon relations of the plant types in the area of interest need to be integrated at different organizational levels in order to effectively test for any links with competitiveness in relation to survival of plants in stressful environments.

Acknowledgments

I thank G. F. MIDGLEY for useful discussion and J. DE W. BÖSENBERG, T. B. A. RUTHERFORD and L. W. POWRIE for sterling work in the production of the manuscript and assistance with literature acquisition.

Bibliography

ASCHMANN, H., 1973, Distribution and peculiarity of mediterranean ecosystems. In: DI CASTRI, F.; MOONEY, H. A. (Eds.), Mediterranean type ecosystems. Origin and structure, Chapman and Hall, London, pp. 11–19.

AUSTIN, R. B., 1989, Genetic variation in photosynthesis.
J. Agric. Sci., Camb. **112**, 287–294.

BOND, P.; GOLDBLATT, P., 1984, Plants of the Cape Flora. A descriptive catalogue,
J. S. Afr. Bot. Supplementary volume **13**.

BOTHA, D. J.; VAN DER SCHIJFF, H. P.; VAN TONDER, E. M. A., 1972, The position, structure and ontogeny of the stomata in the stems of *Elegia vaginulata*

Mast. (In Afrikaans),
Tydskrif vir Natuurwetenskappe **12**, 193–199.

BRINCKMANN, E.; WARTINGER, M.; VON WILLERT, D. J., 1985, Turgoränderungen in Blasenzellen von Mesembryanthemaceen.
Ber. Deutsch. Bot. Ges. **98**, 447–454.

CODY, M. L.; MOONEY, H. A., 1978, Convergence versus nonconvergence in mediterranean-climate ecosystems.
Ann. Rev. Ecol. Syst. **9**, 265–321.

COWLING, R. M.; CAMPBELL, B. M., 1980, Convergence in vegetation structure in the mediterranean communities of California, Chile and South Africa.
Vegetatio **43**, 191–197.

CRESSWELL, C. F., 1978, Some features of photosynthesis in Fynbos plants.
In: National Programme for Environmental Sciences. Proceedings of a colloquium on aspects of the ecology of the Fynbos biome, University of Cape Town, Cape Town, pp. 19–24.

DAVIS, G. W.; FLYNN, A. P.; MIDGLEY, G. F., Growth and gas-exchange responses of *Leucadendron xanthoconus* (Proteaceae) seedlings to different nutrient and water regimes (submitted for publication).

DAVIS, G. W.; MIDGLEY, G. F., 1990, Effects of disturbance by fire and tillage on the water relations of selected mountain Fynbos species.
S. Afr. J. Bot. **56** (2), 199–205.

EARNSHAW, M. J.; CARVER K. A.; CHARLTON, W. A., 1987, Leaf anatomy, water relations and crassulacean acid metabolism in the chlorenchyma and colourless internal water-storage tissue of *Carpobrotus edulis* and *Senecio ?mandraliscae*.
Planta **170**, 421–432.

EHLERINGER, J.; MOONEY, H. A., 1983, Productivity of desert and mediterranean-climate plants.
In: LANGE, O. L.; NOBEL, P. S.; OSMOND, C. B.; ZIEGLER, H. (Eds.), Physiological plant ecology **IV**, Ecosystem processes: Mineral cycling, productivity and man's influence, Springer-Verlag, Berlin, pp. 205–231.

ELLER, B. M., 1982, Die Strahlungsabsorption von *Argyroderma pearsonii* (N. E. Brown) Schw. in der Vegetations- und Ruheperiode,
Ber. Deutsch. Bot. Ges. **95**, 333–340.

ELLER, B. M.; BRINCKMANN, E.; VON WILLERT, D. J., 1983, Optical properties and succulence of plants in the arid Richtersveld (Cp., Rep. South Africa).
Botanica Helvetica **93**, 47–55.

ELLER, B. M.; RUESS, B., 1982, Water relations of Lithops plants embedded into the soil and exposed to free air.
Physiol. plant. **55**, 329–334.

ELLER, B. M.; RUESS, B. R., 1984, CO_2 dark fixation in CAM enhances water uptake during the night.
S. Afr. J. Bot. **3** (5), 347–349.

ELLIS, R. P.; VOGEL, J. C.; FULS, A., 1980, Photosynthetic pathways and the geographical distribution of grasses in South West Africa/Namibia.
S. Afr. J. Science **76**, 307–314.

GERBER, J. J., 1987, 'n Ekofisiologiese ondersoek na die invloed van water tekorte op koolstofdioksied wisseling, groei en oorlewing van *Portulacaria afra* saailinge.
B. Sc.–Hons thesis, Department of Botany, University of Port Elizabeth, Port Elizabeth.

GOOD, R., 1964, The geography of the flowering plants.
Third Edition, Longmans, London.

GOUWS, J. B.; AALBERS, J., 1969, Annular curves of the osmotic pressure of certain plants on the cape flats.
Jl. S. Afr. Bot. **35** (2), 113–126.

GROSS, L. J.; CHABOT, B. F., 1979, Time course of photosynthetic response to changes in incident light energy.
Plant Physiol. **63**, 1033–1038.

GURALNICK, L. J.; RORABAUGH, P. A.; HANSCOM, Z., 1984. III, Seasonal shifts of photosynthesis in *Portulacaria afra* (L.) Jacq.
Plant Physiol. **76**, 643–646.

GURALNICK, L. J.; TING, I. P., 1986, Seasonal response to drought and rewatering in *Portulacaria afra* (L.) Jacq.
Oecologia **70**, 85–91.

GURALNICK, L. J.; TING, I. P., 1987, Physiological changes in *Portulacaria afra* (L.) Jacq. during a summer drought and rewatering.
Plant Physiol. **85**, 481–486.

HARDCASTLE, J.; SCHUTTE, K. H., 1983, Aspects of an experimental study on root aerenchyma development and the ecological implications thereof.
Bothalia **14**, (3 & 4), 791–794.

HASTINGS, S. J.; OECHEL, W. C.; SIONIT, N., 1989, Water relations and photosynthesis of chaparral resprouts and seedlings following fire and hand clearing.
In: KEELEY, S. C., The California chaparral. Paradigms reexamined, No **34**, Science series, Natural History Museum of Los Angeles county, pp. 107–113.

HIGGINS, K. B.; LAMB, A. J.; VAN WILGEN, B. W., 1987, Root systems of selected plant species in mesic mountain Fynbos in the Jonkershoek valley, southwestern Cape Province.
S. Afr. J. Bot. **53** (3), 249–257.

JARVIS, P. G.; MORISON, J. I. L., 1981, The control of transpiration and photosynthesis by the stomata.
In: JARVIS P. G.; MANSFIELD, T. A. (Eds.), Stomatal physiology, Cambridge University Press, Cambridge, pp. 247–279.

JEFFERY, D.; MOLL, E. J.; VAN DER HEYDEN, F., 1987, Comparative water potentials of 4-month old and 40-year-old Cape Fynbos plants.
S. Afr. J. Bot. **53** (1), 32–34.

KNIGHT, R. S.; REBELO, A. G.; SIEGFRIED, W. R., 1989, Plant assemblages on Mima–like earth mounds in the Clanwilliam district, South Africa,
S. Afr. J. Bot. **55** (5), 465–472.

KÖSTER, S.; ANDERSON, J. M., 1988, The photosynthetic apparatus of C_3 and CAM–induced *Mesembryanthemum crystallinum* L.,
Photosynthesis Research **19**, 251–264.

KRUGER, F. J.; RICHARDSON, D. M.; SMITH, R. E., 1988, Plant water relations in sclerophyllous trees and shrubs of riparian and hillslope habitats at Swartboskloof, southwestern Cape Province, South Africa.
In: DI CASTRI, F.; FLORET, CH.; RAMBAL, S.; ROY, J. (Eds.), Time Scales and Water Stress – Proc. 5th Int. Conf. on Mediterranean Ecosystems, I. U. B. S., Paris, pp. 575–581.

LAMONT, B., 1982, Mechanisms for enhancing nutrient uptake in plants, with particular reference to mediterranean South Africa and western Australia.
The Botanical Review **48**, 597–689.

LOUW, G. N.; SEELY, M. K., 1980, Exploitation of fog water by a perennial Namib dune grass, *Stipagrostis sabulicola*.
S. Afr. J. Science **76**, 38–39.

LOVEGROVE, B. G.; SIEGFRIED, W. R., 1986, Distribution and formation of Mima–like earth mounds in the western Cape Province of South Africa.
S. Afr. J. Science **82**, 432–435.

LÜTTGE, U.; FISCHER, E.; STEUDLE, E., 1978, Membrane potentials and salt distribution in epidermal bladders and photosynthetic tissue of *Mesembryanthemum crystallinum* L.,
Plant, Cell and Environment **1**, 121–129.

MIDGLEY, G. F. (personal communication), 1990.

MIDGLEY, G. F.; BÖSENBERG, J. DE W., Seasonal and diurnal plant water potential changes in relation to water availability in the western Karoo.
S. Afr. J. Ecology, (In press).

MIDGLEY, G. F.; MUSIL, C. F., 1990, Substrate effects of zoogenic soil mounds on vegetation composition in the Worcester-Robertson valley, Cape Province.
S. Afr. J. Bot. **56** (2), 158-166.

MIDGLEY, G. F.; RUTHERFORD, M. C.; DAVIS, G. W.; BÖSENBERG, J. DE W., Photosynthetic responses of heliophilous *Rhus* species to environmental modification by invasive shrubs (In preparation).

MILLER, J., 1985, Plant water relations along a rainfall gradient, between the succulent Karoo and mesic mountain Fynbos, in the Cedarberg mountains near Clanwilliam, South Africa.
M. Sc. thesis, Dept. of Botany, University of Cape Town, South Africa.

MILLER, J. M.; MILLER, P. C.; MILLER, P. M., 1984, Leaf conductances and xylem pressure potentials in Fynbos plant species.
S. Afr. J. Science **80**, 381-385.

MILLER, J. M.; MILLER, P. M.; MILLER, P. C., Diurnal patterns of leaf conductance and water potentials in selected Fynbos plant species. (Unpublished manuscript, circa 1984).

MILLER, P. C.; MILLER, J. M.; MILLER, P. M., 1983, Seasonal progression of plant water relations in Fynbos in the western Cape Province, South Africa.
Oecologia, **56**, 392-396.

MOLL, E. J.; ROMOFF, N., 1983, Evidence of mist trapping by *Thamnochortus punctatus* Pillans at Pella in the southwestern Cape Province,
S. Afr. J. Science **79**, 432.

MOLL, E. J.; SOMMERVILLE, J. E. M., 1985, Seasonal xylem pressure potentials of two South African coastal Fynbos species in three soil types.
S. Afr. J. Bot. **51** (3), 187-193.

MOONEY, H. A., 1983, Carbon-gaining capacity and allocation patterns of mediterranean-climate plants.
In: KRUGER, F. J.; MITCHELL, D. T.; JARVIS, J. U. M. (Eds.), Mediterranean-type ecosystems, The role of nutrients, Springer-Verlag, Berlin, pp. 103-119.

MOONEY, H. A.; FIELD, C.; GULMON, S. L.; RUNDEL, P.; KRUGER, F. J., 1983, Photosynthetic characteristic of South African sclerophylls.
Oecologia **58**, 398-401.

MOONEY, H. A.; TROUGHTON, J. H.; BERRY, J. A., 1977, Carbon isotope measurements of succulent plants in southern Africa.
Oecologia **30**, 295–305.

OSMOND, C. B.; WINTER, K.; ZIEGLER, H., 1982, Functional significance of different pathways of CO_2 fixation in photosynthesis.
In: LANGE, O. L.; NOBEL, P. S.; OSMOND, C. B.; ZIEGLER, H., Physiological Plant Ecology. II. Water relations and carbon assimilation, Springer, Berlin, pp. 479–547.

PHILLIPS, N. (personal communication), 1990.

RICHARDSON, D. M.; KRUGER, F. J., 1988, Seasonal trends in diurnal patterns of net leaf photosynthesis in *Protea repens* and *Brabejum stellatifolium* (Proteaceae) at Swartboskloof, southwestern Cape Province, South Africa.
In: DI CASTRI, F.; FLORET, CH.; RAMBAL, S.; ROY, J. (Eds.), Time Scales and Water Stress – Proc. 5th Int. Conf. on Mediterranean Ecosystems, I. U. B. S., Paris, pp. 203–207.

RICHARDSON, D. M.; KRUGER, F. J., 1990, Water relations and photosynthetic characteristics of selected trees and shrubs of riparian and hillslope habitats in the south–western Cape Province, South Africa.
S. Afr. J. Bot. **56** (2), 214–225.

RUTHERFORD, M. C.; BÖSENBERG, J. DE W., 1988, Some responses of indigenous western Cape vegetation to the Australian invasive, *Acacia cyclops*.
In: DI CASTRI, F.; FLORET, CH.; RAMBAL, S.; ROY, J. (Eds.), Time Scales and Water Stress – Proc. 5th Int. Conf. on Mediterranean Ecosystems, I. U. B. S., Paris, pp. 631–636.

RUTHERFORD, M. C.; MIDGLEY, G. F.; BÖSENBERG, J. DE W., Photosynthetic responses of indigenous and alien broad–leaved shrub populations establishing on an arid–zone dune system. (Submitted for publication).

RUTHERFORD, M. C.; MIDGLEY, G. F.; BÖSENBERG, J. DE W., unpublished data.

RUTHERFORD, M. C.; WESTFALL, R. H., 1986, Biomes of southern Africa – An objective categorization, Memoirs of the botanical survey of South Africa No 54. Botanical Research Institute, Dept. of Agriculture and Water Supply, South Africa.

RYGOL, J.; BÜCHNER, K.–H.; WINTER, K.; ZIMMERMANN, U., 1986, Day/night variations in turgor pressure in individual cells of *Mesembryanthemum crystallinum* L.,
Oecologia **69**, 171–175.

SCHÜTTE, K. H.; ACHLEITNER, K., 1986, Water relations of the Fynbos. Final report, C. S. I. R., Pretoria, South Africa.

SEELY, M. K., 1981, Desert plants use fog water. Scientific Progress **14** (2), 4.

SNOW, C. P., 1985, Mist interception by three species of mountain fynbos. Department of Environmental and Geographical Science, University of Cape Town, Research report No. **58**, April 1985.

STEUDLE, E.; LÜTTGE, U.; ZIMMERMANN, U., 1975, Water relations of the epidermal bladder cells of the halophytic species *Mesembryanthemum crystallinum*: Direct measurements of hydrostatic pressure and hydraulic conductivity. Planta **126**, 229–246.

STEUDLE, E.; ZIMMERMANN, U.; LÜTTGE, U., 1977, Effect of turgor pressure and cell size on the wall elasticity of plant cells. Plant Physiol. **59**, 285–289.

STOCK, W. D.; LEWIS, O. A. M., 1984, Uptake and assimilation of nitrate and ammonium by an evergreen Fynbos shrub species *Protea repens* L.(Proteaceae). New Phytol. **97**, 261–268.

STOCK, W. D.; MANUEL, T., Gas exchange responses of *Protea lepidocarpodendron* seedlings to an imposed drought when grown at three nutrient availabilities. Submitted for publication.

STOCK, W. D.; SOMMERVILLE, J. E. M.; LEWIS, O. A. M., 1987, Seasonal allocation of dry mass and nitrogen in a Fynbos endemic Restionaceae species *Thamnochortus punctatus* Pill., Oecologia **72**, 315–320.

TING, I. P., 1981, Effects of abscisic acid on CAM in *Portulacaria afra*. Photosynthesis Research **2**, 39–48.

TING, I. P.; HANSCOM, Z., 1977, III, Induction of acid metabolism in *Portulacaria afra*. Plant Physiol. **59**, 511–514.

TREICHEL, S.; BRINCKMANN, E.; SCHEITLER, B.; VON WILLERT, D. J., 1984, Occurrence and changes of proline content in plants in the southern Namib desert in relations to increasing and decreasing drought. Planta **162**, 236–242.

VAN DER HEYDEN, F.; LEWIS, O. A. M., 1988, Photosynthetic carbon-fixation in selected Fynbos growth forms and its variation with season and environmental conditions.

In: DI CASTRI, F.; FLORET, CH.; RAMBAL, S.; ROY, J. (Eds.), Time Scales and Water Stress – Proc. 5th Int. Conf. on Mediterranean Ecosystems, I. U. B. S., Paris, pp. 589–593.

VAN DER HEYDEN, F.; LEWIS, O. A. M., 1988b, Effects of irradiance and temperature on photosynthetic gas exchange of some South African Fynbos species. In: DI CASTRI, F.; FLORET, CH.; RAMBAL, S.; ROY, J. (Eds.), Time Scales and Water Stress – Proc. 5th Int. Conf. on Mediterranean Ecosystems, I. U. B. S., Paris, pp. 595–598.

VAN DER HEYDEN,F.; LEWIS, O. A. M., 1989, Seasonal variation in photosynthetic capacity with respect to plant water status of five species of the mediterranean climate region of South Africa.
S. Afr. J. Bot. **55** (5), 509–515.

VAN DER HEYDEN, F.; LEWIS, O. A. M., Environmental control of photosynthetic gas exchange characteristics of three Fynbos growth forms.
Submitted for publication.

VAN DER WESTHUIZEN, F. G. J.; DE VILLIERS, O. T.; JOUBERT, J. G. V., 1981, Effect of cutting on carbon dioxide absorption by *Osteospermum sinuatum* and *Ehrharta calycina* at different stages of development.
S. Afr. J. Science **77**, 273–275.

VOGEL, J. C.; FULS, A.; ELLIS, R. P., 1978, The geographical distribution of Kranz grasses in South Africa.
S. Afr. J. Science, **74**, 209–215.

VON WILLERT, D. J., 1975, Stomatal control, osmotic potential and the role of inorganic phosphate in the regulation of the crassulacean acid metabolism in *Mesembryanthemum crystallinum*.
Plant Science Letters **4**, 225–229.

VON WILLERT, D. J., 1979, Vorkommen und Regulation des CAM bei Mittagsblumengewächsen (Mesembryanthemaceae).
Ber. Deutsch. Bot. Ges., **92**, 133–144.

VON WILLERT, D. J., 1988, personal communication.

VON WILLERT, D. J.; BRINCKMANN, E., 1985, Kohlenstoff- und Wasserhaushalt von Sukkulenten arider Gebiete.
Ber. Deutsch. Bot. Ges. **98**, 455–464.

VON WILLERT, D. J.; BRINCKMANN, E.; ELLER, B. M.; SCHEITLER, B., 1983, CO_2 exchange of CAM exhibiting succulents in the southern Namib desert in relation to microclimate and water stress.
Photosynthesis Research **4**, 289–298.

VON WILLERT, D. J.; BRINCKMANN, E.; ELLER, B. M.; SCHEITLER, B., 1984, Water loss and malate fluctuations during the day for plants in the southern Namib desert.
Oecologia **61**, 393–397.

VON WILLERT, D. J.; BRINCKMANN, E.; SCHEITLER, B.; ELLER, B. M., 1985, Availability of water controls crassulacean acid metabolism in succulents of the Richtersveld (Namib desert, South Africa).
Planta **164**, 44–55.

VON WILLERT, D. J.; BRINCKMANN, E.; SCHULZE, E.-D., 1979, Ecophysiological investigations of plants in the coastal desert of southern Africa. Ion content and crassulacean acid metabolism.
In: JEFFERIES, R. L.; DAVY, A. J. (Eds.), Ecological processes in coastal environments. The first European ecological symposium and the 19th symposium of the British Ecological Society, Norwich, 12–16 September 1979, pp. 321–331.

VON WILLERT, D. J.; HERPPICH, M.; MILLER, J. M., 1989, Photosynthetic characteristics and leaf water relations of mountain Fynbos vegetation in the Cedarberg area (South Africa).
S. Afr. J. Bot. **55** (3), 288–298.

VON WILLERT, D. J.; KIRST, G. O.; TREICHEL, S.; VON WILLERT, K., 1976, The effect of leaf age and salt stress on malate accumulation and phosphoenolpyruvate carboxylase activity in *Mesembryanthemum crystallinum*.
Plant Science Letters **7**, 341–346.

VON WILLERT, D. J.; THOMAS, D. A.; LOBIN, W.; CURDTS, E., 1977, Ecophysiologic investigations in the family of the Mesembryanthemaceae. Occurrence of a CAM and ion content.
Oecologia **29**, 67–76.

VON WILLERT, D. J.; TREICHEL, S.; KIRST, G. O.; CURDTS, E., 1976b, Environmentally controlled changes of phosphoenolpyruvate carboxylases in *Mesembryanthemum*.
Phytochemistry **15**, 1435–1436.

VON WILLERT, D. J.; VON WILLERT, K., 1979, Light modulation of the activity of the PEP-carboxylase in CAM-plants in the Mesembryanthemaceae,
Z. Pflanzenphysiol., **95**, 43–49.

WALTER, H.; VAN STADEN, J., 1965, Über die Jahreskurven des osmotischen Wertes bei einigen Hartlaubarten des Kaplandes.
J. of S. Afr. Bot. **31**, 225–236.

WERGER, M. J. A.; ELLIS, R. P., 1981, Photosynthetic pathways in the arid regions of South Africa.
Flora **171**, 64–75.

WINTER, K., 1973, NaCl–induzierter Crassulaceensäurestoffwechsel bei einer weiteren Aizoacee (sic): *Carpobrotus edulis*.
Planta **115**, 187–188.

WINTER, K., 1974, NaCl–induzierter Crassulaceen–Säurestoffwechsel bei der salzpflanze *Mesembryanthemum crystallinum*. Abhängigkeit des CO_2-Gaswechsels von der Tag/Nacht–Temperatur und von der Wasserversorgung der Pflanzen.
Oecologia **15**, 383–392.

WINTER, K., 1974, Effect of water stress on phosphoenolpyruvate carboxylase activity in *Mesembryanthemum crystallinum* L. (in German).
Planta **121**, 147–153.

WINTER, K.; LÜTTGE, U., 1976, Malate accumulation in leaf slices of *Mesembryanthemum crystallinum* in relation to osmotic gradients between the cells and the medium.
Aust. J. Plant Physiol. **3**, 653–663.

WINTER, K; LÜTTGE, U., 1979, C_3–Photosynthese und Crassulaceen–Säurestoffwechsel bei *Mesembryanthemum crystallinum* L.,
Ber. Deutsch. Bot. Ges., **92**, 117–132.

WINTER, K.; LÜTTGE, U.; WINTER, E., 1978, Seasonal shift from C_3 photosynthesis to crassulacean acid metabolism in *Mesembryanthemum crystallinum* growing in its natural environment.
Oecologia **34**, 225–237.

WINTER, K.; TROUGHTON, J. H., 1978, Carbon assimilation pathways in *Mesembryanthemum nodiflorum* L. under natural conditions.
Z. Pflanzenphysiol. **88**, 153–162.

WINTER, K.; VON WILLERT, D. J., 1972, NaCl–induzierter Crassulaceen–Säurestoffwechsel bei *Mesembryanthemum crystallinum*.
Z. Pflanzenphysiol. **67**, 166–170.

Chapter 9

The combination of measurements and mathematical modelling for assessing canopy structure effects

U. TAPPEINER and A. CERNUSCA

9.1 Introduction

In assessing ecosystem function and response to stress in mountainous regions the analysis of canopy structure effects is gaining increasing importance (CERNUSCA 1990, in this volume). The study of canopy structure effects often requires the simultaneous measurement of more than 50 different factors, whereby the individual measured data must be acquired and processed with a high degree of accuracy and frequency (CERNUSCA 1987). Computer methods can be successfully applied in such ecosystem research even under difficult field conditions. Using modern computerised measuring systems we have analysed canopy structure effects in a series of differently structured canopies in the subalpine and alpine belt in the

Austrian Alps and in the Caucasus. Our studies also showed that an exhaustive analysis and synthesis of the large volume of experimentally obtained measurements is only possible with the help of mathematical models. The main purposes for which canopy structure models are required can be loosely formulated as follows:

- To quantify the role of individual factors affecting the microclimate, energy regime, water balance and net photosynthesis in a plant stand.
- For the derivation, from the more easily obtainable model values, of important characteristics of a plant stand in cases where their direct measurement is difficult or even impossible.
- To simulate daily patterns of the above–mentioned parameters in a plant stand with the help of a few readily obtainable global values.
- Typification and classification of plant stands of differing structure, using parameters that can effectively be determined by means of mathematical models.
- The analysis of dynamic changes in the ecosystem due to changes in environmental parameters or anthropogenic measures, using mathematical simulation techniques.

Obviously the five possibilities are not independent of one another, and are even to some extent competitive. Nonetheless, it can be said that a satisfactory model should fulfil the following conditions:

- The model should be simple and clearly understandable, and it should be possible to interpret each of its parameters.
- The values entered in the model should be reliable and not unduly difficult to measure.
- The parameters to be deduced from the model should be explicitly represented.

The present article is intended to show how even simple mathematical models can be applied successfully in ecosystem research in order to analyse and synthesise results from large–scale field experiments on canopy structure effects. To begin with, a brief description will be given of the theoretical background and the basic structure of our models on canopy architecture, light extinction, single leaf and canopy net photosynthesis for natural, multispecies plant communities. The second section deals with our methods for selecting and validating appropriate models, and the third section concerns some examples of the application of our modelling techniques.

9.2 Materials and methods

9.2.1 Sites investigated

Canopy structure, environmental and net photosynthesis data used to drive the models were obtained in situ in differently structured plant communities (grassland and shrubs) in the subalpine and alpine belt in the Austrian Alps and in the Caucasus.

The investigations in the Austrian Alps were carried out in the course of the MAB-6-High mountain research program (cf. CERNUSCA 1990, in this volume). The experimental data were obtained at the following sites:

Experimental site "Gastein" valley 47°8′N, 13°32′E; 1960 m a.s.l. (cf. TAPPEINER and CERNUSCA 1989a). The studies were carried out on an almost untouched alpine pasture of the Aveno–Nardetum type, interspersed with *Vaccinium myrtillus* and *Vaccinium uliginosum* L., and a sparsely covered grass area, dominated by *Phleum pratense* and *Festuca rubra*, that had been resown in connection with the levelling of a ski slope. Both of these areas are grazed in summer and are used for skiing in winter.

Experimental site "Guttal" 47°3′N, 12°48′E; 1912 m a.s.l. (cf. TAPPEINER and CERNUSCA 1989b) is an alpine pasture grazed by cows. The dominant plant species are *Nardus stricta*, *Poa alpina* and *Alchemilla vulgaris*. At this site changes associated with the cessation of management were investigated from 1977 to 1980.

Experimental site "Mt. Patscherkofel" 47°13′N, 11°20′E; 1940 m a.s.l. (cf. CERNUSCA 1976, SIEGWOLF and CERNUSCA 1984, GAZARINI 1988). A *Rhododendron ferrugineum* community was studied at this site.

Table 9.1a gives an overview of the most important parameters for the sites investigated in the Austrian Alps.

As part of the scientific cooperation between the Botanical Institutes of the University of Innsbruck and of the Georgian Academy of Sciences, six plant communities were investigated in the subalpine zone of the Central Caucasus (TAPPEINER et al. 1989, TAPPEINER and CERNUSCA 1991). The investigations were carried out in the vicinity of the high-mountain research station Kasbegi (42°48′N, 44°39′E, 1850 m a.s.l.) on the northern slope of the Great Caucasus. The sites investigated were:

Experimental site "Dry grassland" is an area grazed by sheep. The dominant species are *Festuca rupicola*, *Pulsatilla violacea* and *Carex buschiorum*.

Experimental site "Dryadetum" This area is situated on a steep NW slope on which characteristic steps have developed as a result of grazing. This sheep

Table 9.1a: Overview of the most important parameters of the sites investigated in the Austrian Alps. Masses are given in dry matter [g·m^{-2}].

	alpine pasture[1]	*Vaccinium* stand[1]	revegetated grassland[1]	grazed pasture[2]	abandoned pasture[2]	*Rhododendron* stand[3]
Altitude (m)	1960	1960	1960	1912	1912	1940
Inclination of slope	15°	15°	19°	10°	10°	19°
Exposure	ESE	ESE	ESE	SSE-S	SSE-S	WNW
	pasture	dwarf shrub	grassland	pasture	pasture	dwarf shrub
Canopy height (cm)	35	18	15	8.5	60	65
Degree of cover	100%	100%	55%	100%	100%	100%
Phytomass	318	658	87	109	537	4138
Biomass	229	612	62	79	301	3959
Necrotic tissues	89	46	25	30	236	177
Litter	142	190	25	49	98	672
LAI	2.4	2.9	0.6	1.2	4.8	3.5
GAI	2.8	3.4	0.7	1.3	5.8	3.6
PAI	3.8	4.5	1.1	2.1	9.8	5.3
Mean foliage inclination	50°	53°	71°	26°	43°	38°

Experimental sites: [1]"Gastein valley", [2]"Guttal", [3]"Mt. Patscherkofel".

pasture is dominated by *Dryas caucasica*, *Deschampsia flexuosa* and *Daphne glomerata*.

Experimental site "Festucetum" This is a thinly grazed sheep pasture on a steep, convex and stony slope dominated by *Festuca woronowii* and *Carex meinshauseniana*.

Experimental site "Deschampsietum" A mowed meadow on very wet ground. The dominant species are *Deschampsia cespitosa* and *Equisetum palustre*. The stand is interspersed in places with *Phragmites australis*.

Experimental site "Heracletum" This is a tall herbaceous stand situated on a gentle N slope, the soil rich in humus. The community is dominated by

Table 9.1b: Overview of the most important parameters of the sites investigated in the Central Caucasus. Masses are given in dry matter [g·m^{-2}].

	Dry grassland	Dryadetum	Festucetum	Deschampsietum	Heracletum	Rhododendretum
Altitude (m)	2050	2050	2000	1750	2200	2200
Inclination of slope	30°	50°/10°	45°	0°	3°	40°
Exposure	SW	NW	SW	–	N	NW
	pasture	pasture	pasture	meadow	tall herb	dwarf shrub
Canopy height (cm)	18	15	40	50	240	47
Degree of cover	80%	90%	100%	100%	100%	100%
Phytomass	219	310	691	427	768	1893
Biomass	164	182	223	369	674	1817
Necrotic tissues	55	128	468	58	94	76
Litter	92	355	447	44	70	1066
LAI	2.4	2.3	2.5	5.8	6.2	3.8
GAI	2.4	2.4	2.6	6.1	7.2	4.0
PAI	3.1	4.7	8.5	6.7	8.1	4.7
Mean foliage inclination	47°	44°	65°	66°	29°	36°

Heracleum sosnowskyi which attains a height of 2.4 m. The stand is harvested for silo feed.

Experimental site "Rhododendretum" This dwarf shrub stand is dominated by *Rhododendron caucasicum*. A very sparse undergrowth of *Vaccinium myrtillus*, *Vaccinium vitis-idaea* and *Empetrum hermaphroditum* is also present.

Table 9.1b gives an overview of the most important parameters for the different experimental sites in the Central Caucasus.

9.2.2 Experimental ecological methods

In all of the plant stands investigated, detailed measurements were made of the stand structure, light climate, microclimate and CO_2 exchange. All measurements were made during the main flowering stage, when the vegetation was at its maximum.

Canopy architecture. The canopy structure of each stand was analysed by the stratified clipping method (MONSI and SAEKI 1953). In each of the plant stands investigated a homogeneous plot was harvested layer–wise by clipping. To ensure statistically reliable results, the size of the section of the stand to be investigated and the height of the layers were in each case selected according to the type of stand. The thickness of layer chosen varied, depending on the height of the particular stand, between 0.5 cm and 20 cm, and the harvested plot between 50 × 50 cm and 100 × 100 cm. Above–ground phytomass was divided into phanerogams, cryptogams and dead standing plant matter. Subsequently, the phanerogams were separated per species and per layer into stems, leaves of different ages and inflorescences. The area of all plant parts of each species per layer was determined with a leaf area metre (LI–3100, LI–COR Corp., Lincoln USA). Dry weights were determined for all fractions.

Field measurements of leaf and stem inclination of the dominant species were also made. The angle formed with the horizontal was measured with a hand inclinometer, the readings being taken to the nearest five degrees. In each layer approximately 100 measurements were made on the dominant species of the stand. For all stands investigated a total of 26 species was analysed with respect to the inclination of their leaves and axes. Additionally, after measurement of their inclination, the area of each leaf and axis was determined. For each species the mean leaf and stem inclination and the respective relative variance in 10° classes per layer were calculated, each of the angles measured being weighted with the leaf (stem) area. The mean inclination per canopy layer was calculated on the basis of the proportional PAI (plant area index=the area of all plant parts per unit area of horizontal ground) contribution of the constituent species. Our procedure is aimed at allowing for the situation encountered in plant stands consisting of a large number of different species, and even differing in species composition from layer to layer.

Microclimate and light climate. A portable battery–powered data acquisition system (Micromet, A. Cernusca, Innsbruck, Austria) was installed during the main growth period at each investigation site. Profiles of soil, air and leaf temperature, profiles of air humidity, wind speed and CO_2–concentration, within and above the canopy as well as soil heat flux, the incoming global radiation and net radiation were measured. The micrometeorological measurements were made at intervals ranging from 1 min for rapidly changing variables like radiation, temperature, water vapour pressure and CO_2–concentration, to 12 min for soil temperature. This system contains a microcomputer to convert the raw data to appropriate units, calculate energy fluxes, and print out summarized values for each hour. The following sensors were used: Star pyranometer and net radiometer (Schenk, Vienna, Austria), small thermocouples for soil, leaf and air temperature (home made: copper/constantan, 0.08 mm diameter), thermocouple psychrometers (home made), head flux plates (Keithly Instruments, S.A.), cup anemometer (Type AMI, Fa. Rimco, Australia), infrared gas analyser (LCA–2, ADC, Hoddesdon, England

and BINOS 1, Fa. Leybold–Heraeus, Hanau, Germany) for measuring the CO_2–concentration.

Measurements of photosynthetic photon flux density (PFD) were made in all plant stands investigated, using quantum sensors (Lambda Instr. Co., Lincoln, Nebr. USA). One sensor was mounted 1 m above the canopy. A second sensor could be moved vertically and horizontally within the stand to detect spatial variability in the foliage (CERNUSCA 1982). Continuous recording during each pass in different canopy layers provided profiles of actual sunlit and shaded foliage areas and extinction data for direct beam and diffuse PFD. The height of the layer chosen in the individual stands corresponded to that in the stand structure analyses. For measurement of the diffuse PFD the sensors were equipped with shade bands. The instruments were sampled on cloudless and overcast days between 6.00 and 18.00 h, at 2 min intervals using the micro–computer–controlled data acquisition system (CERNUSCA 1987). In all an average of 50 vertical profiles of total and diffuse PFD were measured per plant stand analysed. The profiles of direct PFD (beam fraction) were calculated from the difference between the profiles of total and diffuse PFD.

CO_2–Gas exchange. All measurements were made in the field under the currently prevailing temperature– and light conditions and under different water stress conditions. In each stand analysed, in situ measurements of the photosynthetic responses to light and temperature were made on the dominant species using portable gas exchange systems with single leaf cuvettes. In addition to the measurements of the photosynthetic response curves on single leaves, and mainly for the purpose of validating our canopy photosynthesis model, detailed measurements of the CO_2 exchange of the whole canopy were made with the help of a special whole–canopy chamber (cf. SIEGWOLF and CERNUSCA 1984) and a micrometerological approach, the Bowen–ratio energy–balance method (cf. NORMAN and HESKETH 1980, CERNUSCA 1982).

Single leaf cuvette measurements. Light response curves were measured using the portable ADC–system (ADC Co., Hoddesdon, England), consisting of IRGA and integrated indicator module (LCA–2), an air–supply unit with mass flow control (ASU–M) and a leaf cuvette (PLC–N). The system is an open one (cf. LONG and HALLGREN 1985). Additionally, with a small temperature–controlled glass cuvette (KÖRNER 1977) and a battery–driven IRGA (BINOS 1, Fa. Leybold–Heraeus, Hanau, Germany), the in situ photosynthetic response to temperature was recorded for the different species. For the measurements on *Rhododendron ferrugineum* a fully climatized measuring chamber (Sirigor Siemens, Erlangen, Germany) was used in combination with an infrared gas analyser (Type UNOR 5 Maihak, Hamburg, Germany). Gas exchange rates were calculated according to VON CAEMMERER and FARQUAR (1981) and expressed on a projected leaf area basis. In making the calculations, corrections were made for the water vapour sensitivity of the

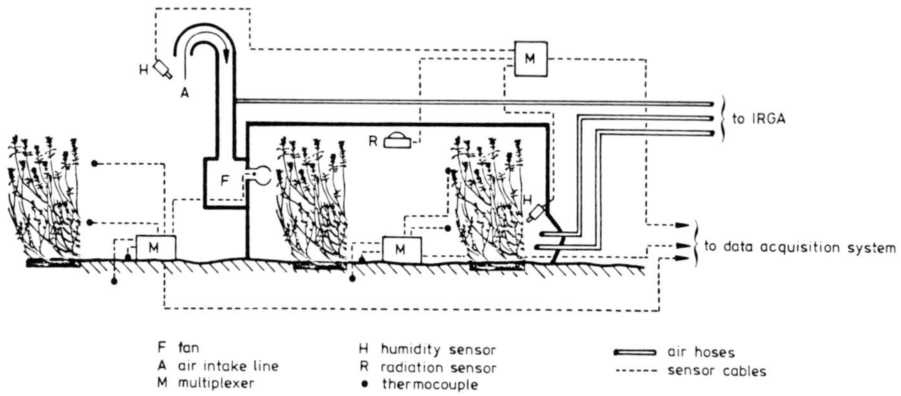

Figure 9.1: Schematic representation of a chamber for measuring the gaseous exchange in a stand of *Rhododendron ferrugineum* (after SIEGWOLF and CERNUSCA 1984). The chamber employed (length 160 cm, width 120 cm, height 70 cm) consisted of a tubular steel frame covered by polyethylene foil (0.16 mm). On each side, steel sheeting to which the foil was firmly attached was inserted about 10 cm into the ground. Air was pumped into the chamber through an aperture in the front of the chamber. The air flow through the chamber was measured continuously with an anemometer. An opening of 15 cm diameter in the back of the chamber served for the escape of air.

IRGA, for local pressure conditions at the various altitudes and for the lower CO_2 concentration in the cuvette due to the use of the differential method for measuring photosynthesis (after DIEMER 1990). Response curves were established for as many individual leaves as possible and for the axes of the dominant species at different heights in the plant stands. In addition, for the measurements on the leaves a rough distinction was made according to age, location within the stand, and sun– or shade characteristics.

Whole–canopy chamber. A special whole–canopy chamber was developed for the investigations on the *Rhododendron ferrugineum* community in the Austrian Alps (cf. Fig. 9.1). The giant cuvette (length 160 cm, width 120 cm, height 70 cm) is inverted over an entire section of the *Rhododendron* community. By measuring the CO_2 difference between the entry and exit of the cuvette using an IRGA (UNCR 5 Maihak, Hamburg, Germany), the total CO_2 input–output of the enclosed ecosystem can be directly obtained.

Bowen–ratio energy–balance method. In the other stands investigated, the photosynthetic uptake of the whole canopy was measured by the Bowen–ratio energy balance method. With this method water vapour, heat flux and CO_2 flux were calculated using profiles of temperature, water vapour pressure and CO_2 concentration measured between the canopy surface and a height 60 cm above it (for details see CERNUSCA 1982). Release of CO_2 from the soil was measured in situ by IRGA techniques (LCA–2, ADC, Hoddesdon, England) using the cuvette method in the "open test gas system" (CERNUSCA and DECKER 1984). The sum of the CO_2 output of the soil and the CO_2 flux in the atmosphere gives the CO_2 exchange of the entire plant stand.

9.3 Model designing

A complex mathematical model of net photosynthesis in a plant stand must take into account the following aspects:

- A mathematical description of the spatial distribution and orientation of phytoelements in a plant stand. The main geometrical characteristics of a plant stand are: its height, the plant area index, the vertical distribution of the area, the inclination and orientation of the surfaces of leaves and stems, as well as the statistical distribution of phytoelements.
- A mathematical description of the spatial distribution of environmental parameters (radiation, carbon dioxide, water, temperature, wind speed, etc.) exerting an influence on net photosynthesis, as well as a mathematical description of mass– and energy–exchange processes between the phytoelements and their microenvironment.
- A mathematical description of the dependence of the net photosynthesis of single leaves and stems of the different species on environmental parameters, taking into account the characteristics of the photosynthetic and respiratory apparatus, the biological age and the adaptivity of plants.

In the following we shall describe the basic structure of a simple model developed by us for the simulation of canopy net photosynthesis of natural, species–rich plant stands. The model consists of a submodel for the canopy structure and the light climate and a single leaf photosynthesis submodel, which can be linked to the canopy photosynthesis model via the measured spatial distribution of structural and environmental parameters in the canopy.

9.3.1 Canopy structure and light climate submodel

Innumerable models for describing the light climate in plant stands have been developed and reported in the literature (for review see LEMEUR and BLAD 1974,

NORMAN 1975, ROSS 1981, MYNENI et al. 1989). The differences between these models depend primarily on the purpose for which they were formulated. We ourselves have developed a series of increasingly complex models for the simulation of the penetration of direct beam solar PFD into a canopy (TAPPEINER and CERNUSCA 1984, 1989a). Table 9.2 gives an overview of such models. In each successive step a different (thus far neglected) parameter influencing the light climate is incorporated in the model as a variable. A hierarchical structure of this type has been used wherever possible in all of our models. This structure has proved to be highly successful, largely because the validation makes possible a specially efficient selection of the particular model that allows a sufficiently accurate simulation of the in situ measurements (cf. TAPPEINER and CERNUSCA 1989a).

The simplest of the light extinction models shown in Tab. 9.2 is based on the classical approach of MONSI and SAEKI (1953) and takes into account only PAI. In order to apply this model, therefore, all that need to be determined experimentally is the total plant area of the canopy. Model 2 requires the additional measurement of the actual distribution of the PAI in layers throughout the depth of the canopy. In model 3 the foliage inclination is included by the factor cos α_i in each layer, where α is the measured frequency distribution of foliage inclination in the layer i. The angle of incidence of incoming radiation is incorporated in model 4 as a variable factor, that is dependent on the ratio foliage inclination (α_i) to solar angle (β). Models 1–4 are based on the assumption of statistically random foliage distribution. In order to obtain an idea of horizontal variability in the stand the effective light–attenuating foliage surface per layer is corrected by a distribution factor, d_i, to give Model 5. A value of $d_i = 0$ means that the entire foliage surface of layer i is in the shade or that its distribution is totally clumped. When $d_i = 1$ the leaf area remains unchanged and distribution is random. If $d_i = 1/(PAI_i \cdot \psi_i)$ the leaf area of this layer would be so arranged as to give complete shade, i.e. distribution would be completely regular. Model 5 can be employed in one of two ways:

- The distribution of a plant stand can be determined experimentally (e.g. using the point quadrat method of WARREN WILSON (1963, 1965), and then inserted in the model in the same way as the other parameters.
- A second possibility is to equate the measured irradiation intensity I_x with the value calculated from the model, from which the only remaining unknown, d_i, can then be derived. This is a useful method for obtaining preliminary information for the distribution of the leaves and was employed by the present authors in the studies reported here, care being taken to employ independent data sets for the calculation of the distribution factor and for the subsequent validation.

For a complete description of the light climate of a plant canopy the diffuse radiation components must of course also be taken into account, in addition to the direct beam PFD. Diffuse PFD from the sky reaching leaves in the different canopy layers

Table 9.2: Overview of the hierarchical structure of our canopy structure and light climate models (cf. TAPPEINER and CERNUSCA 1984). In each step an additional parameter affecting the light climate of the canopy is incorporated in the model as a variable. I_x and I_0 are the respective amounts of direct beam solar PFD in a horizontal plane at the surface of a layer x inside the canopy and above the canopy, n the number of layers, PAI the plant area index, α the frequency distribution of foliage inclination, β the solar altitude and d the distribution factor.

$$I_x = I_0 \cdot (1 - \mathrm{PAI}/n)^{(x-1)} \qquad \text{MODEL 1}$$

$$I_x = I_0 \cdot \prod_{i=1}^{(x-1)} (1 - \mathrm{PAI}_i) \qquad \text{MODEL 2}$$

$$I_x = I_0 \cdot \prod_{i=1}^{(x-1)} (1 - \mathrm{PAI}_i \cdot \cos \alpha_i) \qquad \text{MODEL 3}$$

$$I_x = I_0 \cdot \prod_{i=1}^{(x-1)} (1 - \mathrm{PAI}_i \cdot \Psi_i) \qquad \text{MODEL 4}$$

whereby $\Psi_i = \cos \alpha_i$ for $\alpha < \beta$

$\Psi_i = \cos \alpha_i \cdot (1 + 2 \cdot (\tan \theta_i - \theta_i) / \pi)$ for $\alpha > \beta$

with $\theta_i = \cos^{-1} \cdot (\cot \alpha_i \cdot \tan \beta)$

$$I_x = I_0 \cdot \prod_{i=1}^{(x-1)} (1 - \mathrm{PAI}_i \cdot \Psi_i \cdot d_i) \qquad \text{MODEL 5}$$

was calculated in a manner analogous to that for the direct beam in Model 5. The same canopy geometry is used, but different sky segments are taken as the source of radiation (DUNCAN et al. 1967). The sky hemisphere is treated as nine annular rings of uniform brightness. Penetration of diffuse radiation into the canopy from different sky segments was calculated and then integrated for the whole sky hemisphere.

The models shown in Tab. 9.2 calculate the radiation intensity on a horizontal surface. The light intensity on the inclined foliage was calculated from the light intensity on the horizontal surface. Our procedure was similar to that of ANDERSON and DENMEAD (1969), on the basis of purely geometrical considerations, taking into account the measured frequency distribution of foliage inclination in the single layers.

9.3.2 Single leaf photosynthesis submodel

The literature contains a wealth of examples for the modelling of photosynthetic behaviour, their approach ranging from frank empiricism to a detailed mechanism (e.g. FARQUAR and VON CAEMMERER 1982), depending on the objectives and goals and the level being looked at: e.g. the level of the single chloroplast, leaf level or whole canopy level (cf. THORNLEY 1976). The aim of our study being to analyse effects of the stand structure using the simplest possible model, we employed for the simulation of single leaf photosynthesis a simple empirical model of KÜPPERS and SCHULZE (1985), which takes into account the light– and temperature–dependence of photosynthesis. The photosynthetic response to light and temperature $A_{(I,T)}$ is calculated via an iteration programme, using the following function:

$$A_{(I,T)} = A_{\max(T)} \cdot \left(1 - e^{-a_1 \cdot (I - \Gamma_q)}\right) \tag{9.1}$$

where I is the irradiance and Γ_q the light compensation point (both in μmol photon$\cdot$m$^{-2}\cdot$s^{-1}) and a_1 is a factor determined by an approximation procedure for finding the curve best fitting the measured net photosynthesis and irradiance values. The variation of light–saturated CO_2–uptake ($A_{\max(T)}$) with leaf temperature can be described using a polynomial regression (cf. KÜPPERS and SCHULZE 1985).

The model parameters $A_{\max(T)}$ and a_1 were derived from the in situ CO_2–gas exchange measurements with single leaf cuvettes for different classes of leaves (classified according to the age and the depth of the leaf in the canopy) and stems for the dominant species in the investigated stands. As described above the measurements were made in the field under the currently prevailing temperature– and light conditions and under different water stress conditions. In this way it was possible to calculate response curves for the natural combination of environmental factors most frequently encountered in the stand (for example strong light intensity combined with high leaf temperature and water stress, or weak light intensity and low leaf temperature).

9.3.3 Canopy photosynthesis model

The photosynthesis of a plant canopy is a function of both physiological response characteristics for the individual species of a plant stand and the effects of canopy structure upon environmental variables (PFD, temperature, CO_2, air humidity, wind etc.) Simulation of whole canopy net photosynthesis therefore requires the linkage of the photosynthetic responses and the actual microclimatic situation prevailing in the canopy. Depending on the goals, this kind of canopy photosynthesis model can be employed with varying degrees of complexity in the different sectors as follows:

- the microclimate values can be included as direct measurements or in the form of submodels.
- the photosynthetic responses can be included for a variety of environmental factors which exert a significant effect on photosynthesis.
- the responses curves are species specific as well as coupled to the state of ontogeny (DIEMER 1990) and the location of the leaf in the canopy. It may therefore be necessary to quantify the dependence of CO_2 exchange on environmental factors that exert a significant effect on photosynthesis, for all classes of leaves differing with respect to the above criteria and for every species present.

Since the prediction of canopy photosynthesis is a very complex affair (cf. HESKETH and JONES 1980), requiring great experimental effort to obtain all the necessary data, here again, the simplest model adequate to the aims of the investigation and the structure of the plant stand should be chosen. The approach selected for the present study was based on a combination of the submodels for radiation and single–leaf photosynthesis described on pages 169 and 172. Based on the detailed canopy structure, the beam and diffuse PFD were determined in layers throughout the depth of the canopy with Model 5. Sunlit and shaded foliage at each depth were treated separately. The PFD values entered the photosynthesis submodel. In this submodel the net photosynthesis was determined for different classes of leaves (classified according to age and location in the canopy) and stems of the dominant species in different layers of the stand. Net photosynthesis of these leaf classes was scaled up to the entire canopy by taking into account the canopy structure. Hence the CO_2 uptake of each layer and the whole canopy throughout the course of the day was simulated. This was based on the information obtained concerning the distribution of the surface areas of leaves and stems of the various species in the stand and the calculated net photosynthesis of the individual leaf, as influenced by the daily pattern of the relevant environmental factors.

9.4 Selection and validation of the appropriate model

Before a model can be employed for causal analysis or prediction purposes it must be extensively tested in situ against experimentally obtained data. This process of validation is demonstrated in the following, by a few selected examples. Above all, it also becomes clear that during the process of validation, the use of the mathematical model makes it possible to analyse and quantify important relationships in a plant stand.

9.4.1 Validation of the radiation model

As mentioned above, the stepwise construction of our models makes it possible to check comprehensively the simulation results against experimental field data and thus to select the simplest possible step of the model adequate to the aims of the investigation and the particular plant stand involved. Since in each step an additional structural parameter (the measurement of which is often very laborious) is included in the model, the choice of the simplest adequate model results in a more rational and economical type of research. More important, however, is that validation of the models gives a causal analysis of the stand structure effects on the light climate beyond that provided by the purely experimental data. Assuming that the structural parameters can be additively linked, it is possible to analyse and quantify the effect of individual parameters. Since in every step of the models only one additional parameter is included, an improvement in the predicted values can be exactly attributed to this one parameter. If, additionally, in the validation of the most complex model a large discrepancy is found between the model values and the measured values, this is a clear indication that other parameters not yet taken into consideration play a role in the modelling process. This constitutes an important feedback between theoretical considerations and experimental investigations.

Our procedure for the selection and validation of an appropriate light submodel and the advantages of the hierarchical structure of the models can be convincingly demonstrated by the example of the three structurally different plant communities, *alpine pasture*, *Vaccinium-stand* and the *revegetated grassland* investigated by us in the Gastein valley (cf. TAPPEINER and CERNUSCA 1989a). The three stands differed considerably in their canopy structure. On the resown area a PAI of 1.1 was found, whereas the value measured for the alpine pasture was 3.8, and for the *Vaccinium* stand 4.5 $m^2{\cdot}m^{-2}$. Measurements of foliage inclination revealed for the resown area the erectophilic structure typical of a pure grass vegetation, with a mean leaf inclination of 70° to the horizontal. In the alpine pasture a concentration of flat herbaceous leaves was found in the lowest third of the canopy whereas erect grass predominated in the upper layers. The *Vaccinium* stand was slightly plagiophilic, with leaf inclinations in a middle range, between 40° and 60°. The striking differences in structure and leaf inclination of the three canopies considerably affected light distribution in the stands as well as the descriptive value of the models.

For the validation process the correspondence between the measured and simulated direct PFD intercepted by foliage were determined. An example of measured and simulated PFD with Model 4 for the different stands is shown in Fig.9.2. In order to quantify the deviation of the values measured from the simulated values

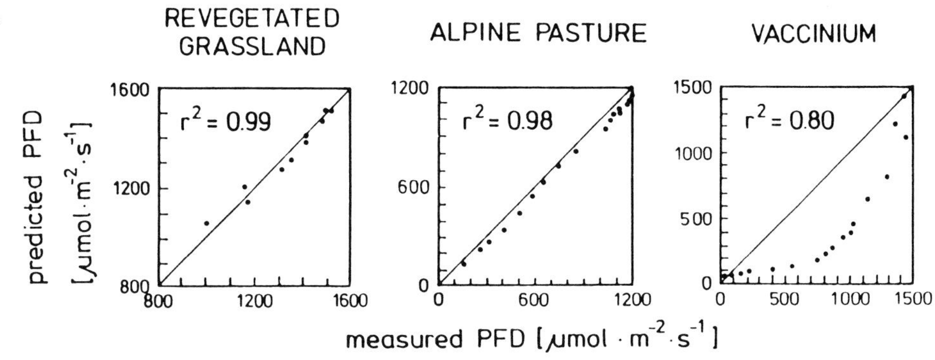

Figure 9.2: Comparison of the distribution of photon flux density (PFD) at different depths in the canopy obtained via direct measurement and calculation of Model 4. Each data point represents the mean of more than 12 profiles measured on a clear day. The coefficient of determination (r^2) based on all observations is included in the upper left corner ($n = 190$, TAPPEINER and CERNUSCA 1989a).

of PFD, a regression line and its coefficient of determination, and thus also the proportion of total variance in the measurements explained by the model, were calculated for each model. The discrepancies between the simulated and measured values can be interpreted as being due to factors not included in the particular model and to stochastic interference. If it is assumed that the increase in explained variance from one model to the next higher step is due solely to the newly introduced parameter, the total variance can be split up as shown in Table 9.3. The results obtained with Model 1 and Model 2, which take into account only the PAI, show that in all stands investigated the PAI is insufficient for the simulation of the distribution of the PFD in the plant canopy. Particularly in the grassy stands the foliage inclination and solar altitude are important for giving a better approximation of the simulated values to those actually measured. Here, however, a difference is revealed between the alpine pasture, with its relatively flat leaf inclination, and the erectophilic revegetated grassland. In the former case an adequate simulation of the light climate is obtained with the simpler Model 3, whereas in the latter case the more complex Model 4 has to be used. This is not surprising, since the position of the sun is a very important factor where the leaves are steeply inclined (cf. ROSS 1975). In addition, further simulations employing Model 4 have shown that the influence of leaf inclination on the light climate is not independent of the stand density, that is to say of PAI. The higher the PAI of a stand, the smaller are the effects of leaf inclination on its light climate.

Table 9.3 and Fig. 9.2 also show that none of the models 1–4 reflects, even approximately, the actual radiation conditions in the *Vaccinium* stand. The ex-

Table 9.3: Absolute values and percent increase in explained variance in the values obtained for the three stands with the models.

model	additional parameters	explained variance revegetated grassland		pasture		*Vaccinium* stand	
		(%)	abs.	(%)	abs.	(%)	abs.
Model 1	*PAI*	79	79	65	65	66	66
Model 2	PAI_i	8	87	26	91	2	68
Model 3	α_i	7	94	6	97	12	80
Model 4	β	5	99	1	98	0	80
Remaining variance (incl. Model 5)		1	100	2	100	20	100

planation may lie in a non-random distribution of the leaves, since models 1–4 assume a statistically random distribution. In the light of this working hypothesis, the next step was to determine approximately the statistical distribution of the leaves by reversing Model 5. Fig. 9.3 shows the results of the determination of the distribution factors for the alpine pasture and the *Vaccinium* stand. In addition to the distribution factor d_i as determined from the model, its confidence interval is also shown. The graph shows at a glance that throughout the entire depth of the pasture, with the exception of the upper layer, no significant deviation from a random leaf distribution could be registered. In the *Vaccinium* stand, however, the upper half of the stand clearly exhibits a significantly clumped distribution corresponding to a reduction in effective projected leaf area by one half to one quarter. On the basis of the simulation results obtained by reversing Model 5, the foliage distribution in the *Vaccinium* stand could be checked experimentally in a subsequent step, using the point–quadrat–method of WARREN WILSON (1959, 1965). The experiments unequivocally confirmed the clumped distribution of the *Vaccinium* leaves in the upper half of the stand. The leaves on the new shoots are seen to be arranged in closely spaced whorls and thus overlap one another.

As in the Gastein valley, a validation of the radiation model employed was carried out for each of the plant stands analysed. With the exception of the *Rhododendron* stands investigated in the Alps and in the Caucasus, very good agreement was found in all cases between Model 4 and the measured values (r^2: 0.93–0.98, cf. TAPPEINER et al. 1989). Simulations with Model 5, however, revealed a slightly clumped leaf distribution in the uppermost third of the stand for both *Rhododendron caucasicum* and *Rhododendron ferrugineum*.

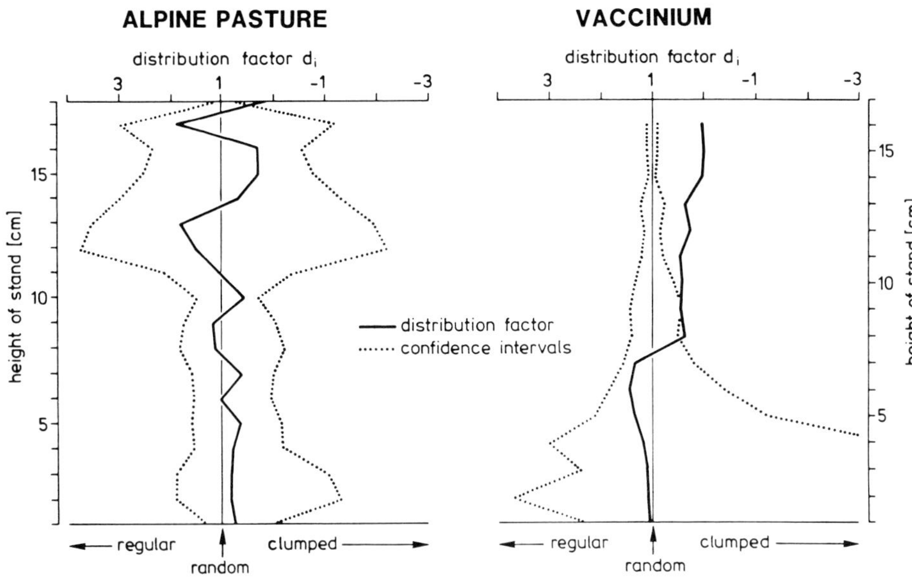

Figure 9.3: Calculated distribution factor (d_i, solid line) and the corresponding confidence intervals (dotted lines) for the different strata of an alpine pasture and a *Vaccinium* stand (TAPPEINER and CERNUSCA 1989a).

9.4.2 Validation of the canopy photosynthesis model

The validation of a canopy photosynthesis model can be carried out on different levels, by different methods. Chamber methods are frequently employed, using either single–leaf chambers or a whole canopy chamber. The single–leaf method is commonly chosen because its application is simple (e.g. CALDWELL et al. 1986, BEYSCHLAG et al. 1990), the performance of the model being checked on individual, selected leaves in the different layers of the canopy. This method has the following disadvantages:

- the validation is achieved by the same method as was used in determining the response curves of net photosynthesis, i.e. the model checks itself.
- in this method the microenvironment of the leaf is disturbed and hence the measurements of exchange processes between leaf and environment are unsatisfactory.

On account of these limitations it is better to validate a canopy photosynthesis model on the basis of measurements of the CO_2 uptake of the whole canopy.

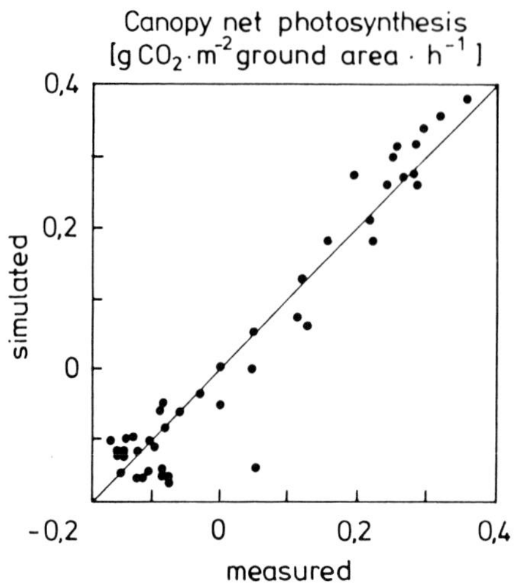

Figure 9.4: The relationship between measured and simulated canopy net photosynthesis of *Rhododendron ferrugineum* for the period Sept. 22 – Sept. 26, 1981, on Mt. Patscherkofel near Innsbruck. The coefficient of determination is 0.93, $n = 56$, (SIEGWOLF 1987).

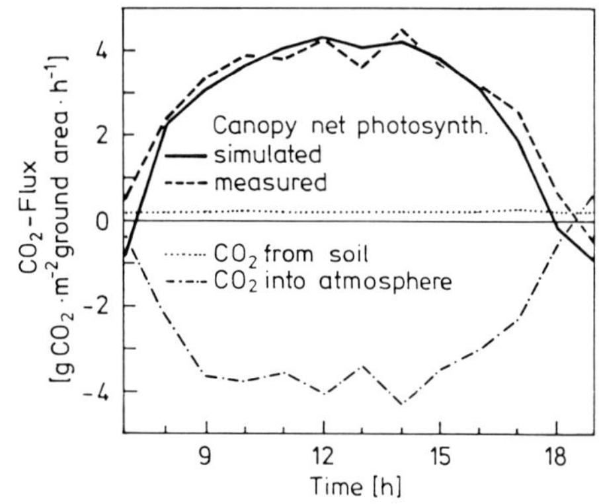

Figure 9.5: Model validation showing the diurnal course of canopy net photosynthesis of a tall herb stand ("Heracletum"). Dashed lines are values of CO_2–fluxes obtained via measurement and calculation of the Bowen–ratio energy balance–method. The continuous line is the result of model calculations based on the structural characteristics of the "Heracletum", the photosynthetic light response of leaves in various layers of the canopy and measured diurnal courses of incoming PFD on a clear day in August 1988.

In using whole–canopy chambers the CO_2 exchange of an entire section of the stand is measured in a giant cuvette. This allows the model to be tested for whole sections of the canopy, which means that the effects of the differences in statistical distribution of leaf inclination, age structure of the leaves and species composition on net photosynthesis can be correctly taken into account. However, in this method, too, the microenvironment in the cuvette is significantly distorted by chamber effects. In spite of this, the method gives good validation results if the artificial environmental conditions measured in the cuvette are used in the simulation.

In order to test the canopy photosynthesis model a whole–canopy chamber of this type was developed and successfully employed on the *Rhododendron* community in the Alps (SIEGWOLF and CERNUSCA 1984). The validation, for the part of the canopy enclosed in the cuvette, showed a very good agreement between measured and simulated canopy net photosynthesis especially under overcast conditions (cf. Fig. 9.4). At higher insolation, which was invariably combined with extremely high temperatures in the chamber, greater differences were found. This was also the case in longer periods of bad weather with consistently low temperatures. Here again, the combination of experiment and model proved its value. From the difference between simulated and measured data the conclusion could be drawn that, in addition to the photosynthetic responses to light, temperature and water vapor included in the model, in extreme situations still other factors play a role. For example the plants are subjected to stress phenomena under the extreme microclimatic conditions of high insolation in the whole–canopy chamber. In addition the plants react to the elevated or extremely low temperature conditions, by modulative adaptations of the temperature optimum of single leaf net photosynthesis. Thus the combined application of the model and specifically planned experimentation gave fundamental information concerning the survival of high–mountain plants (cf. SIEGWOLF 1987). The most sophisticated validation method is a micrometeorological approach for estimating the photosynthetic CO_2 uptake of the whole canopy. The great advantage of such methods is that by their very nature, the values obtained represent averages over substantial areas, that the microenvironment of the photosynthesizing leaves is not significantly disturbed, and that the diffusive resistance of the canopy layer is included (NORMAN and HESKETH 1980).

We have also successfully employed a micrometeorological approach (Bowen–ratio energy balance method) to determine carbon dioxide fluxes in the plant stands investigated. This can be illustrated by the example of a *Heracleum* community in the Central Caucasus. The comparison of measured and simulated canopy net photosynthesis for the "Heracletum" for a clear day is shown in Fig. 9.5. Validation of the model revealed that in the present case only the light climate need be considered in explaining effects due to canopy structure. In this canopy light is quite obviously the dominant environmental factor for photosynthesis. This is not surprising since it is already known from in situ measurements of the microclimate that temperature and water balance are not limiting factors under the conditions obtaining in this habitat.

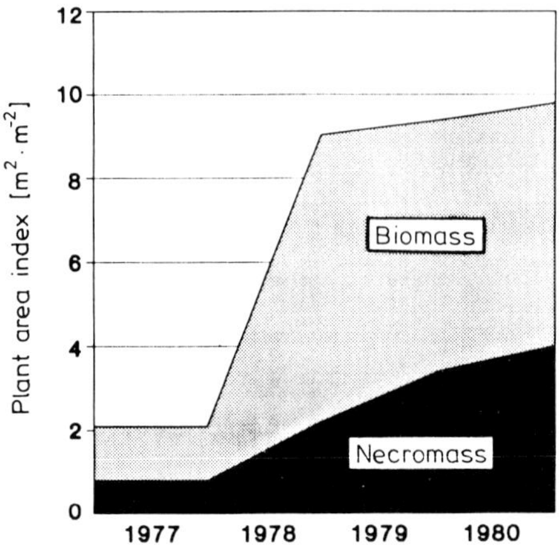

Figure 9.6: Time course of total plant area index (PAI), green area index (GAI), as well as the area index of necrotic matter in the alpine pasture "Guttal" during the investigation. The alpine pasture was last grazed in 1977, after which grazing management ceased (TAPPEINER and CERNUSCA 1989b).

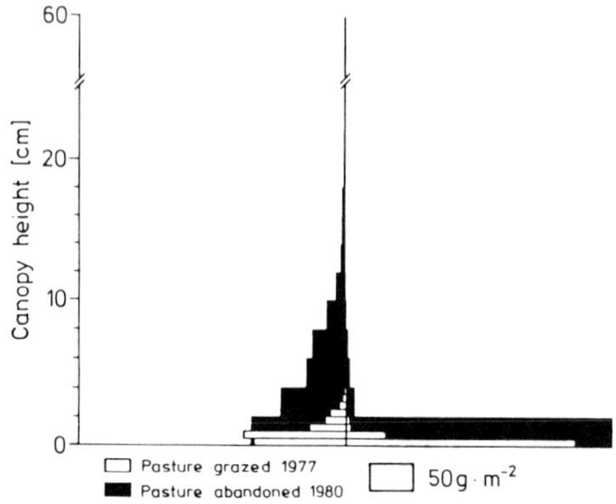

Figure 9.7: Comparison of the stratification of phytomass in the grazed and the abandoned pasture drawn to scale. Dry mass (g·m^{-2}) of photosynthetically–active components is depicted to the left of the ordinate, while mass of photosynthetically–inactive components and litter is shown on the right (TAPPEINER and CERNUSCA 1989b).

9.5 Application of the presented models

9.5.1 Investigation of the dynamic changes in stand structure, light climate, and photosynthesis, following abandonment of an alpine pasture

At the experimental site "Guttal" changes associated with the cessation of management in an alpine pasture were investigated from 1977 to 1980 (TAPPEINER and CERNUSCA 1989b). The studies showed that the plants on the alpine pasture responded very quickly to the new situation, with characteristic changes in the phytomass, stand structure, light climate and plant production:

In the first year the photosynthetically active leaf area (GAI) rose by a factor of 5. The necromass also increased by a factor of 2.8 in the first year and rose even further in the following years (Fig. 9.6). Measurements of leaf inclination showed that the leaves of the grazed area were in every case less steeply inclined than after its abandonment. After three years the aboveground phytomass on

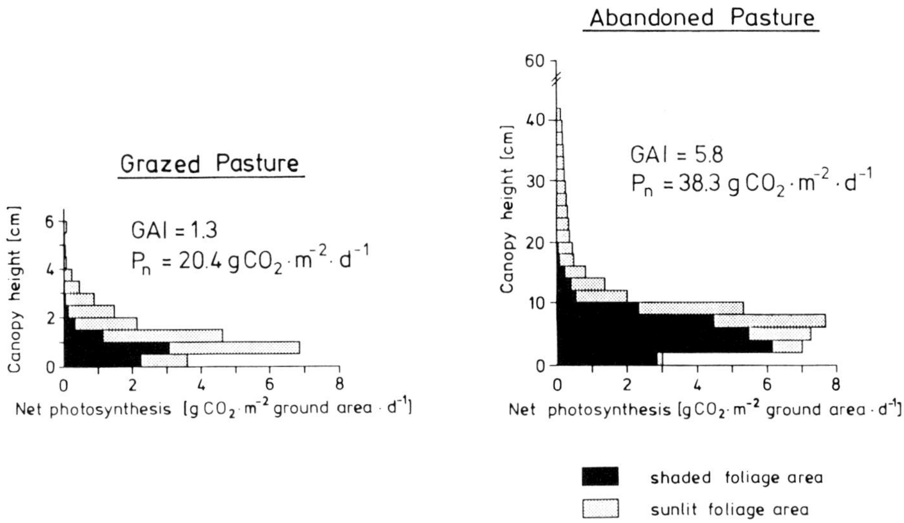

Figure 9.8: Daily net CO_2 uptake of the various layers in the grazed and abandoned pastures "Guttal" in a cloudless situation in mid–August, differentiated between sun and shade foliage.

the abandoned pasture was 5 times higher than at the time when it was still grazed, and the biomass 4 times higher. In Fig. 9.7 the phytomasses of the two canopies are compared on the same scale. Production measurements obtained by the harvesting method (CERNUSCA and JUSSEL, unpublished) revealed very surprising results, however, when compared with stand structure data. Although the photosynthetically active leaf surface of the abandoned pasture was in 1980 4.5 times that of the grazed pasture in 1977, production on the abandoned pasture was only roughly twice as large. The explanation for this was found by employing the radiation and canopy potosynthesis models, as follows:

Using Model 4 for simulating the light climate in the canopy (see pages 169 to 172) at the time of maximum biomass in each of the four years covered by the investigation, it was found that disuse of the alpine pasture results in fundamental changes in the light distribution in the stand. After removal of the grazing pressure, light becomes more and more the decisive limiting factor for net photosynthesis. In the grazed pasture, even in the lowest layer, leaves are exposed to light intensities above light saturation during most of the day. However, in the course of the succession the density of the stand increases and the leaves assume steeper angles in order to obtain a maximum amount of light. Nevertheless on the ungrazed pasture, during the entire day the lowermost four centimetres of the stand only receive diffuse light, so that in this case light is quite obviously the limiting factor. This is confirmed by the results of computer predictions for net photosynthesis, which are shown in Fig. 9.8. With the help of the model it was possible to analyse exactly how much of the daily photosynthetic uptake of CO_2 took place at various levels in the canopy. On a clear day, approximately 60% of the CO_2 uptake occurs through sunlit foliage area. In the abandoned pasture this contribution amounts merely to 30% under the same conditions. The simulations also showed that in the lowermost four centimetres of the ungrazed pasture the leaves operate for almost the entire day close to light compensation points. Although containing 57% of the photosynthetically active leaf surface of the whole canopy, this layer is only responsible for 17% of the total photosynthetic gains. Thus mutual shading is responsible for the surprising result that net photosynthesis on the abandoned pasture is only 1.9 higher than that of the grazed pasture, although the photosynthetically active leaf surface of the former is 4.5 times larger. The greatly increasing losses of net photosynthesis due to mutual shading on the abandoned pasture over the course of the four years of the investigation, as calculated by the model, are shown in Fig. 9.9. In each case the actual canopy CO_2 uptake and the losses due to mutual shading are shown. Whereas on the grazed area the losses amount to only 28%, in the abandoned area they are 68–73%.

9.5.2 Typification of structurally different plant stands in the Central Caucasus

Comprehensive investigations were carried out on six plant communities in the subalpine zone of the Central Caucasus with the aim of documenting the effects of anthropogenic disturbance on canopy structure and of analysing the effects on primary production and water regime (TAPPEINER et al. 1989). Furthermore, the relevance of light as a primary competitive factor within different structurally plant communities was assessed.

The results of the canopy structure analyses are shown in Fig. 9.10. The intensively managed plant communities "Deschampsietum" and "Heracletum" all had a very large biomass and only a small proportion of dead matter and litter. The LAI was on an average twice that of the other stands. The thinly grazed and the unexploited canopies ("Festucetum", "Rhododendretum"), on the other hand, were characterised by a high proportion of necromass and litter. In addition to the differences in phytomass composition, the stands investigated also differed structurally. In the "Heracletum" and "Rhododendretum", for example, canopy structure was characteristically two-layered: the upper third, in which almost the entire photosynthetically active leaf surface was accumulated, and the underlying layer of photosynthetically inactive stems. The arrangement of the leaves in these two stands was more or less horizontal whereas the grassy stands "Deschampsietum" and "Festucetum" were dominated by very erect leaves. In the "dry grassland" and in the "Dryadetum" both types of leaf inclination were represented, the grass

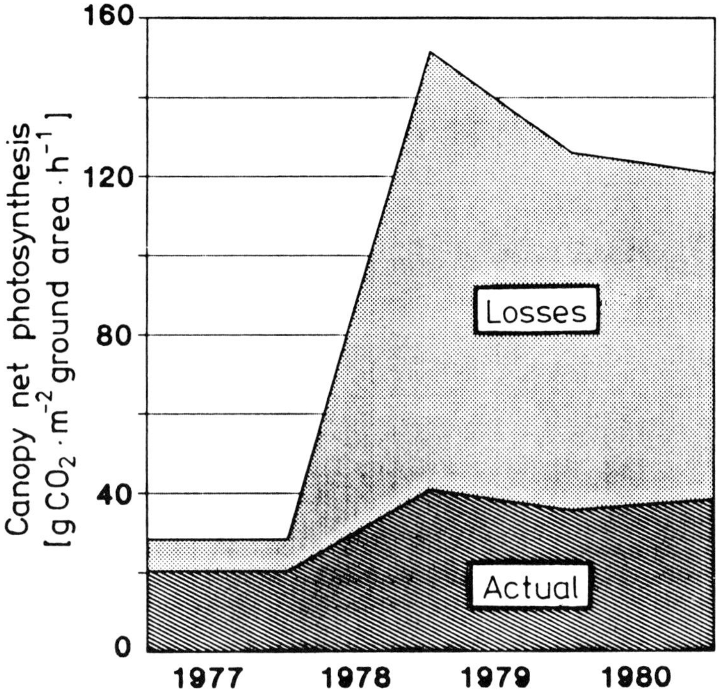

Figure 9.9: Time course of simulated current net canopy photosynthesis, as well as the potential net photosynthesis in the absence of mutual shading, within the alpine pasture "Guttal", at maximum biomass. The alpine pasture was last grazed in 1977, after which grazing management ceased (TAPPEINER and CERNUSCA 1989b).

leaves with a steep angle of inclination being predominant in the upper half of the canopy.

In order to obtain a synthesis of the large number of experimental results and an analysis of the canopy structure effects on light climate we employed the radiation Model 5. The model was used to simulate the light climate in its relation to the spatial structure of the stand in the different layers of the canopy. In this way it was possible for each stand to arrive at an exact analysis of the spatial distribution of the shaded and sunlit foliage area and hence of the degree to which the photosynthetically active components of the variously structured plant stands were able to exploit incoming light. Thus the use of the canopy photosynthesis model also provided information about canopy structure effects on the photosynthesis in the stand.

On the basis of the simulation results it could be shown that the plant canopies

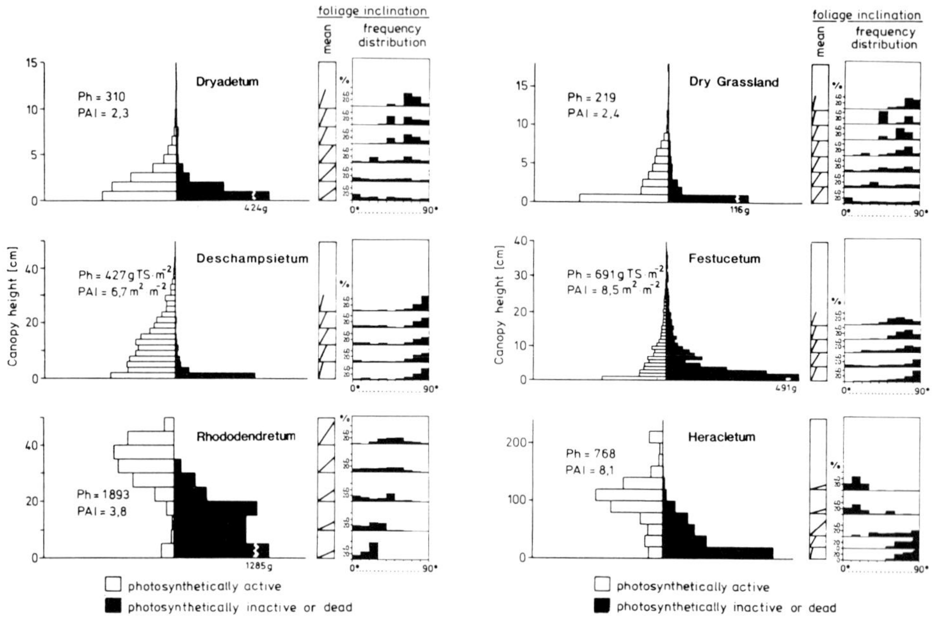

Figure 9.10: Stratification of phytomass, as well as inclination of leaves and stems of the plant communities investigated in the Central Caucasus. Dry mass (g·m^{-2}) of photosynthetically–active components is depicted to the left of the ordinate; the mass of photosynthetically–inactive components and litter is shown on the right. In addition, total phytomass (Ph) and plant area index (PAI) are included in the upper left corner. Depicted on the right side are the means of leaf and stem inclination, as well as their relative frequency in 10 degree classes (TAPPEINER and CERNUSCA 1991).

investigated differ characteristically with respect to absorption of irradiation and photosynthesis, and can be unequivocally assigned to three fundamentally different structural types (cf. Fig. 9.11).

1. Canopies in which leaf inclination is evenly distributed over all classes of angle, and in which there is a concentration of phytomass in the lowest third. Over 50% of the light is absorbed in this layer.

2. Canopies with erectophilic leaves and a gradual increase in phytomass from the canopy surface to the ground. This results in an almost linear decrease in light absorption with depth in the canopy.

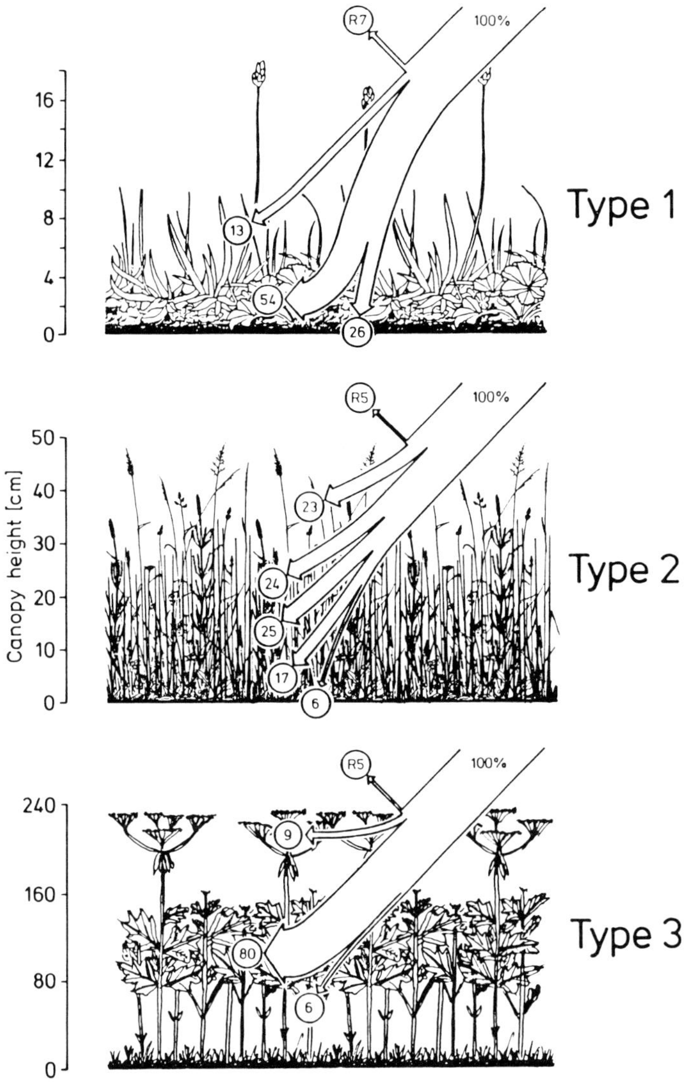

Figure 9.11: The stands analysed in the Caucasus can be assigned to three characteristic types with respect to their canopy structure. **Type 1**: Photosynthetically–active leaf area and maximum radiation interception are concentrated near the soil surface (pastures). **Type 2**: Nearly constant distribution of leaf area and radiation absorption throughout the various canopy layers (meadow, thinly grazed pasture). **Type 3**: Concentration of leaf area in the upper half of the canopy, where 80% of the radiation are absorbed (tall herb and dwarf shrub community). From TAPPEINER and CERNUSCA (1991).

3. Canopies with mainly planophilic leaves and a concentration of leaf area in the upper one–third; 80% of the light is absorbed in this region.

Type 3 includes the "Heracletum" and "Rhododendretum". The photosynthetically active leaf area is concentrated in the upper half of such canopies, and the leaves are almost horizontal. The utilization of radiation is extremely high in this layer, roughly 90% of the incoming PFD being absorbed here by the photosynthetically active components. In the "Heracletum" this layer is between 110 and 130 cm above the ground and with 7 to 10 g $CO_2 \cdot m^{-3} \cdot h^{-1}$ it exhibits throughout the entire day by far the highest net photosynthesis of the canopy (Fig. 9.12). In contrast, in the layer between 80 cm and 20 cm from the ground the net photosynthesis is only 0 to 1 g $CO_2 \cdot m^{-3} \cdot h^{-1}$, whilst below 20 cm, due to lack of light, photosynthesis is no longer possible for *Heracleum sosnowskyi*. In the layer from 0 to 20 cm only diffuse light is registrable, with an average intensity of between 20 and 60 $\mu mol \cdot m^{-2} \cdot s^{-1}$, and only briefly attaining values of up to 150 $\mu mol \cdot m^{-2} \cdot s^{-1}$ in a few places. Apart from the axes of *H. sosnowskyi*, this layer contains only isolated leaves of *Alchemilla* sp., exhibiting extreme characteristics of shade leaves (large leaves, saturation of photosynthesis at low light intensities). The simulation experiments clearly indicate that canopies of type 3 may be dominated by plants that have succeeded in spreading out their leaves in the upper half of the stand, thus effectively exploiting the PFD to suppress competition from species in the lower layers. These results indicate that the position of the leaf area in the upper canopy layers is crucial for success in the competition for light in a mixed species stand (compare also KÜPPERS 1985, CALDWELL 1987, BARNES et al. 1990).

Type 1, to which the "Dryadetum" and "dry grassland" of our study belong, is the opposite of Type 3, in that the dominant factor in competition is not the light. The structure of this type of canopy is merely influenced by grazing. The flat leaves spreading over the ground are especially effective in ensuring that the stability of the canopy is not endangered by excessive grazing losses. The larger part of the phytomass is concentrated in the lower 2 cm of the canopy, a layer that is scarcely affected by grazing sheep (CERNUSCA and NAKHUTSRISHVILI 1983). As the simulation experiments with Model 5 revealed, the disadvantage of this type of structure, however, is the relatively poor utilization of photosynthetically active radiation. Thus in the "dry grassland" only 53%, and in the "Dryadetum" only 63% of the incoming light are absorbed by photosynthetically active components, which means that there was a loss of 20% in the net canopy photosynthesis on these pastures. Since ecophysiological field measurements in the pasture had revealed that, in the early afternoon, water stress due to limiting soil water conditions and high temperatures caused a severe reduction in single-leaf photosynthesis (ABDALAZE 1988), this aspect was subjected to further analysis in the model. Simulation experiments showed that on the pastures lack of water and high temperatures caused a further loss of 15% in canopy net photosynthesis. These simulation model results could also be validated by measurements of photosynthesis using the Bowen ratio energy–balance method (CERNUSCA 1987).

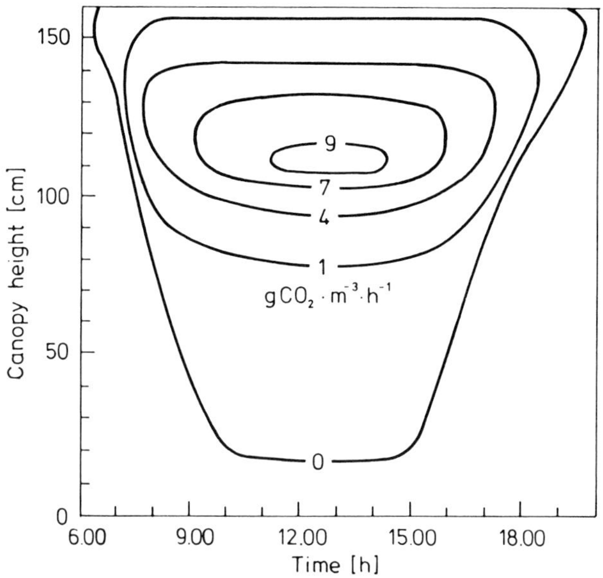

Figure 9.12: Results of the simulation of the vertical distribution of canopy net photosynthesis in the "Heracletum" on a clear day in August. Depicted are the isolines of canopy net photosynthesis expressed on a m^3 area (g $CO_2 \cdot m^{-3}$ ground and canopy height area $\cdot h^{-1}$). The isoline designed by "0" represents the compensation depth of the canopy during the course of the day.

Type 2, to which the grassy stands "Festucetum" and "Deschampsietum" belong, occupies an intermediate position between types 1 and 3. In these stands radiation is almost evenly distributed to all layers. In the "Deschampsietum" the light utilization of 84% provides similarly favourable conditions for plant productivity to those in the "Heracletum" (Type 3), whereas in the "Festucetum" the relatively high proportion of photosynthetically inactive components results in the extremely low figure of merely 30% for light utilization.

9.6 Conclusion

It could be shown that a combination of measurements and mathematical modelling can be applied successfully in ecosystem research for assessing the importance of canopy structure in influencing light climate and net photosynthesis. Whereas the experimental data provide the basis for causal analysis of the most important functional relationships, only mathematical modelling is capable of producing an exhaustive analysis and synthesis of the data from large-scale field experiments.

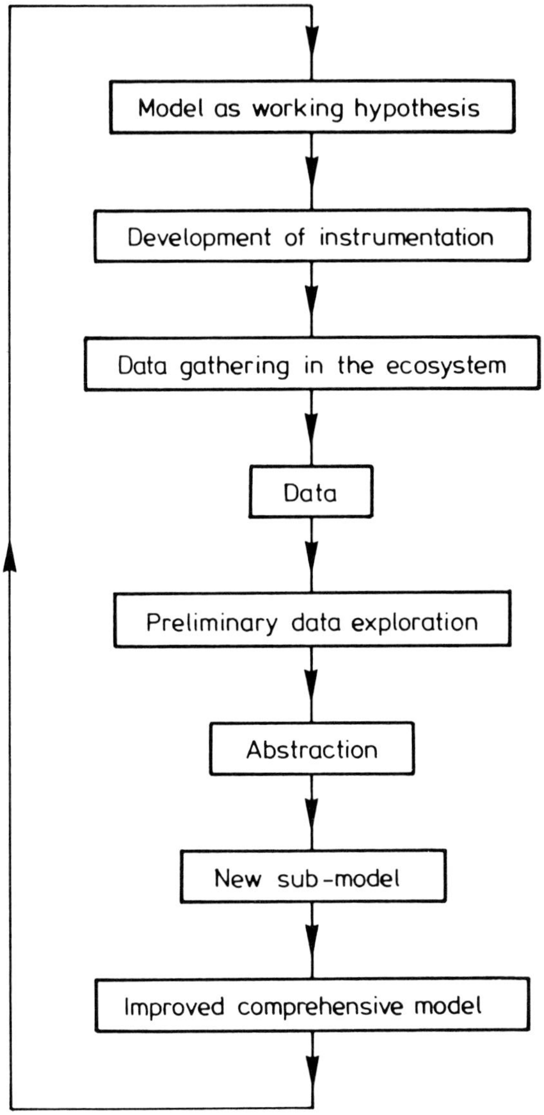

Figure 9.13: Characteristic research schedule of an ecosystem study. On the basis of preliminary model conceptsions ("models as working hypotheses") new scientific information is obtained through experimentation. Such information provides a validation and a means for expansion of the original model, or lead to its revision. This research procedure facilitates a "feedback" between theoretical considerations and experimental investigations and thus guarantees efficient research (CERNUSCA 1990a).

As compared with simple statistical procedures, the essential advantage of models is that the parameters employed in the model allow conclusions to be drawn concerning the interplay of reactions in the ecosystem. The examples provided by our investigations show that with the help of radiation models a classification of plant stands can be made on the basis of the distribution of light within the canopy and its utilization for photosynthesis. Not until simulation experiments on the computer were introduced did it become possible to demonstrate the competitive advantage of a particular stand structure with respect to the light climate.

Due to the complexity of the ecosystem studies, even detailed ecological data obtained in the field frequently only represent the endresult of a large number of interacting processes. The use of mathematical models in this case makes it possible to arrive at a detailed analysis of the causal relationships. The investigations on the abandoned pasture provide a good example: measurements from the abandoned site revealed a much lower CO_2 uptake by the canopy and a lower net primary production that could be explained on the basis of the photosynthetically active leaf area. The detailed analysis of the light climate (experimentally impracticable) and above all of the photosynthesis in the individual layers of the grazed and ungrazed pasture by means of the mathematical model revealed mutual shading to be the decisive effect.

The results presented in this paper show further that particularly the hierarchically structured models provide the following advantages in the phases of validation and application of the model:

- the validation of the individual steps of the model permits an exact quantification of the errors that (would) arise from using a simpler step of the model. It is thus possible to select the model variant that is the most economical of time and labour for the particular problem to be investigated.

- A mathematical model contains current state of knowledge about the interrelationships of the various functions, as well as additional theoretical considerations. Discrepancies between the results of simulation and of field experiments provide an indication of additional factors that have so far been overlooked. Thus, in the *Vaccinium* stand, for example, the marked effect of a statistically clumped leaf distribution was accurately revealed. In addition, taking the *Rhododendron ferrugineum* community as an example, clear indications of the hitherto neglected modulative adaptations of *Rhododendron ferrugineum* were revealed in the validation phase of the model. A feedback of this kind between theoretical considerations and experimental investigations very poignantly illustrates the importance of just such simple and uncomplicated mathematical models.

In conclusion it seems to us to be of fundamental importance for the continued progress of ecosystem research that, in the planning and performance of investigations, a productive interplay between field measurements and modelling of the ecological processes be ensured (Fig. 9.13). Based on specified working hypotheses it is possible to obtain extensive data sets by means of computer–controlled data acquisition systems. From this data, ecological contexts can be derived in a series of deductive abstractions. Various results can then be combined into a functional entity with the aid of mathematical models. Subsequent computer simulations provide a validation of the model, by a comparison of model behaviour with actual ecosystem response. Finally the validated model provides an effective instrument for the causal analysis of the structure and function of the ecosystem subject to investigation. Furthermore via the "inductive method", the model can be used to arrive at non–testable scenarios of the status and reaction of the ecosystem, which have not been directly investigated.

Bibliography

ABDALAZE, O. G.,1988, On field research plant CO_2–gas exchange in high mountains of the Caucasus.
In: NAKHUTSRISHVILI, G. SH. (ed.), The ecology of High mountains. Mezniereba, Tbilisi 1988, p.63–75.

ANDERSON, M. C.; DENMEAD, O. T., 1969, Short wave radiation on inclined surfaces in model plant communities.
Agron. J. **6**, 867–871.

BARNES, P. W.; BEYSCHLAG, W.; RYEL. R.; FLINT, S. D.; CALDWELL, M. M., 1990, Plant competition for light analyzed with a multispecies canopy model. III. Influence of canopy structure in mixtures and monocultures of wheat and wild oat.
Oecologia **82**, 560–566.

BEYSCHLAG, W.; BARNES, P. W.; RYEL, R.; CALDWELL, M. M.; FLINT, S. D., 1990, Plant competition for light analyzed with a multispecies canopy model. II. Influence of photosynthetic characteristics on mixtures of wheat and wild oat.
Oecologia **82**, 374–380.

CALDWELL, M. M., 1987, Plant architecture and resource competition.
In: SCHULZE, E.–D.; ZWÖLFER, H. (eds), Ecological studies, vol **61**. Springer, Berlin, Heidelberg, New York, 164–179.

CALDWELL, M. M.; MEISTER, H.–P.; TENHUNEN, J. D.; LANGE, O. L., 1986, Canopy structure, light microclimate and leaf gas exchange of *Quercus coccifera* L. in a Portuguese macchia: measurements in different canopy layers and simulations with a canopy model.
Trees **1**, 25–41.

CERNUSCA, A., 1976, Bestandesstruktur, Bioklima und Energiehaushalt von alpinen Zwergstrauchbeständen.
Oecol. Plant. **11**, 71–102.

CERNUSCA, A., 1982, Standardmeteorologie und Mikrometeorologie.
In: JANETSCHEK, H. (ed.), Ökologische Feldmethoden, 9–27. Stuttgart: Verlag Eugen Ulmer.

CERNUSCA, A., 1987, Application of computer methods to assess ecosystem function and response to stress. NATO ASI Ser. G15, 158–164.
In: TENHUNEN, J. D. et al. (eds.), Plant Response to Stress. Berlin–Heidelberg–New York: Springer 1987.

CERNUSCA, A., 1990a, Die Ökologie aus der Sicht der interdisziplinären Forschung.
In: RAINALTER H. (ed.), Gespräche der Fakultäten – Interdisziplinarität. Veröff. der Univ. Innsbruck, **174**, 37–52.

CERNUSCA, A., 1990b, Ecosystem research on grassland in the Austrian Alps and in the Central Caucasus.
In this volume.

CERNUSCA, A.; DECKER, P., 1984, Respiratorische Kohlenstoffverluste und Kohlenstoffbilanz einer alpinen Grasheide.
Ver. Ges. Ökol. **12**, 73–88.

CERNUSCA, A.; NAKHUTSRISHVILI, G., 1983, Untersuchung der ökologischen Auswirkungen intensiver Schafbeweidung im Zentral–Kaukasus.
Verh. Ges. Ökol. **10**, 183–192.

DIEMER, M., 1990, Wachstum und Kohlenstoffbilanz von Blättern krautiger Tal– und Hochgebirgspflanzen.
Thesis, Universität Innsbruck.

DUNCAN, W. G.; LOOMIS, R. S.; WILLIAMS, W. A.; HANAU, R., 1967, A Model for Simulating Photosynthesis in Plant Communities.
Hilgardia **38**, 181–205.

FARQUHAR, G. D.; VON CAEMMERER, S., 1982, Modelling of photosynthetic response to environmental conditions.
In: LANGE, O. L.; NOBEL, P. S.; OSMOND, C. B.; ZIEGLER, H. (eds.), Physiological Plant Ecology II, Encyclopedia Plant Physiol., vol **12**B. Springer Berlin, Heidelberg, New York, 549–587.

GAZARINI, L. C., 1988, Bestandesstruktur und Strahlungsextinktion von Zwergstrauchbeständen (*Vaccinium myrtillus* L. und *Rhododendron ferrugineum* L.) an der alpinen Waldgrenze.
Thesis, Univ. Innsbruck.

HESKETH, J. D.; JONES, J. W., 1980, What is Modelling and Why Model Photosynthesis?.
In: HESKETH, J. D.; JONES, J. W. (eds.), Predicting Photosynthesis for Ecosystem Models, vol **I**. CRC Press Inc, Bota Raton, FL, 1–8.

KÖRNER, CH., 1977, Der CO_2 Gaswechsel verschiedener Pflanzen im alpinen Grasheidegürtel. I. Der Einsatz einer neuen teilklimatisierten Meßkammer für in situ Messungen an kleinwüchsigen Gebirgspflanzen.
In: CERNUSCA, A. (ed.), Alpine Grasheide Hohe Tauern, Veröff. Österr. MaB-Hochgebirgsprogrammes Hohe Tauern, Bd **1**, 47–68. Universitätsverlag Wagner, Innsbruck.

KÜPPERS, M., 1985, Carbon relations and competition between woody species in a Central European hedgerow. IV. Growth form and partitioning.
Oecologia **66**, 343–352.

KÜPPERS, M.; SCHULZE, E. D., 1985, An empirical model of net photosynthesis and leaf conductance for the simulation of diurnal courses of CO_2 and H_2O exchange.
Aust. J. Plant Physiol. **12**, 513–526.

LEMEUR, R.; BLAD, B. L., 1974, A critical review of light models for estimating the short wave radiation regime of plant canopies.
Agric. Meteorol. **14**, 255–286.

LONG, S. P.; HALLGREN, J.-E., 1985, Measurement of CO_2 assimilation by plants in the field and the laboratory.
In: COOMBS, J.; HALL, D. O.; LONG, S. P.; SCURLOCK, J. M. O. (eds.), Techniques in bioproductivity and photosynthesis. 2nd ed., 62–94. Pergamon Press, Oxford, New York, Beijing, Frankfurt, Sao Paulo, Sydney, Tokyo, Toronto.

MONSI, M.; SAEKI, T., 1953, Über den Lichtfaktor in den Pflanzengesellschaften und seine Bedeutung für die Stoffproduktion.
Jap. J. Bot. **14**, 22–52.

MYNENI, R. B.; ROSS, J,; ASRAR, G., 1989, A review of photon transport in leaf canopies.
Agric. Forest Meteorol. **45**, 1–153.

NORMAN, J. M., 1975, Radiative Transfer in Vegetation.
In: DE VRIES, D. A.; AFGAN, N. H. (eds.), Heat and Mass Transfer in the Biosphere. I. Transfer Processes in Plant Environment, 187–207. Scripta Book Co., Washington 1975.

NORMAN, J. M.; HESKETH, J. D., 1980, Micrometeorological Methods for Predicting Environmental Effects on Photosynthesis.
In: HESKETH, J. D.; JONES, J. W. (eds.), Predicting Photosynthesis for Ecosystem Models, vol I. CRC Press Inc, Bota Raton, FL, 9–35.

ROSS, J., 1981, The radiation regime and architecture of plant stands.
The Hague–Boston–London: Dr. W. Junk Publishers 1981.

SIEGWOLF, R.; CERNUSCA, A., 1984, CO_2-Gaswechsel von *Rhododendron ferrugineum* L. an der alpinen Waldgrenze.
Verh. Ges. Ökol. **12**, 111–122.

SIEGWOLF, R., 1987, CO_2-Gaswechsel von *Rhododendron ferrugineum* L. im Jahresgang an der alpinen Waldgrenze.
Thesis, Univ. Innsbruck.

TAPPEINER, U.; CERNUSCA, A., 1984, Modellmäßige Beschreibung von Bestandesstruktur und Strahlungsklima alpiner Pflanzenbestände.
Int. Dr. F. Sauberer Gedächtnissymp., Univ. Bodenkultur, Wien, 32–34.

TAPPEINER, U.; CERNUSCA, A., 1989a, Canopy Structure and Light Climate of Different Alpine Plant Communities: Analysis by Means of a Model.
Theor. Appl. Climatol. **40**, 81–92.

TAPPEINER, U.; CERNUSCA, A., 1989b, Veränderung der Bestandesstruktur und der Lichtausnützung nach dem Brachfallen einer Almweide.
In: CERNUSCA, A. (ed.), Struktur und Funktion von Graslandökosystemen im Nationalpark Hohe Tauern. Veröff. Österr. MaB-Hochgebirgsprogrammes Hohe Tauern, Bd. 13, 531–548. Innsbruck: Universitätsverlag Wagner, 1989.

TAPPEINER, U.; CERNUSCA, A., 1991, Charakterisierung subalpiner Pflanzenbestände im Zentralkaukasus anhand von Bestandesstruktur und Strahlungsabsorption.
Verh. Ges. Ökol., in press.

TAPPEINER, U.; CERNUSCA, A.; NAKHUTSRISHVILI, G. SH., 1989, Bestandesstruktur und Lichtklima ausgewählter Pflanzenbestände der subalpinen Stufe des Zentralkaukasus.
Sitzungsber. Österr. Akad. Wiss., Mathem.-naturw. Kl., Abt. I, **197**. Bd., 5. bis 10. Heft, 395–421.

THORNLEY, J. H. M., 1976, Mathematical models in plant physiology.
Academic Press London, New York, San Francisco, 74–91.

VON CAEMMERER, S.; FARQUHAR, G. D., 1981, Some relationships between the biochemistry of photosynthesis and the gas exchange of leaves.
Planta **153**, 376–387.

WARREN WILSON, J., 1963, Estimation of foliage denseness and foliage angle by inclined point quadrats.
Aust. J. Bot. **11**, 95–105.

WARREN WILSON, J., 1965, Stand structure and light penetration. I. Analysis by point quadrats.
J. Appl. Ecol. **2**, 383–390.

Chapter 10

The water balance of deciduous forests: methods and models

B. SAUGIER and J.-Y. PONTAILLER

10.1 Introduction

Water availability is likely to be the main factor limiting plant productivity on continents in the temperate and tropical zones. It is controlled by precipitation, evaporative demand and soil storage capacity, but also by plant factors that regulate transpiration and water uptake by roots. The study of forest water balance is important for at least three reasons:

1. In forests as in crops, water availability strongly affects productivity as is easily seen from the smaller tree rings that form in dry years.

2. Forests play an important role in hydrology at the regional scale.

3. Due to their surface properties (albedo, roughness, evaporation) forests act on climate and may modulate the climatic changes forecast for next century.

Several methods have been used to study the water balance of forests. Precipitation interception, especially important in forests, is deduced from measurements of precipitation above and below the forest canopy. Soil water budget may be monitored weekly by gravimetric or neutron probe measurements. Within certain limits, these techniques allow the calculation of forest water consumption but the timescale of a week, however, is too large for studying the physiological and climatic control of evapotranspiration, which exhibits strong diurnal variations.

Micrometeorological methods have been used to obtain hourly values of water vapour fluxes on selected days, but have rarely been used during a complete growing season due to instrumental difficulties. Recently, sapflow measurements by thermal methods have become routine, filling an important gap between sophisticated micrometeorological measurements and long term soil water budget studies.

Until now, most studies concerned coniferous forests but deciduous and tropical forests play a major role on continental evapotranspiration. Deciduous forests, dominant in temperates latitudes, have their own specificity; water uptake is not only dependent on water availability and evaporative demand but also on leaf area index variations. Four periods can be defined: during leaf development and maturation phase, canopy transpiration progressively increases by the combined effects of leaf growth, stomatal development and often a decrease in leaf pubescence (NIZINSKI et al. 1989). The following phase is the summer leafy period in which canopy transpiration reaches a constant fraction of potential evaporation. In autumn, the foliage senescence and the coming out of absicission layers cause an increase in resistances to water transfers. Later, the winter period is characterized by a very low and stable soil water uptake. This cyclic evolution also has a large impact on water interception and energy balance of the canopy.

10.2 Soil Water Balance

Soil water reserve (R) fluctuates by the effects of precipitation (P), evapotranspiration (E) and drainage (D). In the absence of run-off, the water balance (presented in Fig. 10.2) may be written as:

$$P = E + \frac{\Delta R}{\Delta t} + D \tag{10.1}$$

Regular monitoring of soil moisture is necessary for understanding forest water balance. There is no absolute reference method but the gravimetric determination of soil water content (105 °C up to a constant weight) is generally used to calibrate the other methods (combined with a density measurement to get volumetric moisture).

After the Second World War, a new technique was developed, based on the slowing of fast neutrons (PIEPER 1949). This technique is now widely used in field studies because it allows a correct determination of soil water content without excessive perturbation. A fast neutrons source is exploring an aluminium tubing vertically settled in soil. Elastic collisions occur with the nuclei of the soil elements. Small nuclei like that of hydrogen are the most efficient at slowing neutrons. As a

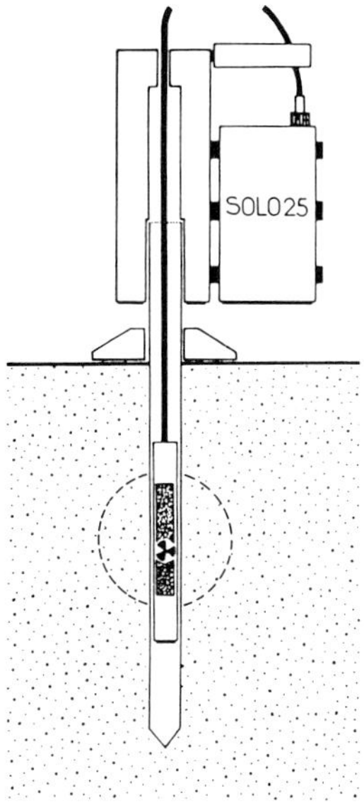

Figure 10.1: Neutron probe in use in the field (from Nardeux).

result, slow (or thermal) neutrons are emitted in all directions. Some of them go in the tubing direction, where a counter is placed, close to the fast neutrons source (Fig. 10.1).

The volume of soil measured by this technique is a sphere or an ellipsoid (depending on source–detector geometry) that ranges from 15 cm in diameter in a wet soil to 50 cm in a dry one (COUCHAT 1974). Neutron probes give a linear response to soil volumetric water content. The slope of this response must be precisely determined for an accurate calculation of variation in water content.

Calibration can be done in the field by comparison with gravimetric measurements made at different times including wet and dry spells. Another method, commercially available, consists of determining the soil neutron constants of absorption and diffusion in a graphite block where two counters analyze the thermal flux perturbation induced by a soil sample. This service is available from the CEA (DPVE, CEN Cadarache, 13108 Saint Paul–lez–Durance Cedex France).

Sampling is an important problem in forest stands where water inputs and uptakes show an irregular spatial pattern. As tubings are small compared to individual tree size, a large number of them (5 to 20) are generally used to get representative results.

Under field conditions, soil water content may vary between two limits, saturation and a wilting point below which no further water can be extracted by plants because it is retained with too much energy by soil microporosity. In practice, soil water available for vegetation is usually calculated between two limits that are very useful even if they are not perfectly defined physically: the upper one is the field capacity and the lower one is the permanent wilting point. Many models consider soil as one or several reservoirs within the root zone. Their levels fluctuate between these two limits according to water supply and demand.

10.2.1 Field capacity

There is an endless discussion of this concept. BETREMIEUX (1975) published an interesting 41 page "actualization and rehabilitation" of it. He quoted 9 different definitions proposed by scientists from 1948 to 1972 and proposed another one: *"water held in a previously highly rehumected soil, well drained and protected from evaporation, after excess water has drained away by gravity only"*. In fact, after a fast drainage phase, soil humidity reaches a sort of plateau with little further decline. In deciduous forests established on well drained soils, the soil surface is protected from direct evaporation by litter and field capacity becomes a measurable variable; soil water content measurements show a high stability over long dry periods, in winter (PONTAILLER et al. 1988).

When possible, *in situ* determination of field capacity appears to be the least perturbative and so the most satisfactory: in winter–time, after an important rainfall event, soil surface is covered with a large waterproof plastic sheet and the soil moisture of one or several profiles is regularly checked with a neutron probe until a relatively stable level is observed (RODE 1960).

10.2.2 Wilting point

Wilting point is generally defined as water content reached at a potential equal to −1.5 MPa. During long dry periods, this level is reached very slowly since vegetation increasingly limits its consumption by stomatal regulation.Thus, *in situ* measurement of the soil wilting point is difficult, so the pressure technique is widely used in the laboratory: humected soil samples are enclosed in a chamber submitted to an air pressure of 1.5 MPa, forcing water to flow out of samples. Moisture is measured when its level stabilizes.

Equation (10.1) shows that evapotranspiration can be deduced from variations in soil water reserve and from rainfall measurements only if drainage is known or neglected. Drainage may be measured by placing water collectors in the soil under the root zone. This procedure is difficult and not very accurate. It may also be calculated as the product of soil hydraulic conductivity (K) and of the water potential gradient (Darcy's law). Water potential is easily measured by tensiometers but the uncertainty on K is often 100 to 200%.

Fortunately, drainage reaches very low values when soil moisture decreases and it may be neglected on sandy soils during a large part of the growing season. Soil water budget studies are quite satisfactory under certain conditions (no drainage,

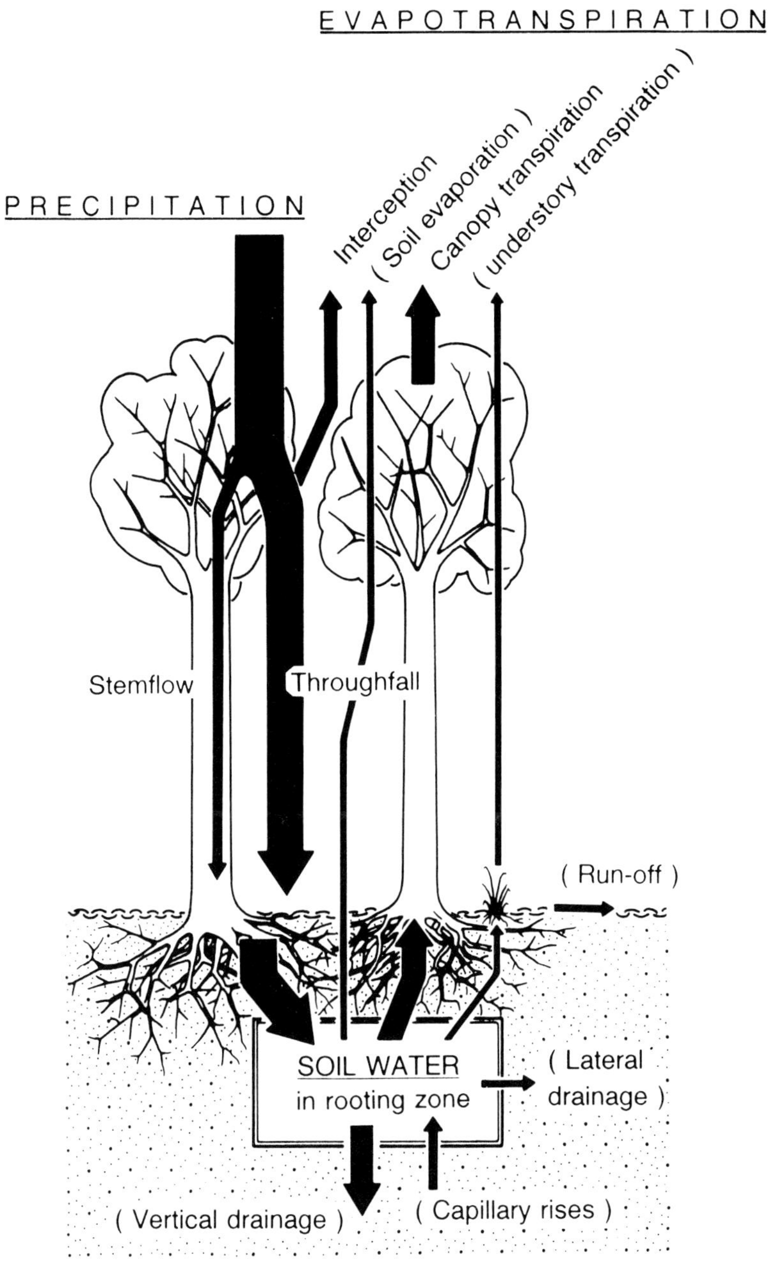

Figure 10.2: Elements of soil water balance in forest. Parameters between parenthesis are often neglected.

no run–off, well defined rooting zone) and using long time steps, a week or more, but in many cases other techniques are required.

10.3 Rainfall distribution pattern

10.3.1 Throughfall, stemflow and interception loss

Incident precipitation over a forest ecosystem may be split into three quantities: the first one, which is usually the largest amount of water for a forest, is the throughfall. It is composed of the rain reaching the soil surface, either directly or indirectly. It starts after humectation of the canopy and can be measured by raingauges placed at ground level under the canopy. A large number of raingauges is necessary since throughfall repartition in forests presents an important heterogeneity. Part of throughfall can be lost by run–off in forests occuring on slopes. In some cases, run–off appears to be a major parameter but measuring or computing it is difficult. Run–off is spatially quite variable but may be neglected on flat terrain.

The second part of incident precipitation is stemflow, which is the water intercepted by leaves and branches that runs down tree trunks. It begins to be important when all of the aerial parts of the trees are saturated (after 2 or 3 mm rainfall according to SCHNOCK 1970 and AUSSENAC and BOULANGEAT 1980). Stemflow is collected in gutters made of mastic, plastic sheets or polyurethane foam and may be monitored with an automatic logging system. A good correlation between individual stemflow and trunks diameter was found by FARDJAH and LEMEE (1980) in a beech forest (*Fagus silvatica*), with minimum values during the leafy period. On the other hand, a good relationship between stemflow fraction and leaf area index was found in Sweden on a young dense willow stand (*Salix viminalis*, LINDROTH personal communication). Stemflow, as expressed as a percentage of incident precipitation, may vary from less than 1% in a mature oak forest (*Quercus petraea*, NIZINSKI and SAUGIER (1989a) to 20% or more in the previously mentioned willow plantation, depending on stand density, tree height and morphology or bark characteristics. The length and the intensity of rainy events also must be taken into account.

The difference between incident precipitation and the sum of stemflow and throughfall is water intercepted by leaves and bark and evaporated. Since the work of HORTON (1919), interception loss is considered as a major component of the water balance of forests. Until the early seventies, most of the results were presented as a regression relating interception loss to incident precipitation. This simple approach assumed an average evaporation rate and an average rainfall intensity during precipitation events and thus only had a local interest, as shown by DOLMAN (1987).

Later, dynamical models were proposed (RUTTER et al. 1971, 1975; GASH 1979; MULDER 1985; CHASSAGNEUX and CHOISNEL 1986) for computing the evaporation of a wet canopy and the changes in intercepted water during and after rain.

10.3.2 Interception by tree litter

Two points are important to consider:

1. Tree litter can be an important component of forest stands in which water is largely unavailable to root systems. NIZINSKI and SAUGIER (1988a) measured the change in water content of an oak–tree litter between two times: one day after a substantial rainfall, and after a three week drying period at ambient temperature. They found a change of 2.9 mm, which is quite a high value. However the evaporation rate of water intercepted by litter is considerably lower than that of water intercepted by foliage because evaporative demand is reduced under a tree canopy by low levels of radiation, wind velocity and saturation deficit. Thus litter interception is often neglected in water balance studies.

2. Litter protects soil from direct evaporation, which can thus be disregarded in forests evapotranspiration models.

Total interception due to canopy and litter may reach a value equal to 5 mm in many woody stands of the temperate zone (oak, beech, sweet chestnut). When rainfall events are small and occur at wide intervals, it may have a considerable impact.

10.4 Evapotranspiration

This term in the water balance equation may be deduced on a hourly basis from micrometeorological measurements by either energy balance or eddy–correlation methods. It may also be computed on a weekly basis from the soil water balance.

10.4.1 Soil water balance method

Careful studies of soil water reserve may allow good estimates of forest transpiration during drying periods when drainage may be neglected. NIZINSKI and SAUGIER (1989b) have studied the seasonal variation of transpiration in a deciduous oak (*Quercus petraea*) forest under these conditions. In the spring, the ratio E_c/E_p of canopy transpiration to potential evaporation (PENMAN 1948) increases from 0 to 0.4 as the leaf area index (L) increases from 0 to its maximal value (4.2). It then takes several more weeks for E_c/E_p to reach its maximum (0.75 in this forest) which is quite surprising and seems to be related to a change in leaf epiderm morphology and particularly to a decrease in leaf pubescence (NIZINSKI et al. 1989). Predictions of the time of budburst and of leaf area growth may be done from current daylength and heat sum over the 10 days prior to budburst as presented by NIZINSKI and SAUGIER (1988b). Once E_c/E_p has reached its maximum value it may decrease either through decreasing soil moisture or through foliage senescence that should also be predicted in a model of forest evapotranspiration.

10.4.2 Energy balance (Bowen ratio) method

The Bowen ratio β of the sensible heat flux H to the latent heat flux λE is calculated as:

$$\beta = \frac{H}{\lambda E} = \frac{\gamma \Delta T}{\Delta e} \tag{10.2}$$

where γ is the psychrometric constant (66 Pa/K at normal pressure), ΔT and Δe are the differences in air temperature and water vapour pressure between two levels above the canopy. Latent heat flux is then computed from the energy balance equation as:

$$\lambda E = \frac{R_n - G}{1 + \beta} \tag{10.3}$$

where R_n is the net radiation and G the heat flux into the ground.

Since a forest is an aerodynamically rough surface, ΔT and Δe are small and thus difficult to measure. Besides this, the layer just above a forest is influenced by individual surface elements (the trees) and the exact similarity between heat and water vapour in this layer has been questioned (LINDROTH 1984).

10.4.3 Eddy correlation method

Thus the eddy correlation method is often preferred to estimate evapotranspiration above forests. It is based on the fact that on average, air rising above the forest carries extra humidity coming from evaporation whereas air going down is drier. Evaporation is calculated as:

$$E = \overline{\rho_e' w'} \tag{10.4}$$

where ρ_e' is the fluctuation of the water vapour density around its average value and w' is the vertical component of the windspeed. This formula has the advantage of making no assumption on the turbulent transfer of either heat or water vapour. ρ_e' is measured with a fast response hygrometer (time constant = 0.5 s approx.) such as an ultra–violet gas analyser, and w' is now measured with sonic anemometers.

Once E is known, canopy resistance may be calculated directly if leaf temperature is measured, or indirectly using the Penman–Monteith's equation given below (10.5). Results for various forest types have conveniently been summarized by SHUTTLEWORTH (1989) and reveal a striking similarity between values of r_c for forests as different as a monoculture of *Pinus pinaster* in the Landes (France) and the Amazonian forest near Manaus in Brazil. This implies that in spite of considerable variations in stomatal resistance of individual leaves and tree species, the forest behaves as an entity that may be easier to study as a whole rather than as a sum of components. As a result, forest transpiration in the absence of rain is a relatively constant fraction of potential evaporation when soil water is not limiting. This fraction, for a closed forest, stays between 0.7 and 0.9 (MONTENY et al. 1985; RAUNER 1976; AUSSENAC and GRANIER 1979; NIZINSKI and SAUGIER 1989a).

10.5 Sapflow

Sap must flow from the soil to the canopy top, which is several tens of meters above the soil surface in a mature forest. Since vacuum pumps can not lift a vertical column of water more than ten meters high, sap ascent in trees has long been intriguing to scientists. We now know that water is under tension (i.e. negative pressure) in the outside cell walls of transpiring leaves. This negative pressure is easily measured using a pressure chamber and may reach −1 to −2 MPa in a transpiring tree. This pressure is transmitted to the sap within the xylem down to the roots and to soil water. Thus, the entire column from soil to leaf is pulled by the transpiration flow that increases tension in the leaf. Why is it not ruptured by such low negative pressures? Because water molecules are firmly bound by cohesion forces thanks to their strong polarity. However the column is occasionally broken by an air bubble under strong dehydratation. When this happens, a cavitation occurs, easily detected by an acoustic detector and the corresponding vessel loses its conducting capacity (TYREE et al. 1984). It is believed that trees have the ability to eliminate these air bubbles at night using positive pressure produced by roots in the absence of evaporation, but no conclusive proof of this ability has yet been given.

Resistances to liquid water flow are important at the root entrance, at the connections between leaf petioles and branches or between branches together or between branches and the trunk. Xylem itself poses little resistance per unit length but it results in a moderately high resistance in tall trees. The mere existence of these resistances leads to a pressure drop between soil water and leaf water, increasing with sapflow and thus with transpiration.

10.5.1 Sapflow measurement

Many methods have been applied to measure sapflow in tree trunks because it is an elegant way to estimate forest transpiration. Complex radioisotope incorporation techniques are not accurate enough. The use of a potometer on small branches causes injury to sapwood and changes the water potential gradient. Besides, data obtained are difficult to extrapolate to a whole tree stand. To allow both better accuracy and minimum perturbation, many methods based on heat transfer were developed. Using the following devices connected to an automatic data logging system, sapflow measurements are possible over the whole growing season with a hourly time step.

- Heat pulse method (HUBERT and SCHMITT 1937; SWANSON 1965): it is the most widely used; a heater is inserted into sapwood and is activated for short periods at regular intervals. A temperature sensor (thermocouple or thermistor), is placed in sapwood at a higher position and detects the heat wave. Generally, a second sensor is placed below the heat source and is used as a reference of xylem conduction. COHEN (1981) improved the efficiency of the

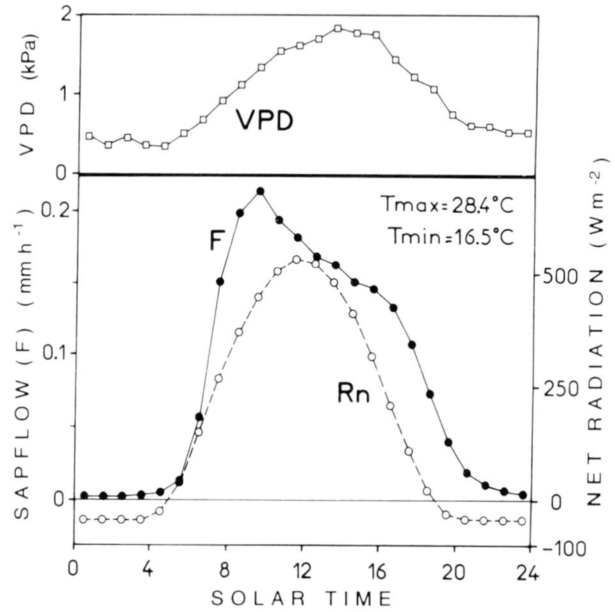

Figure 10.3: Evolution of hourly values of sapflow in a sweet chestnut stand (from BOBAY and SAUGIER, 1989).

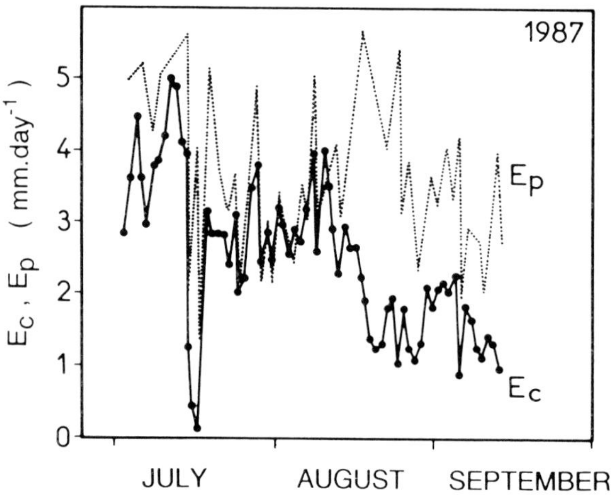

Figure 10.4: Summer evolution of the sapflow in a pine (*Pinus pinaster*) stand. A diminution of the canopy transpiration, due to a soil water reserve decrease, occurs in mid-August (from LOUSTAU et al., 1990).

method by developing probes with several thermistors for the determination of a sap velocity profile in the sapwood.

- Heat dissipation method (GRANIER 1985) is easier to implement; two cylindrical probes are inserted into the conductive xylem, one of them is heated at constant power while the other, at a lower position in the trunk and unheated, is used as a reference. The temperature difference between the two probes is measured by thermocouples and it can be related by an empirical relationship to sapflow density around the heating probe. This relationship is independant of tree species and is also valid for non–living material (GRANIER 1985). Figure 10.3 shows hourly values of sapflow in a sweet chestnut stand (BOBAY and SAUGIER 1989) and Figure 10.4 shows a long term evolution of the sapflow in a pine stand. One can note a severe decrease in transpiration during a dry period (LOUSTAU et al. 1990).

- Heat balance method. With the two first methods, one of the major problems is the difficulty of determining the cross–sectional area of the sap conducting system. Thus, a calibration is necessary to obtain correct sapflow measurements because the probe's presence perturbs nearby sap movements, creating inaccuracy in the estimates. The heat balance method (SAKURATANI 1979) avoids calibration by determining both heat loss due to conduction in the stem and convection into ambient air. If a good insulation is used, the second term can be neglected (NASR 1987). The method is accurate but technically more difficult (it requires five measurements instead of one) and it has only been applied on relatively small stems (diameter < 7 cm).

10.6 Models of forest water balance

Forest evapotranspiration E is the sum of three terms: (1) evaporation of intercepted water during and after rain E_i; (2) evaporation from forest soil and understory E_s; (3) transpiration from forest canopy E_c.

Under a closed forest canopy the second term E_s is usually less than 10% because net radiation and water vapour saturation deficit are small. The first term E_i depends both on foliage interception storage capacity and on rainfall pattern, being higher when rain occurs in long events of low intensity that cause the foliage to be wet for long periods. The last term is the dominant one and is often expressed by the Pennman–Monteith's equation:

$$\lambda E_c = \frac{\Delta(R_n - G) + (\rho c_p / r_a)\delta e}{\Delta + \gamma(1 + r_c/r_a)} \tag{10.5}$$

With:

λ	latent heat of vaporization of water
Δ	derivative of the function relating saturation vapour pressure to temperature
R_n	net radiation
G	heat flux into the ground
ρ, c_p	air density and specific heat at constant pressure
r_a	aerodynamic resistance of forest canopy
δe	water vapour saturation deficit
γ	psychrometric constant
r_c	canopy resistance

The aerodynamic resistance r_a is computed with the following formula:

$$ra = \frac{\ln^2(z-d)/z_0}{k^2 u} \tag{10.6}$$

In a forest, the ratio d/h (displacement height of the wind profile to canopy height) varies from 0.6 in an heterogeneous population to 0.9 in a relatively regular one such as one *Hevea* plantation (BARBIER 1983). Whereas z_0, the roughness length, is of the order of $h/10$, if we take representative values for a deciduous forest ($d/h = 0.63$, $z_0/h = 0.1$, $z = h + 2$m, $h = 30$m) we find $r_a = 7$ s·m^{-1} for a wind speed (u) of 2 m·s^{-1}.

Forests are thus characterized by small values of aerodynamic resistance r_a of the order of 5 to 10 s·m^{-1}. By contrast, they have relatively large values of r_s (typically 100 s·m^{-1}). With these values, the convective term $(\rho c_p/r_a)\delta e$ becomes larger than the radiative term $\Delta(R_n - G)$, thus λE_c tends towards:

$$\lambda E_c' = \rho c_p \delta e / r_c \tag{10.7}$$

Equation (10.7) can be deduced from (10.5) when r_a tends towards zero and it is assumed that for a well ventilated forest canopy, leaf temperature is nearly equal to air temperature. The latter is roughly true for small needles like that of coniferous species but not so much for the larger leaves of deciduous trees.

10.6.1 Evaporation of intercepted water

Evaporation of intercepted water (E_i) occurs directly at the leaf–air interface and may be computed from equation (10.5) with $r_c = 0$ (RUTTER et al. 1971). With the low values that characterize r_a in forest canopies, latent heat fluxes of wet foliage are fairly high and commonly reach 2 or 3 times the net radiation. HALLDIN et al. (1984) as well as NIZINSKI and SAUGIER (1988a) have modelled evapotranspiration as the sum of E_i and of E_c neglecting E_s. During rain, canopy intercepts an amount of water S that is less than or equal to maximum storage capacity S_m. When the canopy is wet ($S > 0$), evaporation is E_i (equation (10.5) with $r_c = 0$) and E_c is

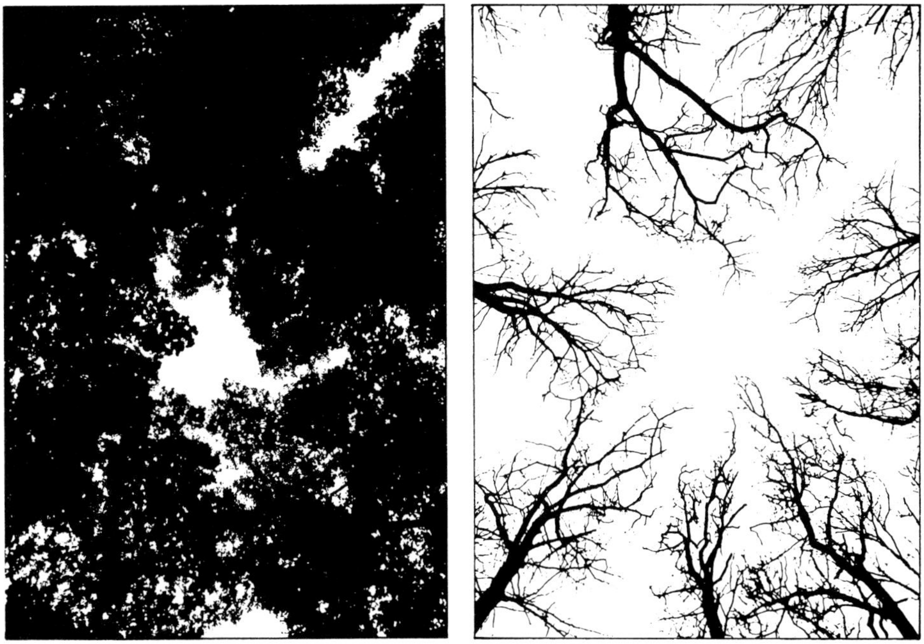

Figure 10.5: Vertical photographs of a canopy in summer and in winter (white oak). These pictures are helpful in the determination of water interception.

assumed to be zero. When the canopy is dry, evaporation is E_c given by equation (10.5). SHUTTLEWORTH (1989) reports average S_m values of 1.2 mm for coniferous forests and of 0.8 mm for deciduous forests when in leaf and 0.3 mm when leafless. It is possible to estimate locally S_m from vertical views of the canopy (Fig. 10.5) and from humectation experiments on leaves, branches and bark. NIZINSKI and SAUGIER (1988a) have measured higher values of S_m in an oak forest; leaves: 0.8 mm, branches: 1.2 mm, trunks: 1 mm, total: 3 mm. Such high values characterize species with thick and corky bark such as oak.

10.6.2 Canopy resistance r_c

Equation (10.5) assumes r_c is known. It may be estimated from measurements of leaf area index L and of stomatal resistance r_s on a sample of leaves, or directly from micrometeorological measurements of evaporation. In the first case r_c is often calculated as r_s/L, where r_s is the average stomatal conductance, or it may be calculated using the light profile through the canopy and the light dependance of r_s (SAUGIER and KATERJI 1991). Stomatal resistance depends not only on light but also on temperature, vapour pressure deficit (VPD) and soil moisture.

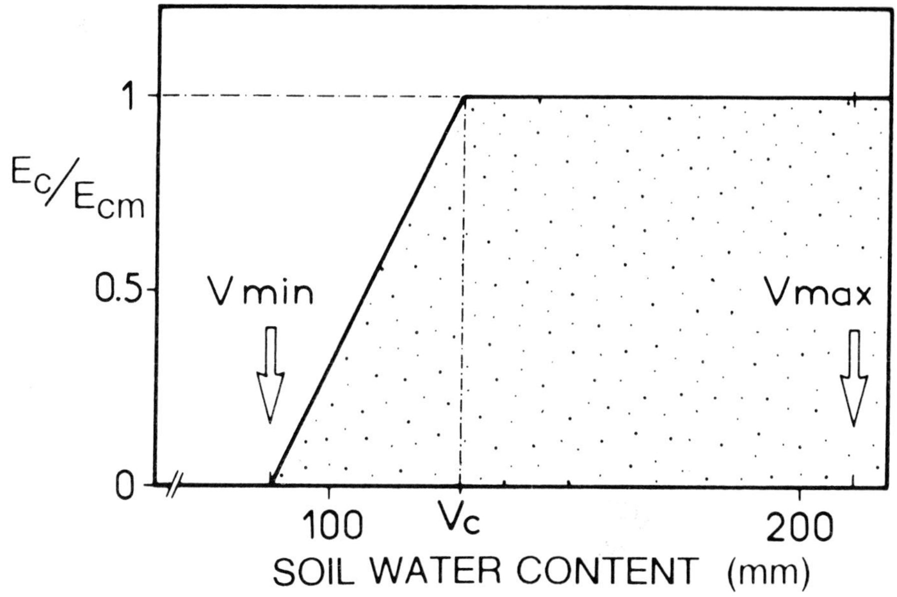

Figure 10.6: Variations of real over maximum canopy transpiration ratio as a function of soil water reserve. V_c is the critical value of this reserve below which E_c/E_{cm} starts to decrease.

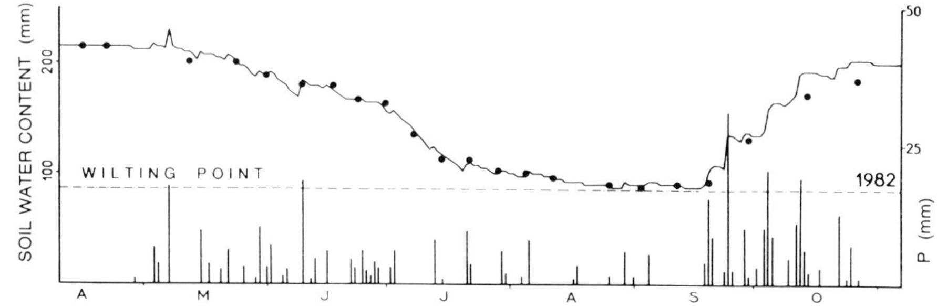

Figure 10.7: Simulation of soil water moisture in a white oak stand (solid line = simulation; dots = measurements), from HALLDIN et al. (1984).

SHUTTLEWORTH (1989) reports similar values of r_c for three pine forests and for the Amazon forest, with minimal values near 50 s·m^{-1}, a doubling of r_c for VPD increasing from 0 to 1 kPa and average day–time values close to 100 s·m^{-1}.

10.6.3 Soil moisture and transpiration

The effect of soil moisture on transpiration is taken into account differently according to the model. Simple models compute values of r_c and of E_c in the absence of stress (E_{cm}) and take the ratio of E_c to its maximum value E_{cm} as a simple function of soil moisture in the rooting zone (Fig. 10.6). More complex models compute the gradient of water potential between soil and leaves, and take canopy resistance as a function of leaf water potential (NIZINSKI and SAUGIER 1989b). In practice, the result is not very different and good simulations of soil moisture dynamics are obtained by both methods. Fig. 10.7 reports such a simulation, obtained for an oak stand in Fontainebleau (France) with a model using routine meteorological data (HALLDIN et al. 1984). There is a good agreement between simulated and measured values of soil water content in the rooting zone. The drying period is especially well simulated, which suggests that model hypotheses on transpiration reduction by soil water deficit are basically correct. However, modelled values overestimate soil water content during rehydration in October, which may result in a crude treatment of drainage.

10.7 Conclusion

The water balance of forests differs considerably from that of crops or grasslands in two ways:

1. High ventilation rates of forest canopies lead to high evaporation rates of intercepted water, E_i. As a result E_i may reach 30 to 40% of incident precipitation and is a major component of water balance.

2. Canopy transpiration, E_c is lower than crop transpiration due to values of canopy resistance around 100 $s{\cdot}m^{-1}$ compared to 30 $s{\cdot}m^{-1}$ for crops. For closed deciduous forests E_c/E_p (E_p = Penman evaporation) reaches a maximum between 0.7 and 0.8. As a result, total evapotranspiration that includes both E_i and E_c may exceed E_p by more than 20% in rainy places such as Wales (SHUTTLEWORTH, 1989).

In deciduous forests as in annual crops, E_c/E_p increases with leaf area index (L). After L reaches its maximum value, E_c/E_p continues to increase for several weeks in an oak forest, as stomatal conductance increases with leaf maturation (NIZINSKI et al. 1989). This phenomenon has no equivalent in crops and requires more investigation in other species.

Sapflow measurements have become a routine method to estimate water uptake. Hourly values of sapflow density lag behind transpiration measurements (1/2 hr to 1 hr delay) but daily values are quite representative of stand transpiration (GRANIER et al. 1990) and allow precise monitoring of transpiration in isolated trees or open forest as well as in closed stands.

Usually, models of forest water balance provide good estimates of soil water reserve on sites in which soil and vegetation parameters are well known. It should be possible to use them in regional estimates of forest water use when the spatial distribution of these parameters is known. L and maximum available soil water are two key parameters in this regard. L may be estimated from a vegetation index derived from satellite observations (NEMANI and RUNNING 1989). Whether forest transpiration can be estimated from surface temperature (T_s), as in crops or grasslands is still an open question, since variations in T_s are relatively small due to the low aerodynamic resistance of forests. RUNNING and HUNT (1989) showed an interesting relationship between surface temperature and the normalized difference vegetation index (NDVI) from NOAA/AVHRR[1] data; surface temperature increases when vegetation cover decreases, and more so on dry soil than on wet soil. They were able to relate the slope of the T_s/NDVI relationship to a simulated canopy resistance with an encouraging accuracy. Thus, it might thus become feasible in the future to monitor canopy resistance and forest transpiration from satellites.

Bibliography

AUSSENAC, G., BOULANGEAT, C., 1980, Interception des précipitations et évapotranspiration réelle dans des peuplements de feuillus (*Fagus silvatica* L.) et de résineux (*Pseudotsuga menziesii* F.).
Ann. Sc. Forest **37**(2), 91–107.

AUSSENAC, G., GRANIER, A., 1979, Etude bioclimatique d'une forêt feuillue (*Fagus silvatica* L. et *Quercus sessiliflora* Salisb.) de l'est de la France. II.- Etude de l'humidité du sol et de l'évapotranspiration réelle.
Ann. Sc. Forest. **36**(4), 265–280.

BARBIER, J.M.,, 1983, Bilan d'énergie et evapotranspiration d'un couvert d'hévéas (*Hevea brasiliensis*) en Côte d'Ivoire.
Thèse de docteur–ingénieur Université Paris–Sud Orsay 162 pp.

BETREMIEUX, R., 1975, Réflexions à propos de la capacité au champ et de sa mesure: actualisation et réhabilitation de cette notion.
GEPPA 22eme réunion Versailles 41 pp.

BOBAY, V., SAUGIER, B., 1989, Hourly and daily variations of xylem sapflow in sweet chestnut coppices using a thermal measurement method.
Ann. Sci. For. **46** (suppl.), 350s–352s.

[1]NOAA: National Oceanic and Atmospheric Administration. AVHRR: Advanced Very High Resolution Radiometer. The NDVI combines the reflected radiances measured in the visible and near-infrared bands and is linearly related to vegetation cover.

CHASSAGNEUX, P., CHOISNEL, E., 1986, Modélisation de l'évaporation globale d'un couvert forestier. I. Principes physiques et conception du modèle.
Ann. Sci. For. **43**, 505–520.

COHEN, Y., FUCHS, M., GREEN, G.C., 1981, Improvement of the heat pulse method for determining sapflow in trees.
Plant, Cell & Env. **4**, 391–397.

COUCHAT, P., 1974, Mesure neutronique de l'humidité des sols.
Thèse, Université Paul Sabatier, Toulouse. 123 pp. et annexes.

DOLMAN, A.J., 1987, Summer and winter rainfall interception in an oak forest: predictions with an analytical and a numerical simulation model.
Journal of Hydrology **90**, 1–9.

FARDJAH, M., LEMEE, G., 1980, Dynamique comparée de l'eau sous hêtraie et dans des coupes nues ou à *Calamagrostis epigeios* en forêt de Fontainebleau.
Bull. Ecol. **11**, 11–31.

GASH, J.H.C., 1979, An analytical model of rainfall interception by forests.
Quart. J. Roy. Met. Soc. **105**, 43–55.

GRANIER, A., 1985, Une nouvelle méthode pour la mesure du flux de sève brute dans le tronc des arbres.
Ann. Sci. Forest. **42**, 81–88.

GRANIER, A., BOBAY, V., SAUGIER, B., GELPE, J., GASH, J.H.C., SHUTTLEWORTH, W.J., 1990, Vapour flux density and transpiration rate comparisons in a stand of maritime pine (*Pinus pinaster* Ait) in Les Landes forest.
Agric. For. Met. **51**, 309–319.

HALLDIN, S., SAUGIER, B., PONTAILLER, J.Y., 1984, Evapotranspiration of a deciduous forest: simulation using routine meteorological data.
Journal of Hydrology **75**, 323–341.

HORTON, R.E., 1919, Rainfall interception.
Mon. Weath. Rev. **47**, 603–623.

LINDROTH, A., 1984, Gradients distributions and flux profile relations above a rough forest.
Quart. J. R. Met. Soc. **110**, 553–563.

LOUSTAU, D., GRANIER, A., EL HADJ MOUSSA, F., 1990, Evolution saisonnière du flux de sève dans un peuplement de pin maritime.
An. Sci. For. (in press).

MONTENY, B.A. AND BERNOS, C.M., 1985, Determination of the energy exchanges of a forest-type culture: *Hevea brasiliensis*.
In: The forest-atmosphere interaction, D. Reidel, Dordrecht.

MULDER, J.P.M., 1985, Simulating interception loss using standard meteorological data.
In: The forest-atmosphere interaction, D. Reidel, Dordrecht.

NASR, Z., 1987, Une nouvelle technique de pilotage de l'irrigation des arbres: mesure de flux de sève par des méthodes thermiques.
Mémoire de fin d'études INA, Tunis 70 pp.

NEMANI, R.R., RUNNING, S.W., 1989, Testing a theoretical climate–soil–leaf area hydrologic equilibrium of forests using satellite data and ecosystem simulation.
Agric. Forest Meteorol. **44**, 245–260.

NIZINSKI, J., SAUGIER, B., 1988a, Mesures et modélisation de l'interception nette dans une futaie de chênes.
Acta Oecol., Oecol. Plant. **9**, 311–329.

NIZINSKI, J., SAUGIER, B., 1988b, A model of leaf budding and development for a mature *Quercus* forest.
J. Appl. Ecol. **25**, 643–652.

NIZINSKI, J., SAUGIER, B., 1989a, Dynamique de l'eau dans une chênaie (*Quercus petraea* (MATT) LIEBL) en forêt de Fontainebleau.
Ann. Sci. For. **46**, 173–186.

NIZINSKI, J., SAUGIER, B., 1989b, A model of transpiration and soil–water balance for a mature oak forest.
Agric. For. Meteorol. **47**, 1–17.

NIZINSKI, J., MORAND, D., SAUGIER, B., 1989, Variation of stomatal resistance with leaf age in *Quercus petraea*: effect on the soil–water balance of an oak forest.
Ann. Sci. For. **46** (suppl.), 429s–432s.

PIEPER, G.F., 1949, The measurement of moisture content of soil by the slowing of neutrons.
Thesis, Cornell University.

PENMAN, H.L., 1948, Natural evaporation from open water, bare soil and grass.
Proc. Roy. Soc. Series A **193**, 120–145.

PONTAILLER, J.Y., NIZINSKI, J., SAUGIER, B., 1988, Bilan de l'eau et évapotranspiration de forêts feuillues.
In: Calvet, R. (ed.), Etudes sur les transferts d'eau dans le système sol — plante — atmosphère.
INRA Paris, 329–355.

RAUNER, J.L., 1976, Deciduous forests.
In: Vegetation and the atmosphere (2), J.L. Monteith (ed.), Academic Press.

RODE, A.A., 1960, Methods of studying the water regimes of soil.
Izd. Akad. Nauk. SSSR Moscow 244 pp. (in Russian).

RUNNING, S.W., HUNT, E.R., 1989, Contributions of plant capacitance and root–xylem flow resistance to ET calculations: are they required for GCMs?.
In: Proc. of workshop on stomatal resistance, J.C. Price (ed.), Penn. State Univ., 23–24.

RUTTER, A.J., KERSHAW, K.A., ROBINS, P.C., MORTON, A.J. , 1971, A predictive model of rainfall interception in forests. I. Derivation of the model from observations in a plantation of Corsican pine.
Agric. Met. **9**, 367–384.

RUTTER, A.J., MORTON, A.J., ROBINS, P.C., 1975, A predictive model of rainfall interception in forests. II. Generalization of the model and comparison with observations in some coniferous and hardwood stands.
J. Appl. Ecol. **12**, 367–380.

SAKURATANI, T., 1979, A heat balance method for measuring water flow in the stem of intact plant.
J. Agr. Met. **37**, 9–17.

SAUGIER, B., KATERJI, N., 1991, Some plant factors controlling evapotranspiration.
Agric. For. Met. (in press).

SCHNOCK, G., 1970, Le bilan d'eau et ses principales composantes dans une chênaie mélangée calcicole de Haute Belgique (bois de Virelles Blaimont).
Thèse d'état, Université de Bruxelles, 352 pp.

SHUTTLEWORTH, W.J., 1989, Micrometeorology of temperate and tropical forest.
Phil. Trans. R. Soc. Lond. B. **324**, 299–334.

SWANSON, R.H., 1965, Seasonal course of transpiration of Lodgepole pine and Engelmann spruce.
Intern. Sympos. Forest Hydrology. Penn. State Univ. USA, 417–432.

TYREE, M.T., DIXON, M.A., THOMPSON, R.G., 1984, Ultrasonic acoustic emissions from the sapwood of *Thuja occidentalis* measured inside a pressure bomb.
Plant Physiol. **74**, 1046–1049.

Chapter 11

Combination effects of water and salt stress on growth, hydration and pigment composition in wheat (Triticum aestivum L.): A mathematical modelling approach

K. H. KREEB and T. CHEN

11.1 Introduction

One aim in modern ecosystem research is to describe and understand in detail the behavior and adaptive mechanisms of organisms under stress. One has to realize that under natural conditions not just one single ecological factor is acting but a combination of many, although one factor may dominate under certain circumstances. Experimenting inside the naturally complicated network of high complexity seems to be impractical in many cases. One way to learn about the interaction between important ecological stressors is to work firstly with only 2 or 3 variables. A mathematical modelling approach is useful to give an exact description of the results of a special combination test. Only then can the type of relationship be clearly characterized and, in the case of only 2 factors, graphically shown.

Plants react very differently to stress conditions. In wheat, it has been shown that the growth is profoundly influenced by a single application of drought or salt stress (e.g. GORHAM et al. 1985, TERMAAT et al. 1985). Several authors present data supporting the possibility of osmotic adjustment (TERMAAT et al. 1985; see also ACKERSON 1985, TURNER and BEGG 1981). VIRGIN (1965) found that the chlorophyll content was decreased by water stress. There is little information available about combination effects (LÜHMANN et al. 1989, WIETSCHORKE et al. 1990), which is very necessary for ecological interpretation, because it is difficult to extrapolate from results gained from single factor experiments to environmental conditions (LICHTENTHALER 1984, KREEB and WEINMANN–KREEB 1987, LÜHMANN 1987, CHEN 1988).

With regard to the quantitative description and modelling of the reactions of plants to external influence, papers and textbooks have been available for nearly a decade (e.g. HALL and DAY 1977, PALLARDY and KOZLOWSKI 1979, THOMAS 1979). HALL (1982) presented mathematical models of plant water loss and plant water relations. PENNING DE VRIES (1983) gave a review on mathematical models and simulations of simple biological processes. SOLIMAN (1984), KREEB and WEINMANN–KREEB (1987), CHEN and KREEB (1989), LÜHMANN (1987), CHEN (1988) (see also LAYZELL et al. 1988), used the multiple regression analysis for testing the types of combination between various experimental factors. Other applications with ecological relevance include multiple regression equations relating leaf resistance to solar irradiance, air temperature, vapor pressure deficit, and xylem pressure potential (HINCKLEY et al. 1975), plant water loss and plant water relations (HALL 1982), and competitive interference and plant performance (JOLLIFFE 1988). There is no doubt that mathematical models are an objective and efficient instrument for integrating and summarizing experimental data, and they are quite necessary for expressing and testing hypotheses (HALL 1982). It is a general assumption that, regarding the multiple linear regression analysis, all values observed have been generated, in part, by random deviations from a linear relationship. The following study represents the advanced attempts to use in addition the statistical method of non-linear multiple regression for testing combination effects, as has been done already in some papers mentioned above.

The two stress factors, drought and salinity, have been investigated into for their relevance to plant growth, hydration and pigment composition of wheat. Both are related phenomena and represent two of the most important environmental stresses influencing the productivity. Although they are natural phenomena, their impact and severity are intensified by the activities of man, which underlines the importance of interaction research as an ecological base, even for agricultural practice.

11.2 Materials and Methods

11.2.1 Plants, experimental layout, and ecophysiological methods

Wheat plants (*Triticum aestivum* L., GIZA 157) were grown in 20 cm diameter and 18 cm high metallic pots containing sand in a controlled environment chamber described earlier (CHEN 1988). The pots were irrigated twice weekly with nutrient solution and deionized water unless otherwise noted (CHEN 1988). Compared with the control (T0) which obtained 400 ml H_2O, the water stressed plants received less irrigation: T1, T2, T3 were treated with 300, 200 and 100 ml H_2O respectively. The salt stressed plants were arranged in three classes (S1, S2 and S3) with the same nutrient solution, to which 0.1%, 0.5% and 1.0% NaCl (weight/volume) were added. The combination stresses of water and salt corresponded to the combinations of the above mentioned treatments; for example, the plants of the treatment S1T1 were irrigated once a week with 300 ml of deionized water only and once a week with the nutrient solution containing 0.1% NaCl (weight/volume). Each treatment was repeated four times. The ecophysiological parameters were measured weekly, i.e. plant height, the potential osmotic pressure of the cell sap (π^*), as described by KREEB (1990), and the chlorophyll and carotinoid content using the techniques given by ZIEGLER and EGLE (1965) and URBACH et al. (1983).

11.2.2 Statistical and regression analysis

General remarks

For calculating the basic data, means and standard errors, and xy–diagrams, the software "LOTUS–123" was used on PC–XT computers. A t–test was accomplished employing a programme "TMATRIX" (BASIC, KREEB/CHEN). 3d–graphical illustrations were designed by the software "OPEN ACCESS" and the programme "3DDMOD" (BASIC, KREEB/CHEN). The mathematical modelling was realized by two repeated runs of the programmes "MODEEDL" (KREEB/CHEN) and, for controlling parts of our results, by the software "SAS" (SAS INSTITUTE INC. 1987) (Tab. 11.1). Because the non–linear multiple regression analysis (SPIEGEL 1972, HARTUNG 1982) had to be adapted for the analysis of our data, we will have to give some notes on that in the following.

Multiple regression equations

Basics (real linear regression). The fundamental principles involved in the problem of multiple correlation are based on the least square method, by which the best fit of a curve for given experimental data can be found (see e.g. HARTUNG

1982, SAS INSTITUTE INC. 1987, DE PETTER et al. 1987). It is most easily done for linear regression. Practically this enables explanation of the values of a variable Z by means of other variables $X_1, X_2, \ldots, X_n$. For example, it may be thought that the growth of a plant is a function of water supply and salts present, as in this study given. In the case of a simple linear relationship, the following equation is postulated, derivable from a matrix arrangement:

$$Z_{(X_1,\ldots,X_n)} = \beta_0 + \beta_1 X_1 + \beta_2 X_2 +, \ldots, +\beta_n X_n + \epsilon \tag{11.1}$$

where $\beta_0, \beta_1, \ldots, \beta_n$ are constants called regression parameters and ϵ is the random error (HARTUNG 1982).

In the case of three variables only, equation (11.1) can be written more simply as

$$Z_{(X_1,X_2)} = \beta_0 + \beta_1 X_1 + \beta_2 X_2 \tag{11.2}$$

where $\beta_0, \beta_1, \beta_2$ are constants, or bearing in mind the 3–dimensional axis system, as

$$Z_{(X,Y)} = a + bX + cY \tag{11.3}$$

This form (11.3) represents a plane fitting a set of n data points $(X_1, Y_1, Z_1), (X_2, Y_2, Z_2), \ldots, (X_n, Y_n, Z_n)$ in a three dimensional scatter diagram.

The constants a, b, and c may be calculated by the extension of the method of least squares. Thus 3 easily solvable so called normal equations corresponding to the least square plane (11.3) are derived by multiplying both sides by 1, X and Y, respectively, and then summarizing:

$$\begin{array}{lcllll} \sum Z & = & aN & + \; b\sum X & + \; c\sum Y \\ \sum ZX & = & a\sum X & + \; b\sum X^2 & + \; c\sum XY \\ \sum ZY & = & a\sum Y & + \; b\sum XY & + \; c\sum Y^2 \end{array} \tag{11.4}$$

Non real linear relations. Mathematical modelling regarding combination effects has meaning only if not just simple linear relations are considered. But calculations of multiple regressions then become more complicated. Besides the real linear relationship, we should be able to test for a (natural) logarithmic, an exponential, and a power function (real non–linear relation). They are basic ones, which could be easily understood, especially in combinations of 2 (see 3d–graphs later showing planes).

Logarithmic and exponential equations of the type

$$Z = a + b\ln X + cY \tag{11.5}$$

$$Z = a + bX + c\ln Y \tag{11.6}$$

$$Z = a + b\ln X + c\ln Y \tag{11.7}$$

(combination of a linear with a logarithmic regression; note the additive way of combining X and Y; multiplications of X and Y are not solvable) can be simply changed to a real linear form by substitution arrangements, and then solved by linear regression procedure as normal. We get

$$Z = a + bX' + cY \qquad (X' = \ln X) \tag{11.8}$$
$$Z = a + bX + cY' \qquad (Y' = \ln Y) \tag{11.9}$$
$$Z = a + bX' + cY' \qquad (X' = \ln X, Y' = \ln Y) \tag{11.10}$$

resp. for

$$Z = a + b\exp(X) + cY \tag{11.11}$$
$$Z = a + bX + c\exp(Y) \tag{11.12}$$
$$Z = a + c\exp(X) + c\exp(Y) \tag{11.13}$$

similarly as with equations (11.5), (11.6), and (11.7)

$$Z = a + bX' + cY \qquad X' = \exp(X) \tag{11.14}$$
$$Z = a + bX + cY' \qquad Y' = \exp(Y) \tag{11.15}$$
$$Z = a + bX' + cY' \qquad X' = \exp(X),\ Y' = \exp(Y) \tag{11.16}$$

This principle works just as well in the case of other combinations, like "ln with ln", "exp with exp", "ln with exp", and "exp with ln".

For regression combinations, together with the power function (real non-linear relationships; SPIEGEL 1972, KREEB and WEINMANN–KREEB 1987, LAYZELL et al. 1988), we need transformations, i.e. from e.g.

$$Z = aX^bY^c \tag{11.17}$$
$$Z = aX^b\exp(cY) \tag{11.18}$$
$$Z = a\exp(bX)Y^c \tag{11.19}$$
$$Z = a\exp(bX + cY) \tag{11.20}$$

(note the multiplicative form of combining X and Y; additions of X and Y are not solvable, except in 11.20).

To use the linear regression method, these multiple non-linear equations must be transformed into linear expressions like

$$\ln Z = \ln a + b\ln X + c\ln Y \tag{11.21}$$
$$\ln Z = \ln a + b\ln X + cY \tag{11.22}$$
$$\ln Z = \ln a + bX + c\ln Y \tag{11.23}$$
$$\ln Z = \ln a + bX + cY \tag{11.24}$$

Table 11.1: Possible regression equations for the modelling of combination experiments with 2 variables using the least square method, as in linear relations (+ = additive, $*$ = multiplicative combination of X and Y: 13 possibilities). S = additional models only to be done with SAS software (altogether then 32 possibilities).

X \ Y	lin	exp	ln	pow
lin	+S	+S	+S	SS
exp	+S	+$*$	+S	$*$S
ln	+S	+S	+S	SS
pow	SS	S$*$	SS	S$*$

and further for determining the constants a, b and c, as above explained, linearizing the logarithmic expressions, with $Z' = \ln Z$, $a' = \ln a$, $X' = \ln X$, and $Y' = \ln Y$, (equation (11.21) for example) to

$$Z' = a' + bX' + cY' \tag{11.25}$$

which has the same form as the equation (11.3). The constants a', b and c of equation (11.25) are obtained by solving simultaneously the 3 normal equations described earlier. The constants b and c arising as a result can be directly used for equation (11.17). But the constant a' must be re–transformed from $\ln a$ into a again. Since $a' = \ln a$, we find $a = \exp(\ln a)$, which can then be used for the equation (11.17), in order to draw the corresponding graph. The other equations (11.18), (11.19), and (11.20) are to be treated similarly.

The result of these derivations is that we are now able to test our experimental data of combination experiments using 2 variables by 13 models, partly of the multiplicative, partly of the additive type, only by actually applying the least square method, as it is taken for linear regression. The models are found in Tab. 11.1.

Significance test. In order to select the best adapted model, we need a significance test. With regard to SPIEGEL (1972) and HARTUNG (1982) the "coefficient of multiple correlation" is a good measure of the explanatory power of the regression. It is to be calculated as

$$R = \sqrt{\frac{\text{explained variation}}{\text{total variation}}} \tag{11.26}$$

$$= \sqrt{\frac{\sum(Z_{est.i} - Z_m)^2}{\sum(Z_i - Z_m)^2}} \tag{11.27}$$

where $Z_{est.i}$ are the values derived through the regression equation and Z_m is the mean of the experimental data Z_i (HARTUNG 1982). It depends on the precalculation of the "standard error of estimate" (measure of the scatter of the regression of Z on X and Y with properties analogous to those of standard deviation) as

Table 11.2: The 13 mathematical functions derived from the measured data of the combination experiment (CHEN 1988). Independent variables: drought (Y), salt stress (X); dependent variable chlorophyll content of wheat leaves (Z). L = linear regression, l = logarithmic, E = exponential, and P = power function.

L+L:	$Z = 2.01 - 46.35 \cdot X$	$+0.14 \cdot Y$	$R = .897$
P · P:	$Z = 1.70 \cdot X^{0.015}$	$\cdot Y^{0.1455}$	$R = .913$
l+l:	$Z = 1.64 - 0.0777 \cdot \ln X$	$+0.69 \cdot \ln Y$	$R = .908$
E · E:	$Z = 2.00 \cdot \exp(-21.03 \cdot X)$	$\cdot \exp(0.0636 \cdot Y)$	$R = .909$
L+l:	$Z = 2.12 - 46.35 \cdot X$	$+0.69 \cdot \ln Y$	$R = .890$
l+L:	$Z = 1.52 - 0.0777 \cdot \ln X$	$+0.14 \cdot Y$	$R = .915$
P · E:	$Z = 1.61 \cdot X^{-0.015}$	$\cdot \exp(0.0636 \cdot Y)$	$R = .919$
E · P:	$Z = 2.11 \cdot \exp(-21.03 \cdot X)$	$\cdot Y^{0.14}$	$R = .903$
L+E:	$Z = 2.20 - 46.35 \cdot X$	$+0.00753 \cdot \exp(Y)$	$R = .885$
E+L:	$Z = 47.01 - 45.06 \cdot \exp(X)$	$+0.16 \cdot Y$	$R = .893$
l+E:	$Z = 1.72 - 0.0777 \cdot \ln X$	$+0.00753 \cdot \exp(Y)$	$R = .903$
E+l:	$Z = 47.60 - 45.52 \cdot \exp(X)$	$+0.81 \cdot \ln Y$	$R = .884$
E+E:	$Z = 47.75 - 45.54 \cdot \exp(X)$	$+0.00734 \cdot \exp(Y)$	$R = .884$

the best adapted function is P·E

$$S_{Z,XY} = \sqrt{\frac{\sum(Z_i - Z_{est.i})^2}{N}} \tag{11.28}$$

Since the ratio of the explained variation to the total variation is always non–negative, we find that the equation (11.27) can be written, disregarding the sign, as

$$R = \sqrt{1 - \frac{\sum(Z_i - Z_{est.i})^2}{\sum(Z_i - Z_m)^2}} \tag{11.29}$$

From the 13 models the best equation (model) can be selected automatically. It is the one with the greatest correlation coefficient. As an example of the computer output after the run of the programme "MODEEDL", we show the results of the combination effects from drought and salt on chlorophyll content of wheat leaves in Table 11.2 (CHEN 1988).

It can be seen, in this case that the values of the "coefficient of the multiple correlation" vary between 0.884 and 0.919. The best model is of the type "power (X) in combination with exponential function (Y)". The corresponding general equation reads $Z = aX^b \exp(cY)$.

11.3 Results and Discussion

11.3.1 The combination effect on plant height

The height of the plants was measured on the 49th day after sowing (Fig. 11.1). Both kinds of stresses had a strong effect on depressing the growth. The result of an overall t–test indicates significant differences (95% level) to the control, and in many cases to the neighbouring treatments, except for S1, S2, and S1T1. The change runs up to 18% and 20% of the control under pure water (T3) and salt (S3) stress conditions respectively. 28% of the control were noted with the maximum combination stress (S3T3). By comparing all experimental data, one can see that the reduction found at a certain combination stress level does not equal the sum of the single effects of water and salt stress, which would mean a pure additive phenomenon: the observed combination effect is, of course, greater than the corresponding single effects with only water or salt stress, but less than their sum. The result gained in our experiments (Fig. 11.2) differs, therefore, from what could be called "a theoretical additive combination effect". The theoretical curve representing the calculated sums of the single effects, runs its course below the curve combining the experimental data. This may be considered as a "dampening effect", certainly not quite in the sense of antagonism. To estimate the role of the combination more quantitatively, we drew our attention to the mathematical modelling of the findings, attempting to describe the complexity of the combination effects quantitatively with regard to the basic functions as linear, logarithmic, exponential and power (see explanations in "methods"). The derived model, showing the effect of combined drought and salt stress on the height of plants is illustrated together with the measured data in Fig. 11.3. We found that an inclined plane exists, which may indicate that an extreme reaction of the used wheat variety to that stress combination does not occur. Its functional characteristics are noted by the formula

$$Z = 38.2 \exp(-11.8X + 0.05Y) \tag{11.30}$$

which means a combination of an exponential characteristic of X = salt treatment multiplied by an also exponential characteristic of Y = drought handling. The calculated coefficient of multiple correlation ($R = 0.89$) indicates a fairly significant result.

11.3.2 The combination effect on the potential osmotic pressure (π^*)

Fig. 11.4 shows that 49 days after sowing, the potential osmotic pressure of the cell sap (π^*) was progressively raised with increased NaCl concentrations. This effect occurs a lot less under single water stress. Compared with the control, the π^* values increased by 49% and 25% under S3 and T3 conditions respectively, but by as much as 112% under the highest stress combination (S3T3) . The systematical changes of the π^* values by the other combinations are shown in the centre of the diagram (Fig. 11.4).

Compared with the curve of theoretical pure addition of the single effects, the observed combination effect on π^*, illustrated in Fig. 11.5, is considerably greater. For the wheat variety investigated, the combinations of drought and salt stress regarding π^* acted clearly towards a synergistic effect.

It is widely recognized that plants have abilities to change their potential osmotic pressure in response to slowly developing water deficits (ZIEGLER and EGLE 1965, TERMAAT et al. 1985). A range of physiological mechanisms enables the plants to adapt to stress. Osmotic adjustment allows the turgor pressure to be maintained as the water content decreases (TURNER and BEGG 1981). One may note that the π^* values of S1 and S2 are significantly higher than those of the control, but the plant height is not (see Fig. 11.1 and 11.3).

The osmotic adjustment, due to high NaCl–concentrations in the nutrient solutions used in this experiment, is remarkable (see Fig. 11.5 and 11.6) and it depends on an active hydrature change by an intensive uptake of Cl^-. A basic understanding of the osmotic adjustment of the plants as a pure physiological effect is well known (TURNER and BEGG 1981, ACKERSON 1985), unlike the complex network interaction necessary for the interpretation of combined effects.

As a result of these treatments (Fig. 11.6), we are able to characterize the combination stress effect on the π^* values in a quantitative manner by searching for the best fitting mathematical model. It corresponds to a fairly complicated combination of a power function for X (salt), indicating a strong non–linear relationship, and an exponential function (Y = drought). The multiplicative property, combining both expressions, underlines the strong effect of the used combination, regarding cell physiological aspects of water relations. The formula appears as

$$Z = 24.75X^{0.03}\exp(-0.102Y) \tag{11.31}$$

The high significance is given by $R = 0.90$. The present experiments demonstrate that NaCl exerts an effect on π^* in response to stress. This enables the plants to maintain leaf expansion under water stress and photosynthetic activity at levels of stress which are not possible in the absence of NaCl (TERMAAT et al. 1985).

11.3.3 The combination effect on chlorophyll content

In addition, the chlorophyll content has been measured from samples which were taken on the 49th day after sowing. As expected, we do find a negative correlation to the applied single stresses and to the combination stress. The experimental data in Fig. 11.7 to 11.9 are presented as previously. As with plant height, the heaviest stress combination S3T3 caused the greatest reduction of chlorophyll content compared with the control.

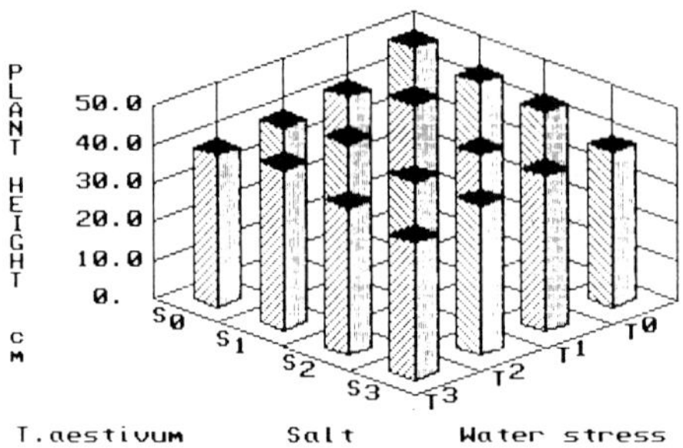

Figure 11.1: The combination effect on plant height at the 49th day after sowing. S salt stress: 0, 1, 2, 3 NaCl concentrations with 0.0%, 0.1%, 0.5%, 1.0% NaCl (weight/volume) in the nutrient solution; T water stress: 0, 1, 2, 3 volume of weekly irrigated water (400, 300, 200 and 100 ml).

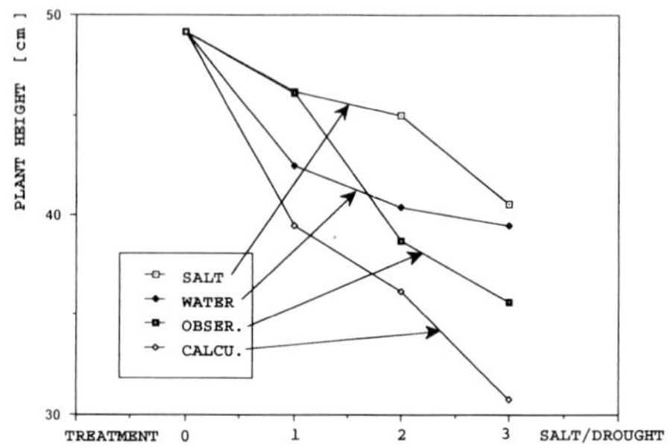

Figure 11.2: A comparison of the observed (OBSER.) and of the calculated (CALCU.) combination effect on plant height according to Fig. 11.1.

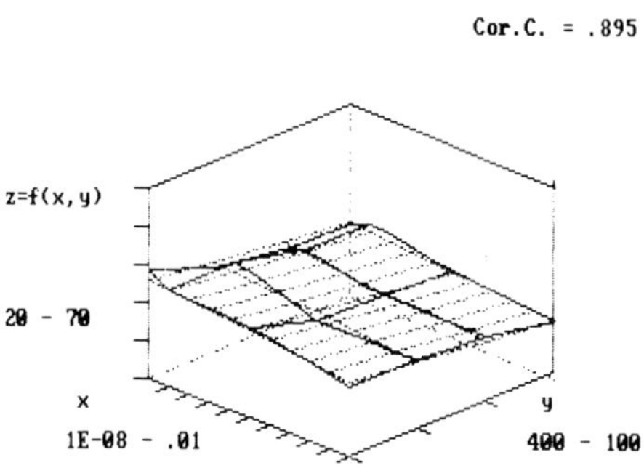

Figure 11.3: Function plot of the combination effect on plant height. Dotted: model; solid: measured data (49 days after sowing). Axes: x concentration of NaCl (weight/volume); y irrigated water (ml); z plant height (cm).

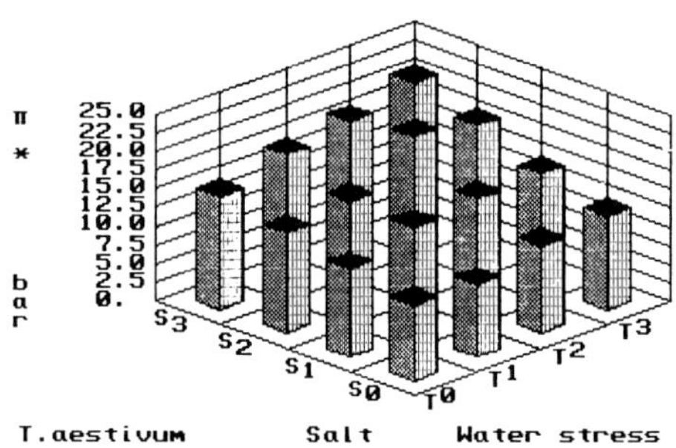

Figure 11.4: The combination effect on potential osmotic pressure of the cell sap (π^*) measured on the 49th day after sowing (abbreviations as in Fig. 11.1).

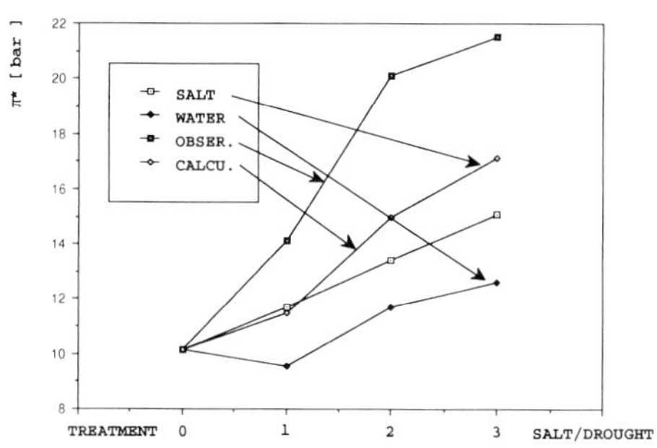

Figure 11.5: A comparison of the observed (OBSER.) and of the calculated (CALCU.) combination effect on potential osmotic pressure of the cell sap (π^*) according to Fig. 11.4.

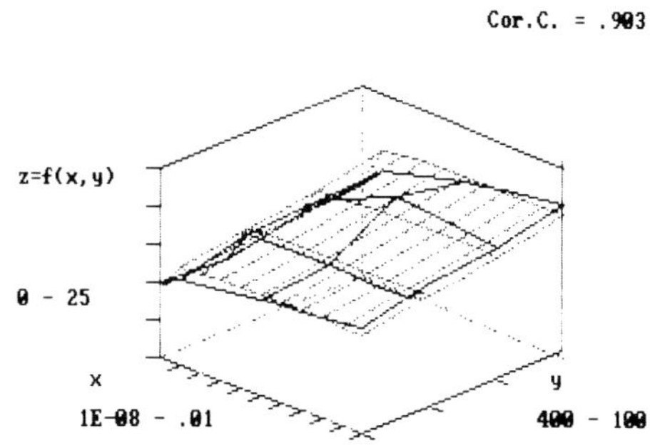

Figure 11.6: The model display of the combination effect on potential osmotic pressure of the cell sap (π^*). The dotted line corresponds to the mathematical model and the solid line to the measured data (49 days after sowing). Axes: x, y as in Fig. 11.3; z potential osmotic pressure (bar).

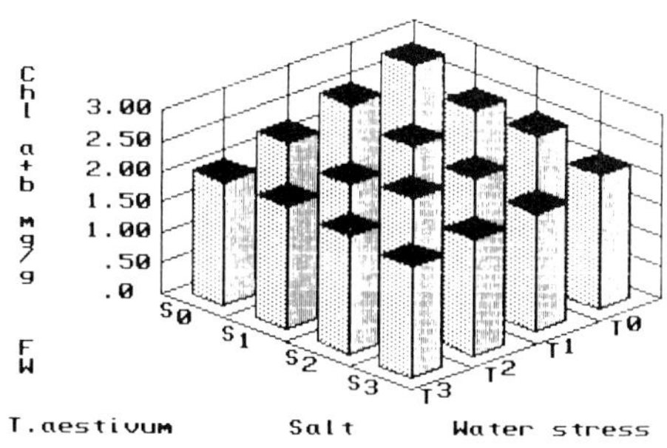

Figure 11.7: The combination effect on chlorophyll measured on the 49th day after sowing (abbreviations as in Fig. 11.1).

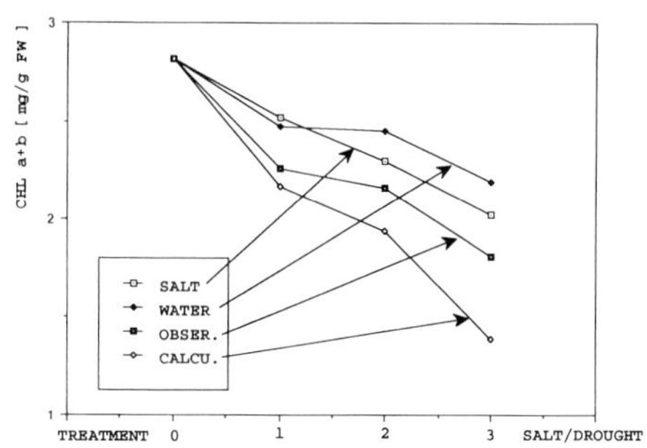

Figure 11.8: A comparison of the observed (OBSER.) and of the calculated (CALCU.) combination effect on chlorophyll according to Fig. 11.7.

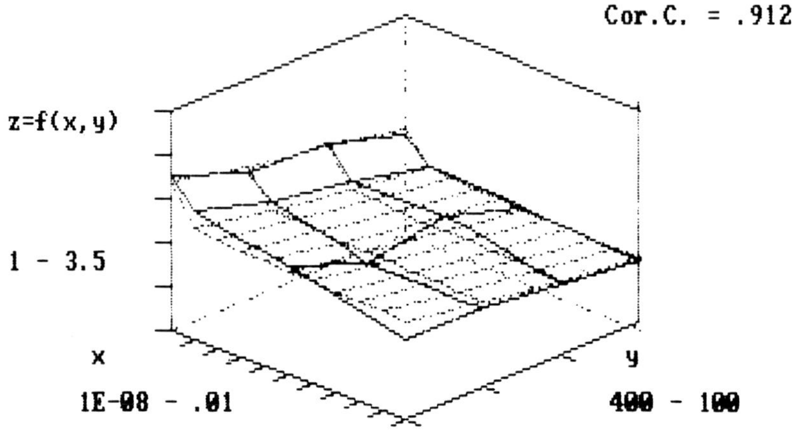

Figure 11.9: The model description of the combination effect on chlorophyll. The dotted line corresponds to the mathematical model and the solid line to the measured data (49 days after sowing). Axes: x, y as in Fig. 11.3; z chlorophyll content (mg·g^{-1} fresh weight).

From Fig. 11.8 it can be seen, that the observed combination effect is less than the theoretical additive effect calculated from the single stress influences on chlorophyll. This corresponds with the result found regarding the height of plants.

The multiple regression analysis leads to the model (Fig. 11.9)

$$Z = 1.61X^{-0.015}\exp(0.063Y) \tag{11.32}$$

which is highly significant ($R = 0.91$). It matches a non–linear multiple function, combined of a power function with a fractional exponent in case of X (salt stress) and an exponential function for Y (drought stress), both connected by multiplication. The carotinoid content, which was also investigated, gave a similar result.

Based on the present experiments and those reported earlier (VIRGIN 1965, BOURQUE and NAYLOR 1971, BENGTSON et al. 1978), the following explanation can be offered to account for the changes in chlorophyll content under the stress conditions. The reduction of the chlorophyll content must have been caused by degradation of chlorophyll and/or inhibition of chlorophyll biosynthesis (DUYSEN and FREEMAN 1974, FREEMAN and DUYSEN 1975). However, it is difficult, without more detailed biochemical studies, to decide, whether the measured decrease in chlorophyll content is due to a decrease in chlorophyll biosynthesis or to a breakdown in chlorophyll. The effect of water stress on chlorophyll formation was first shown by VIRGIN (1965). Nevertheless, due to the very slow rate of chlorophyll biosynthesis (GRONAU 1983), one can assume that the sharp decrease in chlorophyll content, observed in the stressed plants, is a result of a degradation process rather than an inhibition of the synthesis.

11.4 Conclusion

From these results, it can be concluded, that the combined application of stress leads to significant and distinct effects, which are more striking than found with single stresses. To obtain a comprehensible interpretation, a mathematical model is recommended to examine the experimental data. But we believe that this has its limits. It seems, of course, important to test various models, and especially non-linear ones, because the strength of the effects found with different stresses on dissimilar ecophysiological parameters of ecological relevance diverges. If a combination effect is very complicated, the modelling could be done best with the aim of getting the best adaption by polynome expressions possible. SAS software is applicable in this respect. Although the plain of data and the function plot plain in 3-dimensional graph may nearly coincide, the calculated formula does not mean much any more. A comparative analysis showed that models we have applied meet most of the cases found in combination tests up until now, i.e. the use of the logarithmic, the exponential, and the power function as well as linear regression. The suggested model palette includes also the additive and multiplicative version of the combination.

Of course, combination experiments with only 2 factors do not tell us too much about the ecological network interaction. But they are the first and most important step to overcome the monocausal view. Our results (see also SOLIMAN 1984, LÜHMANN et al. 1989, CHEN and KREEB 1989) suggest, that one cannot expect the dominance of a particular model; variations in respect to ecophysiological parameters and the type of stress combination are more often found in practice. It also seems very important to be aware of the fact, that in some cases, we do not have an addition of the single effects, but rather a "dampening effect". Most probably that may not be an antagonistic one in the real meaning of the term, although that could also be of importance.

Another very useful purpose of mathematical modelling with regard to combination experiments, is to obtain arguments for atypical data. LÜHMANN et al. (1989) found e.g. that the uniformity of parameter values (in vivo fluorescence of chlorophyll) in the mid-centre of the experimental area of combinations (Ustinex application — containing Diuron which acts as a photosynthesis blocker — in combination with Pb) can be explained as a real antagonistic phenomenon. Whereas Ustinex diminishes the photosynthesis by blocking the electron transport of the primary electron acceptor Q, and therefore increases the variable fluorescence, Pb, on the contrary, inhibits the electron transport on the donor side of PS2, and this causes a decrease in fluorescence yield.

We can finally conclude that, especially with experiments of stress combinations, the mathematical modelling is a very useful tool to understand and to describe the combination effects quantitatively. Although various models may be acceptable, they are not all predictable. A mathematical model, applied with caution, may be used for the purpose of predicting effects. In any case, it gives exact mathematical information regarding the type of combination effect. The 13 applied models already represent a wide variety of possible regression types for combining 2 factors. Thus the analyzing of data takes only minutes, whereas the also tested complete set of 32 models with SAS (see Tab. 11.1) requires about 2 h for calculation. Nevertheless, it is of interest for expanded combination experiments to include in the regression combinations the square function (being tested) as well, and even trigonometrical functions. For advancing the models to combinations of 3 or more factors (being tested), special software will have to be used.

For further analysis of complex systems, we need, of course, a better understanding of the various static and dynamic features of morphological and physiological adaptive properties, that may enable plants to become more resistant to stress.

Comparing the results of the combination effect of this study on the growth, pigment content, and hydration of wheat, only the hydration shows to have been heavily influenced. However, there are significant and typical effects found also regarding the growth and the chlorophyll content, which effects the rate of net photosynthesis and finally the growth of the plants. Pure linear regressions are not relevant in these experiments. Yet it is of interest that, disregarding the models, drought and salt stress combining had a synergistic effect on π^* but a "dampening effect" on the chlorophyll content and growth in wheat.

Bibliography

ACKERSON, R. C., 1985, Osmoregulation in cotton in response to water stress. III. Effects of phosphorus fertility.
Plant Physiol. **77**, 309–312.

BENGTSON, C.; KLOCKARE, B.; KLOCKARE, R.; LARSSON, S.; SUNDQUIST, C., 1978, The after-effect of water stress on chlorophyll formation during greening and the leaves of abscisic acid and proline in dark grown wheat seedlings.
Physiol. Plant. **43**, 205–212.

BOURQUE, D. P.; NAYLOR, A. W., 1971, Large effects of small water deficits on chlorophyll accumulation and ribonucleic acid synthesis in etiolated leaves of jack bean (*Canavalia ensiformis L. DC.*).
Plant Physiol. **47**, 591–594.

CHEN, T., 1988, Natürlich–anthropogene Kombinationseffekte bei höheren Pflanzen auf Wachstum, Hydratur und Pigmentzusammensetzung unter besonderer Berücksichtigung einer mathematischen Modellierung.
Diss. Universität Bremen.

CHEN, T.; KREEB, K. H., 1989, Combined effects of drought and salt stress on growth, hydration and pigment composition in *Gossypium hirsutum L.*
In: Kreeb K. H.; Richter, H.; Hinckley, T. M. (Ed.), Structure and functional

responses to environmental stresses, Water shortage, SPB Academic Publishing, The Hague, 165–177.

De Petter, E.; van Wiemeersch, L.; Dedonder, A.; Rethy, R.; Fredericq, H.; de Greef, J. A., 1987, Mathematical approach to effects of repeated treatments in the study of very low fluence, low fluence and high fluence germination responses.
Physiol. Plant. **72**, 36–41.

Duysen, M. E.; Freeman, T. P., 1974, Effects of moderate water deficit (stress) on wheat seedling growth and plastid pigment development.
Physiol. Plant. **31**, 262–266.

Freeman, T. P.; Duysen, M. E., 1975, The effect of imposed water stress on the development and ultrastructure of wheat chloroplasts.
Protoplasma **83**, 131–145.

Gorham, J.; Jones, R. G. W.; McDonnel, E., 1985, Some mechanisms of salt tolerance in crop plants.
Plant Soil **89**, 15–40.

Gronau, G., 1983, Untersuchungen zur Chlorophyll–Biosynthese in etiolierten Haferkeimlingen, in vitro Chlorophyll–Synthetase Test, Kinetik und Aufreinigung der Chlorophyll–Synthetase.
Diss. Universität München.

Hall, A. E., 1982, Mathematical models of plant water loss and plant water relations.
In: Encyclopedia of plant physiology, 12B, Springer, Berlin, Heidelberg, New York, 231–261.

Hall, C. A. S.; Day, J. W., 1977, Ecosystem modelling in theory and practice.
John Wiley & Sons, New York, London, Sydney, Toronto.

Hartung, J., 1982, Statistik, Lehr– u. Handbuch d. angewandten Statistik.
Oldenbourg GmbH, München.

Hinckley, T. M.; Schroeder, M. O.; Roberts, J. E.; Bruckerhoff, D. N., 1975, Effect of several environmental variables and xylem pressure potential on leaf surface resistance in white oak.
Forest Sci. **21**, 201–211.

Jolliffe, P. A., 1988, Evaluating the effects of competitive interference on plant performance.
J. theor. Biol. **130**, 447–459.

Kreeb, K. H., 1990, Methoden zur Pflanzenökologie und Bioindikation.
Fischer, Jena, Stuttgart, New York, (vgl. 1. Auflage 1977).

KREEB, K. H.; WEINMANN-KREEB, R., 1987, Pflanzliche Bioindikatoren als streßintegrierende Systeme, Analyse, mathematische Modellierung und Bedeutung für die Erstellung von Schadenskatastern.
VDI Berichte Nr. **609**, 217–248.

LAYZELL, D. B.; GAITO, S. T.; HUNT, S., 1988, Model of gas exchange and diffusion in legume nodules.
Planta **173**, 117–127.

LICHTENTHALER, H. K., 1984, Luftschadstoffe als Auslöser des Baumsterbens.
Naturwiss. Rdsch. **37**, 271–277.

LÜHMANN, H.-J., 1987, Toleranzmechanismen von Flechten gegenüber natürlicher und anthropogener Belastung, in vivo — Chloroplastenfluoreszenz als charakterisierendes Kriterium und Grundlage für ein Streß-Reaktionsmodell.
Diss. Universität Bremen.

LÜHMANN, H.-J.; WIETSCHORKE, G.; KREEB, K. H., 1989, Influences of combined temperature, lead and herbicide stresses on lichen fluorescence and their mathematical modelling.
Photosynthetica **23**, 71–76.

PALLARDY, S. G.; KOZLOWSKI, T. T., 1979, Relationship of leaf diffusion resistance of *Populus* clones to leaf water potential and environment.
Oecologia **40**, 371–380.

PENNING DE VRIES, F. W. T., 1983, Modelling of growth and production.
In: Encyclopedia of plant physiology, **12**D, Springer, Berlin, Heidelberg, New York, 118–150.

SAS INSTITUTE INC., 1987, SAS/STATTM Guide for Personal Computers.
Version **6** Edition, Cary, NC.

SOLIMAN, S. A. F., 1984, Some aspects of drought resistance & salt tolerance of plants.
PH.D. Thesis, Cairo University (Egypt) and University of Bremen (W-Germany).

SPIEGEL, M. R., 1972, Statistics.
Mcgraw Hill Book Company, New York.

TERMAAT, A.; PASSIOURA, J. B.; MUNNS, R., 1985, Shoot turgor does not limit shoot growth of NaCl-affected wheat and barley.
Plant Physiol. **77**, 869–872.

THOMAS, B., 1979, Theorie torgesteuerter Entwicklungsprozesse.
Diss. Universität Köln.

TURNER, N. C.; BEGG, J. E., 1981, Plant-water relations and adaptation to stress.
Plant Soil **58**, 97–131.

URBACH, W.; RUPP, W.; STURM, H., 1983, Praktikum zur Stoffwechselphysiologie der Pflanzen.
Georg Thieme, Stuttgart, New York.

VIRGIN, H. I., 1965, Chlorophyll formation and water deficit.
Physiol. Plant. **18**, 994–1000.

WIETSCHORKE, G. M.; LÜHMANN, H.-J.; KREEB, K. H., 1990, Single and combined effects of lead, herbicide and temperature on net photosynthetic rate of *Hypogymnia physodes*, a corticolous lichen.
Photosynthetica **24**, 102–109.

ZIEGLER, R.; EGLE, K., 1965, Zur quantitative Analyse der Chloroplastenpigmente. I. Kritische Überprüfung der spektralphotometrischen Chlorophyll-Bestimmung.
Beitr. Biol. Pflanzen **41**, 11–37.

Chapter 12

Ecosystem research on grassland in the Austrian Alps and in the Central Caucasus

A. CERNUSCA

12.1 Introduction

As part of the Austrian contribution to the UNESCO research programme "Man and Biosphere", intensive scientific investigations were carried out on the structure and mode of functioning of alpine grassland ecosystems in the "Hohe Tauern" National Park. Fig. 12.1 summarizes the ecosystem studies carried out from 1975 onwards. The two ecosystem studies "Alpine grass heath" and "Altitudinal transect" resulted in important insights into the ecology of various grassland ecosystems along an altitudinal transect from 1000 to 2500 m above sea level on the southern slope of the Großglockner high alpine road. The ecosystem studies "Abandoned pasture" and "Skiing area Schloßalm/Stubnerkogel" were carried out in the Gastein valley. These two projects yielded valuable information concerning the ecological effects of the increasing disuse of the alpine pasture regions by farmers and concerning the ecological damage caused by construction and use of ski runs. The cooperative project "Caucasus" resulted in long-term studies in the Alps and in the Central Caucasus concerning the ecological effects of intensive sheep grazing on mountainous ecosystems.

The results of these ecosystem studies provided a valuable basis for decisions in connection with the ecological consequences of game protection (e. g. restriction of hunting, farming and forestry) as well as with regard to the effects of anthropogenic measures (mining, road-building, harnessing of energy, tourism). Of the many investigations carried out, it should be specially mentioned that for the first time the altitudinal variations in canopy structure, microclimate, energy balance, water regime, plant productivity and soil-microbiology were studied simultaneously along an altitudinal transect ranging from the valley to mountain tops. The present article is a summarizing report of the most important results of the investigations on grassland ecosystems in the Austrian Alps and in the Caucasus. Further informations are given in CERNUSCA (ed. 1989).

12.2 Scope of the studies and methods

Under the direction of the author, an interdisciplinary collaboration between meteorologists, soil scientists, geologists, hydrologists, botanists, zoologists, microbiologists, hygiene specialists, as well as agriculturalists and forestry colleagues was realized (Tab. 12.1).

Macrometeorology: The macrometeorological data were obtained from 14 stations set up for the MAB programme in the Hohe Tauern by Emmerich WEISS (Austrian Academy of Science), in whose charge they remained for the duration of the programme.

Microclimate, Energy budget: Evapotranspiration, sensible heat convection and CO_2-flux between the atmosphere and the vegetated surface were calculated by means of the Bowen-ratio method, using gradients of temperature, water vapor and CO_2-concentration measured between the canopy and a height 120 cm above it (CERNUSCA 1982). Soil heat flux was measured using heat flux plates. For these studies a number of new measuring instruments were developed, including an apparatus for the automatic electrical registration of the water level in an evaporation pan; a recording system, controlled by a pocket computer, for the automatic registration of air temperature, air humidity and CO_2 concentration in different canopy layers; and a variety of battery-driven automatic climate stations. In the youngest generation of these climate stations, Type MIKROMET, all data analyses, including energy balance calculations and mathematical simulation model calculations, can be carried out in the field, under "expedition" conditions. Additionally, the measurements can immediately be expressed graphically by means of a small built-

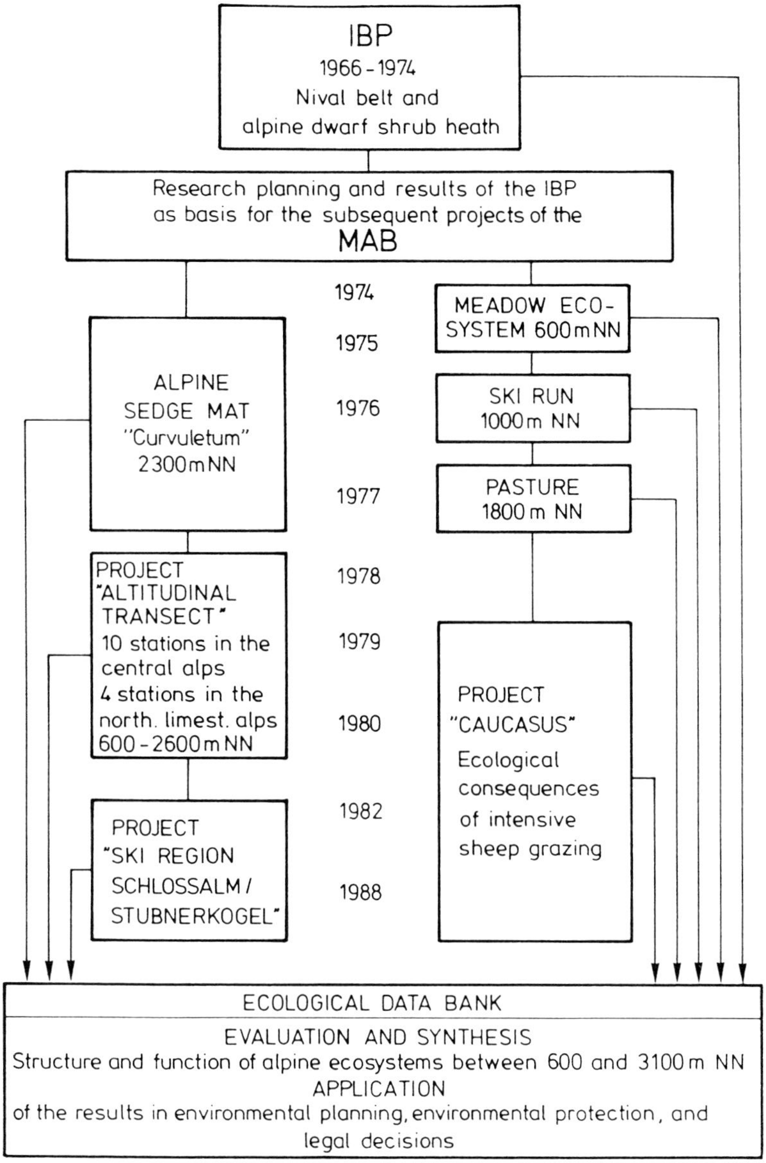

Figure 12.1: Chronological plan of the ecosystem studies carried out in the context of the MaB–program "Hohe Tauern". The present two volumes of the Publications of the Austrian MaB-Program now provide a complete report on the results of the projects "Alpine sedge mat Curvuletum" and "Altitudinal transect".

Table 12.1: Parameters investigated

1. Meteorology Macrometeorology Micrometeorology and energy budget 2. Hydrology Water budget of vegetation and soil Surface run–off and soil erosion 3. Botany Canopy structure Phytomass, primary production Fodder value, digestibility Energy content, minerals Litter production and breakdown Photosynthesis, canopy respiration Stress ecology of the vegetation (ski–slopes, cattle and sheep tracks)	4. Soil biology Soil respiration Soil microbiology Epigaeic arthropods as bioindicators 5. Environmental hygiene Heavy metals and organic waste 6. Data acquisition and modelling Ecological data bank Mathematical system models Development of methods for ecosystem research

in plotter (CERNUSCA 1985, 1987). The application of MIKROMET in measuring microclimate is shown schematically in Fig. 12.2.

Primary production, CO_2 exchange: An important section of the project was devoted to primary production analyses using the harvesting method. This involved the measurement of biomass and necromass throughout the year. The canopy structure of each stand was analysed by the stratified clipping method. Total phytomass was divided into phanerogams, cryptogams and dead standing plant matter. Subsequently, the phanerogams were separated per species and per layer into stems, leaves and inflorescences. The area of all plant parts was determined with a leaf area meter (LI–3100, LI–COR Corp., Lincoln USA). Dry weights were determined for all fractions. Field measurements of leaf and stem inclination of the dominant species were also made using a hand inclinometer. The energy content of the plant samples were determined with an adiabatic calorimeter. For the analysis of the carbon regime of the individual ecosystems in situ measurements of photosynthesis and respiration were carried out. The use of a fully climatized, artificially illuminated cuvette greatly reduced the time required to obtain factor–dependent curves for CO_2 gaseous exchange. CO_2–release from the soil was analysed by means of the cuvette method (open system). The streams of sample gas were controlled by MIKROMET. Further information on the breakdown processes in the ecosystems and the availability of the resulting minerals for the plants was obtained by means

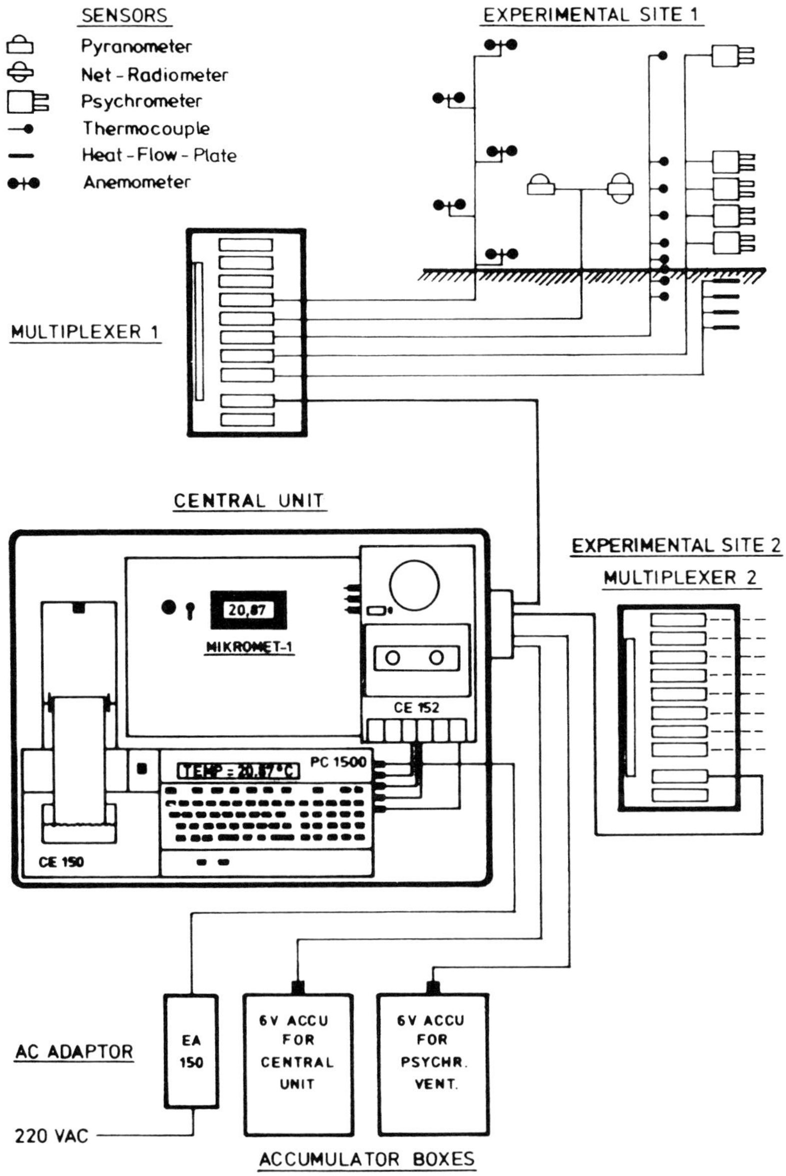

Figure 12.2: Diagram of instrumentation of the MIKROMET used for analyzing CO_2–exchange and the energy balance at two sites by means of the Bowen ratio method.

of microbiological soil analyses.

Hydrology: An important area of the ecosystem studies was devoted to the water balance of individual plants and the ecosystem as a whole. These investigations were carried out using diffusion porometers to measure stomatal opening, the Scholander pressure apparatus to record the water potential of plants, tensiometers for soil water potential measurements, and lysimeters for the continuous recording of evapotranspiration throughout the day and over the course of the year. Here, too, a number of new measuring procedures were devised and successfully put into practice. These included a new type of diffusion porometer, an electrical recording device for tensiometers, and an automatic, electrically-recording lysimeter. These instruments make possible the continuous electrical recording of soil water potential and of evapotranspiration, and thus provide us with the data necessary for calculating mathematical models of the water budget. An important contribution to all of these ecosystem studies was made by the Bavarian Water Authority (Bayerisches Landesamt für Wasserwirtschaft): surface run-off- and erosion experiments carried out by Günther BUNZA using a mobile rain machine proved to be an invaluable supplement to the stationary surface run-off measurements. These experiments provided the basis for an exact analysis of soil erosion caused by catastrophic rain under a wide variety of local conditions.

Data acquisition and modelling: Particular importance was attached to the storage of all measurements in an ecological data bank which at the present time contains 6 million measurements. Additionally, the data bank system contains a number of programmes for evaluation and simulation that not only make possible exhaustive analyses of the stored data but also permit the designing of mathematical ecosystem models based on the entire body of data. Models for soil temperature and soil water content, for the distribution of radiation in a plant canopy, for leaf temperature and canopy photosynthesis are already available.

12.3 Sites investigated

The investigation took the form of an integrated ecosystem analysis of selected sites along an altitudinal transect from 1085 m to 2528 m a. s. l. on the southern slopes of the Hohe Tauern mountain range (Fig. 12.3, Tab. 12.2):

Experimental site "Seppenbauer" (47° 2' 43" N, 12° 15' 22" E; 1612m), is a two-crop hay meadow located above Heiligenblut (Carinthia). The dominant plant species are *Dactylis glomerata* L. and *Trisetum flavescens* (L.) P. B., *Alchemilla vulgaris* L. and *Plantago lanceolata* L.

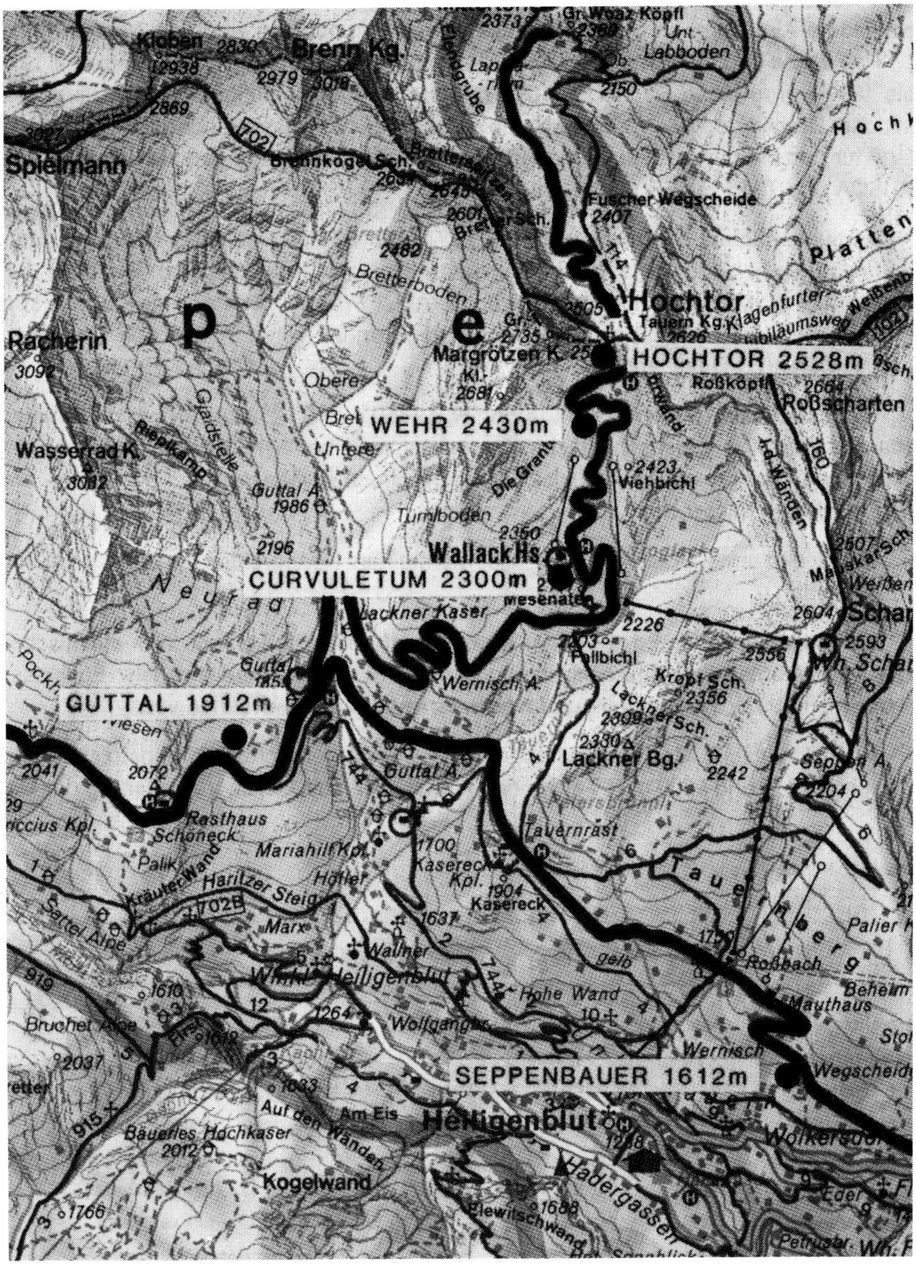

Figure 12.3: Location of the sites of the Austrian MaB–6–High mountain research project "Hohe Tauern" along the Großglockner–high alpine road in the Hohe Tauern range.

Table 12.2: Overview of the most important parameters for the sites investigated.

	Seppen-bauer	Guttal	Seslerietum	Curvuletum	Hochtor Süd
Altitude [m]	1612	1912	2280	2300	2528
Slope	25°	5°	42°	10–15°	10–15°
Exposure	S	SSE–S	E–SE	SW–S	S
Vegetation type	meadow	pasture	sedge mat	sedge mat	cushion plants
Plant cover [%]	100	100	90	95	variable
Canopy height [cm]	100	45	40	14	variable
Canopy (layers)	3	2	2	2	variable
Rooting depth [cm]	25–30	45–50	60–80	85	18–20
Soil type	Para-rendzina	Braun-erde	alpine Pseudogley	alpine Pseudogley	alpine Pseudogley

Experimental site "Guttal" (47° 3' 35" N, 12° 48' 46" E; 1912 m) situated at the timberline, is an alpine pasture grazed by cows. The dominant plant species are *Nardus stricta* L., *Poa alpina* L. and *Alchemilla vulgaris* L.

Experimental site "Curvuletum" (47° 4' 13" N, 12° 50' 38" E; 2300 m), an alpine sedge mat with an abundance of lichens, thinly grazed by sheep. Total plant cover 95%, dominated by *Carex curvula* ALL., *Primula minima* and *Cetraria* spp. On wind-swept ridges, cushions of *Loiseleuria procumbens* (L) Desv.

Experimental site "Hochtor" (47° 4' 56" N, 12° 50' 38" E; 2300 m), at the cushion plant level. The average total cover is 60%; small clumps of *Festuca pumila* develop locally, depending on topography, as well as cushions of *Salix* and *Silene acaulis*.

12.4 Structure and function of alpine grassland ecosystems

This section is intended as a summary of knowledge on the basic structure and function of grassland ecosystems in the Hohe Tauern range, based on the results obtained in the course of the MaB–6–projects "Alpine grass heath" and "Altitudinal transect".

12.4.1 Altitudinal variation of water budget

Evapotranspiration: Using weighing lysimeters, KÖRNER, WIESER and CERNUSCA (1989) measured the diurnal and annual course of evapotranspiration along the altitudinal transect. As shown in Fig. 12.4, the annual total evaporation decreased from 450 mm at 1500 m to 200 mm at 2500 m. The evapotranspiration data obtained from each experimental site were also of value in testing the validity of formulae used for calculating daily and annual evaporation.

Water balance: It could be shown, that with increasing altitude the contribution of evapotranspiration to water losses decreased, whereas run–off became increasingly important (Fig. 12.5): The total annual precipitation rose from 830 $mm \cdot yr^{-1}$ at 1085 m a. s. l. to 1670 $mm \cdot yr^{-1}$ at 2528 m, evapotranspiration dropped from 550 to 210 $mm \cdot yr^{-1}$, and run–off rose from 280 to 1460 $mm \cdot yr^{-1}$. Although this principle has long been recognized, the investigations reported here provided the necessary confirmation for the Hohe Tauern National Park. Furthermore, the scale of the investigation was such that an exact analysis of evaporation and run–off could be made for each ecosystem studied. It is thus now possible, for the first time, to offer an ecologically well–founded judgement on measures resulting in changes in the landscape. This can be illustrated by the case of the dwarf sedge mats in the alpine grass heath, shown in Fig. 12.6. It appears that an annual average of only 7.8% of the annual precipitation leave the ecosystem again via the transpiration of green leaves, and that 60% of the annual precipitation are lost as surface run–off: 57.4% of it as melting snow, over frozen ground, and only 2.5% during the snow–free period. From these results it can be concluded that measures solely affecting the transpiring leaf surface, such as different forms of farming for example, have only a very slight effect on the annual total run–off. Measures that have a direct effect on run–off are a much more serious matter, and become increasingly serious at higher altitude.

Further evaluation of the water budget data showed that the decrease in the annual mean value for evapotranspiration with increasing altitude was chiefly due to the shorter snow–free season at higher altitudes. During the growing season almost identical quantities of water evaporated from valley and mountain ecosystems. Fig. 12.7 shows the altitudinal variation of mean daily sums of evapotranspiration on bright days between July and September at various elevations in the Austrian Alps

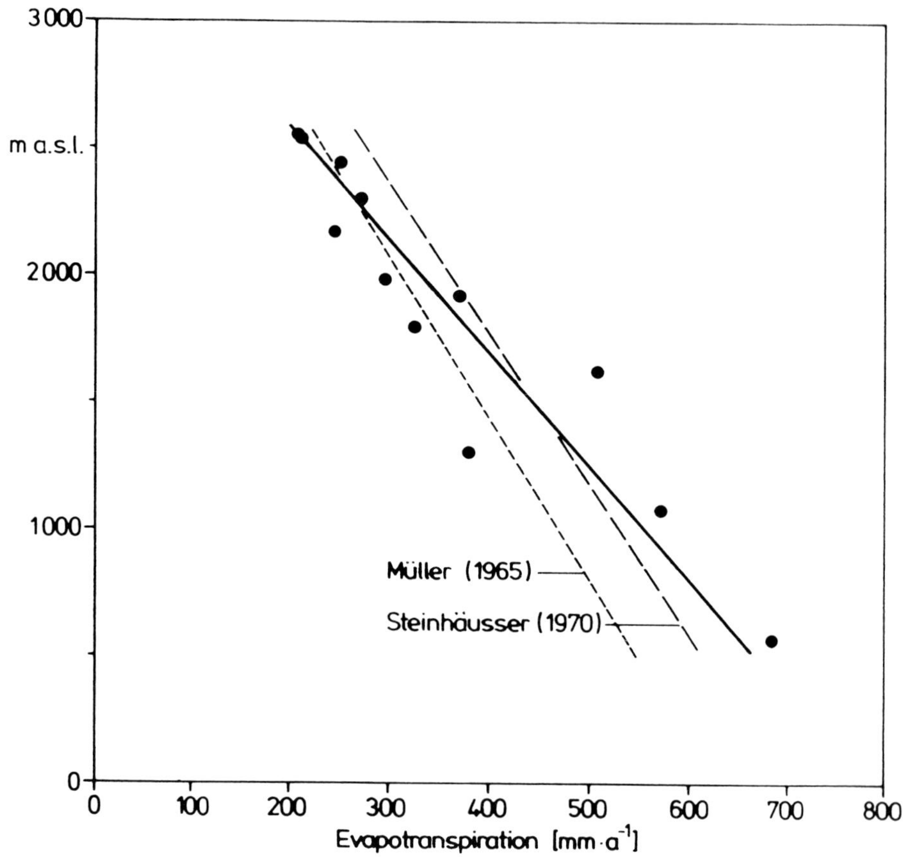

Figure 12.4: Mean annual evapotranspiration of different grassland ecosystems along an altitudinal transect between 1612 m and 2528 m altitude in the Hohe Tauern range (KÖRNER et al. 1989).

and in the Central Caucasus. Average values range from 4.5 mm·d^{-1} in grassland in the valley floor at 600 m to 3.7 mm·d^{-1} in alpine sedge mats at 2550 m.

12.4.2 Altitudinal variation of canopy structure, energy balance and microclimate

Canopy structure: Analyses of canopy structure made by CERNUSCA and SEEBER (1989) revealed that the canopy height decreased along the altitudinal transect, from 90 cm in the hay meadow at 1612 m, to 12 cm on the alpine sedge mats at 2300 m (Fig. 12.8). The above-ground phytomass increased with altitude as follows: 420 g·m^{-2} DM (hay meadow); 534 g·m^{-2} DM(alpine pasture); and 786

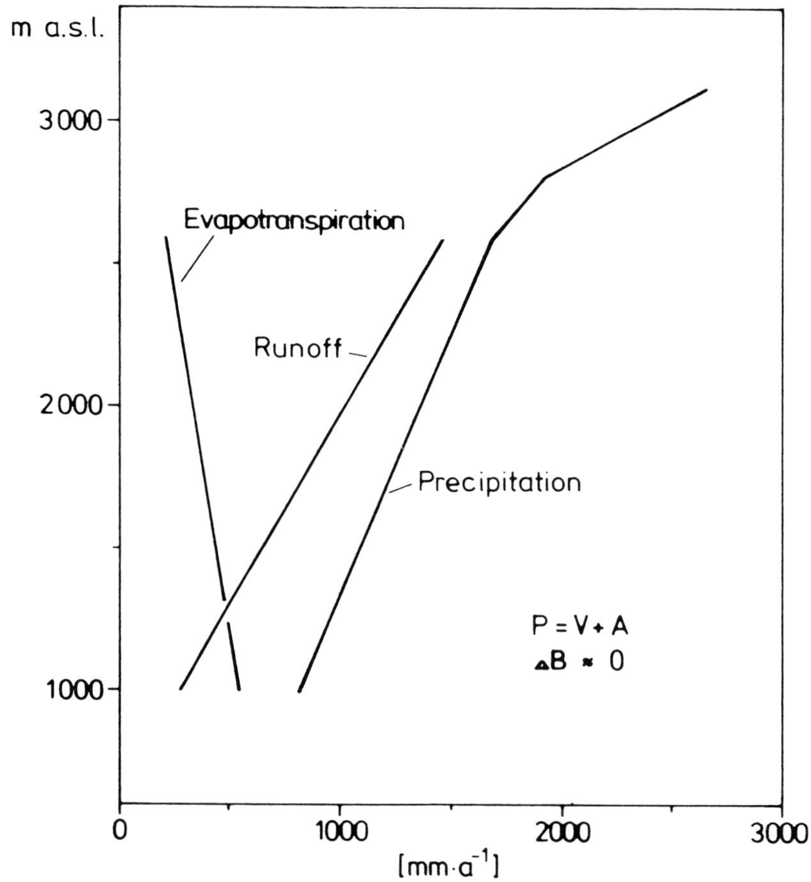

Figure 12.5: Altitudinal variation of the mean annual values of precipitation, evapotranspiration and runoff between 1612 m and 2528 m altitude in the Hohe Tauern range (KÖRNER et al. 1989).

$g \cdot m^{-2}$ DM (Curvuletum). The greater phytomass was due to a large increase in necromass with altitude, although the biomass only differed within the relatively small range of 330 to 460 $g \cdot m^{-2}$. This means an increase in the contribution of necromass to phytomass from 8% on the hay meadow to 42% on the Curvuletum. With increasing altitude the phytomass accumulating near the ground was also seen to increase. The leaf area index (LAI) dropped from 7 to 1.9 $m^2 \cdot m^{-2}$, due to the decreasing specific leaf area (SLA). For canopies of similar biomass, less leaf surface is produced with increasing altitude. Thus the SLA for the canopy of the hay meadow amounted to 0.020, for the alpine pasture 0.019, and for the

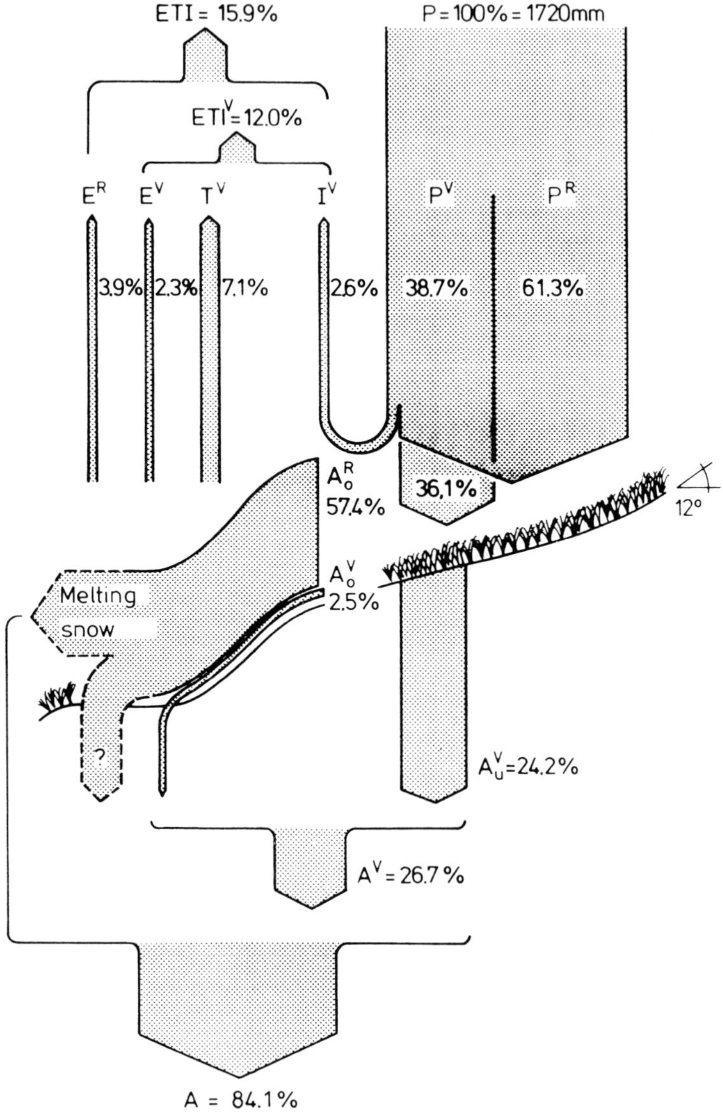

Figure 12.6: The partitioning of annual precipitation to the various runoff and evaporation processes in the sedge mats (Curvuletum 2300 m). Index V indicates events during the periods covered by measurements (more or less the snowfree period); Index R indicates fluxes estimated for the remaining periods. Precipitation (P), run–off (A), interception (I), transpiration (T), evaporation from the soil surface (E) (KÖRNER et al. 1989).

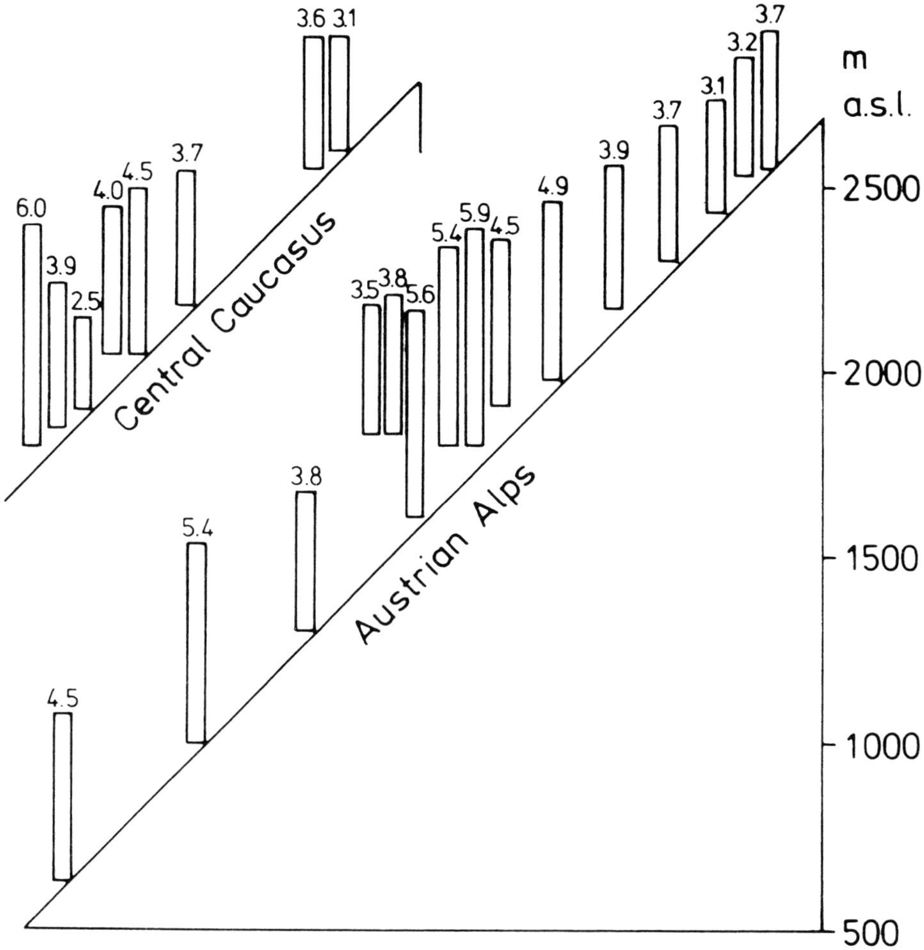

Figure 12.7: Mean daily sums of evapotranspiration on bright days between July and September at various elevations in the Austrian Alps and in the Central Caucasus (KÖRNER et al. 1989).

Curvuletum 0.010 $m^2 \cdot g^{-1}$.

Energy budget: Using the Bowen–ratio method, the altitudinal variation of the energy budget was investigated. The results of the energy balance measurements

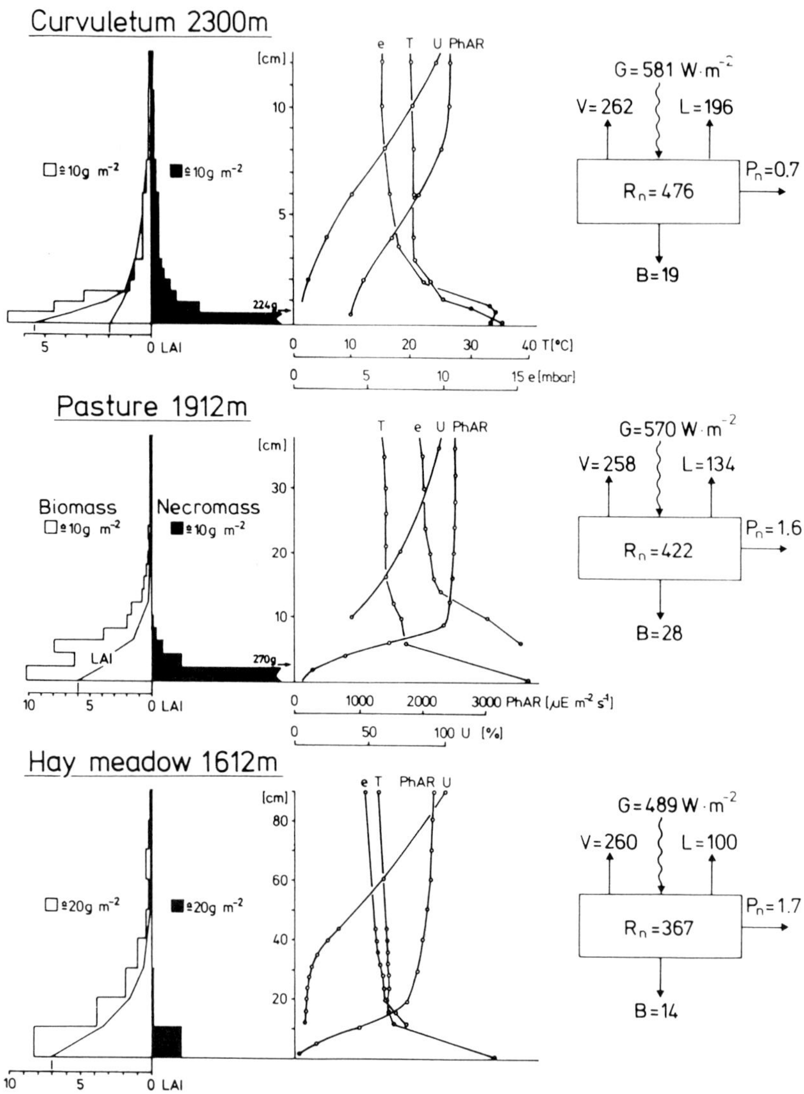

Figure 12.8: Canopy structure, microclimate pofiles and energy budget for the investigated sites. Microclimate profiles are mean values for the time interval 12.00 to 13.00 on a bright day in August. Energy fluxes are mean values for the time interval 8.00 to 18.00 for all bright days in August 1978. Water vapour pressure (e), air temperature (T), windspeed (u), photosynthetically active radiation (PAR), global radiation (G), net radiation (R_n), evapotranspiration (V), sensible heat flux (L), heat flux to ground (B) and net primary production (P_n) (CERNUSCA and SEEBER 1989).

can be summarized as follows (Fig. 12.8):

1. On bright days in August the driving forces for energy transfer between the plant cover and the free atmosphere increased with altitude. Global radiation increased by 20%, net radiation by 27%, and the temperature difference between canopy and screen level by 100%. In contrast to these trends the water vapour pressure difference showed no great change from the meadow to the sedge mats.

2. In agreement with the above-mentioned results of lysimeter measurements, the energy balance method showed for the three sites almost the same evapotranspiration rate of about 260 $W \cdot m^{-2}$. Taking into account the large decrease in transpiring leaf area, this phenomenon cannot be adequately explained by the measured increase of the driving forces. Diffusion porometry provided a further explanation (KÖRNER and MAYR 1980). Measurements of the diffusion resistance have shown a remarkable altitudinal increase in leaf conductance for water vapour for a number of plant species (Fig. 12.9). The increased leaf conductance compensates for the reduction of the green leaf area along the altitudinal transect.

3. As the evapotranspiration rate remained almost constant, the additional net radiation at the higher sites had to be converted into sensible heat flux (C). The sensible heat flux increased from 100 in the meadow to 196 $W \cdot m^{-2}$ in the sedge mats, resulting in an increase in the Bowen ratio (β) from 0.38 to 0.75. The Bowen ratio reflects the ratio of energy consumption due to sensible heat flux and evapotranspiration. In this connection it should be pointed out that the plant communities investigated were always well supplied with water. Plant water potentials hardly fell below −1.5 MPa. Therefore the increase found in the Bowen ratio is not a result of impaired water supply but is due to the drastic decrease in the green leaf area and the large increase in the amount of standing, dead plant parts. The increase in the Bowen ratio along the altitudinal transect in the Hohe Tauern agrees with the results of energy budget measurements carried out by CERNUSCA (1976) at the dwarf shrub level (*Loiseleuria*–stands) on the Patscherkofel (near Innsbruck), where the ratio increased significantly from 0.45 at 1950 m to 0.75 at 2175 m (Fig. 12.10). Of special interest in this connection are also the results obtained by CERNUSCA and NACHUZRISHVILI (1983) in the Central Caucasus (Kasbegi, Georgia, USSR). Their energy budget measurements on a hay meadow at 1850 m (Hordetum; canopy height 50 cm), a sheep pasture at 2150 m (Cobresietum capilliformis; canopy height 5 cm) and a Caricetum tristis at 2600 m (canopy height 3 cm) revealed here, too, a sizable increase in the Bowen ratio with increasing altitude (1850 m: $\beta = 0.20$; 2150 m: $\beta = 0.56$; 2600 m: $\beta = 0.92$). In all ecosystems investigated the changes in energy budget with increasing altitude are associated with a large decrease in leaf area index (LAI) as well as a decrease in the proportion of biomass in the phytomass (b/ph).

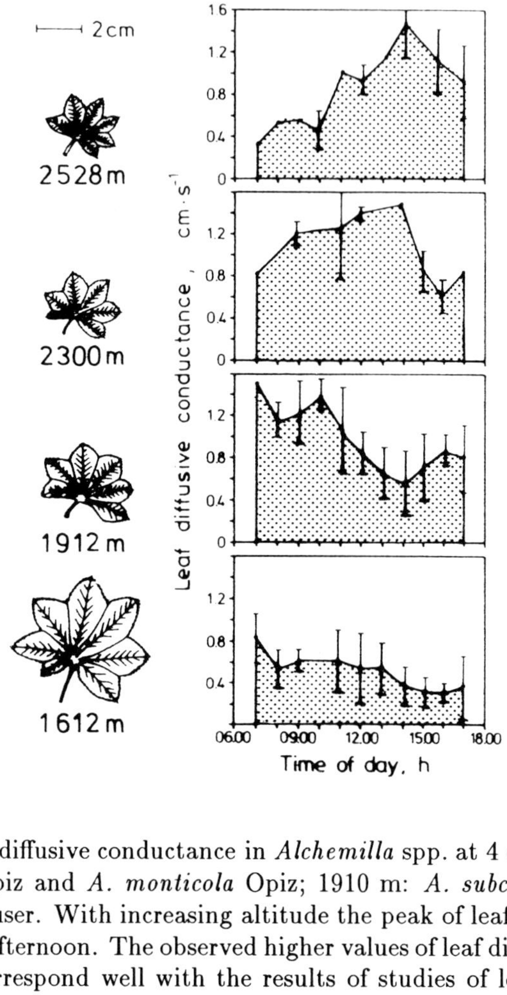

Figure 12.9: Leaf diffusive conductance in *Alchemilla* spp. at 4 different sites. 1630 m: *A. gracilis* Opiz and *A. monticola* Opiz; 1910 m: *A. subcrenata* Buser; 2310 m: *A. colorata* Buser. With increasing altitude the peak of leaf conductance shifts from morning to afternoon. The observed higher values of leaf diffusive conductance at alpine sites correspond well with the results of studies of leaf anatomy. With increasing altitude the frequency of stomata increased with a tendency to increase more in the adaxial (dorsal) leaf surface (KÖRNER and MAYR 1980).

Microclimate: The increase in accumulating dead organic material with rise in altitude also plays a very important ecological role in connection with the microclimate of a habitat. In the higher sites a considerable portion of the available energy is for this reason not consumed by transpiration but is available for sensible heat convection, and therefore contributes towards raising the temperature of the canopy. This means that it is possible for ecosystems at a higher altitude to develop and maintain a climate that partially compensates for the less favourable

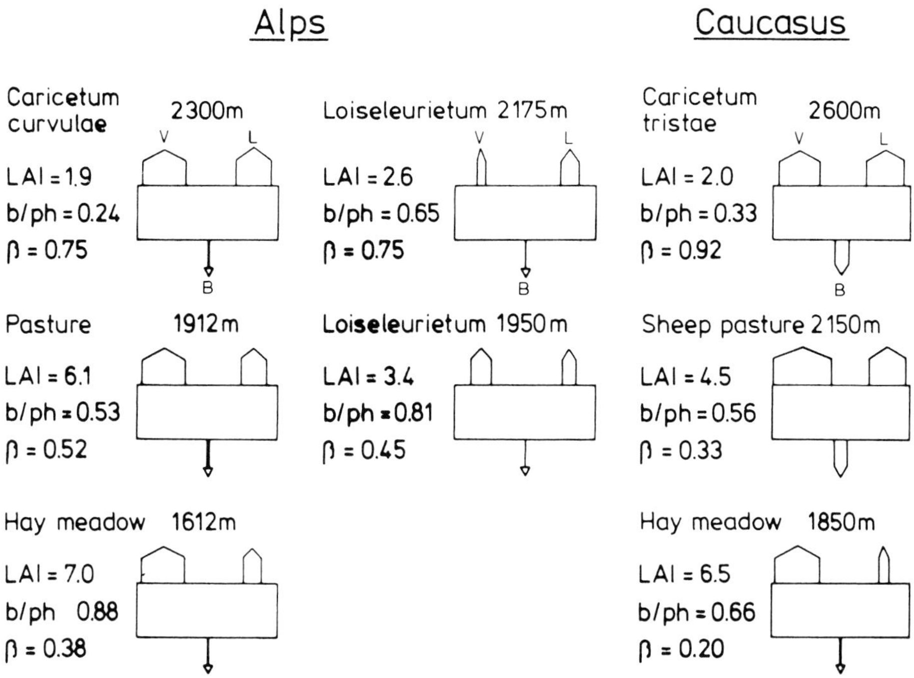

Figure 12.10: Energy budget for the grassland ecosystems investigated in the Hohe Tauern, for two dwarf shrub communities on Mt. Patscherkofel near Innsbruck, and for three grassland communities in the Central Caucasus. Energy fluxes are mean values for the time interval 8.00 to 18.00.

climatic conditions prevailing at such an altitude. In August the mean values of air temperature at screen level decreased from 11 °C in the meadow to 5 °C in the sedge mats, whereas there was no significant difference between the mean values of the temperature within the canopy (in the zone of highest biomass accumulation) at the three sites. On the other hand, the considerable accumulation of dead plant parts at higher sites reduces of course the efficiency of energy utilization for dry matter production. As light extinction measurements have shown, the absorption of photosynthetically active radiation by green plant parts decreases from 82% in the meadow to only 32% in the alpine sedge mats. In a quantitative ecophysiological analysis KÖRNER (1989) showed that in the alpine sedge mats suboptimal quantum flux density restricts yield by 40%.

In conclusion, it should be drawn attention to the fact that the comparative investigations of canopy structure, microclimate and energy budget in the Alps and in the Caucasus have again confirmed the dominant influence of the canopy structure on the microclimate and energy budget of alpine grassland ecosystems. Especially on clear bright days, when irradiation is the main driving force in the energy budget, the canopy structure and its degree of exposure to wind and incoming radiation are decisive factors influencing the energy regime. In addition, an important role is played by morphological adaptations of the plants, including leaf size, stomatal density and so on. The canopy structure may therefore to a considerable extent compensate for differences in altitude and degree of exposure.

12.4.3 Altitudinal variation of phytomass, energy content and fodder value

Phytomass: Fig. 12.11 shows the phytomass on studied stands. A comparison of the results obtained from the Hohe Tauern with values for grass and sedge canopies from different altitudinal levels in different parts of the world revealed good agreement. Detailed comparative measurements of the altitudinal changes in canopy structure, microclimate and energy budget in the Central Caucasus also confirmed the results obtained in the present investigations. The rule of altitudinal adaptation of the canopy structure of plant stands observed in the grassland ecosystems of the Hohe Tauern is probably also of general validity beyond the narrow confines of the region investigated.

Energy content of plant material: CERNUSCA, PIPP and JUSSEL (1989), using an adiabatic calorimeter, determined the energy content of the plant matter in the various canopies investigated. The mean ash-free energy content was 20.7 $kJ \cdot g^{-1}$, which is significantly higher than the values for ash-free energy contents of plants organs from the literature (19.8 $kJ \cdot g^{-1}$). Along the altitudinal transect a distinct increase in the energy contents was registered with increasing altitude. Furthermore, a clear connection was also seen between weather and the energy content of the plant samples. Consequently the annual mean ash-free energy content for the rather cold and wet 1978 amounted to 19.8 $kJ \cdot g^{-1}$, while in 1979, a warm and dry year, this value rose to 20.3 $kJ \cdot g^{-1}$.

Fodder value of the plant canopies studied: SPATZ, MÜHLSCHLEGEL, JUSSEL and WEIS (1989) investigated the fodder value of the various plant canopies. Investigations of the contents of raw protein, raw fibre, lignin, P, K, Ca, Mg and Na, and non-structural carbohydrates led to the recognition of the fact that grazing of these alpine pastures is only practicable if the number of animals are kept to a very low level. Only the current practice of very extensive selective grazing gives the animals the opportunity to provide themselves with adequate quantities of nutritive fodder.

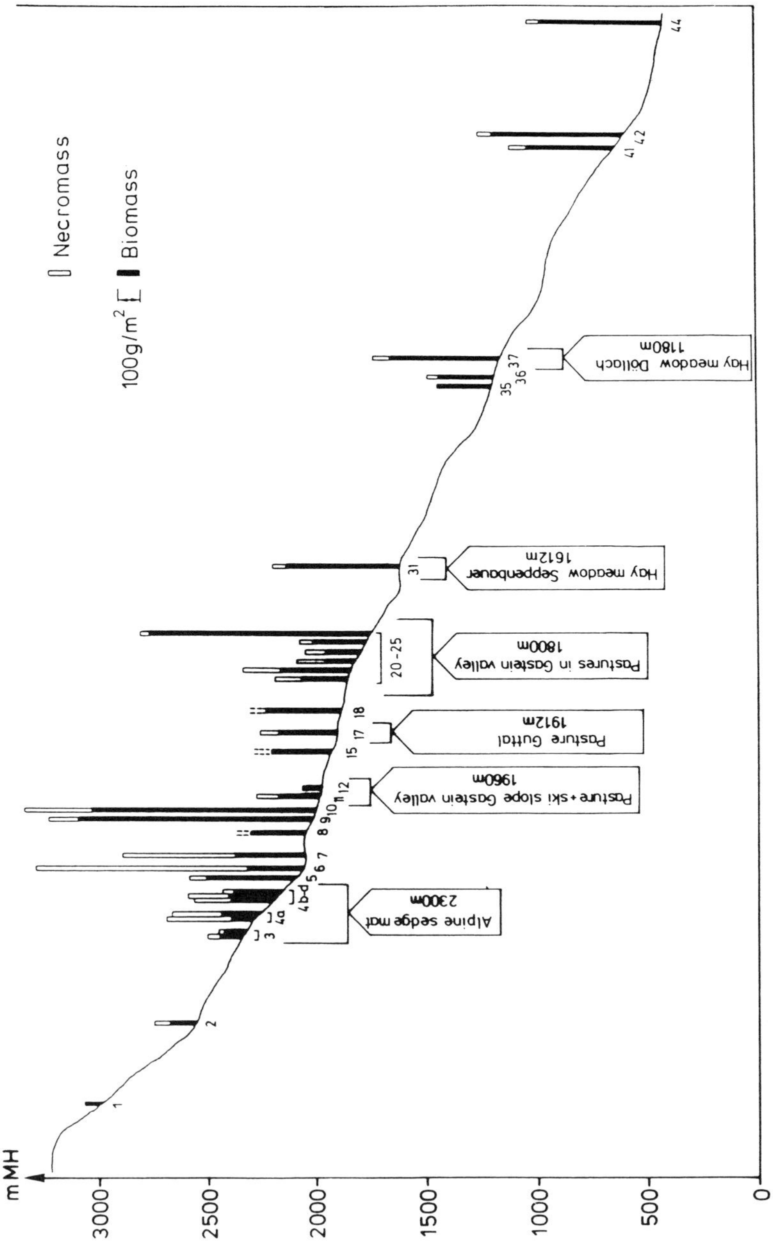

Figure 12.11: Aboveground biomass and mass of necrotic tissues in $g \cdot m^{-2}$ dry matter in various plant communities at different altitudes in the Alps (CERNUSCA and SEEBER 1989).

12.4.4 Altitudinal variation of soil–microbiological processes

In order to gain more information on metabolic processes and decomposition in the soil, the activities of some enzymes involved in nitrogen– and carbon–regimes were measured throughout the year, and the influence of soil temperature and moisture was investigated.

Microbial activities within the nitrogen cycle: SCHINNER F. (1989) analysed the microbial activities involved in the nitrogen cycle. This required the determination of cell numbers of nitrifying–, nitrate reducing–, denitrifying–, aerobic nitrogen fixing–, ammonifying– and ureolytic bacteria. In addition, the urease activity and the nitrite– and nitrate contents of soil samples were also studied. The results revealed considerable differences in cell numbers from site to site, larger numbers being found in the higher than in the lower–lying sites. Nitrification was higher in near–neutral soil samples than in acid soil: greater numbers of denitrifying organisms imply greater nitrogen losses from the water–saturated soil at the alpine level. Since the numbers of nitrogen–fixing, aerobic, free–living bacteria were low, and Leguminosae contributed very little to the vegetation, nitrogen–fixation is probably very slight in the sites investigated. Ureolytic bacteria were invariably found in relatively high numbers wherever animal manure was used; urease activity was also high, but due to high persistence resulting from immobilization it did not show the clear–cut annual pattern of spring and autumn maxima seen in the bacterial microfauna.

Soil–respiration and microbial activities: In order to analyse the altitudinal variation of soil respiration CERNUSCA and DECKER (1989) recorded the daily and annual patterns of respiratory losses of carbon from the soil in situ, using the cuvette method in the "open test gas system". No influence of altitude on soil respiration could be detected (Fig. 12.12). In fact, CO_2 losses from the soil depend to a far greater degree on the composition of the plant canopy, on the dynamics within the soil–canopy complex, and on the local microclimate (temperature and water content of the soil). Soil–respiration rates were relatively low in the cushion plant level at 2528 m (4.8 g $CO_2 \cdot m^{-2} \cdot d^{-1}$) and in the hay meadow at 1612 m (5.3 g $CO_2 \cdot m^{-2} \cdot d^{-1}$), whereas in the alpine sedge mats at 2300 m and the alpine pastures at 1912 m the considerably higher rates of 8.0 and 8.9 g $CO_2 \cdot m^{-2} \cdot d^{-1}$, respectively, were found. The low rates of soil respiration in the cushion plant belt can be attributed to the low soil temperature, and in the hay meadow to the small quantities of decomposable organic matter in the soil. In contrast, the alpine sedge mats and the alpine pasture were characterized by a high content of organic material and by a more favourable combination of soil temperature and soil moisture. Conditions of this kind favour the breakdown activity of microorganisms as the microbiological analyses of SCHINNER F., HOFMANN, NIEDERBACHER and RAINER (1989) revealed: The annual patterns of CO_2 release and activities of dehydrogenase, catalase, phosphatase, saccharase, xylanase, pectinase, and M–cellulase, as well as the numbers of cellulolytic bacteria exhibited that the soil at lower altitudes harbours more efficient microorganisms and offers better reaction conditions for enzymes whereas soils at high altitudes have a high potential for decomposition, although their efficiency is limited by unfavourable environmental conditions.

Mycotrophy in alpine ecosystems: HASELWANDTER (1989) found that mycorrhizal infections occur regularly in the alpine dwarf shrub communities as well as in the alpine sedge mats. It was shown that, for growth and for the provision of mineral nutrients, especially in connection with the phosphorus and nitrogen regime, ericoid mycorrhizal infections are important for Ericaceae, as are vesicular–arbuscular mycorrhiza and fungi with dark, septate hyphae for grasses, sedges and herbs. HASELWANDTER also drew attention to the role of mycorrhizal infections in connection with the recultivation of disrupted ecosystems. Studies on the frequency of occurrence of mycorrhiza in plants in damaged ecosystems as compared with intact ecosystems have shown that disturbances in the ecosystem lead to a drastic reduction in the number of plants harbouring mycorrhiza. This reduces the chance that a plant in a disrupted community will encounter one harbouring mycorrhiza. In this context, HASELWANDTER justifiably emphasizes that future recultivation measures in alpine regions should be so designed as to ensure optimal conditions for the development of symbiosis between fungi and seedlings.

12.4.5 The carbon budget of the alpine sedge mats

A very detailed study was made of the photosynthesis, production of dry matter and CO_2 release by the aboveground components of the canopy and soil in the alpine sedge mats. With the data obtained it has become possible to form a complete picture of the carbon budget of this ecosystem.

Phytomass and primary production: KLUG–PÜMPEL (1989), employing the harvesting method, measured in great detail the phytomass, necromass, primary production and mineral– and energy contents of the alpine sedge mats. In addition to the Curvuletum, a snow hollow (*Salix herbacea, Gnaphalium supinum*) and a wet depression (*Deschampsia cespitosa, Luzula alpino–pilosa*) were also studied. For the Curvuletum an aboveground phytomass of 676 $g \cdot m^{-2}$ DM was found, 40% of which consisted of cryptogams (Fig. 12.13). The underground phytomass amounted to 2440 $g \cdot m^{-2}$ DM. In snow hollows an aboveground phytomass of 120 to 130 $g \cdot m^{-2}$ DM was found, of which 50% was *Salix herbacea* and 20% cryptogams; the underground phytomass was somewhere between 700 and 1200 $g \cdot m^{-2}$ DM. In the wet depression the aboveground phytomass of the *Deschampsia* canopy amounted to 320 and that of the *Luzula* stand 195 $g \cdot m^{-2}$ DM; the underground phytomass values for *Deschampsia* were between 1300 and 1800, and for *Luzula* between 1900

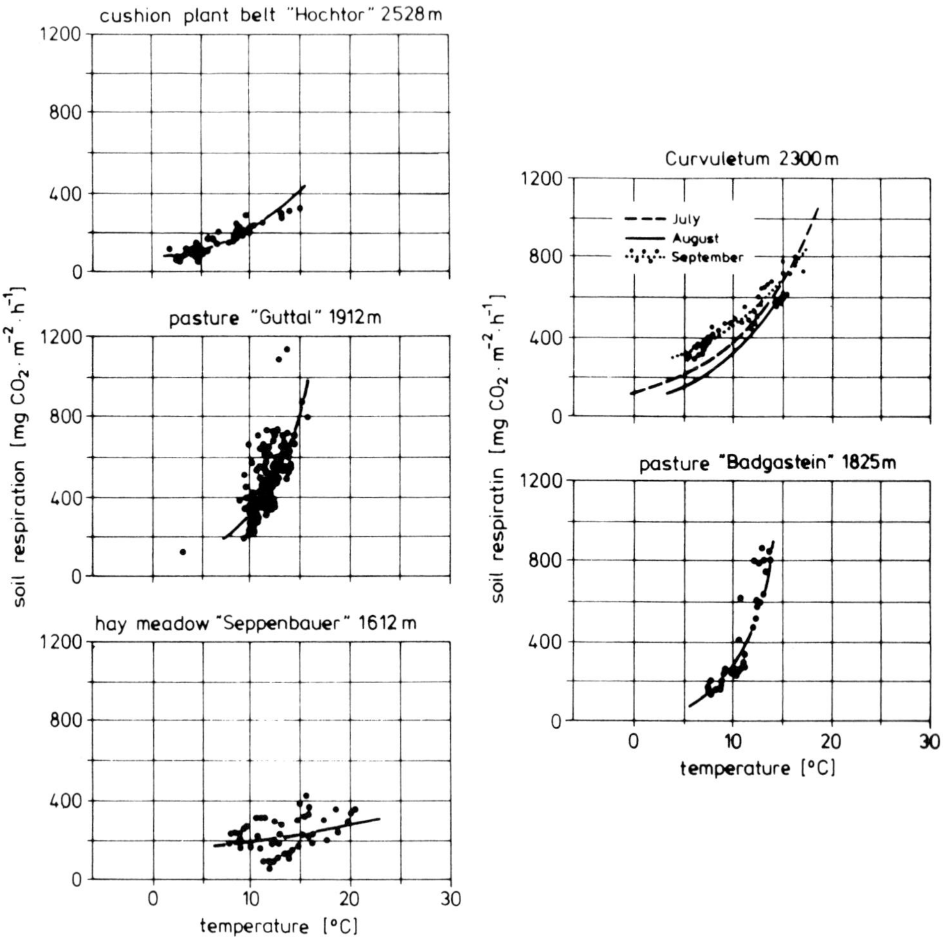

Figure 12.12: Comparison of temperature dependence of soil respiration for the four sites comprising the altitudinal transect. Additionally, data obtained from an abandoned alpine pasture "Badgastein" are presented. The temperature dependence of respiration is based on soil temperatures at 5 cm depth (CERNUSCA and DECKER 1989).

and 2300 $g \cdot m^{-2}$ DM. Depending on the weather conditions during the growth period, the net primary production in the Curvuletum was between 1618 and 2002 $g \cdot m^{-2} \cdot yr^{-1}$ DM, between 880 and 980 $g \cdot m^{-2} \cdot yr^{-1}$ DM in the Salicetum herbaceae, 2325 in the *Deschampsia* stand and 1937 $g \cdot m^{-2} \cdot yr^{-1}$ DM in the *Luzula* stand. From the values for energy content of plant substance and incoming photosynthetically active radiation KLUG-PÜMPEL calculated the efficiency of the primary production, which for the Curvuletum was between 0.32 and 0.51%, for the Salicetum herbaceae 0.36% and for the *Deschampsia* canopy 0.82%.

Gaseous exchange of CO_2 in *Carex curvula*: KÖRNER (1989) carried out a detailed study of the factors influencing gaseous exchange in *Carex curvula*. His investigations revealed *Carex curvula* to be a heliophyte, or in other words a plant that requires a large quantity of light for saturation of its photosynthetic uptake of CO_2. The temperature optimum in mid-summer and at light saturation is 22.5 °C, which is a relatively high optimum for a mountain plant (Fig. 12.14). KÖRNER also showed that CO_2 assimilation in *Carex curvula* and *Geum montanum* per unit leaf surface is greater than that of comparable species from low altitudes. Mitochondrial respiration (dark respiration) in *Carex curvula* in the low temperature range is also higher than in plants from lower altitudes.

Calculations were made with an empirical model revealed a loss of 20% in the annual CO_2 assimilation of *Carex curvula* in its native habitat. This is due to the often very bad weather and in particular to the poor irradiation conditions. Shade conditons within the canopy account for a further loss of 20%, so that inadequate illumination of the leaves is responsible for a loss of about 40% as compared with optimal conditions. The analysis also revealed that, despite the fact that the actual temperature within the Curvuletum stand is clearly lower than the relatively high temperature optimum for photosynthesis, the resulting losses amount to only 7.6% of the annual budget. KÖRNER's computer prediction for photosynthetic yield resulted in a value of 1060 $g \cdot m^{-2} \cdot yr^{-1}$ DM for the entire Curvuletum.

CERNUSCA and DECKER (1989) recorded that in the Curvuletum respiration was clearly dependent on soil temperature and soil water. The rise in soil temperature during the summer months led to a large increase in the amount of CO_2 leaving the soil (Q_{10}: 3.3 to 3.7) but in September the increase was smaller (Q_{10}: 2.3). At the mean soil temperature for the Curvuletum a reduction in soil respiration was already noticeable (Fig. 12.15) if the soil water rose above 30 vol.%. With the help of an approximation polynome calculated from data recorded for soil respiration, soil temperature and soil water content it was shown that at soil temperatures up to 15 °C soil water exerts the primary influence on the intensity of soil respiration and only at higher soil temperatures does temperature take over the decisive role. The overall weather of the preceeding day has little effect on CO_2 release by the soil.

Measurements of dark respiration of aboveground plant organs of the alpine sedge mats revealed that their CO_2 losses are mainly determined by the canopy temperature and the weather of the preceeding day. In autumn, although the average intensity of CO_2 release was lower, a rise in temperature evoked a greater increase (Q_{10}: 2.3) than in summer (Q_{10}: 1.6). After exposure to lower temperatures there was a marked rise in intensity of CO_2 release by the aboveground organs.

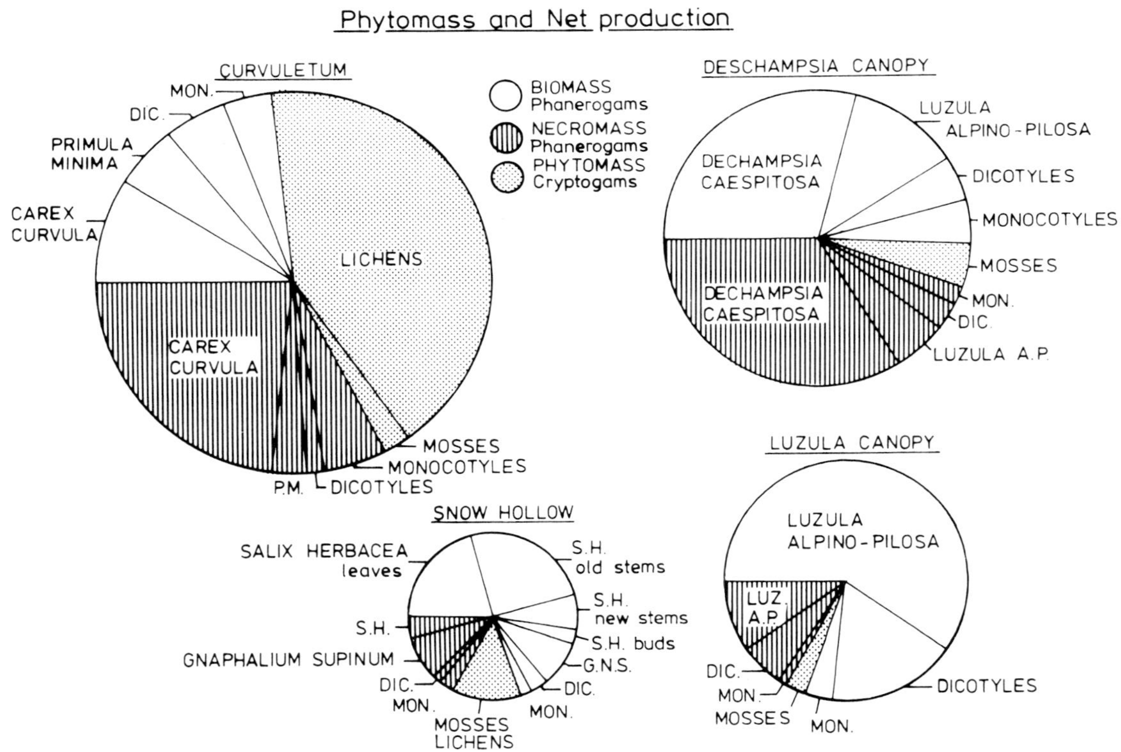

Figure 12.13: Above ground phytomass of the plant communities and its distribution to important components. Circle area corresponding to total above ground phytomass (KLUG–PÜMPEL 1989).

P.m.: *Primula minima*
S.h.: *Salix herbacea*
Gn.s.: *Gnaphalium supinum*

Mon.: Monocotyle
Dic.: Dicotyle
A.: Axes

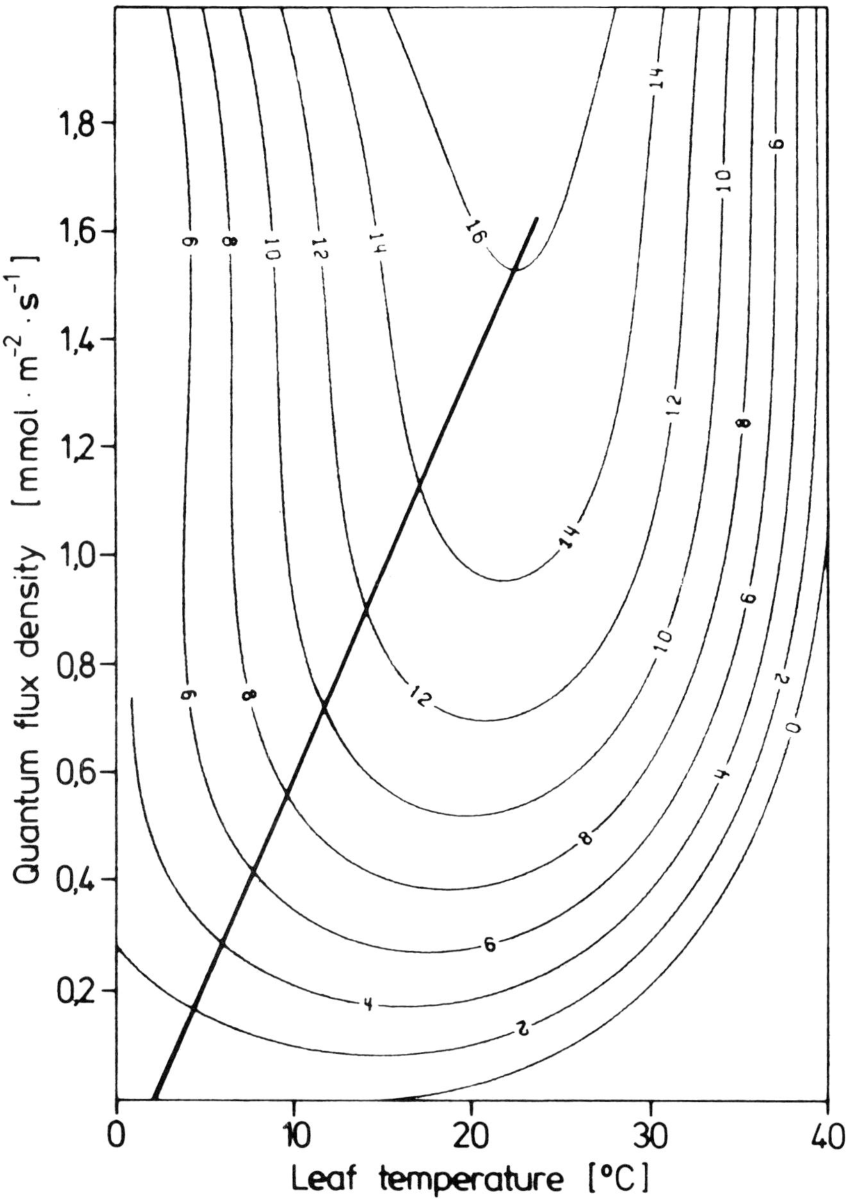

Figure 12.14: Dependency of net CO_2–uptake of *Carex curvula* on leaf temperature and photon flux density. The heavy line shows the actual leaf temperatures measured in the field at the appropriate quantum flux densities (KÖRNER et al. 1989).

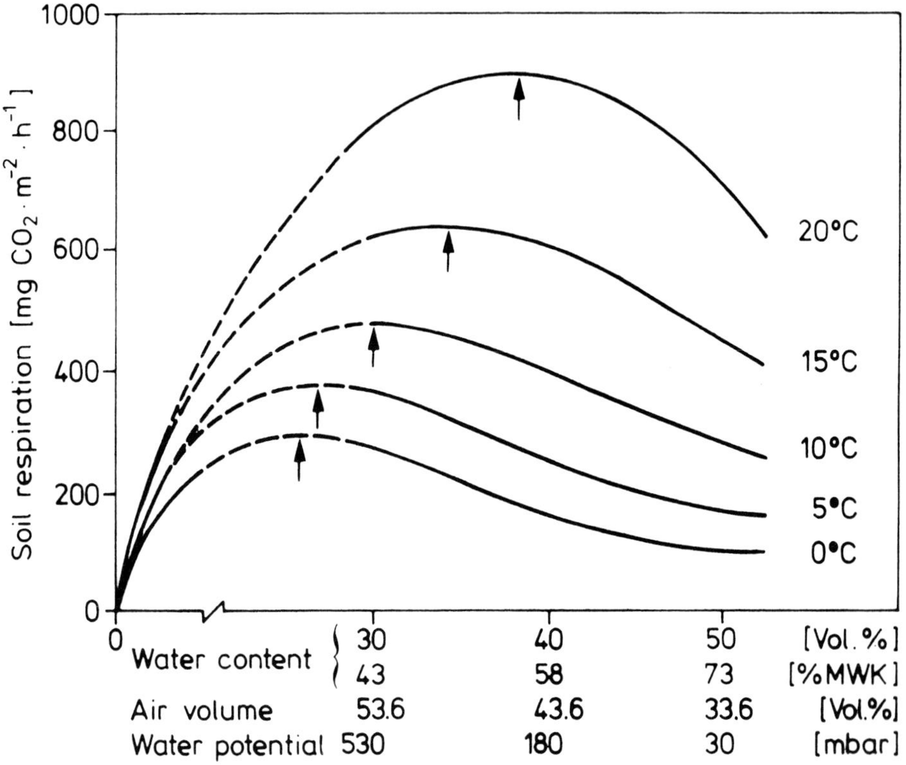

Figure 12.15: Influence of soil moisture on soil respiration over a range of temperatures. The arrows depict the position of the moisture optimum (CERNUSCA and DECKER 1989).

Laboratory and field studies aimed at determining the contribution of the individual components of the plant canopy to total respiratory losses additionally revealed that above- and below- ground organs make roughly the same contribution to the total respiratory losses (Table 12.3). Of the above–ground components, the green plant parts are responsible for most of the carbon lost.

CO_2 budget of the alpine sedge mats: On the basis of the results now available, the annual carbon budget of the alpine sedge mats can be summed up as shown in Tab. 12.3 and Fig. 12.16. In the Curvuletum 1114 $g{\cdot}m^{-2}{\cdot}yr^{-1}$ DM are produced by photosynthesis. Of this, 320 $g{\cdot}m^{-2}{\cdot}yr^{-1}$ DM are utilized (respired)

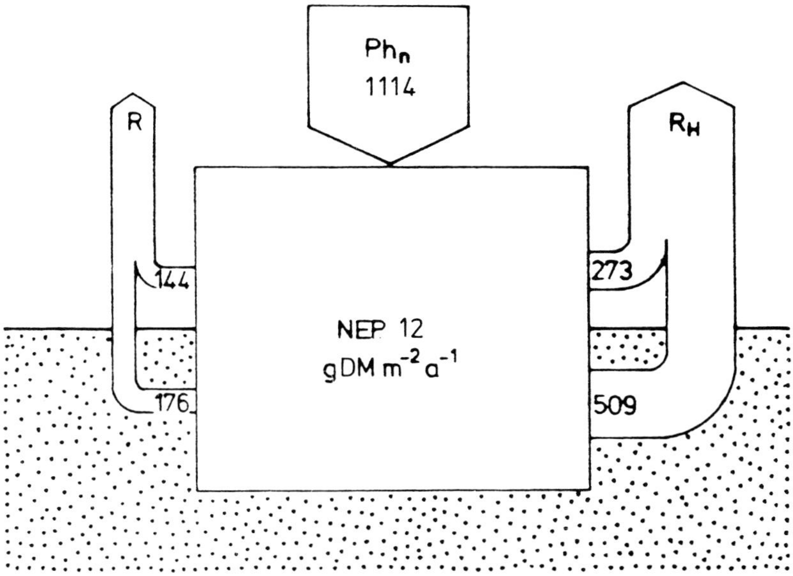

Figure 12.16: Carbon budget of an alpine grass heath for 1978 (all fluxes in g dry matter per m^2) (CERNUSCA 1989).

by the plants themselves for their metabolic processes, and 782 g·m^{-2}·yr^{-1} DM are used up by small animals and microorganisms. The remaining 12 g·m^{-2}·yr^{-1} DM represents the net production of the ecosystem (NEP). The NEP indicates the annual carbon increase within the ecosystem. This is very small, i. e. in the Curvuletum the carbon budget is to a large extent balanced, there is no sizeable increment of carbon and production serves mainly for maintaining the structure and function of the ecosystem. A similar balanced carbon regime was found in an alpine dwarf shrub heath at Mt. Patscherkofel near Innsbruck (LARCHER 1977). As far as the carbon budget is concerned the Curvuletum is a climax ecosystem. However, it should be borne in mind that the biological equilibrium of this alpine ecosystem is maintained by a very small energy flux and is therefore easily disrupted by human interference.

12.5 The effects of anthropogenic measures on alpine ecosystems

The following section is devoted to a description of the effects of anthropogenic measures on alpine ecosystems as revealed by the results of ecosystem studies in

Table 12.3: Release of CO_2 from soil and net primary production of the vegetation of the alpine sedge mats Curvuletum. Net photosynthesis (Ph_n), Net primary production (PP_n), plant respiration (R), changes in plant biomass (δB), losses of standing dead phytomass to animals and micro organisms (V_A), losses of plant biomass to grazers (V_K), production–based economic coefficient of primary production ($\ddot{O}K_p$) and of the entire ecosystem ($\ddot{O}K_s$), net production of the entire ecosystem (NEP), Respiration of animals and microorganisms (R_H). All values are given in $g \cdot m^{-2} \cdot yr^{-1}$ either dry matter (DM) or CO_2.

1. CO_2–Release

	growing season	winter	year
roots	252	23	275
soil without roots	680	114	794
soil respiration	932	137	1069
above ground plant parts during night	225	–	225
attached dead + litter	410	16	426
above ground respiration	635	16	651
plant respiration	477	23	500
heterotrophic resp.	1090	130	1220
entire CO_2-Release	1567	153	1720

2. Net primary production measured by harvesting method

$$\Delta B = 694 \text{ DM} = 1083 \text{ CO}_2$$
$$V_A = 100 \text{ DM} = 156 \text{ CO}_2$$
$$PP_n = 794 \text{ DM} = 1239 \text{ CO}_2$$

3. Net photosynthesis calculated from 1. + 2.

$$Ph_n = PP_n + R = 1739 \text{ CO}_2 = 1114 \text{ DM}$$

4. Net production of the entire ecosystem (NEP) from 1. + 2.

$$NEP = PP_n - R_H = 19 \text{ CO}_2 = 12 \text{ DM}$$
$$\ddot{O}K_p = Ph_n/R = 3.5$$
$$\ddot{O}K_s = Ph_n/(R + R_H) = 1.01$$

the Austrian Alps and in the Caucasus.

12.5.1 The consequences of grazing on the alpine grassland

The effects of grazing on the alpine sedge mats: The alpine sedge mats are nowadays grazed by sheep and young cattle, but at such a low density that even after 10 years no great differences are noticeable as compared with fenced-off areas. Investigations showed that intensive, uniform cattle-grazing can at most deplete the live biomass of the Curvuletum by 46%, whereas sheep can consume 81%. Hence, an uncontrolled increase in sheep grazing presents the very real danger that the critical limiting value for the stability of grazing land (50% according to ODUM 1967) will be exceeded. Additionally it should be borne in mind, that according to the analyses of fodder values the plants of the alpine sedge mats are very poor in minerals and therefore an increase in density of sheep grazing would have to be accompanied by measures aimed at improving the nutritive value (manuring).

Comparative investigations on pastures in the Caucasus: In order to analyse the ecological effects of intensive sheep grazing comparative investigations of microclimate, CO_2-exchange and energy regime of a pasture grazed by sheep and an adjacent meadow (not grazed but mowed once a year) were carried out in the Central Caucasus by CERNUSCA and NAKHUTSRISHVILI (1983). Their results showed that the grazing leads to characteristic adaptations of the plant canopy with respect to its structure, water regime, photosynthesis, and energy budget (Fig. 12.17):

1. Concerning plant community structure, the meadow differs from the pasture in a number of ways. In the pasture the height of the canopy was only 2 cm, as compared with 60 cm in the meadow, whereas phytomass and leaf area index were only half as much as in the meadow. In the pasture, leaf inclination is very plagiophile in contrast to a predominantly erectophile leaf inclination in the meadow. Fifty percent of the phytomass in the pasture is concentrated in the lowermost 10 mm of the canopy. Therefore, it is inaccessible to the sheep.

2. Concerning daily means of the energy budget, there are only slight differences between both communities on bright midsummer days. Thus, although the biomass and leaf area of the sheep pasture were only half those of the meadow, its evapotranspiration was only 18% lower. This is only possible because almost twice as much water evaporates per unit leaf area from the plants of the sheep pasture as from the meadow plants. The elevated rate of transpiration is not elicited by higher gradients of air humidity, but by the lower stomatal resistance to diffusion in the plants on the sheep pasture. Investigations have shown that although the leaves of the pasture plants are smaller, their stomatal density is greater.

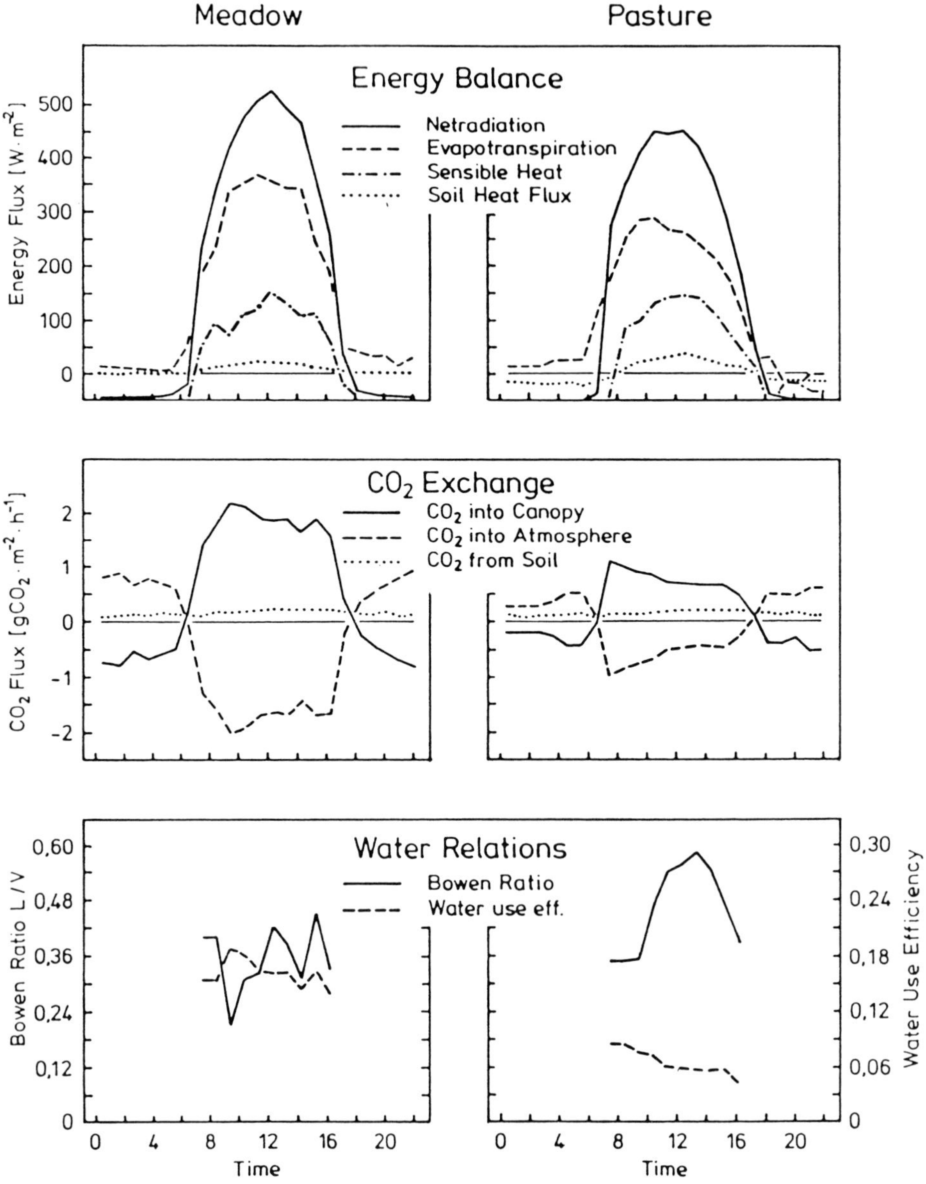

Figure 12.17: Comparative Investigations of energy regime, water relations and CO_2–exchange of a pasture grazed by sheep and an adjacent meadow (not grazed but mowed once a year) in the Central Caucasus (CERNUSCA 1987).

3. In contrast to the relatively small reduction in evapotranspiration the net photosynthesis of the pasture was only 60% of that of the meadow and reflected the large reduction of aboveground plant parts in the pasture. In the early afternoon, the increase in the BOWEN-ratio indicated water stress in the plants due to limiting soil water conditions in the pasture. The decreased efficiency in water use represented an additional limitation to photosynthesis by high temperatures. In contrast to the pasture, photosynthesis in the meadow was limited by the low light intensity in the lowermost 10 cm of the canopy. These results for the whole canopy agree well with results of single leaf experiments of CO_2-exchange.

12.5.2 The consequences of the abandonment of the alpine pastures

For the purposes of ecosystem research, alpine pastures are anthropogenic environments that for centuries have been maintained in a state of artificial equilibrium by the husbandry of alpine farmers. Once the grazing pressure is removed an ecological succession sets in immediately. This could be impressively illustrated by investigations of TAPPEINER and CERNUSCA (1989) for an abandoned alpine pasture near the experimental site "Guttal" at 1912 m. Only one year after abandonment characteristic changes in the phytomass, canopy structure, microclimate and plant production of the pasture were noticeable. In the first year the photosynthetically active leaf area (GAI) of the grasses rose by a factor of 7 and that of the herbs by a factor of 3. The explanation for this is that the "limiting factors" are different. On a grazed pasture the grazing itself is the decisive limiting factor promoting the development of predominantly horizontal leaves near the ground and therefore effectively prevents endangerment of the stability of the canopy from excessive grazing losses. On the abandoned site, however, following removal of the grazing pressure light becomes the decisive limiting factor. Mutual shading is responsible for the surprising result that net photosynthesis on the abandoned pasture is only 1.9 higher than that of the grazed pasture, although the photosynthetically active leaf surface of the former is five times larger. Ecosystem studies in the Gastein valley have shown that in the vicinity of abandoned alpine pastures below the alpine timberline, a characteristic succession sets in, resulting after about a hundred years in a reversion to forest (CERNUSCA 1978). In the course of this succession evapotranspiration increases and surface run-off and soil erosion are considerably reduced. This contributes to the general stabilization of the ground and taken in conjunction with the increasing growth of dwarf shrubs and alder reduces the potential dangers connected with abandonment. Seen in this light, there is nothing to be said against the abandonment of certain alpine pastures and the reestablishment of a natural succession.

12.5.3 The effects of skiing facilities on the water balance of the alpine grassland ecosystems

Water balance measurements made by KÖRNER, WIESER and CERNUSCA (1989) showed that in the alpine sedge mats an average of only 2.5% of the annual precipitation is lost as surface run–off and that the type of farming practiced has little effect on the annual run–off figures. The effects become much more serious, however, if instead of looking at the figures for annual mean run–off, the results of heavy rain are considered, which are the relevant figures for assessing the potential danger from mountain torrents. Experiments with artificial heavy rain showed that in the alpine sedge mats 20% to 40% of a heavy downpour runs off the surface (BUNZA 1989). In similar experiments the effects of ski–runs on the run–off into the Grantenbach in the alpine sedge mats along the Glockner road were estimated. It could be shown that ski–runs on a third of the catchment area would lead to an increase in the runoff–coefficient from 0.51 to 0.67. Further confirmation of these results was provided by detailed investigations in connection with the project "Skiing area Schloßalm/Stubnerkogel" in the Gastein valley (Fig. 12.18). In the Schloßalm area, too, the highest surface runoff is produced by 'levelled' ski–runs. It also could be shown that increased surface run–off from even the smallest catchment areas increases the risk of overflowing streams (see also CERNUSCA 1986).

12.5.4 Susceptibility of alpine ecosystems to disruption, as seen in the case of hiking trails

The susceptibility of alpine ecosystems to disturbance can be illustrated by a brief look at the situation created by tourist trails in the alpine dwarf shrub mats and alpine sedge mats. Fig. 12.19 shows the serious consequences for CO_2 exchange in the Curvuletum caused by the existence of a hiking trail. Soil respiration, which is a measure of the activity of roots, soil fauna and microorganisms, is reduced by half in the vicinity of the trail as compared with the undisturbed surroundings. Soil compaction also causes a strong decrease in bacterial biomass, chiefly in the aerobic bacteria. In addition, a considerable reduction in the infiltration rate of water was noted near the trail. Further analysis showed (KÖRNER 1980) that damage is clearly noticeable in a *Carex curvula* canopy when trampled 3000 times, which is the equivalent of about 50 busloads. *Carex curvula* itself is relatively resistant to trampling and is not so much affected as the "undergrowth" species, such as *Primula minima, Hieracium* sp., *Leontodon* sp. and, to an even greater extent, lichens. Regeneration after trampling proceeds relatively quickly if *Carex curvula* is not damaged, but if it is in fact damaged recovery takes an unusually long time. According to studies made by GRABHERR (1978) in the Ötztaler Alps, bare patches in the *Carex curvula* canopy can only be vegetatively recolonized by rhizomes. The exceptionally slow spread of the rhizomes (0.9 mm·yr^{-1}) means that recolonization of a track of 25 cm width from both sides would take 100 years.

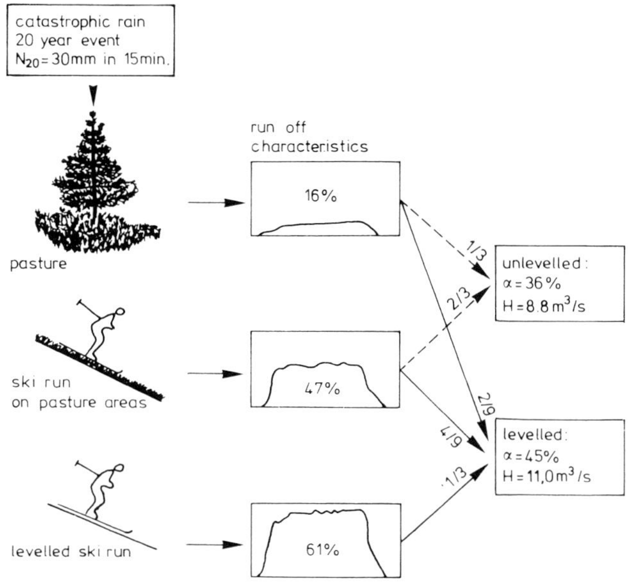

Figure 12.18: Impacts of the ski area Schloßalm on streamflow characteristics (BUNZA 1990). Upperhalf: Study site resembling a portion of the drainage of the Schloßbach (0.96 km^2) including ski lifts. Below: Run–off characteristics associated with a 20–year precipitation event. The ski–runs cause an increase in the runoff–coefficient (a) and the peak discharge (H). The fractions associated with the arrows designate the percentage of area covered by respective vegetation units.

Even more susceptible to disturbance than the alpine sedge mats is the alpine dwarf shrub belt. Investigations on a *Loiseleuria* canopy revealed that here 600 passages along a trail (the equivalent of 10 busloads), cause damage of a largely irreversible nature (KÖRNER 1980). The irreversibility is due to the fact that where the canopy structure of the normally dense carpet of *Loiseleuria* is destroyed, the raw humus on the ground surface dries out almost completely so that the temperature of the soil surface itself may exceed 80 °C on fine days. Such hot and dry patches are hardly ever recolonized by the *Loiseleuria* plants and a process of soil erosion gradually spreads from the trail.

Bibliography

BUNZA, G., 1978, Vergleichende Messungen von Abfluß und Bodenabtrag auf Almflächen des Stubnerkogels im Gasteiner Tal.
In: CERNUSCA, A. (ed.), 1978, Ökologische Analysen von Almflächen im Gasteiner

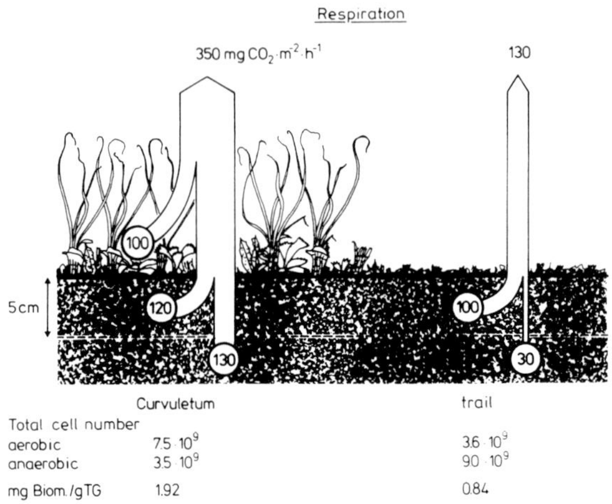

Figure 12.19: Effects of trampling on CO_2–release (CERNUSCA and DECKER 1989). Rates of CO_2–release are more than threefold higher on the undisturbed site (left) than on the trail (right side). Depicted below is a summary of microbial parameters in the soil.

Tal. Veröff. d. Österr. MaB–Hochgebirgsprogramms Hohe Tauern, Bd. 2, Universitätsverlag Wagner, Innsbruck 1978, 315–334.

BUNZA, G., 1984, Oberflächenabfluß und Bodenabtrag in alpinen Graslandökosystemen.
Verh. Ges. Ökologie **12**: 101–109.

BUNZA, G., 1989, Oberflächenabfluß und Bodenabtrag in der alpinen Grasheide der Hohen Tauern an der Großglockner–Hochalpenstraße.
In: CERNUSCA, A. (ed.), Structure and function of grassland ecosystems in the Austrian National Park Hohe Tauern. Publ. Austr. MaB–Programme vol. 13, Universitätsverlag Wagner, Innsbruck, 625 pp., 155–200.

BUNZA, G., 1990, Abfluß und Abtragsmessungen im Gebiet der Schloßalm bei Bad Hofgastein.
In: Veröff. Österr. MaB–Programm, in Vorbereitung.

CERNUSCA, A., 1975, MaB–6–Projekt "Pilotprojekt Alpine Ökosysteme".
Sitzungsber. Österr. Akad. Wiss., Mathem.–naturwiss. Kl., Abt. I, Bd. **184**, 6. bis 7. Heft, 133–141.

CERNUSCA, A., 1976, Bestandesstruktur, Bioklima und Energiehaushalt von alpinen Zwergstrauchbeständen.
Oecol. Plant., **11**, 71–102.

CERNUSCA, A., 1978, Ökologische Veränderungen im Bereich aufgelassener Almen. In: CERNUSCA, A. (ed.), 1978, Ökologische Analysen von Almflächen im Gasteiner Tal. Veröff. d. Österr. MaB-Hochgebirgsprogramms Hohe Tauern, Bd. **2**, Universitätsverlag Wagner, Innsbruck 1978, 7–27.

CERNUSCA, A., 1985, Der Einsatz von Taschencomputern für programmgesteuerte Datenerfassung, Auswertung und Versuchsüberwachung in der Ökologie. Verh. Ges. Ökol. **13**: 369–377.

CERNUSCA, A., 1986, Répercussions de la construction de pistes de ski sur les écosystèmes alpins et lignes directrices concernant la construction de ces pistes. Sauvegarde de la Nature **33**. Strasbourg (Conseil de l'Europe), 170 S.

CERNUSCA, A., 1987, Application of computer methods in the field to assess ecosystem function and response to stress. In: TENHUNEN, J. D. et al. (eds.), Plant Response to Stress. NATO ASI Series, Vol. **G15**, Springer–Verlag Berlin Heidelberg, 157–164.

CERNUSCA, A., 1989, Kohlenstoffbilanz einer alpinen Grasheide (Caricetum curvulae) in 2300 m MH in den Hohen Tauern. In: CERNUSCA, A. (ed.), Structure and function of grassland ecosystems in the Austrian National Park Hohe Tauern. Publ. Austr. MaB-Programme vol. 13, Universitätsverlag Wagner, Innsbruck, 625 pp., 397–404

CERNUSCA, A.; DECKER, P., 1989, Vergleichende Atmungsmessungen an Graslandökosystemen entlang einer Höhenstufenabfolge zwischen 1635 und 2528 m MH in den Alpen. In: CERNUSCA, A. (ed.), Structure and function of grassland ecosystems in the Austrian National Park Hohe Tauern. Publ. Austr. MaB-Programme vol. 13, Universitätsverlag Wagner, Innsbruck, 625 pp., 405–418.

CERNUSCA, A.; DECKER, P., 1989, Faktorenabhängigkeit der respiratorischen Kohlenstoffverluste einer alpinen Grasheide (Caricetum curvulae) in 2300 m MH in den Hohen Tauern. In: CERNUSCA, A. (ed.), Structure and function of grassland ecosystems in the Austrian National Park Hohe Tauern. Publ. Austr. MaB-Programme vol. 13, Universitätsverlag Wagner, Innsbruck, 625 pp., 371–396.

CERNUSCA, A.; NAKHUTSRISHVILI, G., 1983, Untersuchungen der Ökologischen Auswirkungen intensiver Schafbeweidung im Zentralkaukasus. Verh. Ges. Ökologie **10**: 183–192.

CERNUSCA, A.; PIPP, E.; JUSSEL, U., 1989, Energiegehalt der pflanzlichen Substanz von Graslandökosystemen zwischen 1085m und 2528 m ü. NN in den Hohen Tauern. In: CERNUSCA, A. (ed.), Structure and function of grassland ecosystems in the

Austrian National Park Hohe Tauern. Publ. Austr. MaB-Programme vol. 13, Universitätsverlag Wagner, Innsbruck, 625 pp., 497–514.

CERNUSCA, A.; SEEBER, M. C., 1989, Mesoklimatische Hinweise und Beschreibung von Witterungsablauf auf Phänologie während der Ökosystemstudie "Höhentransekt" in den Hohen Tauern.
In: CERNUSCA, A. (ed.), Structure and function of grassland ecosystems in the Austrian National Park Hohe Tauern. Publ. Austr. MaB-Programme vol. 13, Universitätsverlag Wagner, Innsbruck, 625 pp., 311–330.

CERNUSCA, A.; SEEBER, M. C., 1989, Phytomasse, Bestandesstruktur und Mikroklima einer Mähwiese (1635 m), einer Almweide (1912 m) und einer alpinen Grasheide (2300 m) in den Österreichischen Zentralalpen.
In: CERNUSCA, A. (ed.), Structure and function of grassland ecosystems in the Austrian National Park Hohe Tauern. Publ. Austr. MaB-Programme vol. 13, Universitätsverlag Wagner, Innsbruck, 625 pp., 419–462.

CERNUSCA, A.; SEEBER, M. C., 1989, Energiehaushalt einer Mähwiese (1635 m MH), einer Almweide (1912 m MH) und einer alpinen Grasheide (2300 m MH) in den Österreichischen Zentralalpen.
In: CERNUSCA, A. (ed.), Structure and function of grassland ecosystems in the Austrian National Park Hohe Tauern. Publ. Austr. MaB-Programme vol. 13, Universitätsverlag Wagner, Innsbruck, 625 pp., 463–496.

GRABHERR, G., 1978, Schädigung der natürlichen Vegetation über der Waldgrenze durch die Anlage von Skipisten und deren Fähigkeit zur Regeneration (Ergebnisse des MaB-Programmes Obergurgl).
Ber. Bayer. Akad. Naturschutz, 45–51.

GREIF, F., 1989, Geomorphometrische Studien in den Hohen Tauern. Zur Quantifizierung von Geländeeigenschaften auf der Grundlage der amtlichen Österreichischen Karte 1:50.000.
In: CERNUSCA, A. (ed.), Structure and function of grassland ecosystems in the Austrian National Park Hohe Tauern. Publ. Austr. MaB-Programme vol. 13, Universitätsverlag Wagner, Innsbruck, 625 pp., 9–32.

GSTRAUNTHALER, G.; SCHINNER, F., 1989, Bodenpilze in der alpinen, subalpinen und montanen Stufe in den Hohen Tauern (Österreich).
In: CERNUSCA, A. (ed.), Structure and function of grassland ecosystems in the Austrian National Park Hohe Tauern. Publ. Austr. MaB-Programme vol. 13, Universitätsverlag Wagner, Innsbruck, 625 pp., 227–232.

HASELWANDTER, K., 1989, Die Mykotrophie alpiner Pflanzengesellschaften oberhalb der Waldgrenze mit besonderer Berücksichtigung der Mykorrhiza.
In: CERNUSCA, A. (ed.), Structure and function of grassland ecosystems in the

Austrian National Park Hohe Tauern. Publ. Austr. MaB-Programme vol. 13, Universitätsverlag Wagner, Innsbruck, 625 pp., 217-226.

JENEWEIN, I., 1989, Umweltbelastungen durch Straßenoberflächenwasser am Beispiel der "Großglockner-Hochalpenstraße".
In: CERNUSCA, A. (ed.), Structure and function of grassland ecosystems in the Austrian National Park Hohe Tauern. Publ. Austr. MaB-Programme vol. 13, Universitätsverlag Wagner, Innsbruck, 625 pp., 297-310.

KLUG-PÜMPEL, B., 1989, Phytomasse und Nettoproduktion naturnaher und anthropogen beeinflußter alpiner Pflanzengesellschaften in den Hohen Tauern.
In: CERNUSCA, A. (ed.), Structure and function of grassland ecosystems in the Austrian National Park Hohe Tauern. Publ. Austr. MaB-Programme vol. 13, Universitätsverlag Wagner, Innsbruck, 625 pp., 331-356.

KÖRNER, CH., 1980, Zur anthropogenen Belastbarkeit der alpinen Vegetation.
Verh. Ges. Ökologie **8**: 451-461.

KÖRNER, CH., 1989, Der Flächenanteil unterschiedlicher Vegetationseinheiten in den Hohen Tauern: Eine quantitative Analyse großmaßstäblicher Vegetationskartierungen in den Ostalpen.
In: CERNUSCA, A. (ed.), Structure and function of grassland ecosystems in the Austrian National Park Hohe Tauern. Publ. Austr. MaB-Programme vol. 13, Universitätsverlag Wagner, Innsbruck, 625 pp., 33-48.

KÖRNER, CH., 1989, Der CO_2-Gaswechsel verschiedener Pflanzen im alpinen Grasheidegürtel. II. Photosynthetische Kohlenstoffbindung des Bestandes.
In: CERNUSCA, A. (ed.), Structure and function of grassland ecosystems in the Austrian National Park Hohe Tauern. Publ. Austr. MaB-Programme vol. 13, Universitätsverlag Wagner, Innsbruck, 625 pp., 357-370.

KÖRNER, CH.; MAYR, R., 1980, Stomatal behaviour in alpine plant communities between 600 an 2600 meters above sea level.
In: GRACE, J.; FORD, E. D.; JARVIS, P. G. (eds), Plant and their atmospheric environment. Blackwell Scientific Publications, Oxford, London, Edinburgh, Boston, Melbourne, 205-218.

KÖRNER, CH.; WIESER, G.; CERNUSCA, A., 1989: Der Wasserhaushalt waldfreier Gebiete in den österreichischen Alpen zwischen 600 und 2600 m Höhe.
In: CERNUSCA, A. (ed.), Structure and function of grassland ecosystems in the Austrian National Park Hohe Tauern. Publ. Austr. MaB-Programme vol. 13, Universitätsverlag Wagner, Innsbruck, 625 pp., 119-154.

LARCHER, W., 1977, Ergebnisse des IBP-Projektes "Zwergstrauchheide Patscherkofel".
Sitzungsber. Österr. Akad. Wiss. Mathem.-naturw. Kl. Abt. I, **6**, 301-371.

NEUWINGER, I., 1989, Ökologische Kennzeichnung von Boden–Reliefserien längs eines Höhengradienten im Gebiet des Tauernbaches am Großglockner (Hohe Tauern, österreichische Alpen).
In: CERNUSCA, A. (ed.), Structure and function of grassland ecosystems in the Austrian National Park Hohe Tauern. Publ. Austr. MaB–Programme vol. 13, Universitätsverlag Wagner, Innsbruck, 625 pp., 49–94.

ODUM, E. P., 1967, Ökologie.
Reihe Moderner Biologie. Bayerischer Landwirtschaftsverlag, München, Basel, Wien 1967.

SCHINNER, F.; GURSCHLER, A., 1989, Bakterien–, Hefen– und ATP–Gehalte von Böden entlang einem Höhentransekt in den Hohen Tauern.
In: CERNUSCA, A. (ed.), Structure and function of grassland ecosystems in the Austrian National Park Hohe Tauern. Publ. Austr. MaB–Programme vol. 13, Universitätsverlag Wagner, Innsbruck, 625 pp., 233–238.

SCHINNER, F.; NIEDERBACHER, R.; RAINER, J., 1989, Enzymaktivitäten und CO_2–Freisetzung von Bodenmaterialien entlang einem Höhentransekt in den Hohen Tauern.
In: CERNUSCA, A. (ed.), Structure and function of grassland ecosystems in the Austrian National Park Hohe Tauern. Publ. Austr. MaB–Programme vol. 13, Universitätsverlag Wagner, Innsbruck, 625 pp., 239–248.

SCHINNER, F., 1989, Mikrobielle Aktivitäten im Stickstoffkreislauf alpiner, subalpiner und montaner Böden in den Hohen Tauern.
In: CERNUSCA, A. (ed.), Structure and function of grassland ecosystems in the Austrian National Park Hohe Tauern. Publ. Austr. MaB–Programme vol. 13, Universitätsverlag Wagner, Innsbruck, 625 pp., 249–256.

SCHINNER, F.; HOFMANN, J.; NIEDERBACHER, R., 1989, Mikrobielle Aktivitäten des Kohlenstoffmetabolismus in Böden der Alpen, subalpinen und montanen Stufe des Großglocknergebietes (Hohe Tauern).
In: CERNUSCA, A. (ed.), Structure and function of grassland ecosystems in the Austrian National Park Hohe Tauern. Publ. Austr. MaB–Programme vol. 13, Universitätsverlag Wagner, Innsbruck, 625 pp., 257–262.

SCHINNER, M., 1989, Blei–, Cadmium–, Zink– und Kupfergehalte der Vegetation im Bereich der Großglockner–Hochalpenstraße zwischen Heiligenblut und dem Hochtor.
In: CERNUSCA, A. (ed.), Structure and function of grassland ecosystems in the Austrian National Park Hohe Tauern. Publ. Austr. MaB–Programme vol. 13, Universitätsverlag Wagner, Innsbruck, 625 pp., 263–296.

SPATZ, G.; MÜHLSCHLEGEL, F.; JUSSEL, U.; WEIS, G. B., 1989, Zur Futterqualität von Pflanzenbeständen entlang einem Höhengradienten an der Glocknerstraße.

In: CERNUSCA, A. (ed.), Structure and function of grassland ecosystems in the Austrian National Park Hohe Tauern. Publ. Austr. MaB–Programme vol. 13, Universitätsverlag Wagner, Innsbruck, 625 pp., 515–530.

TAPPEINER, U.; CERNUSCA, A., 1989, Auswirkungen des Auflassens einer Almweide auf die Bestandesstruktur und das Lichtklima.
In: CERNUSCA, A. (ed.), Structure and function of grassland ecosystems in the Austrian National Park Hohe Tauern. Publ. Austr. MaB–Programme vol. 13, Universitätsverlag Wagner, Innsbruck, 625 pp., 531–548.

THALER, K., 1989, Epigäische Spinnen und Weberknechte (Arachnida: *Aranei, Opiliones*) im Bereich des Höhentransektes Glocknerstraße–Südabschnitt (Kärnten, Österreich).
In: CERNUSCA, A. (ed.), Structure and function of grassland ecosystems in the Austrian National Park Hohe Tauern. Publ. Austr. MaB–Programme vol. 13, Universitätsverlag Wagner, Innsbruck, 625 pp., 201–216.

WIESER, G.; NEUWINGER, I.; CERNUSCA, A., 1989 Der Bodenwasserhaushalt alpiner Graslandgesellschaften zwischen 1600 und 2600 m Höhe.
In: CERNUSCA, A. (ed.), Structure and function of grassland ecosystems in the Austrian National Park Hohe Tauern. Publ. Austr. MaB–Programme vol. 13, Universitätsverlag Wagner, Innsbruck, 625 pp., 95–118.

Part III

Minerals and vegetation–soils interactions

Chapter 13

Multi–element analysis in plant material

B. MARKERT

13.1 Introduction

Inorganic environmental chemistry is a relatively young and enthusiastic science (HUTZINGER 1980). Studies on the elemental composition of biological systems have produced a lot of analytical data within the last few years (IYENGAR 1989; MCKENZIE and SMYTHE 1988; SANSONI 1987; ZEISLER et al. 1988).

Up to now inorganic environmental chemistry has mainly been related to two objectives:

- Investigations into pollution and toxicity of different elements for various organisms,
- Studies on essentiality or beneficial effects of inorganic substances related to proper growth, development and health of plants, animals and humans.

In particular investigations related to the toxic effects of heavy metals have increased enormously within the last few years (CAROLI et al. 1989; ERNST and JOOSSE VAN DAMME 1983; MARKERT and THORNTEN 1990; NRIAGU and PACYNA 1988). This was promoted by finding direct relationships between trace elements and human health and diseases. F. e. historians have speculated that the decline of the Roman Empire may have been partly caused by a disease affecting the mental capacities of the ruling class, as a result of lead poisoning. The Romans stored wine in pottery that was lined with lead, some of which was leached by the acidic beverage and subsequently ingested (FREEDMAN 1989). Today NEEDLEMAN et al. (1979) found that higher levels of lead in children's milk teeth, a measure of accumulative exposure, correlated with lower scores in tests of intelligence, concentration and other psychological measures. In Minamata, Japan, hundreds died and many more were acutely poisoned as a result of eating mercury–contaminated fish. In this case, an acetaldehyde factory had dumped relatively nontoxic elemental mercury into Minamata Bay. However, microbes in anaerobic sediment converted the elemental mercury into methylmercury. This highly toxic and bioavailable compound of mercury entered into the aquatic foodchain, and caused widespread poisoning of fish–eating birds, cats, and humans (FREEDMAN 1989). Increasing public and medical interest has recently arisen concerning a possible association between human exposure to aluminium from various sources, such as drinking water, and the incidence of Alzheimer's disease in developed countries (SIMPSON et al. 1988). As in most toxicological situations, the dose received by a target organism is not only a function of the concentration of the poison in the environment: it is also a function of the period and duration of exposure.

Therefore, in certain situations a long–term exposure to a small available concentration of some elements may cause a toxic effect. Often this takes place because of a progressive bioaccumulation of the element, until a toxic level is reached (FREEDMAN 1989).

Besides these toxic effects, different chemical elements can play an essential role in the normal growth and development of an organism (BRÄTTER and SCHRAMEL 1988; FIEDLER and RÖSLER 1988; MARSCHNER 1986). After ARNON (1950) an essential element has to fulfil the following demands:

1. the organism can neither grow nor complete its life cycle without it;
2. it cannot be wholly replaced by another element;
3. it has a direct influence on the organism, and is involved in its metabolism.

Up to now about one third of the elements of the periodic table (Fig. 13.1) have achieved the status of essentiality and according to their physiological functions they can be divided into structural, electrolytic and enzymatic elements (Tab. 13.1).

The above mentioned examples may give the impression that all that matters in inorganic chemistry for plant, animal and human nutrition is known. This, however, is a misconception, and should be accepted as such by the total community of scientists involved in biological trace element research (MARKERT 1990). In earlier investigations, a large proportion of the analytical results obtained for various biomaterials were mainly intended to demonstrate the powerful capabilities of the newly emerging methodologies, for example multi–element techniques.

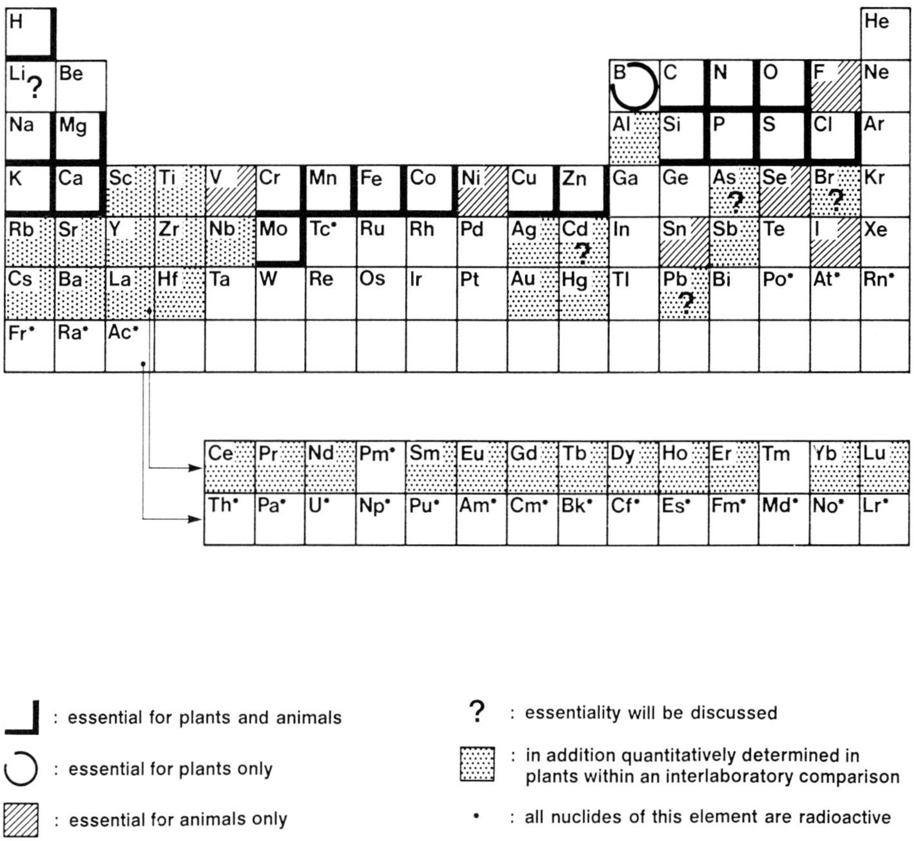

Figure 13.1: The periodic table of the elements with additional information on essentiality and occurence in living organisms.

Unfortunately, during that enthusiastic phase of proving the effectiveness of these extraordinary technical achievements, little or no consideration was given to the biological basis of the problems studied (IYENGAR 1988). For example, a new cleaning procedure involving the removal of the cuticule of spruce needles by chloroform had shown that a main concentration of several heavy metals, which were discussed as toxic substances for the forest decline in central Europe, are located more in the cuticule layer than in the inner part of the needle (WYTTENBACH et al. 1985). Metals located in the cuticule layer of plants have nearly no influence on metabolic processes, so that these results differs extremely from previous analyses of the whole needle. The interpretaion of the older results can lead to completely

Table 13.1: Division of chemical elements according to physiological criteria (changed after SANSONI and IYENGAR, 1978) In brackets: not naturally occurring elements (after HOFMANN and RÜDORFF 1973); ⋆ essential for animals only; ⋆⋆ essential for plants only.

structural elements:	C, H, O, N, P, S, Si, Ca
electrolytic elements:	K, Na, Ca, Cl, Mg
enzymatic elements:	V*, Cr, Mo, Mn, Fe, Co, Ni*, Cu, Zn, B**, Sn*, Se*, F*, I*, Mg

Up to now without biological function:

1st main group:	Li, Rb, Cs, (Fr)	2nd main group:	Be, Sr, Ba, Ra
3rd main group:	Al, Ga, In, Tl	4th main group:	Ge, Pb
5th main group:	As, Sb, Bi	6th main group:	Te, Po
7th main group:	Br, (At)		
8th main group:	He, Ne, Ar, Kr, Xe, Rn		
1st subgroup:	Sc,Y	2nd subgroup:	Ti, Zr, Hf
3rd subgroup:	Tb, Ta	4th subgroup:	W
5th subgroup:	(Tc), Re	6th subgroup:	Ru, Os
7th subgroup:	Rh, Ir	8th subgroup:	Pd, Pt
9th subgroup:	Ag,Au	10th subgroup:	Cd, Hg

Lanthanides:	La, Ce, Pr, Nd, (Pm), Sm, Eu, Gd, Tb, Dy, Ho, Er, Tm, Yb, Lu.
Actinides:	Ac, Th, Pa, U, (Np), (Pu), (Am), (Cm), (Bk), (Cf), (Es), (Fm), (Md), (No), (Lr).

false conclusions. This means inorganic environmental chemistry is an interdisciplinary science which requires hand–in–hand work by biologists, analysts, geologists and physicans.

In this article some ideas for developing an effective concept of modern inorganic environmental chemistry will be given.

13.2 Establishment of baseline values (background concentrations)

The biosphere is the natural environment of living things and is the complex biological epidermis of the earth. Chemical and physical characteristics of the biosphere are determined by the lithosphere, a lower part of the atmosphere, and the hydrosphere. Within evolution relatively constant environments for the existence of living matter in a given ecosystem have been created. The ecosystem is a fundamental environmental system consisting of the community of all living organisms in a given area and having a balanced cycling of elements and energy flow (KABATA–PENDIAS and PENDIAS 1984). In general, the biosphere consists of three main ecosystems—the land environment, the freshwater environment and the marine environment. These fundamental ecosystems include several smaller systems of variable dimensions and conditions (KABATA–PENDIAS and PENDIAS 1984).

The bulk of elements are naturally bound as insoluble inorganic compounds in rocks and sediments. These are unavailable until they are mobilized by weathering processes, like physical breaking, abiotic chemical portioning and biologically mediated dissolution. A multitude of ions can be released from soils or as a consequence of the degradation of organic material and by redox changes (IRGOLIC and MARTELL 1985).

The primary route of entry of trace elements into humans is via the ingestion of foods of plant and animal origin. Thus, it is important to understand the soil–plant–animal relationships. Plants can be considered as intermediate reservoirs through which elements from primary sources are transferred to other organisms (ADRIANO 1986). Elemental uptake can occur via the root pathway or through foliar and stem pathway, or both.

Due to the great influence of plants within the food chain for other organisms a total chemical characterization of plants is of enormous interest. Besides this, only green plants are able to absorb a fraction of solar radiation and store this as chemical energy, which is than used to absorb carbon dioxide from the atmosphere for growth and to produce oxygen essential for all other aerobic reactions.

From the ecochemical point of view it seems to be most interesting to compare the data of the element concentrations in the lithosphere with data in plant systems (Fig. 13.2). The elements which form the earth today were formed by well–defined nuclear processes. It can generally be assumed that uptake from soil solution through roots is the predominant uptake pathway for plants. Uptake of gaseous species from the atmosphere through the leaves is a major pathway for a few constituents (CO_2, O_2, etc.) and may be a common but less widely investigated uptake pathway for other chemicals (BODEK et al. 1988). To get a first impression of the relationship of chemical elements in plant material and the lithosphere, Figure 13.2 compares the concentration found in a specific, well characterized plant species (*Vaccinium vitis-idaea*) with data of the lithosphere extracted from the literature (BOWEN 1979). Fig. 13.2 represents all 88 naturally occurring elements with the

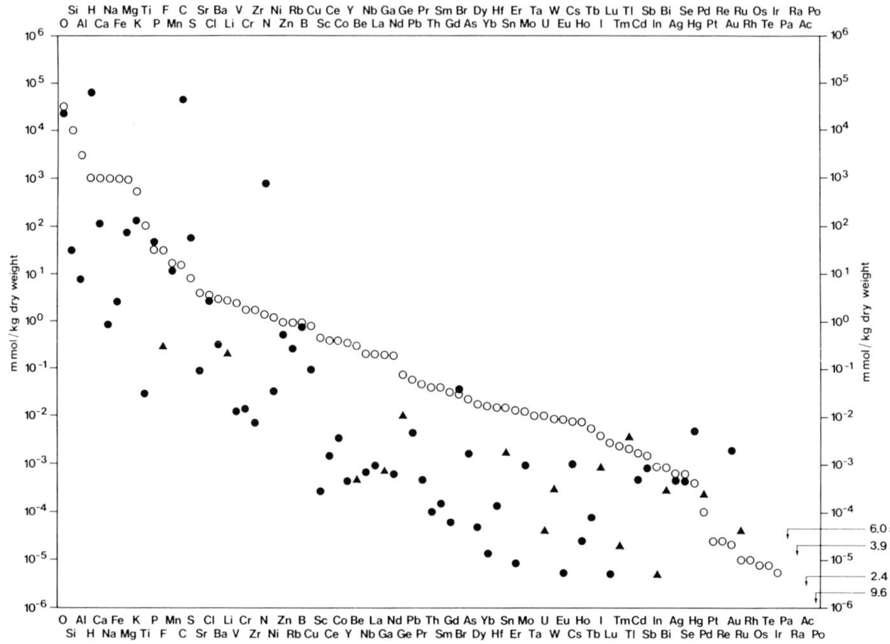

Figure 13.2: Concentration of 88 naturally occurring elements (excluding noble gases) in plant tissues and earth's crust. Elements were ordered according to decreasing occurrence in the earth's crust.
o Values for the earth's crust (extracted from BOWEN, 1979)
• Values for plant material (Leaves of *Vaccinium vitis–idaea*, extracted from MARKERT, 1986)
▲ Additional values for plant material (extracted from BOWEN, 1979).

exception of the noble gases. All nonmetals such as C, H, N, S, and Br appear in higher concentrations and O, P, Cl, and B in similar concentrations as in the earth crust (exception F). Most metals appear in concentrations one or two orders lower than in the earth's crust. Exceptions are K and Mn. The high content of Mn can be explained by the high accumulation rate of Ericaceae, the plant family to which *Vaccinium vitis–idaea* belongs. However, generally nonmetallic elements are present in higher concentrations in plant materials than metallic elements. Irregularities in the behaviour of the remaining third of the elements, which appear in very low concentrations, will be explained later.

Such considerations are a valuable tool in establishing so–called baseline values (background concentrations) in biological samples. Baseline values should reflect the natural concentration of elements in selected organisms of the environment. Especially for elements which are not generally investigated in biological materials due to analytical difficulties, the distribution of elements within the earth's crust can give some indication of the order of magnitude of concentration in which these elements will appear in plant materials. Two examples:

A group of elements which have not been investigated intensively in the past are the rare–earth elements or so–called lanthanides (MARKERT et al. 1989). They are becoming more and more significant in our technical environment. Samarium, praseodymium and cerium are compounds of magnetic materials, lanthanum and yttrium are used for the preparation of the new high–temperature superconductors. Some rare–earth elements are in use as catalysts for chemical reactions or are necessary for the production of laser crystals. In the past little attention has been paid to the concentration and baseline levels of the lanthanides in ecosystems, or to their behaviour and uptake by plants from soil, primarily occasioned by inadequate detection limits of the analytical methods applied, since the concentrations of lanthanides especially in plant tissues are very low. Recently the lanthanides have been determined (Fig. 13.3). Baseline reference values of lanthanide elements in plant material were estimated to fall into the following ranges (all values were given in $mg \cdot kg^{-1}$ dry weight):

La: 0.15–0.2; Pr: 0.030–0.060; Nd: 0.1–0.25; Sm: 0.02–0.04; Eu: 0.005–0.015; Gd: 0.030–0.060; Tb: 0.005–0.015; Dy: 0.025–0.05; Ho: 0.005–0.015; Er: 0.015–0.030; Tm: 0.0025–0.005; Yb: 0.015–0.030 ; Lu: 0.0025–0.005; Y: 0.15–0.25 (MARKERT and ZHANG DE LI 1990).

The measured values of Fig. 13.3 were compared with the expected baseline values which were calculated from the value in the earth's crust divided by factor 100. As can be seen in Fig. 13.4 measured and expected values are in good agreement. It seems that the one hundredth part of the concentration of the earth's crust value is a good indicator of the basic reference values in plant materials for elements distributed in very low concentrations in the environment. This can be especially helpful for about 10 to 20 elements which are not at the moment detectable in plant materials due to the very low content of these elements in natural, unpolluted systems. The platinum metals belong to this group. The platinum metals are a group of 6 metals including platinum, palladium, osmium, iridium, rhodium and ruthenium. The introduction of the catalytic cleaning of motorcar exhaust gases has introduced a new pollution source of, in particular, platinum emission leading to uncontrolled immissions into the environment. Tests made in the U. S. A., where catalytic converters were introduced more than ten years ago, disclosed that, depending on the mode of driving, up to 2 μg platinum and more were emitted per one driven kilometre (ALT 1988; ALT et al. 1988). Due to the low natural concentrations of platinum and the high affinity of this substances to organic molecules, toxic effects can be expected in the future. The analysis of platinum and platinum metals in biological concentrations is just beginning. To get initial information on the natural background concentration of platinum metals and gold

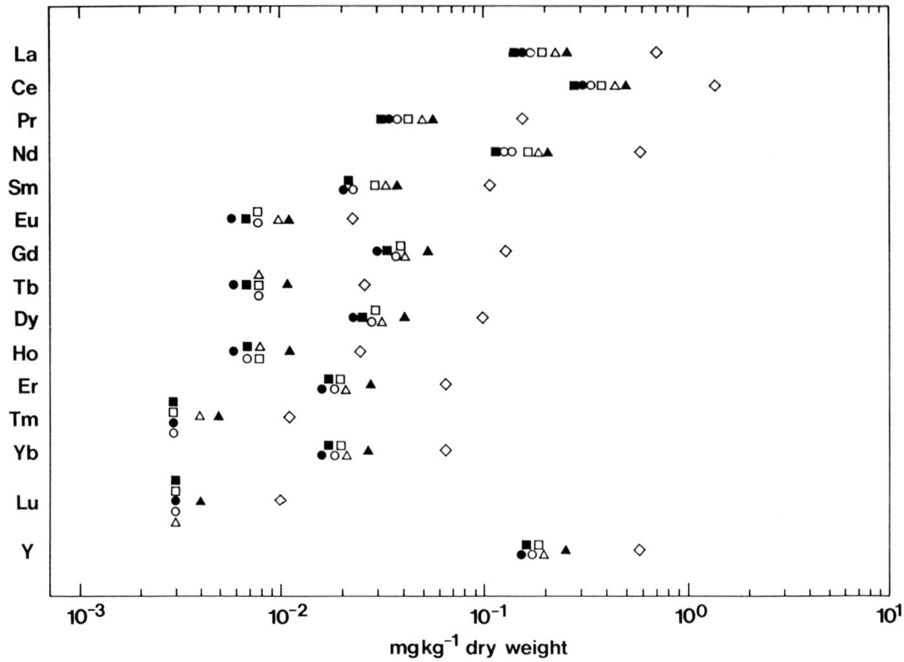

Figure 13.3: Concentration levels of lanthanides found for different plants in a natural forest ecosystem (MARKERT and ZHANG DE LI 1990)
○ *Betula pendula* (leaves); ● *Pinus sylvestris* (needles); □ *Vaccinium vitis-idaea* (leaves); ■ *Vaccinium myrtillus* (leaves); △ *Deschampsia flexuosa* (aboveground parts); ▲ *Polytrichum formosum* (aboveground parts); ◇ *Sphagnum* species (aboveground parts).

in plant materials, the calculated values according to the normalization against the earth's crust described above are given in fig. 13.5. The following background levels were calculated for platinum metals: Pd: 0.0001 mg·kg^{-1} DW, Pt: 0.00005 mg·kg^{-1} DW; Os: 0.000015 mg·kg^{-1} DW; Ir, Rh and Ru 0.00001 mg·kg^{-1} DW.

At the moment analytical data of platinum in plants are available in two reference materials only (Citrus leaves NBS, 1572 and Orchard leaves, NBS 1571). As can be seen in Fig. 13.5, the measured platinum values in citrus leaves are near to the expected value. Orchard leaves represent higher measured values. In addition the same procedure was repeated with gold, and similar good agreement between calculated and measured gold values are observable for citrus leaves. As in the case of platinum in orchard leaves, the gold values determined experimentally are

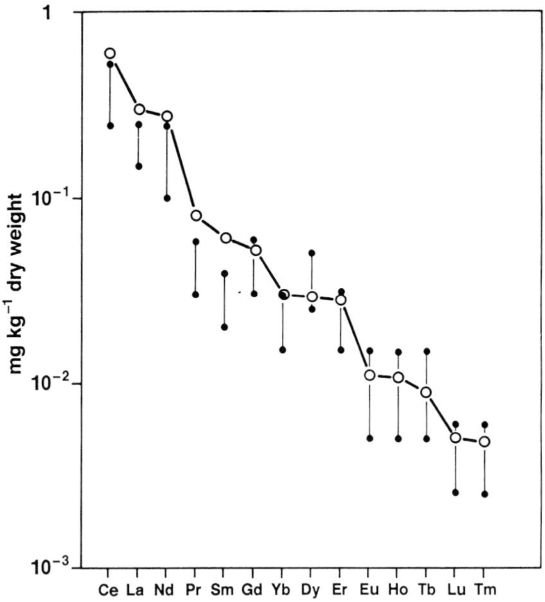

Figure 13.4: Concentration levels of lanthanide elements in different plants of a natural forest ecosystem (•) extracted from Fig. 13.3 compared with baseline values (natural background concentrations) calculated by division of earth's crust values by factor 100 (○).

higher than the expected values, which might be related to a contamination during preparation of this material, a higher accumulation factor for these metals by orchard leaves compared with citrus leaves or that the material was polluted by soil or anthropogenic influences during growth. However, as in the case of lanthanides, the calculation of background values with the help of normalized earth's crust values seems to give a first indication of natural background concentrations for plant materials, especially when no analytical data are available (e. g. for Pd, Os, Ir, Rh and Ru). Coming back to the distribution of the remaining third of the elements in plants of fig. 13.2, it seems that a lot of the latter data do not reflect the natural concentration to be expected in unpolluted or uncontaminated samples. Most of this last group of elements should be expected in lower concentrations (e. g. Tl, Hg, Se, etc.).

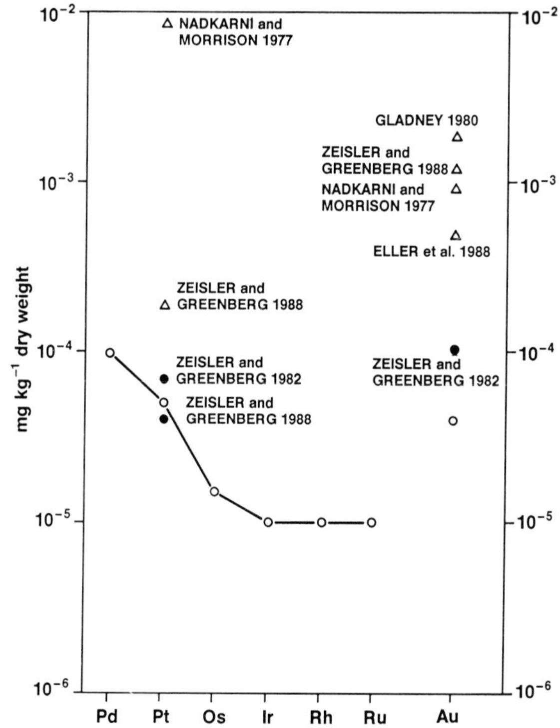

Figure 13.5: Calculated baseline values (natural background concentration) for platinum metals and gold (o)compared with analytical values given in different reference materials of NBS (orchard leaves (•) and citrus leaves (△)).

13.3 Analytical chemistry as an interdisciplinary approach

As can be seen from the examples above, correct and reliable results are the most important guarantee of successful inorganic environmental chemistry. Instrumental techniques available for environmental analysis have reached a high degree of sensitivity for most chemical elements of the periodic table. Especially the ppm and the upper ppb range are open for routine analyses, the lower ppb and ppt range is at the moment a domain for highly specialized laboratories especially equipped with clean room conditions, quartz tools etc. (MARKERT 1990). From the instrumental point of view, optimum equipment have been provided for scientists to perform accurate analytical chemistry. However, in reality results obtained in the past are far from satisfactory. Especially interlaboratory comparisions organized by different national and international institutes demonstrate the analytical disaster of the last three decades (MARKERT 1990). Analytical results obtained in these round robins often differ by some orders of magnitude. In a comparative study HORWITZ (1980) demonstrated that the errors of chemical analysis increase exponentially in the ppb and ppt range. Reasons for errors can be mainly attributed to careless trust in high tech equipment, insufficient care during sample preparation, and unrepresentative sampling. These errors were mainly made by inexperienced persons concusing training in analytical chemistry with a cookery class. This situation is often created by the universities themselves, because analytical chemistry is often taught in a very inadequate and unpopular manner. The most important point of analytical chemistry is, that within an analytical procedure the quality of the results (precision and accuracy) has to be controlled (TÖLG 1989). The precision (reproducibility of single analytical signals) of an analysis can be easily tested by repeating several times the same measurement in the same sample. Only two procedures are possible to check the accuracy of analytical results: a. Use of independent methods for the same sample (JAYASEKERA and MARKERT 1989) or use of certified reference materials available from different organizations (GRIEPINK 1990, International Atomic Energy Agency, 1980). It is most unsatisfactory that for a lot of different matrices and for not commonly investigated elements no reference materials are available at the moment. It is most important that new and well characterized reference materials will be prepared in the near future for all detectable chemical elements (MARKERT 1990).

A second step in environmental analytical chemistry for its reliability is that of sample quality. This means, does the collected sample really represent a portion of the whole; is it selected in such a way as to be truly representative of the whole or the question in mind (KEITH 1988). Since incorrect sampling can introduce an error up to 1000% into chemical analysis, more attention must be paid to this step during the whole procedure of single or multi-element analysis (ALLEN 1974; IYENGAR 1982 and 1989; SANSONI 1985; MARKERT and KLAUSMEYER 1989). Well established sampling procedures developed by experts and intensive discussion (before starting any further activity!)) with experienced operators seem to be the most valuable tools for collecting a representative sample (MARKERT and WECKERT 1989 a and b).

The first holistic approach to determine as many elements as possible in representative plant samples was undertaken in the years 1983 to 1986 by LIETH and MARKERT (1988). The data of chemical analysis were listed in increasing order of magnitudes in so-called element concentration cadasters (ECC) (Fig. 13.6). By comparing different ECC from different plants and various soils interesting environmental results were obtained (LIETH and MARKERT 1988 and 1990). Highly accurate results of multi-element analysis can e. g. be taken for investigating so-called interelement relationships. A chemical balance of inorganic elements in living organisms is a basic condition for their proper growth and development (MARKERT 1988). The interaction of chemical elements is also of similar importance for deficiency and toxicity in the physiology of plants. Interactions between chemical

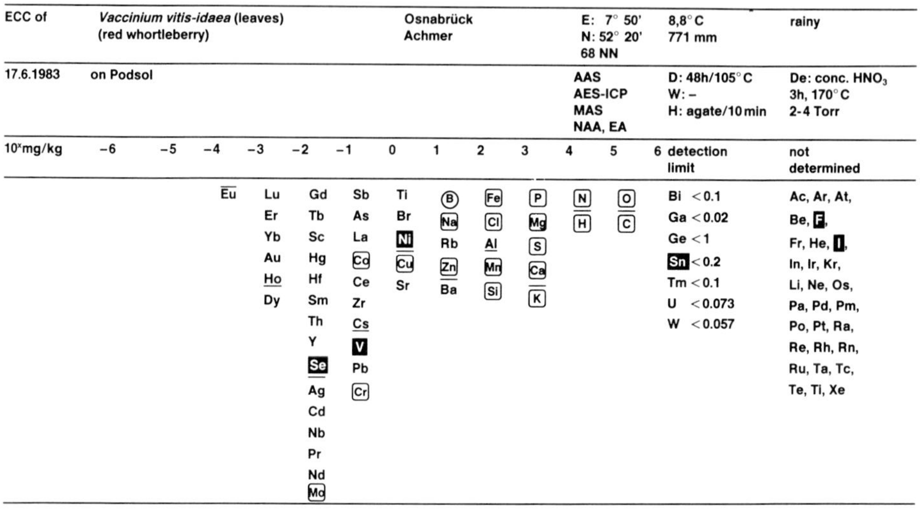

Figure 13.6: Element concentration cadaster of *Vaccinium vitis–idaea* (leaves of red whortleberry); bar: half order of magnitude; D: drying; W: washing; H: homogenization, De: decomposition (modified after LIETH and MARKERT 1990) ○ essential for plants only; □ essential for plants and animals; ■ essential for animals only.

elements may be both antagonistic and synergistic, and their imbalanced reaction may cause a chemical stress in plants. Up to now most of the correlation analysis between different elements in plant materials has been relatively unsatisfying due to obtaining relatively poor correlation. By using the data of plant reference materials, data with high precision and accuracy, a lot of unexpectedly negative or positive correlations were obtained (Fig. 13.7).

However, knowledge of the composition of a sample in terms of total trace element concentrations is a first step towards complete characterization of an environmental sample, but does certainly not provide all or even the most important information. Synergistic or antagonistic interactions of life–sustaining, biologically important molecules in living organisms occur in most cases not with elements or their free ions but with specific compounds of these elements (species). To obtain the necessary knowledge, investigations of the toxic or beneficial effects must be followed by studies identifying the various compounds, their transformations, and their interactions with biomolecules (IRGOLIC and MARTELL 1985). Various methods are available for speciation analysis. Volatile compounds can be determined by gas chromatography with element– or compound–specific detectors such as mass spectrometers, and atomic absorption or emission spectrometers. Non–

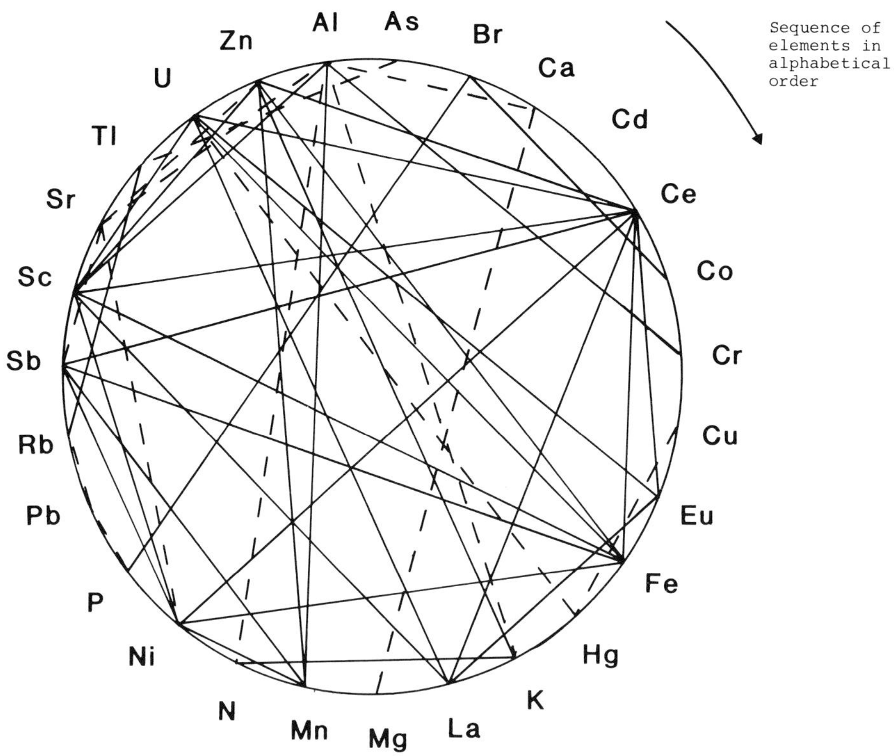

Figure 13.7: Elements which were highly correlated in plant reference materials. Dotted lines: negatively correlated, solid lines: positively correlated. Highly correlated means correlation coefficients $\geq |\pm 0.9|$ (from MARKERT, 1988).

volatile compounds can be separated by high pressure liquid chromatography or ion chromatography and detected by graphite furnace atomic spectrometers or plasma emission spectrometers (IRGOLIC and MARTELL 1985). Analysts are aware of the difficulties associated with the preservation of samples during and after collection for total trace element analysis, and have at their disposal various methods to assure stability of their samples. Little is known about the trace element species in environmental samples and even less about methods for the successful preservation of these compounds during sample storage and sample preparation. It is of vital importance for future work to collect, evaluate, and establish approximate sampling and preservation methodologies for the investigation of natural systems (IRGOLIC and MARTELL 1985). Speciation analysis seems to be the greatest challenge for the next century.

13.4 Conclusions

Inorganic (as well as organic) environmental chemistry is becoming more and more a challenge for scientific research. In the last two decades a trace element has changed its status from being inessential to being essential for living systems approximately every two years, formerly nontoxic elements attained the status of potentially toxic, and recently elements not detectable by instrumental techniques have been found in quantitative amounts in organisms by more sensitive equipment (MARKERT 1990). Besides this a new thinking on trace element research must be established to clarify objectives and interests for biological trace element research in the next century. Perhaps it will then be clarified as HOROVITZ pointed out in 1988 that a distinction between essential and nonessential elements might not exist in nature and each element of the periodic system might play an active role in the physiological chemistry of organisms. From the health point of view it is certain that every single element in the periodic table will produce some detrimental effect on man if the level of exposure is sufficiently abnormal (HAMILTON 1980).

Acknowledgments

Thanks are due to the "Graphische Betriebe" of the Research Centre Jülich for preparing the figures and Mrs. J. Carter–Sigglow for polishing the English of the manuscript.

Bibliography

ADRIANO, D. C., 1986, Trace elements in the terrestrial environment.
Springer–Verlag, Berlin, Heidelberg, New York, 533 pp.

ALLEN, S. E. (ed.), 1974, Chemical analysis of ecological materials.
Blackwell Scientific Publications, Oxford, 557 pp.

ALT, F., 1988, Platinum metals in the environment—A challenge to trace analysis.
in: Brätter, P.; Schramel, P., (eds.), 1988, cited below, 279–298.

ALT, F.; JERONO, U.; MESSERSCHMIDT, J.; TÖLG, G., 1988, The determination of platinum in biotic and environmental materials.
Microchimica acta, **III**, 299–304.

ARNON, D. I., 1950 cited after BOWEN, 1979.

BRÄTTER, P.; SCHRAMEL, P. (eds.), 1988, Trace element analytical chemistry in medicine and biology.
Vol. **5**, Walter de Gruyter, Berlin, New York, 666 pp.

BODEK, I.; LYMAN, W. J.; REEHL, W. F.; ROSENBLATT, D. H., (eds.), 1988, Environmental inorganic chemistry, properties, processes, and estimation methods. Pergamon Press, New York, 1067 pp.

BOWEN, H. J. M., 1979, Environmental chemistry of the elements.
Academic Press, London, 333 pp.

CAROLI, S.; IYENGAR, G. V.; MUNTAU, H., (eds.), 1989, Bioelements: health aspects.
Annali dell'Istituto Superiore di Sanita. Vol. **25**, 373–378.

ELLER, R.; ALT, F.; TÖLG, G.; TOBSCHALL, H. J., 1989, An efficient combined procedure for the extreme trace analysis of gold, platinum, palladium and rhodium with the aid of graphite furnace atomic absorption spectrometry and total reflection x-ray fluorescence analysis.
Fresenius Zeitschrift für Analytische Chemie **334**, 723–739.

ERNST, W. H. O.; JOOSSE VAN DAMME, E. N. G., 1983, Umweltbelastung durch Mineralstoffe.
Gustav Fischer Verlag, Stuttgart, 234 pp.

FIEDLER, H. J.; RÖSLER, H. J., (eds.), 1988, Spurenelemente in der Umwelt.
Ferdinand Enke Verlag, Stuttgart, 278 pp.

FREEDMAN, B., 1989, Environmental Ecology, the impact of pollution and other stresses on ecosystem structure and function.
Academic Press, San Diego, 454 pp.

GLADNEY, E. S., 1980, Elemental concentrations in NBS biological and environmental standard reference materials.
Analytica Chemica Acta **118**, 385–396.

GRIEPINK, B., 1990, Quality and certified reference materials.
In: Lieth, H.; Markert, B., (eds.), cited below, 181–206.

HAMILTON, E. I., 1980, The need for trace element analyses of biological materials in the environmental sciences.
In: IAEA, (eds.), cited below, 39–54.

HOFMANN, U.; RÜDORFF, W., 1973, Anorganische Chemie.
Vieweg Verlag, Braunschweig, 862 pp.

HOROVITZ, C. T., 1988, Is the major part of the periodic system really inessential for life ?
J. Trace Elem. Electrolytes Health Dis. **2**, 135–144.

HORWITZ, W.; KAMPS, L. R.; BOYER, K. W., 1980, Quality assurance in the analysis of foods for trace constituents.
J. Assoc. Off. Anal. Chem. **63**, 1344–1354.

HUTZINGER, O., (ed.), 1980, The handbook of environmental chemistry, Vol. **1**, Part A: The natural environment and the biogeochemical cycles.
Springer–Verlag, Berlin, Heidelberg, New York, 192 pp.

INTERNATIONAL ATOMIC ENERGY AGENCY, 1980, Element analysis of biological materials—Current problems and techniques with special reference to trace elements.
Technical Report Series No. **197**, Vienna, 367 pp.

IRGOLIC, K. J.; MARTELL, A. E. (eds.), 1985, Environmental inorganic chemistry.
VCH–Publisher, Weinheim, 654 pp.

IYENGAR, V. G., 1982, Presampling factors in the elemental composition of biological systems.
Anal. Chem. **54**, 554A–560A.

IYENGAR, V. G., 1988, Biological trace element research: A multidisciplinary science.
The Science of the Total Environment **71**, 1–5.

IYENGAR, V.G., 1989, Elemental analysis of biological systems, Vol. **1**: Compositional and methodological aspects.
CRC Press, Boca Raton, 430 pp.

JAYASEKERA, R.; MARKERT, B., 1989, Multi–laboratory chemical characterization of ecological samples.
Fresenius Zeitschrift für Analytische Chemie **334**, 226–230.

KABATA–PENDIAS, A.; PENDIAS, H., 1984, Trace elements in soils and plants.
CRC Press, Boca Raton, Florida, 315 pp.

KEITH, L. H. (ed.), 1988, Principles of environmental sampling.
ACS Professional Reference Book, American Chemical Society, 458 pp.

LIETH, H.; MARKERT, B., 1988, Aufstellung und Auswertung ökosystemarer Element–Konzentrations–Kataster.
Springer–Verlag, Berlin, Heidelberg, New York, 193 pp.

LIETH, H.; MARKERT, B. (eds.), 1990, Element concentration cadasters in ecosystems, methods and assessment.
VCH–Publisher, Weinheim, 448 pp.

MARKERT, B., 1986, Aufstellung und Auswertung von Element–Konzentrations–Katastern in unterschiedlichen Pflanzenarten und Bodentypen in Deutschland, Österreich und Schweden.
In: Stoeppler, M.; Dürbeck, H. W., (eds.), Beiträge zur Umweltprobenbank, Spezieller Bericht der KFA Jülich, No. **360**, 166 pp.

MARKERT, B., 1988, Interelement correlations in different reference materials. Fresenius Zeitschrift für Analytische Chemie, **332**, 630–635.

MARKERT, B., 1990, Multi–element analysis in ecosystems— analytical tools and biological questions.
Lecture given at the International Conference on Trace Metals in Soils, Plants, Animals, and Waters in Orlando, Florida, U. S., from 30.4 – 5.5.1990. The conference proceedings will be published by CRC–Press, Boca Raton, Florida.

MARKERT, B.; PIEHLER, H.; LIETH, H.; SUGIMAE, A., 1989, Normalization and calculation of lanthanide element concentrations in environmental samples. Radiat. Enviro. Biophysics **28**, 213–221.

MARKERT, B.; WECKERT, V., 1989a, Fluctuations of element concentrations during the growing season of *Polytrichum formosum* (Hedw.).
Water, Air, and Soil Pollution **43**, 177–189.

MARKERT, B.; WECKERT, V., 1989b, Use of *Polytrichum formosum* (moss) as a passive biomonitor for heavy metal pollution (Cadmium, Copper, Lead and Zinc). The Science of the Total Environment **86**, 289–294.

MARKERT, B.; KLAUSMEYER, 1990, Variations in the elemental compositions of plants, and computer aided sampling in ecosystems.
Toxicological and Environmental Chemistry **25**, 201–212.

MARKERT, B.; THORNTON, I., 1990, Multi–element analysis of an English peat bog soil.
Water, Air, and Soil Pollution **49**, 113–123.

MARKERT, B.; ZHANG DE LI, 1990, Baseline levels of rare earth elements in a natural ecosystem.
The Science of Total Environment, in press.

MARSCHNER, H., 1986, Mineral nutrition of higher plants.
Academic Press, London, New York, 762 pp.

MCKENZIE, H. A.; SMYTHE, L. E. (eds.), 1988, Quantitative trace analysis of biological materials.
Elsevier, Amsterdam, New York, Oxford, 791 pp.

NADKARNI, R. A.; MORRISON, G. H., 1977, Neutron activation determination of noble metals using a selective group seperation scheme.
J. Radioanal. Chem. **38**, 435.

NEEDLEMAN, H. L.; GUNNOE, C.; LEVITON, A.; REER, R.; PERESI, H.; MAHLER, L.; BARRET, T. P., 1979, Deficity in psychologica and classroom performance of children with elevated dentine lead levels.
N. Engl. J. Med. **300**, 689–695.

NRIAGU, J. O.; PACYNA, J. M., 1988, Quantitative assessment of worldwide contamination of air, water and soils by trace metals.
Nature **333**, 134–139.

SANSONI, B., (ed.), 1985, Instrumentelle Multielementanalyse.
VCH–Publisher, Weinheim, 782 pp.

SANSONI, B., 1987, Multi–element analysis for environmental characterisation.
Pure & Applied Chemistry **59**, 579–610.

SANSONI, B.; IYENGAR, V., 1978, Sampling and sample preparation methods for the analysis of trace elements in biological materials.
Nuclear Research Centre Jülich, Jül. Spez. **13**, 77 pp.

SIMPSON, A. M.; SOLLARS, C. J.; PERRY, R., 1988, A European overview of Aluminium in drinking water.
In: Thornton, I. (ed.), cited below, 201–228.

THORNTON, I. (ed.), 1988, Geochemistry and Health, Proceedings of the Second International Symposium.
Science Review Ltd., Northwood, 320 pp.

TÖLG, G., 1989, Analytical chemistry and the quality of life.
Kontakte (Darmstadt) **2**, 20–29.

WYTTENBACH, A.; BAJO, S.; TOBLER, L.; KELLER, TH., 1985, Major and trace element concentrations in needles of *Picea abies*; levels, distribution functions, correlations and environmental influences.
Plant and Soil **85**, 313.

ZEISLER R.; GREENBERG, R. R., 1982, Ultratrace determination of Platinum in biological materials via neutron activation and radiochemical separation.
Journal of Radioanalytical Chemistry **75**, 27–37.

ZEISLER, R.; STONE, S. F.; SANDERS, 1988, Sequential determination of biological and pollutant elements in marine bivalves.
Anal. Chem. **60**, 2760–2765.

ZEISLER, R., 1988, Determination of baseline platinum levels in biological materials.
In: Brätter, P.; Schramel, P., (eds.), 1988, p. 297–303, cited above.

Chapter 14

Significance of nutrient relations and symbiosis for the competitive interaction between grasses and legumes in tropical savannas

E. MEDINA and B. BILBAO

14.1 Introduction

Savanna ecosystems in the tropics appear to be controlled by interactions between water and nutrient availability. Strong seasonality in water availability and general low soil fertility are common denominators for all tropical savanna areas. This basic environmental framework is modified by more unpredictable factors such as frequency of fires, intensity of herbivory and land management procedures. Complex interactions among these factors produce a wide range of savanna types differing in species composition and population structure (MEDINA and SILVA 1990).

The species composition of neotropical savannas, excluding the tree components, is constituted mainly by the monocot families Graminae and Cyperaceae, and by Leguminosae (SARMIENTO 1984). Grasses however, are the absolute dominants in terms of space occupation and biomass production. On the other hand, in many savannas diversity of legumes is larger than that of grasses. At first glance it would appear that in an environment where plant productivity is limited by N (MEDINA et al. 1978), particularly in grasslands under the influence of annual fires, legumes would have a competitive advantage against the grasses because of their potential for fixation of atmospheric N. However, savanna soils in the neotropics are also acid, with high mobility of Al, situation leading to a relatively low availability of P.

Knowledge of the nutritional requirements of savanna plants has both scientific and practical significance. Soils of low fertility select for species with low nutrient requirement and therefore low growth potential (CHAPIN 1980), or for species highly efficient in the exploitation of soil nutrient resources (e. g. through the development of a high root/shoot ratio). These characteristics may determine competitive ability and evolution of savanna species. On the other hand, plant matter produced by species from low fertility sites tends to be of low palatability and nutrient content for the herbivores which forage on them. Since the savanna areas are centers of intense cattle raising activities, the question of nutritional relationships of savanna plants acquires considerable economic importance.

In this paper we will explore the subject of grass–legume relationships in natural savannas focusing on their differences in organic matter production capacity and nutritional requirements, as a basis to understand the structure of savanna ecosystems. For this purpose we analyse the species composition and soil physico–chemical properties of eleven *Trachypogon*–savanna sites in Venezuela. We also compare the nutrient content of legume and grasses and evaluate the N_2–fixing capacity and the presence of mycorrhizal associations in selected legume species. Results are then discussed and compared with experimental results dealing with the regulation of organic matter production through nutrient availability both in legumes and grasses. Our goal is to gain insight in the nutritional factors affecting abundance and diversity of legume species in tropical savannas.

14.2 Soil characteristics and vegetation composition

Trachypogon–savanna sites were selected in Venezuela to cover as much as possible the variability in species composition, particularly in regard to the abundance of legume species. General descriptions of species composition and distribution of savanna types in Venezuela may be found in BLYDENSTEIN (1962, 1967) and RAMIA (1967).

The properties of the superficial soil samples (0–30 cm) of the savanna sites differ widely both in chemical composition and texture (Table 14.1). Soil pH ranges from 5.1 to 6.3 and is inversely correlated with organic matter content and the logarithm

Table 14.1: Chemical properties and texture of *Trachypogon*-savanna sites in Venezuela: A, B, C from Cojedes State; D, E, F, G, K from Anzoátegui State; H from Bolívar State; and I, J from Monagas State. Methods of soil analysis described in Anonymous (1967)[1].

Site:	A	B	C	D	E	F	G	H	I	J	K
pH[2]	5.2	5.1	5.5	5.2	5.3	5.6	5.4	6.3	6.3	6.2	5.9
N[3]	1.1	1.3	1.4	1.1	1.8	1.1	1.2	0.7	0.7	0.7	0.6
P[4]	4.8	2.9	5.0	2.4	5.3	4.2	3.2	2.9	2.2	1.7	3.5
OM[5]	2.0	1.6	3.0	1.4	1.3	1.3	1.1	1.1	0.7	0.7	0.8
Ca[6]	0.21	0.12	0.22	0.22	0.26	0.17	0.24	0.36	0.35	0.3	0.4
K[6]	0.22	0.17	0.16	0.06	0.04	0.05	0.07	0.03	0.03	0.03	0.06
Al[6]	0.93	1.95	1.46	1.48	0.5	0.16	0.42	0.21	0.11	0.18	0.06
ΣB[6]	0.62	0.29	0.62	0.46	0.36	0.29	0.37	0.47	0.43	0.37	0.53
EA[6]	1.46	2.77	2.04	1.65	0.75	0.24	0.55	0.30	0.13	0.20	0.10
CEC[6]	2.08	3.06	2.67	2.11	1.11	0.53	0.92	0.77	0.56	0.57	0.63
sand[7]	56.8	43.0	55.8	61.5	73.0	71.5	77.5	72.5	75.0	80.0	75.0
silt[7]	10.5	13.2	11.5	15.0	12.5	15.0	12.5	5.0	5.0	2.5	2.5
clay[7]	32.7	43.7	30.5	23.5	14.5	13.5	10.0	22.5	20.0	17.5	22.5

[1] Kjeldahl N; P and cations extracted with double acid solution; texture measured with Bouyoucos cylinder method. Cations measured with atomic absorption spectrophotometry, and P measured colorimetrically using the method of Murphy and Raleigh.

[2] Soil/water 1:5

[3] $mg \cdot g^{-1}$

[4] available P, $\mu g \cdot g^{-1}$

[5] OM organic matter content (%).

[6] ΣB sum of bases, EA exchangable acidity, CEC cation exchange capacity, in $meq \cdot 100g^{-1}$ air dried soil.

[7] % of dry soil.

of available Al, but directly correlated with the available Ca. Therefore Al and Ca in the soil solution are inversely correlated, although not on a linear scale. The organic matter content is significantly and positively correlated with the contents of N, P and K, and the cation exchange capacity. As expected, the cation exchange capacity is highly and positively correlated with the % clay ($r^2 = 0.68$) and %

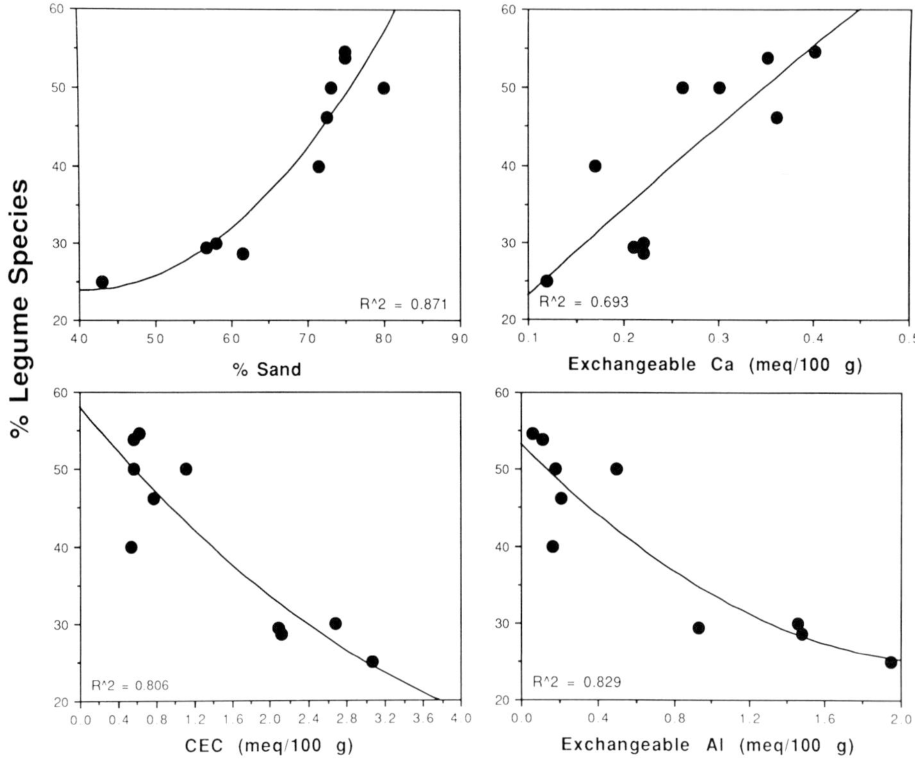

Figure 14.1: Regressions between percentage of legume species in *Trachypogon*–savannas and soil properties. Non–linear regressions correspond to the fitting of second order polynomials. For each graph the determination coefficient is given as the square of the correlation coefficient.

organic matter ($r^2 = 0.61$) and negatively correlated with % sand ($r^2 = 0.88$).

Phosphorus availability (P [μg·g^{-1}]) could be predicted in these savannas with a multiple linear regression using total N (N [mg·g^{-1}]), Σ bases (ΣB [meq·100g^{-1}]) and the cation exchange capacity (CEC [meq·100g^{-1}]); $r^2 = 0.79$:

$$P = 3.392N + 5.91\Sigma B - 0.575CEC - 1.019 \qquad (14.1)$$

A principal component analysis run with the samples of Table 14.1 separates 3 groups based mainly on increasing % sand content and therefore decreasing CEC. The first group is constituted only by the soil B, differentiated by its relatively lower sand content and high content of mobile Al and very low Ca availability and Σbases. The second group is constituted by the soils A, C, D and E, characterized

Table 14.2: Species composition, relative cover and importance value index (IVI = relative density + relative frequency + relative dominance) for plant groups of the *Trachypogon*–savannas sampled.

site:	A	B	C	D	E	F	H	I	J	K
GRASSES										
Aristida riparia Trin.										•
Andropogon brevifolius Sw.	•	•	•							
Andropogon sp.					•	•				
Axonopus canescens (N & T)Pil.		•			•	•	•	•	•	•
A. purpusii (Mez) Chase	•									
Eragrostis acutiflora (HBK)N.	•									
Hackelochloa granularis (L.)K.	•									
Paspalum plicatulum Michx.	•									
Trachypogon plumosus (H & B)N.	•	•	•	•	•	•	•	•	•	•
LEGUMES										
Caesalpinaceae										
Chamaecrista calycioides (DC) Gr.	•							•		•
C. cultrifolia (HBK) Br.& R.				•						
C. flexuosa (L.) Gr.							•		•	
C. rotundifolia (Pers.) Gr.					•					
C. tetraphyla Martyn						•			•	
C. serpens (L.) Gr.		•	•							
Fabaceae										
Aeschynomene brasiliana (Poir.) DC										•
A. histrix Poir.								•	•	
Calopogonium mucunoides Desv.	•									
Centrosema angustifolium (HBX) Benth.	•									
Clitoria guyanensis (Aubl.) Benth.					•	•	•			
Crotalaria stipularis Desv.							•			
Eriosema crinitum (HBK) G.Don	•				•		•	•	•	•
Galactia jussieuana K.	•			•	•		•			
Indigofera pasquorum Benth.	•									
Phaseolus diversifolius Pittier					•					
P. pilosus HBK							•			
Stylosanthes guianensis (Aubl.) Sw.								•	•	•
Stylosanthes sp.		•	•	•						
Tephrosia cinerea (L.) Pers.										•
Zornia curvata Mohlb.							•	•		•

continued next page

Table 14.2, continued

site	A	B	C	D	E	F	H	I	J	K
Mimosaceae										
Mimosa camporum Benth.						•				
M. martensys Br.& Rose		•	•							
SEDGES										
Bulbostylis capillaris (L.)Clarke				•						
B. conifera K.		•						•		
B. junciformis (HBK) Lindm.							•	•	•	•
B. svenzoniana Stey.							•			
Dichronema ciliata Vahl	•	•								
Rhynchospora subplumosa C.B.Cl.		•								
OTHER DICOTS										
Borreria capitata (R & P)DC	•									
Cuphea micrantha K.		•	•							
Diodia teres Walt.		•	•	•					•	
Euphorbia dioica HBK	•	•	•		•					
Evolvulus villosissimus v.Oost.							•	•		
Hyptis suaveolens (L.)Poit.		•								
Pectis elongata HBK							•			
Polycarpea corimbosa (L.) Lamb.										•
Polygala longicaulis HBK				•	•	•		•		
P. monticola HBK		•		•		•				
Ruellia geminiflora HBK	•					•				
Sida linifolia Juss. ex Cav.		•	•							

Summary:

site	A	B	C	D	E	F	H	I	J	K
No. of species	17	16	10	7	10	10	13	13	10	11
Relative cover %										
Legumes	23	15	26	56	12	48	18	21	6	55
Grasses	69	72	64	34	87	37	78	78	94	44
IVI										
Legumes	52	66	112	113	46	106	52	96	82	127
Grasses	192	172	145	105	216	133	193	163	191	132

by higher Σbases and P availability and intermediate levels of mobile Al. The third group, constituted by the soils F, G, H, I, J, K are mostly sandy soils of relatively recent origin near the Orinoco river. Those soils are characterized by high % sand, the lowest mobile Al in the soils sampled, and the lowest values of % organic matter and N contents. In addition, they occur in areas with rainfall values above 1400 mm and are therefore very low in available K.

The phytosociological analysis of the the savanna sites discussed above gave also a large variability among sites regarding species composition, grass dominance and presence of legumes (Table 14.2). The method employed was the counting of individuals along replicated 10 m long transects in each savanna site. From these numbers relative cover and importance value indices were calculated (MÜLLER–DOMBOIS and ELLENBERG 1974).

The relationships between vegetation structure and soil characteristics were not simple. Most savanna sites had 7–13 species with the exception of two sites located in the Cojedes state. The % of legume species ranged from 25–30 % in sites from Cojedes and Anzoátegui and were around 50 % in the other sites. With the exception of savanna A there were more legume species than grasses in each transect. The number of legume individuals or species does not appear to be correlated with any of the soil properties reported in Table 14.1. However, the % of legume species is significantly and positively correlated with % sand and exchangeable Ca, but negatively correlated as would be expected from the discussion of data from Table 14.1, with CEC and exchangeable Al (Fig. 14.1). Per cent of legume species is also negatively correlated with % organic matter and total soil N (not shown).

No direct relationship between legume abundance and P availability could be detected, although the relationship with Ca, Al and CEC could be considered as indirect evidences of the limiting role of P availability. The measurement of extractable P within the root zone does not allow a true evaluation of the availability of P. Measurement of P mineralization rates, or available P below the root zone would be required for a better appraisal of the P nutrition in these grasslands (MEDINA and SILVA 1990).

Our results confirm earlier reports on the species composition and biomass distribution of *Trachypogon*–savannas. Several authors showed that the % cover of legume species is always low, in spite of the fact that on a per area basis legume species frequently outnumber grass species (BLYDENSTEIN 1961, VELÁSQUEZ 1965). In seven sites of *Trachypogon*–savannas in central–eastern Venezuela with low to moderate soil fertility, percentage of grass biomass ranged from 80–100 %, while legumes made up to 3.8 % of total above ground biomass in only one site with the highest pH (ca. 5.5) and relatively high base saturation (81 %) (SAN JOSÉ et al. 1985). Similarly, in a detailed study of the distribution of grassland types along a gradient of increasing higher phreatic level in the northern Amazon Territory in Venezuela, significant representation of legume species was found only on relatively well drained soils of moderate fertility (FÖLSTER and HUBER 1984).

Table 14.3: Allocation of biomass and nutrients to underground organs, and occurrence of nodulation and xylopods in legume species growing under natural conditions in *Trachypogon*–savannas in Venezuela. Nodulation given as total nodule volume per unit root dry weight (number of nodules × nodule average volume divided by the root dry weight).

	Nodules	Xylopod	% allocation to underground organs			
	$mm^3 \cdot g^{-1}$ d.w. of roots	frequency	biomass	N	P	K
Chamaecrista desvauxii	0	0.0	2	1	1	1
Crotalaria stipularis	46	0.0	8	11	2	13
Aeschynomene hystrix	237	1.0	10	9	9	7
Phaseolus pilosus	51	0.0	11	7	8	5
Mimosa martensys	0	0.0	12	10	13	7
Phaseolus diversifolius	9	0.0	22	19	22	37
Zornia curvata	13	0.0	23	40	22	33
Stylosanthes guyanensis	55	1.0	26	29	18	14
Chamaecrista cultrifolia	43	0.0	36	34	34	19
Chamaecrista calycioides	37	0.4	38	27	26	10
Galactia jussieuana	56	1.0	41	36	28	24
Mimosa debilis	7	0.0	46	49	63	10
Chamaecrista serpens	34	0.5	48	46	50	45
Tephrosia cinerea	19	1.0	52	50	43	38
Indigofera pasquorum	0	1.0	54	69	55	29
Chamaecrista flexuosa	4	1.0	57	54	48	34
Chamaecrista rotundifolia	0	0.0	60	38	32	7
Mimosa camporum	0	1.0	60	66	74	51
Eriosema crinitum	42	1.0	61	61	54	57
Aeschynomene brasiliana	30	1.0	63	61	53	29
Clitoria guyanensis	2	1.0	69	63	68	69
Chamaecrista tetraphyla	4	1.0	79	67	74	68

14.3 Biomass distribution and storage of nutrient and mobile carbohydrates

In southamerican savannas the occurrence of enlarged roots, both in herbaceous and woody dicots, is a common phenomenon. Some of these roots constitute xylopodia and have been identified as "tuberous–ligneous, gemmiferous taproots", while other are truly fleshy roots specialized as storage structures (RIZZINI and HERINGER 1961). These organs have been considered to be of ecological significance as endurance structures, providing regeneration of the aboveground biomass after drought or fire. These structures are quite common in legume species, and many species allocate more than 30 % of their total biomass to underground organs (Table 14.3). These organs accumulate also a significant proportion of the plant nutrient reserves. The percentage of N and P accumulated in underground organs is proportional to biomass. However, accumulation of K in underground organs is much more variable.

Non–structural carbohydrates are accumulated in the enlarged roots where they can reach high concentrations (Table 14.4). Measurements included in Table 14.4 were conducted on plants collected in natural savannas during the rainy season, when the reserves may have been remobilized. This probably explains the large variability observed in the samples. These data support the concept that xylopodia and other enlarged underground organs acts as endurance structures allowing regrowth of new branches and leaves after prolongued periods of drought or fire that kill regularly the aboveground biomass in these dicot species.

In contrast to the legume species in the *Trachypogon*–savannas, grasses develop a fibrous, well branched root system able to occupy densely the soil volume in which the plant is growing. Root biomass in *Trachypogon*–savannas oscillate around 1 $kg{\cdot}m^{-2}$ with a variation coefficient of around 25 % (SUSACHS 1980, BULLA et al. 1984). In general the proportion of biomass allocated to roots in grasses (or better underground biomass because rhizomes are also included) is very large and averages 75 % of total biomass in a series of savanna sites in southern Anzoátegui State in Venezuela (BULLA et al. 1984). It is clear that underground biomass in savanna grasses constitute also a large reservoir of non–structural carbohydrates and nutrients, allowing perennial species fast regrowth after fire, drought, or herbivore consumption.

14.3.1 Nutrient content of leaves and roots

The legume species sampled in the eleven sites show intermediate levels of nitrogen, but very low levels of K and particularly P (Fig. 14.2). In general root tissues are lower in N, P and K compared with leaf tissue, although that is not always the case, particularly in herbaceous legumes species. The difference in nutrient requirements of legumes and grasses can be seen from the comparison of nutrient levels in mature leaves of species growing under natural conditions with those which have been grown under different levels of nutrient supply to estimate the critical levels for the occurrence of growth limitation (Table 14.5). For species in their natural environment legumes have considerably higher values of N but similar values of P and K as those measured in grasses of similar sites. The P deficiency of the legume species is particularly striking.

Table 14.4: Concentration of non–structural carbohydrates in underground organs of selected legume species from the *Trachypogon*–savannas in Venezuela. Dry plant material was extracted alternatively with water, ethanol (soluble carbohydrates) or 52 % perchloric acid (insoluble non–structural carbohydrates). The carbohydrates extracted were measured using the anthrone method (HASSID and NEUFELD 1964). The values are expressed as mg of glucose equivalents per g dry weight.

Species	Organ	Soluble	Insoluble	Total
Chamaecrista rotundifolia				
Site F	leaves	28	10	38
	roots	13	243	256
Chamaecrista tetraphyla				
Site F	roots	24	246	270
	xylopodium	58	828	886
Site G	roots	3	39	42
	xylopodium	5	35	40
Clitoria guyanensis				
Site F	leaves	57	193	250
	roots	12	124	136
	xylopodium	75	529	604
Eriosema crinitum				
Site D	leaves	46	74	119
	roots	34	42	77
	xylopodium	84	103	186
Site E	leaves	25	53	78
	xylopodium	35	132	167
Site F	roots	6	290	296
	xylopodium	78	814	892
Galactia jussieuana				
Site E	leaves	11	186	197
	xylopodium	116	466	583
Site G	leaves	4	40	44
	xylopodium	6	78	84
Mimosa camporum				
Site G	leaves	31	55	86
	roots	19	378	397
	xylopodium	54	249	313
Phaseolus diversifolius				
Site G	leaves	11	44	55
	roots	52	19	70

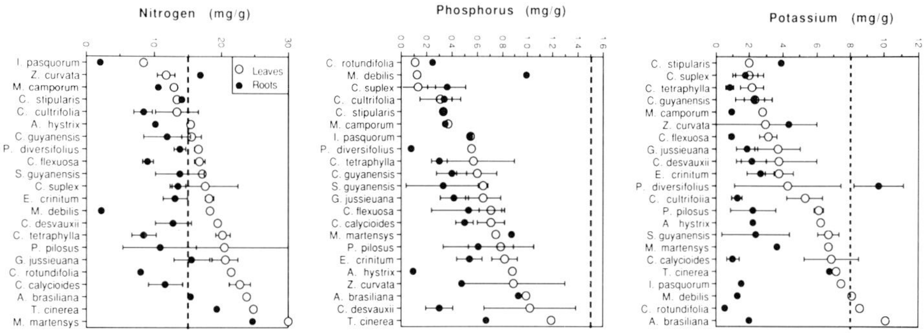

Figure 14.2: Nutrient content of leaves and roots of legume species in *Trachypogon*–savannas. The ranges associated with each point correspond to the standard deviation of the concentration. The number of samples for each point is variable (from 1–5) depending on the frequency of each species. The broken line indicates the assumed level of sufficiency for each element.

14.3.2 Nutrient limitation of productivity of native grass species

Low nutrient levels of N, P, Ca and S are commonly found in South American, West African and northern Australian savanna soils. This condition of low nutrient availability seems to be accentuated by the recurrence of fire, through which most of the N and S remaining in the dead biomass during the dry season is volatilized (MEDINA et al. 1978). It has been suggested that dominant grass species occurring in these soils may have adapted in the course of evolution to grow and reproduce under limited nutrient supply, mainly P (CHRISTIE and MOORBY 1975, MOTT and POPENOE 1977). The fact is that dominant grass species in seasonal, non flooded savannas, at least in South America and northern Australia, have low productive capacity and therefore impose lower demands on soil nutrients (CHAPIN 1980). There is no experimental evidence, however, that they might be more efficient in the use of nutrients as compared with grasses from nutrient rich environments. The capacity of savanna grasses to extract nutrients from poor soils is probably more related to the maintenance of a dense root system and a high root/shoot ratio.

Fertilization experiments in natural grasslands have provided additional insight into the nutritional requirements of tropical savanna grasses. As early as 1938 WEINMAN showed in South Africa that fertilization of a native grassland with N resulted in increased above ground biomass production. Addition of P and K alone did not increase production although the P concentration in forage increased

Table 14.5: Comparative values of leaf nutrient levels of legumes and grasses growing under natural conditions in *Trachypogon*–savannas in Venezuela compared to critical levels of cultivated pastures. Average values followed by standard deviations in parenthesis.

Species	N	P	K	Reference
LEGUMES				
Wild species ($n = 22$)	18.1 (4.9)	0.62 (0.30)	5.1 (2.4)	(1)
Critical levels for cultivated species ($n = 8$)	ca. 20.0	2.11 (0.03)	6.6 (2.3)	(2)
GRASSES				
Trachypogon–savanna ($n = 2$)	8.9 (1.5)	0.76 (0.07)	5.5 (1.8)	(3)
Critical level for cultivated species ($n = 3$)	9.9 (3.2)	0.82 (0.43)	6.8 (0.9)	(4)

References: (1) This paper; (2) ANDREW and ROBINS 1969 a, b; (3) MEDINA et al. 1978; (4) MALAVOLTA et al. 1974.

significantly. In Australian grasslands dominated by *Themeda australis* NORMAN (1966) showed that the addition of neither N or P alone, but only a combination of both, increased above ground biomass production. The experiment also showed that application of P always increased significantly the concentration of this element in forage, although the magnitude of the concentration increase was lower when N was applied simultaneously, probably as a result of stimulated growth. Similar results were reported by MEDINA et al. (1978) and GARCÍA–MIRAGAYA et al. (1983) for *Trachypogon*–savannas in Central Venezuela. Aboveground biomass production was increased by the addition of N or N + P simultaneously, but the increase in N concentration was much lower than that of P indicating the more pronounced dilution of N due to stimulated growth (Table 14.6). MEDINA et al. (1978) showed also that apparent retranslocation of P from biomass at the end of the growing season was very low in all plots with added P.

14.3.3 Biological interactions improving nutrient availability

It is becoming clear that biological interactions between roots and soil microorganisms play an important role in the nutrient relationships of higher plants, partic-

Table 14.6: Effect of fertilization on above ground biomass production in *Trachypogon*–savannas of Central Venezuela (calculated from curves in SAN JOSÉ and GARCÍA–MIRAGAYA, 1981). Fertilizer 100 kg·ha^{-1} of N, P, and K.

Treatment	Peak Biomass g·m^{-2}	Nitrogen mg·g^{-1}	% Increase Biomass	N
Control	222	5.3	–	–
N	250	6.1	13	15
NK	247	6.3	11	19
NP	313	7.4	41	40
NPK	373	7.3	68	38
		Phosphorus mg·g^{-1}	% Increase Biomass	P
Control	223	0.51	–	–
P	263	1.50	18	194
PK	280	1.19	26	133
PN	314	1.88	41	269
PNK	372	1.85	67	263

ularly in soils of intrinsic low nutrient availability. For tropical savannas 3 different associations are of importance: a) *Rhizobium* based symbiosis, b) Vesicular–arbuscular mycorrhizal (VAM) symbiosis, and c) Root associations with free–living microrganisms capable of fixation of atmospheric N such as *Azospirillum*. The first type occurs only in legumes, while the other two systems may be found in association with both legumes and grasses. In addition, free–living blue–green algae can make significant contributions to the N budget of tropical savanna soils.

Legume–*Rhizobium* and mycorrhizal symbiosis. Under the conditions of N deficiency which prevail in most savanna soils, legumes might have a competitive edge against grasses due to the symbiotic fixation of atmospheric N. Nodulation in south american legumes does in fact seem to be widespread and abundant. BARRIOS and GONZÁLEZ (1971) made an extensive survey of nodulated legumes in venezuelan savannas. Among nodulating herbaceous and shruby species the authors listed 9 species of Mimosoids, 11 Caesalpinoid species, and 54 Papilionoid species (Table 14.7). The effectiveness of these nodules for N fixation, however, has not been established. The contribution of the symbiosis to the N economy of the plant may be evaluated by comparing the natural abundance of ^{15}N in

Table 14.7: Nodulation of herbaceous or shrubby Leguminosae in *Trachypogon*–grasslands of Venezuela (extracted from BARRIOS and GONZÁLEZ, 1971).

Genera and subfamily		No. of nodulating species
Caesalpinioideae		
	Chamaecrista	11
Mimosoideae		
	Mimosa	9
Papilionoideae		
	Aeschynomene	7
	Calopogonium	1
	Centrosema	3
	Clitoria	2
	Crotalaria	5
	Desmodium	7
	Eriosema	3
	Galactia	1
	Indigofera	4
	Phaseolus	3
	Rhynchosia	2
	Sesbania	1
	Stylosanthes	6
	Thephrosia	4
	Teramnus	1
	Zornia	4

nodulating legumes and non–nodulating species in the same environment (*Rhizobium*–nitrogenase discriminates little against ^{15}N), or through the enrichment in ^{15}N taken up from added fertilizer (SHEARER and KOHL 1986), or measuring the proportion of ureids and aminoacids to total N in xylem sap (PATE et al. 1980). The analysis of nitrogen compounds in alcoholic extracts from petioles or very young branches in a group of common legume species in the *Trachypogon*–savannas show a large variability within and among species. Three of the 6 species measured did not show higher ureid concentration in the extract than the non–nodulating control (Fig. 14.3). Total aminoacid content is also variable and 2 species showed similar values as the control species. However, all legume species measured, except *Indigofera pasquorum*, showed higher proportion of organic N in the extracts than the non–nodulating control.

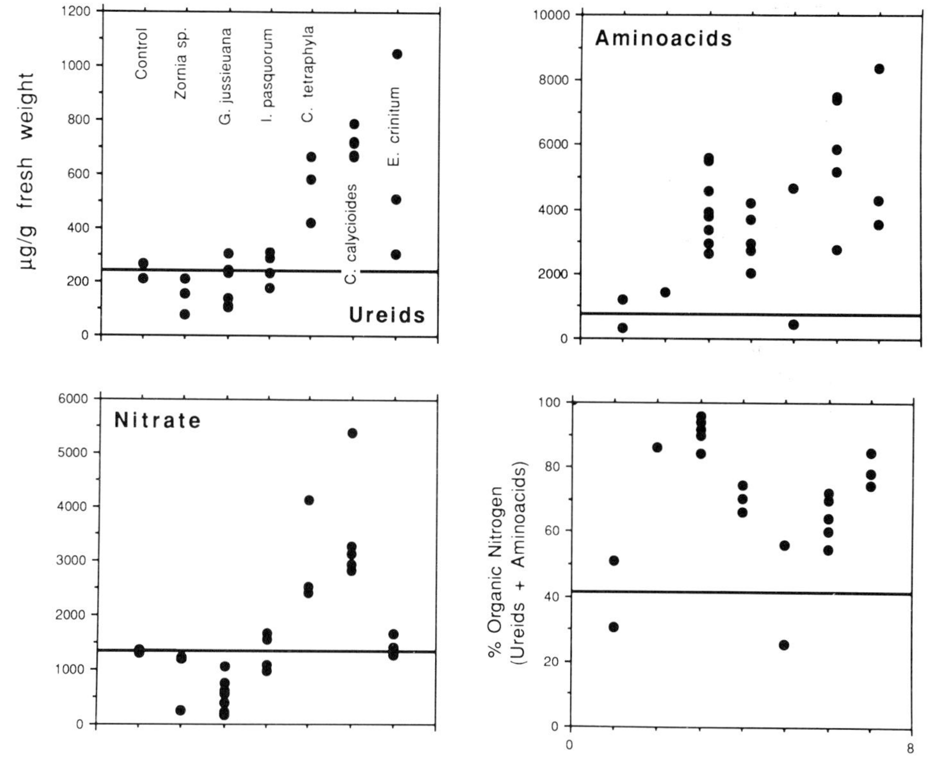

Figure 14.3: Content of ureids (VOGELS and VAN DER DRIFT 1970, PATE et al. 1980), aminoacids (ninhydrin method, PATE et al. 1980) and nitrate (CATALDO et al. 1975) in alcoholic extracts of petioles and young branches of selected legume species from *Trachypogon*–savannas in central Venezuela. The lower right graph shows the percentage of soluble organic nitrogen from total soluble nitrogen extracted. The control species is *Godmania macrocarpa* (*Bignoniaceae*).

Variability in N fixation capacity under natural conditions is also shown by the measurements of $\delta^{15}N$ in the aboveground biomass of legumes in typical *Trachypogon*–savannas. Among 11 common legume species *Eriosema crinitum* and *Chamaecrista fagonoides* appear to be good N fixers, while *Mimosa camporum, Clitoria guyanensis* and *Phaseolus diversifolius* seem to be weak fixers of atmospheric N (Fig. 14.4). These results confirms in general the observation on nodule volume per unit root weight reported in Table 14.3. The other 6 species ocuppy intermediate positions showing $\delta^{15}N$ values ranging from -1 to $-2°/_{oo}$, while the non–nodulating species have $\delta^{15}N$ values around zero or above.

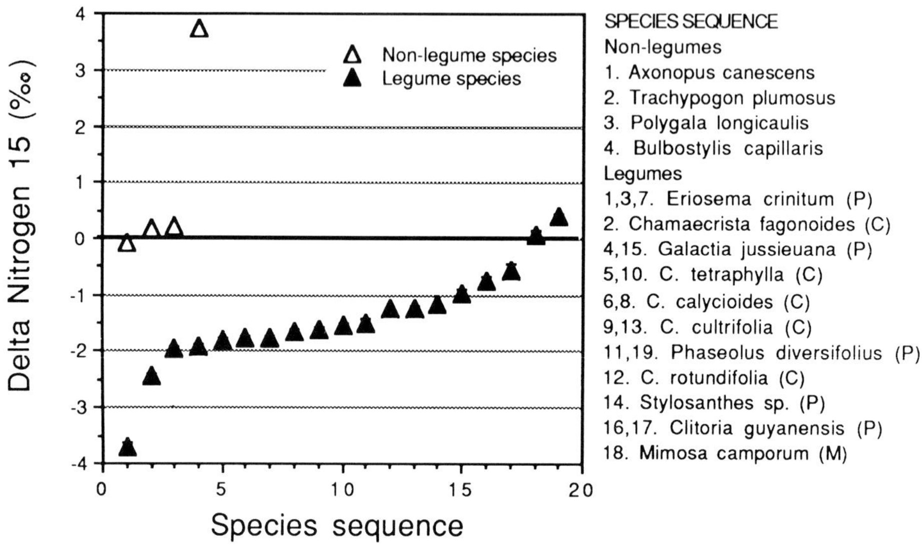

Figure 14.4: Sequential ordination of δ^{15}N values of aboveground biomass of legume species in *Trachypogon*–savannas. Nitrogen isotope ratios were measured in Prof. E.-D. Schulze's Institute in Bayreuth with a differential mass spectrometer (see SHEARER and KOHL 1986). For comparison the δ^{15}N values of two grass species, one sedge and one non–nodulating dicot growing in the same area are included. The solid line is the assumed limit of δ^{15}N value indicating a contribution of the symbiosis to the total nitrogen content of the above ground biomass.

Effectiveness of nodules in N fixation can vary widely depending on soil nutritional conditions. The first stages of the *Rhizobium* infection demand a significant proportion of the plant current photosynthates, an investment returned to the plant at later stages when the nodules begin to export nitrogen compounds. However, plant fitness and competitiveness might be low if the activity of the nodules is reduced by the lack of certain nutritional inputs from soil, such as P and Mo (ANDREWS and ROBINS 1969a).

Many forage species of tropical grasslands can develop mycorrhizal infections which contribute significantly to their P supply and thereby influence their nodulation and N fixation capacity. There appears to be a general agreement on the beneficial effects of mycorrhizal inoculation for agronomic purposes, an issue of particular interest for tropical soils of low nutrient availability (see for instance EZETA and SANTOS 1980). The positive impact of mycorrhizal infection on the overall performance of grasses and legumes has been widely documented and the

Table 14.8: Percentage of root mycorrhizal infection in legume species under natural savanna conditions in Venezuela.

Species	% mycorrhizal infection
Caesalpinioideae (average: 32.3 ± 1.7)	
Chamaecrista flexuosa	30.9
C. cultrifolia	34.7
C. tetraphyla	31.2
Papilionoideae (average: 41.2 ± 10.7)	
Aeschynomene histrix	24.3
A. paniculata	29.7
Clitoria guyanensis	27.6
Crotalaria stipularis	53.9
Eriosema crinitum	44.7
Stylosanthes guyanensis	45.5
Tephrosia cinerea	46.1
Zornia curvata	54.3

obtention of efficient mycorrhizal strains is an active field of research (HARLEY and SMITH 1983).

Mycorrizal infection is also common in neotropical savanna legumes and grasses (G. CUENCA, personal communication), although the percent of infection is not as high as that obtained under cultivation with appropiate inoculation (SAIF 1986) (Table 14.8). There is very little information on the effectiveness of mycorrhiza in improving P uptake by savanna plants under natural conditions.

The occurrence of mycorrhizal symbiosis not only ameliorates the nutrient situation, particularly of P, of a given species, but it may also improve the capacity for uptake of other nutrients and water, thereby increasing transpiration and photosynthesis. However, the presence of mycorrhizal infection constitutes also a cost to the plant which has to be covered by increased photosynthesis. In *Bouteloua gracilis*, a dominant C_4 pasture grass in northamerican grasslands, mycorrhiza inoculation increased chlorophyll content, P content, photosynthesis and transpiration without affecting biomass production compared with control plants supplied with mineral P (ALLEN et al. 1981). The increased photosynthetic rate is the consequence of the increased sink activity of the roots.

In legumes both under natural conditions and under cultivation successful tripartite symbiosis (*Rhizobium*–mycorrhizal fungi–legume) develop easily. This tripartite symbiosis might be of high relevance because of its complementarity. The viability of this arrangement was shown straightforwardly by LINDT et al. (1985) in

Macroptilium atropurpureum. In this species there is a linear relationship between nodule fresh weight and nitrogenase activity per nodule weight. The inoculation with the mycorrhizal fungus *Glomus fasciculatus* increases around 4 times both total nodule fresh weight per plant and the capacity for atmospheric N fixation.

The nature of the interaction between nodulation, mycorrhizal infection and phosphate supply to tropical legumes was elegantly shown by CRUSH (1974). Mycorrhiza overcome the P limitation for effective nodulation and N fixation in these legumes. Similar positive effects of VA–mycorrhiza on nodulation and organic matter production of legumes have been shown for many species under a range of conditions (for example SMITH and DAFT 1977, SAIF 1986).

The tripartite symbiosis increases photosynthetic rates of the inoculated legume more than 40 % above control, but it takes around 17 % of total photosynthates (HARRIS et al. 1985). As expected, there is competition for plant resources between the symbionts *Rhizobium* and the mycorrhizal fungus *Glomus* when they occur simultaneously (BETHLENFALVAY et al. 1985). The sequential inoculation of soybean plants with *Glomus* and *Rhizobium* results in a better development of the symbiont which was first inoculated (BROWN and BETHLENFALVAY 1986).

Another aspect regarding the significance of symbiosis for the nutrient economy of legume in savannas is the efficiency of the *Rhizobium* strain. It has been shown experimentally in *Phaseolus vulgaris* cultivars that the efficiency of *Rhizobium* strains varies within an 8–fold range from about 46 to 383 mg N per gram of nodule (NEVES and HUNGRIA 1987). The inoculation with inefficient strains could result in a drain of carbohydrate resources of the host plant. This argument is also valid for the occurrence of inefficient mycorrhizal strains. Under natural conditions the occurrence of tripartite symbiosis as they do in *Trachypogon*–savannas (Table 14.8) do not necessarily mean an advantage for the energy economy of the legumes.

Furthermore, even in the case that the *Rhizobium*–symbiosis provide a higher N input to the legumes, in mixed communities of perennial plants it would also result in a slowly improvement of the availability of N for the grasses, or in the transfer of N from legume to grasses as observed in experimental mixtures (ISMAILI and WEAVER 1987; BROPHY et al. 1987). Thus the benefit of this type of association for the legumes might not be as effective as it was shown in monoculture experiments.

Nitrogen fixation by free–living bacteria in association with roots and blue–green algae. There is evidence indicating that biological associations between plant roots and microorganisms play a fundamental role in savanna nutritional ecology. Free living nitrogen fixing microorganisms, such as blue–green algae, or the grass–root associations with *Azospirillum*, may supply in sites of low N availability signficant amounts of readily available nitrogen for plant growth (GILLER and DAY 1985).

Blue–green algae can make a substantial contribution to soil N. In Nigerian savannas, with a considerable cover of blue–green algae (ca. 30 %, ISICHEI 1980), the average rate during the rainy season amounts to 2.3 mg N m^{-2} day^{-1} while in the *Trachypogon*–savannas of central Venezuela, with 10–12 % algal coverage during the rainy season, average rates vary between 0.4–0.7 mg N m^{-2} day^{-1} (SANTAELLA 1985). These rates will provide several kg of N per ha during a growing season, an amount which partially offsets potential N losses through fire.

More important contributions to the N supply in savannas may derive from the root–*Azospirillum* associations in tropical grasses (DOBEREINER and DAY 1975). In natural savannas average rates ranging from 1.0–3.7 mg N m^{-2} day^{-1} have been reported (BALANDREAU 1976; SANTAELLA 1985). In cultivated fodder grasses daily rates may be even higher (VAN BERKUM and DAY 1980). Research on the factors regulating the activity of these associations, and the selection of highly efficient bacteria–grass associations, may prove to be important for the management of disturbed savanna areas.

14.4 Concluding remarks

The savannas sites analysed here are growing on soils of general low nutrient status reflected particularly in the small levels of bases in the exchange complex of the soil. This characteristic can be generalized for all *Trachypogon*–savannas in Venezuela and Colombia (BLYDENSTEIN 1962, 1967, SARMIENTO 1984, MEDINA and SILVA 1990). The frequency of legumes was not correlated with these indicators of soil fertility but with those parameters associated with the mobility of Al. Therefore the % of legume species in a given savanna is directly correlated with the % sand and the exchangeable Ca, but negatively correlated with cation exchange capacity. Larger cation exchange capacities are associated in *Trachypogon*–savannas with more exchangeable Al and lower P availability. In contrast, grasses appear not to be as sensitive as legumes to high Al mobility in the soil therefore their importance value indices were less variable than those of legumes.

Most legume species in the savannas are perennial and accumulate a large proportion of their biomass, energy and nutrient resources in enlarged underground organs called xylopodia. These reserves allows the regrowth after fire, herbivory or drought. However, their root system is extensive, with small number of ramifications, therefore the occupation of the soil volume is not as effective as in grasses with their fibrous, highly branched root system. This indicates a higher capacity of grass individuals to extract water and nutrient resources from soil, and suggests a lower competitive capacity of the legumes in this process.

Our analyses show that the nutritional status of legume species in natural savanna environments is particularly deficient in P and K, while N is always around or above the sufficiency level. This contrasts with the situation in grasses where the nutritional status is near the level of sufficiency. The productive capacity of the native dominant grasses appears to be more limited by N than by P. Although fertilization with P might increase aboveground production under certain conditions, there is a pronounced accumulation of this element in the senescing shoots which is not translocated to the roots, indicating that it is being absorbed in larger quantities than those required for maximum growth.

Rhizobium–symbiosis is widespread in savanna legumes but degree of nodulation varies markedly from site to site. Besides, the effectivity of the symbiosis in N supply to the host plant seems to be very variable as judged from the concentration of soluble organic nitrogen (ureids + aminoacids) in alcoholic extract of petioles and young branches, or by the differences of $\delta^{15}N$ values of legumes and non–fixing plants. The potential advantage of the legumes because of their mechanism for N acquisition from the atmosphere can be counterbalanced at least partially by the contribution of bacterial–root associations to the N budget of the savanna. Although the actual rate of fixation is quite low compared with effectively nodulated legumes, accumulation of N through this pathway might be of significance at least to offset the yearly losses of N due to burning. Furthermore, the occurrence of efficient symbiosis might benefit the grasses in the long run because transfer of symbiotically fixed N from legumes to grasses has been shown experimentally.

Mycorrhizal symbiosis is also common in legumes and grasses but we did not find evidences that these symbiosis were effective in reducing the P deficiency in the legume species. Experimental work is necessary to confirm this hypothesis. Tripartite symbiosis (*Rhizobium*–mycorrhizal fungi–legume) were observed in several species but it is not known if the establishment of inefficient symbiosis may constitute a burden rather than an advantage for the competitive capacity of legume species.

Acknowledgments

We wish to thank Drs. María L. Mayoral and Elizabeth Olivares (Instituto Venezolano de Investigaciones Científicas, Caracas) for the determination of nitrogen compounds in plants extracts and to Prof. Dr. E.–D. Schulze (University of Bayreuth, Germany) for the measurement of the $\delta\ ^{15}N$ values. Dr. A. Franco (University of Darmstadt) made valuable comments to an earlier version of this paper.

Bibliography

ALLEN, M. F.; SMITH, W. K.; MORRE JR., T. S.; CHRISTENSEN, M., 1981, Comparative water relations and photosynthesis of mycorrhizal and non–mycorrhizal *Bouteloua gracilis* H. B. K. Lag ex Steud.
New Phytol. **88**, 683–693.

ANDREWS, C. S.; ROBINS, M. F., 1969a, The effect of phosphorus on the growth and chemical composition of some tropical pasture legumes. I. Growth and critical percentanges of phosphorus.
Australian J. Agric. Res. **20**, 665–674.

ANDREWS, C. S.; ROBINS, M. F., 1969b, The effect of potassium on the growth and chemical composition of some tropical pasture legumes. I. Growth and critical percentages of potassium.
Australian J. Agric. Res. **20**, 999–1007.

Anonymous, 1967, Soil Survery Laboratory Methods and Procedures for Collecting Soil Samples.
Soil Conservation Service, U. S. Department of Agriculture, Washington, D. C.

BALANDREAU, J., 1976, Fixation rhizosphérique de l'azote (C_2H_2) en savanne de Lamto.
Revue d'Ecol. Biol. du Sol **13**, 529–544.

BARRIOS, S.; GONZÁLEZ, V., 1971, Rhizobial symbiosis in venezuelan savannas.
Plant and Soil **34**, 707–719.

BARUCH, Z.; LUDLOW, M. M.; DAVIS, R., 1985, Photosynthetic responses of native and introduced C_4 grasses from venezuelan savannas.
Oecologia **67**, 388–393.

BERKUM, P. VAN, DAY, J. M., 1980, Nitrogenase activity associated with soil cores of grasses in Brazil.
Soil Biol. Biochem. **12**, 137–140.

BETHLENFALVAY, G. J.; BROWN, M. S.; STAFFORD, A. E., 1985, *Glycine-Glomus-Rhizobium* symbiosis. II. Antagonistic effects between mycorrhizal colonization and nodulation.
Plant Physiol. **79**, 1054–1058.

BLYDENSTEIN, J., 1962, La sabana de *Trachypogon* del alto Llano.
Bol. Soc. Venezolana Cienc. Nat. **23**, 139–206.

BLYDENSTEIN, J., 1967, Tropical savanna vegetation of the Llanos of Colombia.
Ecology **48**, 1–15.

BROPHY, L. S.; HEICHEL, G. H.; RUSSELLE, M. P., 1987, Nitrogen transfer from forage legumes to grass in a systematic planting design.
Crop Sci. **27**, 753–758.

BROWN, R. H., 1978 A difference in N use efficiency in C_3 and C_4 plants and its implications in adaptation and evolution.
Crop Sci. **18**, 93–98.

BROWN, M. S.; BETHLENFALVAY, G. J., 1986, The *Glycine-Glomus-Rhizobium* symbiosis. III. Endophyte effects on leaf carbon, nitrogen and phosphorus nutrition.
J. Plant Nutr. **9**, 1199–1212.

BULLA, L.; SÁNCHEZ, P.; SILVIO, C.; MALDONADO, A.; DE SOLA, R.; LIRA, A., 1984, Ecosistema sabana.
In: Bases para el diseño de medidas de mitigación y control de las cuencas hidrográficas de los ríos Caris y Paso, Edo. Anzoátegui. Informe final, Proyecto Meneven CAR–33. Facultad de Ciencias, Instituto de Zoología Tropical, Universidad Central de Venezuela. Caracas.

CATALDO, D. A.; HAROO, M.; SCHRADER, L. E.; YOUNGS, V. L., 1975, Rapid colorimetric determination of nitrate in plant tissue by nitration of salicylic acid. Comm. Soil Sci. and Plant Analysis **6**, 71–80.

CHAPIN, F. S. III, 1980, The mineral nutrition of wild plants.
Ann Rev. Plant Physiol. **11**, 233–260.

CHRISTIE, E. K.; MOORBY, J., 1975, Physiological responses of semi–arid grasses. I. The influence of phosphorus supply on growth and phosphorus absorption.
Australian J. Agric. Res. **26**, 423–436.

CRUSH, J. R., 1974, Plant growth responses to vesicular–arbuscular mycorrhiza. VII. Growth and nodulation of some herbage legumes.
New Phytol. **73**, 743–752.

DOBEREINER, J.; DAY, J. M., 1975, Nitrogen fixation in the rhizosphere of tropical grasses.
In: STEWART, W. D. P. (ed), Nitrogen fixation by free living microorganisms. IBP Series. Vol. **6**, pp. 39–56. Cambridge University Press.

EZETA, F. N.; SANTOS, O. M., 1980, Benefício da introdução de endomicorriza eficiente na utilização de nutrientes em latosolos do sul da Bahia.
Rev. bras. Cienc. Solo **4**, 13–17.

FÖLSTER, H.; HUBER, O., 1984, Interrelaciones suelos–vegetación en el área de Galipero, Territorio Federal Amazonas.
Ministerio del Ambiente. (PT) Serie Informes Técnicos DGSIIA/IT/144. Caracas.

GARCÍA–MIRAGAYA, J.; SAN JOSÉ, J. J.; HERNÁNDEZ, J., 1983, Effect of added nitrogen, phosphorus and potassium on aboveground biomass production and nutrient content of *Trachypogon*–savanna grasses.
Trop. Ecol. **24**, 33–42.

GILLER, K. E.; DAY, J. M., 1985, Nitrogen fixation in the rhizosphere.
In: FITTER, A. H. (ed), Ecological interactions in soils. Blackwell Sci. Publ. Oxford. pp. 127–147.

HARLEY, J. L.; SMITH, S. E., 1983, Mycorrhizal symbiosis.
Academic Press. London.

HARRIS, D.; PACOVSKY, R. S.; PAUL, E. A., 1985, Carbon economy of soybean–*Rhizobium–Glomus* associations.
New Phytol. **101**, 427–440.

HASSID, W.; NEUFELD, E., 1964, Whole starches and modified starches. [9] Quantitative determination of starch in plant tissues.
Methods in Carbohydrate Chemistry **4**, 33–36.

ISICHEI, A. O., 1980, Nitrogen fixation by blue–green algal crusts in a Nigerian savanna.
In: ROSSWALL, T. (ed), Nitrogen cycling in West–African ecosystems. pp. 191–198. SCOPE–UNEP International Nitrogen Unit. Stockholm. Royal Swedish Academy of Sciences.

ISMAILI, M.; WEAVER, R. W., 1987, Nitrogen transfer from Siratro to Kleingrass: competition for mineral nitrogen and isotopic fractionation.
Agron. J. **79**, 188–192.

LUDLOW, M. M.; WILSON, G. L., 1971, Photosynthesis of tropical pastures. I. Illuminance, CO_2 concentration, leaf temperature and leaf–air vapour pressure difference.
Australian J. Biol. Sci. **24**, 449–470.

LYND, J. Q.; TYRL, R. J.; PURCINO, A. A. C., 1985, Mycorrhiza–soil fertility effects on regrowth, nodulation and nitrogenase activity of Siratro (*Macroptilium atropurpureum* (DC) Urb.).
J. Plant Nutri. **8**, 1047–1059.

MALAVOLTA, E.; HAAG, H. P.; MELLO, F. A. F.; BRASIL, S., 1974, Nutrição mineral e adubação de plantas cultivadas.
Livraria Pioneira Editora. São Paulo.

MEDINA, E., 1982, Physiological ecology of neotropical savanna plants.
In: HUNTLEY, B. J.; WALKER, B. H. (eds), Ecology of Tropical Savannas. Ecological Studies **42**, 308–335. Springer Verlag. Berlin.

MEDINA, E.; MENDOZA, A.; MONTES, R., 1978, Nutrient balance and organic matter production in the *Trachypogon* savannas of Venezuela.
Trop. Agric. **55**, 243–253.

MEDINA, E.; MOTTA, N., 1990, Metabolism and distribution of grasses in tropical flooded savannas in Venezuela.
J. Trop. Ecol. **6**, 77–89.

MEDINA, E.; SILVA, J. F., 1990, The savannas of northern South America: a steady state regulated by water–fire interactions on a background of low nutrient availability.
J. Biogeog. **17**, 403–413.

MOTT, G. O.; POPENOE, H. L., 1977, Grasslands.
In: ALVIM, P. T.; KOZLOWSKI, T. T. (eds), Ecophysiology of Tropical Crops. pp. 157–186. Academic Press, New York.

MÜLLER-DOMBOIS, D.; ELLENBERG, H., 1974, Aims and Methods of Vegetation Ecology.
John Wiley & Sons, Inc.

NEVES, M. C. P.; HUNGRIA, M., 1987, The physiology of nitrogen fixation in tropical grain legumes.
CRC Critical Rev. Plant Sci. **6**, 267–321.

NORMAN, M. J. T., 1966, Katherine research station 1956–64: a review of published work.
Division of Land Research, Technical Paper No. **28**. CSIRO. Australia.

PATE, J. S.; ATKINS, C. A.; WHITE, S. T.; RAINBIRD, R. W.; WOO, K. W., 1980, Nitrogen nutrition and xylem transport of nitrogen in ureide-producing grain legumes.
Plant Physiol. **65**, 961–965.

PEREIRA, P. A.; DOBEREINER, J.; NEYRA, C. A., 1981, Nitrogen assimilation and dissimilation in five genotypes of *Brachiaria* spp.
Can. J. Bot. **59**, 1475–1479.

RAMIA, M., 1967, Tipos de sabanas en los llanos de Venezuela.
Bol. Soc. Venezolana Cienc. Nat. **27**, 264–288.

RIZZINI, C. T.; HERINGER, E. P., 1961, Underground organs from some southern Brazilian savannas, with special reference to the xylopodium.
Phyton (Argentina) **17**, 105–124.

SAIF, S. R., 1986, Vesicular-arbuscular mycorrhizae in tropical forrage species as influenced by season, soil texture, fertilizers, host species and ecotypes.
Angew. Bot. **60**, 125–139.

SAN JOSÉ, J. J.; GARCÍA-MIRAGAYA, J., 1981, Factores ecológicos operacionales en la producción de materia orgánica de las sabanas de *Trachypogon*.
Bol. Soc. Venezolana Cienc. Nat. **36**, 347–374.

SAN JOSÉ, J. J.; MONTES, R.; GARCÍA-MIRAGAYA, J.; ORIHUELA, B., 1985, Bioproduction of *Trachypogon*-savannas in a latitudinal cross section of the Orinoco Llanos, Venezuela.
Acta Oecologica, Oecol. Plant. **6**, 25–43.

SANTAELLA, S., 1985, Fijación biológica de nitrógeno en sabanas de *Trachypogon* spp.

Trabajo Especial de Grado. Escuela de Biología, Universidad Central de Venezuela. Caracas.

SARMIENTO, G., 1984, The ecology of neotropical savannas.
Harvard University Press. Cambridge.

SHEARER, G.; KOHL, D. H., 1986, N_2-fixation in field settings: estimations based on natural ^{15}N abundance.
Australian J. Plant Physiol. **13**, 699–756.

SMITH, S. E.; DAFT, M. J., 1977, Interactions between growth, phosphate content and nitrogen fixation in mycorrhizal and non–mycorrhizal *Medicago sativa*.
Australian J. Plant Physiol. **4**, 403–413.

SUSACH, F., 1984, Caracterización ecológica de las sabanas de un sector de los Llanos Centrales bajos de Venezuela.
Tesis Doctoral, Universidad Central de Venenzuela, Facultad de Ciencias, Caracas.

VELÁSQUEZ, J., 1965, Estudio fitosociológico acerca de los pastizales de las sabanas de Calabozo, Estado Guárico.
Bol. Soc. Venezolana Cienc. Nat. **25**, 59–101.

VOGELS, G. D; VAN DER DRIFT, 1970, Differential analysis of glyoxylate derivatives.
Ann. Biochem. **33**, 143–157.

WEINMANN, H., 1938, Effect of fertilizer treatment on Transvaal highveld.
South African J. Sci. **35**, 246–249.

Chapter 15

Correlations between the microbial activity, and water, air, temperature and nutrient status of different soils under different land use

H.-P. BLUME, L. BEYER and F. FRIEDRICH

15.1 Introduction

The edaphon and especially the micro-organisms are of utmost importance with regard to terrestrial ecosystems: They break down the litter on and in the soil and in doing so they mineralize nutrients and put these back into the ecological cycle. Some of the micro-organisms are also capable of assimilating the nutrients directly, e. g. nitrogen in the air, and by forming humic substances they create nutrient reserves as well as adsorption places for available water and nutrients. In addition, humus clay complexes are also made, which lead to the development of a loose biotic soil structure and which ensure the plant roots sufficient root penetration and oxygen.

The micro-organisms' activity reflects the water, air, heat and nutrient conditions of a soil. A high biological activity usually stands for favorable growth conditions, taller plants and the higher productivity of cultivated plants, as well as for the favorable transformation of harmful organic substances.

The following paper will report on the comparative investigation of the microbial activity of characteristic North German soil forms (loamy Orthic Luvisols, sandy Haplic Podzols, Eutric and Dystric Gleysols and Histosols) under forest. The influence of intensive and alternative farm management was also determined for the first two soil types.

15.2 Sites and soils

The sites, which were investigated in Schleswig-Holstein, are made up predominantly of glacial and fluvioglacial sediments.

The Eastern Hills in Holstein are moraines of boulder marl. The processes of decalcification, braunification, loamification and clay migration have formed the Orthic Luvisols (Parabraunerden) under *Fagus sylvatica* of this area. These are associated with Stagno-Gleyic Luvisols (Pseudogleye) if rainwater stagnates in the soil. Three loamy Luvisols from glacial till above fluvioglacial sand under forest and arable land on the upper slope of a ridge were investigated. The amount of precipitation is between 560–730 mm and the average temperature is 8.4 °C.

In the Holsteinische Vorgeest region the Weichselian glaciation sandurs dominate. They are extremely poor in nutrients and clays. In former times the forest, which was made up of oak and birch trees was cleared, and since then heather has prevailed. During this time the soils have been podzolized considerably. Three sandy Haplic Podzols from glacial sand under forest and arable land at a hillfoot were investigated. The amount of precipitation here is 750 mm and the average temperature is 8.3 °C.

In Berlin two sandy Gleysols and two Terric Histosols were investigated. The first have developed from fluvioglacial sand without carbonates and the second have developed from fluvioglacial sand containing carbonates. The amount of precipitation is 600 mm and the average temperature is 8.2 °C.

The physical and chemical characteristics of all the sites can be seen in Tab. 15.1. (For further details of the sites, see BEYER 1989, FRIEDRICH 1979, NEUMANN 1976, PETERS 1990, and ZINGK 1988.)

A moderately damp **beech and oak forest** is situated on the **Orthic Luvisol** of Siggen farm. It has been verified that the forest has existed here since the 13th century. The acidification and loss of bases are both very advanced (Tab. 15.1). The trees' roots, however, reach till containing carbonate in the subsoil, which has high to moderate nutrient reserves. The soil vegetation is a typical Melico-Fagetum, as it is characteristically moist. The **Orthic Luvisol under conventional farm management** of Siggen farm is slightly eroded. The crop rotation of wheat, barley and rape is characterized by high yield (wheat 80–90 dt, barley 70–80 dt, rape

Table 15.1: Characteristics of Luvisols, Podzols, Gleysols, and Histosols in Northwest Germany (a–f: BEYER (1989), g–k: NEUMANN (1976)).

hor	depth	d_B	S	U	C	GPV	LK	aWK	k_f	cb	pH	CEC	BS	C_{org}	N_t	C/N	Fe_o
	cm	g/cm³	%			Vol.%			cm/d	%		meq/kg	%	%	%		mg/g
a) Orthic Luvisol under Melico-Fagetum																	
L/O	4-0	0.09									3.9			36.0	1.41	25	
Ah	0-11	1.10	47	37	16	57	22	25	1000	0	3.5	172	13	4.6	0.35	13	6.2
Alv	-47	1.62	41	34	25	37	6	13	710	0	3.8	149	18	0.7	0.10	7	9.5
Bvt	-91	1.67	40	31	29	37	6	13	2.7	0	4.3	158	63	0.3	0.04	8	10.1
BtC	-117	1.71	42	31	27	36	3	13	2.7	7	7.3	166	98	0.2	0.03	6	8.8
Ccv	-135	1.74	42	35	23	34	4	11	1.3	16	7.4	148	100	0.1	0.02	5	6.4
b) Orthic Luvisol under conventional farming, slightly eroded																	
Ap	0-35	1.51	44	32	24	41	12	14	412	0.1	7.0	136	100	1.1	0.16	7	2.4
Bvt	-70	1.74	40	31	29	35	3	12	23	0	6.3	175	81	0.3	0.05	6	1.2
BtC	-86	1.75	40	32	28	33	2	13	23	8	7.6	154	98				1.4
Ccv	-140	1.76	47	27	26	33	4	12	23	19	7.6	134	100		n.d.		0.8
c) Stagni-orthic Luvisol under ecological farming																	
Ap	0-30	1.60	56	30	14	39	4	23	69	0.1	6.6	146	86	1.6	0.26	6	2.9
AB	-50	1.63	55	28	17	39	8	17	22	0	6.5	95	86	0.4	0.08	5	2.7
Bgt	-70	1.73	58	24	18	36	6	15	21	0	6.4	113	84	0.2	0.05	4	3.2
Bvg	-120	1.76	54	29	17	34	5	15	21	0	6.7	106	90	0.1	0.03	4	2.0
Ccv	-160	1.68	64	22	24	38	7	7	9	16	7.3	295	97	0.1	0.02	5	1.3
d) Haplic Podzol under Querco-Betuletum																	
L/O	15-0	0.13									3.0	833	15	44.0	1.51	29	n.d.
Aeh	0-10	0.81	79	17	4	68	14	44	215	0	2.7	298	11	7.6	0.46	16	0.5
Ahe	-20	1.54	94	4	2	45	12	10	226	0	3.1	17	25	1.3	0.05	25	0.1
Bh	-25	1.37	93	2	5	47	30	13	600	0	3.1	289	3	8.0	0.23	35	1.8
Bhs	-30	1.37	91	6	3	47	30	13	609	0	4.0	143	1	2.4	0.09	26	6.3
CBvs	-70	1.59	98	1	1	40	33	5	555	0	4.5	267	3	1.0	0.03	40	0.2
GCv	-160	1.59	97	2	1	40	33	5	519	0	4.7	14	6	0.1	0.01	10	n.d.
e) Haplic Podzol under conventional farming																	
Ap	0-30	1.46	88	8	4	46	18	19	308	0	4.6	139	37	2.6	0.27	9	1.8
Ahe	-38	1.52	92	6	2	42	24	11	204	0	4.8	69	39	1.1	0.07	15	0.4
Bh	-41	1.25	92	5	3	51	21	15	344	0	4.8	226	34	3.8	0.17	22	1.0
Bhs	-47	1.36	93	5	2	48	34	7	744	0	5.0	80	25	1.7	0.09	19	4.6
Bsv	-64	1.44	95	4	1	45	34	6	812	0	5.1	18	19	0.2	0.02	12	0.7
Bv	-127	1.54	96	3	1	42	30	10	685	0	5.1	14	10	0.1	0.01	10	0.1
f) Haplic Podzol under ecological farming																	
Ap	0-30	1.39	85	10	5	46	11	27	320	0	4.7	147	32	3.0	0.18	17	3.5
Aeh	-50	1.35	85	14	0	49	15	30	230	0	4.7	94	35	2.0	0.06	35	0.8
Bhs	-52					n.d.				0	4.2	156	42	1.7	0.04	41	4.8
Bhv	-100	1.49	85	11	4	44	25	16	530	0	4.6	45	24	0.4	0.01	40	0.6
IIBv	-105	n.d.	91	9	0				n.d.	0	4.8	23	15	0.1	0.01	10	0.5
Bbsv	-125	1.64	82	14	4	38	19	17	210	0	4.9	22	23	0.1	0.01	10	0.4

Continued next page

Table 15.1 continued

hor	depth cm	d_B g/cm³	S %	U %	C %	GPV Vol.%	LK Vol.%	aWK Vol.%	k_f cm/d	cb %	pH	CEC meq/kg	BS %	C_{org} %	N_t %	C/N	Fe_o mg/g
g) Sandy Eutric Gleysol under Querco-Betuletum																	
Oh	13-0	0.60				74	9	43	3200	0	3.5	609	30	10.0	0.86	12	1.6
Ah	0-12	0.92	86	10	4	64	13	31	2900	0	6.7	607	95	3.4	0.22	16	2.7
Go	-52	1.63	91	7	2	38	6	30	790	0	6.7	40	95	0.2	0.02	10	0.4
Gor1	-77	1.59	94	5	1	40	12	26	1100	0	6.7	20	95	0.2	0.02	10	0.1
Gor2	-112	1.61	97	2	1	39	4	32	710	0	6.6	15	95		n.d.		0.1
Gr	-140	1.67	91	3	6	37	3	30	340	0	6.6	15	95		n.d.		0.1
h) Sand covered Podzo-dystric Gleysol under Querco-Betuletum																	
Ofh	7-0					75	15	40			2.7	912	10	28.0	1.00	28	2.7
Aeh	0-5	1.18	85	12	3	55	11	37	4500	0	2.9	116	2	5.1	0.15	34	0.9
Bsv	-16	1.44	86	11	3	46	13	29	9500	0	3.5	93	3	2.2	0.07	32	0.3
fAeh	-22	1.35	87	10	3	49	10	23	4300	0	3.5	152	1	3.6	0.10	35	0.8
AeGo	-34	1.47	97	2	1	45	16	27	9800	0	3.6	35	2	7.0	0.02	35	0.3
BhGr	-74	1.45	96	3	1	45	9	30	7300	0	3.7	160	1	3.8	0.08	48	0.3
Gr	-130	1.45	95	4	1	45	5	35	2300	0	3.9	135	1	2.8	0.07	40	0.1
i) Eutri-fibric Histosol under Frangulo-Salicetum-auritae																	
nHhf1	0-20	0.15				92	7	48	5000	0	5.3	1200	58	45.0	0.78	56	6.8
nHhf2	-40	0.10				94	15	55	68000	0	5.1	1140	54	45.0	0.57	79	4.7
nHfh	-80	0.09				94	15	61	37000	0	5.2	1370	59	45.0	0.57	82	3.7
k) Dystri-fibric Histosol under Sphagno-Carietum																	
uHfh	0-5	0.10									3.2	774	8	28.0	1.89	15	4.9
uHhf1	-15	0.54				76	13	28	5000	0	3.7	330	9	9.4	0.55	17	1.2
uHhf2	-47	0.12				92	16	37	63000	0	3.4	1030	24	49.0	1.54	32	5.2
uHfh	-80	0.11				92	12	36	14000	0	3.5	1370	19	50.0	1.51	33	3.0

hor:horizon; d_B:bulk density; S:sand,U:silt,C:clay in % of humus-and carbonate-free fine soil; GPV:total pore volume; LK: >50μm pores; aWK: 50-0.2 μm pores; k_f: water permeability under saturated conditions; cb:carbonate; BS: base saturaction; C_{org}: organic carbon; N_t: total nitrogen; Fe_o: oxalat-soluble iron.
pH in $CaCl_2$; CEC in meq/kg ovendry fine soil; n.d.: not determined

30–40 dt), which is ensured by intensive mineral fertilizing in spring and early summer (225 kg N/ha) and autumn (40 kg P, 100 kg K, 400 kg CaO/ha) and by the preventive application of pesticides. The **Stagni–orthic Luvisol under ecological farming** has been managed biologically–organically since 1979. It has a seven year crop rotation of cash crops and forage, and only organic fertilizer is used here (dung in autumn, liquid manure in spring). In the Luvisols under farm management the low water transpiration sometimes caused water to stagnate and this in turn caused a lack of air.

The **spruce forest** is situated on a **Haplic Podzol**. It was planted at the end of the last century after the heather had been burnt away. The topsoil is very acidified. The natural soil vegetation is a typical Betulo–Quercetum, in which the presence of the *Molinia coerulea* indicates moist conditions. Maize and rye are cultivated on the **Podzol under conventional farm management** using organic (dung in autumn, slurry in spring) and mineral fertilizers (70 kg N in spring, 80 kg P, 200 kg K, 57 kg CaO/ha in autumn). The **Podzol**, which has been under **biological–dynamic farm management** for over 30 years, receives organic fertilizer only (dung and liquid manure in frequent small doses throughout the year). The four to eight year crop rotation consists of the cultivation of vegetables, forage and summer cereals. All three Podzols are gleyed in the subsoil and they have been disturbed considerably by the clearing of heather and the ensuing tilling of the topsoil.

The **Podzo–dystric Gleysol** is situated on the upper slope of a dune exposed to the west and the **Dystri–fibric Histosol** is to be found in a swampy depression of the "Großes Fenn" in the Düppeler forest in Berlin. The Gleysol is covered with colluvial material, and a damp oak and beech forest is to be found here due to the favorable conditions. The forest is recorded as having been present since the 13th century. Nowadays the ground water level fluctuates between 1 to 2 m under the surface and in former times this was approximately 0.5 m higher. The **Dystric Histosol** is a filamentary sedge moor rich in bog moss and containing layers of sand. During our investigations the ground water level fluctuated between 0 m in winter and 0.7 m in summer. In both soils the acidification and the loss of bases are very advanced. The Histosol is very humified in the top 50 cm and otherwise below 50 cm it is moderately humified.

The **Eutric Gleysol** is situated on the upper slope of a dune exposed to the south–west and the **Eutri–fibric Histosol** in a swampy depression "Teufelsbruch" both in the Spandauer Forest in Berlin. In contrast to the soils of the "Großes Fenn" the acidification and loss of bases is only moderate because during the early Holocene chalk accumulated in the ground of the lake "Teufelssee". Today this is dissolved in the groundwater as calcium bihydrogen carbonate. This is why the oak and birch forest on the Eutric Gleysol as well as the alder forest on the Eutric Histosol are made up of more demanding plant species in contrast to the Fenn site, for example *Tilia cordata, Calamagrostis canescens* and *Brachypodium silvaticum* are to be found in the oak and birch forest and *Rhamnus frangula, Iris pseudacorus* and *Peucedanum palustre* are present in the alder forest. In the Gleysol the ground water level fluctuated between 1 and 1.7 m, and in the Histosol between 0.2 and 0.6 m during our investigation period.

At the four Berlin sites the trees were at least able to supply themselves with water from the ground water supply during dry periods.

15.3 Methods

Precipitation: The amount of precipitation was registered at a height of 1 m above the soil in the forest with five parallels and above the field with two parallels.

Amount of water and air: The water content of the soil was determined gravimetrically at a depth of 4–8 cm with three parallels. The water tension was registered at depths of 20 and 50 cm using pressure gauge tensiometers (three parallels). The available water in vol. % (actual water – water content at wilting point) was reconstructed using the correlation between water content and water tension (water desorption of saturated core samples, SCHLICHTING and BLUME 1966). The difference between this and the total pore volume gave the air volume. The amounts of water and air respectively were calculated using the thickness of the horizon.

Temperature: The actual average daily temperature was determined in the soils in Schleswig–Holstein using minimum–maximum thermometers parallel to the measurement of the soil respiration. In the soils in Berlin the temperature was registered hourly and the average daily temperature was then calculated (FRIEDRICH 1979). The determination of the average exponential temperature (as opposed to normal mean arithmetic temperature) was carried out according to PALLMANN et al. (1940) with three parallels at depths of 5, 30 and 50 cm at six week intervals together with the gauze bags used for the determination of the cellulose decomposition. The basis of this method is the inversion of sucrose by acid which is dependent on temperature.

Sum of heat: The heat capacity was calculated from the amounts of water, humus and mineral soil with the aid of known specific heat (SCHEFFER and SCHACHTSCHABEL 1989). The sum of heat was obtained by multiplying this with the average exponential temperature.

Input of N in precipitation: The NO_3^- and the NH_4^+ concentrations in the precipitation were measured by means of Flow–Injection–Analysis (see available nitrogen), and by multiplying this with the precipitation amount. In addition to this, measurements which were obtained in 1978 and 1979 from neighboring forest and arable land sites were used for the soils in Berlin (JAYAKODY 1981).

Available nitrogen: In Berlin the soil was extracted every two to three months with $K(Al(SO_4)_2)$. NO_3^- was determined colorimetrically with Na–salycilate and NH_4^+ with "Neßlerreagenz" according to SCHLICHTING and BLUME (1966). During this process only the organic layer and the uppermost A horizon were investigated. In Holstein NO_3^- and NH_4^+ in the soil solution from depths of 20 and 50 cm were analyzed colorimetrically by means of the Flow–Injection–Method every six weeks. Due to the methodological differences of the methods, a direct comparison of the N dynamic of both sites is not permissible.

Respiration: The output of CO_2 is described as being soil respiration. The output of CO_2 was determined over 24 hours. The CO_2 was absorbed by 1 N

NaOH in Lundegardh cylinders with five parallels every fortnight (WALTER 1952). All the vegetation and the large litter had to be cleared away before measurement, in order to keep the root respiration low and to enable the vessels to be collected cleanly.

Cellulose decomposition: The decomposition of artificially buried cellulose characterizes the decomposition of litter according to UNGER (1960), modified by FRIEDRICH (1979) and differs from the soil respiration. The decomposition of cellulose buried in the soil was investigated at depths of 5, 30 and 50 cm with five parallels. 5 g cellulose fleece (100% cellulose, Firma Hartmann, oven dry) was packed into 12.5×25 cm fly gauze nets with a mesh size of approximately 1 mm. The vegetation which was present was removed in sods and then pressed back into place afterwards, so that this would not be damaged by the precise work. After six weeks (in winter 12 weeks) the remaining cellulose was determined by ashing at 600 °C. The decomposition was calculated in % of the initial weight and the average was determined from three depths.

Assessment of the results of soil dynamics: All calculations were carried out on an area of 1 m^2 and at a depth of 50 cm under the soil surface. This does not apply to available nitrogen in the soils in Berlin due to lack of data. Here calculations could only be carried out at a depth of 30 cm. The curves of Figures 15.1 to 15.4 were polished. See BEYER (1989) and PETERS (1990) for further details about the methods and results of the measurements of the soils in Schleswig-Holstein. Measurements were taken during 1986/87/88. The data for the soils in Berlin were taken from works by NEUMANN (1976), FRIEDRICH (1979) and SCHWIEBERT (1980). Here measurements were taken during 1972/73/74.

15.4 Soil dynamics

15.4.1 Loamy Luvisols under different land use (Fig. 15.1)

Water and air regimes: In the Eastern Hills the highest amount of precipitation falls in the summer, whereas the winters are relatively dry. The Luvisol under the beech and oak forest is reached by much less rain than the Luvisol under farm management due to precipitation interception by the tree canopy. The air content in the soil is therefore correspondingly high. This is also due to intensive transpiration (ZINGK 1988). The high precipitation levels in summer lead to an extreme reduction in the air volume in the forest soil. In spring the stagnation of water regularly occurs for a short period of time. The amount of precipitation at the site of the ecologically farmed Luvisol is approximately 100 mm higher (Tab. 15.2), but the annual cycles correspond with each other. In contrast to the Luvisol under forest, the amounts of air in the arable soils are distinctly lower and the water supply is better. An increase in the stagnating water up to a depth of 30

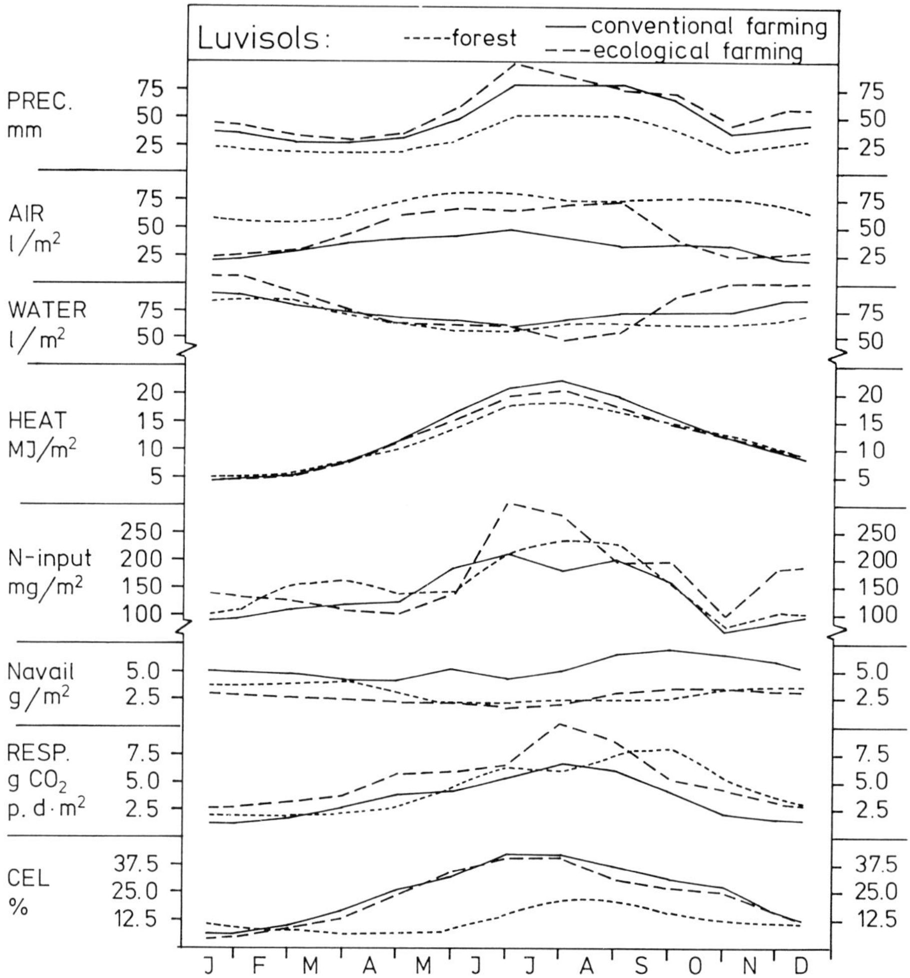

Figure 15.1: Precipitation (PREC., under forest without interception and stemflow) and nitrogen input with precipitation (N–input) on Luvisols in North–Germany, their supply of air, water, heat and nitrogen (at a depth of 0–50 cm) and their annual cycle of respiration (RESP.) and cellulolytic activity (CEL.). Mean values from June 1986 to May 1988.

cm under the surface leads to a worse air supply, above all in autumn and winter. This in turn can have an effect on the biological activity.

Heat regime: The soils warm up gradually in spring. The relatively low temperatures mean that in all three Luvisols the sum of heat is similar for a long period of time. With increasing temperature the larger amount of water in the arable soils warms up and the heat capacity of the water, which is twice as high as that of the minerals, can be used completely. In addition the temperature in the forest does not increase as much due to the shadow cast by the tree canopy, so that the sum of heat i. e. the actual heat supply, is lower there. In winter, however, the litter layer isolates the soil, so that the heat supply here is higher than in the Luvisols under farm management. The ecologically farmed field is covered with weeds or underseed the whole year round, so that the temperature in the soil is lower than in the soil under conventional farm management. The low temperatures, the considerable drying out and the low densities lead to a lower sum of heat in the Luvisol under ecological farm management than in that under conventional management. The larger amount of humus (Tab. 15.2) could not compensate for this.

Nitrogen regime: Approximately 20 kg/ha nitrogen reach the Luvisols in precipitation, 3/5 of this amount in the form of NH_4^+–N and 2/5 in the form of NO_3^-–N. The input in Schleswig–Holstein, which has very little industry, lies at the lower end of the table of nitrogen inputs which have been observed in the Federal Republic of Germany (KREUTZER 1989). In the forest soil this nitrogen input is the only external nitrogen source, apart from the microbial N adsorption in an unknown amount. In addition under ecological farm management 100 kg N reach the soil as organic fertilizer and under conventional management 180 kg N as mineral fertilizer.

In the forest soil the mineralization of litter–N starts in autumn. The average N input in precipitation makes up approximately 5% of the available nitrogen in the soil and in general is of little importance. In summer at the time of high precipitation and high temperatures the proportion increases to approximately 15%. This obviously has a direct effect on the soil respiration.

The high nitrogen input in the soil under conventional farm management results in a high supply of available nitrogen during most of the year. In spring the nitrogen is released from the fertilizer. This effect drops somewhat in summer. As the cereals begin to ripen most of the leaf parts start to die (GEISLER 1980). These parts are then converted in late summer. After the harvest and the mulching of the stubble, the decomposition of the remaining harvest material takes place at an increased rate and nitrogen is released. In winter relatively high amounts of nitrogen are available in comparison with the other Luvisols.

Under ecological farm management more nitrogen is taken up by the organic matter because at 1400 g Nt/m^2 the amount of nitrogen in the soil is a third higher than in the soil under conventional management.

Soil respiration: The distinct annual rhythm with maximum values in summer enables the considerable influence of heat to be recognized (KRZYSCH 1962, FRANKO 1984). In spring the soil life develops slowly, so that the soil respiration gradually increases and in summer reaches its maximum, after which it then drops again. Deviations from this characteristic cycle are caused by the lack or surplus of certain factors. The considerable litter fall in the deciduous forest causes a large increase in the soil respiration in autumn, as long as it is warm enough (EDWARDS 1975).

In summer the sudden fall in the soil respiration in the ecologically farmed soil is put down to a lack of air, as with an improvement in the air supply it increases again immediately. The fall in the soil respiration curve is greater than would be thought in the arable soils in autumn, due to the reduction in the heat supply. This is because harvesting affects the soil animals much more than any other tillage procedure (DUNGER 1983). The removal of the plant covering causes a drastic change in the climate and the food. If the temperature is still high enough and the harvest litter is wet enough, decomposition begins in autumn with a correspondingly high respiration intensity after the population has increased again. In winter the soil respiration drops, however it does not cease altogether. According to TAMM and KRZYSCH (1965) this can be recorded only when a permanent frost of $-7°C$ and lower occurs. The decomposition rates in winter make up approximately 10 % of the C turnover of the summer (TAMM and KRZYSCH 1965, FRANKO 1984, 1986).

Despite having the best heat supply the Luvisol under conventional farm management released the lowest amount of CO_2. In these loamy soils a lack of air hinders an increase in the soil respiration. The soil is very compressed and is hardly loosened at all by earthworms. Despite the high amount of nutrients available, the soil respiration remains lower than that of the forest soil. In winter it almost comes to a standstill, due to the the stagnating water which is less than 50 cm under the soil surface. The relatively high respiration rates in the Luvisol under ecological farm management in winter are put down to the overwintering underseed (clover–grass), which in contrast to the young plant growth of the winter cereals under conventional farm management, has a more positive effect on the soil organisms. The decomposition of organic matter takes place in winter and this is also due to the favorable microclimate. The reduction in temperature, due the thick soil covering of the soil under ecological farm management, does not have a negative effect on the soil respiration. On the contrary the living conditions of the soil organisms were definitely better. The earthworms, *Lumbricus terrestris*, which dig deep into the soil, transport large amounts of organic matter into the depths of the soil. The plant remains and the worm faeces on the other hand prepare the basic food of the smaller bioturbate soil organisms. The lower nitrogen supply, in comparison with the Luvisol under conventional farm management, is of no importance with regard to the soil respiration.

The supply of available nitrogen in general does not have a great influence on the soil respiration. In the Luvisol under ecological farm management the dung is converted in spring, whereas the conversion of the mineral fertilizers is not reflected in the soil. The earthworms' constant bioturbation causes a continuous nutrient supply for the soil organisms. The extensive flora which accompanies the cultivated plants has a shorter life cycle than these plants and is converted during the vegetation period. Under forest a mull–like moder has developed due to the relatively good decomposition performance (Tab. 15.1). This moder becomes mull in large parts of this forest. The litter which falls is of course not completely decomposed within one year, but the accumulation of humus is low in comparison with the Podzol under spruce.

Cellulose decomposition: The maximum and minimum amounts of cellulose decomposition in summer and winter respectively are caused directly by the temperature. In general, under forest decomposition rates are lower than under farm management. In winter the ratios are inverted in exactly the same way as for the soil respiration. Above all at depths of 30 and 50 cm the decomposition was higher than in the arable soils (BEYER 1989), due to the higher temperatures. In comparison with the forest Podzol which is poor in nutrients, the bioturbation and the better nutrient supply have encouraging effects on the cellulose decomposition. The cellulose of the litter was decomposed or converted into hemicellulose respectively very quickly (BEYER, BLUME 1990) so that the products of decomposition (deliquescent sugar) are available quickly. While the soil respiration after the summer maximum values increases immediately after the leaf fall, the cellulose decomposition only reaches a peak in summer. Obviously the litter which is difficult to break down, in contrast to the easily decomposable part is decomposed more during the following year, however, before the new leaf fall begins. The decomposition of the cellulose takes place only when more heat is available, as the increase in the curve occurs later than that of soil respiration.

The living conditions of the micro–organisms which decompose the cellulose improve with increasing pH and nitrogen supply (RUSCHMEYER et al. 1958, SCHMIDT et al. 1958). For this reason the arable soils are characterized by high decomposition rates. In the Luvisol under ecological farm management the cellulose decomposition, despite a much lower nitrogen supply, is just under that of the conventional farm. The dependence of the cellulose decomposition on the nitrogen supply, which has been proved by numerous authors (see SZEGI 1988) can only be seen as being partly the case in the Luvisols. This is possibly slowed down so much by the temporary influence of stagnating water at the conventional farm, so that the high N–supply could not have an effect. Furthermore Kozova (1963) discusses the strong competition between cultivated plants and cellulolytic soil organisms, which in the vegetation period means that the organisms which decompose cellulose do not receive enough nitrogen, because the plants are more competitive. On the other hand the Luvisol under the beech and oak forest has a much lower cellulose decomposition despite having a similar N supply to the Luvisol under ecological farm management. This means that nitrogen can not be the only determining factor.

The fall in the curve after harvest, which was documented for soil respiration, is also to be recorded for the cellulolytic activity. However it is not as distinct. Above all in the Luvisol under conventional farm management the whole population specializes in the decomposition of cellulose, a phenomenon which is caused by straw fertilizing. At the beginning of winter the cellulolytic activity clearly falls once again due to the hindering influence of stagnating water. As almost all the cellulose decomposers are aerobic ones (POKORNA–KOZOVA 1968) the connection between cellulolytic activity and soil moisture is plausible. In winter during the biological rest period the cellulose decomposition however, does not come to a complete standstill (KOZOVA 1963, DUNGER and FIEDLER 1989).

15.4.2 Sandy Podzols under different land use (Fig. 15.2)

Water and air regimes: Even in winter in the Geest large amounts of precipitation fall. The precipitation is approximately 200 mm higher than in the Eastern Hills and causes an improvement in the water supply of the dry Podzols which have a low water capacity. The soil is shielded by the tree canopy in the spruce forest, though this is not as distinct as that of the deciduous forest situated on the Luvisol. Despite this less water reaches the soil here than under arable land. At the site of the Podzol under ecological farm management approximately 100 mm more rain falls (Tab. 15.2).

The difference in the water and air regimes of the two Podzols under farm management is put down to the higher input of water by precipitation (Tab. 15.2) and the higher water capacity of the Ap horizon (Tab. 15.1) under ecological farming, as this generally has a higher water supply. Under spruce the water capacity of the organic layer and the Aeh horizon, which is rich in humus, leads to a large amount of water in the soil, despite the lower amount of water input in comparison with the Podzols under farm management. However, very high precipitation shows only a short term reduction in the high amount of air, which is sometimes 100 liters higher than in the Luvisol under farming. In the Podzols therefore water is assumed to be the restricting factor with regard to the soil organisms.

Heat regime: In winter under spruce the sum of soil heat is clearly higher than under farm management, because the 15 cm thick layer (Tab. 15.1) acts as a better heat insulator than that of the Luvisol under deciduous forest. Therefore the mineral layers warm up slowly in summer. In September and October the sums of heat under farm management reach comparable values, as the fallow fields are subject to a much higher radiation of heat after the harvest than under forest.

The small amount of soil water under farming, in comparison with the Luvisols, reduces the heat capacity of the soil, but the dark color appears to compensate for this. The Podzol under ecological farm management has higher temperatures mostly than that under conventional farming, despite similar air temperatures at

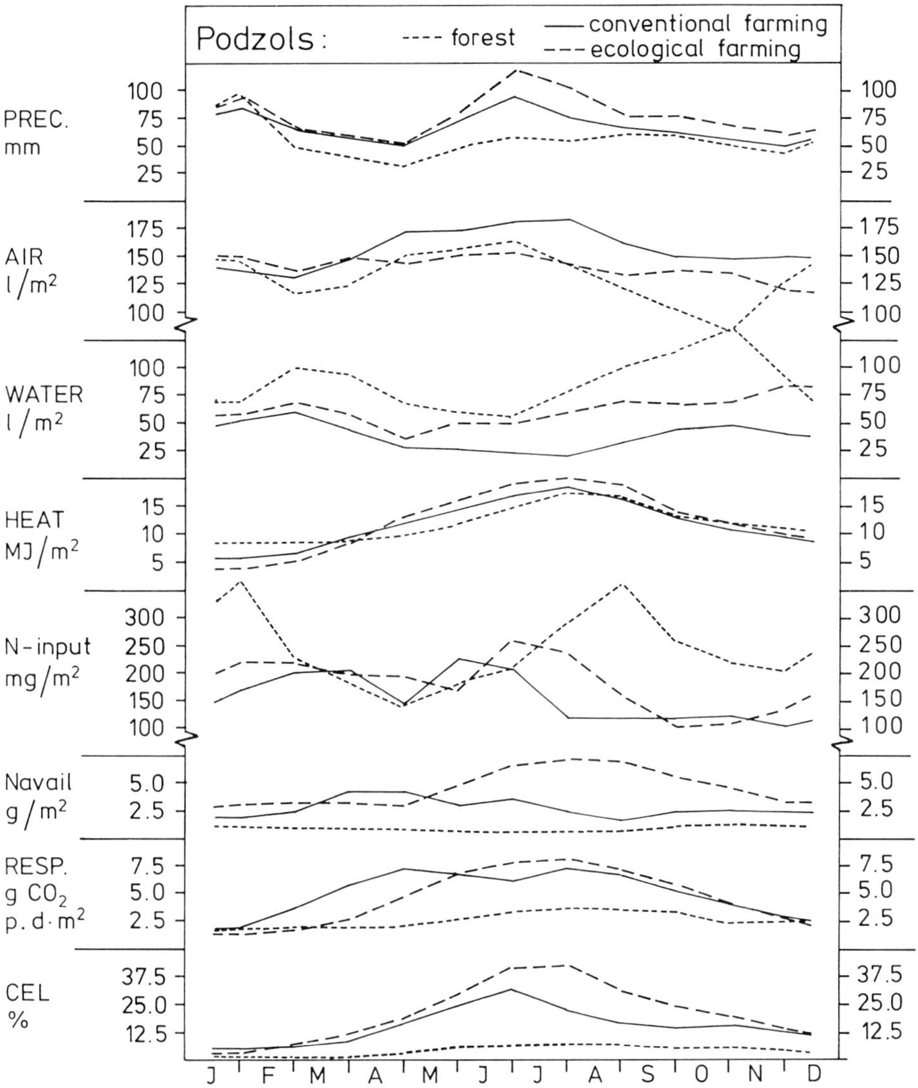

Figure 15.2: Precipitation (PREC., under forest without interception) and nitrogen input with precipitation (N–input) on Podzols in North–Germany, their supply of air, water, heat and nitrogen (at a depth of 0–50 cm) and their annual cycle of respiration (RESP.) and cellulolytic activity (CEL.). Mean values from June 1986 to May 1988.

the sites (BEYER 1989). The large amounts of maize and cereal vegetation at the conventionally farmed site protect the soil extensively from direct sunlight and therefore cause a reduction in the temperature. On the other hand weeds are constantly being hoed in between the vegetables in the Podzol under ecological farm management. This means that the exposed soil warms up much better. This together with the higher amounts of water and humus cause the higher sum of heat during the vegetation period. In winter the soil under conventional farming is still covered with dung, which was applied in autumn and which has a similar effect to that of the humus layer in the forest. The low temperatures in the Podzol under alternative farming are put down to the vegetable plants which do not completely cover the soil in winter.

Nitrogen regime: Under spruce forest the input of nitrogen by precipitation was 30 kg/ha but under farmland 20 kg/ha (Tab. 15.2). 2/3 of this is in the form of ammonium, as conifers filter gases and dust containing NH_4^+ (BLUME et al. 1989) which reach the soil with the precipitation the whole year round. In addition the Podzol under conventional farm management is provided with 160 kg N/ha in anorganic and 120 kg N/ha in organic form. The ecologically farmed Podzol on the other hand is supplied with only 130 kg N/ha in organic form.

In the Podzol under spruce the supply of available nitrogen is, without exception, equally low throughout the whole year. The N input makes up approximately 25% of the available nitrogen, but can reach 30–40% during high amounts of precipitation. In this soil the nitrogen regime is therefore characterized considerably by the input in precipitation accompanied by the mineralization of litter. Under conventional farm management the easily decomposable slurry was applied in March and the anorganic fertilizer in April. The water supply decreases quickly in May, which results in the reduction of available nitrogen. An increasing heat supply on the other hand pushes up the mineralization, but cannot compensate for this. In August at the end of a long period with a low water supply in the soil the release of N almost comes to a standstill. With increasing soil moisture it increases again in autumn. At this time the dung which has been applied is partly converted. Apart from in spring, the nitrogen supply is always higher in the Podzol under ecological farm management. Obviously during the summer months when the temperatures are high and there is sufficient water available a large amount of nitrogen is released from the organic soil substance. The application of frequent small doses of dung during the vegetation period is reflected in the high nitrogen level.

Soil respiration: The dependence of the soil respiration on temperature is even more distinct in the Podzols than in the Luvisols as the similar curves verify.

The soil respiration in the Podzol under spruce takes place at a very low level. In this soil the nutrients restrict the soil respiration, as the amount of water is higher than in the Podzols under farm management. Nevertheless with increasing heat supply the decomposition of the easily decomposable litter increases somewhat during the vegetation period. In contrast to the Luvisol under deciduous forest the litter fall is spread over the whole year, so that the input of litter never reaches a real peak. ALBERTSEN (1979) measured the CO_2 concentration in different depths of the same soil in 1976/77. He calculated 0.3 g CO_2 per day as the mean emission rate for the mineral soil in contrast to the 2.7 g of Tab. 15.2: therefore nearly 90% of the CO_2 should be released from the organic layer. ALBERTSEN (1979) also analyzed higher respiration rates in the B horizons than in the A horizons.

The Podzols under farm management differ more clearly from the Podzol under forest than can be said for the Luvisols. Under conventional farm management a lack of water occurs in summer, which influences the soil respiration. This phenomenon of two peaks in the soil respiration curve, which is determined by moisture could be observed at other places (BUYANOVSKY et al. 1986, MÜLLER 1965, SCHEFFER and SCHACHTSCHABEL 1989), whereas SCHAFFER (1954), KRZYSCH (1962), and BERTRAM (1985) did not find any measurable influence of soil moisture. According to KRZYSCH (1962) moderate soil moisture is sufficient in order to attain higher soil respiration rates, provided that only the soil temperature is high enough. Depressions in the respiration only occur when the soil moisture content reaches the permanent wilting point. In our soil the easily decomposable slurry is converted early in the year and at the same time a high amount of N is released, so that the soil respiration increases unusually quickly. This effect decreases in May, which means that the respiration drops off somewhat once again. This influence of nitrogen can only be recorded in spring, when the temperature is still of secondary importance. Under ecological farm management the soil respiration increases much later, because the dung which has been applied is only decomposed in May as is the case for the Luvisol under ecological farming. The respiration therefore does not drop in summer. In addition the soil has a higher water capacity and the water supply is sufficient. The course of this curve shows clearly the dependency of the soil respiration on the heat supply, because there is no shortage of the other factors (especially water and nutrients).

ALBERTSEN (1979) measured the highest respiration rates in the plough layer of similar Podzols in the neighborhood but relatively high rates in the subsoil too, down to a depth of more than 2 meters.

Cellulose decomposition: Under spruce the decomposition rate is low because of the low nutrient supply and in particular the low N supply. The very low pH, which is below 3, enables only a few types of fungi which decompose the cellulose slowly to have an effect (SZEGI 1988). The cellulolytic activity which is dependent on temperature, increases somewhat despite this during the vegetation period and in contrast to all the other sites runs almost parallel to the soil respiration.

The difference between the Podzols under farm management and under spruce is even more distinct than in the Luvisols, as it is the case for the soil respiration. The differences between the Podzols under farm management are put down to a higher decomposition of cellulose in the topsoil (BEYER 1989), in contrast to the Luvisols.

At a depth of 30 cm both sandy Podzols are similar. In the Luvisol under ecological farm management, in particular at a depth of 30 cm, the rates are clearly lower than those of the conventionally farmed Luvisol because here ploughing enables the nutrients to reach a depth of 30–40 cm and this in turn supports the cellulose decomposition (SZEGI 1988). In the Podzol under ecological farm management the decomposition rates are always significantly higher than those under conventional farming. In contrast to the soil respiration the cellulose decomposition increases with increasing N supply and even when the nitrogen level is high.

15.4.3 Gleysols under forest (Fig. 15.3)

Water and air regimes: During 1972 and 1974 a precipitation deficit of 15% occurred in contrast to the long standing average precipitation. Fine grained sands containing humus which are features of both soils, cause a notable available water capacity of approximately 150 mm in the top 50 cm (including the humus layer; see Tab. 15.1 f and h), which however borders on pores (10–50 $\not{p}$) i. e. voids which are still draining. During the summer months the available water content in the upper layer decreased, without having reached critical values, as the ground water supplied more water. The Eutric Gleysol was more moist the whole year round and therefore poorer in air than the Podzo-dystric Gleysol (Fig. 15.3).

Heat regime: The soil heat of the Eutric Gleysol was usually somewhat higher than that of the Podzo-dystric Gleysol, although the Teufelsbruch as a whole is the larger valley, in which cold air can have a longer and more intensive effect. The south-west exposition (in contrast to the west exposition of the Podzo-dystric Gleysol) and the higher heat capacity as a result of the higher water content increase the soil heat of the Eutric Gleysol.

Nitrogen regime: During the period in which investigations were carried out the input of nitrogen in precipitation was not measured. For this reason measurements which were taken in 1978/79 under arable land and a tree canopy from neighboring sites by JAYAKODY (1981) have been consulted. The input at both sites was probably somewhat below 2 g/m^2 per year; approximately 70% as ammonium and only 30% as nitrate. The average concentration was probably approximately 3.2 mg/l NH_4^--N and 1.2 mg/l NO_3^- (JAYAKODY 1981). With regard to the root space content of dissolved and adsorbed N the Eutric and Dystric Gleysols differ very much from one another: The Eutric Gleysol received on average 80 mg N in the O horizon and 55 mg N in the Ah horizon per l soil. The Dystric Gleysol on the other hand received only 13 and 20. The NO_3^--N content was especially low with only 3 mg. The reason for the larger amount at the eutrophic site is probably due to litter rich in N (compare with C/N ratio in the humus layer in Tab. 15.1) and above all due to microbial N assimilation. At the dystrophic site the nitrification was obviously hindered considerably.

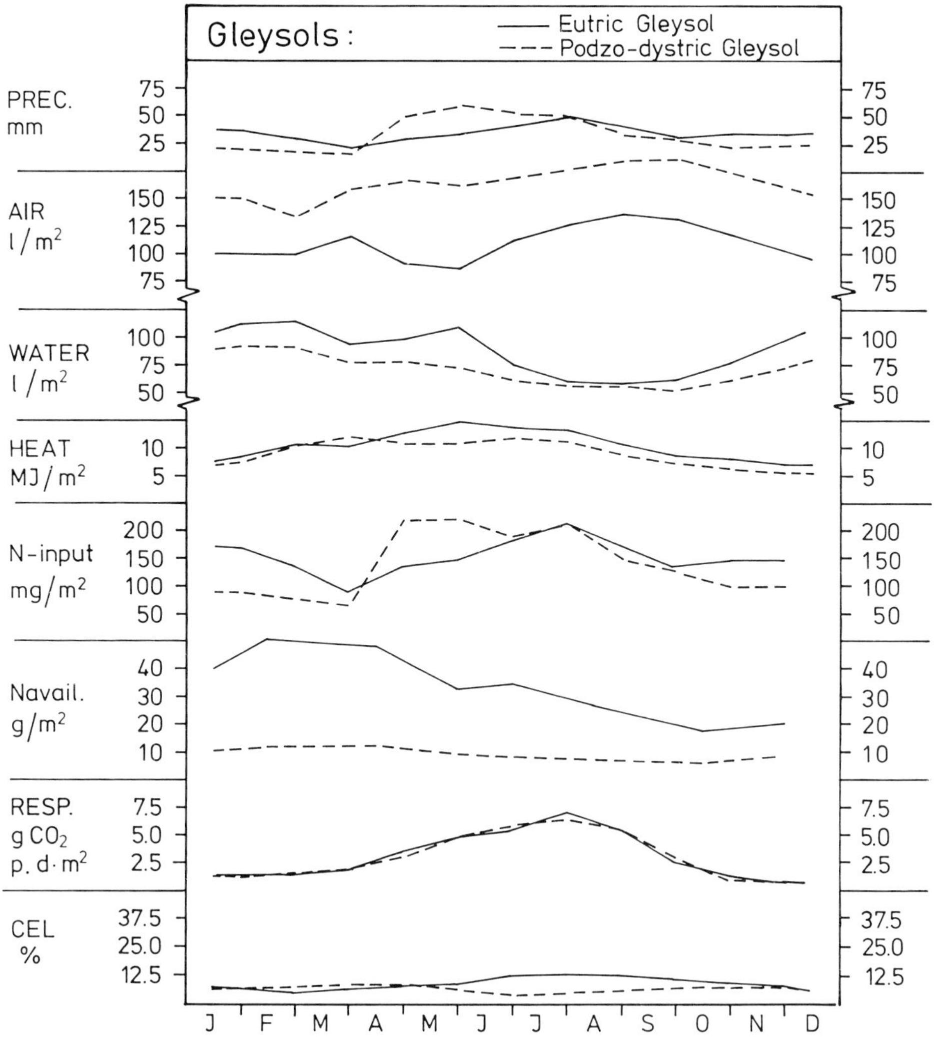

Figure 15.3: Precipitation (PREC.; without interception and stem flow) and nitrogen input with precipitation (N–input) on Gleysols in North–Germany, their supply of air, water, heat and nitrogen (at a depth of 0–50 cm) and their annual cycle of respiration (RESP.) and cellulolytic activity (CEL.). Mean values from May 1973 to September 1974.

Soil respiration: Generally the soil respiration was also determined by the temperature in the Gleysols: At 19 °C during the height of summer it amounted to 6–7 g per m^2 and day of CO_2 in the upper layer; at 1 °C in winter, however, it amounted to only 0.7 g. The eutrophic site surpassed the dystrophic one only slightly as was the case for the somewhat higher heat supply (Tab. 15.2). This can be attributed to the fact that the eutrophic site was continually more wet, so that a lack of O_2 restricted the soil respiration.

Cellulose decomposition: There are clear differences between the eutrophic and dystrophic sites with regard to the cellulolytic activity, the former surpassed the latter by a third. This can probably put down to the much better nitrogen supply.

15.4.4 Histosols under natural vegetation (Fig. 15.4)

Water and air regimes: Both moors naturally showed much higher water contents than the other sites. In comparison with the long standing average approximately 15% less precipitaion fell during the period in which investigations were carried out. This meant that both moors dried out considerably, above all during the vegetation period, and therefore were naturally rich in air. Even during the winter months water saturation did not occur and in turn a lack of air rarely occurred in the upper layer. Considerable differences between the two moors were not found existing.

Heat regime: The higher heat capacity had the effect that both moors contained more heat energy than the neighboring Gleysols. For the same reason the daily and annual fluctuations in the moors were clearly lower. The soil temperature in the eutrophic moor was without exception, slightly lower than that in the dystrophic moor, because the Teufelsbruch which belongs to the former is a larger valley, in which more cold air collects for longer periods of time.

Nitrogen regime: Very large differences were determined in the N contents of both sites, despite the fact that the N input in precipitation was presumed to be approximately the same (1.7 – 1.8 g per m^2 and year) (Tab. 15.2): The eutrophic moor received on average approximately 39 mg NO_3^-–N and 19 mg dissolved and adsorbed NH_4^-–N per liter upper layer; the dystrophic moor, however, received only 3 mg NO_3^-–N and 8 mg NH_4^-–N. Microbial N adsorption and also mineralization do not appear to play a considerable role in the dystrophic moor.

Soil respiration: The soil respiration in the moors was determined by the temperature. An increase in the soil temperature of approximately 10 °C caused the output of CO_2 to be trebled. The rates in the eutrophic moor were similarly high to those of the mineral soil and in summer even surpassed these now and again.

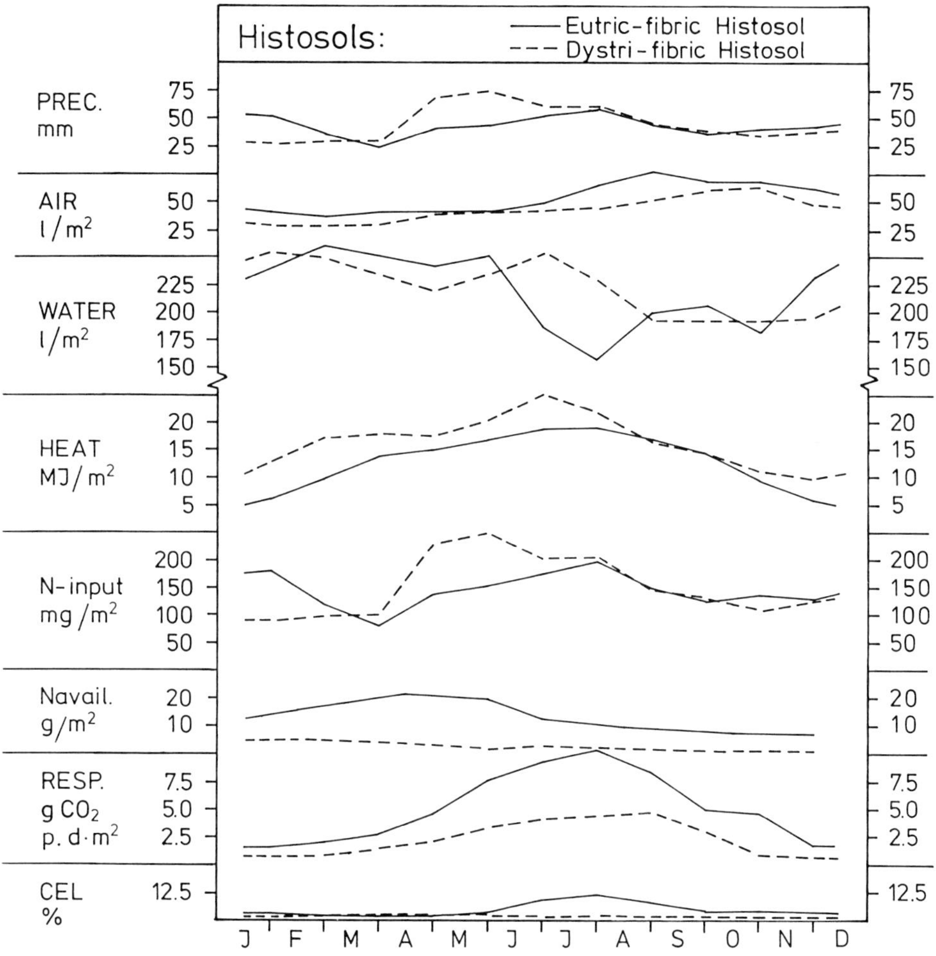

Figure 15.4: Precipitation (PREC.) and nitrogen input with precipitation (N-input) on Histosols in North–Germany, their supply of air, water, heat and nitrogen (at a depth of 0–50 cm) and their annual cycle of respiration (RESP.) and cellulolytic activity (CEL.). Mean values from May 1973 to September 1974.

The relatively low precipitation is probably responsible for the latter, which led to an improvement in the air supply. The dystrophic moor showed, despite low temperatures, clearly lower soil respiration. Here the clearly lower nutrient content and in particular the nitrogen content are reflected.

Cellulose decomposition: The cellulolytic activity was moderate in the eutrophic moor and low in the dystrophic moor. Lack of air and in the dystrophic moor the lack of nutrients were the cause of this.

15.5 Comparative discussion

Mean annual results of the respiration rates activity and the environmental conditions of different soil types in North Germany under different land use were summarized in Tab. 15.2.

It has been shown that the heat supply determines considerably the biological activity of the soils, the CO_2 emission rate as well as the cellulolytic activity. The elimination of this factor "soil temperature" makes a comparison between different soils possible. In addition to this a possibility was found, which enabled the soil respiration and the cellulolytic activity to be compared with reference to the same temperature level (Fig. 15.5). The measurements were assessed correspondingly and the average values were obtained (see columns 11, 13, and 14 of Tab. 15.2). The oxygen supply, next to the temperature, was also found to be an important factor with regard to cellulolytic activity: Both Histosols were characterized by a very low cellulose decomposition which explains the large accumulation of peat.

The soil respiration, on the other hand, was determined more by the nutrient supply, in particular that of nitrogen, than by the oxygen conditions. Haplic Podzol, Podzo-dystric Gleysol and Dystric Histosol, all of which are characterized by low available nutrient contents, show very low respiration values. This podzolization is caused by their low biological activity.

Orthic Luvisols under deciduous forest are richer in nutrients than the above-mentioned soils and also have easily decomposable litter. This led to a higher biological activity (see column 14 of Tab. 15.2). Lumbricidae mix the litter that has fallen with the mineral soil and counteract a translocation in the subsoil. Smaller soil animals then take over the decomposition under the soil surface (DUNGER 1983). The bases pump of the trees alone is not sufficient to prevent the lowering of the pH and the loss of bases (BEYER et al., 1991).

Soil respiration and cellulose decomposition therefore show that great differences exist with regard to the biological activity between the most important soil forms in North Germany.

In addition, the arable soils differ greatly from the forest soils, as the nutrient supply is higher. The soil organisms and the cultivated plants profit from the easily converted fertilizers (see column 7 of Tab. 15.2). In the sandy Podzols under arable land there was no difference in the biological activity between conventional and ecological farm management, as both types of farm management used organic fertilizers. The soil respiration at all the sites with organic fertilizers was high (see column 10 of Tab. 15.2), and therefore it only fell clearly in the Luvisol under conventional farming. But LAVINOVA et al. (1988) did not find any differences in the respiration between Greyzems with and without organic manure, perhaps because

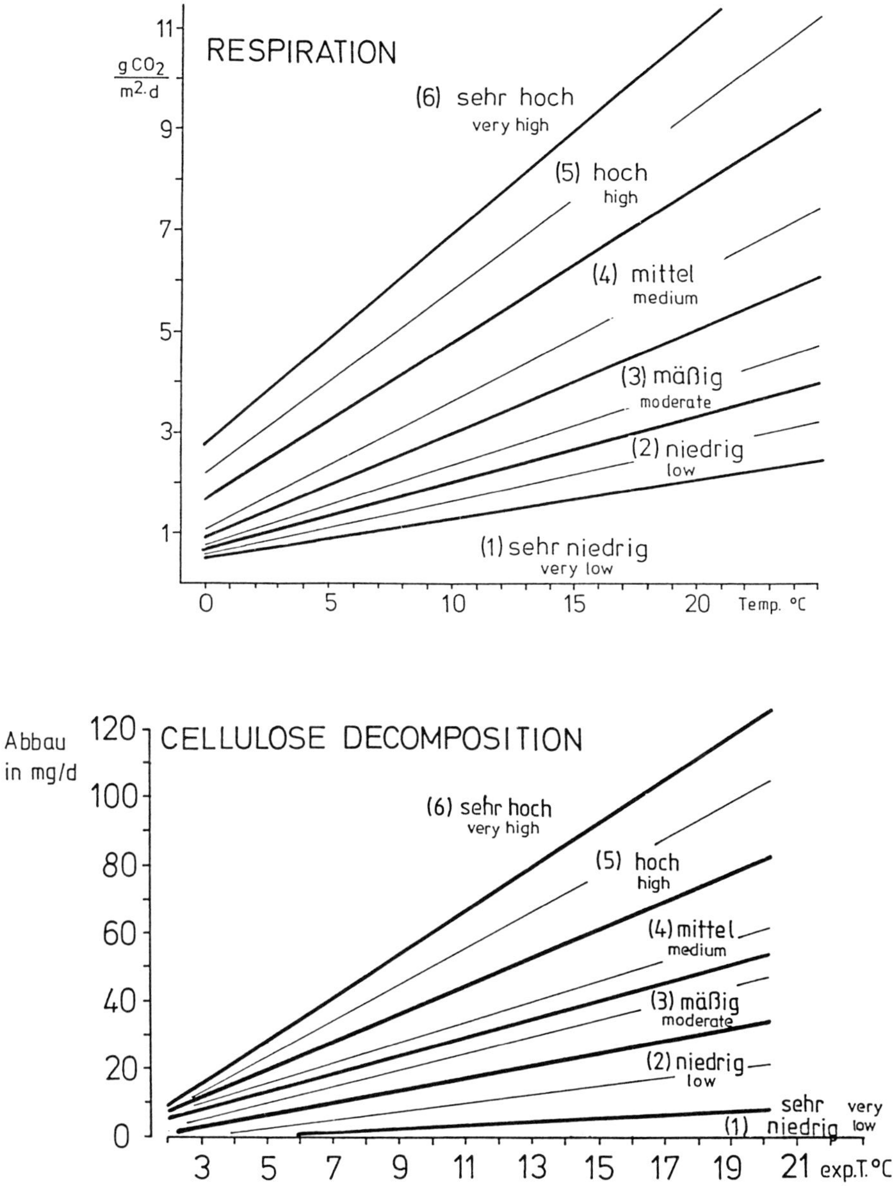

Figure 15.5: Scheme of grouping response of respiration and cellulolytic activity of soils to the soil temperature (from BEYER, 1990).

	(1)	(2)	(3)	(4)	(5)	(6)	(7)	(8)	(9)	(10)	(11)	(12)	(13)	(14)
Luvisols														
a) Beech/Oak	384[1]	71	75	12.1	1.06	0.81	—	0.4	2.8	4.5	3.9	12	3.2	3.6
b) Conv. farming	607	34	77	13.1	1.02	0.68	18.2	0.2	5.1	3.4	3.8	25	4.4	4.1
c) Ecol. farming	707	49	80	12.5	1.17	0.97	10.0	0.2	2.8	5.3	5.2	23	3.9	4.6
Podzols														
d) Spruce	671[1]	134	84	12.0	1.97	1.04	—	0.2	0.8	2.7	3.2	4	1.8	2.5
e) Conv. farming	803	157	38	10.4	1.06	0.82	27.8	0.0	2.7	5.0	4.6	16	3.5	4.1
f) Ecol. farming	922	140	62	12.1	1.20	0.96	12.7	0.1	4.5	4.6	4.1	21	3.8	4.0
Gleysols														
g) Eutric Gl.	411[1]	111	90	10.6	1.32	0.51	—	21.1	11.0	3.2	3.5	9	3.0	3.3
h) Dystric Gl.	392[1]	166	72	9.2	1.17	0.48	—	7.9	1.8	3.1	3.4	6	2.3	2.9
Histosols														
i) Eutric H.	452	53	224	12.7	1.31	0.45	—	5.3	9.0	4.7	4.2	5	1.7	3.0
k) Dystric H.	539	43	224	16.4	1.35	0.46	—	1.6	0.7	2.2	2.6	1	1.3	2.0

(1) Annual precipitation in $l \cdot m^{-2}$, [1]without interception and stem flow.
(2) Mean air filled pores in $l \cdot m^{-2}$ to a depth of 50 cm.
(3) Mean water content in $l \cdot m^{-2}$ to a depth of 50 cm.
(4) Mean sum of heat in $MJ \cdot m^{-2}$ to a depth of 50 cm.
(5) Annual NH_4^+–N input with precipitation in $g \cdot m^{-2}$.
(6) Annual NO_3^-–N input with precipitation in $g \cdot m^{-2}$.
(7) Annual N input with fertilizers in $g \cdot m^{-2}$.
(8) Mean amount of available NH_4^+–N in $g \cdot m^{-2}$ to a depth of 50 cm.
(9) Mean amount of available NO_3^-–N in $g \cdot m^{-2}$ to a depth of 50 cm.
(10) Mean daily respiration in $g \cdot m^{-2}$ CO_2.
(11) Classification of mean (temperature conformed) soil respiration (1 = very low, 6 = very high).
(12) Cellulolytic activity (decomposition in % of 5 g initial cellulose during 6 weeks: mean from 5, 30, 50 cm depth).
(13) Classification of mean (temperature conformed) cellulolytic activity (1 = very low, 6 = very high).
(14) Mean of (11) and (13).

Table 15.2: (See previous page) Environmental and microbial characteristics of typical soils of Northwest Germany (average of two years; a–f: 1986–1988, g–k: 1972–1974).

Greyzems are characterized by high contents of organic matter. The cellulose decomposition on the other hand was influenced more by the oxygen supply, as the Stagni–orthic Luvisol under farm management suffers more from stagnating water, which explains its lower decomposition rates.

Bibliography

ALBERTSEN, M., 1979, Kohlendioxid-Haushalt in der Gasphase der ungesättigten Bodenzone, dargestellt am Beispiel eines Podsols.
Z. Pflanzenernähr. u. Bodenk. **142**, 39–56.

ANDERSON, J. P. E., 1982, Soil respiration.
In: Page, A. L., Methods of soil analysis.
Am. Soc. of Agron., Madison, Wisconsin, USA, 831–871.

BERTRAM, H. G, 1985, Zur Rolle des Bodens im globalen Kohlenstoffkreislauf.
Thesis University Osnabrück.

BEYER, L., 1989 Nutzungseinfluß auf die Stoffdynamik schleswig–holsteiner Böden–Humusdynamik und mikrobielle Aktivität.
Schriftenr. Inst. Pflanzenern. Bodenk. Kiel **6**.

BEYER, L., 1990, Die Bewertung der biologischen Aktivität von Böden über Ermittlung der Bodenatmung und der zellolytischen Aktivität im Feld.
Z. Pflanzenernähr. u. Bodenk. **153**, 261–269.

BEYER, L.; BLUME, H.–P., 1990, Eigenschaften und Entstehung der Humuskörper typischer Wald- und Ackerböden Schleswig–Holsteins.
Z. Pflanzenernähr. u. Bodenk. **153**, 61–68.

BEYER, L.; BLUME, H.–P.; IRMLER, U.; HENSS, B., 1991, Veränderung des Humuskörpers einer Parabraunerde unter Wald in 25 Jahren.
Forstw. Centralbl., in press.

BEYER, L.; IRMLER, U., 1991, The structure of Humus and Dynamic of Litter Composition on a Luvisol and a Podzol under forest.
Pedobiologia, in press.

BLUME, H.-P.; BEYER, L.; PETERS, M.; SCHIMMING, C. G., 1989, Umweltschonende Bodennutzung in der Landwirtschaft.
Publications of the Agric. Faculty of the University of Kiel **71**, 33–47.

BUYANOVSKI, G. A.; WAGNER, G. H., 1983, Annual Cycles of Carbon Dioxide Level in Soil Air.
Soil Sci. Soc. Am. J. **47**, 1139–1145.

BUYANOVSKI, G. A.; WAGNER, G. H.; GANTZER, C. G., 1986, Soil Respiration in a Wheat Ecosystem.
Soil. Sci. Soc. Am. **47**, 338–334.

DUNGER, W., 1983, Tiere im Boden.
Ziemsen, Wittenberg

DUNGER, W.; FIEDLER, H. J., 1989, Methoden der Bodenbiologie.
Fischer, Stuttgart

EDWARDS, C. A., 1975, Earthworms in Soil Formation, Structure and Fertility.
Quaestionaes Entomoligicae **21**, 517–522.

FRANKO, U., 1984, Einfluß niedriger Temperaturen auf die Umsetzung der organischen Substanz im Boden.
Archiv Acker- und Pflanzenbau und Bodenkunde **28**, 533–536.

FRANKO, U., 1986, Die Mineralisierung der org. Substanz im Boden außerhalb der Vegetationsperiode.
Archiv Acker- und Pflanzenbau und Bodenkunde **30**, 391–394.

FRIEDRICH, F., 1979, Humusmetabolik und Wärmedynamik zweier Bodenschaften der Berliner Forsten.
Thesis University Berlin

GEISLER, G., 1980, Pflanzenbau.
Parey, Berlin

JAYAKODY, A. N., 1981, Nitrogen Dynamics of Berlin Soils in Forest, Cultivated Fields and Waste Water Disposal Fields.
Thesis University Berlin

KOZOVA, J., 1963, Mikrobiologische Zellulosetersetzung unter natürlichen Bodenverhältnissen.
Zbl. Bakteriol. II **116**, 459–468.

KRZYSCH, G., 1962, Der Verlauf der Bodenatmung während der Vegetationsperiode und ihre Veränderung durch langjährige Düngungs- und Bearbeitungsmaßnahmen.
Habil. Schrift University Berlin.

LAVINOVA, A.; ROZONOVA, L.; STRUKTOZOWA, V.; SAMOLILOV, T., 1988, The effect of various cultivation methods on the speed of gas exchange in grey forest soil.
Agrokhimiya **9**, 75–87.

MÜLLER, G., 1965, Bodenbiologie.
Fischer, Berlin

NEUMANN, F., 1976, Struktur, Genese und Ökologie hydromorpher Bodengesellschaften West–Berlins.
Thesis University Berlin.

PALLMANN, H.; EICHENBERGER, E.; HASLER, A., 1940, Eine neue Methode der Temperaturmessung bei ökologischen Untersuchungen.
Ber. Schweiz. Bot. Ges. **50**, 337–362.

PETERS, M., 1990, Nutzungseinfluß auf die Stoffdynamik schleswig-holsteinischer Böden — Wasser–, Luft– und Stoffdynamik.
Schriftenr. Inst. Pflanzenern. Bodenk. Kiel **8**.

POKORNA–KOZOVA, J., 1968, Mineralisations–Aktivität der zellulolytischen Makroflora bei verschiedener Düngung.
In: Deutsche Akademie der Landwirtschaftswissenschaften der DDR (ed.) Tagungsberichte Dt. Akad. der Landwirtschaftswissenschaften **98**, Berlin, 59–66.

RUSCHMEYER, O. R.; SCHMIDT, E. L., 1958, Cellulose Decomposition in Soil Burial Beds, 2. Cellulytic Activity as Influenced by Soil Properties.
Appl. Microbiol. **6**, 114–120.

SCHAFFER, G., 1954, Atmungskurven des Bodens in ihren Beziehungen zum Wasser-, P- und K-Haushalt während mehrerer Vegetationsperioden.
Landw. Forsch. **7**, 12–16.

SCHEFFER, F.; SCHACHTSCHABEL, P., eds., 1989, Lehrbuch der Bodenkunde.
Enke, Stuttgart.

SCHMIDT, E. L.; RUSCHMEYYER, O. R., 1958, Cellulose Decomposition in Soil Burial Beds, 1. Soil Properties in Relation to Cellulose Degradation.
Appl. Microbiol. **6**, 108–114.

SCHWIEBERT, H., 1980, Statik und Dynamik des Wasser– und Lufthaushaltes zweier Düne–Moor–Ökotope Berlins.
Thesis University Berlin

SCHLICHTING, E.; BLUME, H.-P, 1966, Bodenkundliches Praktikum.
Parey, Hamburg

SZEGI, J., 1988, Cellulose–Decomposition and Soil Fertility.
Akademiai Kiado, Budapest

TAMM, E.; KRZYSCH, G., 1965, Zur Dynamik der Bodenatmung und des CO_2-Gehaltes der bodennahen Luftschicht während der Vegetationsruhe.
Acker- and Pflanzenbau **117**, 209–224.

UNGER, H., 1960, Der Zellulosetest, eine Methode zur Ermittlung der zellulytischen Aktivität des Bodens in Feldversuchen.
Z. Pflanzenernähr. u. Bodenk. **91**, 44–52.

WALTER, H., 1952, Eine einfache Methode zur ökologischen Erfassung des CO_2-Faktors am Standort.
Ber. Dtsch. Bot. Ges. **65**, 175–182.

ZINGK, M., 1988, Wasser–, Nährstoff– und Schadstoffdynamik charakteristischer Bodentypen Schleswig–Holsteins. Wasser– und Luftdynamik.
Schriftenr. Inst. Pflanzenern. Bodenk. Kiel **3**.

Part IV

Disturbances and management of semi–natural and agricultural systems

Chapter 16

Fire, dry heat and germination of savanna grasses in Botswana

W. H. O. ERNST

16.1 Introduction

Fire, either naturally initiated by lightning or frequently man–made, is one of the seasonal factors, associated with the dry season in savanna ecosystems (for reviews: DAUBENMIRE 1968; TAINTON and MENTIS 1984; TROLLOPE 1984 a, b). In the savannas of Southern Africa fire has been used for a long time to combat shrub encroachment and to favor the development of a grassland vegetation by killing seedlings and juveniles of woody plants and by preventing their development to a taller fire–resistant stage (TROLLOPE 1984a; GERTENBACH and POTGIETER 1979; APRU 1980; SKARPE 1980; NOVELLIE 1987). Although fuel modelling and fire behavior prediction systems have been developed for an increased efficiency of fire management in savanna ecosystems (VAN WILGEN and WILLS 1988), the effect of fire on individual plants and plant's life history are less well documented in Africa (KEELEY 1981) in contrast to some information for Australian species (GILL, GROVES and NOBLE 1981) and Californian species (KEELEY 1981).

Generally, grasses and other herbaceous plants are less adapted to back fires,

i. e. fires burning against the wind, because heat release is close to the soil surface, although there are species-specific differences of established plants (ROBINSON, GIBBS-RUSSEL, TROLLOPE and DOWNING 1979). The impact of fire on the survival of the seed bank is very important on the long-term when perennial grasses senesce or suffer mortality due to fire-based resource depletion (ERNST and TOLSMA 1989) and have to be replaced by juveniles. In addition to fire as a destroying agent for organic matter and thus possibly of disseminules, the fire temperature will also affect the viability and the germination of seeds. In a Venezuelan savanna, seeds of plants from the herb layer could endure 90 °C for 40 sec. (VARESCHI 1962) whereas grasses in the chaparral and in the mediterranean region kept their vitality after a treatment of 5 min. with temperatures ranging from 93 to 127 °C (SAMPSON 1944; NAVEH 1974). Due to the lack of information on the sensitivity of disseminules of savanna grasses in Southern Africa to fire and the importance of seeds in establishing new plants this study reports on the effects of experimental burns on diaspore survival, germination and seedling development.

16.2 Material and Methods

Diaspores of the perennial savanna grasses *Cenchrus ciliaris* L., *Enneapogon cenchroides* C. E. HUBBARD, *Eragrostis rigidior* PILGER and *Panicum maximum* JACQ. and the annual grasses *Setaria verticillata* (L.) BEAUV., *Tragus berteronianus* SCHULTES and *Urochloa trichopus* (HOCHST) STAPF were bulk-collected in *Acacia tortilis* savannas at Gaborone and Kumakwana in Southeastern Botswana between January and June (for site description: ERNST and TOLSMA 1989; TOLSMA, ERNST, VERWEIJ and VOOIJS 1987).

Two experimental burns were performed. In five plots of 30×30 cm^2, consisting of a dominant cover (80%) of 1.20 m tall *Eragrostis rigidior* with approximately 300 g·m^{-2} fuel loading (cf. VAN WILGEN and WILLS 1988) and an assumed heat yield of about 18000 kJ·kg^{-1} (TROLLOPE 1984b), 100 diaspores of each species were deposited per plot on the soil surface. After burning at 13.00 hours and a wind velocity of 2 m·sec^{-1} during the dry season of November 1987, a 1 cm thick surface soil layer with ash from the entire 30×30 cm^2 plot was collected, distributed into petri-dishes and wetted. In total, 85 petri-dishes were exposed to a temperature regime of approximately 25 °/20 °C (maximum/minimum; day/night) under roof shadow (approximately 10% of full sun light). As a site control, two plots of 30×30 cm^2 without addition of grass diaspores were collected for estimation of background seed bank. As a treatment control 3×100 diaspores of each species were sown on soil collected from 30 cm depth of the same site, without any seed bank, and treated in the same manner as the burned soil samples but not burned.

In a second experiment 100 diaspores of each species were sown on a 1 cm thick soil layer on aluminium foil (30×30 cm^2). This foil did not heat up above 60 °C during the burns. After addition of a fuel load of 35 g dry matter of *Eragrostis rigidior* the grass was burned. The burned material was distributed into petri–dishes and controlled for the effect of fire on the caryopsis itself and the disseminule structures such as awns (*Cenchrus ciliaris*), pseudopappus (*Enneapogon cenchroides*) and glumes. The petri–dishes were then wetted with demineralized water and exposed to a temperature regime of 25 °/20 °C (day/night; 12 h/12 h) and a photon flux density of 200 μmol·m^{-2}s^{-1} in a growth chamber. As control 100 diaspores per species were incubated in petri–dishes on wet filter paper under the above described conditions.

For the determination of the effect of short term dry heat, disseminules of each species were exposed to dry heat of 100 °C in preconditioned petri–dishes in an oven for 2 min. In addition caryopses of *Cenchrus ciliaris, Setaria verticillata*, and *Tragus berteronianus* were exposed to a temperature range of 84, 92, 100, and 116 °C during 2, 3, 4 and 5 minutes. Germination was recorded from triplicate samples of 100 caryopses or diaspores of each species kept moist in petri–dishes at 25 °/20C (day/night, 12/12 h) and a photon flux density of 200 μmol·m^{-2}·s^{-1} during day period. The germination experiments lasted for 6 weeks. In addition to total seed germination, the harmonic mean germination day (HMGD) was used as parameter for the germination rate and calculated as follows:

$$HMGD = \frac{100}{\sum N_t/t}$$

N_t = number of germinated seeds at day t, t: time in days.

The viability of not germinated caryopses was tested with the tetrazolium stain, after dissecting the caryopsis in two parts. In the case of *Setaria verticillata*, the lemma and palea were removed (ERNST, KUITERS and TOLSMA 1991).

16.3 Results

16.3.1 Effect of field fire in the savanna

From the disseminules of all 7 grass species, sown in the *Eragrostis rigidior* savanna just before burning, seedlings of all species were registered in the not burnt control, whereas in the burnt plot only *Setaria verticillata* germinated and seedlings were detected up to 4 months after treatment. Compared to the not burnt control the germination of *S. verticillata* was delayed by a factor of nearly 10; the harmonic mean germination day of the caryopses was 33.2 in the control compared to 3.7 in the burnt plot. The final germination percentage of *S. verticillata* was not significantly different in control and burn treatments after 4 months (49.5 ± 3.5% in the control vs. 40.8 ± 6.2%).

Effect of fire on disseminules on bare soil

After exposure of grass disseminules sown on a thin soil layer over aluminium foil, to experimental burning, it was possible to analyze the fire effect on the various

Table 16.1: Mass of disseminules and caryopses (mean±1sd.) of 7 grass species from an *Acacia tortilis* savanna at Gaborone (Botswana). $n = 20$ per species.

Grass species	Disseminule (D) mass (mg)	Caryopsis (C) mass (mg)	Ratio D/C
Cenchrus ciliaris	1.51 ± 0.47	0.790 ± 0.310	1.91
Enneapogon cenchroides	0.33 ± 0.08	0.111 ± 0.042	1.83
Eragrostis rigidior	0.13 ± 0.03	0.076 ± 0.029	1.83
Panicum maximum	0.81 ± 0.15	0.554 ± 0.100	1.51
Setaria verticillata	0.63 ± 0.04	0.455 ± 0.041	1.38
Tragus berteronianus	0.47 ± 0.11	0.220 ± 0.040	2.14
Urochloa trichopus	1.56 ± 0.26	0.730 ± 0.134	2.14

species. All disseminules of *Cenchrus ciliaris*, having long bristles, and *Enneapogon cenchroides*, with a pseudo–pappus, were completely destroyed by fire. The same happened to the disseminules of *Eragrostis rigidior, Panicum maximum, Tragus berteronianus* and *Urochloa trichopus*. In the case of *Setaria verticillata* the glumes were mostly burnt, except the hard palea and lemma. From all investigated species (Table 16.1) the disseminules of *S. verticillata* had the lowest ratio of disseminule mass to caryopsis mass (1.38), whereas this ratio varied in most of the other species between 1.51 (*Panicum maximum*) and 2.14 (*Tragus berteronianus* and *Urochloa trichopus*). As in the previous experiment, the germination of fire–treated caryopses of *S. verticillata* was significantly delayed (P< 0.001) by a factor of 8.

16.3.2 Effect of dry heat

Disseminules below the soil surface are protected from the direct effect of fire and are only exposed to dry heat. After an exposure to a temperature of 100 °C for 2 min. (Fig. 16.1), the final percentage of germination was significantly (P< 0.001) decreased compared to the control.

This effect was independent of the species–specific final germination, which was around 90% in *Cenchrus ciliaris, Enneapogon cenchroides* and *Tragus berteronianus*, 60% in *Urochloa trichopus*, and between 35 and 45% in *Eragrostis rigidior* and *Setaria verticillata* under control conditions. Due to the variation of the amount and duration of heat in the soil profile during a veld fire (BENTLEY and FENNER 1958; Vareschi 1962) caryopses of *Cenchrus ciliaris, Setaria verticillata*, and *Tragus berteronianus* were exposed to temperature between 84 ° and 116 °C for 2-5 minutes (Fig. 16.2). Whereas a dry heat of 84 °C for up to 5 min. did not affect the germination of the caryopses of *C. ciliaris* and *T. berteronianus*, the

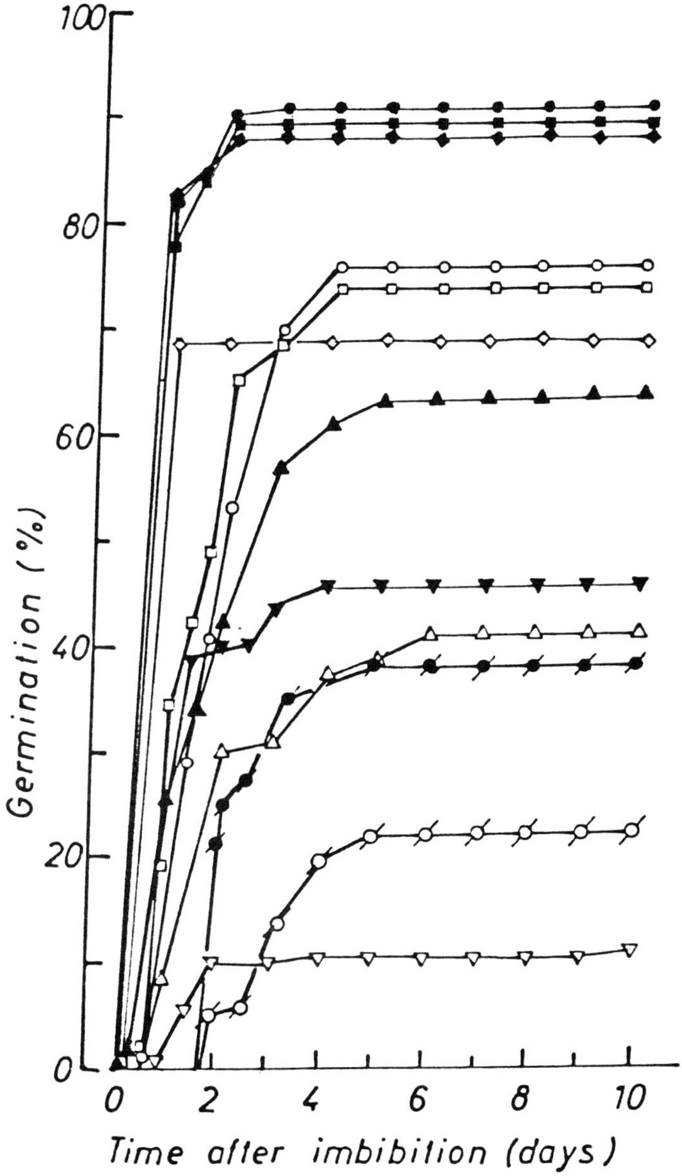

Figure 16.1: The effect of dry heat (100 °C, 2 min.) on the germination percentage of six savanna grass species. The control values at 20 °C are given by filled symbols, that of the heat-treated caryopses by open symbols. ○● *Cenchrus ciliaris*, ◇◆ *Enneapogon cenchroides*, ϕ ɸ *Eragrostis rigidior*, ▽▼ *Setaria verticillata*, □■ *Tragus berteronianus*, △▲ *Urochloa trichopus*.

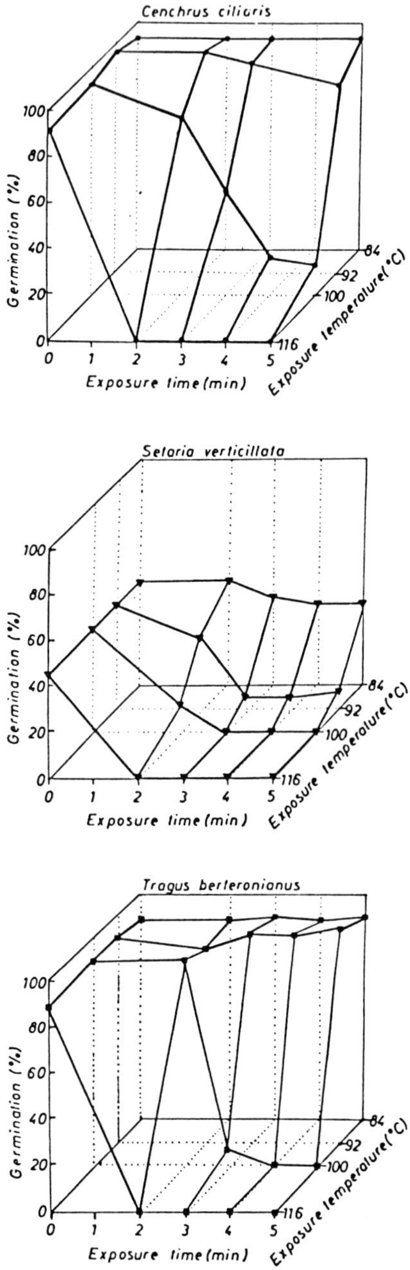

Figure 16.2: The impact of an increasing exposure time to dry heat on the germination of *Cenchrus ciliaris, Setaria verticillata* and *Tragus berteronianus* after six weeks.

Table 16.2: Impact of exposure of caryopses to dry heat (100 °C) for 2 minutes on germination, harmonic mean germination day (HMGD) and seedling survival of 6 savanna grasses one month after germination.

Grass species	Germination (%)	HMGD	Seedling survival (% of germin. seeds)
Cenchrus ciliaris			
control	95.5 ± 3.5	0.78	100.0 ± 0.0
heat	90.2 ± 3.0	0.82	16.9 ± 3.1
Enneapogon cenchroides			
control	89.9 ± 1.4	0.77	97.0 ± 2.6
heat	68.4 ± 7.4	1.46	46.2 ± 10.9
Eragrostis rigidior			
control	34.8 ± 2.4	1.33	100.0 ± 0.0
heat	24.0 ± 2.8	2.51	23.8 ± 6.8
Setaria verticillata			
control	46.0 ± 5.7	1.31	99.0 ± 1.7
heat	11.9 ± 4.2	3.99	9.2 ± 1.3
Tragus berteronianus			
control	94.4 ± 0.8	0.65	100.0 ± 0.0
heat	90.2 ± 0.3	0.94	6.7 ± 2.1
Urochloa trichopus			
control	57.4 ± 2.1	1.05	98.0 ± 1.4
heat	43.4 ± 2.3	1.70	29.2 ± 5.8

germination of *S. verticillata* decreased from 54.0% to 38.5%. An exposure to 92 °C for 5 min. reduced the germinability by 10% in *C. ciliaris*, by 50% in *S. verticillata*, but did not affect caryopses of *T. berteronianus*. At 100 °C an exposure time of 2 min. reduced the germinability of *C. ciliaris* to 75.1 ± 3.8%, that of 3 min. to 32%, and that of 5 min. to 13%, whereas at the same conditions the germination of *T. berteronianus* decreased from 90.2%(2 min.), to 7.2% (3 min.) and zero after 5 min. The strongest effect of dry heat of 100 °C was observed in *S. verticillata*, with only 12% germination after 2 min. exposure, and no germination at all after longer exposure time. A further increase in temperature to 116 °C did not let any caryopsis germinate.

The decreased germination percentage after the exposure of caryopses to dry heat of 100 °C for 2 min. may be due to imposed dormancy or/and to lethal injury. Exposure to 116 °C was lethal to the caryopses of all tested grass species; the same was true for an exposure to 100 °C for 3 min. and more, except the caryopses of *T. berteronianus*, which stayed dormant. The decreased germination of *Ennea-*

pogon cenchroides, Eragrostis rigidior, Setaria verticillata, Tragus berteronianus and *Urochloa trichopus* was due to imposed dormancy.

In addition to the earlier reported decrease of the germination percentage in all species exposure of caryopses to dry heat strongly reduced the vigor of the seedlings (Table 16.2). The strongest impact was on seedlings of *Setaria verticillata* and *Tragus berteronianus* with less than 10% of seedling survival. The most heat resistant caryopses were that of *Enneapogon cenchroides* with a survival mean of nearly 45%. In all species exposure to dry heat damaged either the radicle and/or plumule or gave rise to seedlings with two or three primary leaves, which were either free of chlorophyll or curled within the coleoptile. These deformed seedlings survived for 3 to 14 days.

16.4 Discussion

Fire in a savanna has direct and indirect effects on the viability and germination of grass disseminules. As long as they remain on or above soil surface, they will be destroyed due to the high inflammability of the glumes and their appendixes. Such a destruction of an above-ground seed bank of grass is not restricted to the investigated savanna grasses in Botswana, but has also been reported for disseminules of grasses and other herbs in fire-prone Australian ecosystems (ANDREW 1986; PEART 1984; PURDIE 1977; WESTOBY, RICE, GRIFFIN and FRIEDEL 1988), from Sahelian grassland (BREMAN, CISSÉ, DJITÈYE, and ELBERSE 1982) and recently from common veld grasses in South Africa (ZACHARIAS, TAINTON and OBERHOLSTER 1988).

An escape of the disseminules from the destroying fire can be achieved by various avoidance mechanisms. A low amount of inflammable glume mass together with a hard seed coat, as present in many shrubs and trees of fire-prone ecosystems (THANOS and GEORGHIOU 1988; ZAMMIT and WESTOBY 1988) was found in *Setaria verticillata*. In contrast to other fire-resistant species, the germination of *S. verticillata*, however, is hampered by fire. Another avoidance mechanism of fire is subterranean seed production, as exhibited by the annual grass *Amphicarpum purshii* in fire-prone ecosystems in North America (CHEPLICK and QUINN 1988), but obviously absent in grass species in the savanna in Botswana. A third avoidance mechanism is achieved by active burial of caryopses with the aid of awns due to trampling by ungulates or to transport by harvester ants (ANDREW 1986; MARSH 1987; MOTT and MCKEON 1977), or to washing by precipitation. Such burial in the upper centimeters below the soil surface is very common in the investigated savanna (E. M. VEENENDAAL, pers. commun.) and it has also been reported from Sahelian grassland (BREMAN, CISSÉ, DJITÈYE and ELBERSE 1982).

Burial below the soil surface did not only protect the disseminules against direct effects of fire, but it diminished the exposure to dry heat (VARESCHI 1962; BENTLEY and FENNER 1958; BISWELL 1956). Although heat resistance of caryopses of grass species from other ecosystems has been reported in a range from

82 to 116 °C at 5 min. exposure (SAMPSON 1944), caryopses of savanna grasses in Botswana are only partly resistant to dry heat of 100 °C. A temperature of 116 °C at a exposure period from 2 to 5 min. has devastating effects on the caryopses of *Cenchrus ciliaris, Setaria verticillata*, and *Tragus berteronianus*. In every case, dry heat did not stimulate germination of these grasses, as observed for the tall grass *Heteropogon contortus* in a South African savanna (ZACHARIAS, TAINTON and OBERHOLSTER 1988). The difference of the effects of dry heat may be correlated with the dormancy state of the caryopses which may explain the contrasting results of stimulation (WEST 1965) and depression (ZACHARIAS, TAINTON and OBERHOLSTER 1988) of the germination of *Themeda triandra* in South Africa. At the end of the dry season, when burning is normally initiated by herdmen in Botswana and our experiments were carried out, caryopses of most grass species are ready for germination (ERNST and TOLSMA 1988).

More important than the reduction of germination percentage is the impact of dry heat on the further development of the seedlings. The strong reduction of the survival of the seedling may be due to a deformation of the embryonic tissue, resulting in a lack of root growth and often in the development of more than one primary leaf. In addition the impact on chlorophyll synthesis makes dry heat above 92 °C very devastating for the rejuvenation of grasses after fire.

These reactions of grasses to fire and dry heat may help to explain the decrease of many grass species in the encroached savanna of southeastern Botswana (ERNST and TOLSMA 1989; TOLSMA, ERNST and VERWEIJ 1987). During and after bush encroachment a bush fire causes higher temperature at and just below soil surface than grass fires (KOZLOWSKI and AHLGREN 1974; GILL, GROVES and NOBLE 1981). Therefore the use of fire in savanna management will not only affect the recruitment of shrubs and trees (APRU 1980; SKARPE 1980), but also the long term diversity of grasses in the savanna. In West African and Australian savannas tall grasses were favored in fire-protected sites or sites exposed to moderate fire intensity (BOWMAN, WILSON and HOOPER 1988; RAMSEAY and ROSE-INNES 1963), whereas intense late season fires let shorter grasses dominate that have a lower fuel load. Therefore the disappearance of tall grasses in many Botswana ecosystems during the last decades will not only be due to overgrazing (ERNST and TOLSMA 1989; TOLSMA, ERNST, VERWEIJ and VOOIJS 1987), but also to the frequency of fire.

The lack of fire resistance of savanna grasses in Botswana may be the result of a short history of frequent fires and the low annual contribution of seedlings to the maintenance of perennial grasses. The increase of annual grasses during drought and overgrazing (ERNST and TOLSMA 1989) may therefore not only be explained by a better competition due to a reduced vegetation cover, but also by a lack of fire due to a low standing biomass.

Acknowledgments

I thank Dr. D. J. Tolsma for the help in disseminule collection, the National Institute for Development Research and Documentation, University of Botswana, for hospitality, and Miss D. Hoonhout for typing the manuscript.

Bibliography

ANDREW, M. H., 1986, Granivory of the annual grass *Sorghum intrans* by the harvester ant *Meranoplus* sp. in tropical Australia.
Biotropica **18**, 344–349.

APRU, 1980, Ten Years of Animal Production and Range Research in Botswana. Animal Production Research Unit. Ministry of Agriculture, Gaborone.

BENTLEY, J. R.; FENNER, R. L., 1958, Soil temperatures during burning related to postfire seedbeds on woodland range.
J. For. **56**, 737–774.

BISWELL, B. H., 1956, Ecology of California grassland.
J. Range Manage. **9**, 19–24.

BOWMAN, D. M. J. S.; WILSON, B. A.; HOOPER, R. J., 1988, Response of Eucalyptus forest and woodland to four fire regimes at Mummarlary, Northern Territory, Australia.
J. Ecol. **76**, 215–232.

BREMAN, H.; CISSÉ, A. M., DJITÈYE, M. A.; ELBERSE, W. TH., 1982, La potential botanique des pâturages.
In: Penning de Vries, F. W. T.; Djitèye, M. A. (eds.), La productivité des pâturages sahéliens, Pudoc, Wageningen, 98–132.

CHEPLICK, G. P.; QUINN, J. A., 1988, Subterranean seed production and population response to fire in *Amphicarpum purshii* (*Gramineae*).
J. Ecol. **76**, 263–283.

DAUBENMIRE, R., 1968, Ecology of fire in grassland.
Adv. Ecol. Res. **5**, 209–266.

ERNST, W. H. O.; TOLSMA, D. J., 1988, Dormancy and germination of semi-arid annual plant species, *Tragus berteronianus* and *Tribulus terrestris*.
Flora **181**, 243–251.

ERNST, W. H. O.; TOLSMA, D. J., 1989, Mineral nutrients in some Botswana savanna types.
In: Proctor, J. (ed.) Mineral Nutrients in Tropical Forest and Savannas, Blackwell Scientific Publ., Oxford, 97–120.

ERNST, W. H. O., KUITERS, A. T.; TOLSMA, D. J., 1991, Dormancy and germination of annual and perennial grasses from a savanna in southeastern Botswana (submitted).

GERTENBACH, W. P. D.; POTGIETER, A. C. F., 1979, Veldbrandnavorsing in de struikmopanieveld van de nationale Kringerwildtuin.
Koedoe **22**, 1–28.

GILL, A. M., GROVES, R. H.; NOBLE, J. R. (eds.), 1981, Fire and the Australian Biota.
Australian Academy of Science, Canberra.

KEELEY, I., 1981, Reproductive cycles and fire regimes.
In: Mooney, H. A.; Bonnicksen, T. M.; Christensen, N. L.; Lotan, J. E.; Keiners, W. A. (eds.) Fire Regimes and Ecosystem Properties. USDA Forest Service, Gen. Techn. Rep. WO-**26**.

KOZLOWSKI, T. T.; AHLGREN, C. E. (eds.), 1974, Fire and Ecosystems.
Academic Press, New York.

MARSH, A. C., 1987, The foraging ecology of two Namib Desert (South Africa) harvester ant species.
S. A. J. Zool. **22**, 130–136.

MOTT, J. J.; MCKEON, G. M., 1977, A note on the selection of seed types by harvester ants in Northern Australia.
Austr. J. Ecol. **2**, 231–235.

NAVEH, Z., 1974, Effects of fire in the mediterranean region.
In: Kozlowski, T. T.; Allgren, C. E. (eds.) Fire and Ecosystems, Academic Press, New York, 401–434.

NOVELLIE, P., 1987, Interrelationships between fire, grazing and grass cover at the Bontebok National Park.
Koedoe **30**, 1–17.

PEART, M. H., 1984, The effects of morphology, orientation and position of grass diaspores on seedling survival.
J. Ecol. **72**, 437–453.

PURDIE, R. W., 1977, Early stages of regeneration after burning in the dry sclerophyll vegetation. II. Regeneration by seed germination.
Austr. J. Bot. **25**, 35–46.

RAMSAY, J. M.; ROSE–INNES, R., 1963, Some quantitative observations on the effects of fire on the Guinea savanna vegetation of Northern Ghana over a period of eleven years.
African Soils **8**, 41–84.

ROBINSON, E. T.; GIBBS-RUSSEL, G. E.; TROLLOPE, W. S. W.; DOWNING, B. H., 1979, Assessment of short-term burning treatments on the herb layer of False Thornveld of the eastern Cape.
Proc. Grassl. Soc. S. A. **14**, 79–84.

SAMPSON, A. W., 1944, Effect of chaparral burning on soil erosion and soil moisture relations.
Ecology **25**, 171–191.

SKARPE, C., 1980, Observations on two bush fires in the Western Kalahari, Botswana.
Acta Phytogeog. Sue. **68**, 131–146.

TAINTON, N. M.; MENTIS, M. T., 1984, Fire in grassland.
In: de V. Booijsen, P.; Tainton, N. M. (eds.) Ecological Effects of Fire in South African Ecosystems. Springer Verlag Berlin, 115–147.

THANOS, C. A.; GEORGHIOU, K., 1988, Ecophysiology of fire-stimulated seed germination in *Cistus incanus* ssp. *creticus* (L.) Heywood and *Cistus salvifolius* L.
Plant Cell Environ. **11**, 841–850.

TOLSMA, D. J., ERNST, W. H. O., VERWEIJ, R. A.; VOOIJS, R., 1987, Seasonal variation of nutrient concentrations in a semi-arid savanna ecosystem in Botswana.
J. Ecol. **75**, 755–770.

TOLSMA, D. J.; ERNST, W. H. O.; VERWEIJ, R. A., 1987, Nutrients in soil and vegetation around two artificial waterpoints in eastern Botswana.
J. Appl. Ecol. **24**, 991–1000.

TROLLOPE, W. S. W., 1984a. Fire in savanna.
In: de V. Booijsen, P.; Tainton, N. M. (eds.) Ecological Effects of Fire in South African Ecosystems. Springer Verlag Berlin, 149–175.

TROLLOPE, W. S. W., 1984b, Fire behavior.
In: de V. Booijsen, P.; Tainton, N. M. (eds.) Ecological Effects of Fire in South African Ecosystems. 199–217.

VAN WILGEN, B. W.; WILLS, A. J., 1988, Fire behavior prediction in savanna vegetation. S. A. J. Wildlife Res. **18**, 41–46.

VARESCHI, V., 1962, La quema como factor ecologico en los Llanos.
Soc. Venez. Cienc. Nat. Bol. **101**, 9–26.

WEST, O., 1965, Fire in vegetation and its use in pasture management with special reference to tropical and subtropical Africa.
Commonwealth Bureau of Pastures and Field Crops, Hurley, Berkshire, **1**, 1–53.

WESTOBY, M.; RICE, B.; GRIFFIN, G.; FRIEDEL, M., 1988, The soil seed bank of *Triodia basedowii* in relation to time since fire.
Austr. J. Ecol. **13**, 161–169.

ZACHARIAS, P. J. K.; TAINTON, N. M.; OBERHOLSTER, C., 1988, The effect of fire on germination of five common veld grasses.
J. Grassl. Soc. S. A. **5**, 229–230.

ZAMMIT, C.; WESTOBY, M., 1988, Predispersal seed losses and the survival of seeds and seedlings of the two serotinous *Banksia* shrubs in burnt and unburnt heath.
J. Ecol. **76**, 200–214.

Chapter 17

Does intermediate disturbance increase species richness within deciduous forest understory?

R. J. READER, K. C. TAYLOR and D. W. LARSON

17.1 Introduction

Some plant communities are species-rich while others are species-poor. The factors controlling this variation in species richness (=number) are not completely understood but the intensity of disturbance, that is, the event causing loss of plant biomass is thought to be an important determinant of species richness (GRIME 1973, CONNELL 1978, HUSTON 1979, SOUSA 1979, YODZIS 1986, PETRAITIS et al. 1989). Possible relationships between disturbance intensity and species richness proposed in the studies cited above are shown in Figure 17.1.

At a low disturbance intensity, insufficient biomass is lost to affect the persistence of either sparse or common species. At a high disturbance intensity, so much biomass is lost that only a few tolerant species can persist. At an intermediate disturbance intensity, species richness should *increase* if the following three conditions

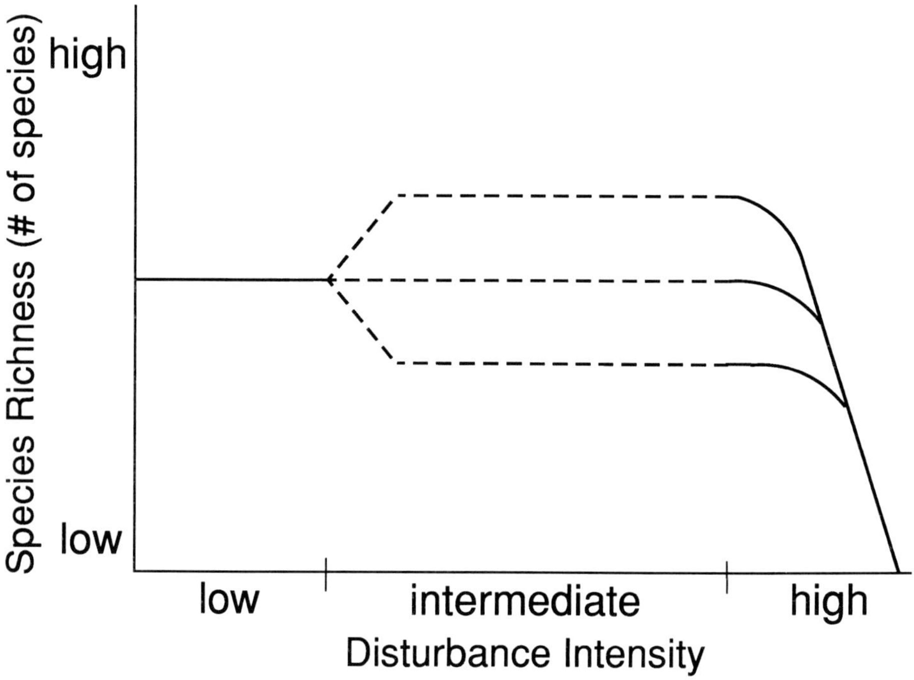

Figure 17.1: Relationship between species richness and disturbance intensity. The dotted lines indicate three alternative effects of intermediate disturbance on richness.

are met:

1. Disturbance reduces the abundance (=density or frequency) of common species without eliminating either common or sparse species.

2. Abundance is reduced sufficiently to release resources (e. g. light, space).

3. Propagules of species not already present utilize the newly released resources.

If these three conditions are not met species richness should either *stay the same* or *decrease* (YODZIS 1986, FUENTES and JAKSIC 1988, PETRAITIS et al. 1989).

The fact that species richness may either increase, decrease or not change following intermediate disturbance makes it difficult to decide whether disturbance should be encouraged or discouraged in communities where maintaining high species richness is an important management objective. To be able to make the correct management decision about disturbance it would help to know whether the three conditions required for increased richness following disturbance are likely to be met. Such information is unavailable for most, if not all, plant communities. Here we consider whether conditions required for increased species richness are likely to be met in managed deciduous forests.

Tree harvesting and trampling are two important sources of disturbance in deciduous forest. The intensity of disturbance ranges from low (e. g. felling a single tree, trampling by deer) through intermediate (e. g. selective tree harvesting, recreational hiking) to high (e. g. clear cutting, trampling by livestock). In this chapter we focus on the effects of intermediate levels of tree harvesting or trampling applied to a medium sized area (e. g. < 100 ha).

It is well documented that an intermediate level of tree harvesting can increase both resource availability and tree regeneration (e. g. MINCKLER et al. 1973, CANHAM and MARKS 1985). It is less clear whether shrubs and herbs (excluding early successional species) also benefit from tree harvesting and therefore, whether the number of forest species present in the understory increases, decreases or is unaffected by intermediate disturbance. Results given in the few studies available (e. g. EHRENFELD 1980, MCCOMB and NOBLE 1982, METZGER and SCHULTZ 1984, MOORE and VANKAT 1986) are especially difficult to interpret because the effects of tree harvesting on shrubs or herbs were potentially confounded by predisturbance differences among plots. The same potentially confounding effect makes it difficult to interpret results of trampling studies done in deciduous forest (MAKHOUM and KHORASANI 1988, HALL and KUSS 1989).

To avoid confounding predisturbance and disturbance effects, we set up permanent plots and surveyed vegetation before and after applying either experimental trampling or tree harvesting. Here we use our experimental results to determine whether the three conditions required for species richness to increase following disturbance are met in deciduous forest understory.

17.2 Expected effects of intermediate disturbance on deciduous forest understory

Tree harvesting and trampling both damage above ground parts of plants in the understory. Consequently, the abundance of common species is likely to be reduced and some sparse species eliminated as well.

Tree harvesting obviously increases the amount of light reaching the understory and tree skidding creates patches of bare ground. Both light and exposed mineral soil are often important limiting resources in deciduous forest (BEHREND and PATRIC 1969, BEATTY 1984, HICKS et al. 1985). Resource availability should also increase following trampling, but to a lesser extent. Any bare ground created by trampling is likely to be compacted, making rooting difficult, and any increase in light is likely to be minor compared to tree harvesting. Consequently, fewer new species should be added following trampling than following tree harvesting.

Table 17.1: Summary of variables measured to determine whether conditions required for increased species richness following tree harvesting were met. Within each row, values with the same small letter do not differ significantly (Tukey test, $P > .05$).

Variable	% of tree basal area harvested		
	0	33	66
Condition 1: reduced abundance without increased species loss			
Mean(±1SE) abundance of the 20 most common understory species immediately after harvesting (% of preharvest abundance)	96 ± 4^a	95 ± 4^a	94 ± 5^a
Mean(±1SE) loss of understory species (% of species censused preharvest)	2 ± 1^a	12 ± 1^b	11 ± 1^b
Condition 2: increase in resource availability			
Photosynthetically active radiation at 1m height (% of incoming)	2 ± 1^a	30 ± 4^b	43 ± 3^c
Quadrats with mineral soil exposed (% of 36 quadrats per plot)	0 ± 0^a	21 ± 6^b	18 ± 4^b
Condition 3: new species added			
Mean(±1SE) number of species added per plot over 3 years	7 ± 1^a	13 ± 2^b	11 ± 1^{ab}
Mean(±1SE) understory species richness 3 years after harvesting (% of preharvest richness)	103 ± 3^a	112 ± 5^a	115 ± 3^a

17.3 Selective tree harvesting effects on deciduous forest understory

Experimental disturbance. Deciduous forest managed for timber production in southern Ontario (Canada) is usually selectively cut, with one–third to two–

thirds of tree basal area harvested. To determine whether this intermediate level of tree harvesting increases the number of forest species present in the understory, we monitored changes in resource levels, population size, species composition and species richness in the understory of 0.2 ha experimental plots ($n = 5$ per cutting treatment) from which zero, one–third or two–thirds of tree basal area (> 5 cm diameter) was harvested. Two tracts (50 and 100 ha) of deciduous forest located in southern Ontario (42° 30' N, 80° 30' W) were used for the experiment. Neither tract had been logged for at least sixty years. Oaks (*Quercus alba* L., *Q. borealis* MICHX.) and maples (*Acer rubrum* L., *Acer saccharinum* L.) dominated the canopy and the understory contained a diverse mixture of woody and herbaceous species (mean of 48 species per plot). Trees were cut in midwinter and stems were removed using a mechanical skidder. Tree tops were then cut for firewood and removed by hand.

Condition 1: reduced abundance without increased species loss. To determine whether harvesting operations reduced a species' abundance in the understory, we compared species' abundance in 1 × 2 m permanent quadrats ($n = 36$ per 0.2 ha plot) immediately before and after tree harvesting. The number of 20 × 20 cm squares occupied per quadrat was compared for each herbaceous species present and stem density was compared for woody species.

The abundance of the twenty most common species (accounting for 83% of total understory abundance) decreased by about 5%, on average, in both cut and uncut plots (Table 17.1). In contrast, the percentage of sparse species eliminated from the understory increased significantly from 2% in uncut plots to 11–12% in cut plots (Table 17.1).

These results do not satisfy the first of three conditions for richness to increase following intermediate disturbance.

Condition 2: increase in resource availability. To determine whether harvesting operations reduced plant biomass sufficiently to release resources we measured both the percentage of incoming photosynthetically active radiation reaching the understory in the first year postharvest and the amount of newly exposed mineral soil.

Radiation reaching the understory increased from about 2% in uncut plots to an average of 30–43% in cut plots (Table 17.1). The amount of mineral soil exposed also increased significantly, from none in uncut plots to 18–21% in cut plots (Table 17.1). These results are consistent with the second of three conditions required for species richness to increase following intermediate disturbance.

Condition 3: new species added. To determine whether species not already present in plots utilized the newly released resources we recorded the number of forest species added to the 1 × 2 m permanent quadrats during the first three years postharvest. Short lived species normally found in open fields were not counted.

The number of species added was greater in cut plots than in uncut plots (Table 17.1), but this difference was only statistically significant in the 33% removal treatment. These results are only partly consistent with the third condition required for species richness to increase following disturbance.

Increase in species richness. Three years after tree harvesting, the number of forest species present in cut plots was slightly greater than in uncut plots but this difference was not statistically significant (Table 17.1). Intermediate disturbance had no significant net effect on species richness because about equal numbers of species were gained and lost from the understory in cut plots.

17.4 Trampling effects on deciduous forest understory

Experimental disturbance. To determine the effect of trampling by hikers on understory vegetation, we monitored changes in species composition, abundance and richness in 4 m long by 2 m wide experimental plots ($n = 5$ per trampling treatment) that received either no trampling, 50 passes or 500 passes during one three month period (May to July). Each pass started and ended at the midpoint of the two ends of the 2 m wide plot.

Plots were set up in an undisturbed section of a mature deciduous forest in southern Ontario (43° 35' N, 79° 55' W). Sugar maple (*Acer saccharum* L.) dominated the canopy and the understory contained a mixture of herbs and woody plants (mean of 24 species per plot). Hiking trails in an adjacent park received about 100 passes per year on secondary trails and up to 26,000 passes per year on main trails (McLEAN 1989). The level of experimental trampling imposed here (i. e. , 50 or 500 passes) simulated an intermediate level of trampling on previously undisturbed vegetation. Our objective was to evaluate the initial effect of trampling rather than the effect of repeated trampling along a preexisting trail.

Condition 1: reduced abundance without increased species loss. To determine whether trampling reduced the abundance of common species in the understory, we measured their abundance before trampling and one year later. Abundance was calculated by summing the number of 17 × 17 cm squares occupied per plot for each of the twenty most common species (accounting for 86% of total abundance).

Their abundance decreased by about 12% in trampled plots, compared with about 8% in untrampled plots (Table 17.2). The difference between trampled and untrampled plots was not statistically significant (Table 17.2). The percentage of species eliminated from the understory was also slightly greater in trampled plots than in untrampled plots (Table 17.2). Again this small difference was not

Table 17.2: Summary of variables measured to determine whether conditions required for increased species richness following trampling were met. Within each row values with the same small letter do not differ significantly (Tukey test, $P > .05$).

Variable	Number of passes per year		
	0	50	500
Condition 1: reduced abundance without increased species loss			
Mean(±1SE) abundance of the 20 most common understory species (% of pretrampling abundance)	92 ± 4^a	88 ± 5^a	88 ± 4^a
Mean(±1SE) loss of understory species (% of species censused before trampling)	12 ± 2^a	14 ± 2^a	17 ± 5^a
Condition 2: increase in resource availability			
Mean(±1SE) weight of litter on path (mg per cm^2)	47 ± 21^a	40 ± 18^a	37 ± 21^a
Condition 3: new species added			
Mean(±1SE) number of species added per plot after trampling	3 ± 2^a	4 ± 1^a	5 ± 2^a
Mean(±1SE) understory species richness 1 year after trampling (% of pretrampling richness)	100 ± 2^a	96 ± 3^a	100 ± 5^a

statistically significant. These results only partly satisfy the first condition for richness to increase following disturbance.

Condition 2: increase in resource availability. To determine whether trampling released space by reducing litter accumulation, we collected and weighed litter present after trampling was applied. Litter was collected from eight, evenly spaced 79 cm^2 areas within each 2 × 4 m plot.

Litter weight decreased slightly on the path created by trampling but this change in litter weight was not statistically significant (Table 17.2). This suggests that trampling released little new space.

Condition 3: new species added. To determine whether new species were added to plots following trampling, we recorded the number of new species present in plots one year after trampling.

The number of forest species added was slightly greater in trampled plots than untrampled plots (i.e., 4–5 versus 3, respectively) but this small difference was not statistically significant (Table 17.2).

Increase in species richness. One year after trampling, species richness was slightly less in trampled plots than in untrampled plots (i. e., 96–100% of pre-trampling richness versus 100%, respectively). This difference was not statistically significant (Table 17.2). Trampling did not increase species richness significantly because about equal numbers of species were added and lost in the trampled plots.

17.5 Does forest species richness increase following intermediate disturbance?

Applying an intermediate level of trampling or tree harvesting did not result in a significant net increase in the number of forest species present in the understory.

While trampling did not significantly affect the number of species lost (condition 1 partly met), neither did trampling increase the number of species added (condition 3 not met). Apparently, resource availability did not increase sufficiently (condition 2 not met) to allow a large number of new species to be added. As a result, species richness did not increase significantly.

Following tree harvesting, the number of forest species added increased significantly in one of the two harvesting treatments (condition 3 partly met) but so did the number of species lost (condition 1 not met). The increased number of species lost during harvesting operations offset the subsequent benefit of increased resource availability (condition 2 met). As a result, species richness did not increase significantly.

We imposed only a small number of possible intermediate disturbances and we still need to determine whether higher or lower levels of intermediate disturbance give comparable results. In the case of trampling, we expect that species richness will be unaffected at higher trampling levels because species richness did not differ significantly on and off existing trails receiving up to 5,500 passes per year in a park adjacent to our study plots (McLEAN 1989). In the case of tree harvesting, removing more than 66% of tree basal area is likely to transform the forest understory into one dominated by early succesional, nonforest plants. Species richness may increase but only because many shade intolerant, nonforest species have been added (BORMANN and LIKENS 1979).

Our experimental results only reflect the short term response of plants to trampling (one year) and tree harvesting (three years). Most forest species are long lived perennials and some species may respond only slowly to increased resource availability following disturbance. To detect possible longer term responses we will continue to survey plots on a regular basis. In the case of tree harvesting, we expect to see no further increase in species richness because Bricker (1989) found no significant change in herb abundance in cut plots between 2 to 9 years after tree harvesting. In the case of trampling, so few resources were released that it is unlikely that many more species will be added.

17.6 Conclusions

Based on our experimental results, we conclude that the number of deciduous forest species present in the understory does not increase significantly following either intermediate tree harvesting or trampling. Therefore, it is unnecessary to apply an intermediate disturbance to help maintain high forest species richness in managed stands. Presumably forest species richness did not increase significantly because some species were lost during tree harvesting operations (condition 1 not met) and insufficient resources were released by trampling for many new species to be added (condition 2 not met).

Bibliography

BEATTY, S. W., 1984, Influence of microtopography and canopy species on spatial patterns of forest understory plants.
Ecology **65**, 1406–1419.

BEHREND, D. F.; PATRIC, E. F., 1969, Influence of site disturbance and removal of shade on regeneration of deer browse.
J. Wildl. Manage. **33**, 394–398.

BORMANN, F. H.; LIKENS, G. E., 1979, Pattern and process in a forested ecosystem.
Springer Verlag, New York.

BRICKER, B., 1989, Response of herb guilds to canopy removal in deciduous forests.
MSc thesis, University of Guelph, Guelph.

CANHAM, C. D.; MARKS, P. L., 1985, The response of woody plants to disturbance: patterns of establishment and growth.
In: Pickett, S. T. A., White, P. S. (eds.) The ecology of natural disturbance and patch dynamics. Academic Press, Orlando, p. 197.

CONNELL, J. H., 1978, Diversity in tropical rainforests and coral reefs.
Science **199**, 1302–1310.

EHRENFELD, J. G., 1980, Understory response to canopy gaps of varying size in a mature oak forest.
Bull. Torrey. Bot. Club **107**, 29–41.

FUENTES, E. R.; JAKSIC, F. M., 1988, The hump-backed species diversity curve: why has it not been found among land animals?
Oikos **53**, 139–143.

GRIME, J. P., 1973, Control of species density in herbaceous vegetation.
J. Environ. Mgmt. **1**, 151–167.

HALL, C. N.; KUSS, F. R., 1989, Vegetation alteration along trails in Shenandoah National Park, Virginia.
Biol. Cons. **48**, 211–228.

HICKS, D. J.; CHABOT, B. F., 1985, Deciduous forest.
In: Chabot, B. F.; Mooney, H. A. (eds.) Physiological ecology of North American plant communities. Chapman and Hall, New York, p. 257.

HUSTON, M., 1979, A general hypothesis of species diversity.
Amer. Nat. **113**, 81–101.

MAKHOUM, M. F.; KHORASANI, N., 1988, Differences between environmental impacts of logging and recreation in mature forest ecosystems.
Environ. Conserv. **15**, 137–142.

MCCOMB, W. C.; NOBLE, R. E., 1982, Response of understory vegetation to improvement cutting and physiographic site in two mid-south forest stands.
Castanea **47**, 60–77.

MCLEAN, K. C., 1989, Effects of trampling disturbance on the richness of forest vegetation: a model and its conservation implications.
MSc thesis, University of Guelph, Guelph.

METZGER, F.; SCHULTZ, J., 1984, Understory response to 50 years of management of a northern hardwood forest in upper Michigan.
Amer. Midl. Nat. **112**, 209–223.

MINCKLER, L. S.; WOERHEIDE, J. D.; SCHLESINGER, R. C., 1973, Light, soil moisture, and tree reproduction in hardwood forest openings.
USDA For. Serv. NC For. Exp. Stn. Res. Pap. NC-89.

MOORE, M. R.; VANKAT, J. V., 1986, Responses of the herb layer to the gap dynamics of a mature beech-maple forest.
Amer. Midl. Nat. **115**, 336–347.

PETRAITIS, P. S.; LATHAM, R. E.; NIESENBAUM, R. A., 1989, The maintenance of species diversity by disturbance.
Quart. Rev. Biol. **64**, 393–418.

SOUSA, W., 1979, Disturbance in marine intertidal boulder fields: the non-equilibrium maintenance of species diversity.
Ecology **60**, 1225–1239.

YODZIS, P., 1986, Competition, mortality and community structure.
In: Diamond, J.; Case, T. J., (eds.) Community ecology. Harper and Row, New York, p. 480.

Chapter 18

Response of a Bromus erectus grassland (Mesobromion) to abandonment and different cutting regimes

H. DIERSCHKE and M. ENGELS

18.1 Introduction

Since 1972 experimental succession studies have been carried out in an abandoned secondary (seminatural) grassland, on calcareous soil, dominated by *Bromus erectus*. The main interest was in the restoration of a species–rich grassland and its maintenance by different cutting regimes. The vegetation, formerly probably a Gentiano–Koelerietum as a result of longtime extensive grazing without manuring, today has developed into a dense, tall–growing meadow–like stand of *Bromus erectus*, ca. 30 years after the end of grazing. High vitality and litter production of this grass have suppressed many low–growing plants of this vegetation type. Still in the thirties, however, the upright brome was one of the rarest plants around Göttingen, due to prevalent sheep grazing (ELLENBERG 1988, p. 487). The table of Gentiano–Koelerietum of BORNKAMM (1960) shows *Bromus erectus* already with higher constancy, sometimes with high cover degree too (see also BORNKAMM 1961).

Some results of succession (1972–84) have already been published (DIERSCHKE 1985). In 1988 a final, more detailed investigation took place (ENGELS 1989). Besides the succession, special emphasis was given to the response of *Bromus erectus* to different frequency of cutting. Some results are presented and discussed here.

18.2 Research area and methods

The research area is situated on the Feldbornberg near Göttingen at the edge of a limestone plateau (ca. 290 m above sea level). On a slight SSW–exposed slope (3–5°) on Rendzina soils, a mosaic of grassland and shrubs (especially *Crataegus, Prunus spinosa, Rosa canina*) has developed. The grassland belongs to the slightly arid swards of Mesobromion, a suboceanic alliance of calcareous xerothermic grasslands of Central Europe (Festuco–Brometea). In spite of the dominance of *Bromus erectus*, the total species combination shows more affinities to the Gentiano–Koelerietum than to the Mesobrometum.

In 1972 a research area of 128 m^2 was established, divided into plots of 2 × 2 m (see also DIERSCHKE 1985). At the beginning the vegetation was a dense *Bromus* grassland not invaded but surrounded by shrubs. The following treatments began in 1972 and continued until 1988, each in four parallel plots:

C1/2 cutting once every two years
C1 cutting once every year
C2 cutting twice every year
C3 cutting three times every year
M mulching every year (mulch being left on the ground)
A control (abandoned, without treatment)

The first cut was at the end of June respectively beginning of July (C2,C3), the second was one month later for all plots, and the third cut was in September (C3). In 1988 all plots were cut in July.

From the harvested above–ground phytomass the dry matter for every plot was determined. In 1988 the phytomass of 3–4 representative plot sections (0.1 m^2 each) was sorted into fractions of wood, grasses, forbs, mosses and litter.

Before the first cut every year a vegetation survey took place with exact estimation of cover degree of all species, in percent. Since 1978 the number of stalks of *Bromus erectus* have been counted in 3 quadrats of 0.1 m^2 of every plot. The length of 30 individual stalks has also been measured.

18.3 Results

18.3.1 Comparison of species combination in one year

The vegetation of different plots in 1987 (15 years after beginning of the investigation) is shown in Table 18.1. For every treatment the average cover degree, in percent, of the four analogical quadrats is given for almost all species (except some very scattered ones). Mostly the plots can be distinguished more by different cover degree than by their own species or species groups. The results are similar to those of 1984 (DIERSCHKE 1985).

Table 18.1 considers seven species groups of different behavior:

Group 1 contains young shrubs, frequently suckers of the neighbouring old shrubs with subterranean runners, forming polycorms. Since 1972, especially in the eighties, shrubs have become established more and more in the control plots (A) forming now a loose low thicket up to a height of 1.5 m. In the other plots only small saplings can be found, due to the cutting every one or two years.

Group 2 contains species that decrease under cutting, especially *Bromus erectus*. Though the grass is present in every plot, high vitality can be recognized only in M, C1/2 and C1. Here *Bromus* forms large tufts with long leaves covering large areas of the plots and overtopping many other species. In order to remove *Bromus*, repeated cutting within a year is necessary. In plot A the grass is tall–growing too but is more limited in space by the shrubs. The other species of group 2 often have their sociological optima in the herbaceous fringes ("Saum") of adjacent shrubs and woods, belonging to the Trifolio–Geranietea. *Brachypodium pinnatum* is more a grass of extensive pastures, but it often is concentrated also to fringe situations.

Group 3 shows two fringe plants with optimal growth in C1/2 and C1. Both can attain high coverage by low branches creeping on the soil surface.

Group 4 contains species almost absent in A but already spreading in plots with slight cutting or mulching, evidently shown by *Trisetum flavescens, Plantago lanceolata* and *Linum catharticum*. Most of them have their optimum in meadows and pastures of Molinio–Arrhenatheretea.

Species of group 5 have only very small coverage in A and M, the plots with litter accumulation. Group 6 shows the clearest correspondence to one or more cuttings every year. Most of its species are low–growing small plants not very competitive with taller meadow species. Especially *Hieracium pilosella*, a rosette plant of open vegetation with quickly growing above soil runners with daughter plants, can reach high cover degree (in one plot of C3 up to 60%). Most of these species are characteristic for calcareous grasslands of Festuco–Brometea.

Species of group 7 are more or less indifferent to the treatments.

Especially the plants of groups 5–6, forming a low–growing sward, contribute much to the many–colored flowering aspects all through the growing season which are typical for calcareous grassland. So plots A and M show the lowest blooming of all. Cutting makes the grassland more attractive for insects and other animals.

Table 18.1: Comparison of the different 2×2 m plots in 1987 (Average cover percent of four 2 × 2 m plots).

Group	Plot variant	A	M	C1/2	C1	C2	C3
	Coverage (%)	97	98	99	99	99	99
	Number of species	25	27	34	34	34	34
Group							
1	*Ligustrum vulgare*	1	·	·	·	·	·
	Rhamnus cathartica	2	·	·	·	·	·
	Cornus sanguinea	1	+	+	+	+	+
	Crataegus spec.	13	+	+	·	+	+
	Prunus spinosa	18	3	4	1	2	2
	Rosa canina	+	+	+	·	+	·
2	*Bromus erectus*	58	69	63	48	43	39
	Brachypodium pinnatum	10	+	4	·	+	+
	Viola hirta	2	2	1	1	+	·
	Galium album	2	6	+	+	·	·
	Poa angustifolia	2	2	2	1	+	1
	Agrimonia eupatoria	1	1	1	+	+	·
	Hypericum perforatum	1	+	1	·	·	·
3	*Medicago falcata*	4	9	24	17	5	2
	Astragalus glycyphyllos	3	7	3	14	1	·
4	*Trisetum flavescens*	+	3	3	5	2	3
	Plantago lanceolata	+	2	1	2	2	2
	Linum catharticum	+	2	+	3	2	2
	Prunella vulgaris	·	+	+	1	+	1
	Luzula campestris	·	+	1	1	1	1
	Hippocrepis comosa	·	+	+	1	+	1
	Cerastium holosteoides	·	+	+	+	+	1
	Plantago media	·	+	+	+	+	1
	Senecio jacobaea	·	+	·	+	1	+
5	*Pimpinella saxifraga*	1	1	2	2	2	2
	Sanguisorba minor	+	+	1	1	1	1
	Briza media	+	+	1	2	2	2
	Polygala comosa	+	+	1	1	2	2
	Carex caryophyllea	·	·	+	+	1	1

continued next page

Table 18.1, continued.

	Plot variant	A	M	C1/2	C1	C2	C3
	Coverage (%)	97	98	99	99	99	99
	Number of species	25	27	34	34	34	34
6	*Festuca ovina* agg.	3	3	3	4	6	5
	Thymus pulegioides	1	+	1	2	5	5
	Trifolium pratense	+	1	1	2	2	2
	Trifolium repens	·	+	·	1	2	3
	Scabiosa columbaria	+	+	+	1	1	1
	Potentilla neumanniana	·	·	+	1	2	2
	Carex flacca	+	+	+	+	1	1
	Medicago lupulina	+	·	1	1	3	5
	Hieracium pilosella	+	1	+	4	13	40
	Ranunculus bulbosus	·	1	+	1	3	3
	Trifolium dubium	·	·	1	1	5	7
	Carlina vulgaris	·	·	·	·	+	+
	Veronica arvensis	·	·	·	·	·	1
7	*Cirsium acaule*	10	3	7	15	13	11
	Lotus corniculatus	1	+	1	1	1	1
	Taraxacum officinale	+	2	1	1	+	1
	Knautia arvensis	1	1	1	+	+	1
	Rhinanthus minor	+	1	1	1	1	1
	Fragaria vesca	1	+	+	1	+	·
	Koeleria pyramidata	1	·	+	+	+	+
	Veronica chamaedrys	·	1	+	+	+	·

Though Table 18.1 shows several differences in species combination, the total coverage is almost equal in all plots. On the other hand, there is a difference of species diversity between A, M and the C plots. Accumulation of litter and overtopping (shading) by shrubs and tall–growing grasses and forbs diminish the number of species markedly. Apparently mulching is not an appropriate treatment for the restoration of such a grassland type.

18.3.2 Changes of species combination over several years

Table 18.2 shows successional trends of four plots that have been abandoned until 1980 and afterwards were cut once a year. At the beginning the vegetation was a meadow–like *Bromus* stand. Between its large tufts, a deep (5–10 cm) layer of loose litter, mainly of dead leaves of *Bromus*, has accumulated. After cutting and

Table 18.2: Successional trends after the beginning of cutting once every year from 1980 onwards (Average cover percent of four 2 × 2 m plots).

	Year	80	81	82	83	84	85	86	87
	Coverage (%)	99	90	97	98	97	97	99	99
	Number of species	21	30	36	35	34	35	39	36
Group									
1	*Bromus erectus*	78	69	65	59	60	59	60	64
	Avenochloa pratensis	+	+	+	·	+	·	·	·
2	*Medicago falcata*	4	7	13	14	13	10	11	16
	Trisetum flavescens	1	2	1	3	2	3	3	3
	Sanguisorba minor	+	1	1	1	1	1	1	1
	Lotus corniculatus	+	1	1	1	1	1	1	1
	Plantago lanceolata	+	1	1	1	1	1	1	1
	Trifolium pratense	+	1	4	3	1	4	4	2
	Medicago lupulina	+	+	2	1	2	6	1	2
	Linum catharticum	+	+	2	+	1	2	4	3
	Briza media	+	+	1	+	1	1	1	1
	Polygala comosa	+	+	1	1	1	1	1	1
	Taraxacum officinale	+	+	1	1	1	1	1	+
	Prunella vulgaris	+	+	1	1	1	1	2	2
	Cirsium acaule	7	7	8	12	14	12	10	9
	Pimpinella saxifraga	1	1	1	2	2	2	2	2
	Thymus pulegioides	1	1	1	2	2	2	2	2
	Scabiosa columbaria	+	+	+	1	1	1	1	1
	Ranunculus bulbosus	+	+	+	1	1	1	1	1
3	*Veronica chamaedrys*	·	+	+	+	+	+	+	+
	Carex caryophyllea	·	+	1	+	+	1	+	+
	Luzula campestris	·	+	1	+	+	1	1	1
	Trifolium dubium	·	+	1	1	+	+	1	+
	Hieracium pilosella	·	+	1	1	1	2	3	2
	Trifolium repens	·	+	+	+	+	+	1	1
	Cerastium holosteoides	·	·	+	1	1	1	+	+
	Veronica arvensis	·	·	1	1	1	1	+	+
	Thlaspi perfoliatum	·	·	+	1	1	+	·	·
	Rhinanthus minor	·	·	·	·	·	+	+	+
	Dactylis glomerata	·	·	·	·	·	·	+	+
	Taraxacum erythrospermum	·	·	·	·	·	·	+	+

continued next page

Table 18.2, continued.

	Year	80	81	82	83	84	85	86	87
	Coverage (%)	99	90	97	98	97	97	99	99
	Number of species	21	30	36	35	34	35	39	36
4	*Festuca ovina* agg.	5	4	4	5	5	6	5	4
	Astragalus glycyphyllos	5	3	3	2	3	5	3	6
	Poa angustifolia	1	2	2	2	1	2	1	1
	Knautia arvensis	1	1	2	1	1	1	1	2
	Brachypodium pinnatum	1	1	1	1	1	2	1	1
	Potentilla neumanniana	1	1	1	1	1	1	1	1
	Viola hirta	1	1	1	1	1	1	1	1
	Agrimonia eupatoria	+	1	+	+	+	1	+	+
	Koeleria pyramidata	+	+	+	+	+	+	+	+
	Hippocrepis comosa	+	+	+	+	+	+	+	+
	Achillea millefolium	+	·	+	+	+	+	+	+
	Daucus carota	+	+	+	+	·	+	+	+
	Carex flacca	+	+	+	+	+	+	+	1
	Galium album	+	+	+	+	+	+	+	+
	Hypericum perforatum	+	1	1	1	1	+	·	+
	Fragaria vesca	+	+	+	+	1	1	+	1
5	*Prunus spinosa*	1	1	1	1	1	1	1	1
	Crataegus spec.	1	1	+	1	1	+	1	+
	Rosa canina	+	1	1	+	+	+	1	1
	Cornus sanguinea	+	+	+	+	+	+	+	+

removing the phytomass, the soil surface showed large bare patches.

In 1981 *Bromus* grew very well again. The structure of single tufts and open spaces remained. Only some young plants of new species could be found. In 1982/83 a new low-growing layer of plants with many small gaps developed, though still poor in blossoms. The tufts of *Bromus* became more and more sparse. The stalks of grasses were lower and more loosely distributed. Since 1984 the new structure of a markedly two-layered grassland can be recognized, with a dense ground layer (15–20 cm high) and sparse taller stalks of grasses and some forbs. The typical flowering aspects of calcareous grassland have also developed during that time.

Summarizing, one can ascertain that species diversity of this grassland type depends mainly on the growing capacity and litter accumulation of *Bromus erectus* as the most important competitor. Reduction of its vitality and litter removal are essential for the existence of many other species.

Table 18.2 consists of five species groups. The first one is mainly represented by *Bromus* itself. Group 2 contains many species already present in the abandoned area but with increase after the first cut. A very quick response is shown by *Medicago falcata, Trisetum flavescens, Trifolium pratense, Medicago lupulina, Linum catharticum* and *Cirsium acaule*, invading the open places.

Group 3, however, shows species absent in the abandoned grassland but with quick regeneration from the neighbourhood or from the soil seed bank. Almost all species are low–growing herbaceous perennials, but therophytes could also become established for a shorter time in open places. Remarkable is the appearance of *Veronica arvensis* and *Thlaspi perfoliatum* in some years. The latter could not be found near the research area.

Moreover there are many species more or less indifferent to cutting, mostly taller–growing plants of grassland and fringe communities (group 4). In 1980 shrubs (group 5) were not important. Only in one plot small shrubs could be found, up to a height of 80 cm. Their portion remained low while shrubs in abandoned plots developed (see Table 18.1, A).

The average cover degree of the plots shows no striking differences. Strong differences, however, can be seen in species diversity, which arose quickly after the first cut. The average number of species has almost doubled over several years. A new floristic balance has developed about 3 years after the beginning of cutting, similar to a species–rich Mesobromion community.

18.3.3 Differences of above–ground phytomass production

At the end of the succession research in July 1988, after 16 years, all plots were cut and the phytomass was sorted into different fractions. The comparison of their mean dry weights confirms the general physiognomic impression and the results of the vegetation survey. Figure 18.1 and Table 18.3 show the data for the different treatments.

Table 18.3: Above–ground phytomass of different plots in 1988 (Average dry weight in kg·ha^{-1} of 4 plots).

	1	2	3	4	5	2–4	2–5
Treatment	Wood	Grasses	Forbs	Mosses	Litter		
A	4,173.6	1,923.4	505.8	822.8	4,300.5	3,252.0	7,552.4
M		1,294.7	197.4	116.7	1,385.6	1,608.8	2,994.5
C1/2	331.5	1,734.8	377.6	95.7	2,566.2	2,208.1	4,774.3
C1		1,816.6	465.3			2,281.9	
C2		1,426.2	331.1			1,757.3	
C3		1,288.2	395.7			1,683.9	

All plots contain grasses and forbs. In contrast to Table 18.1, with an increase of the species number with cutting, the differences in the phytomass of forbs are relatively small. The reason is the low phytomass of many species. So the dry weight of forbs is highest in the species–poor plot A (more than 500 kg·ha^{-1}), resulting only from some tall plants. The further sequence of forb phytomass is C1–C3–C1/2–C2. The mulched plots have the lowest value (197 kg·ha^{-1}).

More striking are the differences in the grass fraction, mainly controlled by *Bromus erectus*. The highest value appears again in the abandoned control (A), with more than 1,900 kg·ha^{-1}, followed by C1/2 and C1. Markedly diminished is the biomass of C2, C3 and M.

Interesting is the appearance of mosses, mainly consisting of *Plagiomnium affine* and *P. undulatum*. Only within dense litter layers with relative high humidity mosses can form their own layer. Therefore the sequence of moss biomass is A–M–C1/2. In A their dry weight is even higher than that of the forbs. In the C1–C3 plots mosses (and lichens) are very sparse and could not be separated.

Clear differences are shown in the litter accumulation. In plot A litter has a higher value than the living phytomass (without wood) of one year (4,300 to 3,250 kg·ha^{-1}). The decomposition of new litter needs about two years. This result is confirmed by the values of C1/2, which was cut before in 1986. After two years again, more than half of the litter of A has accumulated (2566 kg·ha^{-1}). On the contrast the green mulch of M shows a faster decomposition.

The trends of differentiation between the treatments can be seen better in Table 18.4 showing the percentage composition of different herbaceous fractions: the percentage of forbs increases with cutting intensity up to 23.5% in C3. Plot A has only 15.5% but an additional 25.3% of mosses.

Table 18.4: Percentage composition of living phytomass in 1988.

	Grasses	Forbs	Mosses
A	59.2	15.5	25.3
M	80.5	12.3	7.2
C1/2	78.6	17.1	4.3
C1	79.6	20.4	
C2	81.1	18.9	
C3	76.5	23.5	

In 1979 near the research area, an extra plot of abandoned grassland was cut. Table 18.5 shows the dry weight in comparison with 1988. While the living phytomass decreased, the litter increased from 38.6 to 63.4% of total phytomass. This shows the rapid increase of abandonment effects.

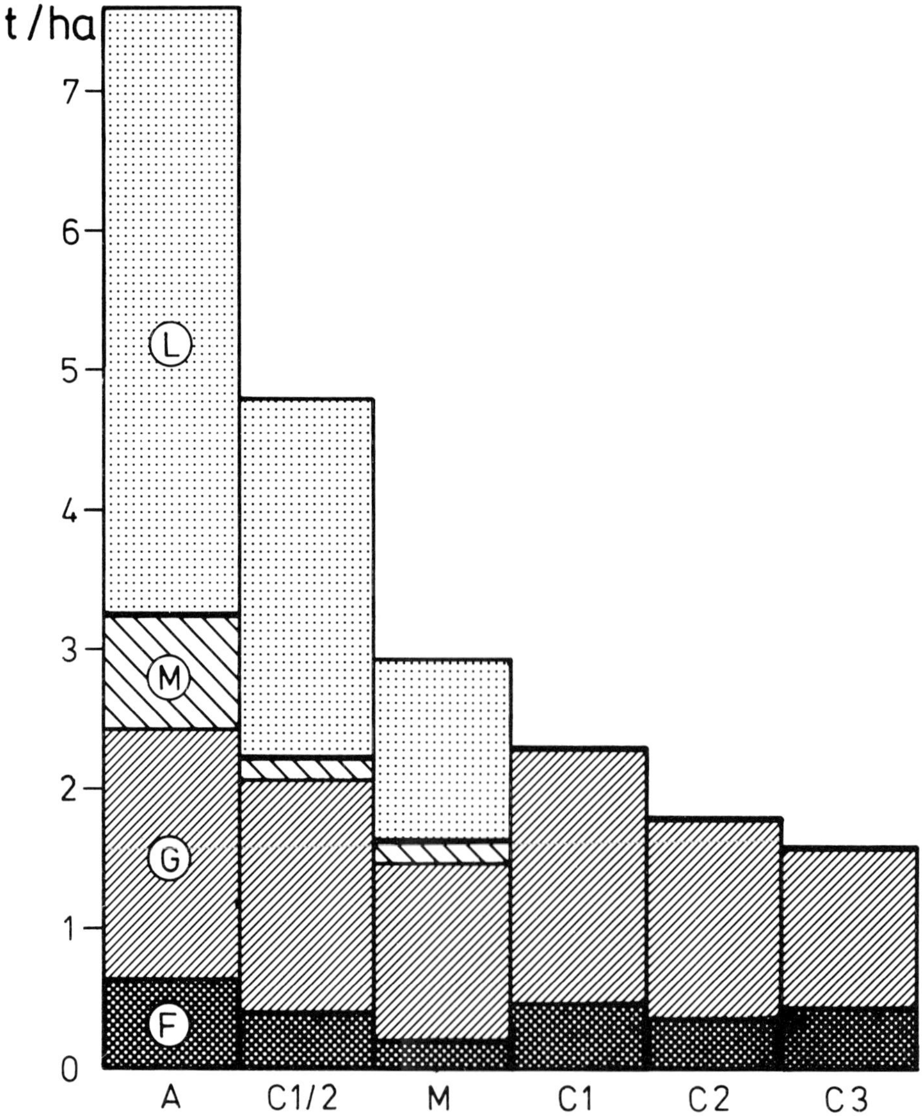

Figure 18.1: Above-ground phytomass of different plots in 1988. F: forbs; G: grasses; M: mosses; L: litter (Average dry weight of 4 plots).

Table 18.5: Comparison of phytomass of abandoned plots in 1979/88.

		1979		1988	
		kg·ha^{-1}	%	kg·ha^{-1}	%
1	Grasses (% 1 + 2)	2,055.0	73.6	1,923.4	79.2
2	Forbs (% 1 + 2)	737.0	26.4	505.8	20.8
1 + 2		2,792.0		2,429.2	
3	Litter (% 4)	1,755.0	38.6	4,300.5	63.4
4	Total	4,547.0		6,729.7	

18.3.4 Response of *Bromus erectus* to different cutting regimes

The described structure of the grassland and its changes are mainly influenced by the vitality of *Bromus erectus*. For that reason special emphasis is given to this species. As an expression of vitality, the number and length of stalks have been measured. The results of 1987 are summarized in the Tables 18.6–18.7.

Table 18.6: Average number (3 quadrats of 0.1 m^2) and length (30 measurements) of *Bromus* stalks of different plots in 1987.

	A	M	C1/2	C1	C2	C3
Number	5.5	21.0	30.1	15.5	10.6	8.0
Length (cm)	97.6	97.4	91.6	85.2	77.3	73.0

The length of stalks shows close correlations to other results. The maximum was measured in plot A (97.6 cm), the minimum in C3 (73.0 cm). High stalks could be found in M and C1/2 too. All together, the results show a loss of vitality parallel to cutting intensity.

The number of stalks, however, shows other tendencies. In A only a few (5.5) stalks were counted, the maximum was found in C1/2 (30.1), followed by M (21.0). Extensive cutting stimulates the development of stalks, whereas more intensive cutting reduces the vitality.

In Table 18.7 longtime results are given for the plots cut every two years (the year of cut is underlined). In order to avoid the influence of annual fluctuations, results are shown in percentage of the annual maximum (for the length in A, for the number mostly in M). In the years following a cut (1979, 1981, ...) the number of stalks is markedly higher than in the second year. This supports the opinion above that occasional cutting stimulates the development of many stalks. The length measurements do not show so clear correlations but a tendency to lower values in years after cutting (see 1983, 85, 87).

Table 18.7: Average number (3 parcels of 0.1 m^2) and length (30 measurements) of *Bromus* stalks of C 1/2 during the period 1978–88 (in percentage of the annual maximum). Cutting years underlined.

	1978	79	80	81	82	83	84	85	86	87	88
Number %	93	100	53	100	66	88	32	92	31	100	31
Length %	97	91	84	85	96	79	96	80	99	87	89

18.4 Discussion and conclusions

Mesobromion communities belong to the most characteristic semi–natural vegetation types of calcareous areas under extensive land use (ELLENBERG 1988, WILLEMS 1984 etc.). During the 20th century these once widespread swards on nutrient–poor soils disappeared more and more due to the cessation or increase of man's influence. They were replaced by manured grassland of the class Molinio–Arrhenatheretea, by arable fields or by afforestations. The remnants have often been abandoned in the last 30–40 years or more, changing their structure or being suppressed by shrubs (HAKES 1987).

Calcareous grasslands belong to the most diverse plant communities of Central Europe. Many of their species can be found in the red data books. In Western Germany (FRG) 41% of the species of dry grassland are endangered (KORNECK & SUKOPP 1988). Therefore a number of such grassland areas has been set up as a nature reserve. To maintain the high floristic diversity a specific management is necessary. Today there are several botanical and zoological investigations in this species–rich vegetation. Emphasis is given mainly to secondary succession by long–term observation of permanent plots as well as to regeneration by different types of management experiments (e. g. cutting, grazing, controlled burning).

18.4.1 Secondary succession of abandoned calcareous grassland

The basic trends of secondary succession of calcareous grassland after abandonment today are known fairly well (e. g. ELLENBERG 1988, SCHREIBER 1980, WILMANNS 1989). Some types of tendencies can be observed, often beginning simultaneously. As a rule, the communities become frequently poorer in species.

Expansion of graminoids ("Vergrasung"). After the end of mowing or grazing the percentage of tall–growing broad–leaved graminoids increases, especially of *Brachypodium pinnatum* with creeping underground rhizomes or *Bromus erectus* with large tussocks. Low–growing, light–loving forbs are overshadowed and suppressed or can survive only with some small remnants. As a consequence the performance of many Festuco–Brometea species is restricted.

Invasion and expansion of fringe (hem) plants ("Versaumung"). Fringe plants are mostly tall–growing or climbing forbs, tolerant to semi–shade but not very resistant against cutting or grazing. They often show a high capacity of vegetative propagation and can invade the abandoned swards of neighbouring communities of the Trifolio–Geranietea very quickly.

Accumulation of litter. The cessation of withdrawing the phytomass in combination with the increase of graminoids (see above) often leads to a thick litter layer within some years. The dead material can form a felted mat, resulting in a distinct decrease of low–growing species. Moreover the invasion or regeneration of plants by seeds is more or less inhibited. Litter accumulation leads also to changes of microclimatic conditions near to the soil surface and can promote some species of Molinio–Arrhenatheretea.

Invasion and expansion of shrubs ("Verbuschung"). A thick litter layer can inhibit the spread of woody species by seeds for a long time. Shrubs with subterranean runners, however, can expand relatively quickly, being effective pioneers of woodland. Especially *Prunus spinosa* forms large colonies of stems (polycorms) by root suckers. In its shadow other woody plants can establish. At the beginning a mosaic of swards and small groups of shrubs develops by and by. Afterwards a long–time succession to dense thickets of Prunetalia takes place, almost without light–loving herbaceous plants.

Reversion to near natural woodland. The potential natural vegetation of slightly arid calcareous grassland of Central Europe often is a beech wood (Carici–, Hordelymo–Fagetum). In dense scrubs, however, tree species cannot invade for a long time because of the heavy shade near to the soil surface. Therefore Prunetalia scrubs can establish very stable permanent communities ("Dauergesellschaften"). Today a succession to woodland can scarcely be recognized. The return of climax communities will perhaps take several centuries.

18.4.2 Own results of secondary succession in comparison with the literature

The succession of abandoned areas depends mainly on the initial species combination, the productivity and the decomposition rate of dead plant material (SCHIEFER 1981, 1982). In comparison with some other non–manured grasslands, the phytomass production of our *Bromus erectus* stand is rather high. The plot with cutting once a year in June/July produces more than 2 t$\cdot$ha^{-1}.

KLAPP (1971) quotes mean values for the Mesobromion (1.2 t) and for an Arrhenatheretum of dryer soils (3.5 t). HARTMANN & OERTLI (1984) measured within a submontane Mesobrometum with cut in July 1.8 t. ELLENBERG (1988) quotes 1.3 t for a *Bromus erectus* sward near Göttingen. SCHIEFER (1982), however, measured on mulched plots of Mesobromion 1.3–3.7 t. KLAPP (1931) mentioned about 3 t for a *Bromus* grassland, BOBBINK (1989) up to 4 t for a *Brachypodium pinnatum* sward.

Altogether our *Bromus erectus* stand shows a relatively high but not unusual productivity. Its vertical structure looks more meadow–like than like a sward.

In many publications the expansion of graminoids at the cost of low–growing forbs in abandoned grassland has been reported. Mostly *Brachypodium pinnatum* becomes the dominant species that can spread with extensive rhizomes rather quickly and overtop all the other plants (e. g. BOBBINK 1989, BOBBINK & WILLEMS 1988, BORSTEL 1974, DIERSCHKE 1985, DURING & WILLEMS 1984, HAKES 1987, 1988, KIENZLE 1979, REICHHOFF 1974, SCHIEFER 1981, WILLEMS 1985).

Succession to dense, species–poor *Bromus erectus* stands has been described more scarcely. BORNKAMM (1961) investigated the spread of *Bromus* in small quadrats where all or almost all plants have been removed at the beginning. Within 5 years *Bromus* reached a coverage of 55%, forming dense clumps, whereas small–growing plants were suppressed more and more. KIENZLE (1979) and KRÜSI (1981) describe the expansion of *Bromus* in an abandoned Mesobrometum of Switzerland. In later phases of succession, however, the brome has been displayed by *Brachypodium pinnatum* (see also REICHHOFF & BÖHNERT 1978). ROTHMEIER (1976) and SCHIEFER (1981) describe similar grasslands of Southern Germany, WILLEMS (1978) from England.

Bromus erectus and *Brachypodium pinnatum* are able to establish a long–term stable succession stage without any trees or shrubs that may persist for decades (SCHIEFER 1981, 1982, SCHREIBER & SCHIEFER 1985 etc.). KIENZLE (1979) describes such a community which remained for more than 22 years, ROTHMEIER (1976) for up to 36 years. The consequence of the expansion of the tall grasses is a notable change in the structure. Special emphasis to the change of growth forms is given by BOBBINK (1989), SCHIEFER (1981, 1982), SCHREIBER & SCHIEFER (1985), and WILLEMS (1985). In particular stolon– and rosette–hemicryptophytes and therophytes are disappearing, whereas tall–growing or climbing hemicryptophytes and geophytes become dominant.

The main factor for this displacement is shading by overgrowing plants and accumulated litter. The majority of Festuco–Brometea species are light–demanding plants and their disappearence is a consequence of shortage of light. BOBBINK (1989) measured the light (PAR) penetration inside the canopy in summer. In a grassland dominated by *Brachypodium pinnatum* the 50% light transmission point in the control plot (mowing at the end of the vegetation period) was 13–15 cm above soil surface compared with 4–5 cm in a plot cut twice a year. The light portion 2 cm above the soil was 4–8% in the control and 14–28% in the other plot (see also BOBBINK & WILLEMS 1988, HAKES 1987, 1988, WILLEMS 1985). Apart from light quantity also its quality should be regarded. The red:far red ratio is higher in open stands and may stimulate the germination especially of short–living species (VERKAAR & SCHENKEVELD 1984). In our own experiment this can be one reason for the almost absence of *Linum catharticum* in the abandoned plot A (see Table 18.1).

Contrary to the rhizome grass *Brachypodium pinnatum* that can develop very dense uniform stands, the tussock grass *Bromus erectus* gives more space between its dense clumps to other species (see DURING & WILLEMS 1984). In abandoned areas, however, both grasses equally form a thick litter layer up to 10 cm with a very slow decomposition rate. The litter can be pressed down by snow in winter and covers all the ground. The grasses themselves are able to penetrate the felted mat by the production of adventitious shoots, but many other plants are suppressed (see page 377). If the litter is removed, between the tussocks of *Bromus* is enough space for a relatively quick regeneration or invasion of other species (see page 379).

Production of phytomass and its decomposition have been investigated in some areas (e. g. HARTMANN & OERTLI 1984, KIENZLE 1979, SCHIEFER 1981, 1982, 1983, SCHREIBER 1980, STÖCKLIN & GISI 1985, WILLEMS 1985, WILMANNS 1989). Most of the above-ground phytomass turns into litter at the end of the vegetation period and takes more than one year for decomposition. Litter accumulation is favored by the increase of tall grasses with slow decomposition of dead material due to its low protein but high cellulose and lignin contents. In addition, summer drought can hinder the activity of microorganisms.

SCHIEFER (1982, 1983) measured in abandoned Mesobromion grassland 2–3 $t{\cdot}ha^{-1}$ of litter. The annual phytomass production was 2–2.5 t. In our experiment (Table 18.3) the mean litter amount is 4.3 t, the annual production (forbs and graminoids only) in the abandoned plots A is 2.4 t, in C1 2.3 t, in C1/2 2.1 t. STÖCKLIN & GISI (1985) investigated the litter decomposition experimentally in litter bags and showed its dependence of the litter quality. Within 12 months 54% of litter of an abandoned Mesobrometum was decomposed compared with 61% of mown plots due to the higher percentage of graminoids. During 90 days, in a meadow about 49% of the litter of *Bromus* and *Brachypodium* was decomposed, contrary to 57% of *Lotus corniculatus*. These results coincide well with our own conclusion, that litter decomposition of *Bromus erectus* needs about two years (see page 383).

SCHIEFER (1982) calculated the ratio of living and death above-ground phytomass for several vegetation types of abandoned areas: Mesobromion 1.0–1.2, Arrhenatheretum 1.4–2.9. In our *Bromus* plots the average value is 0.8 (A), 0.9 (C1/2), and 1.2 (M). These values illustrate the very slow litter decomposition in our research area and can explain the very species-poor abandoned plots.

The expansion of fringe species is also reported in many publications as a general trend of succession (e. g. BÜRGER 1984, DIERSCHKE 1974, 1985, HAKES 1987, 1988, KIENZLE 1979, KNAPP & REICHHOFF 1975, REICHHOFF 1974, REICHHOFF & BÖHNERT 1978, SCHIEFER 1981, 1982, WILMANNS 1975). In our research area this tendency can be observed too (see p. 377, 379). But only a few of Trifolio-Geranietea species are present (*Agrimonia eupatoria, Astragalus glycyphyllos, Hypericum perforatum, Medicago falcata, Viola hirta*). WILMANNS

(1989) pointed out, that *Brachypodium pinnatum* behaves like a skirt species too. It shows optimal growth in half-shaded areas around shrubs (HAKES 1987).

In Southern Germany, especially in the Kaiserstuhl area, positive effects of abandonment have been ascertained (BÜRGER 1984, KRATOCHWIL 1983, 1984, WILMANNS 1989). The invasion of many fringe species into a Mesobrometum leads to higher diversity. Many fringe plants flower in the second half of the vegetation period and are an important nutrient source for flower-visiting insects. In our research area only a few of such species are present, similar to other regions with a sub-atlantic climate. Mostly abandonment of calcareous grassland of Northwestern Germany has negative effects.

The invasion and expansion of woody species in abandoned areas of calcareous grassland has been reported in many publications and can be seen almost everywhere in Central Europe. More detailed results are available especially of *Prunus spinosa* that forms polycorms. The invasion rate into grassland often is 0.25–1.0 m per year (BÜRGER 1984, HAKES 1987, HARD 1975, REICHHOFF & BÖHNERT 1978, SCHREIBER 1980, WILMANNS 1975). In our experiment no special emphasis is given to this question. All woody species which have established in the *Bromus* grassland within 17 years were already present in the neighbourhood. At the beginning of the experiment (1972), only some young shoots could be found. About ten years later the development of a more dense low shrub layer began (DIERSCHKE 1985). This is shown also by the following mean cover values of 4 abandoned plots:

	1972	75	80	82	85	88
Prunus spinosa	1	1	3	5	12	33 %
Crataegus spec.	–	1	1	3	11	25 %

Before the beginning of the experiment, the area had probably been abandoned for 10 years or more. So it needs more than 20 years of abandonment for evident shrub development. BORNKAMM (1975) investigated two small plots in the same area as already mentioned. After 20 years, a dense *Bromus* sward had developed, but shrubs could not establish. From the Rhön mountains, however, BORSTEL (1974) reported a shrub invasion partly after 10 years.

18.4.3 Regeneration of Mesobromion communities

The best method to regenerate species-rich calcareous swards is the re-introduction of the former management, e. g. of extensive grazing or hay making without manuring. The remnants of such grassland, however, are often only small and scattered and therefore not suitable for sheep grazing. So long-term regeneration experiments have been carried out mostly by different cutting regimes. The effects of these treatments on vegetation structure are very similar everywhere (e. g. BOBBINK 1989, BOBBINK & WILLEMS 1988, BÜRGER 1984, DIERSCHKE 1985, DURING & WILLEMS 1984, KRÜSI 1981, REICHHOFF & BÖHNERT 1978, RODI

& KÖNIG 1987, SCHIEFER 1981, SCHREIBER & SCHIEFER 1985, WILLEMS 1985, WILMANNS & KRATOCHWIL 1983). Most important is the retention or suppression of overgrowing shrubs and vigorous herbaceous plants as well as the removal of accumulated litter and the prevention of further accumulation.

After some (3–5) years a species–rich low–growing sward can be regenerated by cutting once a year or more. However, the mobility of many species is very low. Several species that occur, have already been present in the surroundings. Other species were previously present in the plots themselves. The change is often caused more by clonal growth than by seed reproduction (BORNKAMM 1985). ELLENBERG (1988), however, mentions the surprisingly quick spread of many species. In our experiment area, within 13 years 18 species could invade for a longer time, 10 of them only in single plots (DIERSCHKE 1985).

Cutting once a year in summer favors already many low–growing light–loving plants. Cutting several times a year leads to swards with a structure similar to pastures. In contrast, however, the selectivity of grazing by sheep or cattle, favoring pasture weeds, cannot be imitated. Species growing close to the soil surface, e. g. plants with rosettes or prostrate forms as well as plants with storage capacity of nutrients in the soil gain an advantage from this treatment. In our plots, especially the rosettes of *Hieracium pilosella* show a striking spread in plot C3 (see Table 18.1). Mowing leads to a distinct decrease of above–ground phytomass and influences also the ratio of grasses and forbs in favor of the latter (Table 18.3, 18.4). This is a common effect of cutting (ELLENBERG 1952). In our plots, especially *Bromus erectus* is reduced. The higher portion of forbs promotes the decomposition of litter.

Summarizing mowing can be regarded as a suitable management for the regeneration of species–rich calcareous grassland and its maintenance.

Investigations of SCHIEFER (1981, 1983) and SCHREIBER & SCHIEFER (1985) show a positive result of mulching too. During a 8 year experiment in Southern Germany the character of Mesobromion has been preserved. In our experiment, however, the decomposition of mulch being left on the ground is very slow (Table 18.3). Only some tall–growing species (e. g. *Galium album, Medicago falcata, Astragalus glycyphyllos, Trisetum flavescens*) have possibly been favored by better moisture conditions. *Bromus erectus* shows the highest cover degree (69 %). The portion of forbs is very low (Table 18.4) and the species number only scarcely higher than in the abandoned plots (27:25). Therefore mulching cannot be recommended as a suitable management of calcareous grassland in general.

18.4.4 Vitality and competitiveness of *Bromus erectus*

Comparable investigations of vitality and competitiveness of dominant species in abandoned areas are scarcely available. It would be interesting to know more about the behavior of fringe species of Trifolio–Geranietea as well as other tall–growing plants in abandoned grassland.

About *Bromus erectus*, however, there are several ecological investigations as well as experiences from grassland management, well summarized by ELLENBERG (1988). *Bromus* is a perennial tall hemicryptophyte forming dense tussocks. It is only moderate tolerant to cutting and can be eliminated completely by more intensive sheep grazing. In comparison to abandoned areas the grass is favored by cutting once a year but weakened by several cuts (ELLENBERG 1952, KLAPP & OPITZ VON BOBERFELD 1990). After the first cut *Bromus* produces little new material and almost no new inflorescences in the same year (KRÜSI 1981). For hundreds of years it has been weakened by grazing. Today it can spread unhindered and conquer the remnants of calcareous grassland.

Invading areas with open soil, *Bromus* at first develops slowly. As soon as larger tussocks have been established (after 2–3 years), the grass can occupy all the ground and becomes absolutely dominant (see BORNKAMM 1961, 1962, 1988). Afterwards *Bromus* forms a very stable community (BORNKAMM 1975, 1985). This is the case in our plots, too.

There are almost no well comparable results to our investigations in the reaction of *Bromus* to different cutting regimes. In general, the development of stalks is controlled by physiological processes (promotion or inhibition) and by environment (KLAPP 1971). With the increase of radiation the number of stalks can be enhanced. Shading on the contrary diminishes the development of stalks. Cutting can promote the sprouting of lateral buds. KRÜSI (1981) observed an increase of stalks of *Bromus erectus*, cut once in June (30–65 per m^2), in comparison with abandoned plots (7–8). JACQUARD (1984) investigated the influence of density of pure cultures of sown *Bromus* and found a decreasing development of stalks with increasing density. WITTWER (1985) transplanted clonal units of *Bromus* into an experimental garden. In the next year the length of stalks was evidently longer of plants from abandoned areas than from mown plots.

In our own experiment (Table 18.6, 18.7) the abandoned plots (A) show only a very small number of stalks, too (mean value 5.1). The sprouting of buds possibly is inhibited by the shade of shrubs and the thick litter layer. The highest number of stalks (30) has developed in the plots cut every two years (C1/2). Already in C1 the number is diminished to one half. With increasing cutting frequency the vitality of *Bromus* is markedly diminished, also to be seen in the decreasing length of stalks.

Bibliography

BOBBINK, R., 1989, Brachypodium pinnatum and the species diversity in chalk grassland.
Proefschrift Rijksuniv, Utrecht 141 pp.

BOBBINK, R.; WILLEMS, J. H., 1988, Effects of management and nutrient availability on vegetation structure of chalk grassland.

In: H. J. During et al. (eds.), Diversity and Pattern in Plant Communities, The Hague, pp. 183–193.

BORNKAMM, R., 1960, Die Trespen–Halbtrockenrasen im oberen Leinegebiet. Mitt. Flor.-soz. Arbeitsgem., N.F. **8** ,181–208.

BORNKAMM, R., 1961, Zur Konkurrenzkraft von *Bromus erectus*. Ein sechsjähriger Dauerversuch.
Bot. Jahrb. **80**, 466–479.

BORNKAMM, R., 1962, über die Rolle der Durchdringungsgeschwindigkeit bei Klein–Sukzessionen.
Veröff. Geobot. Inst. ETH, Stiftg. Rübel **37**, 16–26.

BORNKAMM, R., 1975, Zwanzig Jahre Vegetationsentwicklung in einem Mitteleuropäischen Halbtrockenrasen.
In: W. Schmidt (ed.), Sukzessionsforschung. Vaduz, pp. 535–537.

BORNKAMM, R., 1985, Vegetation changes in herbaceous communities.
In: J. White (ed.), The Population Structure of Vegetation, Handbook of Veg. Sci. **3**, Dordrecht etc., pp. 89–109.

BORNKAMM, R., 1988, Mechanisms of succession on fallow lands.
Vegetatio **77**, 95–101.

VON BORSTEL U., 1974, Untersuchungen zur Vegetationentwicklung auf ökologisch verschiedenen Grünland- und Ackerbrachen hessischer Mittelgebirge (Westerwald, Rhön, Vogelsberg).
Dissert. Univ. Gießen.

BÜRGER, R., 1984, Successional limestone grassland–communities of the Kaiserstuhl — with regard to their conservation management.
Coll. phytosoc. **11**, 405–419.

DIERSCHKE, H., 1974, Saumgesellschaften im Vegetations- und Standortsgefälle an Waldrändern.
Scripta Geobot. **6**, 1–246.

DIERSCHKE, H., 1985, Experimentelle Untersuchungen zur Bestandesdynamik von Kalkmagerrasen (Mesobromion) in Südniedersachsen. 1. Vegetationsentwicklung auf Dauerflächen 1972–1984.
In: K. F. Schreiber (ed.), Sukzession auf Grünlandbrachen. Münstersche Geogr. Arb. **20**, 9–24.

DURING, H. J.; WILLEMS, J.H., 1984, Diversity models applied to a chalk grassland.
Vegetatio **57**, 103–114.

ELLENBERG, H., 1952, Wiesen und Weiden und ihre standörtliche Bewertung. Landwirtschaftliche Pflanzensoziologie **2**, 1–143.

ELLENBERG, H.,1988, Vegetation ecology of Central Europe. Cambridge University Press.

ENGELS, M., 1989, Auswirkungen langzeitiger Pflegemaßnahmen auf Vegetation und Boden von Kalkmagerrasen. Dipl. Arb. Göttingen.

HAKES, W., 1987, Einfluß von Wiederbewaldungsvorgängen in Kalkmagerrasen auf die floristische Artenvielfalt und Möglichkeiten der Steuerung durch Pflegemaßnahmen. Dissert. Bot. **109**, 1–151.

HAKES, W., 1988, Vergleich der Pflanzenbestandsstruktur genutzter und brachliegender Kalk–Halbtrockenrasen in Nordhessen. Phytocoenologia **16**, 289–314.

HARD, G., 1975, Vegetationsdynamik und Verwaldungsprozesse auf den Brachflächen Mitteleuropas. Die Erde **106**, 243–276.

HARTMANN, J.; OERTLI, J. J., 1984, Produktivität und Produktionsdynamik von Mähwiesen und Brachen des Nordwestschweizer Jura. Acta Oecol., Oecol. Plant. **5**, 255–264.

JACQUARD, P., 1984, Influence de la densité sur la morphogénèse et la production de *Bromus erectus* Huds. Acta Oecol., Oecol. Plant. **5**, 15–37.

KIENZLE, U., 1979, Sukzessionen in brachliegenden Magerwiesen des Jura und des Napfgebietes. Dissert. Univ. Basel.

KLAPP, E., 1931, Wiesen und Wiesenpflanzen in Mitteldeutschland. I. Die Trockenwiesen der Aufrechten Trespe (*Bromus erectus* L.). Landwirtschaftl. Jahrb. **74**, 673–828.

KLAPP, E., 1971, Wiesen und Weiden. Eine Grünlandlehre. Paul Parey, Berlin–Hamburg.

KLAPP, E.; OPITZ VON BOBERFELD, W., 1989, Taschenbuch der Gräser. Paul Parey, 12. Aufl., Berlin–Hamburg.

KNAPP, H. D.; REICHHOFF, L., 1975, Die Vegetation des Naturschutzgebietes "Leutratal" bei Jena. Arch. Natursch. Landschaftsforsch. **15**, 91–124.

KORNECK, D.; SUKOPP, H.,1988, Rote Liste der in der Bundesrepublik Deutschland ausgestorbenen, verschollenen und gefährdeten Farn- und Blütenpflanzen und ihre Auswertung für den Arten- und Biotopschutz.
Schriftenr. Vegetationskd. **19**, 1–210.

KRATOCHWIL, A., 1983, Zur Phänologie von Pflanzen und blütenbesuchenden Insekten (*Hymenoptera, Lepidoptera, Diptera, Coleoptera*) eines versaumten Halbtrockenrasens im Kaiserstuhl — ein Beitrag zur Erhaltung brachliegender Wiesen als Lizenz–Biotope gefährdeter Tierarten.
Beih. Veröff. Natursch. Landschpfl. Baden–Württ. **34**, 57–108.

KRATOCHWIL, A., 1984, Pflanzengesellschaften und Blütenbesucher–Gemeinschaften: biozönologische Untersuchungen in einem nicht mehr bewirtschafteten Halbtrockenrasen (Mesobrometum) im Kaiserstuhl (Südwestdeutschland).
Phytocoenologia **11**, 455–669.

KRÜSI, B., 1981, Phenological methods in permanent plot research. The indicator value of phenological phenomena. A study in limestone grassland in Northern Switzerland.
Veröff. Geobot. Inst. ETH, Stiftg. Rübel **75**, 1–115.

REICHHOFF, L., 1974, Untersuchungen über den Aufbau und die Dynamik des Orchideen–Halbtrockenrasens im Naturschutzgebiet "Leutratal" bei Jena.
Mitt. Sekt. Geobot. u. Phytotax. Biol. Ges. DDR, 115–125.

REICHHOFF, L.; BÖHNERT, W., 1978, Zur Pflegeproblematik von Festuco–Brometea–, Sedo–Scleranthetea– und Corynephoretea–Gesellschaften in Naturschutzgebieten im Süden der DDR.
Arch. Natursch. Landschaftsforsch. **18**, 81–102.

RODI, D.; KÖNIG, G., 1987, Die Entwicklung von Trockenrasengesellschaften in Abhängigkeit von Mahd und Brand an ausgewählten Versuchsflächen des Naturschutzgebietes Bargauer Horn bei Schwäbisch Gmünd.
In: R. Schubert; W. Hilbig (eds.), Erfassung und Bewertung anthropogener Vegetationsveränderungen, Teil 3, Halle, pp. 3–23.

ROTHMEIER, I., 1976, Untersuchungen über die natürliche Vegetationsentwicklung von Brachland bei Dietldorf (MTB 6737 Schmidmühlen).
Hoppea **35**, 235–277.

SCHIEFER, J., 1981, Bracheversuche in Baden–Württemberg. Vegetations– und Standortsentwicklung auf 16 verschiedenen Versuchsflächen mit unterschiedlichen Behandlungen (Beweidung, Mulchen, kontrolliertes Brennen, ungestörte Sukzession).
Beih. Veröff. Natursch. Landschpfl. Baden–Württ. **22**, 1–325.

SCHIEFER, J., 1982, Einfluß der Streuzersetzung auf die Vegetationsentwicklung brachliegender Rasengesellschaften.
Tuexenia **2**, 209–218.

SCHIEFER, J., 1983, Wirkungen des Mulchens auf Pflanzenbestand und Streuzersetzung. — Ergebnisse der Landschaftspflegeversuche in Baden-Württemberg.
Natur u. Landschaft **58**, 295–300.

SCHREIBER, K.-F., 1980, Brachflächen in der Kulturlandschaft.
Daten u. Dokum. Umweltschutz Univ. Hohenheim, Sonderreihe Umwelttagung **30**, 61–93.

SCHREIBER, K.-F.; SCHIEFER, J., 1985, Vegetations- und Stoffdynamik in Grünlandbrachen — 10 Jahre Bracheversuche in Baden-Württemberg.
In: K. F. Schreiber (ed.), Sukzession auf Grünlandbrachen, Münstersche Geogr. Arb. **20**, 111–153.

STÖCKLIN, J.; GISI, U., 1985, Bildung und Abbau der Streu in bewirtschafteten und brachliegenden Mähwiesen.
In: K.-F. Schreiber (ed.), Sukzession auf Grünlandbrachen, Münstersche Geogr. Arb. **20**, 101–109.

VERKAAR, H. J.; SCHENKEVELD, A.J., 1984, On the ecology of short-lived forbs in chalk grasslands: seedling development under low photon flux density conditions.
Flora **175**, 135–141.

WILLEMS, J. H., 1978, Observations on North-West European limestone grassland communities: Phytosociological and ecological notes on chalk grasslands of Southern England.
Vegetatio **37**, 141–150.

WILLEMS, J. H., 1984, Preservation and management of chalk grassland in western Europe.
Coll. phytosoc. **11**, 497–509.

WILLEMS, J. H., 1985, Growth form spectra and species diversity in permanent grassland plots with different management.
In: K.-F. Schreiber (ed.), Sukzession auf Grünlandbrachen, Münstersche Geogr. Arb. **20**, 35–43.

WILMANNS, O., 1975, Junge Änderungen der Kaiserstühler Halbtrockenrasen.
Daten u. Dokum. Umweltschutz Univ. Hohenheim **14**, 15–22.

WILMANNS, O., 1989, Zur Entwicklung von Trespenrasen im letzten halben Jahrhundert: Einblick-Ausblick-Rückblick, das Beispiel des Kaiserstuhls.
Düsseldorfer Geobot. Kolloqu. **6**, 3–17.

WILMANNS, O.; KRATOCHWIL, A., 1983, Naturschutz-bezogene Grundlagen-Untersuchungen im Kaiserstuhl.
Beih. Veröff. Naturschutz Landschpfl. Baden-Württ. **34**, 39-56.

WITTWER, A., 1985, Untersuchungen zur Mikrodifferenzierung von *Bromus erectus* Huds. und *Dactylis glomerata* L. in verschieden bewirtschafteten Wiesen.
Ber. Geobot. Inst. ETH, Stiftg. Rübel **52**, 66-73.

Chapter 19

Phytosociology in vineyards — results, problems, tasks

O. WILMANNS and A. BOGENRIEDER

19.1 Introduction

19.1.1 Framework

The European wine-grape, *Vitis vinifera* L., is cultivated worldwide, on all continents, in a multitude of cultivars that is difficult to survey. Native to the Mediterranean region, its major areas of cultivation are in Mediterranean climates, Zone IV of WALTER and LIETH (1960–67). Where viticulture extends further, into cooler (temperate) regions or into the dry-hot sub-tropics, the vines require special habitats and special handling, for example, special forms of "upbringing". These factors both balance out the climatic hindrances and achieve ecological compensation. Thus on the warm, but windy Macaronesian Islands the plants are drawn close to the ground, occasionally branched spiderlike, and protected from the wind with low walls or with strips of *Arundo donax*; in the laurel-forest climate of Japan whole plots are temporarily covered with plastic sheets to protect them from excess precipitation. This is just to mention two extremes.

When we speak of "Weinrebe", "Weinbau" or "Weinberg" in German, these words (with "Wein-", from Latin *vinum*) are most appropriate in that some 7/8 of the world production from grapevines actually is intended for the wine press, to become must for the production of wine. But a good part will be served as table grapes or be dried into raisins; this is the custom in all Moslem countries. Worldwide, between 55 and 60 million metric tons of grapes are produced yearly on somewhat more than 100,000 km^2 (cf., VOGT and GOETZ 1987).

As well and diversely investigated as the cultivated plant *Vitis vinifera* is (and it continues to be investigated further), the accompanying undergrowth is given relatively little attention outside Central Europe. The reason for this may lie, among other things (but certainly not exclusively), in the fact that in contrast to field crops, the economically used plant is woody. As a rule, woody plants grow higher than the herb layer. This distinct layering of the vine cultures on the one hand makes it easy to control the undergrowth, and on the other hand means there is surely less competition above and below ground level than, for example, in a grain field with its "weed grasses". It should certainly not be denied that many fundamental investigations towards understanding the vegetation, the mosaic of plant communities, and the individual important species as well, have been carried out with practically oriented questions in mind, and this will continue. Recently, the view has been gaining ground that a certain "tamed" herbal undergrowth in the vineyard offers more advantages than disadvantages; this is a stimulus to become better acquainted with the co-existence of the cultivated plant with all the other autotrophs, with the animals and the miccorganisms, that is, with the whole ecosystem, the vineyard.

A key factor here is the type of management practiced. In a macroclimatically homogeneous area this determines the spontaneous herb vegetation in the vineyard; crucial here are the strategies they have developed in their original environment, i.e., genetically fixed patterns of characteristics and behavior which make certain essential vital functions possible. These strategies decide whether a particular sequence of manipulations, which have the character of "disturbances", can be dealt with and favor a given plant relative to the others in the competitive struggle. Thus the type of management is reflected most clearly in the herb undergrowth (see also WILMANNS 1989). This is also true for a change in soil treatment, as has transpired in the last few decades, and is at present in progress.

We would like to illustrate this change in the following with a concrete example. To a case study, in which we fortunately are able to evaluate earlier data material from HÜGIN (1956), we should add a few basic remarks to show the many levels of phytosociological approaches. Thereby we want to present "Tasks for Vegetation Science", as Helmut LIETH so aptly calls them in the series he edits.

19.1.2 A short overview of previously described vineyard wild-herb communities

The classical plant community here, as it were, is the Geranio rotundifolii-Allietum vinealis, or vineyard-leek community (cf., Tables 19.2 and 19.3). Above all, it is distinguished from all other current weed communities (or better: wild-herb communities of Stellarietea mediae) through the many growth forms of its species and resultant textural (sensu BARKMAN 1979) heterogeneity. It claims a series of bulbous geophytes. Besides *Muscari racemosum* are included *Allium, Gagea* and *Ornithogalum* species and, as special embellishment, the wild tulip, *Tulipa sylvestris*. It is important to note that in addition to the morphological diversity of these species, their diverse phenology enters in. In 1989 we characterized 11 different morphological types. This species diversity allows the maintenance of the association even under different management programs, to be sure with drastic quantitative shifts for individual members and not without limits! The main developmental period lies in Spring; a few warm germinators may mark the summer aspect, but disturbances from management and the dry period have a restraining effect. It is important that in the mild vineyard climate many species also are able to assimilate in late Autumn and Winter. The community developed over centuries under the influence of frequent hoeing or shallow plowing. Through the application of herbicides, together with reduced soil movement, shifts in species composition have occurred. The basal constitution of the association has nonetheless been maintained, because the main application of herbicides takes place in late Spring, and hence at a time when the geophytes and evergreen therophytes have already completed most of their development.

Mulching, which has come into practice in recent years and is currently being strongly propagated, is causing a sociological reorganization and in the end a disorganization. Hereby, one mows 3–6 times a year, depending on weather and growth rate; the mown material is left lying on the spot. As a result, a group of trivial grassland species takes over. Continued mulching over several years, especially in cooler places, leads to the development of a *Poa trivialis*–Lolio–Potentillion community, i.e., "Kriechrasen–Gesellschaft"("Kriechrasen" is a special term for grassland mainly composed of plants carrying rooting stolons on the soil surface). Future practice and research are needed to tell how much its advantages (especially long–term erosion protection) outweigh any possible disadvantages (especially root felting near the surface and high water consumption).

19.2 Case study "Markgräfler Hügelland"

19.2.1 Geographical situation

"Markgräfler Hügelland" is understood to be the hilly country forming the southwest border of the "Naturraum" (= NR, "natural area" in a geographical sense) "Oberrhein–Tiefland" (= Upper Rhine Lowland) between Basel and Staufen, south of Freiburg; it encompasses the strip of hills lying just in front of the Black Forest between about 250 and 500 m above sea level. It is made up of Mesozoic and Tertiary rocks, partly covered with loess. Its western part has a colline climate and correspondingly has supported vast areas of vineyards for as long as history records. Starting from the Upper Rhine Plain ("Markgräfler Rheinebene", NR 200)

Figure 19.1: A typical view of the "Markgräfler Hügelland": the vineyards of the village Niedereggenen (July 1990).

in the west and going up toward the peaks of the Black Forest ("Hochschwarzwald", NR 155) to the east, precipitation gradually increases and temperature decreases. As one climbs, the proportion of orchards, cultivated fields, meadows and forests increases. For the wine–growing region as a whole we can take the following approximate values: Annual average precipitation 750–900 mm, mean annual temperature 9.5–10.0 °C .

The emphasis of our presentation pertains to the vineyards of the village of Niedereggenen, i.e. of the Niedereggenen "Gemarkung" (= district), in the eastern part of the wine–growing area, the part closest to the Black Forest (Photo 19.1). Additional ecological measurements were taken in Mauchen and Auggen, which, in that order, lie increasing closer to the Rhine. Niedereggenen offers a number of advantages. To be sure, in 1956/57 it underwent a land consolidation ("Flurbereinigung", a reform to simplify ownership and land use). However, the total extent of the vineyards has not significantly changed since that time, as can be seen by comparing maps from 1965 and 1988 (and, incidentally, 1889!). (We have not included several fresh pieces of ground with new plantings and sowings with *Lolium perenne*.) The vineyards involved cover a single, 2 km long, SE to S exposed slope.

The special geological map also shows rather uniform conditions, though they are occasionally locally difficult to interpret for ecological purposes; this slope encompasses "scree from the middle and upper Dogger mixed with loam", "especially on Opalinuston", peripherally bordering on loess–loam in one place. (The vineyards of the Mauchen and Auggen Districts lie predominantly on loess and loess–loam, respectively.) So these are loamy soils, in places weakly sandy or clayish. As easy as it is to establish this, it is just as difficult to learn anything directly about underground conditions, which become especially relevant when impeded vertical water–flow causes subterranean accumulation above the Loess layer. Indirect inferences, for example as to the occurrence of *Convolvulus sepium*, do not lead to a definitive logical–ecological conclusion.

Livestock–keeping has dramatically decreased since the Second World War, and with it the availability of self–provided stall manure. Consequently, the addition of organic material to the vineyard has decreased. Of course, today the provision of mineral fertilizer is very good.

19.2.2 Available material, methods

To make a comparison on the basis of relevés from different authors at different times, and therefrom further to make a causal interpretation of any possible successions, is only possible in a very favorable situation. This is rarely the case for a community like the Geranio–Allietum, with its pronounced seasonally varying aspects and its strong and direct dependence of its appearance on single management measures. Fortunately, in our case a documentation of outstanding quality is available; this is the doctoral dissertation of GERHARD HÜGIN (deceased 1988) — alas unpublished, but available in Freiburg in the original. He presented 21 relevés each for two Gemarkungen, namely those of the villages Niedereggenen and Mauchen, of the "weed communities of the vineyards" (i.e., of Geranio–Allietum, under the name Muscari–Allietum) for the years 1950–1954, using the BRAUN–BLANQUET method. Also included are a smaller number of relevés from several other Gemarkungen. Almost all of these were made both in Spring and then again in either Summer or Autumn, each in separately presented relevés with the cover–abundance of each species indicated. Although no vegetation mapping was included, systematic notations concerning the frequency of the different community types are given. Thereby, no "atypical" relevés were eliminated.

For about half a decade we have also surveyed vineyard areas in Markgräfler Land ourselves. Specifically for a comparison, we obtained material from Niedereggenen in March, 1990 (with supplementary data from late April). We also sought to encompass the full local diversity and did not reject any relevés as atypical. So individual plots cannot be compared, but the range of community types can. The material from HÜGIN, as necessary for our comparison, is given in Table 19.3 and Figure 19.2 in a condensed, simplified form. We have omitted one relevé of his set with only 6 species in Spring, one of the two surveys marked by him as

Table 19.1: Phytosociological abbreviations and scales used in Tables 19.2, 19.3, 19.4.

Syntaxonomical code for species resp. taxa:

- Ch: character species of the association
- D: differential species
- V: character species of the alliance
- O: character species of the order
- K: character species of the class

Scales (cf., WESTHOFF and VAN DER MAAREL 1973):

We used the slightly modified Braun–Blanquet scales of cover–abundance (= Artmächtigkeit; first number), of sociability (second number), and of presence (Roman number).

Cover–abundance:

- r: 1 individual in the sample plot and not in the near surroundings
- +: 2–5 individuals in the sample plot, cover < 5% of total sample plot area
- 1: 6–50 individuals in the sample plot, cover < 5% of total sample plot area
- 2: very abundant and < 5% cover or 5–25% cover; (m) > 50 individuals in the sample plot; (a) cover 5–15%; (b) cover 16–25%; (a) and (b) irrespective of number of individuals
- 3: cover 26–50% irrespective of number of individuals
- 4: cover 51–75% irrespective of number of individuals
- 5: cover 76–100% irrespective of number of individuals

Sociability:

- 1: growing solitarily
- 2: growing in small groups or small tussocks
- 3: growing in small patches or large tussocks
- 4: growing in extensive patches, carpets, or broken mats
- 5: growing in great crowds or extensive mats

Usually soc. 1 is omitted in combination with Artmächtigkeit r and +

Presence degree: number of relevés of a syntaxon in which a species is noted; given absolute respective in % respective % classes; those are:

- V: 81–100%
- IV: 61–80%
- III: 41–60%
- II: 21–40%
- I: 11–20%
- + 6–10%
- r: 1–5%

Table 19.2: (Pages 406–409) Examples of vineyard herb communities (Reb–Wildkrautgesellschaften) in the "Markgräfler Land". For explanation of the symbols see Table 19.1.

Block	A:	Geranio–Allietum
	Aa:	*Stellaria media*–Agroform
	Ab:	*Bromus sterilis*–Agroform
	Ba:	penetration (Durchdringung) of G.–A. and *Poa trivialis*–Lolio–Potentillion community
	Bb:	*Poa trivialis*–Lolio–Potentillion community
Gemarkung (district)		Au Auggen; Ba Ballrechten; I Istein; Ma Mauchen; W Weil/Rh.; Wo Wollbach;
Bewirtschaftung	F:	frequent soil movement by rotary tillage
(management)	H:	frequent herbicide treatment
	M:	mulching treatment
	N:	young planting on old parcel

Aufnahmefläche: area of the relevé ($\times 10$ m^2)
Zeile: row of grape–vines; Gasse: alley between them; both: +
Deckung: cover (of Cormophytes, Bryophytes, total) +: > 0–$< 1\%$;
dead material is only declared if $> 10\%$ could be identified directly
Artenzahl: species number (further see Deckung)
Rottezeiger: indicators of tilth (proper development of soil organic material);
Spritzbetriebszeiger: indicators of herbicide treatment;
†: only dead shoots present at collection time.

"incomplete"; we can compare the remaining 20 relevés to 20 current–day relevés (see Table 19.3).

Some details as to methods! For low–level representation HÜGIN distinguished between s, r and +; we have grouped these all together as simply +. We have sub–divided his level 2 into 2m, 2a and 2b, as is standard practice today. For the anyway critical sociabilities, we have only given the highest in each case in our tables, unless it happened to be too atypical. In general, the resultant loss of information is bearable, although sociability is of importance for the bulbous plants (see below). In order to be able to compare quantities and not just the presence values, we have introduced a new coefficient, the "Bauwert"–number or "structural value". This is similar to the usual cover value, but easier to calculate. It is calculated rather freely, but not senselessly, as the point total for each species summed up over the 20 relevés; to do this, non–linearly weighted point values were assigned to the levels of cover–abundance (= Artmächtigkeit) in each relevé as follows:

	A							B						
	Aa			Ab				Ba			Bb			
Current number	1	2	3	4	5	6	7	8	9	10	11	12	13	14
Date of relevé: Year	86	86	83	88	86	83	90	83	90	90	86	86	83	90
Month	VI	V	III	III	IV	III	III	III	III	III	V	V	III	III
Day	28	8	20	26	15	17	17	17	9	10	8	8	20	17
District	Au	Ma	Au	I	Ba	W	I	W	Ma	Ma	Au	Au	Au	Wo
	N	N	F	H	H	H	H	M	M	M	M	M	M	M
Exposition	WSW	SW	S	S	SSW	SW	S	SW	SO	WSW	W	NO	W	SO
Slope (°)	15	15	12	20	28	6	20	6	20	5	5	5	3	6
Area of relevé (× $10m^2$)	6	8	·	5	5	3	7	3	7	10	5	5	5	8
Zeile Z, Gasse G, Both +	+	+	+	G	+	Z	+	G	G	G	G	G	G	G
Coverage (%):														
Cormophytes	95	80	98	80	80	65	70	90	80	20	98	95	95	100
Mosses	–	–	3	+	+	3	70	+	50	15	+	+	–	+
Total	95	80	98	80	80	65	98	90	90	99	98	95	95	100
Number of species:														
Cormophytes	19	29	18	23	23	13	24	11	30	24	15	17	21	27
Mosses	–	–	1	1	1	4	4	1	4	4	2	4	–	2
Total	19	29	19	24	24	17	28	12	34	28	17	21	21	29
Ch/D Geranio–Allietum														
Allium vineale	·	+	1 . 2	2m2	2m2	1 . 2	2m2	+ . 2	+ . 2	+ . 2	+ . 2	·	·	·
Ornithogalum umbellatum	·	2m1	·	2m2	·	1 . 2	1 . 2	1 . 2	2m1	2 a 1	·	·	·	·
Muscari racemosum	·	(+)	·	2m1	·	·	2m2	·	2m1	·	·	·	·	·
Tulipa sylvestris	·	·	·	·	·	2m3	·	2m2	·	·	·	·	·	·
Gagea villosa	·	·	·	1 . 1	1 . 2	·	·	·	·	+ . 2	·	·	·	·
D *Ficaria verna*	·	·	·	·	2 b 3	+ . 2	·	·	·	·	·	+ . 2	·	·
D *Corydalis solida*	·	·	·	·	·	1 . 2	·	+ . 2	·	·	·	·	·	·
Ornithogalum nutans	·	·	2m3	·	·	·	·	·	·	·	·	·	·	·
Calendula arvensis	·	·	·	1 . 1	·	·	·	·	·	·	·	·	·	·
D *Anemone nemorosa*	·	·	·	·	2 a 3	·	·	·	·	·	·	·	·	·
D *Corydalis cava*	·	·	·	·	·	·	2 a 2	·	·	·	·	·	·	·
D *Allium scorodoprasum*	·	·	·	·	·	·	·	·	·	·	·	·	·	2m3

"Rottezeiger"? (see text)														
K *Stellaria media*	1.1	1.3	4.4	2m2	·	·	·	·	·	1.2	+.2	·	+.2	·
V *Fumaria officinalis*	2a1	2b2	1.1	·	+	·	+	·	·	·	·	·	·	·
V *Mercurialis annua*	2m1	·	2†	1.1	·	·	·	·	·	1†	·	·	·	·
V *Euphorbia helioscopia*	2b1	2m1	·	+	·	·	·	·	·	·	·	·	·	·
O *Solanum nigrum*	+†	·	·	·	·	·	·	·	·	·	·	·	·	·
K *Urtica urens*	+.2	·	·	·	·	·	·	·	·	·	·	·	·	·
O *Lamium amplexicaule*	·	+.2	·	·	·	·	·	·	·	·	·	·	·	·
Indicators of herbicide treatment														
Polygonum aviculare agg.	·	1.1	·	·	2m2	·	1.1	·	+.2	+.2	·	+	·	·
Galium aparine	·	+.2	·	+.2	+.2	+.2	·	·	·	1.1	·	·	·	·
K *Bromus sterilis*	2m2	·	·	2b3	2a3	·	4.4	·	·	·	·	·	·	·
Cardamine hirsuta	·	·	·	·	·	·	·	·	2b4	·	1.1	2m2	1.2	·
V *Valerianella carinata*	·	+	·	2a2	·	·	·	·	1.2	·	·	·	·	·
Epilobium parviflorum	·	·	·	·	†	·	·	·	+.2	·	·	·	·	(+)
Epilobium tetragonum agg.	·	·	·	·	·	·	·	·	·	·	·	·	·	+
Hypericum perforatum	·	·	·	·	†	·	·	·	+.2	·	·	·	·	·
Rosa cf. canina	·	·	·	·	+.2	·	·	·	·	·	·	·	·	·
Hedera helix	·	·	·	·	·	·	·	·	·	+	·	·	·	·
Further Ch V/O/K Stellarietea mediae														
Veronica persica	1.1	1.2	2a3	2m2	1.2	1.1	2m2	1.1	2m2	+.2	1.1	+	2m2	+
Veronica hederifolia	·	1.2	+.2	1.2	2m2	3.4	+	1.1	1.2	+.2	+	+	·	1.1
Convolvulus arvensis	+°	1.1	+.2	·	†2	†2	1.2	·	2m2	+.2	·	·	·	+.2
Lamium purpureum	·	·	2a3	2m2	·	+	·	·	(+)	2m2	1.1	+	+	+
Senecio vulgaris	2m1	1.2	+	+	·	·	2m2	·	·	·	+	·	+	·
Sonchus oleraceus	+	2m1	1.1	·	·	·	+	·	+	·	+	·	·	·
Malva neglecta	·	1.2	1.1	·	·	+	+.2	1.1	·	1.2	·	·	·	·
Capsella bursa-pastoris	·	+	·	1.1	·	·	+°	·	·	·	·	·	·	·
Amaranthus retroflexus	5.5	·	·	·	·	·	2†	·	·	·	·	·	·	·
Sonchus asper	1.1	·	·	·	·	·	+	·	·	·	·	·	·	·
Lathyrus aphaca	·	+.2	·	·	·	·	·	·	·	·	·	·	1.2	·
Geranium dissectum	·	·	·	·	·	·	·	·	·	+.2	·	·	·	+.2

	A							B						
	Aa			Ab				Ba			Bb			
Lactuca serriola	·	+	·	·	·	·	·	·	·	1 . 2	·	·	·	·
Chenopodium album	1 †	·	·	·	·	·	·	·	·	·	·	·	·	·
Matricaria inodora	·	+ . 2	·	·	·	·	·	·	·	·	·	·	·	·
Veronica polita	·	·	·	·	·	·	+	·	·	·	·	·	·	·
Setaria cf. viridis	·	·	·	·	·	·	†	·	·	·	·	·	·	·
Characteristic species of the Poa tr.–Lolio–Potentillion community														
Poa trivialis	+ . 2	1 . 2	·	+ . 2	·	2 a 2	2m2	5 . 4	2 a 3	2 b 4	4 . 4	4 . 4	5 . 4	3 . 4
Lolium perenne	+ . 2	+ . 2	+ . 2	·	·	·	·	·	+ . 2	·	·	+ . 2	1 . 2	+ . 2
Agrostis stolonifera	·	·	·	·	·	·	·	·	2m2	·	2m3	1 . 2	+ . 2	2 b 3
Trifolium repens	·	·	·	2 a 2	·	·	·	·	1 . 2	1 . 2	+ . 2	+ . 2	+ . 2	·
Rumex obtusifolius	·	·	·	·	·	·	·	+	·	+ . 2	1 . 2	+ . 2	·	1 . 2
Dactylis glomerata	·	·	·	·	·	·	·	+ . 3	+ . 2	·	·	·	+ . 2	+ . 2
Plantago lanceolata	·	·	·	·	·	·	·	·	+ . 2	+ . 2	·	·	+ . 2	·
Festuca arundinacea	·	·	·	·	·	·	·	·	·	+ . 2	·	·	1 . 2	·
Rumex crispus	·	·	·	·	·	·	·	·	·	+ . 2	·	·	·	1 . 2
Ranunculus repens	·	·	·	·	·	·	·	·	·	·	+ . 2	3 . 4	·	·
Glechoma hederacea	·	·	·	·	·	·	·	·	·	·	·	2 . 4	+ . 2	·
Holcus lanatus	·	·	·	·	·	·	·	·	2m2	·	·	·	·	·
Poa pratensis	·	·	·	·	·	·	·	·	+2	·	·	·	·	·
Solidago gigantea	·	·	·	·	·	·	·	·	+ . 2	·	·	·	·	·
Ranunculus acris	·	·	·	·	·	·	·	·	·	·	·	1 . 1	·	·
Potentilla reptans	·	·	·	·	·	·	·	·	·	·	·	·	1 . 3	·
Other, giving evidence species														
Acrocarpi	·	·	2 . 3	1 . 2	2m3	2m2	2m2	·	1 . 2	+ . 2	·	+ . 2	·	·
Eurhynchium swartzii	·	·	·	·	·	+ . 2	4 . 3	1 . 2	3 . 3	1 . 2	+ . 2	+ . 2	·	+ . 2
Brachythecium rutabulum	·	·	·	·	·	1 . 2	·	·	3 . 3	2 a 4	2m3	3 . 4	·	+ . 2
Amblystegium serpens	·	·	·	·	·	2m3	1 . 2	·	1 . 2	+ . 2	·	·	·	·
Cynodon dactylon	·	·	·	·	·	·	+ . 3	·	3 . 5	·	·	·	·	·
Cardaria draba	·	·	·	·	·	2 b 5	·	·	·	·	·	·	·	·
Equisetum arvense	·	·	·	·	·	·	·	·	·	·	·	·	·	+ . 2

Further concomitant species														
Taraxacum officinale	+	1.2	2m2	2a2	+	1.2	1.2	2a2	+	4.5	2b4	2a2	+	2a3
Poa annua	1.1	·	2m2	2b2	·	·	2m2	·	1.2	·	1.2	·	1.2	·
Agropyrum repens	·	1.2	1.2	·	·	·	·	·	+.2	+.2	3.4	2m3	2m2	1.3
Cirsium arvense	·	1.2	1.1	+	·	·	·	·	·	2m2	·	·	·	1.1
Vicia angustifolia	+	+.2	·	·	2m2	·	·	·	1.2	·	·	·	·	·
Crepis capillaris	·	·	+	+	·	·	·	·	1.1	+	·	·	·	·
Urtica dioica	·	+.2	·	·	+.2	·	·	·	·	·	·	·	·	1.2
Convolvulus sepium	·	·	·	·	·	·	·	·	·	·	·	·	·	+.2

In addition, twice each: Veronica arvensis: + in 2. 1.2 in 5; Erodium cicutarium: + in 4. 1.2 in 5; Vicia sepium: 2m2 in 13. 2m2 in 14; Holcus lanatus: cf. +.2 in 13, +.2 in 14;

Once each: Poa nemoralis: 1.2 in 5; Alliaria petiolata: +.2 in 5; Trifolium resupinatum: cf. +.2 in 7; Geum urbanum: +.2 in 9; Veronica serpyllifolia: +.2 in 9; Cerastium holosteoides: +.2 in 13; Fraxinus excelsior: Klg. 2m1 in 12; Brachythecium salebrosum: cf. +.2 in 12; Aegopodium podagraria: 2b4 in 14; Heracleum sphondylium: + in 14; Rubus fruticosus: $+^{\circ}$ in 14; cf. Lolium multiflorum: +.2 in 14.

cover–abundance	points
s, r, +	1
1	3
2m, 2a, 2b	10
3	40
4	60
5	80

Species with large quantities presumably have a much greater importance, also with respect to the animals present, than would be indicated by the arithmetic relationship of numbers of cover–abundance. On the other hand, isolated single individuals of certain species can easily survive in the vineyard communities with their low degree of organization; they do not bear any great significance in the ecosystem and receive a correspondingly low Bauwert–number. The Bauwert–numbers were calculated in each case for springtime conditions, whereby the presence values of HÜGIN for the two different dates were combined.

One may see it as a shortcoming in our data that they only include springtime conditions; but today this is not taken too seriously anymore, because of the general lack of soil movement in Autumn and Winter. As a consequence thereof, warm germinators such as *Amaranthus retroflexus, Solanum nigrum*, etc., are still recognizable as stalks († in the tables), if not always quantitatively. Where relevant, this must be remembered. The average size of our sample plots was a little bit smaller than HÜGIN's. However, species immediately adjacent (+ in the tables) were included in the comparison calculations; so there are no systematic errors here.

The collection of soil samples for the soil analyses carried out in conjunction with this investigation was made on March 15, 1990, at depths of 1 to 5 cm. After sieving and drying (80 °C), the material was pulverized with a ball mill for biochemical analysis. Total–N was determined by the Kjeldahl method (BREMNER 1960; NELSON and SOMMERS 1980; WAGNER 1970) and total–C with a Woesthoff apparatus (BARTSCHER and SCHMIDTS 1963). Determination of limestone–bound carbon was made in a separate procedure (also in the Woesthoff apparatus) through the release of CO_2 with HCl. The amount of organically bound carbon is then calculated as the difference between the two measured carbon values.

As the phytosociological relevés given for Niedereggenen in Table 19.3 may at first glance be confusing, we shall first discuss a few clear, in part extreme, community types for the whole of Markgräfler Land; these are presented in Table 19.2. They need to be interpreted with the aid of our experience from many other regions.

19.2.3 Overview of the community types in the "Markgräfler Land"

Table 19.2 includes relevés from different years; except for Nr. 1, they are all springtime records. Only those from 1990 (March) were supplemented a few weeks later, which may be the explanation for the greater numbers of species here. This notwithstanding, the characteristics of the three types presented remained unaffected. Incidentally, for presentation we have consciously chosen preferentially those plots on which floristic treasures of the southwest German Geranio–Allietum grew, namely *Tulipa sylvestris, Ornithogalum nutans, Corydalis* spp., or *Calendula arvensis*. With the expansion of mulching operations, they all will be further decimated, and on this account they are here documented.

Indeed the most recent operation(s) on the vegetation can be seen. Moreover, earlier measures, repeated with regularity, can still be traced in the present combination of species. Evaluation is made difficult by the fact that the soil treatment and weed control methods often changed every few years, and frequently even within the same year. Hence, a certain procedure "stamps" most of the stands, without one being able to rule out with certainty the application of another in the past.

The character of the Geranio–Allietum as a "weed" communitity with a wealth of spring geophytes is most strikingly shown in relevés 4–7. Their often dominating aspect–determining appearance in the vineyard may be understood from their life cycle. The character–geophytes are species with their main occurrence in sub–Mediterranean–Mediterranean areas. Appropriate to the climate there, they sprout in late Summer (e. g., *Muscari racemosum*), Autumn (e. g., *Ornithogalum umbellatum*) or Winter (e. g., *Tulipa sylvestris*), so as to utilize optimally the mild, moist winter months and the especially favorable spring phase. Before flowering, bulbs will already have formed, to mature in April or May, but this remains to be examined more closely. The *Corydalis* spp., *Ficaria verna* and *Anemone nemorosa*, are deciduous–forest inhabitants and correspondingly are only evaluated here as differential species. (We found the latter in only a single plot). They sprout first in Spring, and thus noticeably later, and their center of distribution is not submediterranean–mediterranean! In any case, they reach the still light, warm places, so to speak on "the edge of the forest" in the "fringe" (= Saum) of the "mantle" (= Mantel) formed by *Vitis*. Here they produce seeds or, in the case of *Ficaria*, bulblets. Quite striking is the frequency of myrmekochoreous species, including species of the genera *Ornithogalum, Corydalis, Anemone* and *Ficaria*. Incidentally, this mode of dispersal is also common among the annuals such as *Veronica, Fumaria, Lamium* and *Euphorbia*.

The concentration of geophytes comes indirectly from the application of herbicides, in that this is in general associated with decreased soil movement and destruction of plants during their vegetation period. This can also be seen in the sociability of the individual plants: the daughter bulbs are not separated and propagated, and whole clusters form. Of course, with frequent enough application and alternation of chemical agents, all plants can be killed, but the herbicides are usually first applied in late April or May just before or after the end of the species–specific vegetation period. And when in Summer, before the vintage, competitors are sprayed again, and then in Winter a turning of the soil is left out, these geophytes are favored relatively. Of course, the action of the various types of herbicides on different species is more differentiated and must be considered in more detail.

In the same group of relevés (4–7) are presented further species as "spray–treatment indicators". For interpretation, the reader is referred to WILMANNS (1989). Briefly, species are involved which either use Winter for assimilation or else, after leaf–less from contact herbicides, can easily resprout thanks to their storage organs. The latter group includes woody plants.

We have proposed the term "Agroform" for such vegetation types, which react quickly to cultivation changes and are not necessarily very sharply definable (on occasion only by quantitatve differences). As in the Kaiserstuhl, with this unit (Ab in Table 19.2) one can speak of a *Bromus sterilis*–Agroform of the Geranio–Allietum.

The three relevés of the unit Aa, which we want to designate as the *Stellaria media*–Agroform, represent the "old" Geranio–Allietum; this type is becoming ever scarcer in Markgräfler Land, as elsewhere, e. g., in the Kaiserstuhl and on the slopes of Steigerwald (WILMANNS 1990). It develops on those parcels cultivated with active soil movement. Previously, this was accomplished by hoeing and plowing, and nowadays usually with a rotary tiller or grubbing cultivator. The geophytes are sparser, generally occurring as isolated individual plants. This in fact on account of the separation of mother and daughter bulbs, insofar as these are less than 10 cm under the surface: with *Tulipa* and *Ornithogalum nutans* they are clearly deeper. These usually also come to light during new plantings on old parcels and are distributed a bit. It is revealing that the "spray indicators", or better yet "spraying–management indicators" appear in this vegetation type (as well as in B), or for that matter that the typological delimitations (in contrast to the man–made stand–boundaries!) tend to be unsharp. Spraying is just one procedure in an overall management program. None of these species is really absolutely dependent on one particular cultivation program or the other. Moreover, as mentioned, cultivation methods are changed often; and finally the expansion into neighboring stands and germination from the seed bank must be considered.

A particular species group is presented in Table 19.2 under the heading "rotting indicators" (= Rottezeiger, Rotting is defined as the partial decomposition of organic material with a good oxygen supply and dry substratum). One also finds these species overproportionally often in young plantings on old vineyard land, but also when individual stocks are replaced on older plots. They also appear regularly where it is evident that manure, grape husks or similar organic material has been worked in. As a rule, manure is still added today during new plantings on

Table 19.3: (Pages 414–421) The span of Geranio–Allietum with its penetrations (Durchdringungen) in the vineyards of Niedereggenen (1950–1990). Synthetic characteristics:

a, d:	Presence degree absolute
b, e:	Presence degree class
c, f:	Bauwert (structural value see text)
g:	Indicator value for nitrogen given by ELLENBERG (1979); "entf." means "no number available" (mosses, cultivated plants, or seedlings of woody plants); ~ means number declared to be tentative by ELLENBERG

For explanation of the symbols see also Table 19.1.

---

old spots. Conversely, many of these species were very strongly reduced after the major land consolidations in the Kaiserstuhl. These species are known as nitrogen indicators and most of them also as "tilth indicators" (= Garezeiger) (OBERDORFER 1983). A soil shows "Gare" when it is loose, crumbly and humus–rich, in other words when it is richly "alive". *Stellaria media*, "chickweed", is regarded by the vintagers as the hallmark of a "good soil". The observation that it particularly flourishes in places where humus has been generously applied agrees with the experimental findings of LYRE (1957). Confronted by the strong nitrogen, as well as nitrogen–phosphate fertilizers commonly used, one must draw the conclusion that organic and inorganic additives are not equivalent. This finding leads to the question as to how far bio–indicators go and how valid even quantitative manifestations, such as indicator–value numbers are. We have begun to investigate this question, so far without any useful results (see the section "Investigations on the problem of nitrogen and tilth", page 426).

Relevés 8–14 represent stands affected by mulching practices. Here we are dealing with a "Kriechrasen" (see page 401) of trivial species; in Spring a multitude of golden yellow bands are formed from the hordes of *Taraxacum officinale*, later to become monotone green stripes. The rapidity of their development depends on whether the appropriate species such as *Poa trivialis, Trifolium repens*, and *Agrostis stolonifera* are already present in the plot. In dry plots *Agropyron repens* or even *Cynodon dactylon* may appear in this role. The strategies of the dominating grassland–species are clear: They form runners, above or below the ground, which escape destruction during mowing. As long as the carpet still has open patches or later is torn open anew, species taking root in the gaps have good chances. Thereto comes the fact that the soil becomes compacted from the frequent use of machines, at least in the tire tracks. The absence of soil movement is expressed in the presence of patches or even blankets of mosses, which additionally are herbicide–resistant (cf., Table 19.2). Today one still finds most frequently mixed stands of Geranio–Allietum and species of the Agrostietea (Relevés 8–13); we classify

	A							B			
Current Number	1	2	3	4	5	6	7	8	9	10	11
Exposition	SSO	SO	SSO	SSO	SO	SO	SO	S	S	SO	SO
Slope (°)	20	20	20	20	20	10	20	25	10	5	20
Zeile Z, Gasse G, Both +	Z	+	G	Z	+	+	+	+	+	Z	G
Area of relevé (× 10m^2)	5	7	8	8	10	10	5	6	10	6	10
Coverage (%):											
Cormophytes	10	50	30	80	70	90	15	97	65	50	15
Dead material										20	
Mosses	+	+	+	+	–	–	–	+	–	+	–
Total	10	50	30	80	70	90	15	97	65	70	15
Number of species:											
Cormophytes	15	25	30	22	20	25	26	33	35	29	22
Mosses	1	1	1	1	–	–	–	3	–	2	–
Total	16	26	31	23	20	25	26	36	35	31	22
Ch/D Geranio–Allietum vinealis											
Allium vineale	2m3	2m2	2m4	2m2	2m2	2m3	1.2	2m2	+	2m3	+
Muscari racemosum	2a4	2a2	2m4	2m2	·	·	+.2	2m3	·	1.2	·
D *Allium scorodoprasum*	·	·	·	·	·	2a4	·	·	·	·	·
D *Allium oleraceum*	·	·	·	·	·	·	·	1.2	·	·	·
D *Ficaria verna*	·	·	·	·	·	·	·	·	1.1	·	·
D *Ornithogalum umbellatum*	·	·	·	·	·	·	·	·	·	·	·
Ch Fumario–Euphorbion											
Veronica polita	·	1.2	·	1.2	2m2	+.2	+	+.2	·	1.2	1.2
Euphorbia helioscopia	+	+	+	1.1	·	+	1.1	·	+	+	·
Valerianella carinata	2m2	2a2	2m1	·	·	·	·	1.1	1.2	·	+
Fumaria officinalis	·	(+)	+	·	·	·	1.2	·	+.2	·	·
Geranium dissectum	·	+	·	·	·	·	·	1.1	1.2	·	+
Mercurialis annua	·	·	†	†	†	†	·	·	·	·	·
Thlaspi arvense	·	·	·	·	·	·	·	·	(+)	+	·
Euphorbia peplus	·	·	·	·	·	·	·	·	·	·	·
Ch O/K Stellarietea mediae											
Lamium purpureum	1.1	2m2	2m2	2a2	2a2	2a3	1.1	1.1	2m2	1.2	1.1
Veronica persica	1.1	2a2	2m2	1.3	2m2	2m2	2m2	2m2	2b4	·	·
Convolvulus arvensis	1.2	1.2	·	1.2	2m2	2m2	1.2	+.2	1.2	2a2	v
Veronica hederifolia	2m1	2m1	2m1	·	1.2	1.2	+	·	1.2	2m2	2m2
Senecio vulgaris	·	1.1	1.1	1.1	1.1	+	1.1	1.1	1.1	·	1.1
Sonchus asper + sp.	+	+	1.1	+	+	1.1	1.1	+	1.1	·	·
Stellaria media	1.1	1.2	2m2	2a3	2a3	·	1.2	+	+.2	·	1.2
Sonchus oleraceus	·	+	+	·	+	1.1	+	·	1.1	·	+
Lactuca serriola	+	+	1.1	·	(+)	·	+	·	·	·	·
Malva neglecta	·	(+)	·	1.2	·	·	1.2	+.2	·	·	·

									Synthetic characteristics						
									1990			1950/1954			
				B					a	b	c	d	e	f	g
12	13	14	15	16	17	18	19	20							
S	SO	SW	S	S	SSO	S	SSO	S							
20	5	5	10	10	20	10	20	10							
+	G	+	Z	G	+	+	G	G							
7	7	10	3	5	8	9	10	7							
65	40	80	85	70	55	20	85	95							
					10										
+	–	–	2	+	25	+	+	–							
65	40	80	85	70	90	20	85	95							
27	24	24	19	20	35	35	21	25							
2	–	–	3	1	3	1	3	3							
29	24	24	22	21	38	36	24	28							
2m2	2m1	1 . 2	·	·	1 . 2	2m3	2m2	1 . 2	18	V	134	19	V	55	7
+	1 . 2	·	·	·	2m2	·	2m2	·	11	III	78	16	IV	65	2
·	·	·	·	·	·	·	·	·	1	r	10	–	–	–	$\underset{\sim}{7}$
·	·	·	·	·	·	·	·	·	1	r	3	–	–	–	$\underset{\sim}{4}$
·	·	·	·	·	·	·	·	·	1	r	3	4	I	5	7
·	·	·	·	·	·	·	·	·	–	–	–	2	+	1°	$\underset{\sim}{5}$
·	1 . 2	1 . 2	·	·	·	·	1 . 1	+ . 2	12	III	35	17	V	37	$\underset{\sim}{7}$
·	·	·	+	·	+	·	·	·	9	III	13	19	V	10	$\underset{\sim}{7}$
1 . 1	·	·	·	·	1 . 1	·	·	·	8	II	43	1	r	1	X
·	·	+ . 2	·	·	·	1 . 2	·	·	6	II	10	12	III	8	$\underset{\sim}{7}$
·	·	·	·	·	·	·	·	·	4	I	8	10	III	13	$\underset{\sim}{5}$
·	·	·	·	·	·	·	·	·	4	I	D	15	IV	2	8
·	·	·	·	·	·	1 . 2	·	·	3	I	5	–	–	–	6
·	·	·	·	·	·	·	·	·	–	–	–	2	+	–	8
1 . 2	1 . 2	2a3	1 . 2	1 . 2	2m2	1 . 2	1 . 1	1 . 1	20	V	116	20	V	95	X
3 . 4	2a2	3 . 4	2m2	2a2	2m2	1 . 2	1 . 2	2m2	18	V	212	15	IV	26	$\underset{\sim}{7}$
·	v	2m2	2m2	2m2	2 . 2	2a2	†	+ . 2	18	V	D	18	V	8	X
·	2m2	2m2	1 . 2	2 . 1	·	+ . 2	+ . 2	2m2	16	IV	105	16	IV	23	$\underset{\sim}{7}$
1 . 1	+	1 . 1	1 . 1	1 . 1	2m3	·	·	+	16	IV	49	16	IV	16	8
+	·	$\underset{\sim}{†}$	+	1 . 1	2m1	·	·	$\underset{\sim}{+}$	15	IV	34	20	V	28	$\underset{\sim}{7}$
·	+ . 2	·	+ . 2	+ . 2	·	·	·	·	12	III	47	20	V	534	8
+	+	·	1 . 1	2m1	·	+	·	·	12	III	27	20	V	12	$\underset{\sim}{8}$
1 . 1	·	·	·	·	+	2m1	·	·	8	II	21	9	III	11	$\underset{\sim}{4}$
·	·	·	·	·	+ . 2	1 . 2	·	·	6	II	12	2	+	1	$\underset{\sim}{9}$

	A							B			
Bromus sterilis	·	·	·	·	·	2m2	·	·	1 . 2	1 . 3	·
Chenopodium album	·	·	·	†	·	†	·	·	†	·	·
Solanum nigrum	·	·	†2	1 . 1	·	·	·	·	·	·	·
Amaranthus retroflexus	·	·	·	·	·	·	·	†	†	·	·
Polygonum convolvulus	·	·	·	·	·	+ . 2	·	·	·	·	·
Geranium columbinum	·	·	·	·	·	·	·	+ . 2	·	·	·
Capsella bursa-pastoris	·	·	+	·	·	·	·	·	·	·	·
Sonchus arvensis	·	·	+	·	·	·	·	·	·	·	·
Galinsoga ciliata	·	·	+ . 2	·	·	·	·	·	·	·	·
Digitaria sanguinalis	·	·	·	†	·	·	·	·	·	·	·
Matricaria inodora	·	·	·	·	·	·	+	·	·	·	·
Sisymbrium officinale	·	·	·	·	·	·	·	+	·	·	·
Alopecurus myosuroides	·	·	·	·	·	·	·	·	1 . 2	·	·
Papaver rhoeas	·	·	·	·	·	·	·	·	·	·	·
Lathyrus aphaca	·	·	·	·	·	·	·	·	·	·	·
Vicia hirsuta	·	·	·	·	·	·	·	·	·	·	·
Geranium pusillum	·	·	·	·	·	·	·	·	·	·	·
Setaria viridis	·	·	·	·	·	·	·	·	·	·	·
Amaranthus lividus	·	·	·	·	·	·	·	·	·	·	·
Characteristic species of the Poa tr.–Lolio–Potentillion community											
Poa trivialis	·	2b3	+ . 2	+ . 2	·	1 . 2	2m2	·	2m2	1 . 2	1 . 2
Lolium perenne	·	·	·	+ . 2	·	·	1 . 2	1 . 2	+ . 2	·	+ . 2
Crepis capillaris	·	+	1 . 2	·	+	·	+	1 . 2	+	·	+
Trifolium repens	·	+ . 2	+ . 2	+ . 2	·	+ . 2	·	2m3	·	·	·
Agrostis stolonifera	·	·	·	·	·	1 . 3	·	·	·	·	+ . 2
Ranunculus repens	·	·	·	·	·	·	·	2a3	+	·	·
Ranunculus acris	·	·	·	·	·	·	·	+ . 2	·	+ . 2	·
Dactylis glomerata	·	·	·	·	·	·	·	·	1 . 2	+ . 2	·
Aegopodium podagraria	·	·	·	·	·	·	+ . 2	·	·	1 . 3	·
Galium album	·	·	·	·	·	·	·	+ . 3	·	(+)	·
Trifolium pratense	·	·	·	·	·	·	·	·	·	·	·
Glechoma hederacea	·	·	·	·	·	·	·	·	+ . 2	·	·
Poa pratensis	·	·	·	·	·	·	·	·	·	2a3	+ . 2
Urtica dioica	·	·	·	·	·	·	·	·	·	1 . 3	·
Rumex obtusifolius	·	·	·	·	·	·	·	·	·	·	·
Rumex crispus	·	·	·	·	·	·	·	·	+	·	·
Potentilla reptans	·	·	·	·	·	·	·	·	·	·	1 . 2

									Synthetic characteristics						
									1990			1950/1954			
B									a	b	c	d	e	f	g
2m3	1.3	·	·	·	·	·	·	+.2	6	II	30	3	I	7	$\underset{\sim}{5}$
·	†	†	·	·	·	†	·	·	5	II	D	–	–	–	7
·	·	·	·	·	†	·	·	·	3	I	D	12	III	1	8
·	·	·	·	·	·	1.3	·	·	3	I	D	1	r	–	$\underset{\sim}{9}$
·	·	·	·	·	·	+	·	·	2	+	2	1	r	1	X
·	·	·	·	·	1.2	·	·	·	2	+	4	5	II	5	entf.
·	·	·	·	·	·	·	·	·	1	r	1	1	r	1	5
·	·	·	·	·	·	·	·	·	1	r	1	4	I	3	X
·	·	·	·	·	·	·	·	·	1	r	1	–	–	–	$\underset{\sim}{7}$
·	·	·	·	·	·	·	·	·	1	r	D	–	–	–	4
·	·	·	·	·	·	·	·	·	1	r	1	–	–	–	6
·	·	·	·	·	·	·	·	·	1	r	1	–	–	–	$\underset{\sim}{7}$
·	·	·	·	·	·	·	·	·	1	r	3	–	–	–	7
+	·	·	·	·	·	·	·	·	1	r	1	2	+	2	6
·	·	·	·	·	·	·	·	·	–	–	–	2	+	2	X
·	·	·	·	·	·	·	·	·	–	–	–	2	+	2	$\underset{\sim}{3}$
·	·	·	·	·	·	·	·	·	–	–	–	1	r	1	$\underset{\sim}{7}$
·	·	·	·	·	·	·	·	·	–	–	–	1	r	–	$\underset{\sim}{7}$
·	·	·	·	·	·	·	·	·	–	–	–	1	r	–	8
2m2	1.2	2a3	2a3	3.4	1.2	1.2	2a3	2m2	17	V	140	3	I	3	7
1.2	+.2	+.2	·	1.2	+.2	1.2	+.2°	1.2	14	IV	26	3	I	3	$\underset{\sim}{7}$
2m1	·	+	·	·	1.2	·	·	1.2	11	III	28	13	IV	11	3
·	·	·	·	·	·	1.2	2m3	·	7	II	27	8	II	8	$\underset{\sim}{7}$
·	+.2	1.3	·	1.3	1.3	·	·	·	6	II	14	4	I	5	$\underset{\sim}{5}$
·	+.2	+.2	·	·	·	·	2m3	2a3	6	II	33	3	I	11	X
·	·	·	·	·	+.2	+.2	+	·	5	II	5	1	r	–	X
1.2	+.2°	·	·	·	·	·	·	(+)	5	II	11	4	I	4	$\underset{\sim}{6}$
2m2	·	·	·	·	·	·	2a4	·	4	I	24	3	I	44	8
·	·	·	·	·	·	1.2	·	(+)	4	I	6	1	r	D	X
·	·	·	·	·	+.2	+.2	+.2	+	4	I	4	4	I	4	X
·	·	·	·	·	·	+.2	·	1.2	3	I	5	–	–	–	$\underset{\sim}{7}$
·	·	·	·	·	·	1.3	·	·	3	I	14	1	r	D	$\underset{\sim}{7}$
·	·	·	·	·	·	1.3	+.2	·	3	I	7	3	I	2	$\underset{\sim}{8}$
·	·	·	1.3	+.2	·	·	·	+.2	3	I	5	4	I	4	9
·	·	·	·	+	·	·	·	·	2	+	2	17	V	20	$\underset{\sim}{5}$
·	·	·	·	·	1.3	·	·	·	2	+	6	1	r	D	$\underset{\sim}{5}$

	A							B			
Further concomitant species											
Taraxacum officinale	2m1	2a2	2m2	4.4	3.3	3.4	1.1	3.4̰	+	2b4	+
Poa annua ssp. annua	·	1.2	1.2	+	1.2	1.2	2m2	+.2	2a2	2m2	2a2
Cardamine hirsuta	·	1.2	·	+.2	·	2m2	2m2	·	2m2	2m3	2m2
Vicia angustifolia	+	(+)	+	·	+.2	·	·	1.1	+.2	+.2	·
Acrocarpi	+.2	+.2	+.2	+.2	·	·	·	2m3	·	1.2	·
Agropyron repens	·	·	·	1.2	·	+.2	+.2	·	+.2	1.4	+.2
Convolvulus sepium	·	·	+	·	·	·	2m1	·	2m2	2m1	·
Brachythecium rutabulum	·	·	·	·	·	·	·	+.2	·	+.2	·
Cirsium arvense	·	·	·	·	+	·	·	1.1	·	·	·
Galium aparine	·	·	·	·	·	+.2	·	+.2	·	2m3	·
Eurhynchium swartzii	·	·	·	·	·	·	·	+.2	·	·	·
Polygonum aviculare agg.	·	·	·	·	·	†2	·	·	·	+.2	·
Torilis arvensis	·	·	+	·	·	·	·	·	·	·	·
Clematis vitalba	·	·	·	·	·	·	·	+	·	r	·
Medicago lupulina	·	·	·	·	·	·	·	1.3	·	·	·
Quercus spec.	r	r	·	·	·	·	·	·	·	·	·
Cerastium glomeratum	(+)	·	·	·	·	·	·	+	·	·	·
Erigeron annuus	·	·	+	·	·	·	·	+.2	·	·	·
Prunus avium + sp.	·	·	±̰	·	·	·	·	·	·	·	·
Lolium multiflorum cf.	·	·	·	·	+.2	·	1.2	·	·	·	·
Veronica arvensis	·	·	·	·	+.2	·	·	·	·	·	·
Lathyrus pratensis	·	·	·	·	·	+.2	·	·	·	·	·
Epilobium parviflorum	·	·	·	·	·	·	·	·	+.2	·	·
Festuca arundinacea	·	·	·	·	·	·	·	·	·	1.3	·
Epilobium tetragonum agg.	·	·	·	·	·	·	·	·	·	·	+
Arrhenatherum elatius	·	·	·	·	·	·	·	·	·	·	·
Arenaria serpyllifolia	·	·	+.2	·	·	·	·	·	·	·	·
Daucus carota	·	·	+	·	·	·	·	·	·	·	·
Vicia cf. cracca	·	·	·	·	+.2	·	·	·	·	·	·
Erodium cicutarium	·	·	·	·	·	·	·	1.2	·	·	·
Picris hicracioidcs	·	·	·	·	·	·	·	[illegible]	·	·	·
Erucastrum gallicum	·	·	·	·	·	·	·	·	1.1	·	·
Diplotaxis muralis	·	·	·	·	·	·	·	·	1.2	·	·
Vicia sepium	·	·	·	·	·	·	·	·	·	+.2	·
Carex muricata agg.	·	·	·	·	·	·	·	·	·	+.2	·
Hypochoeris radicata	·	·	·	·	·	·	·	·	·	·	+
Lapsana communis	·	·	·	·	·	·	·	·	·	·	·
Sedum album	·	·	·	·	·	·	·	·	·	·	·
Sedum reflexum	·	·	·	·	·	·	·	·	·	·	·
Acer pseudoplatanus Klg.	·	·	·	·	·	·	·	·	·	·	·
Symphytum officinale	·	·	·	·	·	·	·	·	·	·	·
Triticum aestivum	·	·	·	·	·	·	·	·	·	·	·
Armoracia rusticana	·	·	·	·	·	·	·	·	·	·	·

									Synthetic characteristics						
									1990			1950/1954			
B									a	b	c	d	e	f	g
1.2	2a1	2a2	2a3	1.2	3.3	2a2	4.5	5.5	20	V	451	20	V	265	7̰
1.2	2m2	2m2	2m2	3.4̰	2m3	+.2	·	2m2	18	V	148	1	r	1	8̰
+	1.2	+.2	1.2	·	+	+	·	2m2	14	IV	74	–	–	–	7
+	+	·	+.2	+.2	+.2	1.1	·	·	13	IV	17	14	III	14	X
·	·	·	2m2	1.2	2a3	+.2	+.2	1.2	12	III	42	–	–	–	entf.
·	+.2	+.2	3.4	2a3	·	(+)	+.2	·	12	III	64	8	II	26	8̰
1.1	V	1.1	2m1	2m1	1.1	·	·	·	10	III	>60	14	IV	1	9̰
+.2	·	·	+.2	·	2m3	·	+.2	+.2	7	II	16	–	–	–	entf.
·	·	+	+	1.1	+	·	·	·	6	II	10	16	IV	34	7
·	1.2	·	1.2	·	·	2m4	·	·	6	II	28	2	+	2	8
+.2	·	·	+.2	·	2m3	+.2	·	+.2	6	II	17	–	–	–	entf.
·	·	†+	·	·	†+	†+	·	·	5	II	18	3	I	–	X
+	·	·	·	·	+	·	·	·	3	I	3	1	r	1	4̰
·	·	·	·	·	r	·	·	·	3	I	3	–	–	–	7
·	·	·	·	·	(+)	1.2	·	·	3	I	7	2	+	2	X
·	·	·	·	·	·	·	·	·	2	+	2	–	–	–	entf.
·	·	·	·	·	·	·	·	·	2	+	2	4	I	6	5̰
·	·	·	·	·	·	·	·	·	2	+	2	–	–	–	entf.
·	·	·	·	·	·	+	·	·	2	+	2	–	–	–	entf.
·	·	·	·	·	·	·	·	·	2	+	4	–	–	–	X
·	+	·	·	·	·	·	·	·	2	+	2	3	I	3	X
·	·	·	·	·	·	1.3	·	·	2	+	4	–	–	–	6
·	·	·	·	·	+.2	·	·	·	2	+	2	–	–	–	5̰
·	·	·	·	·	·	·	1.2	·	2	+	6	–	–	–	4̰
·	·	·	·	+	·	·	·	·	2	+	2	–	–	–	6̰
1.2	·	·	·	·	·	·	1.2	·	2	+	6	1	r	D	7̰
·	·	·	·	·	·	·	·	·	1	r	1	2	+	2	X
·	·	·	·	·	·	·	·	·	1	r	1	–	–	–	4̰
·	·	·	·	·	·	·	·	·	1	r	1	1	r	D	X
·	·	·	·	·	·	·	·	·	1	r	3	2	+	4	X
·	·	·	·	·	·	·	·	·	1	r	1	–	–	–	4̰
·	·	·	·	·	·	·	·	·	1	r	3	–	–	–	4
·	·	·	·	·	·	·	·	·	1	r	3	–	–	–	5̰
·	·	·	·	·	·	·	·	·	1	r	1	–	–	–	5̰
·	·	·	·	·	·	·	·	·	1	r	1	–	–	–	6̰
·	·	·	·	·	·	·	·	·	1	r	1	–	–	–	3̰
1.2	·	·	·	·	·	·	·	·	1	r	3	5	II	5	7̰
1.2	·	·	·	·	·	·	·	·	1	r	3	–	–	–	1̰
(+)	·	·	·	·	·	·	·	·	1	r	1	–	–	–	1̰
r	·	·	·	·	·	·	·	·	1	r	1	–	–	–	entf.
·	·	+.2°	·	·	·	·	·	·	1	r	1	–	–	–	8
·	·	(+)	·	·	·	·	·	·	1	r	1	–	–	–	entf.
·	·	·	+.2	·	·	·	·	·	1	r	1	–	–	–	9̰

	A							B			
Geranium molle	·	·	·	·	·	·	·	·	·	·	·
Solidago gigantea	·	·	·	·	·	·	·	·	·	·	·
Chelidonium majus	·	·	·	·	·	·	·	·	·	·	·
Plantago major	·	·	·	·	·	·	·	·	·	·	·
Hedera helix	·	·	·	·	·	·	·	·	·	·	·
Geum urbanum	·	·	·	·	·	·	·	·	·	·	·
Anthriscus sylvestris	·	·	·	·	·	·	·	·	·	·	·
Rumex acetosa	·	·	·	·	·	·	·	·	·	·	·
Amblystegium serpens	·	·	·	·	·	·	·	·	·	·	·
Plantago lanceolata	·	·	·	·	·	·	·	·	·	·	·
Prunella vulgaris	·	·	·	·	·	·	·	·	·	·	·
Potentilla sterilis	·	·	·	·	·	·	·	·	·	·	·
Cardamine pratensis	·	·	·	·	·	·	·	·	·	·	·
Valerianella olitoria	·	·	·	·	·	·	·	·	·	·	·
Fagus sylvatica Klg.	·	·	·	·	·	·	·	·	·	·	·
Ranunculus bulbosus	·	·	·	·	·	·	·	·	·	·	·
Crepis vesicaria	·	·	·	·	·	·	·	·	·	·	·
Festuca pratensis	·	·	·	·	·	·	·	·	·	·	·
Erophila verna	·	·	·	·	·	·	·	·	·	·	·
Cardaria draba	·	·	·	·	·	·	·	·	·	·	·

A = Geranio–Allietum vinealis
B = Penetration of G.–A. and Poa trivialis–Lolio–Potentillion community

									Synthetic characteristics						
									1990			1950/1954			
B									a	b	c	d	e	f	g
·	·	·	·	·	1 . 3	·	·	·	1	r	3	4	I	3	$\underset{\sim}{4}$
·	·	·	·	·	+ . 2	·	·	·	1	r	1	–	–	–	6
·	·	·	·	·	+	·	·	·	1	r	1	–	–	–	8
·	·	·	·	·	+	·	·	·	1	r	1	1	r	D	$\underset{\sim}{6}$
·	·	·	·	·	·	+ . 2	·	·	1	r	1	–	–	–	X
·	·	·	·	·	·	+ . 2	·	·	1	r	1	–	–	–	$\underset{\sim}{7}$
·	·	·	·	·	·	·	+ . 2	·	1	r	1	–	–	–	8
·	·	·	·	·	·	·	+ . 2	·	1	r	1	–	–	–	$\underset{\sim}{5}$
·	·	·	·	·	·	·	+ . 2	·	1	r	1	–	–	–	entf.
·	·	·	·	·	·	·	·	+ . 2	1	r	1	3	I	2	X
·	·	·	·	·	·	·	·	+ . 2	1	r	1	–	–	–	X
·	·	·	·	·	·	·	·	+ . 2	1	r	1	–	–	–	$\underset{\sim}{6}$
·	·	·	·	·	·	·	·	+	1	r	1	–	–	–	X
·	·	·	·	·	·	·	·	·	–	–	–	12	III	14	X
·	·	·	·	·	·	·	·	·	–	–	–	6	II	6	entf.
·	·	·	·	·	·	·	·	·	–	–	–	5	II	5	$\underset{\sim}{3}$
·	·	·	·	·	·	·	·	·	–	–	–	4	I	4	entf.
·	·	·	·	·	·	·	·	·	2	+	4	3	I	3	6
·	·	·	·	·	·	·	·	·	–	–	–	2	+	2	$\underset{\sim}{2}$
·	·	·	·	·	·	·	·	·	–	–	–	1	r	1	$\underset{\sim}{4}$

these as "penetrations" (= Durchdringungen). The conceptual threshold to a pure *Poa trivialis*–Potentillion community (Nr. 14) must be defined pragmatically. The pure grassland–aspect is not sufficient since the annual *Poa annua* ssp. *annua* co-determines it. Durchdringungen occur primarily when the carpet is still open at the time of a change in cultivation practices; then the Geranio–Allietum annuals can maintain themselves. In dry places this can last many years. Such mixed stands can also arise secondarily when the lawn is torn up for any reason, such as to improve the aeration of the soil, to introduce manure, or to smooth out irregularities caused by the activity of *Arvicola* (voles). Then the seed bank is activated. How long the geophytes can hold out is open; it is definitely several years with *Allium vineale, Ornithogalum umbellatum* and *Tulipa sylvestris*; it also depends in part on the intensity of mulching and must be more carefully investigated. Nonetheless the gradual weakening of the plants is unmistakable. Until recently, usually only the "Gassen" (the alleys between the rows of stocks) have been mulched, while under the "Zeilen" (the rows of stocks) herbicides were sprayed sparingly. Now under-stock mulching equipment is coming into use, yielding a closed grass carpet. This allows one to fear an enhanced regression of the Geranio–Allietum.

19.2.4 The development of the vineyard undergrowth in the "Gemarkung Niedereggenen" (Table 19.3 and Figure 19.2)

Community classification

Our attempt initially to organize our recent survey material according to the classification scheme used by HÜGIN in the 1950s failed. The tables presented here permit the reader to make his own attempt! It turns out that the *Erodium cicutarium* subassociation of the Geranio–Allietum, then frequent and evaluated as locally "dry to very dry", is absent today. Our relevés Nr. 1 and 5 could possibly be assigned to the middle Typicum without one readily being able to notice an ecological difference from the rest. To be sure, it is not discernible, for example, where Opalinuston (Dogger clay) just under the ground hinders drainage. For all that, nearly all our material can be assigned to the *Convolvulus sepium* subassociation or else as Durchdringung of the same by the "Kriechrasen" (see page 401) community. In the 1950s this subassociation was "sparsely developed" on the Niedereggenen Gemarkung. Indeed, the differentation was not especially good back then, besides which it was only represented by a few species in low quantity.

The indicator value of the *Erodium* group is no longer valid in this form. The differential species are generally rather rare in Markgräfler–Land vineyards today, with the exception of *Cardaria draba*. But just this species is found abundantly in the wettest flanks of neighboring Districts; according to statements from vintagers, it is favored by rotary tilling. Thereby, its strong storage roots are chopped up and their propagation is favored. The thick roots, able to advance straight down into

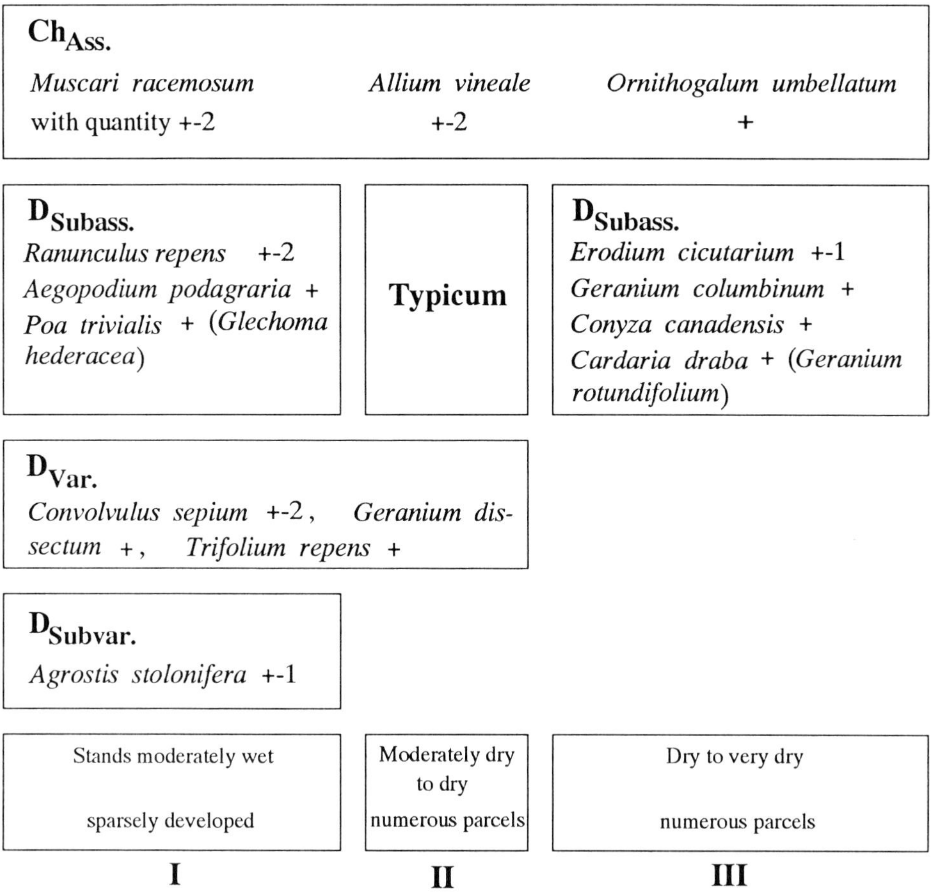

Figure 19.2: Diagram of the vineyard communities described by HÜGIN for Niedereggenen, simplified from his tables.

clay, should have a positive effect for the edaphon in the "Kriechrasen" (see page 401).

In interpreting this shift, mesoclimatic factors can be excluded, because the modification of the terrain morphology as a result of the land consolidation was insignificant in this area. The following factors seem much more plausible:

1. intensive fertilizing, which can have an effect as in the farmer's proverb "nitrogen replaces water";

2. compacted and less permeable soils with light surface accumulation of wa-

ter, as a consequence of increased usage of machines and decreased organic fertilizing;

3. luxuriantly flourishing competitors; in general, today one finds *Erodium*, as well as *Erophila*, almost exlusively in vegetation–poor border strips of parcels.

A whole array of "moisture indicators", most importantly *Poa trivialis*, but also others with above–ground runners, are quite likely favored relatively by mulching and more strongly impaired by soil movement, as compared to annuals and species with underground storage organs. *Glechoma hederacea* is quite resistant to spraying as well. All relevant species were already to be found in the 1950s in the immediate or a broader investigated area.

Number of species and homotoneity ("Homotonität")

The average number of species per 20 relevés is practically identical at 26.6 for HÜGIN's material and 26.85 for that of WILMANNS. If one only considers the material of Group A, at 23.86 today, there is a slight reduction. But it should be recalled that here only springtime relevés are included. The total number of species has, in contrast, increased: 77 vs. 109! Evidently the heterotoneity is somewhat higher today. This is also apparent in comparing the number of species in the individual presence degree ("constancy") classes (= Stetigkeitsklassen):

Presence degree class	1950/54	1990
V	10	7
IV	8	6
III	6	9
II	6	16
I (incl. r & +)	47	71

Nevertheless, the homotoneity at present is still good. One can test this by determining the "Stetenanteil" according to PASSARGE (e. g., 1985); the number of species in presence degree classes V and IV is divided by the mean number of species in the community, the result being expressed as a percentage. Thus one obtains a "Passarge–coefficient" of 67.7 for the 1950s, vs. 48.4 for the present. But since the 20 current relevés involve two basically different communities belonging to different classes, one should consider just the 7 relevés of Geranio–Allietum. The result for them is still a coefficient of 67.1! A value of around 50 is typical for Geranio–Allietum (WILMANNS 1990). Thus the values for Niedereggenen are pleasantly high.

The higher number of species is caused almost completely by "chance" trivial species from meadows and fringes; *Erucastrum gallicum* and *Diplotaxis muralis* should be noted as positive. Nonetheless, it is remarkable how well developed the vineyard flora still is, especially in view of the serious impoverishment of our cultivated fields!

Changes in presence and quantity of species

Of course, the above summarized finding does not mean that no quantitative changes have occurred. Quite the contrary, as can be seen in Table 19.4. Here are presented the pairs of numbers which were the most convincing to us, even if not all are statistically significant. Species are arranged into groups according to the regularity of their occurrence, i.e., their presence, and then additionally according to their quantities in the relevés. An interpretation can be made based on the dependencies on management methods, as discussed above.

Species which have decreased:

Group A1: *Rumex crispus* and *Cirsium arvense* are obvious, as species provided with underground storage organs and thus easy to drive out with purposeful application of systemic herbicides. We have no explanation for the fact that *Valerianella locusta* (still present in the Gemarkung today if scarce) has been replaced by the previously rare sister–species, *Valerianella carinata*. In fact, its phenological rhythm is at least very similar. The increase of *V. carinata*, however, fits well into our model of things (see below). Ecologically, *Stellaria media* resembles the rotting indicators of the next group.

Group A2: Here are to be found 4 character species of the alliance Fumario–Euphorbion, together with numerous tilth indicators. Again, we encounter here the question of the relevance of the input of organic material, discussed in section "Overview of the community types in Markgräfler Land" on page 411 and to be taken up again in "Investigations on the problem of nitrogen and tilth" on page 426. For *Muscari racemosum*, the frequently overdone rotarytilling of the 1960s may have played a role; thereby, clumps of soil were smashed up, and with them the bulbs. Geophytes also suffer as a consequence of the land consolidation, since they do not produce any seed bank worth mentioning (Fischer 1983). Where they had maintained themselves, they could increase because they could reproduce under the influence of spraying policies, as explained above; this makes the slight increase in the structural value (Bauwert) of *Muscari* understandable. Back then in the 1950s, the *Fagus* seedlings (presumably close to the forest) had no chance of developing, in contrast to other, present–day woody species.

Group B: Several "universal" mosses are placed here. (Under Acrocarpi are included all acrocarpous small mosses of the genera *Barbula, Bryum*, etc.) They may be considered "soil–rest" indicators. *Malva neglecta* and *Bromus sterilis* also owe their increase to the reduction in plowing and rotary tillage and to the increased application of herbicides and mulching (see also Wilmanns 1989).

As to the cormophytes, species like *Poa trivialis, Lolium perenne* and *Agrostis stolonifera, Ranunculus acris, R. repens* and *Trifolium repens* are mulch–favored. For *Lolium* and *Trifolium* a general advantage may intervene through seeding–in, also in neighboring parcels. Clear spraying indicators are present in the form of *Cardamine hirsuta, Valerianella carinata, Galium aparine, Clematis vitalba, Allium vineale* and *Polygonum aviculare* (see above). The causal relationship is especially clear for *Cardamine hirsuta*. On the basis of older floristic reports, this

Table 19.4: (Page 427) Species with substantial quantitative changes in the synthetic characteristics. Presence degree and "Bauwert" (based on springtime material): per 20 relevés in the "Gemarkung Niedereggenen". Decrease means decreasing number of species, increase means increasing number of species. AP absolute presence degree, PC presence degrees class, BW "Bauwert". For the explanation of the other symbols see Table 19.1. From G. HÜGIN and O. WILMANNS, respectively.

---

species was certainly present in this region, but was much less common than at present. This is also true of other rather distant areas, for example the Lucerne hinterlands (BRUN–HOOL 1981). The first individuals of this plant develop hesitantly from Summer, then the great majority follow in the winter halfof the year. If it is not eliminated at this time by turning the soil, as was done previously; its seeds will be dispersed already in March. If one brushes the flocks in April, with a crackle they spew out seeds in all directions; a turgor mechanism in the pods can throw the seeds reportedly as far as 1.4 m (MÜLLER–SCHNEIDER 1977). The quite extraordinary increase in *Poa annua* in its annual sub–species can also be attributed to the very early ripening of the fruit. In addition, thanks to its low growth, the meristem cone — indeed nearly the whole biomass — is not directly affected by mowing. *Taraxacum officinale* is known to be tolerant of mulching and contact herbicides, as well as occasional damage to the roots, in that it is rich in storage–materials and resprouts. Moreover, it can take advantage of the stronger fertilizing.

19.2.5 Investigations on the problem of nitrogen and tilth

Currently, the question of fertilizing with mineral nitrogen is especially crucial, particularly in grape–growing areas. This is because it has been recognized that excessive application has led to ground–water pollution. Thus efforts are being made to inform the vintagers, on the basis of local measurements, as to the current available nitrogen in springtime and the appropriate quantity of manure fertilizer to use. This situation and the evidence outlined above concerning the behavior of the so–called tilth indicators led us to ask the following question: Is there a correlation between the occurrence of *Stellaria media*, on the one hand, and the current amount of available nitrogen and a parameter for humus content and quantitity on the other? To answer this question we collected 40 samples of mixed soil in vineyards from places with and without *Stellaria media*. Using the methods described above in section "Available material, methods" (page 403), we determined the organic carbon content (as total–C minus carbonate–C) and the nitrogen content, and then calculated the C/N ratio.

The results are presented in Table 19.5. For plots where relevés were available, the average N–number has also been calculated from ELLENBERG's (1979)

	AP	PC	BW	AP	PC	BW
Group A: decrease		material of 1950/54			material of 1990	
A_1: according to presence and "Bauwert"						
Rumex crispus	17	V	20	2	+	2
Valerianella locusta	13	IV	14	–	–	–
Cirsium arvense	16	IV	34	6	II	10
Stellaria media	20	V	534	12	II	47
A_2: according to presence only						
Euphorbia helioscopia	19	V	10	9	III	13
Solanum nigrum	12	III	1	3	I	·
Sonchus oleraceus	20	V	12	12	III	27
Fumaria officinalis	12	III	8	6	II	10
Geranium dissectum	10	III	13	4	I	8
Veronica polita	17	V	37	12	III	35
Muscari racemosum	16	IV	65	11	III	78
Fagus sylvatica	6	II	6	–	–	–
A_3: according to "Bauwert" only						
Veronica hederifolia	16	IV	105	16	IV	23
Group B: increase						
B_1: according to presence and "Bauwert"						
Poa annua ssp. *annua*	1	r	1	18	V	148
Poa trivialis	3	I	3	17	V	140
Cardamine hirsuta	–	–	–	14	IV	74
Acrocarpi	–	–	–	12	III	42
Brachythecium rutabulum	–	–	–	7	II	16
Eurhynchium swartzii	–	–	–	6	II	17
Lolium perenne	3	I	3	14	IV	26
Valerianella carinata	1	r	1	8	II	43
Malva neglecta	2	+	1	6	II	12
Ranunculus acris	1	r	–	5	II	5
Galium aparine	2	+	2	6	II	28
Bromus sterilis	3	I	7	6	II	30
Ranunculus repens	3	I	11	6	I	26
Clematis vitalba	–	–	–	3	I	3
B_2: according to "Bauwert" only						
Taraxacum officinale	20	V	265	20	V	541
Veronica persica	15	IV	26	18	V	212
Allium vineale	19	V	55	18	V	134
Trifolium repens	8	II	8	7	II	27
Polygonum aviculare	3	I	–	5	II	18
Agrostis stolonifera	4	I	5	6	II	14

Table 19.5: (Page 429) Soil analyses from vineyards of 3 districts (Gemarkungen) of the "Markgräfler Land": Niedereggenen (N), Mauchen (M), and Auggen (A) (as noted in column 1).
Date of sampling was everywhere March 15, 1990. The order of the samples here is according to increasing C/N ratio. Columns 2–5: All data are expressed in $mg \cdot g^{-1}$ dry weight. Column 7: Location of sampling was in the immediate vicinity of *Stellaria media* (+) or, respectively, of other tilth indicators (Gare–Zeiger, GZ), of *Urtica dioica* (Ud), or none of these species nearby or even present on the sampling plot (-) (see text). Column 9: Notes: GA = Geranio–Allietum, PLGD = *Poa trivialis*-Lolio–Potentillion community with penetrations (Durchdringung). Column 10: a1/a2, b1/b2, etc.: either sample was taken from the same parcel. Note: In m1 amd m2, *Stellaria media* was present at some distance from the point of sampling. "·"means: no statement.

---

indicator–value table. This was done, notwithstanding the fact that these values were, for the most part, marked by the author with justification as still being tentative. (The numbers for the individual species in Table 19.3 are listed in column 9.) (In the sample pairs j1/j2 and k1/k2 the absence of *Stellaria* at that particular point may have been purely coincidental; naturally, not every species grows at every suitable point in the survey plot. In each of the other cases it was clear whether, over a larger homogeneous area, tilth indicators were present or absent.)

The results contradicted our expectations and leave us without definitive relationships! This pertains to the amount of organic–C, total N, the C/N ratio and — certainly no one would have guessed — the indicator–value numbers. In summary it can be said: The humus levels of 3–10% are not bad (rounding off and assuming a C–component of 50%), with C/N ratios of 10–20; thus our values lie in the normal range for agriculturally used soils. The numerical values on one and the same parcel may be very similar, but must not be. The humus content can be higher in the mulched alley than in the row, but it must not be. In the cases g1/g2 and l1/l2, where *Stellaria* grew only in the immediate vicinity of the individual young stocks (g1 and l2) in the midst of older ones (g2 and l1), we found less N right under the *Stellaria* than in the surroundings.

Any interpretations of the situation still seem too speculative at this time; we can only call attention to the problem.

19.3 Discussion and prospects

19.3.1 Facets of Plant Sociology

In non–European countries, with the exception of Japan (!), Plant Sociology or Phytosociology, if it is actually practiced at all, is still in the "inventory" stage.

1	2 C total	3 C $CaCO_3$	4 C organ.	5 N	6 C/N	7	8	9	10
M	43.8	35.1	8.7	1.19	7.3	–	·	·	a1
A	50.3	33.4	16.9	2.09	8.1	–	·	·	b1
A	58.6	37.0	21.7	2.27	9.5	+	·	·	·
N	49.0	20.3	28.7	2.95	9.7	+	6.3	GA	·
M	51.0	41.3	9.7	1.00	9.7	–	6.3	LPGD	·
A	52.4	35.2	17.2	1.74	9.9	+	·	·	c1
M	54.7	35.1	19.6	1.96	10.0	+	6.8	GAd1	·
A	58.8	32.3	26.5	2.62	10.1	+	·	·	·
N	23.3	0.3	23.0	2.27	10.1	–	·	·	e1
A	51.4	29.7	21.7	2.10	10.3	–	·	·	·
N	24.6	2.1	22.5	2.17	10.4	+	·	·	·
N	17.6	0.1	17.5	1.68	10.4	+	6.5	LPGD	f1
M	55.4	35.8	19.6	1.85	10.6	Ud	·	·	d2
N	28.1	1.4	26.7	2.51	10.6	–	7.0	LPGD	e2
A	56.9	31.7	25.2	2.37	10.6	+	·	·	·
N	59.7	32.1	27.6	2.57	10.7	–	6.5	GA	·
N	45.3	35.4	9.9	0.92	10.8	+	6.7	LPGD	g1
N	21.1	1.0	20.1	1.86	10.8	+	6.8	GA	h1
N	60.0	10.2	49.7	4.58	10.9	+	·	·	i1
N	45.5	9.5	36.3	3.33	10.9	+	7.0	GA	i2
N	22.3	1.4	20.9	1.90	11.0	+	·	·	h2
A	47.9	33.6	14.3	1.30	11.0	+	·	·	c2
N	20.6	4.1	16.5	1.49	11.1	GZ	·	·	·
A	69.2	30.9	38.3	3.41	11.2	+	·	·	·
A	57.3	30.5	25.8	2.36	11.3	+	·	·	·
M	43.0	30.5	12.5	1.11	11.3	–	·	·	·
N	34.7	1.9	32.8	2.85	11.5	–	6.5	LPGD	·
N	26.1	0.2	26.0	2.24	11.6	+	·	·	j1
N	25.5	0.1	25.4	2.18	11.7	–	·	·	j2
A	50.3	26.5	23.8	2.03	11.7	+	·	·	b2
N	47.2	0.4	46.7	4.00	11.7	–	·	·	k1
M	53.0	38.7	14.3	1.21	11.8	–	6.4	LPGD	a2
N	34.8	0.4	34.5	2.87	12.0	–	6.5	LPGD	f2
N	30.3	0.4	30.0	2.49	12.0	+	·	·	k2
N	44.4	4.1	40.3	3.20	12.6	–	·	·	l2
N	36.4	21.5	14.8	1.14	13.0	–	·	·	g2
N	45.2	6.1	39.1	2.67	14.6	+	·	·	l2
M	55.9	34.4	21.5	1.27	16.0	+	6.3	GA	m1
N	67.8	2.9	64.9	3.24	20.0	GZ	·	·	·
M	60.6	34.8	25.8	0.96	26.9	+	·	GA	m2

This involves making the first comprehensive surveying of stands (phytocoenoses), typing (to phytocoena, that is to defined vegetation units according to their species combinations) and the classification of the units into an "organized whole based on principles", as IMMANUEL KANT defined a system. Following BRAUN–BLANQUET, we also want to include the theory of formations in the conception of Phytosociology and put it on an equal status with Vegetation Science and Vegetation Ecology. It does not seem appropriate to us to speak of Phytosociology in a highly restrictive sense, i.e., only when one means a certain working method or just a school or, as regretably often occurs, only the primary description of communities and syntaxonomy. Why should one, without any sufficient grounds, perform a conceptual amputation and thereby delimit a term, here a whole science, in a different sense than that which the originator intended?!

At least in Central Europe a certain consolidation can be seen with respect to the "inventory", insofar as this pertains to an overview of higher syntaxa. This manifests itself, for example, in the fact that OBERDORFER and his co-workers were able to publish "Süddeutsche Pflanzengesellschaften" in four volumes (three have appeared so far, in 1977, 1978 and 1983). It is also made evident by PREISING's "Pflanzengesellschaften Niedersachsens" (alas, still unpublished), "Rote Liste der Pflanzengesellschaften Schleswig–Holsteins" of DIERSSEN (1983, 1988), and by the working groups under the aegis of the Reinhold–Tüxen–Gesellschaft, which are working out a comprehensive survey of the plant communities of the western part of the Federal Republic of Germany. Similarly comprehensive and detailed syntaxonomical work has been carried out recently respectively is in progress presently (e.g. in Austria, the ČSFR, Japan, the Netherlands). Conversely, progress on the "Prodromus der europäischen Pflanzengesellschaften" has not been as rapid as Reinhold TÜXEN had hoped when he initiated his project in 1970. It seems that it was contentually and geographically too comprehensive and really too ambitious, and it so far still only covers four of at least 60 classes (probably more).

It goes almost without saying, from the beginning many other facets of phytosociology have been investigated both qualitatively and analytic–quantitatively; to recall just the highlights, we should think of synecology and syndynamics, for example the studies in the Swiss National Park by BRAUN–BLANQUET and the soil scientist JENNY (1926), or any of the early, still strongly speculative considerations on succession and climax by CLEMENTS (1916), as well as the synchorological overviews, and — last but not least — the strong emphasis on the application of phytosociology for practical applications. Its support right up to the present, perhaps its very establishment, has been thanks to this last aspect. It is not just by chance that there have been two different, independent series with the same title, "Angewandte Pflanzensoziologie" (published in Stolzenau/Weser in W. Germany and Klagenfurt in Austria, respectively).

These research directions naturally need to be extended and deepened with modern methods. But it is also reasonable and necessary to take up more actively certain directions which have received little attention so far and can only be furthered interdisciplinarily. Examples here include the study of the development of communities in the course of the Earth's history and of cultural history as synevolution; further, population biology and its consequences for vegetation; and especially biocoenology, i.e., the interconnections of plants and animals, and ultimately also microorganisms in biocoenoses and ecosystems. What should be investigated here are not just the fluxes of matter and energy, but more the qualitative relationships which follow from the "requisite character" of the plant communities. Indeed they offer that multiplicity and variety of specific sites, specific sources of nutrition and — only recently brought to attention — specific signals, all of which permit the existence of the just as varied animal species (cf., WILMANNS 1987).

In the following section we would like to point out some of these fields for research, using examples from the vineyards, without trying to be complete. The division into different research directions must be somewhat arbitrary.

19.3.2 Symmorphology and symphenology

Symmorphology, according to BARKMAN (1979), is understood to mean the study of the horizontal, vertical and temporal arrangement of individuals in a stand (phytocoenosis). By building on this basic description one can also add the formation of types such as taxa, life and growth forms, as well as pollination and dispersal types, and finally their study in the communities (phytocoena). Thanks to BARKMAN (1979), we have a review of terms and concepts and of Dutch studies on the topic "vegetation texture and structure". According to him, we should understand texture to mean "the qualitative and quantitative composition of the vegetation with respect to different morphological elements ... regardless of their arrangement, whereas structure is concerned with the spatial (horizontal and vertical) arrangement (the architecture) of these elements". Thereby, the temporal periodicity inherent to phytocoenoses may also be brought in (note that successional steps are not intended to be included here). Depending on intepretation, one may also speak of texture and structure in a very broad sense by including functional relationships, such as competition. (In the zoological literature one even finds the term of structure fully freed from the relationship network with expressions such as "taxonomic structure" in the sense of "spectrum of taxa").

Thus symphenology is often incorporated, and rightly so. Indeed the close connection becomes especially clear in the vineyard herb community. We have (1989) set up a list of 11 morphological–phenological types for the modern Geranio–Allietum and its Durchdringung with the *Poa trivialis*–Lolio–Potentillion community, and it may well be possible to increase this number. Important parameters here are:

- Lifetime of individuals as a whole, lifetime of an individual leaf, lifetime of the individual in an assimilation–competent state, lifetime of the population in a green state; *Veronica persica*, for example, fruits after a few months. Thanks to overlapping generations, it is present the year around as an assimilation–

competent population, hence as an "evergreen". Such therophytes are known to be able to maintain themselves for decades as a seed bank. In contrast, *Veronica hederifolia* is restricted to the cool winter half–year, because it is a cold germinator.

- Position of the major part of its development in Winter or Summer. This is not necessarily coupled with any particular germination–temperature requirements, as shown by *Bromus sterilis*.
- Mode of survival, simultaneously the manner of restitution of the population: generative with a seed bank or vegetative with an underground bud bank, which can be provided by bulbs, rhizomes, tubers or roots.
- The regenerative capacity of the individual plays a decisive role in frequently disturbed places; the increase of "mulch species" on the one hand and "spraying indicators" on the other serve to illustrate this. If both assaults can be withstood, through a positioning of the buds close to the ground and through mobilization of reserve materials, the pertinent species become modern "facies–formers", e.g., *Convolvulus arvensis* and *Bromus sterilis*.

In all these cases phytosociology touches very close to population biology.

The super– and the subterranean structures also play an important role in practical problems. Here, as mentioned, the root conditions, as regards form and function, are still largely unclear.

The height of the plants influences the microclimate between the vines; high growing plants such as *Amaranthus retroflexus* increase the danger to the vines of molds. The short growth height of the cover–forming "classic" vineyard herbs, e. g., *Stellaria, Veronica* and *Lamium*, even with excessive fertilizing (in contrast to *Bromus sterilis, Galium aparine* and *Rumex*), has always been advantageous, besides which they also act as a ground protector at critical times.

19.3.3 Ecology

Phytosociology has given less interest to the usual research objects, soil and microclimate, in the case of the Geranio–Allietum than for studies of communities closer to nature. Competition problems have also been neglected. This is understandable: The intensive soil preparation, the generally high intensity of fertilizing, the more or less uniform mesoclimatic special situation (which, to be sure, has disappeared as a consequence of the expansion of grape cultivation since the 1960s), all have created a relatively homogeneous habitat for the wild plants. The highly bred grape cultivars are known to react with great sensitivity to the finest variations (cf. also MAY, BECKER and BOGENRIEDER 1985); is this also true for the wild plants? Without a doubt, some of them also show reactions, as shown by the example of the ever so "common" *Stellaria media*, which is evaluated as a class character species; these reactions often occur in vitality, which is influenced by so many factors and is thus not easy to interpret. It is an interesting question whether, in concrete cases, an individual of a wild plant population reacts just as finely to particular factors as one from a cultivated plant population consisting of genetically uniform individuals; their maintainance in a variety of locations might then be attributable to genetic polymorphism, thus to the flexibility of the population.

Bio–indication is one of the current problems of ecology; this is the utilization of living beings as indicators. That plants indicate certain environmental factors and their alteration has been known since pre–scientific times; certainly, without the ability to read this "language" no *Homo sapiens* could have developed! Admittedly, if one bores scientifically, many questions are still open, even if one does not reach the ecophysiological–cellular level.

In Central European agriculture the avoidance of over–fertilizing plays a major role currently. Here the herb layers can not only take on an indicator role, but may have a buffering function as well, since they can utilize temporary ion surges. It has been observed that in the vineyard the so–called "Stiellähme", a metabolic disturbance of the vine which can arise from strong fertilizing, is reduced by a well–developed herb layer (ARNETH 1979).

If one wants to let the wild herbs "speak" meaningfully, there are, to be sure, several questions to be answered, not only in principle, but also specifically for each species and individual factor. For example:

- Where in principle do the indication boundaries lie for a particular factor (e. g., threshold concentration of a substance)?
- How do the reactions of a species differ in different communities, i.e., in different background situations (including the competition)?
- How long does it take before a reaction becomes visible, and how long until a new equilibrium can be established in the communities?
- What role do the reproductive rate and possibilities for dispersal and establishment play thereby?
- How long can a population survive, although its habitat has changed and a new settlement is impossible?
- Conversely, how quickly does the restitution of a population or even the reconstitution of a community take place? In our case of the vineyard, where the possibilities for new settlement are favorable, this may occur much more quickly than in closed communities. As for "disturbance" and "recruitment", compare the critical and more extensive analysis of GRUBB (1988).

Clearly, a stronger connection of Phytosociology with zoological and microbiological soil biology is urgently called for.

19.3.4 Temporal change: syndynamics (succession theory) and synevolution

Floristic changes occur rapidly in the vineyard, because the frequent rigorous measures of the cultivator lead continually to a chance for new settlement; r–strategists find exceptionally good starting conditions. For the classical Geranio–Allietum we do not even want to speak of "biological instability", since the species combination is adapted even to the frequent mechanical disturbance; the "disturbance", i.e., the removal of biomass or even killing, pertains to the plants. However the habitat, as the totality of all factors acting on the plants, is not "disturbed"; rather, frequent movement of the soil is specific for it.

The vegetation is completely in a steady state with this anthropogenic habitat, as confirmed as well by the high proportion of species of high presence (see section "Overview of the community types in Markgräfler Land" on page 411). The vineyard vegetation and the natural vegetation in the strict sense (in fact mostly Asperulo–Fageten and Carici–Fageten) are really quite divergent; they are so different from each other that they share virtually no species in common. (*Corydalis* spp. and *Ficaria verna* would be those exceptions that prove the rule.) In analyzing successions several phenomena should be taken more into consideration and better distinguished than heretofore:

- A population newly appearing on the ground surface may have arisen either from an in situ seed bank or else from propagules from outside which have newly reached the plot (cf. GRUBB 1988). The second of these two processes is the pre–condition for a gradual enrichment of the flora and vegetation in destroyed areas. It probably comes into play more infrequently than is generally assumed, as shown by the observations and especially the experiments of FISCHER (1982).

- One must distinguish between the new appearance of a population (= recruitment), an enlargement (= expansion), and a strengthening (= densification, i.e., more intensive settlement as quantitative increase in an already settled spot). (The distinction between expansion and densification depends, of course, on the definition of the term population). Individual species differ from one another in this ability to a remarkable degree. This also plays a role in the evaluation of management measures and thus for nature conservation. *Cardamine hirsuta*, for example, today obviously overcomes expansion, as well as densification, with ease. *Tulipa sylvestris*, on the other hand, can deal with densification in that it forms daughter bulbs on short stolons. It cannot achieve new settlement at all; and expansion only over distances of decimeters — and that by vegetative means. New settlements are favored by r–strategies; densification can be coped with by r– and k–strategists. These demographic strategies (WHITTAKER and GOODMAN 1979) have a close causal relationship with dispersal strategies.

The short–term conversion of phytocoenoses of already existing phytocoena through succession can be compared to synevolution, as a mostly long–term formation of new types. (To be sure, the Agroforms arise exceptionally rapidly!) Synevolution consists, on the one hand, of the evolution of new taxa, mostly species, and on the other hand, in the novel constellation of already existing species in communities. The fact that, in the adaptation to agriculture, new taxa have arisen is shown by the well–known crop mimetics, weeds with camouflage (such as *Silene linicola* and *Echinochloa crus–galli* var. *oryzicola*); that they are still arising today is demonstrated by the notorious acquisition of herbicide resistance (for example, by *Chenopodium album*).

From which habitats and from which communities do the individual weed species originate? For the field weed *Tripleurospermum inodorum* the original biotope can be established: The diploid cytotype (ssp. *maritimum*) lives in the Cakiletea communities on maritime shores; the tetraploid (ssp. *inodorum*) has been shown from sub–fossils to have arisen in cultivated fields (review by WILLERDING 1986). It may have happened similarly in the case of the hexaploid *Veronica hederifolia* var. *hederifolia* found in our Geranio – Allietum, because the tetraploid var. *lucorum* belongs to the fringe inhabitants, for which living space was surely available in the pre–neolithic landscape.

The original habitats for the sub–Mediterranean geophytes still remain to be investigated more closely. Escaped and naturalized Medieval decorative and medicinal plants play a role in the vineyard land in the form of *Tulipa sylvestris, Ornithogalum umbellatum, O. nutans* and *Aristolochia clematitis. Arenaria serpyllifolia* and *Sedum album*, among others, come from thin–soil, rocky habitats. Viewing a short lifetime as a predisposition from the natural original habitat is illuminating; the ability to bloom and bear fruit already as a plantlet a few cm high (*Amaranthus retroflexus, Mercurialis annua, Senecio vulgaris*) is an advantage under xerothermic conditions. For the *Sedum* species, in contrast, herbicide tolerance plays the decisive role in the vineyard; it would be interesting to learn its physiological basis and determine whether there was also a predisposition here.

19.3.5 Sigmasociology: vegetation complexes

Spatial patterns within the stands and communities fall under the heading symmorphology. But there is a principally different pattern formation: its components are the stands (phytocoenoses), therefore, respectively, the communities (phytocoena). They form "mosaics", which we more generally call vegetation complexes. We can typify them like the communities, and they may form combinations of a higher order and scale (complexes of higher degrees). We give the following examples.

The strips so characteristic of modern vineyards (not in all areas, but in many) depend on management methods: mulch strips, accordingly with "grassland" in the alleys, vs. spray–induced Agroforms in the rows. It is not like this everywhere. In the drier Steigerwald, for example, one is more cautious with mulching, and correspondingly the individual parcels are much more homogeneous (WILMANNS 1990). Since the owners cultivate differently – not necessarily in principle but certainly in detailed practice – the parcels, as such, again form patterns of a second degree; to be sure, these are more irregular.

The contact communities in the vineyard regions are quite different. (Here vineyards are meant in a broader sense, as being a piece of the landscape). The geologically determined landscape forms determine the kind of the "embedding" of the vineyard plots in the surroundings. Slopes in areas with hard rocks required terracing in order to reduce the sloping of the cultivated ground proper, hence for example, the low walls in the Muschelkalk region on the Neckar River, with its xerothermic flora and fauna. In loess areas inclination of the cultivated ground should only be maximally 5°; so terraces were set up here. The slopes were overgrown with mown Mesobrometa, and after mudslides Diplotaxi–Agropyreta developed, at least in the Kaiserstuhl (FISCHER 1982). In the course of the major land consolidation by the responsible state agencies, low walls and small terraces were destroyed on a large scale. In Keuper areas the land took revenge through enorm erosion the length of the broken–up, no longer articulated slopes. The overall geomorphological and biological impoverishment of such regions is notorious.

It is possible to record this phytosociologically quite well in that one registers the communities and their fragments, including living, as well as non–living, individual structures (such as trees, vintagers' huts, etc.) analogously to the way one records the species of a stand. Then these relevés must be worked up and thereby assigned to types of community complexes. (Sigma or Σ is meant in the sense of the sum of the communities; for a comprehensive presentation, see TÜXEN 1979.) The vegetation complexes so defined are quite well suited for use as a basic framework for faunistic investigations; this was demonstrated for the birds of the vineyard regions by SEITZ (1981, 1988).

Finally, we can go one step further and ask about the "embedding" of the whole vineyard terrain in the countryside, i.e. about vegetation complexes of higher order. The bordering forests are, as a rule, meso– to xerophytic; still they often bear witness to a previous use by the farmers, for here one used the litter (fallen leaves) as fertilizer for the vines, there one cut vine posts of *Castanea vesca* and *Robinia pseudacacia*. The wood shortage in the anciently and heavily settled, dry regions tended to be so general and great that the forests were terribly overused (for the Kaiserstuhl, see WILMANNS and BOGENRIEDER 1986). A modern characteristic of the vineyard landscape in the vicinity of large cities is the transformation of small lots for recreational use. Still another embedding pattern arose, for example on the middle Rhine, where many poorly accessible small plots, managed only in the ways of our forefathers, have gone fallow. They have been taken over by Prunetalia thickets. There are still many sigmasociological investigations here, awaiting study in conjunction with historical, geographical and zoological problems.

19.3.6 Biocoenology

Basically, this is the specifically biological aspect of ecosystem research. Considering its great scientific and practical significance, this research direction has thus far received far too little attention. But this situation appears to have improved recently; the establishing of a working group for biocoenology (under the aegis of the "Gesellschaft für Ökologie", through the initiative of A. KRATOCHWIL) is indeed more than a symbol (see, esp., KRATOCHWIL 1987).

In the biocoenological consideration of the vineyard, one can only with difficulty avoid the question of beneficial and detrimental animals. Granted, there have been some claims regarding "the birds", but these stand on weak footing, as it is only since SEITZ's studies (1981, 1988) that we even know anything about the avicoenosis of vineyards, as regards the species composition and utilization of this living space. They depend critically on the imbedding pattern of the plots present. This is because the vines themselves are the actual breeding habitat for only one species of bird, the linnet (*Carduelis cannabina*); in contrast, the scattered shrubs, the *Clematis* vines and high forbs on slopes play an important role as breeding and feeding habitats. Where they take up a major portion of the vineyard land, as they do in the Kaiserstuhl, they affect the biological situation quite differently than they do in undivided areas. Swarms of starlings (*Sturnus vulgaris*), and to a minor extent thrushes (*Turdus* spp.) in the vicinity of forests, on occasion cause economic damage through berry consumption. Here it is not the breeding population itself, rather migratory birds or autumnal birds of passage, that gorge themselves on the berries. The starling question offers a typical example of biocoenological and sigmasociological interrelationships. The birds prefer non–articulated open landscapes free of bushes or trees, where they can practice their enemy–confusing flight behavior of swooping down to skim just above the ground. And such areas, at least in the Kaiserstuhl, are just the recently consolidated large tracts.

It can not be definitively stated just how great the contribution of the vineyard birds to a biological regulation of pests is or can be. But at the moment, serious attempts are being made to reduce the application of insecticides, for example by inhibiting the egg deposition of the "Traubenwickler" (= grape–leaf roller, 2 spp. of Lepidoptera) with the aid of pheromones. Thus one should also encourage bird wild–life, especially summer–breeding birds. Quite independently of this, species like the red–back shrike (*Lanius collurio*), and the stonechat (*Saxicola torquata*), among others, require protection and support anyway. From SEITZ (1981) we know about the differentiated requirements for structural bush types; the space requirements for strips of such plantings should be easily acceptable, given the current overproduction of winee.

It is known that predatory mites are important helpers in the fight against pest arthropods. As a nutritional supplement, they also require pollen. The question arises here, as to how far one can further their survival by influencing the kind of undergrowth. This would involve favoring as much as possible a multi–seasonally flowering herb layer, not just one that covered the ground vegetatively. Finally, a finding of ENGLERT (1978) should be mentioned as promising for the future, namely, that the larvae of pest noctuids only go over onto the vine stocks when they have no herbal plants available (such as *Senecio vulgaris*, and the like).

The demand for a reduction in the application of biocides can be taken just as well as a demand for the intensification of biocoenological research.

Acknowledgment

We thank Dipl. Biol. W. H. Müller once more for his reliable and capable technical help and Dr. R. Cassada for the translation of our text.

Bibliography

ARNETH, A. G., 1979, Untersuchungen über die Eignung von dürreresistenten Pflanzen und von Mulchschichten im Weinbau.
Diss. Gießen.

BARKMAN, J. J., 1979, The investigation of vegetation texture and structure.
In: WERGER, M. J. A. (ed.): The study of vegetation, p. 125–160. The Hague–Boston–London, Junk.

BARTSCHER, W.; SCHMIDTS, W., 1963, Beitrag zur relativkonduktometrischen Kohlendioxidbestimmung.
Z. analyt. Chem. **203**, 168–178.

BRAUN–BLANQUET, J.; JENNY, H., 1926, Vegetationsentwicklung und Bodenbildung in der alpinen Stufe der Zentralalpen (Klimaxgebiet des Caricion curvulae).
Denkschr. Schweiz. Naturf. Ges. **63**, 183–349.

BREMNER, J. M., 1960, Determination of nitrogen in soil by the Kjeldahl method.
J. Agric. Sci. **55**, 11–33.

BRUN–HOOL, J., 1981, Pflanzenschmuck an Bauernhaus und –hof im Luzerner Hinterland.
Heimatkd. Wiggertal **39**, 113–196.

CLEMENTS, F. E., 1916, Plant Succession: An analysis of the development of vegetation.
Carnegie Inst. Washington **242**, 1–512.

DIERSSEN, K., 1983, 1988, Rote Liste der Pflanzengesellschaften Schleswig–Holsteins.
Schriftenr. Landesamt Naturschutz Landsch. pfl. Schl.–H. **6**, 157 pp. + append. (2nd edition)

ELLENBERG, H., 1979, Zeigerwerte der Gefäßpflanzen Mitteleuropas.
2nd edition Scripta Geobot. **9**, 122 pp.

ENGLERT, W. D., 1978, Die Bedeutung von Weinbergsunkräutern als Nährpflanzen für Noctuidenlarven.
Mitt. dtsch. Ges. allg. angew. Entomol. **1**, 29–30.

FISCHER, A., 1982, Mosaik und Syndynamik der Pflanzengesellschaften der Lößböschungen im Kaiserstuhl (Südbaden).
Phytocoenologia **10**, 73–256.

FISCHER, A., 1983, Wildkrautvegetation der Weinberge des Rheingaus (Hessen): Gesellschaften, Abhängigkeit von modernen Bewirtschaftungsmethoden, Aufgaben des Naturschutzes.
Phytocoenologia **11**, 331–383.

GRUBB, P. J., 1988, The uncoupling of disturbance and recruitment, two kinds of seed bank, and persistence of plant populations at the regional and local scales.
Ann. Zool. Fennici **25**, 23–36.

HÜGIN, G., 1956, Wald–, Grünland–, Acker– und Rebenwuchsorte im Markgräflerland.
Diss. Freiburg i. Br.

KRATOCHWIL, A., 1987, Zoologische Untersuchungen auf pflanzensoziologischem Raster — Methoden, Probleme und Beispiele biozönologischer Forschung.
Tuexenia **7**, 13–51.

LYRE, H.–H., 1957, Beiträge zur Biologie und Ökologie der Vogelmiere, *Stellaria media* L.
Diss. Hohenheim.

MAY, T.; BECKER, N.; BOGENRIEDER, A., 1985, Vergleich verschiedener Methoden zur Beurteilung der lokalklimatischen Wärmegunst von Reblagen.
Die Wein–Wissenschaft **6**, 371–393.

MÜLLER–SCHNEIDER, P., 1977, Verbreitungsbiologie (Diasporologie) der Blütenpflanzen.
2nd edition Veröff. Geobot. Inst. ETH, Stiftg. Rübel, Zürich **61**, 226 pp.

NELSON, D. W.; SOMMERS, L. E., 1980, Total nitrogen Analysis of Soil and Plant Tissues.
J. Assoc. Off. Anal. Chem. **63** (4), 770–779.

OBERDORFER, E., 1983, 1990, Pflanzensoziologische Exkursionsflora.
5th edition 1051 pp., 6th edition 1050 pp., Stuttgart, Ulmer.

OBERDORFER, E. (ed.), 1977, 1978, 1983, Süddeutsche Pflanzengesellschaften. 2nd edition part I–III, 1097 pp., Jena, Fischer.

PASSARGE, H., 1985, Gagea pratensis–Allium oleraceum–Ass. Tuexenia **5**, 107–112.

SEITZ, B.–J., 1982, Untersuchungen zur Koinzidenz von Vegetationskomplexen und Vogelgemeinschaften im Kaiserstühler Rebgelände. Tuexenia **2**, 233–255.

SEITZ, B.–J., 1988, Zur Koinzidenz von Vegetationskomplexen und Vogelgemeinschaften im Kulturland — Untersuchungen im südwestdeutschen Hügelland. Phytocoenologia **16**, 315–390.

TÜXEN, R., 1979, Sigmeten und Geosigmeten, ihre Ordnung und ihre Bedeutung für Wissenschaft, Naturschutz und Planung. Biogeographica **16**, 79–92.

VOGT, E.; GÖTZ, B., 1987, Weinbau. 7th edition 452 pp., Stuttgart, Ulmer.

WAGNER, R., 1970, Neue Aspekte zur Stickstoffanalytik in der Wasserchemie. Vom Wasser, **37**, 263–318.

WALTER, H.; LIETH, H., (1960–1967), Klimadiagramm–Weltatlas. Jena, G. Fischer.

WESTHOFF, V.; VAN DER MAAREL, E., 1973, The Braun–Blanquet approach. In: Whittaker, R. H. (ed.), Ordination and classification of communities. Hb. Veget. Sci. **5**, 617–726, Junk, The Hague.

WHITTAKER, R. H.; GOODMAN, D., 1979, Classifying species according to their demographic strategy. I. Population fluctuations and environmental heterogeneity. Amer. Nat. **113**, 185–200.

WILLERDING, U., 1986, Zur Geschichte der Unkräuter Mitteleuropas. 382 pp., Neumünster, Wachholtz.

WILMANNS, O., 1975, Wandlungen des Geranio – Allietum in den Kaiserstühler Weinbergen? — Pflanzensoziologische Tabellen als Dokumente. Beitr. naturk. Forsch. Südwest–Dtld. **34**, 429–443.

WILMANNS, O., 1987, Zur Verbindung von Pflanzensoziologie und Zoologie in der Biozönologie. Tuexenia **7**, 3–12.

WILMANNS, O., 1989, Vergesellschaftung und Strategie–Typen von Pflanzen mitteleuropäischer Rebkulturen. Phytocoenologia **18**, 83–128.

WILMANNS, O., 1990, Weinbergsvegetation am Steigerwald und ein Vergleich mit der im Kaiserstuhl.
Tuexenia **10**, im Druck.

WILMANNS, O.; BOGENRIEDER, A., 1986, Veränderungen der Buchenwälder des Kaiserstuhls im Laufe von vier Jahrzehnten und ihre Interpretation – pflanzensoziologische Tabellen als Dokumente.
Abh. Landesmus. Münster **48**, 55–79.

Chapter 20

The dissipation of energy through soil invertebrates in wheat field and meadow

L. RYSZKOWSKI

20.1 Introduction and area description

Agroecosystems are managed to provide food supplies for man. In order to obtain this goal man controls or modifies many effects of the natural driving variables in agroecosystems. This is usually accomplished by simplification of the system structure in order to focus, as far as possible, only on food–producing components. Thus, a lot of energy subsidies are spent by man to control the composition of primary producer species and their yield efficiencies, as well as to protect cultivated plants against pests, severe climatological conditions, and so on.

The effects of these agricultural measures on the functioning of the total agroecosystems are still poorly understood. This situation especially concerns basic functional characteristics as energy flow in different types of cultivated field soils.

Interdisciplinary studies on energy and matter fluxes in agroecosystems (RYSZKOWSKI 1974, 1975, 1979) carried out in Turew situated about 40 kilometers south of Poznan (Western Poland) by the research Centre for Agricultural and Forest Environment Studies of the Polish Academy of Sciences, provided long–term data on primary production as well as estimates of density, biomass and respiration of the entire animal community. In order to examine the changes in the pattern of energy flow evoked by agriculture, through soil animals a cultivated field was compared with a meadow, where the soil had not been ploughed. Thus, soil animals in the meadow were living under conditions of much less agricultural activity than the soil animals in the cultivated field.

The soil in cultivated fields is mainly composed of light, loamy sand, whilst that of meadows is mainly hydromorphic. The organic carbon content ranges from 0.55% (in fields) to 1.95% (in meadows). The difference in elevation between slightly inclined cultivated fields and flat meadows does not exceed 2 meters. The areas of the field and meadow are 13.4 ha and meadow 3.6 ha respectively. The ground water table depth is related to the elevation and ranges from 0.5 m to 3.0 m below the terrain surface.

The area climate is the warmest in Poland, with a mean annual temperature of 8 °C. The vegetation growing season (with an air temperature of over 5 °C) lasts for 225 days (from 21 March to 30 October). The climate is also one of the driest in Poland. The mean annual precipitation is 527 mm, of which 375 mm falls in the vegetation growing period (RADOMSKI, LYKOWSKI and MADANY 1974).

The region in which the studies were carried out is located about 40 km south of Poznan and is renowned for its old and superior traditions of agriculture. The average dose of mineral fertilizer is about 290 kg NPK per hectare. The obtained yields are high and, in the case of four cultivated cereals (rye, wheat, barley and oats) amount to 3.2 – 4.0 $t \cdot ha^{-1}$.

20.2 Primary production

20.2.1 Methods

The annual primary production of the wheat field was estimated by the summation of the biomass increments; including the production of the shed parts, weeds, wheat regrowth from scattered seeds after harvesting, catch–crop (yellow lupine — *Lupinus luteus* L. and white mustard — *Sinapis alba* L.) cultivated for green manure, all of which were produced from the beginning of September to the end of October. Roots were sampled to a depth of plough layer (0.3 m). Details of the sampling procedure are described by KUKIELSKA (1973 and 1975). The above–ground meadow primary production was estimated by the summation of the biomasses harvested in three cuttings. The increment of biomass from the live standing crop at the end of winter (16 March), which had been produced the previous year, was added to the annual yield of hay and the biomass remaining on the ground that had not been cut during harvest (stubble to a height of 0.1 m). The autumn increase of biomass, after the third cutting, was also included. Root production, to a depth of 0.3 m, was estimated by subtraction of the biomass detected at the end of winter (March 16) from the peak value achieved on June 26. The meadow production was

Table 20.1: Annual primary production of wheat cultivation.

Elements of production	production [g·m^{-2}] dry weight
Above–ground:	
grain	373.0
harvested straw	488.5
stubble	134.4
shed stems and leaves	46.1
weeds	65.3
Below–ground:	
wheat	202.6
weeds	11.3
Above and below–ground:	
wheat regrowth and catch crop	138.0
Total	1,459.2
Total reduced by the weight of sown seeds	
for cultivation of wheat	–25.0
catch crop	–14.0
Total primary production	1,420.2
removed yield	861.5
input to soil	597.7

under–estimated, because the rate of plant decomposition (especially the roots) was not determined. Thus, the changes in the root biomass indicated the gross trend to either increase or decrease (after June 26) resulting from the balance of production and decomposition processes (WIEGERT and EVANS 1964).

20.2.2 Values of annual primary production and inputs of plant biomass for energy requirements of heterotrophs

The largest components of primary production in wheat cultivation were straw and grain, removed from field as yield, which amounted to 60% of total production (Table 20.1).

The additional production of catch–crop was equal to 9.7% of the annual field production, increasing the input of plant biomass into the soil. The plant biomass, plowed into the soil after the removal of the yield, amounted to 558 g dw$\cdot$m^{-2}, but if the catch-crop had not been cultivated and the field was laid fallow during autumn, then the input of organic matter into the soil would have been 420 g dw $\cdot$m^{-1}.

The production of weeds was equal to 75.9 g dw$\cdot$m^{-2}, of which the above-ground production amounted to 65.3 g dw$\cdot$m^{-2} per square meter. The most abundant species of weeds found in the wheat field were: *Apera spica-venti* L., *Agropyron repens* L., *Stellaria media* VILL., *Matricaria chamomilla* L., *Centaurea cyanus* L. and *Poa annua* L.. The contribution of weeds to the total primary production amounted to 5.3%.

The total biomass produced during the year amounted to 1,459.2 g dw$\cdot$m^{-2} (Table 20.1). But it should be kept in mind that, to raise cultivation, wheat seeds were sown in the soil. In order to obtain the net annual production, therefore, the weight of seeds equal to 25 g dw$\cdot$m^{-2} should be subtracted from the produced biomass. The same applies to the catch-crop. Thus corrected, the annual net primary production of the wheat field was equal to 1,420.2 g dw$\cdot$m^{-2}. The estimated value of the primary production is similar to the result obtained by PASTERNAK (1974) who estimated the annual primary production of a wheat field near Warsaw at 1,537 g dw$\cdot$m^{-2}. The value of the plant biomass input to the soil (597.7 g dw$\cdot$m^{-2}) is very similar to the value estimated by PASTERNAK (1974) which was equal to 575.8 g dw per square meter.

The dominant species of plants in the meadow were: *Lolium multiflorum* LAM. showing a frequency of 30%, *Festuca pratensis* HUDS.) – 25% and *Lolium perenne* L. with a frequency of 22%. Frequency values for *Trifolium pratense* L., *Dactylis glomerata* L. and *Poa pratensis* ranged from 8 to 5% (KOZLOWSKI, S. personal communication). The annual primary production in the meadow was equal to 1,336.2 g dw$\cdot$m^{-2} (Table 20.2) and was comparable to the estimates obtained for similar meadows (COUPLAND 1979; STEEN 1985).

In the studied meadow, the below-ground production was 52% of the total production. In the yield (hay) 40% of the primary production was removed in three cuttings. Frequent mowing of the above-ground growth prevented losses from elimination of shoots or leaves by different natural causes leading to the decomposition of plant material. One can assume, therefore, that the above-ground meadow production was more accurately estimated than that for the below-ground production. The method of estimating the production of the peak biomass used in this study did not allow for the detection of root production in late summer or early autumn. The total amount of plant material remaining in the ecosystem for use by heterotrophs, was approximately 805.2 g dw$\cdot$m^{-2}.

In order to estimate the energy value of the plant biomass in kilojoules (kJ) left in the studied ecosystems after the removal of yields a factor of 16.72 was used (GOLLEY 1961). Thus, the plant biomass left in the meadow, out of the total primary production, was 13,462.9 kJ$\cdot$m^{-2} and, in the field with wheat cultivation,

Table 20.2: Annual primary production in meadow.

Elements of production	production ($g \cdot m^{-2}$ dry weight)
Sum of 3 swaths	$210.0 + 179.4 + 141.6 = 531.0$
Difference between biomass of stubble left after mowing and biomass of live grass at the beginning of spring	$83.2 - 33.1 = 50.1$
Autumn increase after third cutting	$143.5 - 83.2 = 60.3$
Difference between root peak biomass (June 26) and root biomass at the beginning of spring (March 16)	$1,460.5 - 765.7 = 694.8$
Total production	1,336.2
Biomass remaining in ecosystem	805.2

9,993.5 $kJ \cdot m^{-2}$. Thus, the energy input for maintaining the heterotrophs metabolic activity was higher in the meadow than in the cultivated field by 135%.

20.3 Total community of soil invertebrates

20.3.1 Methods of sampling and evaluating energy flow

Information on the biomass structure of the soil–animal community, under wheat cultivation and in meadows, was gathered in 1979 not only for all animal groups concerned but also in the same period as when the primary production was estimated. These studies were made by several scientists involved in the project on the ecological effects of intensive agriculture coordinated by the Research Centre for Agricultural and Forest Environment Studies (RYSZKOWSKI 1974, 1975, 1979). Here, reference will only be made to the main source of information on the season–long estimation of standing crops for various animal groups, as well as a short description of the sampling methods used. The average density of *Protozoa* for the vegetation season was obtained by PAPROCKI (1985), using the extinction dilution method of CUTLER (1929). Their biomass was evaluated on the basis of the mean specific gravity and the average cell volume of the dominant groups (HEAL 1971).

The Nematoda were sampled by a soil corer and extracted from the soil by centrifuge and flotation techniques (WITKOWSKI 1966). The mean density was estimated by WITKOWSKI (personal communication) and ZAMSZYN (1986). The biomass was approximated by ANDRASSY'S (1956) method.

The *Enchytraeidae* were sampled by soil corers by RYL (1980, 1985), using the O'CONNOR (1965) extraction method. The biomass was estimated by the DUNGER (1968) method.

The *Lumbricidae* density was estimated by RYL (1984) using the method of ejection from soil using formalin solution (SATCHELL 1971).

The *Acarina* density was estimated by SELL (1985) using soil corers and the Murphy split-funnel method of extraction (WOOD 1965). The biomass was determined on the basis of body measurements using the conversion factors of MACFADYEN (1963).

The *Collembola* mean density was estimated by CZARNECKI (1979), using soil corers and were extracted by the Murphy split-funnel method. The biomass was estimated as in the case of *Acarina*.

The density of soil-living insects was estimated by KARG (1985) and NABIALCZYK-KARG (1985) taking soil monoliths and then subjecting them to the flotation method (NABIALCZYK-KARG 1980). The body weight of invertebrates larger than *Enchytraeidae* was estimated by using suitable balances. The energy cost of maintenance was approximated, using the equations showing the relationship between live body weight and respiration, developed by the following authors: HEMMINGSEN (1960) for *Protozoa*, O'CONNORS (1967) for *Enchytraeidae*, BERTHET'S (1964) for *Acarina* and for the rest of the invertebrates, the equation of RYSZKOWSKI (1975) was used. Corrections for different thermal conditions of animal life were introduced by assuming $Q_{10} = 2$.

20.3.2 Biomass structure and energy cost of soil maintenance

The total soil fauna in a cultivated field is almost five times lower than that in an adjacent meadow (Table 20.3).

A similar large difference was observed when the biomass of soil animals in a meadow steppe soil was compared with that of a closely situated cultivated wheat field (FRENCH et al. 1979). RYSZKOWSKI (1985) comparing biomasses of soil fauna (without *Protozoa*) has shown that by analysing 15 available estimates, the biomass of soil animals is larger in forests and grassland ecosystems than in cultivated fields. Thus, one can conclude that the biomass difference observed between total communities of soil animals in cultivated fields and meadows, reflects the process of fauna impoverishment caused by agricultural activity (RYSZKOWSKI 1985).

The most striking impoverishment of soil animals is observed amongst insect larvae which means that the season-long biomass is more than 78 times higher in meadows than in cultivated fields (Table 20.3). Tillage prevents the development of a rich insect larvae community in the soil. Large size larvae, whose development is interrupted by ploughing, are often eliminated (TISCHLER 1971, EDWARDS and LOFTY 1975, RYSZKOWSKI 1981). In the case of the studied cultivated field,

Table 20.3: Mean standing biomass (mg dw$\cdot$m^{-2}) and annual dissipation of energy (kJ$\cdot$m^{-2}) of soil animals

Taxon	wheat field		meadow	
	biomass	dissipation of energy	biomass	dissipation of energy
Protozoa	440.0	1,190.0	525.0	1,360.0
Nematoda	557.4	314.6	816.6	499.5
Lumbricidae	990.0	56.0	3,840.0	254.2
Enchytraeidae	50.9	26.5	198.6	88.7
Acarina	11.2	1.3	305.7	22.6
Collembola	47.4	20.5	65.6	21.1
Winged insects mainly larvae	62.2	14.6	4,883.0	223.7
Total	2,159.1	1,623.5	10,634.5	2,469.8

ploughing eliminated large larvae of *Coleoptera* similar to *Elateridae* as *Agriotes* sp. and among *Diptera*: *Tipulidae, Bibionidae, Muscidae* and large larvae from other taxons (NABIALCZYK–KARG 1985, KARG 1985). Thus, the biomass of *Diptera* larvae, living in the meadow, was on average, 195 times greater than that of those in the wheat cultivation soil (KARG 1985). Tillage and the lack of litter covering the soil is a reason for an almost four–fold decrease of earthworms in cultivated field soil in comparison with that in meadows (RYL 1984). The great difference between the mean biomasses was also detected amongst *Acarina* (Table 20.3). This group of animals is also effectively eliminated by tillage (MILLS and ALLEY 1973, EDWARDS and LOFTY 1975, HENDRIX et al. 1987).

The biomasses of *Protozoa, Nematoda* and *Collembola* showed smaller differences between cultivated fields and meadows as the animal groups discussed above (Table 20.3). The smaller biomasses in cultivated fields can therefore be explained mainly by the lower input of plant biomass to the system providing food for their metabolic activity. It seems that their existence is threatened less by tillage than insect larvae, earthworms and mites.

Beside the impoverishment of the soil animal community RYSZKOWSKI (1985) shows another important change in the functional characteristics of soil fauna, due to agricultural impacts. Smaller body sizes were detected in many groups of soil animals when the body weight distribution of animals living in cultivated fields was compared with those of meadows. The differences in the mean body sizes were particularly noticeable in the winged insect larvae in meadow soil, being about

7.4 times heavier than the mean body weights of larvae found in wheat field soil (RYSZKOWSKI 1985). Among particular taxonomic groups, the differences in the mean body sizes were even larger. E. g. the mean body weight of *Muscidae* larvae found in meadow soil was equal to 12.5 mg dw per specimen, whilst that of *Muscidae* larvae sampled in wheat field soil amounted to 0.54 mg dw per specimen (NABIALCZYK–KARG 1985). After analysing the distribution of the mean body weights in 14 families of Diptera the bigger larvae were always found in meadow soil rather than in cultivated field soil (NABIALCZYK–KARG 1985). Thus, species showing bigger body sizes are eliminated with increased tillage. Comparing biomasses of soil animals in ploughed and unploughed soils, HENDRIX et al. (1987) detected much larger differences in macroarthropods and earthworms than in Protozoa and nematodes. Thus, one can assume that ploughing is the main factor in eliminating soil macrofauna. Only large plant pests, such as *Agrotis* sp. (*Lepidoptera*) feeding on plant roots and *Cantharidae* (mobile predators) show no smaller body sizes in cultivated fields.

The body sizes of *Nematoda, Enchytraeidae, Acarina* and *Collembola* living in cultivated field soil were found to be smaller than those living in meadow soil, although the differences were not as great as in the case of the winged insect larvae. The opposite distribution of mean body weights in specimens were detected in earthworms (RYSZKOWSKI 1985). Thus, species living in meadows are characterized by larger body sizes in the majority of soil animal taxons. One can conclude, therefore, that the agricultural activity of man creates conditions which can be survived by smaller and rapidly developing animals. It can be said, in other words, that agriculture is favourable for r–species, in the terminology of MACARTHUR and WILSON (1967).

Because of the well–known inverse relationship between body weight and the rate of metabolism, the amount of energy dissipated for maintaining living processes in soil animals (energy losses in respiration) does not follow the pattern of biomass distribution in various distinguished groups of fauna. The highest contribution to the dissipation of energy was shown by *Protozoa* (Table 20.3).

Second in rank, but much lower than in *Protozoa*, were the energy amounts dissipated by *Nematoda*. Winged insects living in meadow soil made up 45% of the total soil community biomass structure, but their contribution in terms of dissipated energy amounted to only 9%. The same was true for earthworms. In terms of biomass, the contribution to the total community living in meadow soil was 36%, whilst in terms of dissipated energy, it was only 10%.

The biomass of the total soil–animal community, being equal to 2,159.1 mg dw$\cdot$m^{-2}, was almost five times lower than the biomass of soil–animals in the meadow, which amounted to 10,634.5 mg dw$\cdot$m^{-2}. But because the distribution of body sizes in soil–animals is altered by tillage, in favour of smaller animals, the amount of energy dissipated by soil animals was equal to 1,623.5 kJ$\cdot$m^{-2} and was only 1.5 times lower than the energy dissipated in the meadow (2,469.8 kJ$\cdot$m^{-2}). Thus, to maintain one unit of biomass in the cultivated field, much more energy is

Table 20.4: Utilization of available energy by soil animals in $kJ \cdot m^{-2} \cdot a$.

Ecosystem	wheat field	meadow
Primary production	23,745.7	22,341.2
Input into soil	9,993.5	13,462.9
Energy dissipated by soil animals	1,623.5	2,469.8
Ratio of energy dissipated to energy input	0.16	0.18

needed than in the meadow.

20.4 Dissipation of energy by soil–animals

The farmer left 39.4% of the annual primary production in the studied cultivated field and in the meadow he left 60.3%. A more abundant soil–animal community was established in the meadow ecosystem, because of the higher input of organic matter and the lack of soil disturbances caused by ploughing. The mean season-long biomass of soil–animals was almost five times higher in the meadow than in the cultivated field. Despite this great difference in biomasses, the amount of energy dissipated for the maintenance of living processes in animals differed only by a factor of 1.5 in two studied ecosystems. The ratio of energy dissipated by the soil–animals in the cultivated field to the energy value of the phytomass left in the agroecosystem (energy available for heterotrophs) was equal to 0.16 (Table 20.4).

Thus, 16% of the energy available was dissipated in metabolic processes of the soil–animals. The remainder was used by microbes and animals living above ground or stored in newly formed humus. The contribution of animals living above-ground relative to the total animal community is small in the studied agricultural landscape (Ryszkowski 1975, 1979, 1981). A similar ratio of dissipated energy to available energy, equal to 0.18, was found in the meadow (Table 20.4). Thus, the efficiency of energy used by soil–animals was the same, despite the enormous difference in the biomasses of soil fauna in these two ecosystems. The results of the analysis indicate that the input of available energy in the studied ecosystems is more directly correlated with the amount of energy dissipated by soil–animals than with biomasses. The ratio of the biomass of soil–animals in the cultivated field (2,159.1 mg $dw \cdot m^{-2}$) to the available plant biomass (597,700 mg $dw \cdot m^{-2}$)

was equal to 0.0036, whilst in the meadow this ratio was equal to 0.0132. The complicated relationship between the input of organic matter and the heterotroph energy dissipation, the energy demands of humification and the various types of tillage are still badly understood. The results of reported studies indicate that, although there is a positive relationship between the input of organic matter to soil and the biomass of soil fauna, the cause is not direct. The distribution of body sizes in the animal community can efficiently moderate this relationship. The lower input of organic matter to cultivated field soil, coupled with ploughing, not only conditions a lower biomass but also enhances this effect by altering the body size distribution causing higher energy costs of biomass unit maintenance.

Acknowledgments

This study was carried out within the "Natural bases for protection and management of agricultural landscapes" project (CPBP 04.10.03), supported by the Ministry of Education in Poland.

Bibliography

ANDRASSY, J., 1956, Die Rauminhalts- und Gewichtbestimmung der Fadenwürmer (Nematoden).
Acta Zool. Acad. Sci. Hung. **11**, 1–15.

BERTHET, P., 1964, L'activité des Oribates d'une chemie.
Mem. Inst. Roy. Sc. Nat. Belg. **152**, 1–152.

COUPLAND, R. T., 1979, in: Coupland, R. T. (ed.), Grassland ecosystems of the world. Conclusion.
Cambridge University Press, Cambridge, pp. 335–355.

CUTLER, D. W.,1920, A method for the estimation of the number of active protozoa in the soil.
J. Agric. Sci. **10**, 135–143.

CZARNECKI, A., 1979, Density of soil mesofauna (Acari Collembola) in the fields under normal and simplified plant rotation.
(in Polish), XII Conference of Polish Zoological Society, Poznan, Poland, 3 – 6 September 1979, 30–31.

DUNGER, W., 1968, Die Entwicklung der Bodenfauna auf rekultivierten Kippen und Halden des Braunkohlentagebaues.
Abh. Ber. Naturk. Mus. Gorlitz **43**, 1–256.

EDWARDS, C. A.; LOFTY, J. R., 1975, The influence of cultivations on soil animal populations.
in: Vanek, J. (ed.) Progress in soil zoology, Academia, Prague, 399–407.

FRENCH, N. R.; RYSZKOWSKI, L.; TITLYANOVA, A. A.; ZLOTIN, R. J., 1979, Comparison of organic matter structure in diverse ecosystems.
(in Russian), Izvestya Sibirskovo Oddielenya Akademii Nauk SSSR. **2**, 8–22.

GOLLEY, F. B., 1961, Energy values of ecological materials.
Ecology **42**, 581–584.

HEAL, O. W., 1971, in: Philipson, J. (Ed.) IBP Handbook No. **18**, Methods of Study in Quantitative Soil Ecology. *Protozoa*. Blackwell, Oxford, 51–57.

HEMMINGSEN, A. M., 1960, Energy metabolism as related to body size and respiratory surface and its evoluting.
Rep. Steno Hosp., Copenhagen, **9**, 7–110.

HENDRIX, P. F.; CROSSLEY, D. A.; COLEMAN, D. C.; PARMELEE, R. W.; BEARE, M. H., 1987, Carbon dynamics in soil microbes and fauna in conventional and no-tillage agroecosystems.
Intecol Bulletin **15**, 59–63.

KARG, J., 1985, Impact of crop rotation on soil insect larvae.
Intecol Bulletin **12**, 95–101.

KUKIELSKA, C., 1973, Primary productivity of crop fields.
Bull. Acad. Polonaise Sci. **21**, 109–115.

KUKIELSKA, C., 1975, The effect of fertilization on the primary production of rye and potato fields.
Polish Ecological Studies **1**, 17–26.

MACARTHUR, R. H.; WILSON, E. O., 1967, The theory of island biogeography.
Princeton Univ. Press, Princenton N. Y.

MACFADYEN, A., 1963, Animal ecology.
Pitman and Sons, London.

MILLS, J. T.; ALLEY, B. P., 1973, Interactions between biotic components in soil and their modification by management practices in Canada.
Can. J. Plant Sci. **53**, 425–441.

NABIALCZYK–KARG, J., 1980, Density and biomass of soil inhabiting insect larvae in a rape field and in a meadow.
Polish Ecological Studies **6**, 305–316.

NABIALCZYK-KARG, J., 1985, Differentiation of soil insect larvae in agroecosystems.
(in Polish), Prace Komisji Naukowych Polskiego Towarzystwa Gleboznawczego **90**, 30–36.

O'CONNOR, F. B., 1965, Extraction of enchytraeid worms from a coniferous soil. Nature **175**, London, 815–816.

O'CONNOR, F. B., 1967, The Enchytraeidae.
In: Burges, A. and Row, F. (eds.), Soil Biology, Academic Press, London,

PASTERNAK, D., 1974, Primary production of field with winter wheat.
Ekol. Pol. **22**, 369–378.

RADOMSKI, CZ.; LYKOWSKI, B.; MADANY, R., 1974, Preliminary climatological and agrometeorological characteristics of cultivated fields in the neighbourhood of Turew.
In: Ryszkowski, L. (ed.), Ecological effects of intensive agriculture, Polish Scientific Publishers, Warszawa, 7–17.

RYL, B., 1980, Enchytraeid (*Enchytraeidae, Oligochaeta*) populations of soils of chosen crop fields in the vicinity of Turew (Poznan Region).
Polish Ecological Studies **6**, 277–291.

RYL, B., 1984, Comparison of communities of earthworms (Lumbricidae) occuring in different ecosystems of agricultural landscape.
Ekol. Pol. **32**, 155–165.

RYL, B., 1985, Qualitative and quantitative differentiation of enchytraeids in agroecosystems and shelterbelt.
(in Polish), Prace Komisji Naukowych Polskiego Towarzystwa Gleboznawczego **90**, 36–43.

RYSZKOWSKI, L., (ed.), 1974, Ecological effects of intensive agriculture.
Polish Scientific Publishers, Warszawa.

RYSZKOWSKI, L., 1975, Energy and matter economy of ecosystems.
in: van Dobben W. H., Lowe-McConnell, R. H., (eds.), Unifying concepts in ecology, Junk, Hague, 109–126.

RYSZKOWSKI, L., 1979, Croplands.
in: Coupland, R. T., (ed.), Grasslands ecosystems of the world, Cambridge University Press, Cambridge, pp. 301–331.

RYSZKOWSKI, L., 1981, Effects of agriculture intensification on fauna.
Zeszyty Problemowe Postepow Nauk Rolniczych **233**, 7–38.

RYSZKOWSKI, L., 1985, Impoverishment of soil fauna due to agriculture. Intecol Bulletin **12**, 7–17.

SATCHELL, J. E., 1971, Earthworms. In: Phillipson, J., (ed.), IBP Handbook **18** Methods of study in Quantitative Soil Ecology, Blackwell, Oxford, 107–127.

SELL, D., 1985, Differentiation of biomass numbers and respiration of soil *Acarina* in agroecosystems. (in Polish), Prace Komisji Naukowych Polskiego Towarzystwa Gleboznawczego **90**, 43–50.

STEEN, E., 1985, Root and rhizome dynamics in a perennial grass crop during an annual growth cycle. Swedish J. agric. Res. **15**, 25–30.

TISCHLER, W., 1971, Agrarökologie. (Polish translation), Panstwowe Wydawnictwa Rolnicze i Lesne, Warszawa.

WIEGERT R. G.; EVANS, F. C., 1964, Primary production and the disappearance of dead vegetation on an old field in southeastern Michigan. Ecology **45**, 49–63.

WITKOWSKI, T., 1966, Structure of nematoda communities living in soil of cultivated fields. (in Polish), Studia Societatis Scientiarum Torunensis **8**, 75–127.

WOOD, T. G., 1965, Comparison of a funnel and a flotation method for extracting Acari and *Collembola* from moorland soils. Pedobiologia **5**, 131–139.

ZAMSZYN, Z., 1986, Numbers and biomass of soil nematodes on alfalfa, meadow and shelterbelt. (in Polish), Zeszyty Problemowe Postepow Nauk Rolniczych **322**, 311–332.

Part V

Eutrophication and pollution in terrestrial systems

Chapter 21

Lead tolerance of annuals at roadsides

U. HELLMUTH and W. SCHMIDT

21.1 Introduction

The increase of anthropogenically caused heavy metal contamination leads to serious toxic effects in exposed plants. A few plant species show a particular ability to establish themselves on such sites, either on the basis of a fundamentally existing insensitivity ("inherent tolerance", MCNAUGHTON et al. 1974), or by means of evolving heavy metal tolerant ecotypes. Several plant groups seem incapable of achieving either the one or the other of these possibilities. These groups include gymnosperms and tree–like angiosperms. Therefore, the lack of many annual plants on heavy metal sites and ore slag heaps, for example, could be caused by their inability to evolve any tolerance (ANTONOVICS et al. 1971, ERNST 1985).

However, the vegetation and soil close to high traffic volume motorways have a high heavy metal content, in particular lead, but nevertheless produce a good stock of annuals (STOTTELE and SCHMIDT 1988, ULLMANN and HEINDL 1989, SCHMIDT 1990). Perennial plants and mosses which have evolved lead tolerance have also recently been found on such sites (BRIGGS 1972, WU and ANTONOVICS 1976, HELMING and RUNGE 1979). Thus, it seemed to be of interest to test whether populations of annual plants from such habitats differ with regard to lead tolerance from control populations grown on unaffected places.

21.2 Materials and methods

Three annual plant species were investigated: *Poa annua* L., *Polygonum aviculare* L. (= *P. aequale* LINDM. = *P. arenastrum* BOR.), and *Anagallis arvensis* L. These are typical for roadside vegetation (STOTTELE and SCHMIDT 1988, ULLMANN and HEINDL 1989) and can be found growing on the sites investigated. Their concept of growth, continuously and alternately bringing forth vegetative and reproductive organs, ensures a low influence of the ontogenetic stage of development on the results of the experiment.

Four habitats were chosen in the Land of Hessen in the Federal Republic of Germany:

- a motorway with high traffic density (70,000 motor vehicles/24 h) and eight lanes in the industrial agglomeration of Southern Hessen (A–l),
- a four lane motorway with moderate traffic density (15,000 motor vehicles/24 h) in an agricultural area of Northern Hessen (B–l),
- and two uncontaminated areas on a track across the fields far from the motorways with comparable soil properties (A–n, B–n).

The total lead content in the soil on these sites (HNO_3–extract, high pressure method) and the results of 1n NH_4–acetate–(pH 4.8)–extracts are summarized in Table 21.1. The latter can be regarded as an approximate reflection of the ability of the plants to take up this element from the soil solution.

All plants were collected as seedlings. In the case of the heavy traffic sites the seedlings originated up to a distance of 20 cm away from the edge of the road surface (shoulders). Tolerance tests were carried out by means of water cultures. The thoroughly rinsed plant material was first cultivated in pure quartz sand until the root length was sufficient for transference to hydroculture vessels. Groups of four plants, of approximately the same size, were put into 1.5l–vessels of a synthetic substance. The nutrient solution corresponded to the instructions of BAUMEISTER and ERNST (1978), using the method of SHIVE at a concentration of 25% for the first week. At the beginning of the actual tolerance test, the plants were transposed every other day from a 50% concentrated SHIVE nutrient solution to a solution containing lead nitrate. Calcium nitrate was added to the latter to ensure that the ionic strength was about the same in each solution. The test was carried out in five different regular stages (0, 15.6, 31.2, 46.8, and 62.4 mg Pb^{2+}/l). Each stage was represented by three parallel test vessels, i.e. twelve plants. They were grown in a growth chamber at 15–20°C with a 14h day period; the pH was adjusted to 5.6; all solutions were aired.

The experiment was continued as long as there was enough space for the root system of those plants showing the highest biomass production and as long as

Table 21.1: Lead content of the soil in the growth habitats at 0–5cm and 5–10cm soil depth. Habitats:

A Industrial agglomeration of Southern Hessen
B Agricultural landscape of Northern Hessen
l Habitats bordering the motorways
n Habitats on field tracks far away from the motorways

growth habitat	total lead content [mg Pb^{2+}/kg dry soil]				NH_4–Acetate–extract [mg Pb^{2+}/kg dry soil]			
	0–5 cm		5–10 cm		0–5 cm		5–10 cm	
	$\overline{x}$	$\pm s$	$\overline{x}$	$\pm s$	$\overline{x}$	$\pm s$	$\overline{x}$	$\pm s$
A–l	1578	198	897	128	69.8	2.8	42.1	7.3
A–n	21.5	2.8	17.8	7.9	2.5	0.5	2.1	0.8
B–l	483	34	119	19	16.1	3.3	5.8	0.9
B–n	11.2	4.4	9.1	4.4	1.0	0.7	1.8	0.5

the differences between the separate lead concentration stages could be clearly distinguished. *Polygonum aviculare* was harvested after 36 days, *Poa annua* after 44, and *Anagallis arvensis* after 48 days. The material was separated into roots and shoots. Then the plants were dried at 105°C for at least 48h, separately packed for each culture vessel. The increase in root and shoot dry weight in the culture was calculated.

After the material was crushed, it was broken down by acid digestion under high pressure with concentrated HNO_3. Lead contents were determined via atomic absorption spectroscopy. Both the flame and the flameless methods were used, employing standard addition.

21.3 Results

21.3.1 Lead concentration in the soil

The measured concentrations of lead in the soil are primarily correlated with traffic congestion in the growth habitats. The highest concentration of lead was found in the upper layer from 0 to 5cm of the soil bordering the very busy motorway in Southern Hessen. In the adjoining layer of soil from 5 to 10cm, the lead concentration was halved.

Compared with Southern Hessen, the lead content of the upper soil layer of the motorway in Northern Hessen was about a third. This is in accordance with the traffic density at this site which is about a fourth. With increasing soil depth, the lead concentrations here decrease more clearly than in Southern Hessen. Besides the lower traffic volume, this may also reflect the lower age of this motorway. It was opened to traffic in 1971 whilst the motorway in Southern Hessen has been in use since 1935.

The lead contents of the soil from the sites far away from the roads lie close together and are typical for less contaminated soils in Central Europe. The measurements indicate a higher lead burden for the industrial agglomeration in Southern Hessen.

The soil extractant NH_4–acetate could dissolve 3 to 5% of the total lead content in the case of the sites close to the motorways, and in the case of the uncontaminated sites, about 10%, so that a range of 0 to 70mg Pb^{2+}/kg soil dry weight available for the plants could be postulated. These concentrations correspond to the lead content used for the culture solutions.

21.3.2 Plant growth and lead content

The values of the dry weight results are listed in Table 21.2. As a rule, the increase of shoots and roots was equal. The plants answered the increasing level of lead with a decrease in growth. The most sensitive criterion for lead tolerance was shown to be root growth. Thus, the dry weight increase of roots is used to explain the results in detail. As it could be shown that, obviously independent of lead tolerance, specific ecotype differences in growth were present, an index of Pb tolerance (relative growth index) was estimated as the mean dry weight produced in a (lead containing/leadless) concentration stage (Fig. 21.1). The smaller the value of the index, the higher the sensitivity of a plant to the lead burden. This enables direct comparison of plants from different sites within a concentration stage.

With increased levels of lead in the solution, both *Polygonum* and *Poa* showed a more or less continuous decrease in dry weight. The levels of lead in both plant species increased almost regularly, although the concentration in the roots was much higher than in the shoots (Fig. 21.2). The plants originating from both the contaminated and non–contaminated sites behaved in the same manner; differences in lead tolerance were not apparent.

However, *Anagallis* clearly reacted differently, depending on its origin: whilst the root tolerance index of the first lead–containing stage from plants far away from motorways fell to a value of 0.32 and 0.25 respectively, it remained at 0.83 in the case of the moderately frequented motorway (Fig. 21.1). The plants originating from the roadside with the highest lead contamination produced more than double their root dry weight in comparison to the control value. This fact indicates a clearly distinct lead tolerance of *Anagallis arvensis* on both roadsides. This was directly visible in the constitution of the plants, especially the roots. In the low lead concentration stage II (Table 21.2), they seemed to be completely (A–l) or almost completely (B–l) unaffected. However, the roots of the two reference habitat plants showed the typical marked damages under lead influence. First brown discolouration of the root tips and cortex area could be observed, followed finally

Table 21.2: Dry weight increase of shoots and roots during lead tolerance tests (values in mg dry matter/culture vessel, mean values and standard deviations). Description of habitats A–l through B–n: see Table 21.1.

	Pb	A–l		A–n		B–l		B–n	
	[mg/l]	$\bar{x}$	±s	$\bar{x}$	±s	$\bar{x}$	±s	$\bar{x}$	±s
				Polygonum aviculare					
shoots	0	9287	538	6740	698	6784	1027	5025	817
	15.6	6957	867	4940	1029	6281	1078	4817	805
	31.2	4596	1679	3588	688	5173	629	3570	690
	46.8	6128	612	3317	171	3757	136	3345	412
	62.4	2581	365	2681	104	2557	197	2433	820
roots	0	2101	258	1504	62	2124	328	1504	62
	15.6	1320	196	1097	268	1676	304	1097	268
	31.2	1260	324	996	214	1501	40	996	214
	46.8	1435	230	848	161	817	140	848	161
	62.4	823	107	569	72	791	204	569	72
				Poa annua					
shoots	0	4180	559	3656	354	2772	261	3240	556
	15.6	3144	823	2448	249	1772	342	2280	346
	31.2	1268	464	916	257	616	128	1064	430
	46.8	1236	146	660	245	480	340	700	395
	62.4	876	105	548	126	212	65	496	189
roots	0	971	147	763	186	389	143	792	213
	15.6	946	255	709	180	452	54	669	147
	31.2	427	127	384	197	207	33	474	242
	46.8	415	53	287	90	145	71	244	120
	62.4	310	57	213	103	106	38	244	94
				Anagallis arvensis					
shoots	0	1461	360	2687	831	2825	174	3608	767
	15.6	2097	649	828	257	2234	598	1772	232
	31.2	360	282	6	30	495	217	208	22
	46.8	375	141	34	68	265	161	134	260
roots	0	143	33	235	80	240	65	383	83
	15.6	302	66	76	29	200	62	95	34
	31.2	44	46	−2	7	22	16	5	23
	46.8	18	19	20	39	28	11	16	20

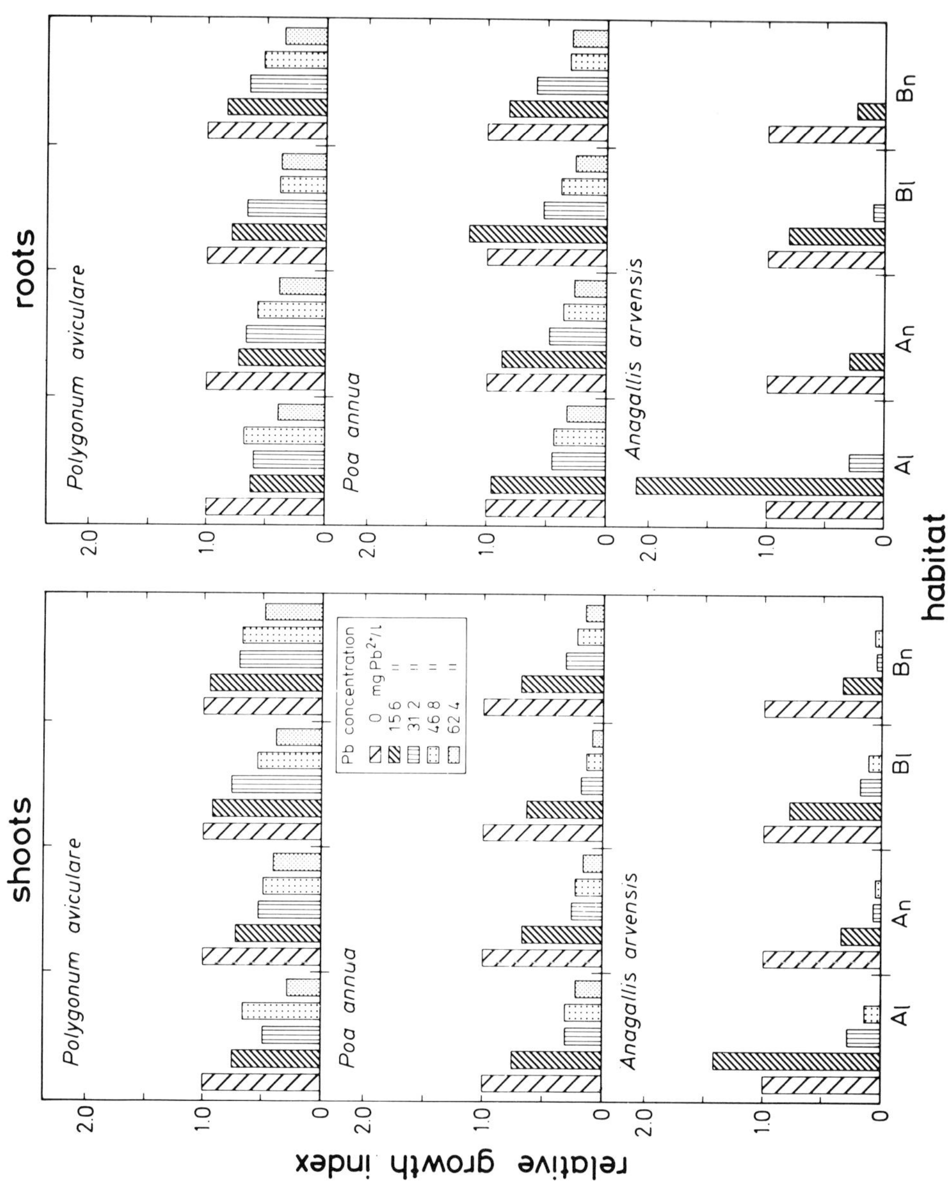

Figure 21.1: Relative growth index of shoot and root weight increase alter lead tolerance tests. Habitats:

A Industrial agglomeration in Southern Hessen

B Agricultural landscape in Northern Hessen

l Habitats bordering the motorways

n Habitats on field tracks far away from the motorways

by a white fluffy decomposition of almost the entire root system. This was also the case for the other two examined plant species under lead influence.

Whilst in the highest lead concentration stages, the level and grading of the lead content in the roots of *Anagallis* are similar to *Polygonum* and *Poa*, the shoots show a different reaction: with increased lead concentration in the solution the lead contents of *Anagallis*–shoots decrease significantly in all cases, independent of their origin (Fig. 21.2).

21.4 Discussion

A central question arises from the fact that only one of the three examined plant species — on both growing sites — evolved lead tolerance. In fact *Anagallis* differs yet further from *Poa* and *Polygonum*, and these striking features could be fundamentally related to lead tolerance:

1. The shoot/root–quotient of *Anagallis* has a higher value in the control and in the lead containing stages (see Table 21.2). Consequently, a given root mass has to supply comparatively more plant tissue; the destruction of an equal root part accordingly leads to more drastic effects.

2. Lead content of the shoots decreases in the higher concentration stages. That of *Poa* and *Polygonum*, however, shows an increase (Fig. 21.2). This could point to a reduced ionic transport into the green plant parts in general which in turn could be caused by the destruction of root tissue and by decreased water permeability.

3. Lead content in the roots of the first lead containing stage are at a level which is first reached by *Poa* and *Polygonum* in the following concentration stage (Fig. 21.2). It seems likely that the roots of *Anagallis* have an exceptional affinity for lead ions.

4. There is almost no increase in the biomass during cultivation in the higher concentration stages (Table 21.2, Fig. 21.1). In concentration stage V, most *Anagallis*–plants died after a few days of cultivation. The ability of *Anagallis* to bear high lead concentrations is therefore comparatively low.

It is to be expected that the consequence of the observed sensitivity is that *Anagallis* is exposed to a particularly high selection pressure whilst populating lead contaminated sites. This is also indicated by the fact that the *Anagallis*–plants grown at the roadway–sites died towards July/August after sufficient seed production, and a second generation grew in the same year. Again, this may promote the rapid development of adapted ecotypes (SYMONIDES 1988). One can assume that a strong selection pressure would support evolution of tolerance towards the stressing factor (CONARD and RADOSEVICH 1979). This conclusion, evolution of lead tolerance of plants with low inherent tolerance, has a remarkable similarity

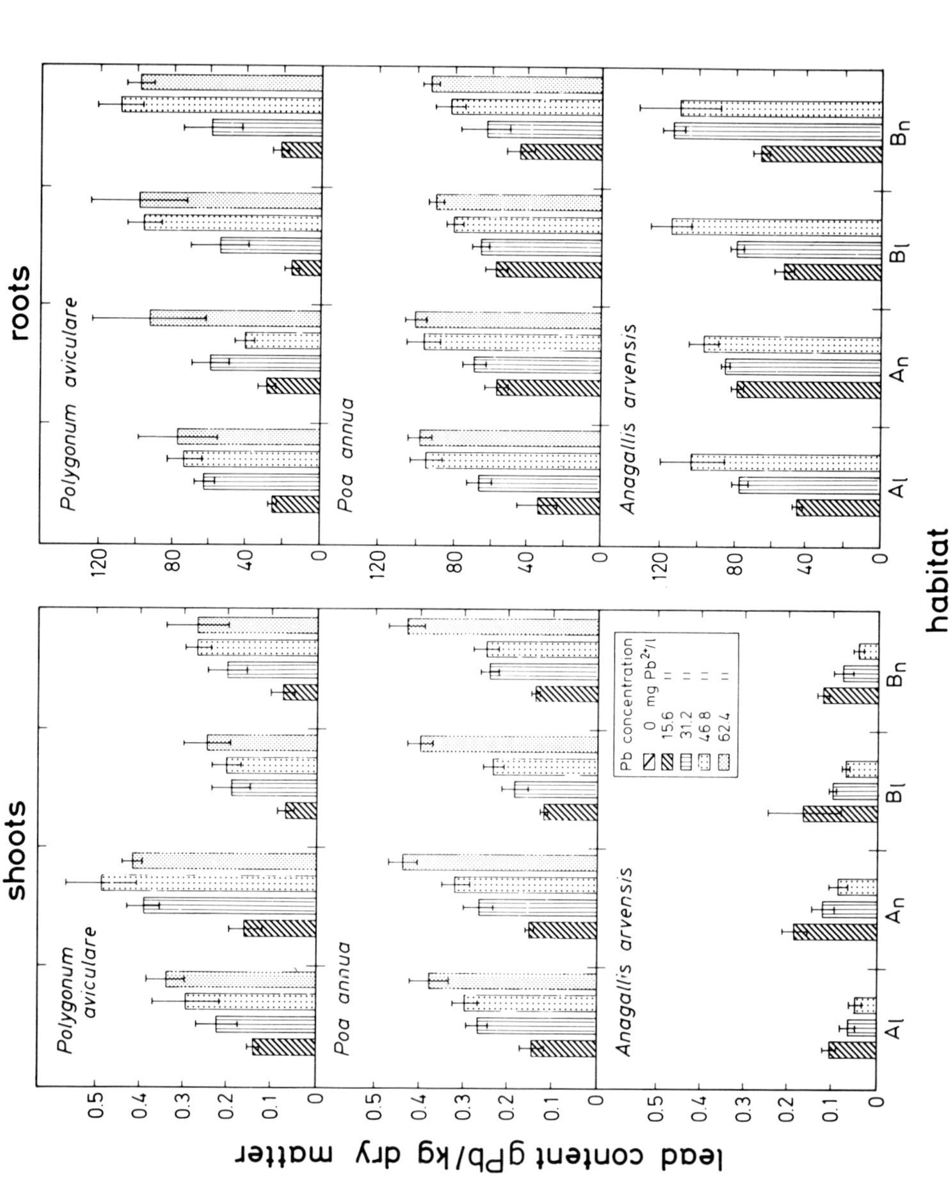

Figure 21.2: Lead concentration of shoots and roots at the end of lead tolerance tests (mean value and standard deviation). Habitats:

A Industrial agglomeration in Southern Hessen

B Agricultural landscape in Northern Hessen

l Habitats bordering the motorways

n habitats on field tracks far away from the motorways

to the observations of WU and ANTONOVICS (1976), who tested lead tolerance of perennial plants from roadsides.

It may be quite correct that, in the case of annuals as compared with perennials, an evolutionary process directed towards lead tolerance only starts when selection pressure is extremely high, i.e. probably only with such a sensitivity as shown by *Anagallis*. One reason could be the short life span and the phenotypic plasticity of this plant group (MAHN 1989) which may put it in a position to adapt to heavy metal content without any changes in the gene pool. Especially the short life span seems to be a phenomenon convergent to the growth concept that rosette and cushion plants growing on heavy metal mines continuously shed leaves or organs. In both cases plants or parts of them die after producing progenies, before metabolism breaks down as a result of critically increased heavy metal accumulation.

However, one result of this examination definitely disproves the postulate of the inability of annual plants to evolve heavy metal tolerance. In fact, on closer examination it seems that especially many properties of annuals could favour the genesis of such a feature. Their characteristics as "r–strategists" belong particularly to these attributes: high genotypic plasticity, large gene pool resulting from wide distribution, high rates of reproduction and generation, dispersal of seeds through the air, i.e. factors favouring a high rate of recombination (CONARD and RADOSEVICH 1979, MAHN 1987, 1989). One has to bear in mind the fact that heavy metal tolerance can evolve via the combination of a few genes which only exist in non–adapted populations with a low frequency (URQUART 1971). The site by the roadside with a sparse vegetation cover especially gives an invading plant species the chance to establish themselves again and again without limiting interspecific competition.

Furthermore, attention must be paid to the special growing conditions at the roadside. In fact, some factors such as high levels of pH and high contents of sodium chloride or calcium immobilize a high percentage of lead (BRÜMMER et al. 1986). The existence of further heavy metals, namely zinc and cadmium, seems to intensify lead toxicity, this phenomenon being more distinctive for lead sensitive lines than for tolerant ones (COUGHTREY and MARTIN 1979).

Especially on verge sites, plants are often exposed to serious water stress, as a result of wind, compressed soil, high levels of salt in the soil, dust coating on leaves and the open vegetation cover (ELLER 1977, FLÜCKIGER et al. 1978, ODZUCK 1979, ELLENBERG et al. 1981). A considerable factor in lead toxicity for plants is based on water stress. This can be explained by lead precipitation on root surfaces, thereby decreasing water permeability (ENGENHART 1980). It is possible that plant species which survive extreme water balance conditions are hardly sensitive to lead contamination as well. This could be true of comparatively drought resistant plant species, such as *Polygonum aviculare* and *Poa annua*, for which specific ecotype differences with respect to lead tolerance could not be found.

It has to be emphasized that drought and heavy metal resistance are not synonymous. As the findings of numerous authors show (f. e. POULTER et al. 1985,

WIERZBICKA 1987), an essential mechanism of lead tolerance seems to be the inactivation of lead through immobilization in the cell wall. Consequently, direct heavy metal intrusion into metabolism occurs less frequently. This model even gives an explanation for the fact that lead tolerant plants of *Anagallis* in the leadless control solution showed a lower dry weight than under low lead influence. Such "need of lead" could be caused by not quite specific catching reactions. They result in a deficiency situation in the plant on such occasions when a lack of the toxic substance facilitates the binding of other metabolic ions to the binding sites.

It would be interesting to know how far the distribution areas of lead tolerant annuals exist beyond the shoulder sites. On soils with a low heavy metal content such individuals are unable to withstand intraspecific competition with normal plants (COOK et al. 1972). For areas bordering the roadside, the problem of lead pollution will remain for decades to come, in spite of the extensive limitation of lead content in fuel, since the movement of lead in soil is very slow (HILDEBRAND 1974).

Another aspect of interest, especially with regard to annuals, arises from the necessity to revegetate heavy metal slag heaps or mine sites, which is most difficult. Lead contents there can be tenfold higher, or more, than at the sites examined in this work (ERNST 1974, SHAW et al. 1984). It remains to be seen whether such a high lead contamination is able to evolve a higher lead tolerance in plants with a certain inherent tolerance, such as *Polygonum* or *Poa*. Perhaps productive annual ecotypes will be found, as in the case of perennials, which can support recolonization and revegetation at such contaminated sites (JOHNSON et al. 1977).

Bibliography

ANTONOVICS, J.; BRADSHAW, A. D.; TURNER, R. G., 1971, Heavy metal tolerance in plants.
Adv. Ecol. Res. **7**, 1 85.

BAUMEISTER, W.; ERNST, W., 1978, Mineralstoffe und Pflanzenwachstum.
3. ed. Fischer, Berlin

BRIGGS, D., 1972, Population differentiation in *Marchantia polymorpha* L. in various lead pollution levels.
Nature **238**, 166–167.

BRÜMMER, G. W.; GERTH, J.; HERMS, U., 1986, Heavy metal species, mobility and availability in soils.
Z. Pflanzenernähr. Bodenk. **149**, 382–398.

CONARD, G.; RADOSEVICH, S. R., 1979, Ecological fitness of *Senecio vulgaris* and *Amaranthus retroflexus* biotypes susceptible or resistant to atrazine.
J. Appl. Ecol. **16**, 171–177.

COOK, S. C. A.; LEFEBVRE, C.; MCNEILLY, T., 1972, Competition between metal tolerant and normal plant population on normal soil.
Evolution **26**, 366–372.

COUGHTREY, P. J.; MARTIN, M. H., 1979, Cadmium, lead and zinc interactions and tolerance in two populations of *Holcus lanatus* L. grown in solution culture.
Environ. Exp. Bot. **19**, 285–290.

ELLENBERG, H.; MÜLLER, K.; STOTTELE, T., 1981, Straßenökologie. Auswirkungen von Autobahnen und Straßen auf Ökosysteme deutscher Landschaften.
In: Deutsche Straßenliga und Hessisches Landesamt für Straßenbau (eds.) Ökologie und Straße. Broschürenreihe der Deutschen Straßenliga **3**, 19–116.

ENGENHART, M., 1980, Hemmung des Wasserdurchsatzes in Wurzelsystemen von *Phaseolus vulgaris* L. unter Bleieinfluß.
Diss. Univ. Wien

ERNST, W., 1974, Schwermetallvegetation der Erde.
Geobotanica selecta **5**, Fischer, Stuttgart, 194 pp.

ERNST, W. H. O., 1985, Schwermetallemmissionen — ökophysiologische und populationsgenetische Aspekte.
Düsseldorfer Geobot. Kolloq. **2**, 43–57.

ELLER, B. M., 1977, Road induced increase of leaf temperature.
Environ. Pollut. **13**, 99–107.

FLÜCKIGER, W.; FLÜCKIGER–KELLER, H.; OERTLI, J. J., 1978, Der Einfluß von Straßenstaub auf den stomatären Diffusionswiderstand und die Blatt–Temperatur — ein antagonistischer Effekt.
Staub–Reinhalt Luft **38**, 502–505.

HELMING, W.; RUNGE, M., 1979, Zum Auftreten bleitoleranter pflanzlicher Populationen an der Autobahn.
Verh. Ges. Ökol. **7**, 407–412.

HILDEBRAND, E., 1974, Die Bindung von Immissionsblei in Böden.
Freiburger Bodenkundl. Abhandl. **4**, 147 pp.

JOHNSON, M. S.; MCNEILLY, T.; PUTWAIN, P. D., 1977, Revegetation of metalliferous mine spoil contaminated by lead and zinc.
Environ. Poll. **12**, 261–277.

MAHN, E. G., 1987, Bedeutung anthropogenen Einflusses auf Entwicklungszyklus und Stoffproduktion bei annuellen Arten.
Wiss. Z. Univ. Halle **36**, 61–68.

MAHN, E. G., 1989, Anpassungen annueller Pflanzenpopulationen an anthropogen veränderte Umweltvariable.
Verh. Ges. Ökol. **18**, 655–663.

MCNAUGHTON, S. G.; FOLSOM, T. C.; LEE, T.; PARK, F.; PRICE, C.; ROEDER, D.; SCHMITZ, J.; STOCKWELL, C., 1974, Heavy metal tolerance in *Typha latifolia* without the evolution of tolerant races.
Ecology **55**, 1163–1165.

ODZUCK, W., 1979, Effects of railway and road traffic pollution on meadow vegetation (Molinio–Arrhenatheretea) — comparison and attempt of evaluation.
Verh. Ges. Ökol. **7**, 331–336.

POULTER, A.; COLLIN, H. A.; THURMAN, D. A.; HARDWICK, K., 1985, The role of the cell wall in the mechanism of lead and zinc tolerance in *Anthoxanthum odoratum* L.
Plant Science **42**, 61–66.

SCHMIDT, W., 1990, Struktur und Funktion von Straßenrändern in der Agrarlandschaft.
Verh. Ges. Ökol. **19** (in press).

SHAW, S. C.; RORISON, I. H.; BAKER, A. J. M., 1984, Solubility of heavy metals in lead mine spoil extracts.
Environ. Poll. Ser. B **8**, 23–33.

STOTTELE, T.; SCHMIDT, W., 1988, Flora und Vegetation an Straßen und Autobahnen der Bundesrepublik Deutschland.
Forschung Straßenbau Straßenverkehrstechnik **529**, BMV, Bonn–Bad Godesberg, 191 pp.

SYMONIDES, E., 1988, On the ecology and evolution of annual plants in disturbed environments.
Vegetatio **77**, 21–31.

ULLMANN, I.; HEINDL, B., 1989, Geographical and ecological differentiation of roadside vegetation in temperate Europe.
Botanica Acta **102**, 261–269.

URQUART, C., 1971, Genetics of lead tolerance in *Festuca rubra* L.
Heredity **26**, 19–33.

WIERZBICKA, M., 1987, Lead accumulation and its translocation barriers in roots of *Allium cepa* L. — Autoradiographic and ultrastructural studies.
Plant Cell Environ. **10**, 17–26.

Wu, L.; Antonovics, J., 1976, Experimental ecological genetics in *Plantago* II. Lead tolerance in *Plantago lanceolata* and *Cynodon dactylon* from a roadside. Ecology **57**, 205–208.

Chapter 22

Environmental mutagenesis: mutational load in natural populations of *Eisenia fetida*?

W. NAU and W. KÖHLER

22.1 Introduction

More than at any previous period in history, scientific and industrial developments have enabled man today to change nature. Many of the subsequent effects on the natural environment are merely attendant phenomena of interventions that had other objectives, or side effects of industrial activity. In the field of chemistry, for example, new substances are being developed at an ever increasing rate. Many of these newly engineered chemicals are being produced in greater amounts and are ultimately released into the environment. These chemicals, in turn, have the potential to give rise, through decomposition or transformation, to other substances not occuring in nature.

In nature this inundation of chemical compounds comes into contact with organisms which have been evolving over a long period of time and changes their environment. One possible result of the interaction between these organisms and the newly engineered substances in the environment is a genotoxic effect. Newly induced mutations, which could be damaging to the affected individual and its descendants, can persist through many generations. Even the minutest alterations in genetic make–up (point mutation) can accumulate and lead in the short or long run to severe genetic load of the populations (ASHBY 1985, DE RAAT et al. 1985).

AUERBACH and ROBSON (1946) already pointed out alterations of genetic material by chemicals. Since that time mutagenic effects have been confirmed for a variety of chemicals. Therefore new chemicals are tested in the laboratory before being introduced into the marketplace. Due to the large number of substances to be tested, fast and cost–effective procedures must be in place to screen for genotoxic compounds (HARPER and MORRIS 1984, TASSINGTON 1985). These laboratory tests can only examine a single given substance on isolated test organisms under standardized conditions. As a result, organisms to be tested for toxicity and mutagenesis are chosen on the basis of their ability to be maintained in the laboratory. They are not exposed to natural conditions of stress, like competition with other species. In natural ecosystems several different genotoxic substances influence all members of the system.

Environmental mutagens include not only naturally occurring metabolism products from plants, bacteria, and fungi, but also synthetic substances and some heavy metal compounds from industrial emissions (HSIEH 1982, GAYATHRI and KRISHNAMURTHY 1985, RAY 1982, DE RAAT et al. 1985, WARE 1989), as well as a number of pesticides used in agriculture and gardens (BAO et al. 1974, CROSBY 1982, WARE 1988). These anthropogenic compounds are often more dangerous than those substances which have always been present in nature (DOMSCH 1984). Their effect on individual organisms or even on whole populations cannot be extrapolated from laboratory data on mutagenesis, because under natural conditions different substances act simultaneously on organisms. A whole range of potential interactions makes a reliable prediction about the resulting effect impossible. Therefore it is imperative that in situ tests on natural ecosystems be carried out in addition to the laboratory test systems (ANDERSON 1985, ASHBY 1983, 1985, MA and HARRIS 1985, VANPARYS and MARSBOOM 1985).

In such ecotoxicological studies (biomonitoring) it is not possible from a financial and time standpoint to test all the organisms of an ecosystem for genetic alterations. Rather we must generally limit ourselves to one species as a bioindicator for the stress on the system. This species, whether animal or plant, should possess sufficient tolerance to stress factors, so as to make quantitative statements feasible. Bioindicators allow for an understanding of acute, accumulative, direct, and indirect effects of environmental stressors (SCHUBERT 1985, DUMPERT 1981).

Among other species, earthworms are suitable for bio–monitoring terrestrial systems. As soil organisms they have intensive contact to all hazardous substances in the soil. Not only by direct uptake, but also by way of epidermal diffusion processes, chemicals penetrate into the bodies of these animals and influence their metabolism. Earthworms take up many substances, some of which, for example mutagens like cadmium, mercury, PCBs and phenol chlorines, are accumulated. The burrowing and mixing habits of earthworms, on the one hand, and their active propagation in confined spaces on the other hand, makes the earthworm a good indicator for the environmental stress of a limited area (IRELAND 1983, HEIMBACH 1984, EDWARDS 1979, DIERCXENS et al. 1985, BEYER et al. 1987, CONNELL et al. 1988, VAN GESTEL and MA WEI–CHUN 1988, CURL et al. 1987, TARRADELLAS et al. 1982).

Household waste sites are an especially interesting terrain in relation to mutational stress. These sites are terminals for materials no longer used nor wanted, among which can be found remnants of pesticides, varnishes, paints, solvents, or pharmaceuticals and heavy metals. In the waste itself other critical compounds such as organic metallic salts, chlorinated aromates and hydrocarbons like trichlorethane or trichlorethylene are produced by decomposition and transformation of the deposited substances. In solid waste as well as in sewage sludge a considerable concentration of harmful and potentially genotoxic substances can be found (DE RAAT et al. 1985, USEPA 1982, ROWLEY et al. 1983, RACKE and FRINK 1989, MCBRIDE et al. 1989, BEYER et al. 1982, HOLZBECK and VENTULETT 1987, HOINS 1986). Earthworms living in waste sites, then, are exposed to a wide variety of anthropogenic hazardous materials. Thus it is to be expected that in an ecosystem in which a multitude of potential mutagens act on organisms, environmental stresses on natural populations caused by increased mutational pressure due to environmental chemicals become more evident.

These considerations form the premise for this paper. In it the exemplary genetic load of an animal organism in a terrestrial ecosystem is examined. The household waste site at Hasenbühl in Schwäbisch–Hall County was chosen as the presumably stressed environment. There were a large number of earthworms available as bioindicators of the environmental stress (COLLINS and SPILLMANN 1985).

For the investigation of hereditary genotype–altering effects on the waste site the procedure of Ferguson plots after polyacrylamide gel electrophoresis and enzyme assay was used, which allows a clear identification of protein molecules as gene products in relation to charge and structure. Most DNA alterations of the appertaining structural gene, even simple base substitutions or base alkylations (point mutations) can be proven by this method (HAHN 1989, JOHNSON 1977). It was developed by HAHN (1989) under laboratory conditions using tube gels on samples of *Drosophila melanogaster* and for this experiment adapted to slab gels and *Eisenia fetida* as the test organism. To validate this method for a species other than *Drosophila*, a laboratory population treated with a mutagenic substance and an untreated control population were investigated. The compost worm *Eisenia fetida* was chosen for breeding and maintaining in the laboratory. This species is especially suitable for these purposes due to its relatively rapid development and easy maintenance (GRAFF 1983, HEIMBACH 1985, EDWARDS 1985).

22.2 Material and Methods

We got a loboratory culture ($n = 130$) of *Eisenia fetida* from the Institute of Phytopathology, University of Gießen, which initially came from the Forschungsanstalt für Landwirtschaft (FAL) in Braunschweig (Germany). After multiplication of this founding population a sample of 300 animals were injected in the genital area with 15 micrograms each of the gene mutation–inducing agent ethyl–nitroso–urea (ENU) (VOGEL 1984). The first filial generation of these animals was later examined for mutations as validation of the test systems. The treated animals and their offspring then continued as separate populations. The untreated control population and the treated ENU populations were kept in the laboratory at 25–27 °C in plastic pails containing a compost substrate.

A natural population of earthworms of the species *Eisenia fetida* were collected from the waste site of Schwäbisch–Hall County as random samples. Some areas of the site were heavily populated with *Eisenia fetida*. After removing the outermost layer of waste, the animals could simply be gathered by hand. The earthworms were living in the areas of the waste site not yet compacted, where the waste could decompose aerobically over a period of months before compaction led to anaerobic conditions, which make the continued survival of the worms impossible. This special waste site procedure reduces the space requirements for the deposited wastes (COLLINS and SPILLMANN 1985).

The earthworms come into direct contact with the hazardous materials of the deposited waste. Through spoiling of the seepage water that overflows from the compacted waste site into the non–compacted waste, substances that normally form in the deeper anaerobic layers of the waste piles get into the biosphere of the earthworm and can contaminate them. At the waste site in Hasenbühl, contamination of the solid waste and seepage water with heavy metals could be determined. In addition, several characteristic values of organic contamination were determined (see appendix). However, in accordance with investigations of other waste sites in which the same recurring organic hazardous substances were found, it was presumed that a whole range of potentially mutagenic substances can impact living organisms at Hasenbühl site, too.

The procedure used for the investigation of genotype–altering effects at the waste site was a special electrophoresis. Ferguson plots after polyacrylamide gel electrophoresis and enzyme specific staining was employed in order to clearly identify certain protein molecules as gene products with relation to charge and structure. A Ferguson plot is the linear relation of the logarithm of the relative mobility versus gel concentration (log–linear regression line). It allows the determination of two parameters independently from each other: the retardation coefficient (a measure of particle size) and the free mobility, which is related to protein surface net charge (TIETZ 1987). In this way most of the DNA alterations of appertaining structural genes, including simple base substitutions and alkylations (gene mutations) can be verified (HAHN 1989, JOHNSON 1977). This method was adapted to slab gel electrophoresis using four different gel concentrations. For the detection and correction of experimental errors hemoglobin as a standard protein was added in each electrophoresis. The separation of proteins was carried out with discontinuous electrophoresis by a modified procedure after ORNSTEIN (1964) with polyacrylamide in stacking and resolving gels (HAHN 1989). For the vertical slab gel electrophoresis the Desaga VA 200 electrophoresis apparatus was used. Separation distances of up to 18 cm in length were thus reached. The electrophoresis was steered in two phases by tension voltage apparatuses with volt/hour integrators. In the first phase a voltage of 100V over 3 hours was applied, then 300V over 9 hours.

Following electrophoresis, enzymes under consideration were stained by means of specific reactions. The following enzymes were assayed:

Acid phosphatase	[ACP]	EC 3.1.3*
Diaphorase II	[DI2]	EC 1.6.2
beta-esterase	[EST]	EC 3.1.1
Glucose-6-phosphate-isomerase	[GPI]	EC 5.3.1
Leucine-amino-peptidase	[LAP]	EC 3.4.11
Triose-phosphate-isomerase	[TPI]	EC 5.3.1
Xanthine-dehydrogenase	[XDH]	EC 1.1.1

(*Enzyme Nomenclature 1973)

The enzymes were stained according to formulas by HARRIS and HOPKINS (1976), SHAW and PRASAD (1970).

The location of the visualized proteins on the gel was ascertained with a densitometer. The measured data were directly transferred to a personal computer and further processed on the mainframe with software developed by the authors (NAU 1990).

The parameters of the log-linear regression line of each allozyme band of the analyzed loci were calculated from four electrophoreses using different gel concentrations:

b = slope, average change in relative mobility,
Y = mean relative mobility in a medium gel concentration.

These parameters characterize charge and structural properties of proteins. Alterations of genetic material influencing charge and/or structure can be proven (JOHNSON 1977). In order that the different experimental conditions in the electrophoresis will not superimpose upon the deviations looked for in the parameters of the Ferguson plot, the regression lines of the investigated proteins were corrected using the accompanying protein standard hemoglobin. In this way most methodical variances could be eliminated (HAHN 1989, JOHNSON 1977, NAU 1990). For every examined individual, the calculated parameters b and Y were recorded in a

Cartesian coordinate system. For the standard protein hemoglobin (Hb) the parameters were listed for each concentration in order to be able to determine the position of the enzyme bands relative to the Hb bands.

For all individuals together a confidence ellipse was then established. If, after applying the χ^2–test, more points lie outside the ellipse than would be expected according to the error probability α, then the variation between the different specimens is not merely random. Some of the proteins clearly deviate in charge and/or structure from the other specimens and are thus the result of mutations. Hence, an increased mutation rate can be proven with statistical reliability.

22.3 Results

Seven enzymes in the three populations (control, ENU–treated and waste site) were investigated:

Acid phosphatase (ACP)
Diaphorase II (NADPH specific) (DI2)
beta–esterase (EST)
Glucose–6–phosphate–isomerase (GPI)
Leucine–amino–peptidase (LAP)
Triosephosphate–isomerase (TPI)
Xanthine–dehydrogenase (XDH)

In all populations the five enzymes DI2, GPI, LAP, TPI and XDH showed only one band apiece with no differences between the populations. For the rest of the enzymes, several bands could be stained. The bands were marked 1a, 1b, ... 2a, 2b ..., whereby all the bands with the same number were grouped to the same gene locus and the bands with lower numbers or letters had run further down on the electrophoresis gel than the other bands.

For ACP, several different bands were found in all three populations. In the lower regions of the gels, five distinct bands were stained in all three populations (see Fig.22.1). The fastest of the five bands appeared in all individuals and displayed a higher activity level than the others. Of the four remaining bands, a maximum of any two appeared together. This suggests the association of the bands to two loci. The fast allele of locus 1 can be found for all individuals homozygously. At locus 2, four distinct alleles can be identified. They are named after their migration distances during electrophoresis 2a to 2d. Allele frequencies could be determined for the control and waste site populations (see Table 22.1). A χ^2–test on deviation from Hardy–Weinberg equilibrium did not furnish any basis for larger alterations of the waste site population by selection, migration, or other evolutionary factors. For ENU–treated populations the bands of locus 2 could only be determined clearly enough for fresh specimens (not older than three weeks). Most of the specimens no longer fit this category. Therefore an evaluation of the allele frequencies for this

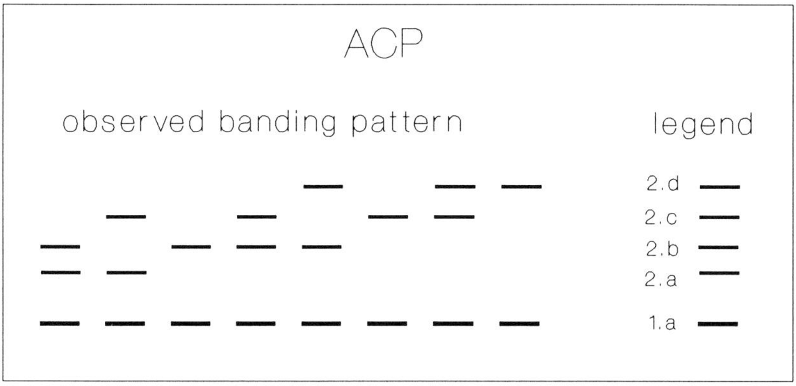

Figure 22.1: Observed banding patterns for ACP for locus 1 (allele a) and locus 2 (alleles a, b, c, d).

Table 22.1: Genotype and allele frequencies for the locus ACP2 (see Figure 22.1).

A genotype frequencies

genotype	aa	ab	ac	ad	bb	bc	bd	cc	cd	dd
control		2				2		126	50	6
waste site		3	4		14	50	2	38	6	

B allele frequencies

allele	a	b	c	d	n
control	0.005	0.011	0.817	0.167	186
waste site	0.030	0.355	0.581	0.034	117

n number of individuals studied

group had to be forgone. For the control population the bands of locus 2 were not marked clearly enough in all gel concentrations. Therefore, no Ferguson plot was calculated for this. In the waste site worm population allele 2a and 2d appeared so rarely, that no Ferguson plot was calculated for this either.

The beta–esterases of the three populations exhibited several bands (see Figure 22.2). Moreover, marked differences in the banding patterns were exhibited between the control population and ENU population on the one hand, and the waste site population on the other. The control and ENU populations had the same banding patterns (EST 2 and EST 3). Out of a total of four distinct bands, the fastest

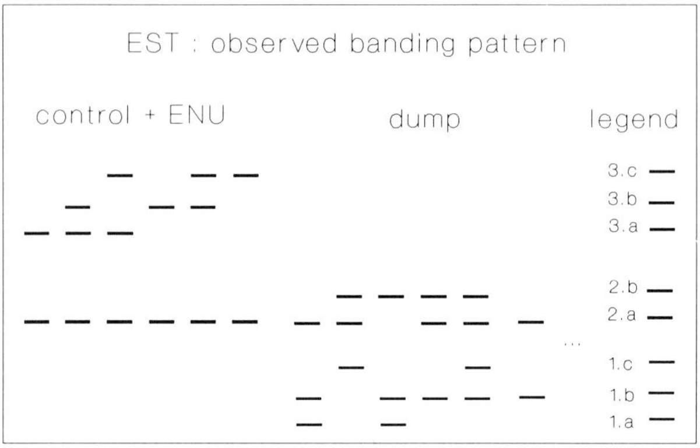

Figure 22.2: Observed banding pattern for EST. Not all of the possible combinations of bands from locus 1 and 2 are represented for the waste site population.

was present in all individuals. This one was assigned its own locus (EST 2). The other three bands were combined at a second locus, which presumably encodes a monomeric enzyme EST 3), because a maximum of two bands of EST 3 appeared together. Each band corresponded to one allele. The frequencies of genotypes and alleles are combined in Table 22.2. The denotation of the alleles conforms to the above-mentioned definition.

A different pattern was observed for the waste site population (EST 1 and EST 2). The slow bands of the other two populations were not present. In their place, however, three fast bands appeared, located ahead of the bands EST 2. Since, of the three bands, at most only two appeared together, a single locus system of a monomeric enzyme can be inferred. However, the bands at this locus (EST 1) generally had a low intensity and could not always be clearly determined for all gel concentrations. Only the middle band stained frequently and clearly enough to be used for the Ferguson plot. An allele frequency for this locus was not estimated. Two other bands ran somewhat more slowly. The band that had been observed in all individuals of the control population (EST 2) was present in only 90% of the worms of this population. In the remaining 10% a slower band appeared, which also occurred together with EST 2a in about 45% of the population. This slower band was designated as a second allele to the same gene locus. No deviation from the Hardy–Weinberg equilibrium was shown for the allele frequency determination in all three populations. The allele frequencies of the ENU population differed not significantly from those of the control population. This was not surprising, as the parental animals of the ENU population investigated had randomly been selected from the control population.

The presumed locus and allele structure was confirmed for both polymorphic enzymes in a cross-breeding experiment (NAU 1990).

Table 22.2: Genotype and allele frequencies for beta–esterases. EST 1 is omitted (see text).

A	genotype frequencies						
population	locus/genotype	aa	ab	bb	ac	bc	cc
control	EST 3	28	42	25	19	25	12
ENU	EST 3	18	21	29	8	13	12
waste site	EST 3	–	–	–	–	–	–
control	EST 2	151					
ENU	EST 2	101					
waste site	EST 2	63	70	15			

B	allele frequencies of polymorphic loci				
population	locus/allele	a	b	c	n
control	EST 3	0.407	0.348	0.245	151
ENU	EST 3	0.426	0.247	0.327	101
waste site	EST 2	0.662	0.338		148

n = number of individuals studied

Of the enzyme bands studied, a Ferguson plot was calculated for all bands which could be well represented and which were observable in at least 30 specimens. In Table 22.3 the mean values for Y and b for the standard protein are listed for the different concentration domains. Y is the negative natural logarithm of the relative mobility of a band for the calculated mean gel concentration of a concentration domain. Y values of 0.5, 1.0, or 2.0 indicate that a band near the mean gel concentration has covered 60, 37, or 13% , respectively, of the total length. Fast bands thus have a low Y value.

The b value describes the reaction of a protein when the gel concentration is increased. A high b value indicates that the relative mobility values for a protein decreases sharply with increasing gel concentrations. A comparison of the bands ACP 1.a and TPI 1.a shows that the ACP 1.a band migrates considerably further and reacts less to changes in gel concentration than the TPI 1.a band. The trend of the results was the same for all enzymes. Therefore, a detailed description of the results for individual enzymes is superfluous. Only graphs for the enzyme Diaphorase II are shown here as an example for the Ferguson plot.

In the following table the results of the respective populations are listed. The number of bands evaluated for the Ferguson plot and the number of points lying outside the 95% and 99% confidence ellipse, respectively, are listed. The results

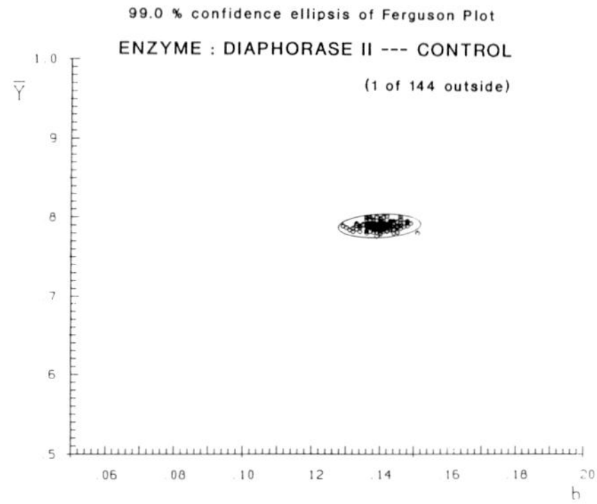

Figure 22.3: Ferguson plot for the enzyme DI2 for the control population, corrected data, $\alpha = 1\%$

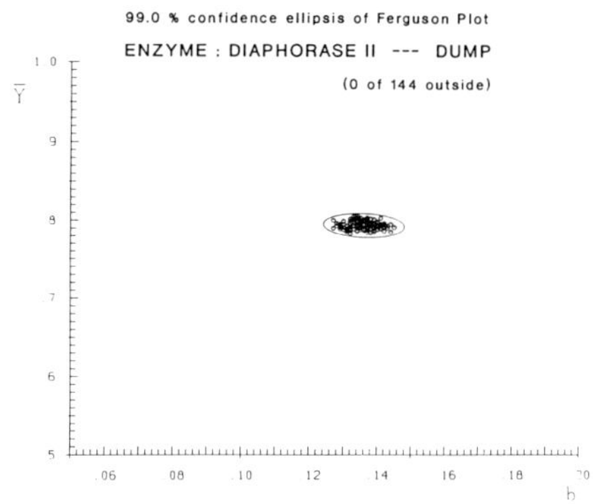

Figure 22.4: Ferguson plot for the enzyme DI2 for the waste site population, corrected data, $\alpha = 1\%$

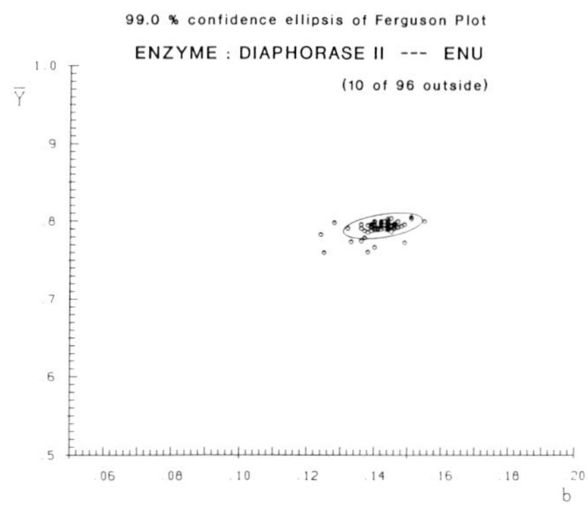

Figure 22.5: Ferguson plot for the enzyme DI2 for the ENU population, corrected data, $\alpha = 1\%$

Table 22.3: Mean mobility Y and slope b of the Ferguson plot calculated for hemoglobin (Hb) and the different enzyme bands.

gel concentrations	band	Y	b
7 – 8 – 9 – 10%	Hb	1.06	0.11
	DI2 1.a	0.79	0.14
8 – 9.5 – 11 – 12.5%	Hb	1.25	0.11
	EST 1.b	0.59	0.13
	2.a	0.63	0.13
	2.b	0.67	0.13
	3.a	0.69	0.11
	3.b	0.77	0.11
	3.c	0.85	0.11
	GPI 1.a	1.45	0.16
	LAP 1.a	0.82	0.16
10 – 11 – 12 – 13%	Hb	1.42	0.11
	ACP 1.a	0.31	0.11
	2.b	0.40	0.12
	2.c	0.47	0.12
	TPI 1.a	1.97	0.16
	XDH 1.a	2.39	0.10

for the bands of one locus are combined in Table 22.4 to determine the mutation rate per locus (see below). The number of points lying outside the confidence ellipse was compared to expected values. Even after correction, the points of the Ferguson plot that were determined for the enzyme bands were still affected by a residual error, e.g., random effects caused by methodical inaccuracies. In addition, there are systematic deviations when the bands of one individual deviate from the others, because the enzyme protein is slightly altered. Theoretically five (one) in a hundred points should lie outside the 95% (99%) confidence ellipse.

χ^2 was used to test whether the observed number of outliers was higher than that expected with a random residual error. Alternately, it must be presumed that a larger number of observed outliers is brought about by systematic alterations of the proteins due to changes in the DNA sequences (point mutations).

The χ^2–test proved that only in the case of the ENU population, in fact, for all the bands investigated, the number of outliers was greater than the number that would have explained methodical inaccuracies ($\alpha < 0.1\%$).

The outliers are attributed to altered proteins, while the protein variants are attributed, in the end, to an altered base sequence of DNA, hence to gene muta-

Table 22.4: Number of bands outside the confidence ellipse, combined for all populations and enzymes.

population	enzyme	locus	number	95%	99%	%
control	ACP	1	141	3	0	
	DI2	1	143	9	1	
	EST	2	112	6	1	
		3	172	9	0	
	GPI	1	120	2	0	
	LAP	1	143	0	0	
	TPI	1	144	0	0	
	XDH	1	132	1	0	
ENU	ACP	1	96	9	5	5.2
	DI2	1	96	13	10	10.4
	EST	2	96	20	12	12.5
		3	132	27	13	9.9
	GPI	1	95	13	5	5.3
	LAP	1	96	15	6	6.3
	TPI	1	120	32	10	8.3
	XDH	1	96	18	15	15.6
waste site	ACP	1	142	2	1	
		2	139	4	0	
	DI2	1	144	4	0	
	EST	1	50	6	1	
		2	184	4	2	
	GPI	1	143	2	0	
	LAP	1	143	2	2	
	TPI	1	144	0	0	
	XDH	1	132	1	0	

number : number of bands analyzed
95% : outliers for $\alpha = 5\%$
99% : outliers for $\alpha = 1\%$
% : percentage of outliers for $\alpha = 1\%$

tion. If we consider only the percentage of protein variants for the 99% confidence ellipse to be reliable mutations, then the percentage of individuals of the ENU population mutated on at least one of the enzyme–coding chromosomes fluctuates between 5.2% for ACP 1 and 15.6% for XDH 1 (see also below). These values do not correspond to the mutation rate per treated chromosome. This cannot be determined directly from the Ferguson plot. It is only in the case of monomeric enzyme proteins for heterozygous individuals at a polymorphic locus that one band corresponds to exactly one chromosome. If the genetic information of the corresponding gene locus on both chromosomes differs in a manner not clearly exhibited by electrophoresis, it is not possible to determine with certainty whether an individual is homozygous or heterozygous. Therefore an extrapolation of the number of mutated chomosomes from the number of observed bands outside the confidence ellipse was omitted.

The concept "mutation rate per locus" will henceforth refer to the percentage of individuals in the population mutated on at least one of the two chromosomes from which an enzyme system is encoded. The percentage of outliers ("%" column in Table 22.4) was computed only for the ENU population, because it is only for this population that the χ^2–test confirmed the significant increase of mutations.

22.4 Discussion

Hardy–Weinberg equilibrium is achieved in random–mating populations in the absence of mutation, migration, selection and genetic drift after only one generation. A deviation from equilibrium is an indication of greater acute alterations in a population. The allele frequencies of polymorphic enzyme systems were used to test deviation from the Hardy–Weinberg equilibrium. Of the seven enzymes investigated in all three populations, it could only be analyzed for two polymorphic systems whether the populations fit in to the Hardy–Weinberg equilibrium. No deviation could be found. This does not mean, however, that no changes whatsoever took place in the populations. As regards the ENU population, the first filial generation of the treated animals that itself was not exposed to any further mutagenes was investigated. Except for a change in allele frequencies at the observed loci no deviation from Hardy–Weinberg equilibrium was found in this case, too.

The test to determine whether the populations are in Hardy Weinberg equilibrium was conducted, since ZADORY (1986), for example, discovered a suspicious alteration of the allele structure in snail populations that lived in contaminated waters. KLEKOWSKI (1978) and THOMPSON et al. (1988) found chromosome aberrations for ferns in contaminated water and rats in a toxic waste site. These alterations are an expression of a severe mutational load which should also be discernible at the enzyme protein level, as in the alteration of allele frequencies in certain enzyme systems, so that a deviation from Hardy–Weinberg equilibrium is observable. These alterations were not found in the populations studied in this project. This was to be expected for the control and ENU populations. For the waste site population it cannot be concluded on the basis of the allele frequencies that a stronger acute mutative pressure was exerted.

Bands which could be clearly differentiated on the gel were used to enumerate allele frequencies. Subtler differences between bands associated with one allele can only be detected by the Ferguson plot. As shown in Table 22.4 for the ENU population there were considerably more smaller alterations of proteins evident than would have been expected for randomly or normally distributed residual deviation. The smaller alterations of proteins are interpreted as a consequence of alterations of the corresponding structural gene due to gene mutation. The ENU chemical used to treat the parents of the investigated animals of the ENU population induces point mutation on a large scale (GOCKE et al. 1983). Base-substituted DNA may produce proteins altered in at least one amino acid. Even a change in a single amino acid can have an effect on the net charge and structure of the enzymes. Ferguson plots help verify this effect. The increased detection of protein variants in the ENU population confirms on the one hand the efficiency of the test used in verifying point mutations, and on the other hand, the efficacy of injecting ENU in the genital region of *Eisenia fetida* in order to produce mutations in the germ cells.

In the population treated with ENU between 5 and 15% outliers were found for the 99% confidence ellipse and between 8 and 26% for the 95% confidence ellipse. If these numbers are corrected to the amount of the expected errors then a mutation rate remains of between 4 and 14%, and 3 and 21%, respectively, per locus. Thus the values for the 95% confidence ellipse show greater fluctuation than for the 99% confidence ellipse. For the 99% confidence ellipse, only points clearly separated from the common cluster of the remaining specimens were defined as outliers. In relation to the enzyme molecules investigated this means the following: for those bands located outside the 99% confidence ellipse, the enzyme proteins are clearly altered compared to the normally occurring protein in the population. The corresponding gene locus is mutated.

Data from the 95% confidence ellipse are subject to greater fluctuation and are less reliable. Therefore in this study the mutation rate per locus is established according to the outlier definition with the 99% confidence ellipse. This does not result in a lower mutation rate than for calculations with the 95% confidence ellipses for all enzymes. Consequently a rate of 4 to 14% detectable mutations results on the different enzyme loci. It is possible for several base substitutions to occur simultaneoulsy at a locus. Using the applied method it cannot be established whether several substitutions cancel each other out in their effect, or whether a mutation has no influence on the charge and/or structure of a protein.

The mutations determined can be characterized as point mutations on the basis of the applied mutagen and the testing procedure. The average rate of mutated loci established in this investigation of 7.5% cannot be directly compared with the values of the average gene mutation rate per locus ascertained by HAHN (1989) in the development of the test procedure. The differences among the test organisms *Eisenia fetida* and *Drosphila melanogaster*, the mutagens ethyl-nitroso-urea (ENU) and ethyl-methane-sufonate (EMS), and the applied quantities and methods used in this study, are too great for this to be the case.

HAHN (1989) established that the dose-response function of the detectable mutation rate does not increase linearly with raising the EMS concentration in the feed, but rather, in some cases, actually decreases. Therefore an expected gene mutation rate at lower or higher mutational load cannot be extrapolated from this investigation. In order to do so, it would be necessary to carry out further studies using *Eisenia fetida* and differing degrees of mutational stress. We therefore deviate from our original considerations and conclude that the results of the ENU population cannot be directly compared with those of the waste site test population. Investigations with the ENU population could, as discussed earlier, only confirm the suitability of the method as verification of point mutations in *Eisenia fetida*.

The load situation of the waste site population can be compared to that of the control population, although investigations showed that the animals of the subspecies *Eisenia fetida andrei* made up the control population, while only the subspecies *Eisenia fetida typica* was observed at the waste site. It was only in the enzyme system beta esterase (EST) that qualitatively different bands could be determined. According to OIEN and STENERSON (1987) these qualitative differences characterize the two subspecies *andrei* and *typica*, respectively, and may not depend on induced mutations in the waste site populations. Therefore we conclude that all animals of the waste site population belong to the subspecies *Eisenia fetida typica*. All the other enzyme systems had proteins with the same electrophoretic properties, as was shown by the mean values of the parameters Y and b of the Ferguson plot. At the high resolving power of the applied electrophoresis technique, the enzyme molecules can be considered to be identical. In light of the very high earthworm population density in a suitable substrate at the waste site, it is unlikely that the mutational stress at the waste site would be so high that no point mutations occur in the next generation nor accumulate since all mutated germ cells or zygotes would die at an early date due to a high load. This was how HAHN (1989) substantiated declining point mutation rates for several enzymes. An increase in mutational load cannot be verified for either of the two populations using this test procedure.

If only the points lying outside the 99% confidence ellipse are estimated as mutations, then, with the applied method, at least 3 points out of 100 specimens, 5 out of 200 and 17 out of 1000 must lie outside the 99% confidence ellipse, in order to verify with statistical reliability ($\alpha = 5\%$) an increased gene mutation rate. If the outlier rate is corrected, as was the case above, then this corresponds to an effective (influencing charge and/or structure) mutation rate of 2%, 1.5% or 0.7% at one locus. Taken together, all investigations performed on all loci in the waste site population yield a specimen number of 1448. At the same time an average effective mutation rate of 0.6% per locus could still be proved significant. Corresponding outlier numbers could not be established at the loci of the control or waste site populations. It cannot be concluded from the results of this study that the increased mutational pressure brought about by environmental chemicals at domestic waste sites have led to an increased mutational load on the *Eisenia fetida* population.

According to the investigations of KLEKOWSKI (1978) and THOMPSON et al.

(1988), a detectable increase in mutation rate was at least potentially expected. Both authors, however, were able to show cytological evidence of chromosomal alterations in populations of water fern in highly contaminated sewage and rats in a toxic waste site. Gene mutations, on the other hand, are smaller alterations of genetic material which do not necessarily affect the fitness of the mutated genotype. Thus they can accumulate in a population. In populations that have been exposed to more severe mutative loads over a long period of time, gene mutations should therefore be detected more frequently. However, this was not the case for the population of *Eisenia fetida typica* (waste site population) investigated in this study. The loads emanating from the chemicals to the waste site are evidently either so slight as to enable the repair mechanisms to reverse DNA alterations, or they are at least sufficient to prevent an increase of the average effective mutation rate per gene locus to over 0.5%. Mutation rates below this threshold cannot be determined with statistical reliability for the specimen numbers used in this investigation. On the basis of the results under consideration, however, neither an increased mutation rate nor an accumulation of mutations (mutational load) could be detected with the method applied. Nevertheless, there is every reason to believe a mutational stress on the earthworm population at the Schwäbisch-Hall domestic waste site. Therefore, alternative approaches which might be more sensitive to changes in the genetic material, like DNA-fingerprinting, may prove suitable.

Acknowledgments

We gratefully acknowledge the significant support of Dr. Hashemi and Prof. Stein at the Hasenbühl domestic waste site and the helpful comments of an anonymous reviewer. We thank S. Teng-Whitehead, Milpitas, CA, for translating and editing the text. We also thank B. Sonntag for her skilful technical assistance. Part of this work was supported by a grant of the German Federal Environmental Agency.

Bibliography

ANDERSON, D., 1985, Genetic toxicology in the food industry.
Food Chem. Toxicol. **23**, 11–18.

ASHBY, J., 1983, The unique role of rodents in the detection of possible human carcinogens and mutagens.
Mutat. Res. **115**, 177–182.

ASHBY, J., 1985, Genetic Toxicology in industry; perspectives and initiatives.
Food Chem. Toxicol. **23**, 43–45.

AUERBACH, C.; ROBSON, J. M., 1946, Chemical production of mutations.
Nature **157**, 302.

BAO, VAN, T.; SZABO', D.; RUZICSKA, P.; CZEIZEL, A., 1974, Chromosome aberrations in patients suffering acute organic phosphate insectizide intoxication. Humangenetik **24**, 3–57.

BEYER, W. N.; HENSLER, G.; MOORE, J., 1987, Relation of pH and other soil variables to the concentration of lead, copper, zinc, cadmium, and selenium. Pedobiologia **30**, 167–172.

BEYER, W. N.; CHANEY, R. L.; MULHERN, B. M., 1982, Heavy metal concentrations in earthworms from soil amended with sewage sludge. Jour. Environ. Qual. **11**, 381–385.

COLLINS, H. J.; SPILLMANN, D., 1985, Forschungsbericht der DFG – Wasser- und Stoffhaushalt von Abfalldeponien und deren Wirkung auf Gewässer. Verlag Chemie Weinheim.

CONNELL, D. W.; BOWMAN, M.; HAWKER, D. W., 1988, Biocentration of chlorinated hydrocarbons from sediment by oligochaetes. Ecotoxicol. Environ. Saf. **16**, 293–302.

CROSBY, D. G., 1982, Pesticides as environmental mutagens. In: Fleck, R. A. and Hollaender, A., (eds.), Genetic toxicology. Basic life sciences, Vol **21**, Plenum Press New York.

CURL, E. A.; EDWARDS, P. J.; ELLIOTT, C.; LEAHEY, J. P., 1987, The conjugation and accumulation of metabolites of cybermehtrin by earthworms. Pestic. Sci. **20**, 207–222.

DIERCXENS, P.; DE WECK, D.; BORZINGER, N.; ROSSET, B.; TARRADELLAS, J., 1985, Earthworm contamination by PCB's and heavy metals. Chemosphere **14**, 511–522.

DOMSCH, K. H., 1984, Effects of pesticides and heavy metals on biological processes in soils. Plant and Soil **76**, 367–378.

DUMPERT, K., 1981, Literaturstudie über die ökologische Bedeutng tierischer Bodenorganismen und die ökotoxikologische Berwertung von Umweltchemikalien. Forschungsbericht **03 7245**, Bundesminister für Forschung und Technologie.

EDWARDS, C. A., 1979, Tests to assess the effects on the uptake of pesticides on beneficial soil organisms. In: Schmidt, E. (ed.), Tests for the ecological effects of chemicals. Springer Verlag, Berlin.

EDWARDS, C. A., 1985, Production of feed proteins from animal waste by earthworms. Philosophical Transactions of the Royal Society of London, Series B **310**, 299–308.

ENZYME NOMENCLATURE, 1973, Recommendations (1972) of the International Union of Pure and Applied Chemistry and the International Union of Biochemistry. Elsevier Amsterdam.

GAYATHRI, M. V.; KRISHNAMURTHY, N. B., 1985, Investigations on the mutagenicity of two organomercurial pesticides, Ceresan and Agallol 3, in *Drosophila melanogaster*.
Environ. Res. **36**, 218–229.

GESTEL, VAN C. A. M.; MA WEI-CHUN, 1988, Toxicity and bioaccumulation of chlorophenols in earthworms in relation to bioavailebility in soil.
Ecotoxicol. Environ. Saf. **15**, 289–297.

GRAFF, O., 1983, Unsere Regenwürmer.
Verl. M. and H. Scharper, Hannover (Germany).

GOCKE, E.; WILD, D.; ECKHARDT, K.; KING, M. T., 1983, Mutagenicity studies with the mouse spot test.
Mutat. Res. **117**, 201–212.

HAHN, E. M., 1989, Nachweis von Punktmutationen mit Hilfe hochauflösender Gelelektrophorese – Entwicklung und Validierung eines Testsystems.
Wissenschaftlicher Fachverlag Gießen.

HARPER, B. L.; MORRIS, D. L., 1984, Implications of multiple mechanisms of carcinogenesis for short-term testing.
Teratogenesis, Carcinogenesis, and Mutagenesis **4**, 483–503.

HARRIS, H.; HOPKINSON, D. A., 1976, Handbook of enzyme electrophoresis in human genetics.
North-Holland Publishing Company Amsterdam.

HEIMBACH, F., 1984, Correlations between three methods for determining the toxicity of chemicals to earthworms.
Pestic. Sci. **15**, 605–611.

HEIMBACH, F., 1985, Comparison of laboratory methods, using *Eisenia fetida* and *Lumbricus terrestris* for the assessment of the hazard of chemicals to earthworms.
Jour. Plant Diseases Protect. **92**, 186–193.

HOINS, G., 1986, Minderung des Gefahrenpotentials alter Hausmülldeponien in bezug auf Gas und Sickerwasser.
Müll und Abfall **18**, 426–432.

HOLZBECK, L.; VENTULETT, G., 1987, Untersuchungen von Behandlungsmöglichkeiten für das Sickerwasser der Zentraldeponie in Düsseldorf.
Müll und Abfall **19**, 206–212.

HSIEH, D. P. H., 1982, Genotoxic agents in the agro–ecosystem.
In: Fleck, R. A.; Hollaender, A., (eds.), Genetic toxicology. Basic life sciences Vol. **21**, Plenum Press New York.

IRELAND, M. P., 1983, Heavy metal uptake and tissue distribution in earthworms.
In: Satchell, J. E. (ed.), Earthworm Ecology Chapmann and Hall, London, 247–265.

JOHNSON, G. B., 1977, Assessing electrophoretic similarity: The problem of hidden heterogeneity.
Ann. Rev. Ecol. Syst. **8**, 309–328.

KLEKOWSKI, E. J., 1978, Detection of mutational damage in fern populations: An in situ bioassay for mutagens in aquatic ecosystems.
In: Hollaender, A.; de Serries, F. J. (ed.), Chemical mutagens – principles and methods for their detection, Vol. **5**, Plenum Press, New York.

MA, T. H.; HARRIS, M. M., 1985, In situ monitoring of environmental mutagens.
In: Saxena, J. (ed.), Hazard assesment of chemicals, Vol **4**, current developments, Academic Press, Orlando.

MCBRIDE, R. A.; GORDON, A. M.; GROENEVELT, P. H., 1989, Treatment of landfill leachate by spray irrigation – an overview of research results from lake Ontario, Canada. I. Site hydrology.
Bull. Environ. Contam. Toxicol. **42**, 510–517.

NAU, W., 1990, Nachweis induzierter Mutationen bei *Eisenia fetida*: Aufbau eines halbautomatischen Analysesystems.
Diss. FB Agrarwissenschaften Justus–Liebig–Universität Gießen.

ORNSTEIN, L., 1964, Disc electrophoresis. I. Background and theory.
Ann. New York Aced. Sci. **121**, 321.

RAAT, DE W. K.; HANSETREIT, A. O.; KREUK, DE J. F., 1985, The role of mutagenicity testing in the ecotoxicological evaluation of industrial discharges into the aquatic environment.
Food Chem. Toxicol. **23**, 33–41.

RACKE, K. D.; FRINK, C. R., 1989, Fate of organic contaminants during sewage sludge composting.
Bull. Environ. Contam. Toxicol. **42**, 526–533.

RAY, V. A., 1982, Approaches to the genetic safety evaluation of agricultural chemicals.
In: Fleck, R. A.; Hollaender, A. (eds.), Genetic toxicology. Basic life sciences, Vol **21**, Plenum Press New York.

ROWLEY, M. H.; CHRITIAN, J. J.; BASU, D. K.; PAWLIKOWSKI, M. A.; PAIGEN, B., 1983, Use of small mammalians to assess a hazardous waste site at love canal Niagara Falls, New York.
Arch. Environ. Contam. Toxicol. **12**, 383–397.

SCHUBERT, R., 1985, Bioindikation in terrestrischen Ökosystemen.
Gustav Fischer Verl. Stuttgart.

SHAW, C. R.; PRASAD, R., 1970, Starch gel electrophoresis of enzymes: A compilation of recipes.
Biochem. Genet. **4**, 297–320.

TARRADELLAS, J.; DIERCXENS, P.; BOUCHE, M. B., 1982, Methods of extraction and analysis of PCB's from earthworms.
Intern. Jour. Environ. Anal. Chem. **13**, 55–61.

TASSINGTON, J. P., 1985, Genetic toxicology in the industrial practice: General introduction.
Food Chem. Toxicol. **23**, 5–9.

THOMPSON, R. A.; SCHRODER, G. D.; CONNOV, T. H., 1988, Chromosomal aberrations in the cotton rat *Sigmodon hispidus* exposed to hazardous waste.
Environ. Mol. Mutagen. **11**, 359–368.

TIETZ, D., 1987, Review. Gel electrophoresis of intact subcellular particles.
Jour. Chromat. **418**, 305–344.

USEPA (US Environment Protection Agency), 1982, Fate of priority pollutants in public owned treatment works.
USEPA 440/1–82/303.

VANPARYS, P.; MARSBOOM, R., 1985, Genetic toxicology in the pharmaceutical industry.
Food Chem. Toxicol. **23**, 19–22.

VOGEL, E. W., 1984, Assessment of genotoxic profiles of monofunctional alkylating agents in relationship to their initial interaction pattern with DNA.
Bull. Acad. Sci. Biol. **32**, 185–196.

WARE, G. W., 1988, Reviews of environmental Contamination, Vol. **104**.
Springer Verl. New York.

WARE, G. W., 1989, Reviews of environmental contamination and toxicology, Vol. **107**.
Springer Verl. New York.

ZADORY, L., 1986, Biomonitoring von Schwermetallen mit Hilfe von Mollusken.
In: Arndt, U. (ed.), Gesellschaft für Ökologie. Verhandlungen Band XIV, Erich Göltze GmbH Göttingen, pp. 445–449.

22.5 Appendix: Contamination of the household waste site at Hasenbühl.

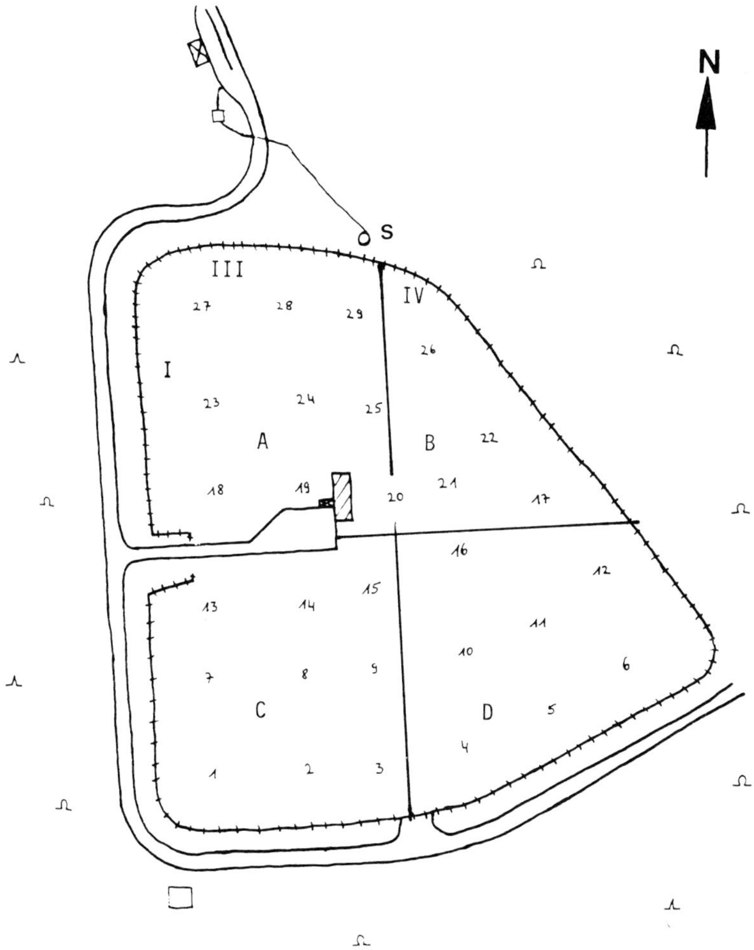

Figure 22.6: General map of the household waste site (7 ha) at Hasenbühl in Schwäbisch–Hall County.

1–29	gas outlets in the 4 sectors A, B, C, and D
I, III, IV	seepage cisterns
S	main seepage reservoir
+++++++++	cockroach fence.

Table 22.5: Contamination of solid waste with heavy metals at Hasenbühl in Schwäbisch-Hall County (COLLINS and SPILLMANN 1985).

substance	decomposing waste, sector A 0–8 mm	decomposing waste, sector A 8–40 mm	waste, rotation drum 0–8 mm	waste, rotation drum 8–40 mm
	% dry matter			
P_2O_5	0.59	0.46	0.74	1.51
K_2O	0.80	0.44	0.74	0.47
MgO	1.08	1.00	0.98	0.66
CaO	6.30	5.50	7.60	6.20
	$\mu g \cdot g^{-1}$ (ppm) dry matter			
Cd	11.9	5.7	5.1	6.3
Pb	390.0	290.0	261.0	157.0
Hg	1.6	1.0	3.2	1.2
As	1.5	2.1	1.5	1.3
Cr	60.0	106.0	158.0	78.0
Ni	44.0	333.0	56.0	190.0
B	39.0	30.0	31.0	26.0
Cu	216.0	283.0	291.0	191.0
Zn	1,310.0	1,840.0	1,100.0	545.0

Table 22.6: Contamination of seepage (1987) determined from random samples out of the cisterns (I, III, IV) of the waste site (personal communication, Dr. H. Hashemi, Schwäbisch-Hall).

date	Nr. cist.	pH	e.c.* $\mu S \cdot cm^{-1}$	BOD_5^*	TOC*	NH_4^+	NO_3^-	PO_4^{3-}	Cl^-	SO_4^{2-}
						$mg \cdot l^{-1}$				
Oct. 11	I	7.7	500	26		0.3				
	III	7.6	840	19		2.8				
	IV	7.5	600	29		2.6				
Oct. 14	I	7.7	510	18		0.3				
	III	7.6	820	26		2.9				
	IV	7.6	610	18		2.2				
Oct. 23	I	8.2	520	27		0.3	1.9		29.3	
	III	7.8	800	16		3.2	4.8		17.7	
	IV	7.7	600	29		2.2	4.0		8.5	
Oct. 24	I	7.8	500	16		0.3	3.9	0.28	35.5	
	III	7.6	740	19		0.2	5.1	0.18	24.1	
	IV	8.0	380	27		0.8	3.8	0.07	5.6	
Nov. 05	I	7.8	540	20		0.3	3.9	0.20	41.1	
	III	7.5	820	22		0.2	5.0	0.20	51.6	
	IV	7.8	420	16		1.0	3.7	0.06	6.1	
Nov. 13	I	7.8	530		125	0.4	6.3	0.50	49.1	30
	III	7.5	660		185	0.2	3.1	0.35	60.1	40
	IV	7.5	430				3.0	0.22	9.1	10

* e.c. electric conductivity, BOD_5 biochemical oxygen demand, TOC total organic carbon.

Table 22.7: Contamination of seepage (1987). Samples were taken from different places of the dump (gas outlets 1 to 25) and from the main seepage reservoir (S), (personal communication, Dr. H. Hashemi, Schwäbisch-Hall).

date	Nr.	pH	e.c.* $\mu S \cdot cm^{-1}$	BOD_5^*	TOC*	NH_4^+	NO_3^- $mg \cdot l^{-1}$	PO_4^{3-}	Cl^-	SO_4^{2-}
Oct. 23	1	8.7	14,580	240		1,500			1,690	
	7	8.5	17,000	220		2,000			1,720	
	15	8.8	12,250	210		2,000			1,380	
	16	7.9	13,700	230		1,500			1,290	
	23	8.2	18,600	360		2,300			2,070	
	S	8.9	14,700	190		700			1,840	
Oct. 28	8	8.1	8,400	130					1,070	
	24	8.2	19,600	730					2,200	
Nov. 05	1	8.2	14,200	220		1,200		9.0	1,700	
	2	7.8	14,700	510		1,280		20.0	1,560	
	3	7.8	9,200	210		900		40.5	1,280	
	7	7.8	13,400	180		2,000		30.0	1,840	
	8	8.3	8,600	130		750		15.0	1,280	200
	15	8.0	12,600	230		2,200		30.8	1,370	
	16	8.3	13,200	250		1,600		20.6	1,420	
	21	8.2	19,200	710		2,800		18.2	2,130	
	23	8.3	21,600	520		2,400		21.0	2,200	200
	24	8.4	20,300	690		2,100		30.1	2,400	250
Nov. 13	1	8.2	14,300			1,200		10.0	1,940	150
	2	7.6	14,700		3,150	1,300	3.1	13.0	1,700	200
	3	7.8	9,750		2,500	1,200	7.3	34.6	2,260	150
	7	8.1	11,700		1,850	1,100	3.8	30.2	2,200	100
	8	7.9	8,400		1,380	1,000	2.7	19.2	1,800	500
	15	7.8	12,400		2,010	1,200	3.8	33.5	2,550	120
	16	8.0	13,400		1,990	1,600	4.3	30.2	1,990	280
	21	8.4	19,700		3,280	2,400	1.7	19.2	3,640	210
	23	7.9	18,500		3,025	3,200	2.5	20.5	4,540	220
	24	8.0	18,700		3,140	2,000	2.7	33.2	4,300	250

* e.c. electric conductivity, BOD_5 biochemical oxygen demand, TOC total organic carbon.

Chapter 23

The sublethal enchytraeid test system: guidelines and some results

W. WESTHEIDE and D. BETHKE–BEILFUSS

23.1 Introduction

As compared to aquatic environments, there is a deplorable lack of appropriate terrestrial ecotoxicological test systems. Whereas a number of aquatic species have become widely used test organisms (STEBBING et al. 1980, PERSOONE and VAN DE VEL 1988), so far the only soil inhabiting test species widely accepted is the mistworm *Eisenia fetida* (SAVIGNY, 1826) used in an acute toxicity test system on artificial soil (EDWARDS 1983, RUDOLPH and BOJE 1986). (In most cases this test may have actually been carried out with the sibling species *Eisenia andrei* BOUCHÉ, 1972, although this cannot usually be discovered from the literature). Disadvantages of this earthworm test are obvious and have been much debated (DEAN–ROSS 1983, HEIMBACH 1986); for instance, both species usually inhabit compost, waste areas and standing manure heaps, and therefore may be of reduced ecological relevance as test species for cropped and forested soil with less organic material.

Emphasis has also been placed on the fact that so far the objective of the

Eisenia test guidelines (OECD 1984; EEC 1984) is the determination of acute toxicity of chemicals by LC_{50} values. Considerations that mortality alone is a rather insensitive parameter in these species under the conditions of the test procedure (DEAN–ROSS 1983) stimulated VAN GESTEL et al. (1989) to develop a reproduction toxicity test with *Eisenia*.

Various attempts, therefore, have been made to work out alternative terrestrial test systems or tests which could supplement the *Eisenia* test, e.g., using species of the highly abundant terrestrial arthropod taxon Collembola (THOMPSON and GORE 1972, SPAHR 1981).

The main difficulty in obtaining new terrestrial test species especially lies in the fact that contrary to the case with many aquatic species terrestrial species generally lack the most essential criterion for a test organism: problem–free cultivation in the laboratory. Consequently, the most convincing reason for developing a test system with enchytraeid annelids was the discovery that species of the genus *Enchytraeus* are easy to culture.

Although RÖMBKE (1989) decided to use soil as a test medium for an acute test with *Enchytraeus albidus* HENLE, 1837, to which he later added a reproductive test procedure, the difficulty of direct observation of life cycle data even in small amounts of soil prompted us to develop and suggest a test system in an artificial medium. Easy observation of the entire set of life cycle parameters of individuals kept in agar makes it possible to use enchytraeids for a reproductive test system on a completely sublethal level.

Several species of the genus *Enchytraeus* HENLE, 1837, have been shown to be potentially useful test organisms (BETHKE–BEILFUSS and WESTHEIDE 1987, WESTHEIDE et al. 1989). They can be easily cultured en masse in petri dishes filled with agar medium for longterm laboratory stocks as well as in low numbers for observation of individual life histories. In our laboratory, during the last six years more than 25 populations of *Enchytraeus* species originating from different habitats and geographic areas and belonging to about 15 different species have been kept in continuous culture on agar. Animals were kept at different temperatures from 15 °C upwards. The most favorable results could be obtained at a temperature of 21 °C, allowing rapid developmental rates and high reproduction, with minimal growth of undesirable fungi on the surface of the agar. At 21 °C, life history characteristics (Figure 23.1) of several *Enchytraeus* species are as follows: As is generally the case in clitellate annelids, eggs (1 to 39 in the species investigated) are spawned in cocoons. The cocoon production remains essentially constant over most of the lifetime, resulting in a cumulative cocoon production curve that is almost linear during the first 40 days (Figs. 23.1, 23.2). Approximately 15 days before death the animals cease to produce eggs and cocoons. Also cocoon production rates are species–specific (0.5 to 1.1 cocoons per day for the first 30 days after maturation). The success of egg hatching is generally high; at least 91% (in other series of experiments 96%) of the eggs are fertile. Egg development time (from spawning to hatching) is from 4 to 8 days; since 8 to 10 days after hatching the animals are mature this results in a considerably short generation time of about two weeks. Mortality rates are almost constant throughout the entire lifespan — there is no adult age-specific mortality (Figure 23.3). Only one species has a lifespan of less than 90 days; mean life spans of other species are between 120 and 160 days under these laboratory conditions.

Enchytraeids provide another test species essential: ecological relevance. The family, including the genus *Enchytraeus*, comprises abundant non-target terrestrial animals in European temperate regions. The species inhabit a large variety of acid soils, including cropped and forested areas, where they occur in high numbers and may represent the dominant taxon (SCHOCH-BÖSKEN and GREVEN 1987, MELLIN 1988, for references). The wide spectrum of food: bacteria, fungi, fresh and dead plant parts, and all kinds of decaying organic material (DOSZA-FARKAS 1982, MELLIN 1988, BROCKMEYER et al. 1990) indicates the complex role of enchytraeids in the soil system; this role also includes litter decomposition by several species. Enchytraeids generally prefer to settle within the surface layer and its wet decomposing material; more than 70% of them may occur in the 0–6 cm horizon when the moisture is sufficient (MELLIN 1988). Thus, they may be highly affected by pesticides and other xenobiotics.

23.2 The test guidelines

Principle of the test (Figure 23.4). Enchytraeids of a given species (for taxonomy see below) are kept in an artificial substrate to which the test substance is added in defined concentrations. (1) The number of cocoons produced in the first 30 days after maturation has started (= **cocoon production**), and (2) the number of cocoons in which the entire set of eggs is fertile (= **hatchability**) are compared with data for untreated individuals. Control survival must exceed 80% of the animals for the entire duration of the test. (Also life span data (see Figure 23.5) can be used as a suitable criterion, but may be considered to be inappropriately time-consuming in routine testing procedures.)

Test organisms. Our investigations have shown that various species of *Enchytraeus* can be successfully used as test organisms and that all give similar results. *Enchytraeus* sp. E (=*E. cf. globuliferus*) is the species of choice because of its particularly easy cultivation together with its relatively high susceptibility to chemicals. The species cannot be separated from *E. variatus* BOUGUENEC and GIANI, 1987 on morphological features. Ultrastructure of sperm, protein data, immunological distance, and DNA data, however, distinctly distinguish it from this species. Since also cross-breeding experiments with *E. variatus* failed, the erection of a new species is envisaged (GRAEFE and WESTHEIDE, in preparation).

The animals are mass cultured in large petri dishes (diameter 15 cm) half filled with 1.5% agar medium that contains a solution of 1 g $Ca(NO_3)_2$, 0.25 g $MgSO_4$ ×

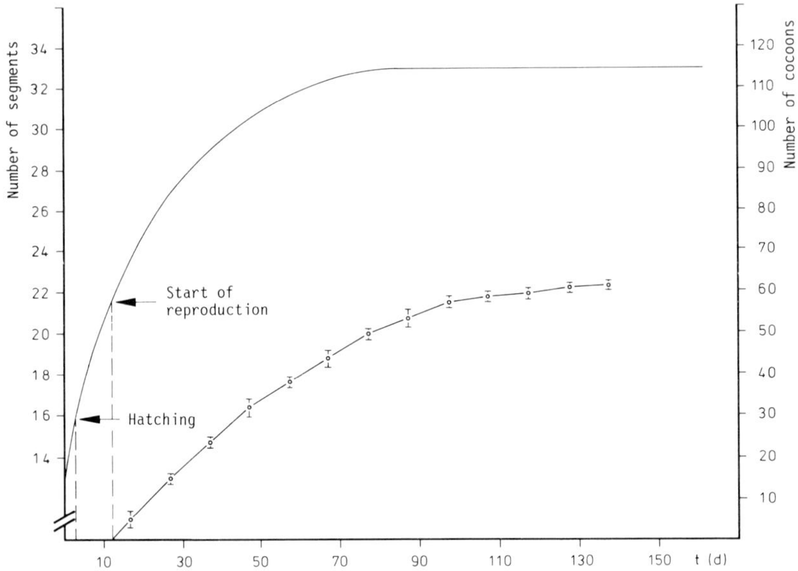

Figure 23.1: Generalized growth curve of *Enchytraeus* sp. I at 21 °C after mean number of segments of 13 specimens cultured in agar, and their cumulative cocoon production.

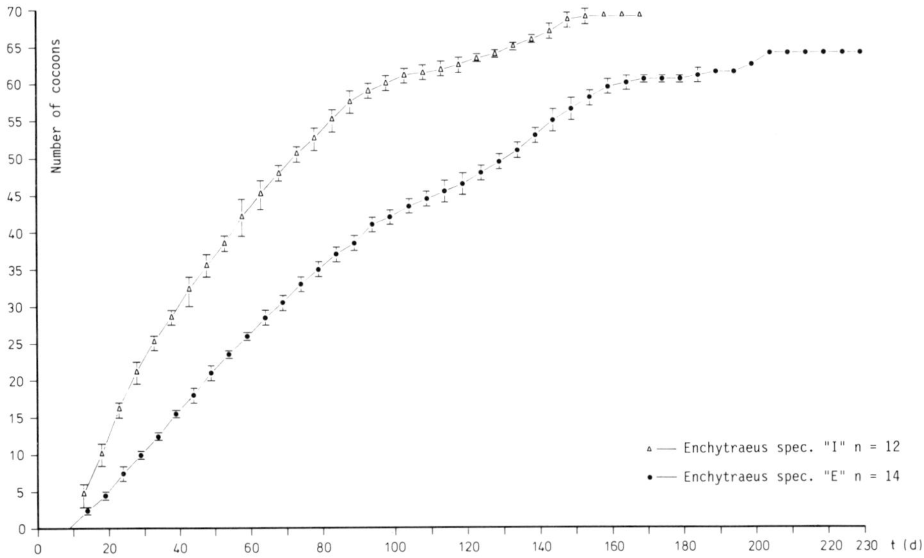

Figure 23.2: Cumulative counts of cocoons of two *Enchytraeus* species in agar at 21 °C, demonstrating almost linear production curves for a considerable time during the reproductive period.

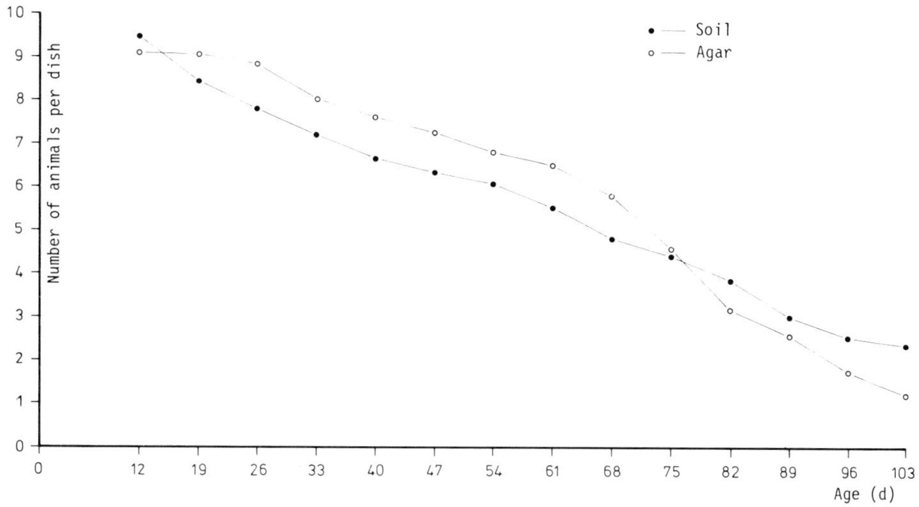

Figure 23.3: Mortality of *Enchytraeus* sp. E in agar plates and small amounts of natural soil. Points indicate number of surviving animals. Mean values of 40 (agar) and 24 culture dishes (soil). Recordings were started with each 10 juvenile individuals of the same age. Food: flaked oats. Temperature: 21 °C.

7 H_2O, 0.25 g KNO_3, 0.25 g KH_2PO_4 and traces of $FeSO_4$ in 1 l deionised water (pH approximately 5.0). The cultures are stored in incubators in the dark at a temperature of 21 °C (humidity 85–90%). Flaked oats are used for food and put in excess on the surface of the agar plates. After about two months the stock organisms must be transferred to fresh agar. Cultures of this species could be shown by isoenzyme and general protein electrophoresis to be genetically completely monomorphic populations (WESTHEIDE and BROCKMEYER, in press; BROCKMEYER, in preparation).

Test conditions and test procedure. Test conditions and the test procedure is pointed out in Figure 23.4. The test ist performed in the same substrate as that used for the culture medium of the stock organisms, except that food and the substance to be tested are added. Ground flaked oats are folded into the agar and autoclaved at 120 °C. The chosen concentrations of the test substances are thoroughly mixed with the agar immediately before it is poured into the petri dishes, i.e. at a temperature of about 40 °C. Non–liquid chemicals must be finely ground, dissolved or thoroughly dispersed in water before they are added to the medium.

Controls contain the basic medium and an amount of deionised water adequate to the amount of the test substance. All test procedures are carried out in glass

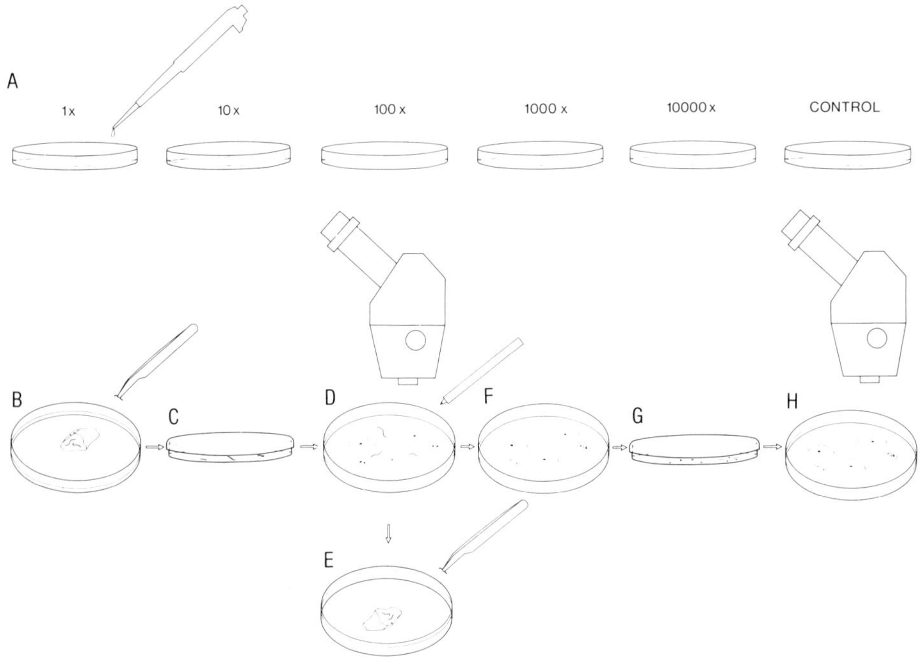

Figure 23.4: Pictorial representation of the sublethal enchytraeid laboratory test procedure. A Addition of different concentrations of the test substance to the agar medium in petri dishes. B Transfer of three test individuals into a small depression of the agar layer. C Storage of the test dish with cover at 21 °C in an incubator for 5 days. D, F Inspection of test dish under a dissecting microscope; counting and marking of the cocoons produced on the bottom of the dish. E Transfer of test specimens to another, fresh test plate. H Inspection of cocoons after 10 days (G) and recording of the cocoons from which all juveniles have been emerged.

petri dishes with a diameter of 5 cm. The covers of the dishes must allow free access of air.

The test organisms should be mature and of the same age (8 days old from 21 °C cultures). Three individuals are carefully transferred by flat–bladed featherweight forceps onto one agar plate. To prevent the worms creeping on the surface of the medium, eventually crawling over to the cover and drying out, they should be placed into a small depression of the agar layer cut out by the forceps and afterwards covered by a piece of agar. The worms will burrow within the agar, their surface being completely in contact with the substance to be tested.

The duration of the test is 30 days, 25 days are needed for counting the cocoons produced and 30 days for evaluating hatchability. After every 5th day the test dishes must be inspected. The cocoons, which are easily visible within the agar, are counted under a dissecting microscope. During each inspection the test organisms are transferred to another, fresh agar test plate, prepared together with the first one. This is repeated five times. The position of the cocoons produced is marked on the bottom of the dishes, from which the worms are removed. These dishes stay in the test run and are inspected ten days later.

The number of cocoons completely abandoned by the juveniles reveals the percentage of hatchability. (It is usually possible to observe the single eggs and their fertility within the cocoons. This procedure, however, is time-consuming. The counting suggested above appears to be justified, because fertility or infertility usually affects the entire set of eggs within an cocoon.)

The test is run with a control and five different non-lethal concentrations of the substance to be tested, comprising five replicates each. Range-finding tests must be carried out in advance, to determine lethal concentrations below 10% in the first five days after application.

23.3 Test results and discussion

Enchytraeid specimens in agar have been exposed to a series of different pesticides and other environmental substances according to the test guidelines presented. These tests demonstrated susceptibility of the enchytraeids to a wide range of chemicals and revealed substance-specific reactions, which made it possible to establish an order of relative toxicity.

The insect growth regulator Dimilin 25WP (Diflubenzuron), for instance, resulted in no effects on the production of cocoons even at very high concentrations (Figure 23.6). Ten times the minimum recommended concentration of the commercially obtained agricultural fungicide DuPont Benomyl (Benomyl) reduced the number of cocoons by more than 50%. Ripcord 10 (Cypermethrin) known as a highly effective pesticide to non-target soil animals in the field (SCHMITT et al. 1989, FUNKE and ROTH-HOLZAPFEL, this volume), was also the most toxic chemical in this comparison. It reduced the number of cocoons produced by more than 50% already at its minimum recommended concentration. The production of cocoons was ceased completely at the 100-fold concentration.

There was a comparable gradation of sublethal effects on hatchability with Dimilin 25WP, Curaterr 500 SC, and DuPont Benomyl. Again, Dimilin showed no effects at any of the concentrations applied (Figure 23.7). Curaterr considerably reduced fertility at concentrations exceeding the minimum recommended concentration. Benomyl produced striking effects at an even lower concentration level. Egg development and hatching completely ceased in agar plates containing this substance's minimum recommended concentration for agricultural application.

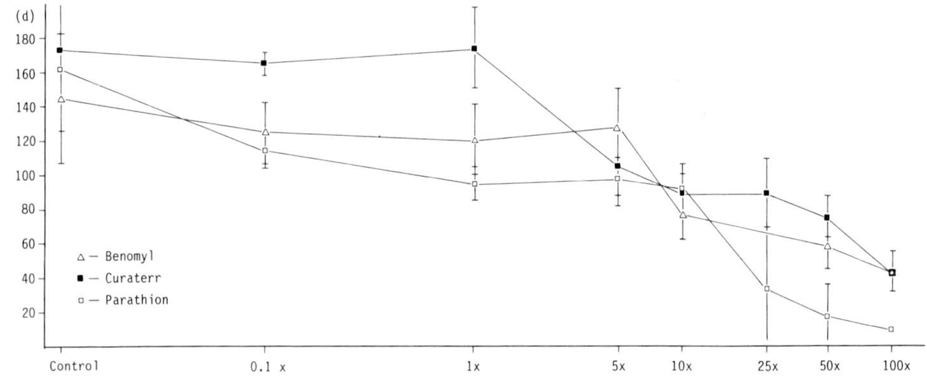

Figure 23.5: Relative toxicity of three pesticides. Lifespan of *Enchytraeus* sp. E in agar plates at 21 °C with different concentrations of DuPont Benomyl (Benomyl) ($1\times = 23.6\ \mu l \cdot ml^{-1}$ substrate), Curaterr 500 SC (Carbofuran) ($1\times = 0.12\ \mu l \cdot ml^{-1}$ substrate), and E 605f (Parathion, $1\times = 0.008\ \mu l \cdot ml^{-1}$ substrate). Bars: standard deviation.

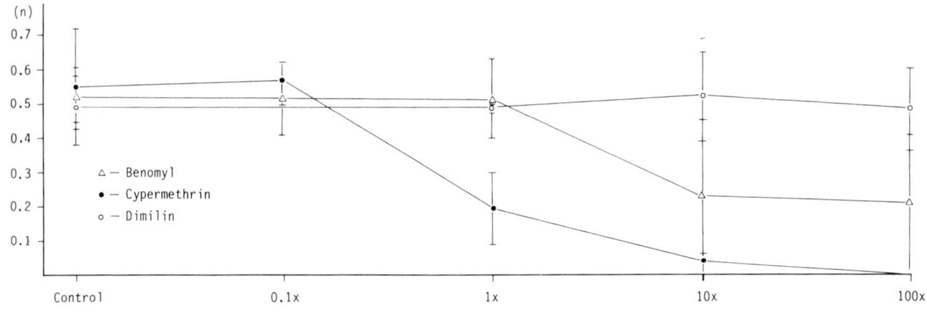

Figure 23.6: Relative toxicity of three pesticides. Number of cocoons produced per day of one *Enchytraeus* sp. E individual in agar plates at 21 °C with different concentrations of DuPont Benomyl ($1\times = 23.6\ \mu l \cdot ml^{-1}$), Ripcord 10 (Cypermethrin) ($1\times = 0.012\ \mu l \cdot ml^{-1}$), and Dimilin 25WP (Diflubenzuron) ($1\times = 11.8\ \mu l \cdot ml^{-1}$). Mean values of 15 individuals. Bars: standard deviation.

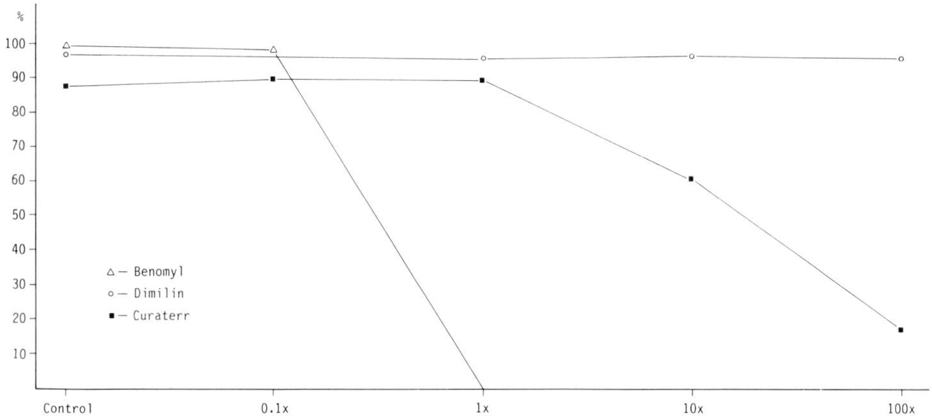

Figure 23.7: Relative toxicity of three pesticides. *Enchytraeus* sp. E. Percentage of cocoons from which all juveniles emerged. Mean values of individuals in agar plates at 21 °C with different concentrations of DuPont Benomyl ($1\times = 23.6\ \mu l \cdot ml^{-1}$), Dimilin 25WP ($1\times = 11.8\ \mu l \cdot ml^{-1}$), and Curaterr 500 SC ($1\times = 0.12\ \mu l \cdot ml^{-1}$).

The sublethal data for these chemicals do not necessarily correspond with their acute toxicity data or with data obtained from lifespan experiments. It could be observed, indeed, that concentrations of benomyl lower than the minimum recommended concentration significantly shortened the lifespan of test organisms (Figure 23.5). But even at a concentration corresponding to 100–fold the recommended application concentration lethality did not exceed normal mortality rates in the controls during the first 30 days after application of the test species recommended (WESTHEIDE et al. 1991). Thus, the impact of benomyl may well be underestimated if its sublethal effects, especially on hatchability, are not considered. Non–acute effects of this kind, however, cause population changes and have high ecological impact. Accordingly, the test parameters cocoon production and hatchability may be considered to be especially meaningful, because they directly affect the populations' dynamics.

Ecotoxicological laboratory test data generally present the problem that they cannot be directly applied to give an exact indication of the chemicals' impairment in the environment. This may be even more problematic if a test system uses an artificial medium, as with the procedure proposed here. Various characteristics of normal soil, e.g., adsorption of chemicals to particles, must be considered to be different from those in agar. The animals contact with the pore water in natural soil and its soluted xenobiotics, however, may be as intimate as with the agar and the test substances which it contains. Exposure of substances to the intestinal tract by ingested food and substrate may also be very similar in natural soil or in the artifical medium. Indeed, the toxicity values of benomyl and other substances obtained from tests carried out in small quantities of soil could be reconciled to a large extent in agar (BETHKE–BEILFUSS and WESTHEIDE 1987, WESTHEIDE et al. 1989), and effects often were of the same order of magnitude. It also could be demonstrated that the concentration of benomyl which stopped hatching in the agar plates resulted in a comparable effect even in larger amounts of soil: considerable decrease of the reproductive capacity (WESTHEIDE et al. 1991).

Thus, the high advantage of direct, easy and exact quantitative observation of test parameters, which is the most obvious characteristic of the test system presented, may well outweigh the general disadvantage of the medium's lack of ecological relevance.

General advantages of using enchytraeids as test organisms and realization of test essentials by the specific system proposed are as follows:

- ecological relevance, high abundance and wide distribution of the taxon
- easy culturing of test species
- unproblematic feeding
- economical maintenance of stock cultures and test dishes
- short generation time
- production "on demand" of test specimens of same age
- genetically monomorphic test specimens
- direct and easy observation of specimens during the test procedure
- quantifiable numerical test parameters not exposed to subjective judgement
- meaningful effect–criteria
- susceptibility to a wide range of chemicals and other xenobiotics
- substance–specific reactions.

Acknowledgments

For obtaining enchytraeids for our cultures, we wish to thank Dipl. Biol. U. Graefe, Hamburg, and Dr. J. Römbke, Frankfurt. We are grateful to Mary E. Petersen, Copenhagen, and Prof. Dr. H. Becker, Berlin, for critically reading the manuscript. The investigations were supported by BMFT grant ♯ PTB 03 7426 7.

Bibliography

BETHKE–BEILFUSS, D.; WESTHEIDE, W., 1987, Subletale Schädigungen terrestrischer Enchytraeiden (*Oligochaeta, Annelida*) durch Pflanzenbehandlungsmittel: Veränderungen der Kokonproduktion und der Eifertilität.
Verhdlg. Ges. Ökol. **16**, 417–422.

BROCKMEYER, V.; SCHMID, R.; WESTHEIDE, W., 1990, Quantitative investigations of the food of two terrestrial enchytraeid species (*Oligochaeta*).
Pedobiologia **34**, 151–156.

DEAN–ROSS, D., 1983, Methods for the assessment of the toxicity of environmental chemicals to earthworms.
Regulatory Toxic Pharmac. **3**, 48–59.

DOSZA–FARKAS, K., 1982, Konsum verschiedener Laubarten durch Enchytraeiden (*Oligochaeta*).
Pedobiologia **23**, 251–255.

EEC, 1984, Directive 79/831, Annex V, Rev. **5**. Toxicity for earthworms artificial soil test.

EDWARDS, C. A., 1983, Report of the second stage in development of a standardized laboratory method for assessing the toxicity of chemical substances to earthworms.
DOC XI/83/700, Commission of the European Communities.

FUNKE, W.; ROTH–HOLZAPFEL, M., 1991, Animal–coenosis in the ecosystem "Spruce Forest" (Protozoa, Metazoa – Invertebrates) — Indicators of alterations in forest–ecosystems. (This volume).

GESTEL, C. A. M. VAN; DIS, V. A. VAN; BREEMEN, E. M. VAN; SPARENBURG, P. M., 1989, Development of a standardized reproduction toxicity test with the earthworm species *Eisenia fetida andrei* using copper, pentachlorphenol, and 2,4–dichloroaniline.
Ecotoxicology and Environmental Safety **18**, 305–312.

HEIMBACH, F., 1981, Methoden zur Bestimmung des Gefährdungspotentials von Pflanzenschutzmitteln für Regenwürmer.
Verdhlg. Ges. Ökol. **14**, 237–241.

MELLIN, A., 1988, Untersuchungen zur Autökologie und Funktion von Enchytraeiden, Tubificiden und Aeolosomatiden (*Annelida, Oligochaeta*) im Ökosystem Kalkbuchenwald.
Ph. D. Thesis, Universität Göttingen, FR Germany.

OECD, PARIS, 1983, Guideline for Testing of Chemicals, 208 "Earthworm, Acute Toxicity Test".

PERSOONE, G.; VAN DE VEL, A., 1988, Cost–analysis of five current aquatic toxicity tests.
EC Report EUR/11432/EN, Commission of the European Communities, Brussels.

RÖMBKE, J., 1989, *Enchytraeus albidus* (*Enchytraeidae, Oligochaeta*) as a test organism in terrestrial laboratory systems.
Arch. Toxicol., Suppl. **13**, 402–405.

RUDOLPH, P.; BOJE, R., 1986, Ökotoxikologie. Grundlagen für die ökotoxikologische Bewertung von Umweltchemikalien nach dem Chemikaliengesetz.
Ecomed, Landsberg.

SCHMITT, G.; FUNKE, W.; KRANIZ, V., 1989, Auswirkungen von Pflanzenschutzmitteln auf die Eklektorfauna eines Fichtenforstes.
Jber. naturw. Ver. Wuppertal **42**, 112–115.

SCHOCH–BÖSKEN, J.; GREVEN, H., 1987, Artenspektrum und Dominanzstruktur von Enchytraeiden einiger Buchen– und Fichtenwälder Nordrheinwestfalens.
Verhdlg. Ges. Ökol. **16**, 79–84.

SPAHR, H. J., 1981, Die bodenbiologische Bedeutung von Collembolen und ihre Eignung als Testorganismen für die Ökotoxikologie.
Anz. Schädlingskde, Pflanzenschutz, Umweltschutz **54**, 27–29.

STEBBING, A. R. D.; AKESSON, B.; CALABRESE, A.; GENTILE, J. H.; JENSEN, A., LLOYD, R., 1980, The role of bioassays in marine pollution monitoring. Bioassay panel report.
Rapp. P.–v. Reun. Cons. int. Explor. Mer **179**, 322–332.

THOMPSON, A. R.; GORE, F. L., 1972, Toxicity of twenty–nine insecticides to *Folsomia candida*: Laboratory studies.
J. Econ. Entom. **65**, 1255–1260.

WESTHEIDE, W.; BETHKE–BEILFUSS, D.; GEBBE, J., 1991, Effects of benomyl on reproduction and population structure of enchytraeid oligochaetes (*Annelida*) — Sublethal tests on agar and soil.
Comp. Biochem. Physiol. (C), (in press).

WESTHEIDE, W.; BETHKE–BEILFUSS, D.; HAGENS, M.; BROCKMEYER, V., 1989, Enchytraeiden als Testorganismen — Voraussetzungen für ein terrestrisches Testverfahren und Testergebnisse.
Verhdlg. Ges. Ökol. **17**, 793–798.

WESTHEIDE, W.; BROCKMEYER, V., 1991, Suggestions for an index of enchytraeid species (*Oligochaeta*) based on protein patterns.
Soil Biology and Biochemistry (in press).

Chapter 24

Transfer mechanisms and deposition rates of atmospheric pollutants

W. Kuttler

24.1 Introduction

In light of the noticeable damage to different environmental elements in the marine and terrestrial ecosystems over the past few decades, the extent of man's influence on the natural and cultural environment is increasingly being brought into question. In this context man-made air pollution with its wide-spread interferences in natural cycles is of great importance.

The atmosphere is being burdened with a great volume and increasing number of airborne pollutants from various sources. Consequently it has become of far-reaching importance for the environment that the mechanisms of distribution of these pollutants, their residence time, the deposition, and in the final analysis, the effects of their deposition on the ground of totally different ecosystems are examined.

Although the causes and effects are well-known in most cases, objective data are often not available to obtain a balanced picture of the whole state of nature from the knowledge of the atmospheric deposition of pollutants and their distribution in

Table 24.1: Emissions of the pollutants nitrogen oxide (NO_x, calculated as NO_2), organic compounds (org.), sulphur dioxide (SO_2), carbon monoxide (CO), and dust, in the Federal Republic of Germany for the period 1966 to 1986 (data in million tons per year, after UBA 1989). The right column gives the change 1966–1986 in percent of the value of 1966.

	1966	1968	1970	1972	1974	1976	1978	1980	1982	1984	1986	%
NO_x	1.95	2.05	2.35	2.50	2.60	2.70	2.85	2.95	2.85	2.95	2.95	151
org.	2.20	2.35	2.60	2.65	2.55	2.55	2.55	2.50	2.40	2.40	2.45	111
SO_2	3.35	3.40	3.75	3.75	3.65	3.55	3.40	3.20	2.85	2.65	2.30	69
CO	12.40	12.90	14.00	14.00	13.70	13.00	12.40	11.70	9.80	9.20	8.90	72
dust	1.75	1.50	1.30	1.10	0.95	0.78	0.70	0.69	0.60	0.59	0.56	32

time and space.

Facts are needed to solve these problems, to find out about the kind of atmospheric transfers, their temporal variability as well as their spatial differences with regard to the different emission patterns.

Furthermore, it is important for any investigation of air purity and ecosystems to know about the physical and chemical changes of the pollutants which are occuring and how they are deposited over a long period of time. First, detailed statements about pollution in the Federal Republic of Germany and the various mechanisms of self-cleaning in the atmosphere will be discussed before the concluding results of typical examinations for single pollutants in different areas are dealt with.

24.2 The emissions of different pollutants in the Federal Republic of Germany

In this report about air pollution in the Federal Republic of Germany only those typical pollutants will be dealt with for which possible trends in the distributed amounts of emissions can be illustrated, because they have been measured over a sufficiently large area and over an adequate long period of time. These consist of the following: nitric oxides, organic compounds, sulphur dioxide, carbon monoxide and dust (Tab. 24.1).

Nitrogen compounds are found in the atmosphere as oxides, acids and in reduced form. The nitric oxides (NO_x), in particular, play an important role in the formation of photo-chemical smog and the acidification of fog and precipitation (FABIAN 1989). These pollutants which result from the reaction of atmospheric nitrogen with atmospheric oxygen are released into the air by nearly all kinds of combustion. Nitric oxides are emitted by traffic on a large scale. Additionally – but mainly in agricultural areas – nitrogen is released by nitrogenous fertilizers.

In 1986 2,950,000 tons of nitric oxides were emitted; compared to 1966 (= 100%) this means an increase to 151%. This can mainly be explained by the enormous increase in the number of automobiles, reaching 27,000,000 cars in 1986, twice the number of, for example, 1960. The largest shares of these emissions in 1986 were from traffic (52%) and power stations (24%).

The **organic** compounds are mainly substances which are derived from solvents (41%) or are released by combustion (45%). They are additionally released by transshipment and evaporation of petrol, e.g. during refuelling of automobiles. The emissions of organic compounds have hardly changed since 1966 having reached an amount of 2,450,000 tons in 1986.

95% of the **sulphur** released into the atmosphere by man-made processes is in the form of sulphur dioxide (KELLOG et al. 1972). Power stations produced the largest share of these emissions with 63%; industry had an 18% share. The remainder is made up of emissions from domestic households and, to a lesser extent, traffic. Since 1966 the SO_2 emissions have decreased from 3,350,000 tons to 2,300,000 tons (1986) thus falling to only 69% of the former amount. There are several reasons for this decrease. For example, the increasing consumption of low-sulphur fuels in power stations and the application of filters in smoke stacks.

The emissions of **carbon monoxide** – 80% of which are produced by incomplete combustion – have decreased from 12,400,000 tons (1966) to 8,900,000 tons (1986). This is due to the decreased combustion of solid fuels and the emission controls for automobiles.

The **dust** emissions have also been lowered in the same period from 1,750,000 tons (1966) to 560,000 tons (1986), dropping to less than a third of the former amount. This could be achieved by improved modern technology of dust filtration. However, industry still emits the greatest share of the dust, and traffic 10%. The remainder is caused by domestic households and low-consumption units.

All these emissions are bound up by transmission into long-range transport and are removed from the atmosphere by the different mechanisms of deposition.

24.3 Mechanisms of atmospheric self-cleaning

The atmospheric pollutants released by natural and man-made processes are removed from the atmosphere sooner or later. The important factors for the residence time are the height of the polluting sources, the speed of the chemical changes, the intensity of the ground-layer exchanges as well as the surface characteristics of the affected area and its affinity towards the pollutant on the ground (JACOBSEN et al. 1982).

We distinguish between three groups of gaseous pollutants because of their differing residence time in the atmosphere (Tab. 24.2).

While nitrogen and oxygen have residence times of some thousand, or even million, years gaseous pollutants of the second group (i.e. hydrogen, carbon dioxide, methane) have periods between two and eight years. In the third group we have

Table 24.2: Residence times of different gases in the atmosphere (from MESZAROS 1981)

group 1	
Helium	10^7 yr
Nitrogen	$\sim 10^6$ yr
Oxygen	5×10^3 yr
group 2	
Hydrogen	6–8 yr
Methane	4–7 yr
Carbon dioxide	5–6 yr
Ozone	2 yr
group 3	
Water vapor	10 d
Nitrogen dioxide	8–10 d
Ammonia	$\sim$ 5 d
Sulfur dioxide	$\sim$ 2 d
Carbon monoxide	5–12 d
Hydrogen sulfide	$\sim$ 12 d

the short-lived gases such as water vapour, sulphur dioxide and carbon monoxide with a duration of days or some hours.

The time that a substance remains in the atmosphere is expressed by JUNGE (1987) in the following equation:

$$\frac{dM}{dt} = Q - S \tag{24.1}$$

The term dM/dt describes the change in concentration of a substance in the atmosphere with respect to time; Q is the change in intensity at the source and S is the increase in intensity.

Ignoring the chemical changes, there are three mechanisms of pollutant reduction. These are dry, humid and wet deposition. A description of the theoretical fundamentals of dry and wet deposition is given by HERBERT (1987). JAESCHKE (1987) and SCHMITT (1987) discuss the mechanisms of humid deposition. While dry deposition occurs continually, humid and wet deposition are limited to the duration of fog and precipitation. For wet deposition to take place it is necessary that the rate of fall of raindrops or snowflakes be greater than any possible upwinds and that the drops do not evaporate on their way to the ground.

24.3.1 Dry deposition of pollutants

Dry deposition occurs at interfaces like the ground, vegetation or water surfaces. We distinguish between the following mechanisms of dry deposition:

- Sedimentation, especially with particles having radii of more than 0.5 μm,
- Impaction, the deposition at obstacles in the stream of the air according to the law of inertia; this can also be the collection of particles by raindrops falling through a polluted atmosphere,
- Diffusio- and thermophoresis, pollutants are deposited because of prevailing water vapour in the atmosphere or temperature differences in the air,
- Turbulent and molecular diffusion, steering the process of deposition through Brownian motion and convective or turbulent motion.

The precise determination of dry deposition is still considerably difficult. The possibilities for the determination of dry deposition are listed in the following table; the capital letters show whether the quoted method is suitable for gases, G, or aerosols, A. The brackets indicate that the given method is less suitable for this form.

Profile method/Gradient method	G,	(A)
Bowen-ratio method	G	
Eddy correlation	G,	(A)
Eddy accumulation	G,	A
Variance method	G,	(P)
Chamber method	G,	(P)
Tracer method	(G),	A
Mass-stream balance	G,	A
Exposition of artificial surfaces	(G),	A

It is impossible to describe all these individual methods within this report; instead the reader is referred to the survey of various works by Höfken and Bauer (1985) from which the above information was also drawn.

The deposition rates of pollutants from the atmosphere onto exposed surfaces can be calculated with equation (24.2).

$$F = v_d \cdot C \tag{24.2}$$

Knowing the deposition velocity (v_d) and the concentration of pollutants (C) one can determine the rate of deposition (F).

For atmospheric gaseous pollutants, which are absorbed onto the ground relatively well, a mean value for the deposition velocity of 1 cm·s^{-1} (Beilke 1975) is used. Examples for different deposition velocities have been summarized in Tab.

Table 24.3: Deposition velocities (v_d) of SO_2 over different surfaces (after MEYERS and BALDOCCHI 1988, FÄHNRICH and GEORGII 1989; from FÄHNRICH and GEORGII 1989 (No. 1–8), GEORGII 1989 (No. 9, 10))

No.	surface	method	v_d ($cm{\cdot}s^{-1}$)	author	year
1	water	Gradient	0.41	GARLAND	1977
2	calcareous-soil	Gradient	1.20	GARLAND	1977
3	grass	$^{35}SO_2$-Tracer	0.80	OWERS	1974
4	wheat	Gradient	0.74	FOWLER	1976
5	pine	$^{35}SO_2$-Tracer	0.10–0.60	GARLAND	1978
6	oak–hickory	Eddy-Correlation	0.21–1.14	MEYERS	1986
7	spruce	Filter sampling	0.11–0.86	FÄHNRICH	
8	beech	Filter sampling	0.25–0.70	FÄHNRICH	
9	snow	–	0.10	GEORGII	1989
10	forest (10 m tall)	–	0.10–2.00	GEORGII	1989

24.3. SO_2-deposition velocities of up to 8 $cm{\cdot}s^{-1}$ were determined for urban areas (compare survey in SEHMEL 1980).

A mean value of $v_d = 0.8$ $cm{\cdot}s^{-1}$ is often used for general calculations in Europe (OECD 1977), which together with the mean concentration of pollutants is taken to estimate deposition of SO_2 in areas with varying pollution patterns. As dry deposition is considerably influenced by vegetation we need precise facts for cause-and-effect research.

A calculation of the deposition rates of atmospheric sulphur e.g. areas of vegetation lead to errors in the estimates if a deposition velocity for SO_2 of $v_d = 0.8$ $cm{\cdot}s^{-1}$ is used to the part of the dry deposition. It would be better to take the so called "critical loads" into consideration when calculating the total pollutant input in an ecosystem (NILSSON and GRENNFELT 1988). This will be dealt with again later.

24.3.2 Humid deposition of pollutants

Humid deposition is the process of pollutant decrease whereby droplets of fog and clouds act as "carriers" of the pollutants. This process ends with the settling of condensed vapour on the different surfaces – e.g. by the "sweep-out" of vegetation.

Pollutants can easily be absorbed during the formation of droplets in fog or cloud. We distinguish between three different forms of fog: radiation fog, upslope fog and frontal fog.

Radiation fog develops by radiation from the ground and tends to form in morphological depressions. It consists of small droplets, is not moist and contains predominantly local pollutants which have mixed with the cold air in valleys or basins. Fog in valleys is mostly separated from a clear air mass above by a strong temperature inversion located at the height of the valley rim. This kind of fog is most prevalent in weather conditions with low wind velocities.

The formation of *upslope fog* is connected with expansive anticyclonic weather conditions. This fog is generated at inversions due to the presence of radiation together with sufficient air humidity. Since upslope fog usually covers relatively large areas, the pollutants in this fog are not only limited to those from local sources, but also consist of pollutants from other regions.

Frontal fog covers mountainous regions. As the occurance of this fog is tied to the existence of frontal systems, the incorporated pollutants result from long distance transport. In the winter months, because of the lower level of the clouds larger areas are exposed to the humid deposition of pollutants than in the summer months. A considerable part of the annual precipitation on the luffslopes of highlands, in particular, is due to the "sweep-out" by vegetation. According to LINKE (1916) the additional moisture from dripwater is 66% in the Taunus region and 150% at the Rhön relative to precipitation in open spaces.

Previous works (e.g. HOUGHTON 1955) have already referred to the fact that the concentration of pollutants in fogdroplets was in some cases substantially higher than that in comparable samples of rainwater. Furthermore, there are considerable differences between the pollutants contained in the numerous droplets. The increased concentration of pollutants in fog is mainly caused by the favourable surface-to-volume ratio of fog, which consists of smaller droplets than rain, and by the higher concentration of pollutants near to the ground causing greater rates of absorption than at higher altitudes.

24.3.3 Wet deposition of pollutants

Wet deposition takes place through the precipitation of rain and snow. According to VDI RDL (1983) the mechanisms leading to the incorporation of gaseous pollutants and pollutant particles in clouddroplets or raindrops are :

- The formation of clouddroplets using sulphurous and nitrogenous aerosols as the nuclei of condensation; 80% of the sulphur and nitrogen compounds incorporated in raindrops are "embodied" in this way.
- The dissolution and oxidation of pollutants (SO_2 and NO_x) in raindrops. Because of the great surface-to-volume ratio and lower rate of fall small raindrops have a relatively high rate of absorption compared to large raindrops (LANDSBERG 1954, GEORGII 1965). This process is assumed to account for between 10% and 30% of the total absorption of pollutants.
- Diffusiophoresis, whereby smaller particles have a higher absorption rate than larger ones; on the whole this accounts for only about three per cent of the total absorption of pollutants.

- Brownian motion; this contributes only about one per cent to the absorption of pollutants.
- Impaction; this incorporation of pollutants by falling raindrops depends on the size of the falling drops and the intensity of precipitation, for smaller rain quantities (less than 2 mm per hour) this process contributes to about 30% of the total absorption of pollutants.

In wet deposition a distinction is made depending on where the pollutants have been absorbed by the raindrops. Incorporation into the falling raindrop which occurs below the cloud – that means between the base of the cloud and the ground – is called "washout" or "below cloud scavenging". Should the absorption occur within the cloud, then the expressions "rainout" or "in cloud scavenging" are used.

Wet deposition can be calculated according to (24.3)

$$D_w = r_w \cdot C_{aq} \tag{24.3}$$

(D_w = wet deposition of pollutants, r_w = rate of precipitation, C_{aq} = Concentration of pollutants in the liquid phase).

In addition to this, the intensity of precipitation determines the washout rate: a high intensity provides a more efficient reduction of air-pollution; drizzle with a constant velocity is more efficient than rain with larger drops as it has a higher number of drops and, thus, a greater total surface area available for absorption (GEORGII and WÖTZEL 1970).

24.3.4 Filtration of pollutants by vegetation

Together with the settling of the atmospheric pollutants on the ground the filtration and ensuing absorption of pollutants by vegetation play an important role. If plants are seen as filtering elements in the environment it can be concluded that the total deposition of pollutants at forest locations below the canopy of treetops, either evergreen or deciduous, derives from humid, wet and dry depositions.

The humid and wet deposition by rain, snow or fog transfer the pollutants to the ground by way of an interdeposition on the plants.

A part of the pollutant particles or gaseous pollutants deposited by the mechanism of dry deposition are sedimented on to the different surfaces, another part is absorbed or adsorbed by the surfaces. This will be explained in detail later on. In addition, numerous pollutants are oxidized; this occurs either on the surface of the plant (e.g. by dew, interceptive or phytogenic water) or after the absorption by the plant into the cell.

The combined pollutants are gathered in the dripwater consist of according to their origin: the pollutants carried by rain (sum of "rainout" and "washout"), the pollutants deposited during dry periods which are washed off, thus excluding those particles remaining on the part of plants above the ground (according to HÖFKEN et al. (1981), referring to HULL (1970), this is, with the exception of heavy metals and nitrogen components, only a negligible percentage) and those phytogenic substances derived from the washout of the plants ("leaching"). Leaching of these substances depends on the species of plant, the supply of nutrients, the pollutants involved as well as on the season (ULRICH et al. 1981). Furthermore, the leaching of Mg, Ca und K is intensified by the influence of various pollutants – such as SO_2 and/or O_3 (GUDERIAN et al. 1987). The leaching of sulphate–sulphur and nitrate–nitrogen could not be proved (HÖFKEN et al. 1981), and so the increase in these pollutants can only be due to the added dry deposition in the dripwater.

Dry deposition, which is interpreted by ULRICH et al. (1981) as the interception which is comprised of the adsorption of aerosols and fog and the absorption of gases at the plant surfaces, can be determined, for those pollutants which do not derive from leaching, from the difference between the concentrations in precipitation and the run-down from trunks (and stalks) as well as dripping.

The sum of all single depositions yields the total deposition in the region of the trunk and in the soil below. An admittedly smaller share of pollutants that sticks to the leaves of the trees can pass back into the atmosphere (re-suspension or re-insertion), by way of evaporation or dry mechanisms such as removal by the wind. Pollutants can also be taken up and integrated in the metabolism.

A certain portion of substances in the region of the tree trunk finds its way to the ground via the litter of needles and leaves. There they decompose and return to the natural soil and plant cycle.

However, when atmospheric pollutants are taken up by the plant two basic methods of uptake beside the rootsystem can be distinguished: cuticular and stomatic uptake. The cuticles do not put up the same resistance to diffusion at every point, possibly for species-specific reasons based on the different structures, but also because of differing sunshine conditions (leaves in the shadow or in the sun; ZIEGLER 1984). Age-dependent influences on the gas transport must also be considered. Gaseous pollutants are taken up by the stomata to a considerably higher degree than by the cuticles. The resistance which the gaseous pollutants have to overcome in the organs of assimilation consists of stomata resistance and resistance brought against the entering gases by the mesophyllic cells.

The influence open and closed stomata have on deposition velocity of SO_2 are described by the results of JACOBSEN et al. (1982). While the deposition velocities lie between 0.8 and 1.0 $cm \cdot s^{-1}$ when the stomata are open, this value decreases below 0.4 $cm \cdot s^{-1}$ at night, when the stomata are closed.

24.3.5 Methods of measurement of dry, wet and humid deposition

A distinction is made between two different collectors for rain-and-snow samples, the "bulk-collector" and the "wet-only collector" (dry–wet collector).

The bulk collectors are permanently exposed instruments consisting of funnels which are connected to the collecting bottles by tubes. Since the funnels are permanently open they receive humid, wet and even dry depositions as a total integrated sample for the exposed period. A distinction between the three different deposition components cannot be made with this method. A separate collection of dry and wet depositions can be realized with a dry-and-wet collector. This apparatus basically consists of two collecting units with funnels and one lid, which closes the one or the other unit depending on the weather conditions. The samples are usually taken after each shower or, regardless of the number of rain or snowfalls, after one day.

The samples of dry depositions are collected over a period of some days or often one week in order to obtain a sufficient amount of collected pollutants for an analysis. In the last few years several different instruments have been used to collect dry and wet depositions in different projects. As the instruments were produced by different firms, they have different designs which can lead to changes in flow and, thus, cause diverging results. Comparing the different instruments in field studies it was found that obtaining similar results mainly depended on the shape of the funnel and the sensitivity of the precipitation sensor (WINKLER et al. 1989).

This is a potential source of error which has to be taken into consideration in the interpretation and comparison of measured values when different apparatuses are used.

Considerable difficulties in taking an exact and comprehensive sample of the range of droplet sizes still exist.

Fog collectors should be constructed in such way that the fog droplets can be sampled and sorted according to sizes ranging from 1 μm to 50 μm (PRUPPACHER and KLETT 1978). Furthermore for analytical reasons the sampling time should be kept as short as possible, in order to be able to analyze single fog occurences.

The fog sampler should, therefore, have as large a collecting surface as possible to collect sufficient samples for chemical analysis in a very short time. There are different procedures to obtain fogwater samples:

- Passive (stationary) collectors using the Grunow method (GRUNOW 1953) and
- active rotating collectors and suction (impaction) collectors.

The advantage of active collectors is that the samples can be taken in relatively short time intervalls whereas this is only the case for passive collectors, if the wind velocity during the fog is great enough. This is usual for frontal fog.

Suction collectors are preferable in weather conditions with prevailing radiation fog (JAESCHKE 1987, SCHMITT 1987).

Due to the method of collection passive samplers probably represent the best approximation to the natural interception of vegetation.

24.4 Comparative analysis of pollution in an industrial agglomeration and a clean-air area

In the following the rates of deposition of pollutants will be discussed comparing two locations of measurement which differ with respect to their altitudes as well as their air quality. The first location is in the south of the City of "Bochum" (145 m a.s.l.) in the central Ruhr-District and is an urban location influenced by industry; the other location is far from industrial influence in the well-wooded "Sauerland" on the highest mountain, the "Kahler Asten", 845 m a.s.l., in the eastern Rhenish Slate Mountains. At both stations measurements were taken with automatic electrically driven dry-and-wet collectors.

24.4.1 Concentrations of pollutants

For a comparative analysis of the precipitation contents the diurnal concentrations of "wet-only samples" from both stations will be contrasted (Tab. 24.4).

The precipitation volume weighted mean values showed similar pH-values at both stations with a pH about 4.1. However, if the "Kahler Asten" values are used as a reference (= 100%) for the "Bochum" values to obtain an improved analysis then the values for the industrial location are:

152% for electrical conductivity,
130% for sulphate, 117% for nitrate,
178% for ammonium, 198% for chloride,
214% for calcium and 222% for lead.

This means an extraordinarily heterogeneous increase in the concentration of certain pollutants at the industrial location – quite low for nitrate (117%) and high for the heavy metal lead (222%) – with respect to the "Kahler Asten"-values.

24.4.2 Anion equivalents

In order to find out which of the examined ions contribute most to the acidity in precipitation the proportional shares of anion equivalents in precipitation were determined at both stations. While the amounts of SO_4 hardly showed any differences between both stations differences in anion equivalents of NO_3 were considerable (Tab. 24.5).

At the "Kahler Asten" the NO_3 value is higher than at the "Bochum" station, whereas the chloride value is lower. The higher NO_3 value at the less polluted location in the "Sauerland" could possibly arise from the higher degree of fertilization of agricultural acreage in the so-called "Soester Börde" and "Warburger Börde"; the higher chloride value in "Bochum" may be due to industry.

A seasonal pattern of the anion equivalents at both stations shows (Fig. 24.1) that the amount of SO_4 at the "Bochum" station increases in summer while at the

Table 24.4: Means of "wet-only" trace concentrations in rainfall samples ($mg \cdot l^{-1}$), pH-values, and electric conductivity ($\mu S \cdot cm^{-1}$) at "Bochum" station (Jan. 1983–Feb. 1984) and at "Kahler Asten" station (Nov. 1982–Nov. 1984, in italics). After KUTTLER (1986).

	weighted mean	arithm. mean	min.	max.	number of values
pH-value	4.16	4.23	3.45	5.50	124
	4.09	*4.22*	*3.30*	*6.65*	*276*
H^+-ions	0.069	0.0588	0.00316	0.3548	124
	0.080	*0.0593*	*0.00022*	*0.5010*	*276*
electric conductivity	67.20	69.40	14.10	299.9	124
	44.38	*44.07*	*6.50*	*166.5*	*276*
SO_4^{2-}-S	2.09	2.73	0.41	10.99	124
	1.61	*1.82*	*0.27*	*6.31*	*276*
NO_3^--N	0.62	0.79	0.15	4.04	114
	0.53	*0.61*	*0.06*	*3.28*	*263*
NH_4^+-N	1.64	1.59	0.31	4.91	96
	0.94	*1.11*	*0.24*	*6.75*	*216*
Cl^-	2.18	3.07	0.21	22.76	84
	1.10	*1.28*	*0.15*	*12.25*	*158*
Ca^{2+}	1.33	1.75	0.20	11.30	102
	0.62	*0.73*	*0.09*	*6.64*	*218*
Pb^{2+}	0.040	0.036	0.001	0.283	62
	0.018	*0.019*	*0.001*	*0.117*	*89*

"Kahler Asten" station it is higher in spring and fall than in summer or winter. The clearest seasonal dependency was determined for chloride; at both stations considerably higher values were obtained in winter and fall than in summer and spring. It should also be mentioned that the seasonal differences are higher at the top of the "Kahler Asten" (winter 31.5%, spring 12.8%) than at the station

Table 24.5: Percentage of relative anion equivalents in rain water at "Bochum" station and "Kahler Asten" station (Jan. 1983–Feb. 1984). After KUTTLER (1986).

	SO_4^{2-}	NO_3^-	Cl^-
Bochum	55.5	18.5	26.0
Kahler Asten	57.6	23.1	19.3

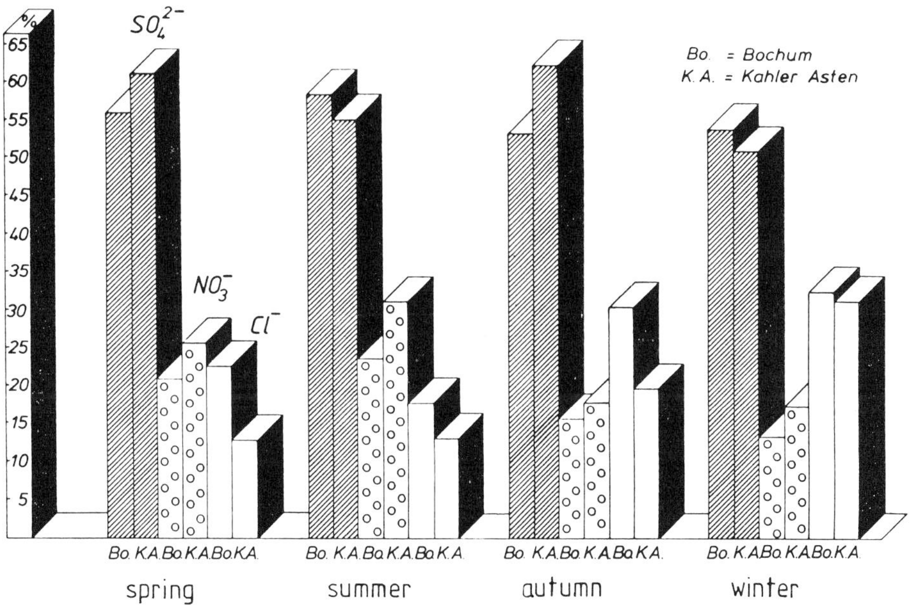

Figure 24.1: Percentage of anion equivalents in rain water samples at "Bochum" station and "Kahler Asten" station for different seasons (Jan. 1983–Feb. 1984). After KUTTLER (1986).

in "Bochum" (winter 32.7%, summer 17.8%). The more homogeneous pattern for chloride at the station in "Bochum" can be explained by the influence of industry. The consistently higher values at both stations in winter are caused by the predominance of advective weather conditions with more frequent intervalls of "sea-spray" depositions than in other seasons.

Clear seasonal differences are also found in the case of NO_3: for summer and,

Table 24.6: Means of "wet-only" pollutant deposition at industrial area station (Bochum, 145 m a.s.l.; Jan. 1983–Feb. 1984) and of remote highland area station (Kahler Asten, 845 m a.s.l.; Nov. 1982–Nov. 1984). Values are in $mg \cdot m^{-2} \cdot d^{-1}$). After KUTTLER (1986).

	H^+	SO_4^{2-}-S	NO_3^--N	NH_4^+-N	Cl^-	Ca^{2+}	Pb^{2+}
Bochum	0.165	4.47	1.32	3.23	5.04	2.98	0.067
Kahler Asten	0.322	6.28	1.94	3.71	4.78	2.50	0.071

to a lesser extent, for spring considerably higher values were calculated at both stations than for winter at both stations which may possibly be due to agricultural fertilization in the "Soester Börde" in these seasons. The seasonal differences are again greater at the "Kahler Asten" station than at the city station.

These results compare well with those determined by PERSEKE (1982) for the location "Deuselbach" (Hunsrück) and those by GEORGII et al. (1982) for precipitation samples on the "Kleiner Feldberg" in the Taunus Mountains.

24.4.3 Deposition of pollutants

The deposition rates at a particular location are determined from the concentration of a pollutant in precipitation and the amount of precipitation during the period of measurement. In order to obtain a better comparison the total amounts of precipitation were standardized to a period of one year: the total amount of precipitation for "Bochum" was 935 mm while the "Kahler Asten" received 1,724 mm, 1.8 times the amount of "Bochum". The mean values of "wet-only depositions" at both stations are arranged in Tab. 24.6. They show the following values for the "Kahler Asten" relative to "Bochum" (= 100%): H-ion = 195%, SO_4–S = 140%, NO_3–N = 147%, Cl = 95%, Ca = 84%, Pb = 106%. With the exception of chloride and calcium consistently higher rates of deposition were determined at the top of the "Kahler Asten" in spite of the lower concentration of pollutants (see Tab. 24.4).

The rates of dry deposition of the pollutants sulphate–sulphur, nitrate–nitrogen and chloride were also determined. Relative to total deposition (dry and wet) the values for dry deposition on the "Kahler Asten" were consistently lower.

The influx of aerosol depositions is therefore of less importance at the top of the "Kahler Asten" than in the industrial region. The standard deviations of the mean values of dry depositions at the mountain station shown in the third column of Tab. 24.7 are considerably lower than those in "Bochum". As these standard deviations are calculated from the values of each month they reflect the monthly fluctuations in the levels of dry deposition which are more homogeneously distributed on the "Kahler Asten" than in "Bochum".

Table 24.7: Total deposition (td), the percentage of dry deposition (dp), and its standard deviation (sd), of trace constituents at "Bochum" station and "Kahler Asten" station (Jan. 1983–Feb. 1984). After KUTTLER (1986).

			Bochum	Kahler Asten
SO_4^{2-}–S	td	$[mg \cdot m^{-2} \cdot d^{-1}]$	6.60	7.26
	dp	%	34.1	21.8
	sd	%	16.1	9.4
NO_3^-–N	td	$[mg \cdot m^{-2} \cdot d^{-1}]$	1.93	2.35
	dp	%	26.8	20.3
	sd	%	18.2	11.8
Cl^-	td	$[mg \cdot m^{-2} \cdot d^{-1}]$	5.14	6.18
	dp	%	29.2	15.5
	sd	%	15.5	12.9

This means that the lower amounts of dry deposition on the "Kahler Asten" show a more uniform distribution over the period of measurement than those at the station in the polluted area in "Bochum".

In this context the dependence of total deposition on the amounts of dry and wet deposition was also examined. It could be proved that high amounts of total deposition were caused, in general, by the predominating amounts of wet deposition at both stations. In order to show the dependence of the total deposition on the wet and dry depositions it was necessary to relate the total deposition to the percentage shares of dry and wet depositions at both stations. In the analysis shown for "Bochum" it can be seen that high depositions of the pollutants sulphate, nitrate and chloride occur when the ratio of wet deposition to dry deposition is much greater than 1. This relation is described shown by the regression line R 1 (Fig. 24.2).

For the three pollutants wet deposition amounts to between 80 and 95% when there are higher values of total deposition. In contrast, the course of regression line R 2, which describes the ratio of wet to dry deposition in July and August 1983, is completely contrary to that of regression line R 1. Although both months were extremely dry (July 25%, August 30% of the long-term mean values) relatively high amounts of total deposition, which were mainly caused by dry depositions – not by wet depositions which otherwise caused the main influx of these pollutants could be determined. This can be attributed to the extraordinarily high proportion of dry hours (98% of all hours in the month) causing an accumulation of the aerosols sulphate and nitrate in the layer of the air at ground level. The same analysis

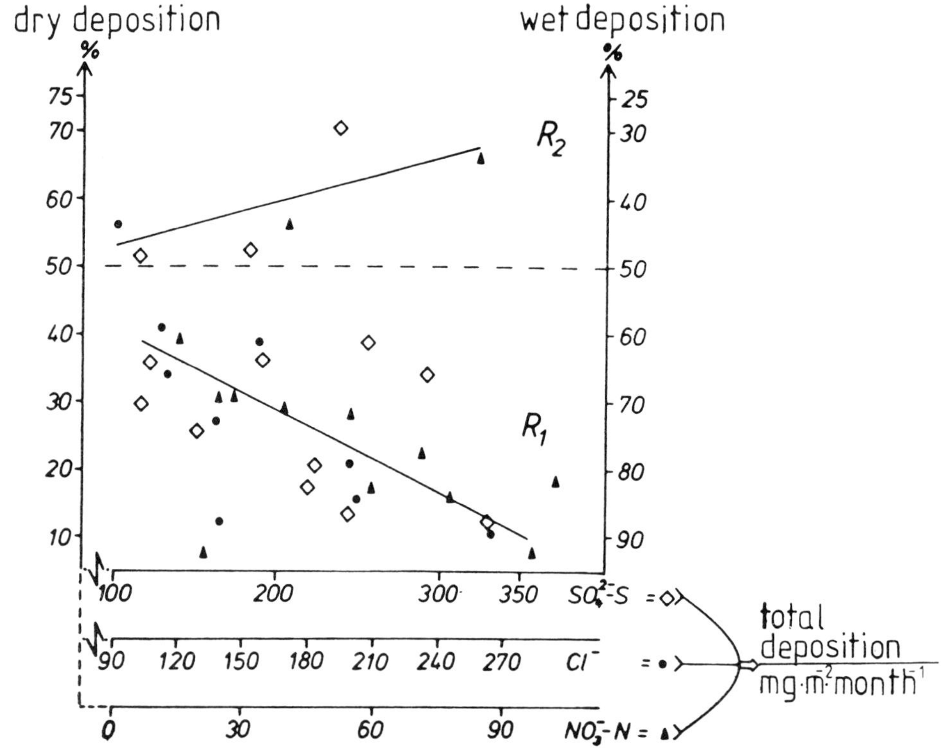

Figure 24.2: Dependence of total deposition on dry and wet deposition at "Bochum" station (Jan. 1983–Feb. 1984). After KUTTLER (1986).

shows a totally different situation at the "Kahler Asten" location (Fig. 24.3).

On the one hand there is no clear regression of the values of the pollutants to a straight line as was the case for "Bochum" – this is especially the case for chloride which shows the greatest deviations. On the other hand there is no dominance of dry deposition in relation to wet deposition in July and August although these months were also extremely dry in the mountain area.

Dry deposition is substantially less important at the top of the "Kahler Asten", even during dry months, than in the industrial region where the dry deposition of the aerosols sulphate and nitrate exceeds wet deposition when the amount of precipitation is low.

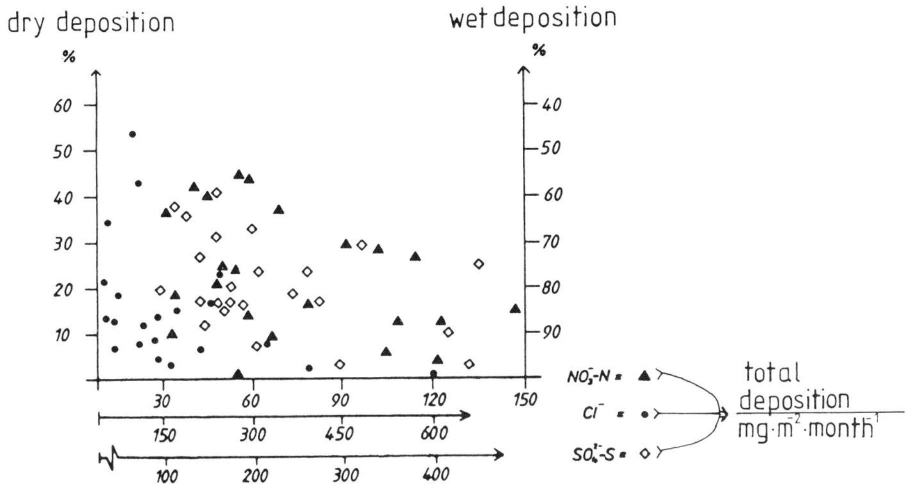

Figure 24.3: Dependence of total deposition on dry and wet deposition at "Kahler Asten" station (Nov. 1982–Nov. 1984). After KUTTLER (1986).

24.5 Results of investigations of pollutant transfer into individual tree stands

It is a significant aim of research to obtain information about the filtration of trees or forests in the analysis of ecological cycles in order to ascertain the effects of pollution on ecosystems covered with vegetation.

In this respect it is not only necessary to determine the concentration of pollutants in the air, but also the resulting influx in the soils and, consequently the amounts of pollutants per unit ground area.

In the following the extent of filtration of H-ion, SO_4–S and NO_3–N on the basis of the monthly mean values of pollutant concentrations in the dripwater from tree tops as well as pollutant deposition will be presented, with the aid of Tables 24.8 and 24.9, for different trees at a site in "Bochum" (the Botanical Gardens of the "Bochum" University).

The mean pH-value of precipitation was 4.1 during the measurement period of 13 months. Slightly increased values were found under a canopy of beech and holly. Under spruce and pine as well as oak, however, the values were lower. The sulphate–sulphur concentrations were generally higher in the dripwater; under spruce a value of 38 mg of SO_4–S per litre, nearly 16 times the concentration outside the canopy, was obtained.

Considerably increased concentrations of nitrate – nitrogen were also found in dripwater and were again the highest under spruce (as was the case for sulphate–

Table 24.8: Monthly means of pH-values and trace concentrations [$mg \cdot l^{-1}$] of SO_4^{2-}–S and NO_3^-–N of open space rain samples and dripwater of different trees. The symbols in column 1 mean: S: SO_4^{2-}–S; N: NO_3^-–N. After KUTTLER (1986).

	Jan.	Feb.	Mar.	Apr.	May	Jun.	Jul.	Aug.	Sep.	Oct.	Nov.	Dec.	mean
Open space:													
pH	4.78	4.41	4.15	4.75	4.04	3.28	4.66	5.31	6.63	4.46	4.18	4.17	4.10
S	1.91	2.18	1.51	3.10	2.15	2.71	3.00	3.43	–	2.24	1.78	2.67	2.42
N	0.45	0.54	0.68	0.93	1.06	0.72	0.61	0.86	–	0.45	0.54	0.36	0.65
Beech (*Fagus sylvatica*):													
pH	4.56	6.29	4.05	4.43	4.92	4.32	4.62	5.01	6.38	4.21	3.78	4.08	4.33
S	3.17	4.13	3.63	4.06	5.38	2.61	5.61	4.32	16.50	4.88	4.12	5.14	5.30
N	0.18	0.61	1.67	1.31	1.76	1.69	2.08	1.63	6.64	0.88	1.51	1.04	1.80
Holly (*Ilex aquifolium*):													
pH	4.88	4.53	4.87	4.78	4.19	4.75	5.18	5.17	5.91	4.92	5.64	5.01	4.79
S	5.38	4.61	6.77	6.73	6.40	4.76	8.58	4.81	18.05	5.61	5.84	9.76	7.23
N	1.56	1.38	2.19	1.92	3.61	1.74	2.57	1.69	5.48	0.27	2.76	0.95	2.18
Spruce (*Picea abies*):													
pH	3.64	3.90	3.48	3.54	3.78	3.91	3.65	5.18	4.84	3.68	3.18	3.57	3.66
S	34.91	23.99	49.30	52.20	35.80	16.04	21.19	14.38	75.34	44.74	35.01	52.99	38.00
N	7.54	5.13	15.06	7.72	12.87	9.84	9.13	5.47	32.95	17.03	12.95	10.43	12.17
Black pine (*Pinus nigra*):													
pH	4.16	4.05	3.89	3.77	3.70	3.75	4.05	4.25	4.07	3.58	3.50	3.60	3.80
S	10.49	6.50	14.12	10.92	8.31	4.19	5.94	4.62	29.40	6.96	5.77	12.05	9.94
N	2.60	1.94	4.00	2.28	4.13	1.89	2.82	2.05	11.14	2.42	1.54	2.73	3.30
Scotch pine (*Pinus sylvestris*):													
pH	4.01	4.06	3.77	3.90	3.62	3.79	3.95	4.54	3.86	3.56	3.52	3.53	3.77
S	10.13	8.18	12.01	11.41	7.95	3.40	9.77	5.48	31.45	9.70	8.09	13.04	10.88
N	1.90	1.85	4.54	2.39	3.68	2.12	2.64	2.55	18.20	3.47	2.50	3.09	4.08
Oak (*Quercus robur*):													
pH	4.24	3.87	3,80	8.81	4.12	4.71	4.89	5.12	6.02	4.07	3.25	3.69	3.91
S	5.47	5.91	7.16	5.64	7.26	7.00	10.92	6.76	26.69	9.07	7.72	6.93	8.88
N	1.08	0.79	2.01	1.45	2.30	1.71	1.21	2.15	9.31	1.42	0.72	0.97	2.09

Table 24.9: Monthly sums of trace depositions of open space rain samples and drip water of different trees. The symbols in column 1 mean: S: SO_4^{2-}–S; N: NO_3^-–N. Values are in [mg·m^{-2}·month^{-1}]. After KUTTLER (1986).

	Jan.	Feb.	Mar.	Apr.	May	Jun.	Jul.	Aug.	Sep.	Oct.	Nov.	Dec.	mean
Open space:													
H^+	2.00	2.32	4.94	0.74	6.14	66.30	1.51	0.77	0.005	3.90	3.85	0.55	7.75
S	232.00	130.30	103.90	129.70	144.60	271.30	208.00	358.10	–	252.80	103.30	218.50	195.70
N	53.10	32.20	46.70	38.20	72.20	92.40	41.90	134.40	–	51.10	31.60	29.20	56.60
Beech (*Fagus sylvatica*):													
H^+	2.50	0.025	3.90	1.28	0.48	3.58	0.91	1.13	0.004	3.36	5.55	4.72	2.90
S	296.30	202.60	158.50	139.50	213.90	194.90	212.50	501.10	171.70	266.10	212.00	292.40	230.90
N	75.30	30.10	73.70	35.50	70.70	126.10	78.60	184.80	69.00	47.90	50.50	58.80	75.40
Holly (*Ilex aquifolium*):													
H^+	0.96	0.95	0.58	0.41	3.35	1.43	0.29	0.93	0.017	1.50	0.066	0.44	0.91
S	393.90	134.50	289.50	164.60	322.70	384.40	374.00	665.40	254.90	700.40	168.00	261.90	343.70
N	113.00	45.40	93.90	47.40	188.30	140.80	112.40	232.90	70.40	33.80	79.40	42.90	100.60
Spruce (*Picea abies*):													
H^+	14.40	3.34	12.21	2.97	2.62	10.37	5.19	0.93	0.045	8.64	6.00	6.78	6.12
S	2196.70	636.20	1819.50	535.20	509.80	1352.80	491.60	2045.90	233.50	1852.70	318.50	1335.30	1110.60
N	474.90	135.90	555.60	158.20	203.30	830.80	211.70	777.30	102.20	705.60	117.20	263.30	378.00
Black pine (*Pinus nigra*):													
H^+	7.20	4.03	7.31	3.82	7.88	19.80	4.61	7.96	1.26	18.90	10.70	14.90	9.03
S	1096.00	297.40	810.60	246.20	332.60	470.60	311.60	658.90	439.30	503.90	197.40	719.80	507.00
N	269.10	87.60	227.10	102.30	163.20	210.80	146.40	290.90	164.80	172.90	66.90	161.90	172.00
Scotch pine (*Pinus sylvestris*):													
H^+	13.10	4.20	15.60	3.80	13.50	19.30	4.40	4.30	1.50	22.10	13.80	22.80	11.60
S	1374.70	397.20	1113.20	326.90	450.70	409.40	475.10	826.00	352.40	786.90	372.80	1016.70	658.50
N	340.30	108.50	417.80	143.80	206.30	253.60	127.60	382.30	202.10	280.30	114.80	238.60	234.70
Oak (*Quercus robur*):													
H^+	9.70	10.60	15.70	7.50	4.10	2.10	0.50	1.12	0.01	8.10	38.90	22.20	10.00
S	933.50	467.50	705.70	275.30	392.30	750.20	404.20	1007.60	355.60	874.10	540.50	836.80	628.60
N	183.20	61.70	198.20	70.00	123.10	181.50	44.70	318.50	122.50	134.90	49.40	104.40	132.70

sulphur). The highest concentrations occured when there was rain after a dry period of several days. The characteristic months for these results were April and September for which the highest concentrations of pollutants in the dripwater were found.

It was generally evident that the variability of the monthly mean values of pollutant concentrations was relatively high during the year in open spaces as well as under vegetation canopy. In particular, the dripwater under spruce showed considerable monthly dependent deviations which lead to the intermittent influx of pollutants in soil-ecosystems.

The distribution and level of pollutant concentrations in Tab. 24.9 show that the influx rates of H-ions at sites with holly, beech and spruce were, in some cases, considerably lower than at sites without vegetation. For pine and oak, however, higher values were obtained.

The influx rates for sulphate–sulphur and nitrate–nitrogen were generally higher under all the measured trees. Values for sulphate – sulphur from 1.2 (under beech) up to 5.7 (under spruce) times the value for open space was measured. For nitrate–nitrogen a range from 1.3 (beech) to 6.7 (spruce) times the value for open space was determined.

Very high monthly deposition rates for all sites were obtained during the rainy months January, March, August and December and mainly low values for the dry months April and September.

Referring to the accumulation of pollutants NO_3 and SO_4 one can basically say that spruce and scots pine have the greatest filtration capacity while the beech has the lowest (compared to open space).

Not only the absolute amount but also the temporal distribution of the depositions during the measuring period, as was the case with the concentrations in the dripwater are relevant when considering the ecological consequences of the pollutants.

As a comparison of the monthly totals within the measurement period for the different locations shows there are extraordinarily large fluctuations between the months (Tab. 24.9). These can be considered in terms of the proportional standard deviation calculated on the basis of the monthly total values with reference to the total mean value of the respective measurement station. These values, for the pollutants sulphate and nitrate (excluding the location under spruce), varied from 40% to 62%. Under spruce trees, however, they reached a higher standard deviation of between 66% and 73%. It can be concluded that tree stand are exposed to a high influx of pollutants over various periods of time resulting in the corresponding impact on the given ecosystem.

24.6 Conclusion and discussion

In this report results of measurements of wet and dry depositions of atmospheric pollutants have been presented for an air polluted industrial site and a location far from industrial influence in the Rhenish Slate Mountains ("Kahler Asten"). It is not only their geographical location which is of significance to their air quality, but also their orographic situation: "Bochum" at an altitude of 145 m a.s.l., the "Kahler Asten" 845 m a.s.l.

The evaluation of the dry and wet depositions clearly shows the different air qualities of both areas. Comparing these values to those found by other measuring networks (Tab. 24.10) the weighted wet-only concentrations of precipitation for the trace substances sulphate and lead show typical increases in the industrial areas, while there are considerably lower values at those sites far from industry which are mainly exposed to long-range transport.

However, there is not such a clear dependence on industry observable for the trace substance nitrate. This is probably due to the fact that this trace sub-

Table 24.10: Weighted means of "wet-only" trace concentrations in rainfall samples at different stations in the Federal Republic of Germany. After GEORGII et al. 1982 (No. 1–6), KUTTLER 1986 (No. 7, 8); from KUTTLER (1986).

No.	Location	SO_4^{2-}-S	NO_3^--N ($mg \cdot l^{-1}$)	Cl^-	Pb^{2+} ($\mu g \cdot l^{-1}$)
1	Schleswig	1.78	0.76	3.51	12.8
2	Braunschweig	1.96	0.87	1.30	23.2
3	Essen	2.19	0.88	2.03	47.2
4	Kl. Feldberg/Ts.	1.73	0.72	1.15	24.0
5	Frankfurt/M.	1.89	0.76	1.14	32.3
6	Hohenpeißenberg	1.35	0.72	3.64	11.6
7	Bochum	2.09	0.62	2.18	41.0
8	Kahler Asten	1.51	0.53	0.85	10.7

stance is released into the atmosphere by agricultural utilization nearly everywhere. Chloride shows increased values at coastal stations and in industrial regions. The comparatively high value measured in precipitation on the mountain "Hohenpeißenberg" may have been caused by a local source.

With regard to the proportions of the most important acidic anions in precipitation for both stations sulphate ions account for more than 50% followed by chloride ions (26%) at the "Bochum" station and nitrate ions (18.5%). On the mountain "Kahler Asten", however, nitrate accounts for a higher share than chloride. "Bochums" higher chloride level can be explained by its somewhat shorter distance from the sea and the industrial release of HCl. Results obtained by the UBA-network in the Federal Republic of Germany show that sulphate participates to 50–60%, nitrate to 23–27% and chloride to 4–10% in the process of acidification.

Increases in the values for chloride by up to 40% have been found in precipitation samples originating from industrial agglomerations and coastal stations (MÜLLER et al. 1982). The analysis of the annual distribution of the anion contents showed a high level of sulphate ions for both stations with a summer maximum in "Bochum" and highest values in fall and spring on the "Kahler Asten". The clearest seasonal dependence, however, was determined for chloride anions whose part in acidification was greater in winter and fall than in summer and spring. When rain coming from the north-western direction was collected at both stations it could indeed be determined that the levels of chloride ions were enhanced at both stations proving the unmistakable influence of the sea.

The different altitudes of "Bochum" and the "Kahler Asten" also allow statements about the absorption of the trace substances in the raindrops to be made. By being able to distinguish between the different kinds of rain the proportion of "rainout" ("in cloud scavenging") and "washout" ("below cloud scavenging") could be calculated.

Comparative measurements (for "Bochum" and "Kahler Asten") showed that the total amount of precipitation at "Bochum" was 57% of the "Kahler Asten" level. However, in "Bochum" the trace substance levels were, for sulphate 77%, for nitrate 71% and for each of ammonium and chloride 91% of those levels found on the "Kahler Asten". Relative to the lower amount of precipitation a higher influx of pollutants was clearly observed in "Bochum". Assuming that the greatest share of wet deposition on the less polluted "Kahler Asten" originates from "rainout" and – among other things because of the increase of the duration of rainy periods – , it can be concluded that the higher levels in "Bochum" mainly result from the "washout" from the approximately 700 m high air-layer between the altitudes of "Kahler Asten" and "Bochum". On this basis the following levels of below cloud scavenging were calculated: for sulphate 40%, for nitrate 29% and for ammonium and chloride 65% – all relative to the determined total deposition for "Bochum".

These levels were confirmed not only by comparative measurements in an earlier investigation in Frankfurt/M. and on the nearby mountain "Kleiner Feldberg" (800 m a.s.l. in the Taunus Mountains) carried out by GEORGII (1965), but also by investigations into the "washout" phenomenon in the Soviet Union in which levels of 45–78% for below cloud scavenging were quoted (PETRENCHUK and SELEZNEVA 1970).

The measured values of depositions at both stations clearly showed that dry deposition makes up considerable share of total deposition in the industrial region while it is less important in the clean-air locations.

The results of our measurements are compatible with the data available for several regions in the FRG. If the sulphur deposition is taken (Tab. 24.11) the converted value of 1.6 g $S \cdot m^{-2} \cdot yr^{-1}$ for the industrial city "Bochum" compares well with the values of other industrial cities. The annual "Kahler Asten" value of 2.0 g $S \cdot m^{-2} \cdot yr^{-1}$ corresponds to those found at other hill stations. Furthermore, the proportional shares of dry deposition fit well to the distributional pattern of the FRG with high levels in urban and industrial regions and lower levels at clean-air locations.

Our data reveals that wet deposition reaches values of between 1 and 2 g $S \cdot m^{-2} \cdot yr^{-1}$ in the FRG while dry deposition only makes up 0.2 to 0.8 g $S \cdot m^{-2} \cdot yr^{-1}$. When calculated over the total area of the FRG (250,000 km^2) annual dry and wet depositions of 300,000 to 700,000 tons of sulphur are obtained. If this is compared with the total SO_2–S emissions in 1984 (see Tab. 24.1) proportions of at least 23% to a maximum of about 54% were deposited in this way. In these calculations of total sulphur deposition the deposition rates of gaseous sulphur (e.g. SO_2) have not been considered. In addition, the influx of trace substances by fog was not determined.

In the last few years more studies into the influx of pollutants by fog have been carried out at several different locations in the FRG (e.g. SCHMITT 1987, SCHELL and GEORGII 1989, MUNGER et al. 1983). They confirm earlier investigations

Table 24.11: Wet and dry sulphate–sulphur deposition in the Federal Republic of Germany (results of measurements No. 1 to 10 after GEORGII 1989; No. 11 and 12 after KUTTLER 1986).

No.	location	precipitation ($mm \cdot yr^{-1}$)	deposition wet ($gS \cdot m^{-2} \cdot yr^{-1}$)	deposition dry ($gS \cdot m^{-2} \cdot yr^{-1}$)	dry part of total deposition %
1	Frankfurt/M.	752	1.4	0.4	22
2	Essen	1,018	2.0	0.6	23
3	Jülich	735	1.4	0.5	26
4	Braunschweig	745	1.2	0.4	25
5	Hamburg	1,011	1.8	0.4	18
6	Hof	811	1.3	0.4	23
7	Schleswig	1,091	1.7	0.4	19
8	Deuselbach (Hunsrück)	914	1.1	0.2	15
9	Kl. Feldberg (Taunus)	1,192	2.0	0.5	25
10	Hohenpeißenberg	1,201	1.4	0.3	18
11	Bochum	863	1.6	0.8	34
12	Kahler Asten (Rhenish Slate Mnts.)	1,498	2.0	0.6	22

(e.g. MROSE 1966) showing that the concentrations of various trace substances were considerably higher in fog water than in rainwater.

KROLL and WINKLER (1989) recently published the results of their comparative measurements at several stations in clean-air areas (highlands), including mountain "Kahler Asten". In Tab. 24.12 the measured concentrations and calculated depositions from KROLL and WINKLERS fog study are compared with the results of examinations of the rainwater at the same location. The values from the fog water analysis have been converted to $mg \cdot m^{-2} \cdot d^{-1}$ for the sake of comparison. The results show considerably higher concentrations in fog water than in rainwater, especially for nitrate, ammonium and lead. With respect to depositions via fog and rain the deposition of sulphate and chloride is higher in rain, whereas nitrate is deposited to a greater extent by dripwater in the tree stands due to fog.

The deposition values for ammonium via fog and rain are in the same order of magnitude. There is no doubt that the depositions by fog, especially in the foggy highlands, for example there are up to 100 foggy days in the Rhenish Slate Mountains annually (SCHIRMER 1976), play an important part in pollution analysis of any ecosystem which have to be considered in examinations of ecological balances. This is true on the one hand because of the higher influx into forest-soil ecosystems,

Table 24.12: Mean values of trace substance concentrations in fog water and pollutant deposition rates via fog interception (after KROLL and WINKLER 1989) compared with trace substance concentrations in rain water and wet-only deposition rates via rain (after KUTTLER 1986) at the remote highland area station "Kahler Asten" (845 m a.s.l.; Rhenish Slate Mountains).

	SO_4^{2-}–S	NO_3^-–N	Cl^-	NH_4^+–N	Pb^{2+}	reference
Concentrations:						
		$(mg \cdot l^{-1})$			$(\mu g \cdot l^{-1})$	
fog	3.5	4.7	2.1	4.5	86	KROLL & WINKLER (1989)
rain	4.1	1.6	1.1	0.9	18	KUTTLER (1986)
Deposition:						
		$(mg \cdot m^{-2} \cdot d^{-1})$				
fog	3.5	4.8	2.2	4.5	–	KROLL & WINKLER (1989)
rain	6.3	1.9	4.8	3.7	–	KUTTLER (1986)

on the other hand because of the fact that increases in acidity of up to a factor of 1,000 have been found, due to the reduction in volume by evaporation of the fog droplets which settled onto the leaves of the trees (FREVERT and KLEMM 1984). Of course, this can cause lethal damage to the epidermal cuticles.

The dominating filtration of atmospheric pollutants by trees could be clearly demonstrated. However, the values under spruce were considerably higher in "Bochum" than at other locations in Northrhine–Westphalia (see Tab. 24.13). This can be explained by the industrial influence in "Bochum", but also by the measurements under single trees. HÖFKEN et al. (1981) mentioned the same fact concerning their investigations. They calculated a factor of increase of 1.3 for nitrate compared to measurements under forest canopies.

In recent discussions about the specification of emission limits for the protection of vegetation and ecosystems the emphasis on the concentration of pollutant in the air is no longer seen as a sufficient mean of control. The so-called "critical loads" which indicate those limits under which no damage to the ecosystem will occur have increasingly become the point of discussion. Hitherto, "critical loads" were only published for sulphur and nitrogen compounds (NO_x and NH_4^+) (NILSSON and GRENNFELT 1988). The critical loads for forest soils for sulphur range from 3 to 32 $kg \cdot ha^{-1} \cdot yr^{-1}$. The relatively large range is due to the number of different soils with various buffer capacities. If the measured values for "Bochum" are converted, a sulphate–sulphur influx of 24 $kg \cdot ha^{-1} \cdot yr^{-1}$ is obtained; under spruce the influx rate reached as much as 135 $kg \cdot ha^{-1} \cdot yr^{-1}$ – exceeding the critical load many times.

For the influx of nitrogen an exceeding of the critical load by up to a factor of 20 can be assumed from the results of this investigation and other surveys in Northrhine–Westphalia (e.g. BARTELS and GEHRMANN 1990).

Table 24.13: Total Deposition of SO_4^{2-}–S and NO_3^-–N at open space and in throughfall of beech and spruce at different locations in Northrhine–Westphalia (NRW), Federal Republic of Germany. Values are in $mg \cdot m^{-2} \cdot d^{-1}$.

SO_4^{2-}–S	NO_3^-–N	location	period	reference
Open space:				
8.8–12.8	3.2–4.1	Burgholz/Solingen	5/83–10/83 5/84–10/84	KUTTLER (1987)
5.8	2.5	Paderborn	11/82–10/83	BLOCK & BARTELS (1985)
4.2	1.6	Monschau	11/82–10/83	BLOCK & BARTELS (1985)
6.5	1.9	Bochum	4/82–4/83	KUTTLER (1986)
–	1.6–2.8	10 stations at non-industrial sites of NRW	1982–1988	BARTELS & GEHRMANN (1990)
Beech:				
16.3–15.8	3.3–4.2	Burgholz/Solingen	5/83–10/83 5/84–10/84	KUTTLER (1987)
7.0	2.8	Paderborn	11/82–10/83	BLOCK & BARTELS (1985)
6.0	1.9	Monschau	11/82–10/83	BLOCK & BARTELS (1985)
7.7	2.5	Bochum	4/82–4/83	KUTTLER (1986)
–	2.7–4.1	10 stations at non-industrial sites of NRW	1982–1988	BARTELS & GEHRMANN (1990)
Spruce:				
15.4–23.7	3.3–6.2	Burgholz/Solingen	5/83–10/83 5/84–10/84	KUTTLER (1987)
9.1	2.6	Paderborn	11/82–10/83	BLOCK & BARTELS (1985)
7.4	2.2	Monschau	11/82–10/83	BLOCK & BARTELS (1985)
37.0	12.6	Bochum	4/82–4/83	KUTTLER (1986)
–	3.0–5.8	10 stations at non-industrial sites of NRW	1982–1988	BARTELS & GEHRMANN (1990)

As any exceeding of the critical load values leads to an extraordinarily severe destabilization of ecosystems in forests and lakes, these critical deposition values – together with the air-pollution limits – become more and more important in discussions about the protection of our environment.

Bibliography

BARTELS, O.; GEHRMANN, J., 1990, Sind naturnahe Ökosysteme durch Stickstoffeinträge gefährdet?
In: Landesanstalt für Ökologie, Landschaftsentwicklung und Forstplanung, Nordrhein–Westfalen, Jahresbericht 1989, 38–42.

BEILKE, S., 1975, Die Abscheidungsprozesse der Spurenstoffe aus der Atmosphäre. Promet **5,2/3**, 35–38.

Block, J.; Bartels, U., 1985, Ergebnisse der Schadstoffdepositionsmessungen in Waldökosystemen in den Meßjahren 1981/82 und 1982/83.
Forschung und Beratung, Reihe C, Wissenschaftliche Berichte und Diskussionsbeiträge Heft **39**, Landesausschuß für landw. Forsch., Erz. u. Wirtsch.-Beratung b. MURL, NRW.

FABIAN, P., 1989, Atmosphäre und Umwelt.
3. edn. Springer, Berlin.

FÄHNRICH, W.; GEORGII, H. W., 1989, Experimental estimation of the SO_2-deposition on an artificial surface.
In: Georgii H. W. (ed.), Mechanisms and effects of pollutants transfer into forest, Proceedings of the meeting on mechanisms and effects of pollutants transfer into forests, held in Oberursel/Taunus FRG, November 24–25, 1988, 69–76.

FREVERT, T.; KLEMM, O., 1984, Wie ändern sich die pH-Werte im Regen und Nebelwasser bei Abtrocknen auf Pflanzenoberflächen?
Archiv für Meteorologie, Geophysik und Bioklimatologie, Serie B, **34**, 75–81.

GEORGII, H. W., 1965, Untersuchungen über Ausregnen und Auswaschen atmosphärischer Spurenstoffe durch Wolken und Niederschläge.
Berichte des Deutschen Wetterdienstes **14**, 100 pp.

GEORGII, H. W., 1989, Schwefelverbindungen in der Atmosphäre.
Promet **3/4**, 65–72.

GEORGII, H. W.; WÖTZEL, D., 1970, On the relation between dropsize and concentration and trace elements in rain water.
Journal for Geophysical Research **75**, 1772–1731.

GEORGII, H. W.; PERSEKE, C.; ROHBOCK, E., 1982, Feststellung der Deposition von sauren und langzeitwirksamen Spurenstoffen aus Belastungsgebieten. Umweltforschungsplan des Bundesministers des Innern, Luftreinhaltung, Forschungsprojekt 10 40 26 00 sowie Datenband. Im Auftrag des Umweltbundesamtes.

GRUNOW, J., 1953, Kritische Nebelfroststudien. Archiv für Meteorologie, Geophysik und Bioklimatologie, Serie B **4**, 389–419.

GUDERIAN, R.; KLUMPP, A.; KÜPPERS, K., 1987, Gehalte und Leaching von Magnesium, Calzium und Kalium bei Fichte (*Picea abies* KARST.) nach Einwirkung von Ozon und Schwefeldioxid. Verh. Ges. Ökologie, Band **XVI**, 311–322.

HERBERT, F., 1987, Mechanismen der nassen und trockenen Deposition. In: Jaenicke, R. (ed.) (1987) Atmosphärische Spurenstoffe, DFG Sonderforschungsbereich, 241–269.

HOUGHTON, H. G., 1955, On the chemical composition of fog and cloud water. Journal of Meteorology **12**, 355–357.

HÖFKEN, K. D.; BAUER, H. (eds.), 1986, Bestimmung der trockenen Deposition. Gesellschaft für Strahlen- und Umweltforschung, München, Bereich Projektträgerschaften. BPT – Bericht 6–85.

HÖFKEN, K. D.; GEORGII, H.; GRAVENHORST, G., 1981, Untersuchungen über die Deposition atmosphärischer Spurenstoffe an Buchen- und Fichtenwald. Berichte des Institutes für Meteorologie und Geophysik der Universität Frankfurt/Main, Nr. **46**.

JACOBSEN, J.; MÜLLER, E.; HEESE, M.; BETZ, M., 1982, Untersuchungen zur großräumigen Ausbreitung von Luftbeimengungen, Umweltforschungsplan des Bundesministers des Innern, Luftreinhaltung, Forschungsprojekt 10404105, Deutscher Wetterdienst, Abteilung Forschung.

JAESCHKE, W., 1987, Physikalische Chemie des Niederschlags. In: Jaenicke, R. (ed.), Atmosphärische Spurenstoffe, DFG Sonderforschungsbereich, 31–76.

JUNGE, C., 1987, Kreisläufe von Spurengasen in der Atmosphäre. In: Jaenicke, R. (ed.), Atmosphärische Spurenstoffe, DFG Sonderforschungsbereich, 19–30.

KELLOG, W. S.; CADLE, R. D.; ALLEN, E. R.; LAZARUS, A. L.; MARTELL, E. A., 1972, The sulfur cycle. Science **175**, 587–597.

KROLL, G.; WINKLER, P., 1989, Trace substance input to coniferous forest by cloud interception.

In: Georgii, H. W. (ed.), Mechanisms and effects of pollutants transfer into forest, Proceedings of the meeting on mechanisms and effects of pollutants transfer into forests, held in Oberursel/Taunus FRG, November 24–25, 1988, 205–211.

KUTTLER, W., 1986, Raum-zeitliche Analyse atmosphärischer Spurenstoffeinträge in Mitteleuropa.
Bochumer Geographische Arbeiten **47**, 240 pp.

KUTTLER, W., 1987, Zur Filterkapazität und zum Bestandsklima eines Buchen- und Fichtenforstes im Bergischen Land. Teil II (Abschlußbericht).
Jber. naturwiss. Ver. Wuppertal **40**, 117–133.

LANDSBERG, H. E., 1954, Einige Beobachtungen zum pH-Wert des Niederschlags – Some Observations on the pH of precipitation.
Archiv für Meteorologie, Geophysik und Bioklimatologie Serie A, **7**, 219–226.

LINKE, F., 1916, Niederschlagsmessung unter Bäumen.
Meteorologische Zeitschrift **33**, 140–141 und Meteorologische Zeitschrift **38** (1921), 277.

MESZAROS, E., 1981, Atmospheric chemistry.
Studies in Environmental Science, Vol. **11**, Amsterdam.

MEYERS, T. P.; BALDOCCHI, D. B., 1988, A comparison of models for deriving dry deposition fluxes of O_3 and SO_2 to a forest canopy.
Tellus **40**, B, 270–284.

MROSE, H., 1966, Measurements of pH and chemical analysis of rain-, snow- and fog water.
Tellus **18**, 266–270.

MUNGER, J. W.; JACOB, D. J.; WALDMAN, J. M.; HOFFMANN, M. R., 1983, Fog water chemistry in an urban atmosphere.
Journal of Geophysical Research **88**, 5109–5121.

MÜLLER, J.; REUVER, H. J.; JOST, D., 1982, Measurements of F^-, Cl^-, NO_3^- and SO_4 – Ions in rainwater and particulate matter by aid of ionic chromatography.
In: Proceedings of the second European symposium on physicochemical behaviour of atmospheric pollutants, Varese, Italy, 29.9.–1.10.1982, 440–448.

NILSSON, S. J.; GRENNFELT, P., (eds.), 1988, Critical loads for sulfur and nitrogen.
Report from a workshop held at Skokloster, Sweden 19–24 March, Nordic Council of ministers, Store Strandstraede 18, 1255 Kopenhagen/ Dänemark, 15 pp.

OECD, 1977, The OECD Program on the long-range transport of air pollution, Organization for economic cooperation and development.
Paris.

PERSEKE, C., 1982, Die trockene und feuchte Deposition säurebildender atmosphärischer Spurenelemente.
In: Berichte des Institutes für Meteorologie und Geophysik der Universität Frankfurt/Main, Nr. **48**.

PETRENSCHUK, O. P.; SELEZNEVA, E., 1970, Chemical composition of precipitation in regions of the Soviet Union.
Journal of Geophysical Research **75**, 3629–3634.

PRUPPACHER, H. R.; KLETT, J. D., 1978, Microphysics of clouds and precipitation.
Reidel Publishing Company Dordrecht, Holland.

SCHELL, D.; GEORGII, H. W., 1989, Design and operation of a two stage fog water collector.
In: Georgii, H. W. (ed.), 1989, Mechanisms and effects of pollutants transfer into forest. Proceedings of the meeting on mechanisms and effects of pollutants transfer into forests, held in Oberursel/Taunus FRG, November 24–25, 1988 221–229.

SCHIRMER, H., 1976, Klimadaten.
Bd. 1: Nordrhein–Westfalen, Lieferung 7, Deutscher Planungsatlas.

SCHMITT, G., 1987, Methoden und Ergebnisse der Nebelanalyse.
Berichte des Instituts für Meteorologie und Geophysik der Universität Frankfurt/Main, Nr. **72**.

SEHMEL, G. A., 1980, Particle and gas dry deposition.
Atmospheric Environment **14**, 983–1011.

ULRICH, W.; MAYER, R.; KHANNA, P. K., 1981, Deposition von Luftverunreinigungen und ihre Auswirkungen im Waldökosystem im Solling.
Schriften aus der Forstlichen Fakultät der Universität Göttingen und der Niedersächsischen Forstlichen Versuchsanstalt, 58, 2. Aufl.

UMWELTBUNDESAMT, 1989, Daten zur Umwelt 1988–1989,
E. Schmid Verlag, Berlin.

VDI RDL = VEREIN DEUTSCHER INGENIEURE, Kommission Reinhaltung der Luft (ed.), 1983, Säurehaltige Niederschläge / Entstehung und Wirkungen auf terrestrische Ökosysteme.
Düsseldorf.

WINKLER, P.; JOBST, S.; HARDER, C., 1989, Meteorologische Prüfung und Beurteilung von Sammelgeräten für die nasse Deposition.
Gesellschaft für Strahlen- und Umweltforschung, München, Projektträger Umwelt- und Klimaforschung, BPT – Bericht 1–89.

ZIEGLER, H., 1984, Weg der Schadstoffe in die Pflanze.
In: Hock, W.; Elstner, E. (eds.), Pflanzentoxikologie, Der Einfluß von Schadstoffen und Schadwirkungen auf Pflanzen, 35–46.

Chapter 25

Gaseous air pollutants and forest floor vegetation — a case study at different levels of integration

L. STEUBING and A. FANGMEIER

25.1 Introduction

During recent years the air quality over wide areas of the globe has considerably changed due to the input of emissions into the atmosphere caused by human activities. Fuel combustion is releasing dusts and gases into the atmosphere, the most important of which are CO_2, SO_2 and NO_x. Beside these gases, chlorofluoromethanes are produced and released which contribute to the depletion of the stratospheric ozone layer possibly having severe impacts on terrestrial life (WORREST and CALDWELL 1986). CO_2 is known as "greenhouse gas", the release of which may influence the global carbon cycle and cause a global climate change.

Most attention concerning air pollutant effects on terrestrial plant life, however, has been paid to the effects of SO_2 (GUDERIAN 1977), NO_x (GUDERIAN and TINGEY 1987), O_3 (GUDERIAN 1985), and, more recently, NH_3 (NIHLGARD 1985;

ASMAN and DIEDEREN 1987). Several studies have shown that these gases are toxic for plants even when occurring at relatively low levels. Beside their direct effects on plants, gases like SO_2 and NO_x are the precursors of acids formed during their atmospheric transport. These acids can directly affect plants or act via acidification of the soil (VDI 1987). NH_3 is another compound contributing to soil acidification by formation and deposition of NH_4^+–ions (NIHLGARD 1985; ASMAN and DIEDEREN 1987).

Most studies investigating the effects of gaseous air pollutants on terrestrial vegetation focus on forest tree species (MCLAUGHLIN 1985; KRAUSE et al. 1986; KLEIN and PERKINS 1988) or on agricultural crops (HECK et al. 1984; HEAGLE 1989). Only very few studies have been carried out which evaluated the effects of gaseous pollutants on native plants, and nearly no experimental studies (at all) have tested the impact of gases like SO_2, NO_2, and O_3 on intact plant communities in their natural habitat. Only one experimental study (TRESHOW and STEWART 1973) investigated the reaction of forest floor species to ozone in their natural environment. Thus, information on the reaction of the forest floor vegetation to gaseous air pollutants based on fumigation experiments with known fumigant concentrations is poor.

To fill this gap, fumigations with SO_2, SO_2+NO_2, and SO_2+NO_2+O_3 were carried out in open–top chambers in a beech forest located near Gießen for a period of five years to test the reaction of the occuring understorey species. In this study, measurements of the plant reaction were undertaken at different levels of integration. Pollutant effects on biochemical and physiological parameters were detected as well as effects on the micromorphology of selected species. Visible injuries were estimated to detect the fumigant stress at plant species level, and estimations of growth parameters provided integrating information on the reaction at population level.

25.2 Gaseous air pollutants and their effects on forest floor vegetation — what is known?

Some of the studies concerning the effects of gaseous air pollutants on forest floor vegetation are listed in Table 25.1. Obviously, nearly all of these studies are descriptive, and in most cases, the exact pollution dose which might have caused the observed effects is not known.

The descriptive investigations listed in Table 25.1 can be divided into studies examining a gradient in time (comparison between older and recent plant distribution maps) and studies examining a gradient in space (comparison of stands with differing pollution).

In addition to the investigations cited in Table 25.1, several studies have been carried out concerning the impact of acid stemflow in the trunk area of beech trees on the ground vegetation. These studies can be considered as investigations at a micro-gradient in space. They have shown that the soil pH in the trunk area of beeches is much lower than in the interstem areas and that the plant community composition in the trunk area is characterized by a lower diversity and an increase of acid indicators (GLAVAC, KRAUSE and WOLFF-STRAUB 1970; GLATZEL et al. 1983; JOCHHEIM 1986; FALKENGREN-GRERUP 1989; WITTIG et al. and WITTIG and NEITE 1985).

Most of the descriptive studies listed above and in Table 25.1 are dealing with impacts of air pollutants on the forest floor vegetation *via the soil*. In moderately polluted areas, the species composition is thought to be influenced mainly by the deposition of nitrogen and hydrogen. Nothing is known about the direct effect of moderate concentrations of gaseous air pollutants from these studies. Those effects were only discussed in case of high SO_2-concentrations (FREEDMAN and HUTCHINSON 1980; GUDERIAN and KÜPPERS 1980). Only one (experimental) study investigated the direct effect of a gaseous air pollutant (ozone) at a known concentration on native plant species of the forest floor (TRESHOW and STEWART 1973).

25.3 The experiment

25.3.1 Study area

The experiment was carried out in a melick-beech forest (Melico-Fagetum) located about 5 km south-east of Gießen. The herb layer is characterized by a sharp delimitation between two subassociations of the plant community. The dominating forest floor species in the Melico-Fagetum typicum is *Melica uniflora* Retz. (wood melick), whereas the Melico-Fagetum allietosum is dominated by *Allium ursinum* L. (broad-leaved garlic). Beeches (*Fagus silvatica* L.) are building the tree layer in both subassociations. Ash trees (*Fraxinus excelsior* L.), however, are building a second tree layer in the Melico-Fagetum allietosum and, thus, cause a lower PAR level at the forest floor in this community.

25.3.2 Fumigations

Fumigations were carried out at plots of 1 m^2 each. Transportable open-top chambers as described earlier (FANGMEIER, GNITTKE and STEUBING 1986) were used to apply SO_2 at 300 $\mu g \cdot m^{-3}$, SO_2+NO_2 at 300 + 100 $\mu g \cdot m^{-3}$, and $SO_2+NO_2+O_3$ at 300 + 100 + 200 $\mu g \cdot m^{-3}$ for four consecutive hours a week. These concentrations represent the highest ambient air peaks measured in the study area during the vegetation periods. Thus, the fumigations did not mean higher pollution peaks for the investigated plant communities, but a higher frequency of such peaks. Controls were treated with ambient air. The resulting concentration means at control and fumigation plots are listed in Table 25.2. Fumigant concentrations and microclimatic parameters were monitored continuously both in the chambers and in

Table 25.1: Results of previous studies on the effects of air pollutants on herb layer species of forests in their natural environment. The columns of the table mean: (1) type of study (descriptive, experimental), (2) location, (3) type of pollution (see footnote to table), (4) pollution measured ?, (5) method used (see footnote to table).

(1)	(2)	(3)	(4)	(5)	Results and Author(s)
d	Germany	1	no	5	increase of acidophilic species from 1976 to 1983 WITTIG & NEITE (1985)
d	Germany	1	no	5	dependent on the investigated plant communtity ROST–SIEBERT (1986)
d	Germany	1	no	5	decrease of species number; increase of acidophilic species during the past 40 years STEUBING & FANGMEIER (1986)
d	Germany	1	no	5	increase of nitrophilic species during the past 10 to 40 years BÜRGER (1988)
d	Sweden	1	no	5	increase of nitrophilic species; decrease of species requiring high soil pH FALKENGREN–GRERUP (19889)
d	Poland	1	no	5	increase of nitrophilic species during the past 20 years; decrease of species indicating stands poor in nitrogen SIENKIEWICZ (1989)
d	Germany	1	no	5	decrease of species number; no definite results concerning the indicator values for N and pH WILMANNS (1989)

continued next page

Table 25.1, continued

(1)	(2)	(3)	(4)	(5)	Results and Author(s)
d	Germany	1	no	6	increase of nitrophilic species; decrease of species indicating stands poor in nitrogen TRAUTMANN et al. (1970)
d	USA (Ohio)	2	yes *	6	species richness, evenness, and diversity decreased with increasing pollution MCCLENAHEN (1980)
d	Canada (Sudbury, Ontario)	3	no	6	cover, frequency, species number, and diversity severely decreased FREEDMAN & HUTCHINSON (1980)
d	Germany	3	no	6	diversity severely decreased with increasing pollution GUDERIAN & KÜPPERS (1980)
d	Sweden	1	no	7	more nitrophilic and less "nitrophobic" species in the more heavily polluted area TYLER (1987)
e	USA	4	yes	8	at 150 ppb visible injury in five, at 250 ppb in 20, at 300 ppb in 44, and at 400 ppb in 65 (of 70 investigated) species TRESHOW & STEWART (1973)

Numbers in columns (3) and (5) mean:
1 = probably moderate N– and acid deposition
2 = moderate SO_2–, NO_x–, Cl^-– and F^-–pollution (several sources)
3 = acute SO_2–pollution (from a point source)
4 = O_3–pollution (fumigations)
5 = comparison between recent and older plant distribution maps
6 = comparison of stands with differing pollution
7 = comparison of areas with differing pollution
8 = ozone fumigations at 150, 250, 300 or 400 ppb for two hours
* SO_2 dose derived from Huey sulfation plates; F^-– and Cl^-–content measured in leaves of selected species.

ambient air and logged as half hourly means.

SO_2 treatments commenced in 1984 both in the Melico–Fagetum typicum and the Melico–Fagetum allietosum. Combined fumigations were carried out in the Melico–Fagetum allietosum only and started in 1986. Nine plots were established for each treatment, thus, altogether 54 plots were under investigation. Five of nine plots of each treatment served as permanent plots which were investigated with non–destructive methods only. Harvests for biochemical analyses occurred at the remaining plots only. Three additional harvest plots were established in the Melico–Fagetum allietosum. These were fumigated with SO_2, SO_2+NO_2, and SO_2+NO_2+O_3 at the same concentrations as mentioned above. The duration of the fumigations, however, was 48 hr weekly (SO_2 and NO_2) or 16 hr weekly (O_3, from 9^{00} to 17^{00} on two consecutive days).

25.3.3 Vegetation analyses

Measurements at the population level were carried out non–destructively. Leaf area index (LAI) and visible injuries were estimated as described earlier (FANGMEIER 1989). The developmental stage of the plants was estimated according to a modified key (ENGELBACH and FANGMEIER 1989) first developed by DIERSCHKE (1972).

The wettability of the cuticle was used as a micromorphological parameter. Selected species of the herb layer (*Allium ursinum, Melica uniflora, Viola reichenbachiana* Jord., wood dog violet, and *Oxalis acetosella* L., wood sorrel) were harvested both from the four hr weekly and the 48 hr weekly harvest plots for these measurements. The method is described elsewhere (STEUBING et al. 1989). Neutralizing capacity of *Allium ursinum* leaves was measured according to STEUBING et al. (1989) to gain additional information about the pollutant effects on the cuticles.

A transportable field porometer (Walz, Germany), combined with an IRGA (Leybold–Heraeus, Germany), was used to measure transpiration and net photosynthesis of *Allium ursinum, Melica uniflora, Oxalis acetosella* and *Viola reichenbachiana* from the four hr weekly treatments at 1–15 (ambient irradiation), 100, and 400 $\mu E \cdot m^{-2} \cdot s^{-1}$. Details are given elsewhere (STEUBING et al. 1989).

Biochemical measurements comprised starch and D–glucose content, pigment content, sulphur content, glutamate dehydrogenase activity, peroxidase activity, and ethylene evolution. The methods are described elsewhere (FANGMEIER 1989; STEUBING et al. 1989; BOTH et al. 1987).

25.4 Results and discussion

25.4.1 Pollutant effects at the population level

Leaf area index (LAI) estimations were used to monitor the pollutants effects on the above–ground development of the occurring species. Fig. 25.1 shows the LAI

Table 25.2: Mean gaseous air pollutant concentrations [$\mu g \cdot m^{-3}$] at control (ambient air) and fumigation plots (fumigated with 300 μg SO_2 m^{-3}, 100 μg NO_2 m^{-3}, and 200 μg O_3 m^{-3} for 4 hr weekly). Based on half-hourly means.

Year		1984	1985	1986	1987	1988
Fumigation period (from — to)		Apr. 16 Oct. 19	Mar. 04 Oct. 18	Mar. 31* Oct. 10	Mar. 30 Oct. 23	Mar. 21 Sep. 30
Number of weekly 4 hr fumigations		27	33	28/22*	30	28
a) Annual mean:						
SO_2	control	15.8	24.2	35.7	31.8	14.7
	fumigated	19.3	28.3	39.1	35.5	18.4
NO_2	control			30.8	27.5	31.2
	fumigated			31.5	28.5	32.6
O_3	control			28.5	28.3	19.4
	fumigated			30.2	30.6	25.1
Increase of pollutant dose at fumigation plots compared with control plots (%)						
	SO_2:	+22	+17	+10	+12	+25
	NO_2:			+2.2	+3.6	+4.5
	O_3:			+6.0	+8.3	+29
b) Mean during the fumigation period:						
SO_2	control	11.3	19.5	22.1	8.0	10.3
	fumigated	18.2	26.2	28.7	15.0	17.2
NO_2	control			31.7	27.9	27.4
	fumigated			33.3	29.7	29.1
O_3	control			33.7	37.2	21.7
	fumigated			37.7	41.1	25.9
Increase of pollutant dose at fumigation plots compared with control plots (%)						
	SO_2:	+61	+34	+30	+88	+67
	NO_2:			+5.1	+6.2	+6.3
	O_3:			+12	+10	+19

* Combined fumigation in 1986 started at May 12.

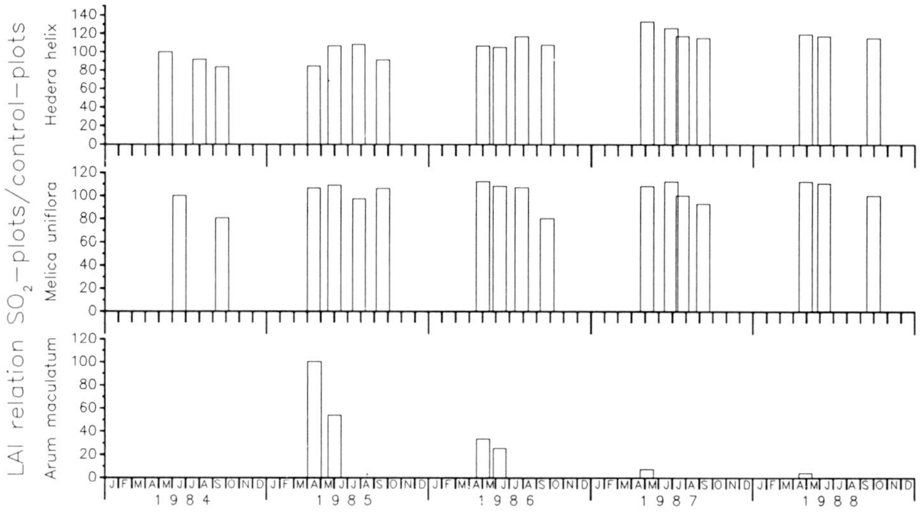

Figure 25.1: LAI relations of selected herb layer species between SO_2 (300 $\mu g \cdot m^{-3}$ for four hr weekly) and control (ambient air) plots in a Melico–Fagetum typicum. Expressed as percentage of the initial LAI relation at the first recording date.

relations for three selected species between SO_2 fumigated and control plots in the Melico–Fagetum typicum. *Hedera helix* L. (common ivy) did not show any LAI decrease at the SO_2 plots. *Melica uniflora* turned out to be nearly insensitive too. However, a decreased LAI relation at the last recording date was observed in all but one years. The population of *Arum maculatum* L. (common arum) was severely reduced at the SO_2 plots.

Examples for the above–ground development of selected species from the Melico–Fagetum allietosum are given in Figs. 25.2–25.5. LAI of the *Allium ursinum* population (Fig. 25.2) was depressed at the last recording date in 1984 to 1987 by SO_2. SO_2+NO_2 had a similar effect. In both treatments, these LAI depressions at the end of the vegetation period of *Allium ursinum* did not lead to a worse development in the consecutive year. The addition of ozone, however, led to a decrease of the LAI of 28% from 1986 to 1987, and the population did not recover in 1988. Most of the observed decrease was due to depressions of older, flowering plants (5–8 years old), while younger *Allium ursinum* plants, especially the smallest ones (one year old and seedlings), could at least partially profit from the decrease of the "overstorey" older plants and the resulting increase in irradiation and, thus, showed a better development at the fumigation plots.

LAI of *Melica uniflora* (Fig. 25.3) was depressed at the end of 1984, 1985 and

1986 by SO_2. The development in the consecutive years was not influenced. A slight depression of the LAI relation occurred at the SO_2+NO_2 plots from 1986 to 1988. This depression was more severe at the SO_2+NO_2+O_3 plots. *Oxalis acetosella* (Fig. 25.4) was severely influenced by SO_2 and SO_2+NO_2+O_3. Surprisingly, SO_2+NO_2 had nearly no effect in 1986 and 1987 on LAI. Only in 1988, a decrease of the LAI relation was observed.

Obviously, the fumigations influenced the above–ground development of sensitive species. This is not only demonstrated by decreases of the LAI relations shown in Figs. 25.1–25.4, but also by the amount and time of occurrence of visible injuries at fumigation plots compared with control plots. The amount of visible injuries observed in 1987 is shown in Fig. 25.5. For *Anemone nemorosa*, the fumigations, especially the combination of all three pollutants, severely enhanced the yellowing of the leaves. At the beginning of June, already 90% of the leaf area of *Anemone nemorosa* was injured at the SO_2+NO_2+O_3 plots, while this high degree of injury was achieved only six weeks later at the control plots. Yellowing of *Melica uniflora*, however, was not definitely influenced by the fumigations. *Viola reichenbachiana* was affected by the pollutants, however, the progress of the injuries was much slower than for *Anemone nemorosa*.

The results shown above suggest that the pollutants mainly influenced the senescence of sensitive understorey species. This is proved by estimations of the developmental stage which were carried out at three day intervals in 1987. The results of these estimations, shown in Fig. 25.6, demonstrate that mainly the senescence phase was influenced by the fumigants. The time of emergence of new leaves differed not or only to a minor degree between the treatments, and phase 6 (fully developed) started at the same time. The beginning of senescence (phase 7 in Fig.6), however, was earlier for all species shown in Fig.6 at least in the SO_2+NO_2+O_3 treatment.

Since the vegetative development of sensitive species was influenced by the fumigations, the question arises whether the generative properties of the understorey vegetation were also affected. Fig. 25.7 shows the fumigation effects on number of flowers and seeds, and on seed weight of *Allium ursinum*. Obviously, the treatments, especially with ozone, were destructive for the generative abilities of this species. Since the survival of *Allium ursinum* is dependent on seed development (generative propagation via bulbs contributes only to a minor part of the survival of the species), these fumigant effects on seed number and quality (weight) have a severe meaning for the fitness of the population.

Taking into account all the results of estimations at population level, the occurring species can be ordered according to their pollutant sensitivity as shown in Table 25.3. Obviously, there is a correlation between leaf morphology and life–form and the sensitivity to the fumigations. The most sensitive species belong to the group of vernal geophytes with hygromorphic or hygromorphic to mesomorphic leaves. Species with mesomorphic leaves were much less sensitive than the former ones, while the only xeromorphic species (*Hedera helix*) did not show any reaction

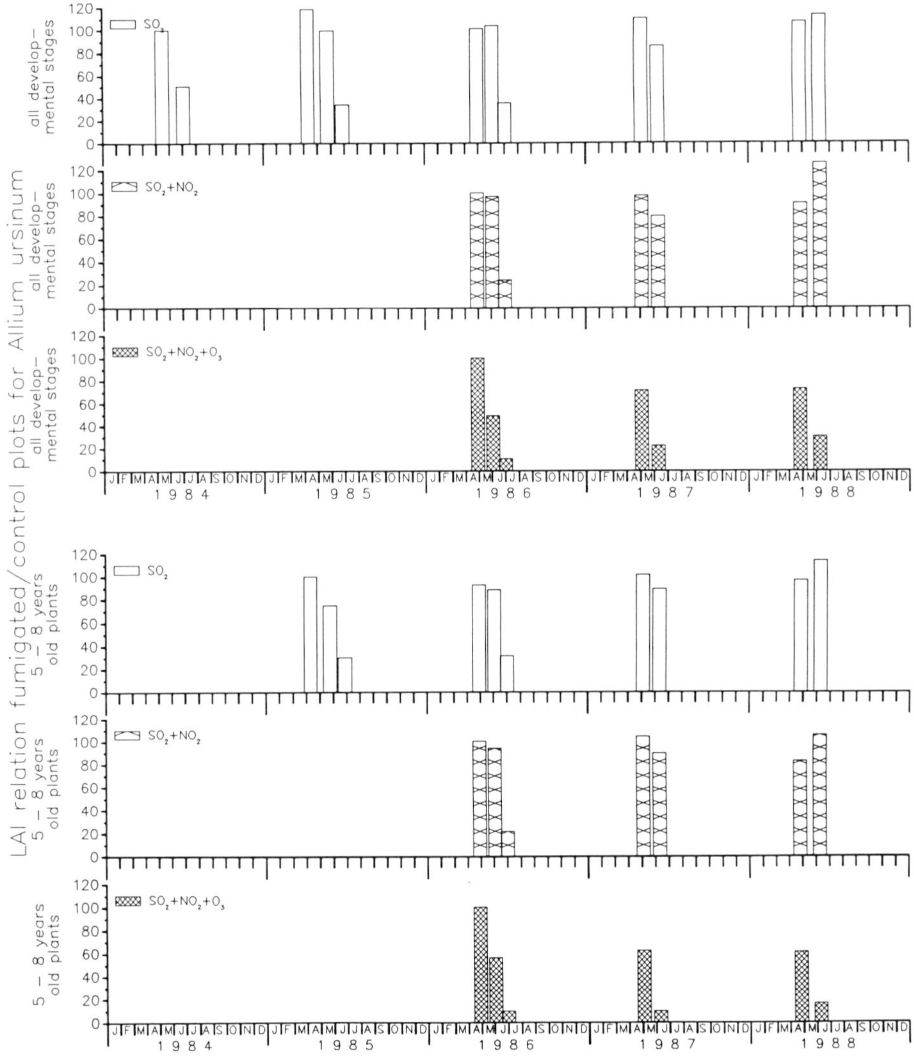

Figure 25.2: LAI relations of *Allium ursinum* between fumigated and control plots in the Melico–Fagetum allietosum. Fumigations were carried out at 300 $\mu g\ SO_2\ m^{-3}$, 100 $\mu g\ NO_2\ m^{-3}$, and 200 $\mu g\ O_3\ m^{-3}$ for 4 hr weekly. Controls received ambient air.

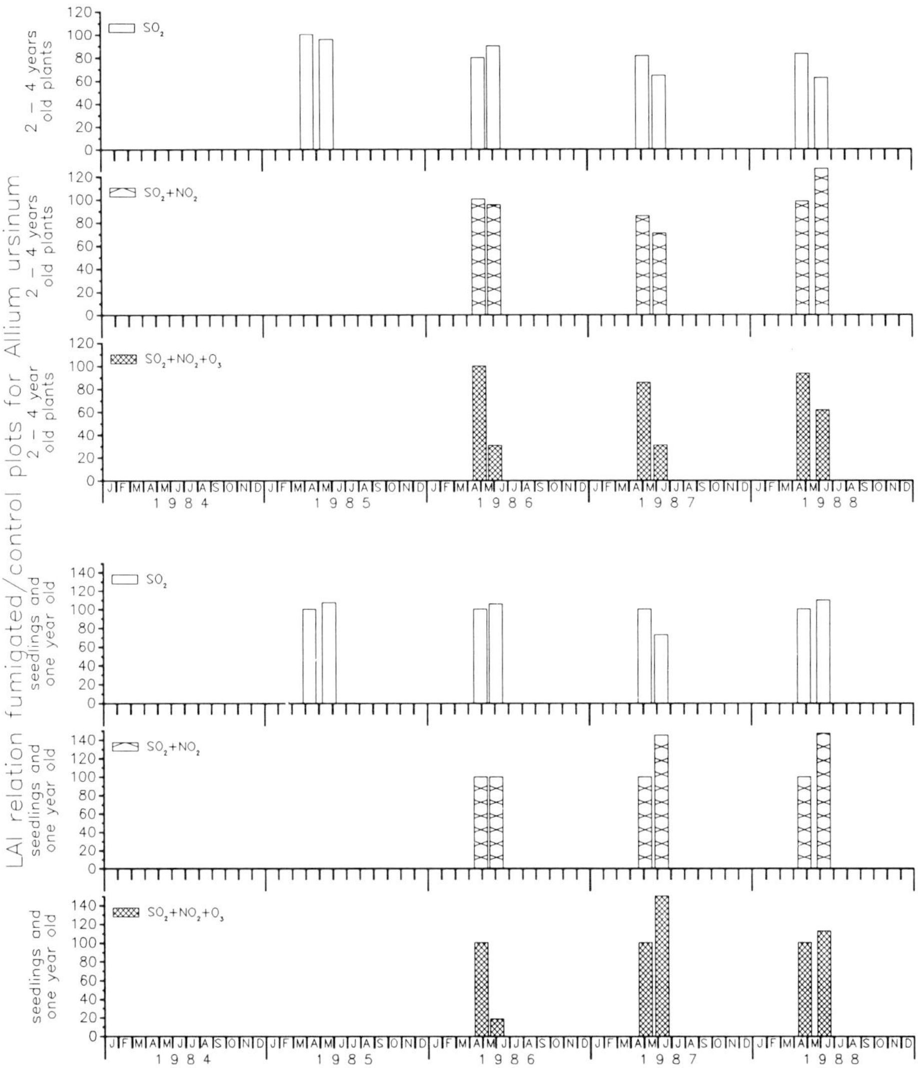

Figure 25.2, continued.

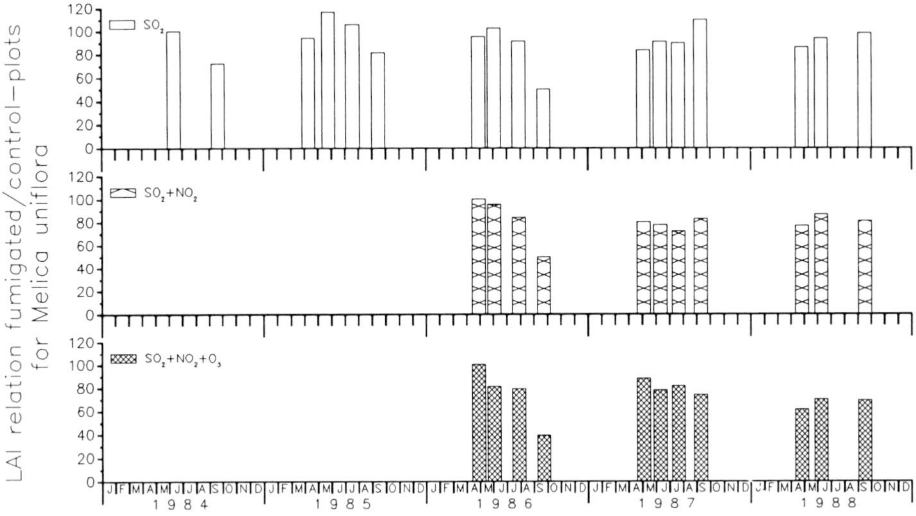

Figure 25.3: LAI relations of *Melica uniflora* between fumigated and control plots in the Melico–Fagetum allietosum. Fumigations as described in Fig. 25.2.

to the fumigations.

Pollutant effects at population level are integrating all the pollutant attacks at morphological, physiological, and biochemical level. Some of these parameters were measured in selected herb layer species to test their suitability as biomarkers to detect pollution stress before the occurrence of effects at the population level and to gain information on the mode of action of the applied pollutants.

25.4.2 Pollutant effects at the micromorphological level

The cuticle is the barrier between the plant and its aerial environment. Air pollutants are known to be destructive for the cuticle (BARNES and BROWN 1990) by eroding the epicuticular wax layer. Measurements of the wettability of leaf surfaces of selected herb layer species (Fig. 25.8) clearly demonstrate that the epicuticular wax layer was one point of attack of the applied pollutants. The contact angles of 10 μl droplets applied to the leaf surface decreased, i. e. the wettability increased significantly in all investigated species with increasing contamination, indicating erosion of the waxes. The effects were most pronounced in *Melica uniflora* and *Viola reichenbachiana* and less pronounced in *Oxalis acetosella* and *Allium ursinum*.

Another parameter providing information on pollutant effects on the cuticle and on the biomembranes is the neutralizing capacity of leaf surfaces. High neutralizing capacities were reported in damaged leaves, where Ca^{2+}, Mg^{2+} and K^{+} can easily

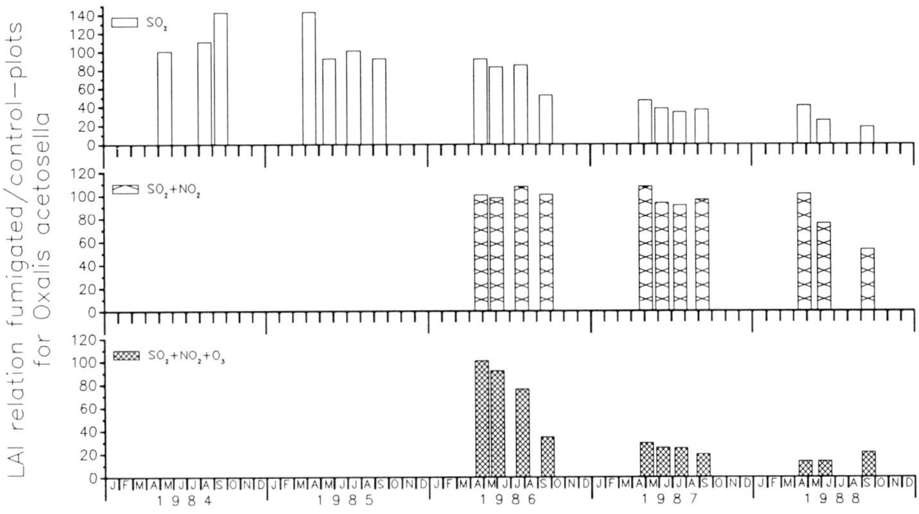

Figure 25.4: LAI relations of *Oxalis acetosella* between fumigated and control plots in the Melico–Fagetum allietosum. Fumigations as described in Fig. 25.2.

be exchanged with H^+ (HUTCHINSON and ADAMS 1987).

The neutralizing capacity was severely increased in fumigated leaves of *Allium ursinum* as shown in Fig. 25.9, whereas *Viola reichenbachiana* did not show significant reactions. Measurements of the ion content of the leachate water, which were published earlier (BOTH et al. 1987), indicated that an efflux of Ca^{2+}, Mg^{2+} and, to a smaller amount, of K^+ took place in leaves of *Allium ursinum* in the SO_2 and SO_2+NO_2 treatments, whereas a strongly increased efflux of K^+ in the SO_2+NO_2+O_3 treatment indicated severe membrane damages.

Thus, species sensitivity was quite different concerning the reaction of the cuticle and of the membranes. The epicuticular waxes were most affected in *Melica uniflora* and *Viola reichenbachiana* and less affected in *Allium ursinum*. The latter species, however, was most sensitive concerning the leakiness of the leaves measured as neutralizing capacity.

25.5 Pollutant effects at the physiological level

The most important parameter for the productivity of plants is the photosynthesis. Gaseous air pollutants are known to have severe impacts on net photosynthesis (HÄLLGREN 1984), either by affecting the stomates or the mesophyll metabolism or both (BLACK 1982). Photosynthesis, transpiration, and stomatal conductance

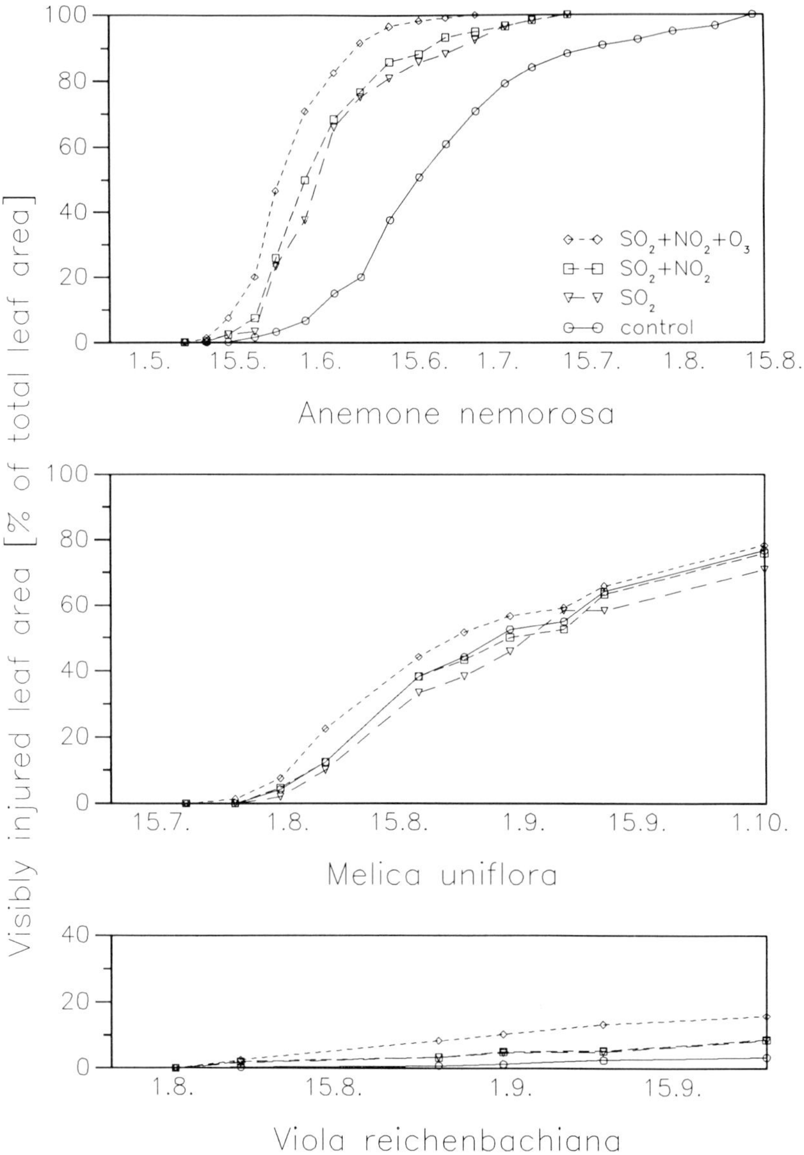

Figure 25.5: Development of visible injuries of selected species in the Melico–Fagetum allietosum at fumigation and control plots. Fumigations as described in Fig. 25.2.

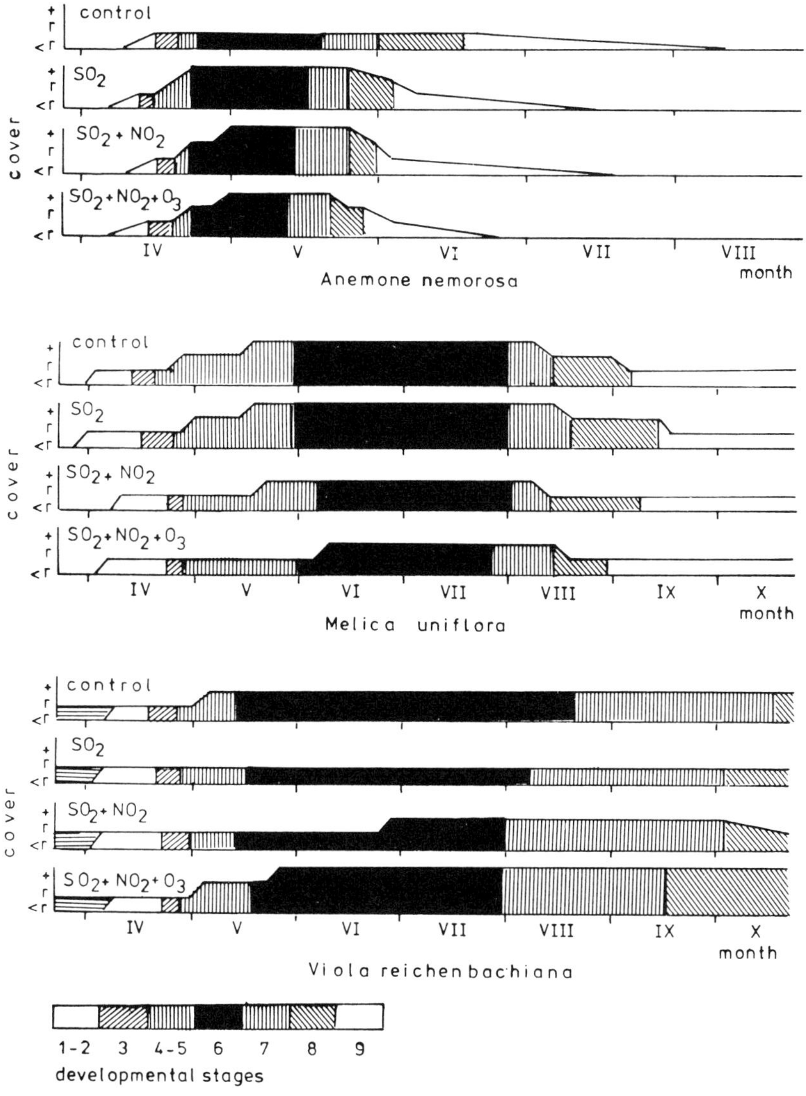

Figure 25.6: Developmental stages of selected species in the Melico–Fagetum allietosum at fumigation and control plots. Fumigations are described in detail in Fig. 25.2.

1–2 = development of new leaves starts
3 = 50% of new leaves developed
4 = 75% of new leaves developed
5 = nearly fully developed
6 = fully developed
7 = beginning of yellowing
8 = 50% of leaves injured
9 = > 50% injured

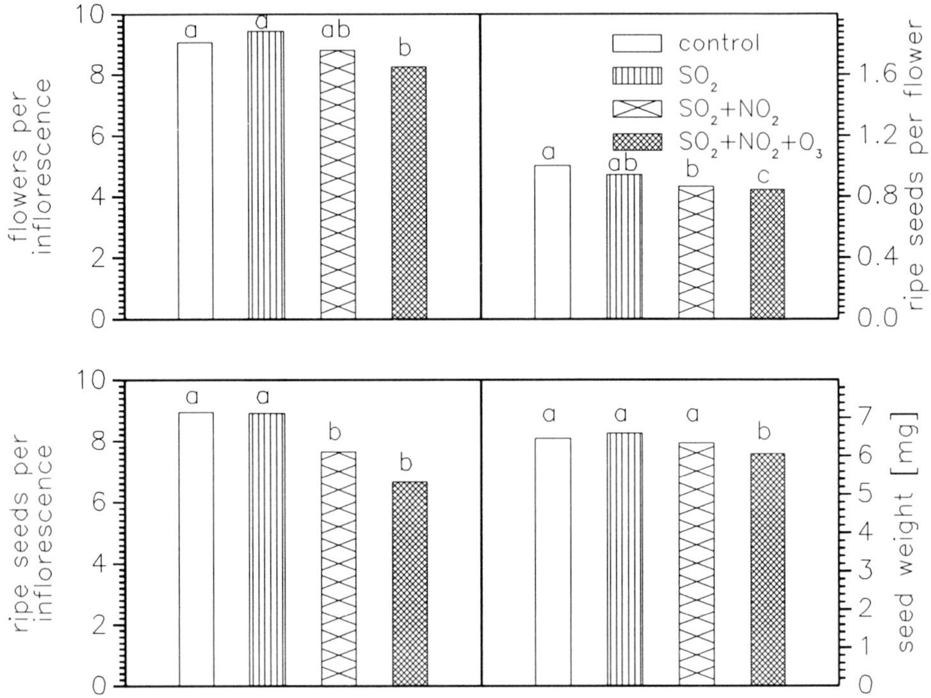

Figure 25.7: Effects SO_2, NO_2, and O_3 on generative properties of *Allium ursinum*. Fumigations as described in Fig. 25.2. Different indices at bars indicate significantly different treatment means ($P < 0.05$).

were measured in selected species of the forest floor throughout their growing season at different PAR levels. Some of the results were published earlier (STEUBING et al. 1989). The fumigation effects on photosynthesis were dependent on species (Table 25.4). SO_2 and SO_2+NO_2+O_3 significantly reduced photosynthesis in *Allium ursinum* at 100 and 400 $\mu E \cdot m^{-2} \cdot s^{-1}$, however, no significant effect was observed in the SO_2+NO_2 treatment. Photosynthesis was depressed in *Melica uniflora* by SO_2+NO_2+O_3 independent of PAR and by SO_2 at 100 $\mu E \cdot m^{-2} \cdot s^{-1}$. Effects on *Oxalis acetosella* were most pronounced in the SO_2+NO_2 treatment. *Viola reichenbachiana* reacted to SO_2+NO_2 and SO_2+NO_2+O_3 at 100 and 400 $\mu E \cdot m^{-2} \cdot s^{-1}$.

The effects of the pollutants on stomatal conductance are listed in Table 25.5. Photosynthesis and stomatal conductance were influenced in a very similar manner by the pollutants. Thus, effects on photosynthesis can mainly be explained by

Table 25.3: Sensitivity of the investigated forest floor species to gaseous air pollutants (fumigated with 300 $\mu g \cdot m^{-3}$ SO_2, 100 $\mu g \cdot m^{-3}$ NO_2, and 200 $\mu g \cdot m^{-3}$ O_3 for 4 hours weekly) according to their reaction at population level (0, ..., 6 = insensitive to highly sensitive).

Species	leaf morphology	life–form	SO_2	SO_2 NO_2	SO_2 NO_2 O_3
Allium ursinum	h	G	4	4	6
Arum maculatum	h	G	6	3	5
Ranunculus ficaria	h	G	2	2	2
Oxalis acetosella	h	G/H	4	2	6
Anemone nemorosa	h/m	G	5	6	6
Viola reichenbachiana	h/m	H	4	4	4
Fraxinus excelsior (seedlings)	h/m	(P)	3	4	4
Lamium galeobdolon	m	C	4		
Melica uniflora	m	G/H	1	1	2
Milium effusum	m	H	1		
Galium odoratum	m/sk	H	3		
Hedera helix	sk	Z	0		

leaf morphology: h=hygromorphic; m=mesomorphic; sk=xeromorphic;
life–form: G=geophytes; H=hemicryptophytes; P=phanerophytes; C=herbaceous chamaephytes; Z=woody chamaephytes
According to ELLENBERG (1990)

stomatal mechanisms. In some cases, significant reductions of conductance were detected when no significant effects on photosynthesis occurred. However, this hold true at ambient PAR (1–30 or 1–15 $\mu E \cdot m^{-2} \cdot s^{-1}$) only, when variation in photosynthesis was too high to gain significant results.

A depression of the stomatal conductance caused by the pollutants may have two antagonistic effects for the sensitivity of the plant. Stomatal closure decreases the CO_2 flux into the leaves and, thus, contributes to the depression of photosynthesis, which causes a diminished productivity of the plant. On the other hand, stomatal closure prevents the pollutant flux into the leaves and, thus, acts as a protective mechanism (TINGEY and TAYLOR 1982; MANSFIELD and FREER–SMITH 1984). Since fumigations were carried out for four hours weekly only, a depression

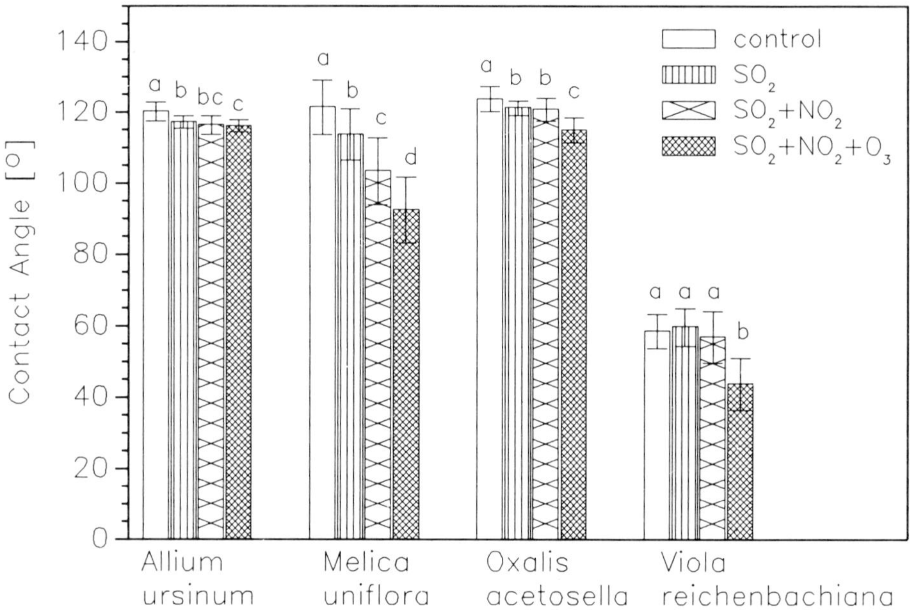

Figure 25.8: Wettability of leaf surfaces of selected herb layer species expressed as contact angle within distilled water droplets of 10 μl applied to the leaf surface (means of 40 estimations each). *Allium ursinum* was harvested after eight four hr weekly fumigations had taken place, while the other species were harvested after 16 fumigations. Fumigant concentrations as described in Fig. 25.2. Different indices at bars indicate significantly different treatment means ($P < 0.05$).

of the net photosynthesis within these four hours will cause small but not severe growth reductions. The recovery period after the pollutant stress has stopped is of major importance for the productivity of the species and provides information on the degree of injury at biochemical level and on the need for repair mechanisms within the mesophyll.

Therefore, additional measurements of photosynthesis, transpiration, and stomatal conductance were carried out. During the four hr weekly fumigations, plants from the fumigation plots were exposed to ambient air applied via the measuring cuvette of the porometer in short term experiments (up to 30 minutes). These experiments clearly demonstrated that recovery times were differing with species. *Melica uniflora* and *Oxalis acetosella* were able to recover from the fumigations within five to ten minutes, while no recovery was observed in *Allium ursinum* within 30 minutes (BOTH 1990). It may be concluded that the fumigations caused

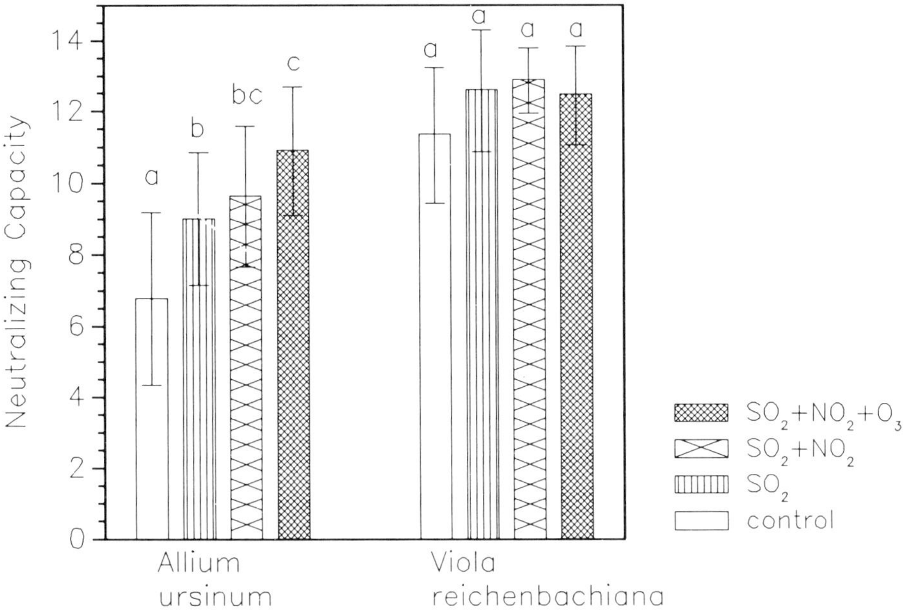

Figure 25.9: Neutralizing capacity of leaves of *Allium ursinum* and *Viola reichenbachiana* measured as neutralized amount of H^+–ions (nmol) of droplets of pH 3.5 applied to the leaf surface within 12 hr. *Allium ursinum* was harvested after eight four hr weekly fumigations had taken place, and *Viola reichenbachiana* after 13 fumigations. Fumigant concentrations as described in Fig. 25.2. Different indices at bars indicate significantly different treatment means ($P < 0.05$).

more severe destructions in the mesophyll in *Allium ursinum* than in *Melica uniflora* and *Oxalis acetosella*. This may prove as an important factor contributing to species sensitivity.

25.6 Pollutant effects at the biochemical level

Gaseous air pollutants may influence the carbohydrate metabolism in plants not only by affecting the primary production processes (photosynthesis), but also by disturbing transport processes, probably the loading of the phloem (KOZIOL 1984). Starch and D–glucose content were measured in *Allium ursinum* leaves. The results indicate a starch accumulation and a decrease of the glucose concentration in the leaves (Fig. 25.10) which can be explained with fumigant effects on the sugar transport. Starch accumulation took place in the SO_2+NO_2 and SO_2+NO_2+O_3

Table 25.4: Effects of 300 μg SO_2 m^{-3}, 100 μg NO_2 m^{-3}, and 200 μg O_3 m^{-3} on net photosynthesis of selected herb layer species in 1987. Expressed as percentage above or below photosynthesis of controls. Values for fumigated and control plants were compared with Student's t–test: ns no significant difference; * $0.01 < P < 0.05$; ** $0.001 < P < 0.01$; *** $P < 0.001$. From STEUBING et al. (1989).

Species	PAR [$\mu E \cdot m^{-2} \cdot s^{-1}$]	SO_2	SO_2 NO_2	SO_2 NO_2 O_3
Allium	1–30	−19% ns	+60% ns	−41% ns
ursinum	100	−37% **	+6% ns	−31% ***
	400	−40% ***	+2% ns	−32% ***
Melica	1–15	−117% ns	+117% ns	−134% **
uniflora	100	−49% **	−25% ns	−54% ***
	400	−24% ns	−11% ns	−34% ***
Oxalis	1–15	−116% ns	−72% *	−250% ns
acetosella	100	−22% ns	−46% **	−47% *
	400	−20% **	−48% ***	−42% *
Viola	1–15	−95% ns	−43% ns	−44% ns
reichenbachiana	100	±0% ns	−27% **	−40% ***
	400	−11% ns	−38% ***	−39% ***

treatments only, while glucose reduction was also found in the SO_2 treatment.

The fumigant effects on pigment content differed with species, as shown in Table 25.6. In *Allium ursinum*, SO_2 alone had no effect on pigment contents. $SO_2+NO_2+O_3$ reduced chlorophylls and carotinoids. Since chlorophyll a was more sensitive than chlorophyll b and chlorophylls were more sensitive than carotenoids, the chlorophyll a/b ratio and the chlorophyll/carotenoid ratio were significantly reduced. Pigments of *Melica uniflora*, however, were as sensitive to SO_2 as to $SO_2+NO_2+O_3$. Fumigations caused pigment reductions in this species, the pigment ratios, however, were not influenced. No significant fumigation effect at all was detected in pigments of *Viola reichenbachiana*.

Pigment content is often used to detect pollution stress at biochemical level (JÄGER 1982; DARRALL and JÄGER 1984). The results obtained in this study, however, indicate that the pollutant effect on pigments is — besides several other environmental factors — depending on species. Without knowledge of the specific reaction of the investigated plant material it will be difficult to draw conclusions

Table 25.5: Effects of 300 μg SO_2 m^{-3}, 100 μg NO_2 m^{-3}, and 200 μg O_3 m^{-3} on on stomatal conductance of selected herb layer species in 1987. Expressed as percentage above or below conductance of controls. Values for fumigated and control plants were compared with Student's t-test: ns no significant difference; * $0.01 < P < 0.05$; ** $0.001 < P < 0.01$; *** $P < 0.001$. From STEUBING et al. (1989).

Species	PAR [$\mu E \cdot m^{-2} \cdot s^{-1}$]	SO_2	SO_2 NO_2	SO_2 NO_2 O_3
Allium	1–30	−49% ***	−30% ns	−40% **
ursinum	100	−47% ***	−32% ns	−22% ns
	400	−56% ***	−30% ns	−37% *
Melica	1–15	−28% ns	−25% ns	−40% *
uniflora	100	−49% *	−25% ns	−21% ns
	400	−15% ns	−8% ns	−19% ns
Oxalis	1–15	−46% ns	−57% **	−26% ns
acetosella	100	−22% ns	−55% **	−46% ns
	400	−18% ns	−43% *	−46% *
Viola	1–15	+28% ns	−42% *	−33% *
reichenbachiana	100	−13% ns	−48% **	−41% *
	400	−4% ns	−50% **	−36% ns

concerning the pollution situation from those measurements.

Another parameter often used to detect pollution stress, especially when SO_2 is present, is the sulphur content of needles or leaves (WEIGEL and JÄGER 1985; ARNDT, NOBEL and SCHWEIZER 1987). The sulphur content increased slightly, but not significantly in *Allium ursinum* and *Melica uniflora* after fumigations with SO_2 (Table 25.7). A significant increase was found in *Viola reichenbachiana*. Fumigations with SO_2+NO_2+O_3, however, led to significantly reduced sulphur contents in *Allium ursinum* and *Melica uniflora* (compared with controls) and *Viola reichenbachiana* (compared with the SO_2 treatment). Probably, the enhanced leakiness of the membranes caused by O_3 led to an efflux of sulphate (and other sulphur compounds) and caused the lowered sulphur contents. It can be concluded that a biochemical indication of SO_2 pollution in an area by sulphur content measurements of leaves only makes sense if the pollution with photooxidants is known (FANGMEIER 1989).

Table 25.6: Effects of 300 μg SO_2 m^{-3}, 100 μg NO_2 m^{-3} and 200 μg O_3 m^{-3} applied for 4 hr weekly on the pigment content [mg·g−1 dry weight] of selected herb layer species. *Allium ursinum* was harvested after 10 fumigations had been applied and the remaining species after 17 fumigations. Different indices in lines indicate significantly different treatments means ($P < 0.05$).

Species	control	SO_2	SO_2 NO_2	SO_2 NO_2 O_3
Allium ursinum				
chlorophyll a	11.57 a	10.98 ab	10.70 b	9.22 c
chlorophyll b	4.08 a	3.96 a	3.80 a	3.48 b
total chloro. (a+b)	15.65 a	14.94 a	14.50 a	12.70 b
carotinoids	2.18 a	2.13 a	2.08 a	1.88 b
chloro. a/b ratio	2.84 a	2.79 a	2.82 a	2.64 b
total chl./carot.	7.16 a	7.05 ab	7.00 ab	6.77 b
Melica uniflora				
chlorophyll a	13.96 a	11.78 b		11.16 b
chlorophyll b	4.96 a	4.13 b		3.95 b
total chloro. (a+b)	18.92 a	15.91 b		15.11 b
carotinoids	2.59 a	2.23 b		2.12 b
chloro. a/b ratio	2.82 a	2.85 a		2.83 a
total chl./carot.	7.31 a	7.11 b		7.12 ab
Viola reichenbachiana				
chlorophyll a	9.25 a	9.08 a	8.26 a	8.68 a
chlorophyll b	3.01 a	2.96 a	2.68 a	2.79 a
total chloro. (a+b)	12.26 a	12.04 a	10.94 a	11.47 a
carotinoids	1.58 a	1.55 a	1.47 a	1.51 a
chloro. a/b ratio	3.08 a	3.07 a	3.08 a	3.12 a
total chl./carot.	7.76 a	7.77 a	7.42 a	7.61 a

Several other biochemical parameters were tested in selected herb layer species of the forest floor. Results were published earlier (FANGMEIER 1989; STEUBING et al. 1989; BOTH et al. 1987; FANGMEIER and STEUBING 1987). The reaction to the pollutants was dependent on species. Glutamate dehydrogenase activity was significantly increased by SO_2 in *Lamium galeobdolon* L. (yellow dead–nettle) and decreased in *Melica uniflora* and *Milium effusum* L. (spreading millet grass).

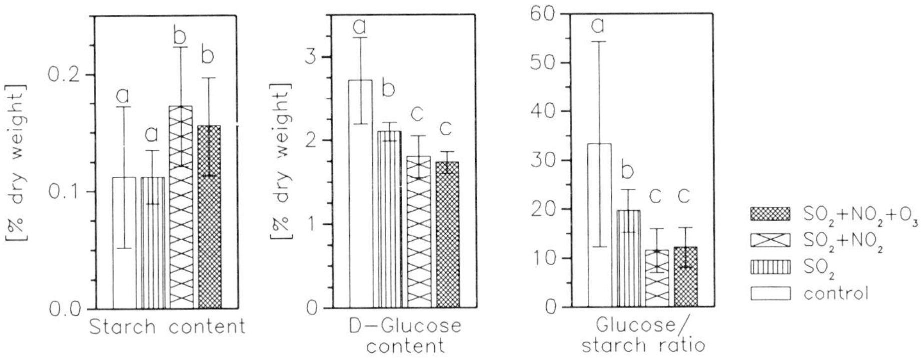

Figure 25.10: Starch and D–glucose content in leaves of *Allium ursinum* after nine four hr weekly fumigations had taken place. Fumigant concentrations as described in Fig. 25.2. Different indices at bars indicate significantly different treatment means ($P< 0.05$).

Table 25.7: Effects of 300 μg SO_2 m^{-3}, 100 μg NO_2 m^{-3}, and 200 μg O_3 m^{-3} applied for 4 hr weekly on the sulphur content [% dry weight] of selected herb layer species. *Allium ursinum* was harvested after 10 fumigations had been applied and the remaining species after 17 fumigations. Different indices in lines indicate significantly different treatments means ($P< 0.05$).

Species	control	SO_2	SO_2 NO_2	SO_2 NO_2 O_3
Allium ursinum	1.70 a	1.77 a	1.71 a	1.42 b
Melica uniflora	0.256 a	0.260 a		0.225 b
Viola reichenbachiana	0.296 a	0.365 b	0.350 b	0.333 b

Peroxidase activity was increased by fumigations in *Melica uniflora* and *Viola reichenbachiana* and depressed in *Allium ursinum* (at least in the SO_2+NO_2 and SO_2+NO_2+O_3 treatments). Ethylene evolution was enhanced in all investigated species. Analysis of variance showed that harvest date and harvest plot had more influence on the measured parameters than the treatment (FANGMEIER 1989). Besides harvest plot and harvest date influences, the reaction of the species to the pollutants was often different in the Melico–Fagetum typicum and the Melico–Fagetum allietosum.

Thus, some conclusions can be drawn concerning the detection of air pollution stress by biochemical investigations:

1. The specific reaction of the investigated plant species must be known.
2. Environmental factors such as climate and soil, and biotic factors such as competition have to be similar if different stands are to be compared.
3. The pollution with different pollution types such as acid deposition and photooxidants has to be investigated since the presence or absence of one pollution type may influence the biochemical reaction to another pollution type.

25.7 Conclusions

The habitat at the forest floor is characterized by buffered climatic extremes, relatively high CO_2 concentrations caused by soil respiration, and, most important as a "stress" factor limiting plant growth, low PAR levels, and a special spectral composition of the irradiation (CRAWFORD 1989). The plants living at the forest floor have developed different strategies to meet the requirements to survive at these conditions (GRIME 1981). Vernal geophytes such as *Allium ursinum*, *Anemone nemorosa*, and *Arum maculatum* can be considered as stress–avoiders since these species have their main above–ground development before bud break of the tree layer takes place. Summer green species such as *Melica uniflora* and seedlings of *Fraxinus excelsior*, however, are stress tolerators.

Within the investigated forest floor species, vernal geophytes represented the group most sensitive to gaseous air pollutants. This high sensitivity must be based on the absence of sufficiently working protection mechanisms at biochemical and physiological level. It is improbable that resistance mechanisms at biochemical level could evolve in the relatively short time span during which man–made gaseous air pollutants influenced terrestrial vegetation in such long–living species as occurring at the forest floor, since adaptation to new stresses will require several generations (ROOSE, BRADSHAW and ROBERTS 1982).

One major mode of action of gaseous air pollutants such as SO_2, NO_2, and O_3 is the formation of free radicals (ALSCHER 1989). Free radicals, however, are also formed within the tissues in case of high irradiation and blocking of the calvin cycle, which may occur if the internal CO_2 level is low (caused by stomatal closure in water–stressed plants), or if the carbohydrate metabolism is disturbed for some reason (for example fungal infection) (ELSTNER and OSSWALD 1984). Plants have developed several defense systems (antioxidants) to protect the tissues against free radical attack (ALSCHER 1989; ELSTNER and OSSWALD 1984; HALLIWELL and GUTTERIDGE 1989). These defense systems are able to compensate for air pollutant stress at least to a certain degree.

Vernal geophytes as investigated in this study obviously were not able to protect their tissues against free radical attack. These species are adapted to relatively high PAR levels. Additionally, they have to stand relatively low temperatures (up to

−10 °C occurred in the study area when *Allium ursinum* leaves were already developed to nearly 50%), and they have short-living leaves with a high production rate during a short span of time. It is unlikely that these species are adapted to gaseous air pollutants forming free radicals within the mesophyll, since high concentrations of such pollutants are unlikely in unpolluted atmospheres in spring-time (ozone is present to a certain degree in unpolluted atmospheres, the highest concentrations, however, are found in summer). Thus, the introduction of pollutants into the habitat of vernal geophytes of the forest floor may, together with the relatively high PAR levels experienced by these species during their above-ground development, lead to an overflow of the antioxidant systems in the tissues with free radicals to which these species are not able to respond adequately. This theory is partly supported by the fact that *Allium ursinum* was not able to respond to the fumigations with increased activities of peroxidase (a detoxifying enzyme group), whereas this parameter was increased in *Melica uniflora* and *Viola reichenbachiana*. Thus, the strategy of avoiding the "stress" of low PAR levels by an early, rapid, and short lasting above-ground development may turn out to be disadvantageous to face the additional stress of anthropogenic air pollutants.

However, the knowledge about the physiological and biochemical properties of plants adapted to the forest floor habitat and their reaction to gaseous air pollutants is still poor, and it is difficult to draw final conclusions. The results in this study, however, suggest that the investigated forest floor communities are more sensitive to gaseous air pollutants than crops or trees. This should be taken into account if air quality guidelines are developed to protect the vegetation.

There is not only a gap in the knowledge of the reaction of forest floor communities to man-made stresses such as air pollutants, but also a gap concerning the reaction of native plants and plant communities to such stresses in general. Further attention should be paid to the problem of reactions to stress at different levels of integration.

Acknowledgments

This study was supported by the German Environmental Agency (Umweltbundesamt), Berlin (contract No. 10607046-07) and the Commissions of the European Communities (contract No. ENV-894-D(B)).

Bibliography

ALSCHER, R. G., 1989, Biosynthesis and antioxidant function of glutathione in plants.
Physiol. Plant. **77**, 457–464.

ARNDT, U.; NOBEL, W.; SCHWEIZER, B., 1987, Bioindikatoren: Möglichkeiten, Grenzen und neue Erkenntnisse.
Ulmer, Stuttgart.

ASMAN, W. A. H.; DIEDEREN, S. M. A., (Eds.), 1987, Ammonia and acidification.
Symposium of the European Association for the Science of Air Pollution (EURASAP), Bilthoven, The Netherlands, 13–15 April 1987, Bilthoven.

BARNES, J. D.; BROWN, K. A., 1990, The influence of ozone and acid mist on the amount and wettability of the surface waxes in Norway spruce [*Picea abies* (L.) Karst.].
New Phytol. **114**, 531–536.

BLACK, V. J., 1982, Effects of sulphur dioxide on physiological processes in plants. In: Unsworth, M. H.; Ormrod, D. P. (Eds.), Effects of gaseous air pollution in agriculture and horticulture, Butterworth, London, 67–91.

BOTH, R., 1990, Wirkungen gasförmiger Immissionen auf den CO_2/H_2O-Gaswechsel von Waldpflanzen eines Melico–Fagetum.
Wissenschafts–Verlag Dr. Wigbert Maraun, Frankfurt.

BOTH, R.; FRANKENFELD, M.; MARKGRAF, U. W.; SAURE, H.; STEUBING, L., 1987, Immissionswirkungen in einem Melico–Fagetum. II: Ökophysiologische und biochemische Effekte bei diskontiunierlicher Begasung in Open–top Kammern.
Verh. Ges. Oekologie **16**, 101–106.

BÜRGER, R., 1988, Veränderungen der Bodenvegetation in Wald- und Forstgesellschaften des mittleren und südlichen Schwarzwaldes.
KfK–PEF **52**, Karlsruhe.

CRAWFORD, R. M. M., 1989, Studies in plant survival.
Ecological case histories of plant adaptation to adversity. Blackwell Scientific Publications, Oxford.

DARRALL, N. M.; JÄGER, H. J., 1984, Biochemical diagnostic tests for the effect of air pollution on plants.
In: Koziol, M. J.; Whatley, F. R. (Eds.), Gaseous air pollutants and plant metabolism, Butterworth, London, 333–349.

DIERSCHKE, H., 1972, Zur Aufnahme und Darstellung phänologischer Erscheinungen in Pflanzengesellschaften.
In: Grundfragen und Methoden in der Pflanzensoziologie. Ber. Int. Symp. IVV Rinteln, Junk, Den Haag, 291–311.

ELLENBERG, H., 1990, Zeigerwerte der Gefäßpflanzen Mitteleuropas.
Scripta Geobotanica **9**.

ELSTNER, E. F.; OSSWALD, W., 1984, Fichtensterben in "Reinluftgebieten": Strukturresistenzverlust.
Naturwiss. Rundschau **37**, 52–61.

ENGELBACH, G.; FANGMEIER, A., 1989, Wirkungen gasförmiger Immissionen im Gießener Stadtwald auf die saisonale Entwicklung der Krautschicht unter besonderer Berücksichtigung der Seneszenz.
Verh. Ges. Oekologie **18**, 379–385.

FALKENGREN–GRERUP, U., 1989, Effect of soil acidification on beech forest vegetation in southern Sweden.
J. Appl. Ecol. **26**, 341–352.

FALKENGREN–GRERUP, U., 1989, Soil acidification and its impact on ground vegetation.
AMBIO **18**, 179–183.

FANGMEIER, A., 1989, Effects of open–top fumigations with SO_2, NO_2, and ozone on the native herb layer of a beech forest.
Env. Exp. Bot. **29**, 199–213.

FANGMEIER, A.; GNITTKE, J.; STEUBING, L., 1986, Transportable open–tops for discontinuous fumigations.
In: Air Pollution Research Report **5**: Microclimate and plant growth in open–top chambers, Commission of the European Communitites, 102–112.

FANGMEIER, A.; STEUBING, L., 1987, Physiological responses of native plants fumigated with SO_2, NO_2, and ozone in their natural environment.
GSF–Ber., 9/87, 39–41.

FREEDMAN, B.; HUTCHINSON, T. C., 1980, Long–term effects of smelter pollution at Sudbury, Ontario, on forest community composition.
Can. J. Bot. **58**, 2123–2140.

GLATZEL, G.; SONDEREGGER, E.; KAZDA, M.; PUXBAUM, H., 1983, Bodenveränderung durch schadstoffangereicherte Stammablaufniederschläge in Buchenbeständen des Wienerwaldes.
Allg. Forst Zeitschr. **38**, 693–694.

GLAVAC, V.; KRAUSE, A.; WOLFF–STRAUB, R., 1970, Über die Verteilung der Hainsimse (*Luzula luzuloides*) im Stammfußbereich der Buche im Siebengebirge bei Bonn.
Schriftenr. Vegetationsk. **5**, 187–192.

GLAVAC, V.; KOENIES, H., 1986, Kleinräumige Konfiguration wichtiger bodenchemischer Meßgrößen in dem vom Stammablaufwasser beeinflußten Bodenbereich alter Buchen.
Verh. Ges. Oekologie **14**, 293–298.

GRIME, J. P., 1981, Plant strategies in shade.
In: Smith, H. (Ed.), Plants and the daylight spectrum, Brit. Photobiol. Soc. Intern. Symp. Leicester, 159–186.

GUDERIAN, R., 1977, Air pollution.
Ecological Studies **22**. Springer, Berlin/Heidelberg/New York.

GUDERIAN, R., (Ed.), 1985, Air pollution by photochemical oxidants. Formation, transport, control and effects on plants.
Ecological Studies **52**, Springer, Berlin/Heidelberg/New York/Tokyo.

GUDERIAN, R.; KÜPPERS, K., 1980, Response of plant communities to air pollution.
In: Proc.of Symp.on Effects of Air Pollutants on Mediterranean and Temperate Forest Ecosystems, June 22–27, Riverside, California, USDA Forest Service, Berkeley, California, 187–199.

GUDERIAN, R.; TINGEY, D. T., 1987, Notwendigkeit und Ableitung von Grenzwerten für Stickstoffoxide.
Umweltbundesamt Berichte 1/87. Erich Schmidt Verlag, Berlin.

HÄLLGREN, J. E., 1984, Photosynthetic gas exchange in leaves affected by air pollutants.
In: Koziol, M. J.; Whatley, F. R. (Eds.), Gaseous air pollutants and plant metabolism, Butterworth, London, 147–159.

HALLIWELL, B.; GUTTERIDGE, J. M. C., 1989, Free radicals in biology and medicine.
Clavendon Press, Oxford.

HEAGLE, A. S., 1989,. Ozone and crop yield.
Ann. Rev. Phytopathol. **27**, 397–423.

HECK, W. W.; CURE, W. W.; RAWLINGS, J. O.; ZARAGOZA, L. J.; HEAGLE, A. S.; HEGGESTAD, H. E.; KOHUT, R. J.; KRESS, L. W.; TEMPLE, P. J., 1984, Assessing impacts of ozone on agricultural crops: I. Overview.
JAPCA **34**, 729–735.

HUTCHINSON, T. C.; ADAMS, C. M., 1987, Comparative abilities of leaf surfaces to neutralize acidic rain drops. I. The influence of calcium nutrition and charcoal filtered air.
New Phytol. **106**, 169–183.

JÄGER, H. J., 1982, Biochemical indication of an effect of air pollutants on plants.
In: Steubing, L.; Jäger, H. J. (Eds.), Monitoring of air pollutants by plants, Junk, The Hague, 99–107.

JOCHHEIM, H., 1986, Einfluß des Stammablaufwassers auf den Pflanzenbewuchs und den chemischen Bodenzustand im Oberboden von Altbuchenbeständen. Verh. Ges. Oekologie **14**, 299–308.

KLEIN, R. M.; PERKINS, T. D., 1988, Primary and secondary causes and consequences of contemporary forest decline. Bot. Review **54**, 1–43.

KOZIOL, M. J., 1984, Interactions of gaseous pollutants with carbohydrate metabolism. In: Koziol, M. J.; Whatley, F. R. (Eds.), Gaseous air pollutants and plant metabolism, Butterworth, London, 251–273.

KRAUSE, G. H. M.; ARNDT, U.; BRANDT, C. J.; BUCHER, J.; KENK, G.; MATZNER, E., 1986, Forest decline in Europe: development and possible causes. Water Air Soil Pollut. **31**, 647–668.

MANSFIELD, T. A.; FREER–SMITH, P. H., 1984, The role of stomata in resistance mechanisms. In: Koziol, M. J.; Whatley, F. R. (Eds.), Gaseous air pollutants and plant metabolism, Butterworth, London, 131–146.

MCCLENAHEN, J. R., 1978, Community changes in a deciduous forest exposed to air pollution. Can. J. For. Res. **8**, 432–438.

MCLAUGHLIN, S. B., 1985, Effects of air pollution on forests. A critical review. JAPCA **35**, 512–534.

NIHLGARD, B., 1985, The ammonium hypothesis — an additional explanation to the forest dieback in Europe. AMBIO **14**, 2–8.

ROOSE, M. L.; BRADSHAW, A. D.; ROBERTS, T. M., 1982, Evolution of resistance to gaseous air pollutants. In: Unsworth, M. H.; Ormrod, D. P. (Eds.), Effects of gaseous air pollutants in agriculture and horticulture, Butterworth Scientific, London, 379–409.

ROST–SIEBERT, K., 1986, Feststellung von Veränderungen in der Bodenvegetation und im chemischen Oberbodenzustand während der letzten Jahrzehnte. In: Querschnittsseminar Bioindikation, 28.–29.11.85, UBA–Texte 18/86, Berlin, 246–256.

SIENKIEWICZ, J., 1989, Forest community changes as bioindicators of contamination. In: Bucher, J. C.; Bucher–Wallin, I. (Eds.), Air pollution and forest decline. Proc. 14th Int. Meeting for Specialists in Air Pollution Effects on Forest Ecosystems, IUFRO P2.05, Interlaken, Switzerland, Oct. 2–8.1988,Birmensdorf, 245–248.

STEUBING, L.; FANGMEIER, A., 1986, Immissionssituation der Waldbodenvegetation: Immissionsbelastung.
Allg. Forst Zeitschr. **41**, 469–471.

STEUBING, L.; FANGMEIER, A.; BOTH, R.; FRANKENFELD, M., 1989, Effects of SO_2, NO_2, and O_3 on population development and morphological and physiological parameters of native herb layer species in a beech forest.
Environ. Pollut. **58**, 281–302.

TINGEY, D. T.; TAYLOR, G. E., 1982, Variation in plant response to ozone: a conceptual model of physiological events.
In: Unsworth, M. H.; Ormrod, D. P. (Eds.), Effects of gaseous air pollution in agriculture and horticulture, Butterworth, London, 113–138.

TRAUTMANN, W.; KRAUSE, A.; WOLFF–STRAUB, R., 1970, Veränderungen der Bodenvegetation in Kiefernforsten als Folge industrieller Luftverunreinigungen im Raum Mannheim–Ludwigshafen. Schriftenr. Vegetationsk. **5**, 193–207.

TRESHOW, M.; STEWART, D., 1973, Ozone sensitivity of plants in natural communities.
Biol. Conserv. **5**, 209–214.

TYLER, G., 1987, Probable effects of soil acidification and nitrogen deposition on the floristic composition of oak (*Quercus robur* L.) forest.
Flora **179**, 165–170.

Verein Deutscher Ingenieure VDI–Kommission Reinhaltung der Luft, 1987, Acidic precipitation. Formation and impact on terrestrial ecosystems.
VDI, Duesseldorf.

WEIGEL, H. J.; JÄGER, H. J., 1985, Physiologische und biochemische Verfahren zum Nachweis von Schadstoffwirkungen.
Staub Reinh. Luft **45**, 269–271.

WILMANNS, O., 1989, Zur Frage der Reaktion der Waldboden–Vegetation auf Stoffeintrag durch Regen — eine Studie auf der Schwäbischen Alb.
Allg. Forst- u. J.- Ztg. **160**, 165–175.

WITTIG, R.; NEITE, H., 1985, Acid indicators around the trunk base of *Fagus silvatica* in limestone and loess beechwoods: distribution pattern and phytosociological problems.
Vegetatio **64**, 113–119.

WITTIG, R.; BALLACH, H.-J.; BRANDT, J. C., 1985, Increase of number of acid indicators in the herb layer of the millet–grass–beech forest of the Westfalian Bight.
Angew. Bot. **59**, 219–232.

WORREST, R. C.; CALDWELL, M. M., (Eds.), 1986, Stratospheric ozone reduction, solar ultraviolet radiation and plant life.
NATO ASI Series G, Vol. **8**, Springer.

Chapter 26

Eutrophication in forest ecosystems

P. JAKUCS

26.1 Introduction

Nowadays, eutrophication is one of the most important problems which nature conservation and landscape management have to deal with. Among the scientists concentrating on the matter, the circulation of fresh water eutrophication is a well-known and widely studied ecological phenomenon, which can be defined as "a process resulting in the abundance of plants in the surface water as a biological response to the enrichment of plant nutrients, i.e. the increase of water trophity due to natural and artificial effects. During the process the water quality changes drastically and deteriorates as far as human use is concerned" (FELFÖLDY 1984).

In studies focussing on the new type of tree decline in deciduous forests, the deforested areas and the "monoculture" forests consisting of non-native tree species (e.g. black locust), many signs of a process being analogous with the eutrophication of water bodies can be recognized. This can be called terrestrial eutrophication. In habitats where eutrophication has arisen or may arise, similar changes in certain environmental factors can be detected. Thus, the radiation reaching the soil surface significantly increases and there is a surplus of the main nutrients in the soil. As a biological-ecological reaction, large abundances of plants may develop, i.e. the degree of trophity endures significant changes which may start with a marked species turnover and the decrease in original diversity. Changes in the leaf area index, the phytomass and the chlorophyll content of the herb layer can also be recognized.

26.2 Indicator populations of eutrophication

Environmental factors resulting in terrestrial eutrophication are normally indicated by populations and population assemblages of species occurring in the herb and shrub layers of forests which were already earlier regarded as nitrophyllous organisms, e.g. *Sambucus nigra, Urtica dioica, Galeopsis tetrahit, Chelidonium majus, Torilis japonica, Galium aparine, Rumex obtusifolius, Parietaria officinalis, Impatiens parviflora*. These herbs can generally only be found on rubbish heaps, manure heaps and ruderal places. As a consequence of eutrophication they may become very abundant in forests and suppress the native plant species. In connection with the biondication, the problem of terrestrial eutrophication emerged earlier as a result of over-fertilization of agricultural land (SCHUBERT 1985, REIF et al. 1985).

26.3 Examples of terrestrial eutrophication in forest stands

26.3.1 Terrestrial eutrophication as an accompanying phenomenon of the new type of forest decline

The results shown in part A of Figure 26.1 were obtained from sessile oak–Austrian oak forest stands in two research areas. One of them is the sampling plot of the "Sikfőkut Project" described by JAKUCS (1985) and the other is situated on the hilly region south of the "Sajo–valley" industrial district in the Northeastern part of Hungary (JAKUCS 1985, JAKUCS 1984, JAKUCS 1986, BERKI and HOLES 1988).

Since 1979 a large–scale decline of sessile oak has been observed in both areas where now 40–60% of the trees are dead. The dead trees remained standing for 4 or 5 years and then fell. The microenvironments of healthy, diseased and dead trees were studied. It was stated that the radiation arriving at the soil surface through the foliage of healthy trees (thus in a closed forest with full canopy) amounted to 120.45 J cm^2/day whilst through the foliage of dead trees 2190 J cm^2/day reached the forest floor. A considerable decrease occurred in the species' richness of shrub and herb layers in the environs of dead trees compared to the healthy forest. There were great differences in the nutrient content of the soil between the healthy and dead trees. It was especially noticeable in the case of nitrate–N content, which was 13 times higher in the soil under the dead trees than under the healthy ones. It seems likely that this nitrogen surplus was mainly attributed to the unhampered input of NO_x through wet and dry deposition from the atmosphere. A slighter

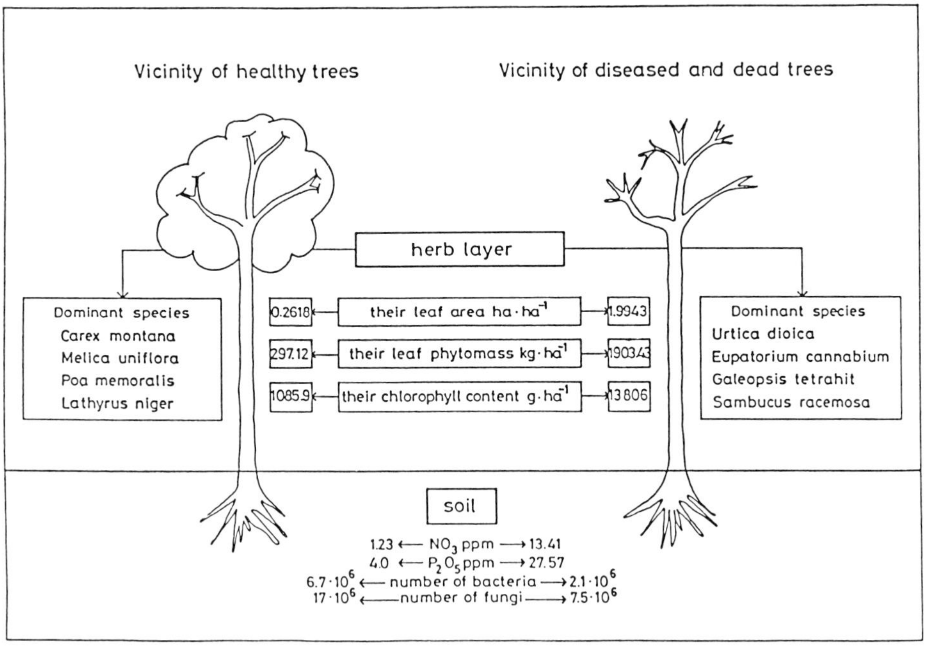

Figure 26.1: Some signs of terrestrial eutrophication due to tree decline in sessile oak forest stands

increase was observed in the phosphorus and potassium content of the soil in the environs of the dead trees, due to the lack of absorption of tree roots. The LAI and the phytomass of herb layer species under the dead trees increased twofold, whilst the content of chlorophylls (a+b) in their biomass was 12 times higher than under the healthy trees. In the environs of dead trees the threefold lowering in the number of bacteria and fungi, in comparison with the healthy trees, reflects the inhibition of litter decomposition.

26.3.2 Terrestrial eutrophication as an accompanying phenomenon of clear–cutting

The results shown in part A of Table 26.1 were obtained during the studies on the research area of the "Rejtek Project" in the Bűkk Mountains (N. Hungary) (Jakucs 1987, Katona and Tóthmérész 1985, Bodnár 1990, Holes 1990, Mészáros 1988). Until January 1981 the area was covered by an old contagious stand of sub–montane beech forest (Melico–Fagetum). But then an area of 4.3ha of it was deforested for experiments. Since then many different parameters have

Table 26.1: Some indications of eutrophication after clear–cutting of a beech forest (A) and after the plantation of black locust (B)

	(A)		(B)	
treatment	clear–cutting		Plantation of black locust	
location	Rejtek		"Great Forest" Debrecen	
	clear–cut area	beech forest	black locust forest	oak forest
soil properties:				
NO_3^- ppm	4.060	2.680	39.400	4.150
P_2O_5 ppm	53.800	19.800	71.450	84.500
K ppm	136.500	88.300	—	—
number of bacteria (10^6)	1.342	1.237	0.585	0.471
number of fungi (10^5)	6.331	6.483	0.441	0.266
characteristics of the stand:				
LAI	2.500	0.760	4.820	1.480
phytomass ($kg \cdot ha^{-1}$)	703.500	58.000	349.700	112.400
chlorophyll (a+b) content ($g \cdot ha^{-1}$)	15,392.000	978.400	3,545.400	775.300
dominant species in the herb layer:	*Atropa belladonna, Urtica dioica, Alliaria petiolata, Solanum nigrum, Cirsium arvense, Rubus idaeus,*	*Asperula odorata, Viola silvatica, Lathyrus vernus, Melica uniflora, Geum urbanum,*	*Sambucus nigra, Alliaria petiolata, Impatiens parviflora, Urtica dioica, Chelidonium majus, Chenopodium album,*	*Polygonatum latifolium, Myosotis silvestris, Ficaria verna, Gagea lutea,*

been measured in the clear–cut area and in the remaining section of the forest.

It was stated that by removing the trees 94–95% of the global radiation can reach the soil surface which contains the whole light spectrum. In the closed forest, however, only 5–6% of the global radiation can get through the canopy layer to the herb layer and there is a shift in the spectral composition towards the green light.

The soil type on the research area is rocky black Rendzina. Shortly after the deforestation, a rapid alteration of the nutrient content in the soil took place. The total nitrogen content doubled and the concentration of the available phosphorus increased threefold. A significant increase of available potassium also occurred. The rapid increase in the supply of important nutrients may be attributed to two main reasons. First, during the deforestation a significant amount of thin twigs and other organic residues fell onto the soil surface and their decomposition resulted in the surplus of nutrients in the soil. Secondly, the accumulation of nutrients in the clear–cut area may also be explained by the fact that more individuals of large game (deer and moufflons) from the environs visit the area for longer periods than before and as a result, there has been an increase in the amount of faeces falling down onto the soil surface. In winter, a year after the clear–cutting, the quantity of faeces collected from the snow surface and dried at 98 °C amounted to 167.9 kg/ha, which considering its effects on the area, would be similar to organic farmyard manure of the same quantity being spread. A slight decrease in the number of soil bacteria and fungi occurred in the clear–cut area in comparison with the forest stand which was mainly due to the lower moisture content.

A drastic and rapid production of species and the decrease in the diversity of species could be observed in the herbaceous vegetation. As early as in the summer following the deforestation, the large–scale appearence of nitrophyllous species (*Atropa belladonna, Alliaria petiolata, Urtica dioica, Solanum nigrum, Cirsium arvense, Rubus idaeus,* etc.) was recorded. The phytomass, the leaf area index and the amount of chlorophylls in the herbaceous vegetation has reached values three times higher than when the clear–cutting took place.

26.3.3 Terrestrial eutrophication in black locust forest stands

It is well–known that in black locust forests the soil is covered by a large density of nitrophyllous herb layer species. This phenomenon can also be regarded as an eutrophication process. Part B of Table 26.1 shows the results of a comparative study which aimed to describe a planted black locust stand and a native peduncular oak forest stand in the "Great Forest" of Debrecen (N. Hungary). The stands of different forest types are next to each other, and their climatic and soil conditions are similar (Jakucs 1988, Nagy and Nagy 1981, Nagy and Tóth 1983).

Whilst in a native oak forest stand the proportion of light transmitted through the foliage of trees and shrubs and reaching the soil surface amounts only to 1–2% of the incoming radiation. In a stand of black locust it ranges between 40 and 50%. This may be explained partly by the structural characteristics of the canopy layer of the black locust stand (composite leaves with many gaps among them) and partly by the the thickness of the leaves. Whilst the average thickness of peduncular oak leaves is 200μ, the thickness of black locust leaves is 120μ. The surplus of light energy reaching into the interior of the black locust stand and the richer light spectrum may be responsible for the significant increase of above-ground shoot density in the herb layer in comparison with the oak forest.

Our studies indicate high concentrations of N, P and K in the soil of the black locust stand. Compared with the oak forest stand, the surplus of the nutrients is the result of the rapid litter decomposition and nitrification. In one hectare of the black locust stand 377 trees produced more than 7 million composite leaves with a dry weight of over 3 tons per year, which provide an abundant nutrient supply for the herb layer through the annual litter fall.

In the case of the nitrophyllous species dominating in the herb layer and indicating the eutrophication (*Sambucus nigra, Alliaria petiolata, Impatiens parviflora, Urtica dioica, Chelidonium majus, Cannabis sativa, Chenopodium album*) in the black locust forest stand, three or fourfold increases were observed in comparison with the herb layer of the native oak stand (Table 26.1/B).

The terrestrial eutrophication seems to develop especially strongly in the monocultures of black locust. 17% of the Hungarian forests are black locust stands and in Europe the so-called culture black locust forest is spread most widely in Hungary. In its original native habitat the black locust (N. America, Appalache Mountains) grows only in mixed deciduous forests.

26.3.4 Eutrophication in forests as the deterioration in the quality of the forest

Eutrophication processes taking place in forest stands result in the permanent deterioration in the quality of the forest. The aestethic and health–recreation value of forests decreases significantly. It is almost impossible to stay and relax in these forests because of the unpleasant (sticky, stinging) outside of nitrophyllous species, which occur in large masses. These processes can also cause serious problems and the solution requires high financial expenses in the silvicultural management. The large quantity of organic matter may become compact under the snow in winter, resulting in anaerobic conditions and partial litter decomposition, and consequently the large-scale propagation of toxic microorganisms. The reforestation or the regeneration of forests in these habitats may become almost impossible; the occasional treatment with herbicides against weeds may also induce newer deterioration processes.

26.3.5 Terrestrial eutrophication in semi–natural forest communities

In maple forests covering the ravines (Phyllitidi–Aceretum) and in lime–ash forests (Tilio–Fraxinetum) of our Central Range of Mountains there have long been signs that point to the eutrophication process. In the former forest stands the surplus of N and other nutrients originated from the large amount of fallen wood. In the lime–ash forest stands however, the favourite habitats of our large game, the nutrient surplus resulted from the faeces and the favourable light conditions. The large-scale appearance of species indicating the eutrophication can often be observed in the vicinity of forest wild-feeders and springs which are frequently visited by the game.

26.3.6 Natural and artificial terrestrial eutrophication

Some signs of terrestrial eutrophication or the change in trophity can always be recognized in our forests during biotic and secular succession (e.g. grove forests). The disturbance of the natural matter circulation, the formation of centres with longer residence periods in the cycle of individual nutrients, result in the alteration of the degree of trophity. In the case of the above-mentioned examples, the processes of eutrophication differ in some respects. The interruption of matter circulation is induced by direct human intervention (clear-cutting) or the indirect effects of his activity (tree decline, plantation of black locust). If a poplar forest stand is supplied by sewage water or is grazed by domestic animals, or in other words, if its soil is fertilized, we can unanimously speak about the artificial method of terrestrial eutrophication, as well as the eutrophication of water bodies.

Bibliography

BERKI, I.; HOLES, L., 1988, Lokale industrielle Emission und Waldschaden in Nordungarn II.
Acta Bot. Hung. **34**, 25–37

BODNÁR, T., 1990, Tarvágásos fakitermelési mód hatása sekély rendzina talajokra. (Effects of clear-cutting on shallow Rendzina soils.)
Mscr.

FELFŐLDY, L., 1984, Hidrobiológia – szavakban. Hidrobiológiai értelmező szótár (Hydrobiology – in words.
Hydrobiological explanatory dictionary) VGI, Budapest

HOLES, L., 1990, Rendzina talaju bűkkős tarvágásának hatása a talaj mikrobiális aktivitására. (Effects of clear-cutting of a beech forest on the microbial activity of rendzina soil.)
Mscr.

JAKUCS, P. (ed.), 1984, Komplex ökológiai vizsgálatok erdőink állapotjavitása érdekében. (Komplex ecological studies for improving the state of our forests.)
Final Research Report, Debrecen

JAKUCS, P. (ed.), 1985, Ecology of an oak forest in Hungary. Results of "Sikfőkut Project".
Akadémiai Kiadó, Budapest

JAKUCS, P., 1987, Rejtek Project. Ecological researches for favourable regeneration of forests covering shallow soil derived from limestone in Hungary.
Acta Biol. Debrecina **19**, 5–12

JAKUCS, P.; BERKI, I; HOLES, L.; TÓTHMÉRÉSZ, B., 1988, Lokale industrielle Emission und Waldschaden in Nordungarn I.
Acta Bot. Hung. **34**, 11–24

KATONA, É.; TÓTHMÉRÉSZ, B., 1985, Szubmontan erdők lágyszáru növényzetének változása tarvágás után. (Change in herbaceous vegetation of submontane forests after clear-cutting.)
Botanikai Közlemények **72**, 17–25.

MÉSZÁROS, I., 1988, Strukturális és ökofiziológiai jellemzők másodlagos erdőszegélyekben. (Structural and ecophysiological characteristics of secondarily induced forest edges.)
Ph. D. Dissertation, Debrecen

NAGY, M.; NAGY, J., 1981, Akácerdő aljnövényzetének diverzitása. (Diversity of herb layer of black locust forest.)
Acta Biol. Debrecina, 18

NAGY, J., 1981, Homoki akácos struktura és fitomassza vizsgálata. (Study on structure and phytomass of black locust forest on sandy habitat.)
Mscr.

NAGY, M.; TÓTH, J. A., 1983, Akác lombavar bomlása egy Debrecen környéki akácosban. (Litter decomposition in a black locust forest near Debrecen.)
Mscr.

REIF, A.; TECKELMANN, M.; SCHULZE, E.-D., 1985, Die Standortamplitude der Großen Brennessel.
Flora, **176**, 365–382

SCHUBERT, R. (ed.), 1985, Bioindication in terrestrischen Okoszisztemen.
G. Fischer, Jena

Chapter 27

Animal-coenoses in the "spruce forest" ecosystem (Protozoa, Metazoa-invertebrates): Indicators of alterations in forest-ecosystems

W. FUNKE and M. ROTH-HOLZAPFEL

27.1 Introduction

The structure and function of animal coenoses in Central European forest-ecosystems are still largely unknown. The present studies should shed light on some of these deficiencies. The investigations emphasize the suitability of *Protozoa* and *Metazoa* (invertebrates) as indicators of extreme alterations in spruce forests (i.e. especially functional disturbances) and of the effects of forestry management (for detailed information see FUNKE 1986; FUNKE and JANS 1989).

For statistics see the studies of co-workers (FUNKE 1989; ROTH-HOLZAPFEL and FUNKE 1989). Some studies are in preparation.

27.2 Study sites, material and methods

The objects under investigation are 50 to 100-year-old spruce stands in southern West Germany. The research areas are situated in the Black Forest (near Freiburg i. Br. and Freudenstadt), in the Swabian-Franconian Forest (near Stuttgart/Welzheim), in the southeastern part of the Swabian Jura (near Ulm), in Upper Swabia (near Ochsenhausen) and in the "Fichtelgebirge" (near Bayreuth/Weissenstadt). In order to deal with special questions forests in the northern Harz Mountains (near Bad Harzburg) were also taken into consideration. The stands must contain different degrees of forest decline.

Most of the investigations are carried out at U1 (university forest near Ulm) and Ux (near Ochsenhausen). At both of these stands several sub-areas have been distinguished:

1. control areas (NF),
2. humid area (FF), which has been irrigated since 1979 during the summer in addition to the natural rainfall, whenever the siphon tension of the soil (measured in a depth of 5 cm) reaches a –70 cm column of water,
3. limed area (DF) in Ux limed in December 1983 (200 g $\cdot m^{-2} CaCO_3$),
4. limed area (DF) in U1 which was limed in March 1984 and fertilized in May 1984 with 5 $Ca(NO_3)_2$: NH_4NO_3 (50 $g \cdot m^{-2}$).

Studies at U1NF began already in 1977 and at U1FF in 1980; at the other sub-areas they began in 1983/84. Weather data are provided by the meteorological service of West Germany (Ulm/Stuttgart). The pH measurements (KCl) of the upper soil layers (0–4 cm) yielded mean values of 2.5 to 3.5 at the control areas U1NF and UxNF, of 4.4 to 4.9 at U1DF, of 3.2 to 4.1 at UxDF and of 2.9 to 3.3 at the irrigated area U1FF. In deeper soil layers of the areas U1 and Ux, the values varied between 2.3 and 3.4. A pedological characterization exists only for U1 and, to some extent, for Ux (FUNKE 1986; ROTH 1986a). Within animal coenoses about 1,200 taxa, and among them about 800 species, are taken into consideration by several cooperators (FUNKE 1986, in prep.). Seven species (including six species of *Protura*, s. STUMPP 1987) were unknown until now.

Detailed information is given, mostly on group level, for *Testacea*, *Ciliata*, *Nematoda*, *Lumbricidae*, *Enchytraeidae*, *Araneae*, *Acari*, *Protura*, *Collembola*, *Thysanoptera*, *Hymenoptera* and *Diptera*.

Most animals are selected by (microscopic) examination of small samples of litter and soil, partly using cultivation methods (s. BUITKAMP 1979 or FOISSNER 1982 in LEHLE 1990); by the extraction of litter and soil samples (according to

BAERMAN in BALOGH 1958; O'CONNOR 1962; MACFADYEN 1962; HEALEY and RUSSELL-SMITH 1971; KEMPSON et al. 1963, all cited in FUNKE 1986; BORIN and HERLITZIUS 1987; RATAJCZAK 1989) and by ground and arboreal photoeclectors (FUNKE 1971, 1986). The decomposition of litter was determined by the deposition of hazel leaf as test-material in special litter containers (HERLITZIUS 1983). Element concentrations of plant material and animals were measured by instrumental neutron activation analysis (iNAA), induced-coupled-plasma atomic emission spectroscopy (ICP–AES) and graphite furnace atomic absorption spectrometry (ROTH 1986b; ROTH-HOLZAPFEL and FUNKE 1989).

27.3 Results and discussion

27.3.1 Soil animals as sensitive indicators of forest decline

The *Protura*, a small group of apterygote insects, are of special significance here, because it seems that they predominantly feed on the ectotrophic mycorrhiza of trees (STURM 1959; STUMPP 1987). Thus, the species number and population density, especially of larvae, are related to the vitality and development of mycorrhiza (Fig. 27.1, FUNKE et al. 1987; STUMPP 1987). If there is a correlation between the vitality of mycorrhiza and the physical condition of trees (OBERWINKLER et al. 1986; SCHLECHTE 1986) the population data of *Protura* may yield valuable information about the physical condition of single trees and/or whole stands (Tab. 27.1). But the influence of interspecific trophic relationships between *Protura* and other arthropods, especially predaceaous *Acari*, also have to be taken into consideration.

27.3.2 Fluctuations of population density and litter decomposition as indicators of alterations in forest-ecosystems (caused by climatic conditions)

Fluctuations of population density of soil arthropods can partly be explained by interspecific activities, especially by predator–prey relationships or by competition (s. e.g. spiders and beetles, esp. *Staphylinidae*, Fig. 27.2). The stark reduction of several groups of arthropods and less clearly of *Enchytraeidae* at U1 (Fig. 27.2) between 1979 and 1982 (especially in 1981 and partly in 1986) may be attributed to extreme climatic events (FUNKE 1986). Thus, 1980 will be a key year, as it was particularly cool and dry at the same time. This may be responsible for the strong decline in the population density at U1, especially of the most important decomposers, such as *Enchytraeidae* and *Diptera* larvae, and consequently, due to the assumed reduced microbial activities, the litter decomposition decreased (Fig. 27.2, HERLITZIUS et al. 1986). This correlation is supported by results obtained at U1FF (irrigated area). There, already in 1981, an increase in abundance was observed in a number of populations, and this trend continued partly to an extraordinary extent

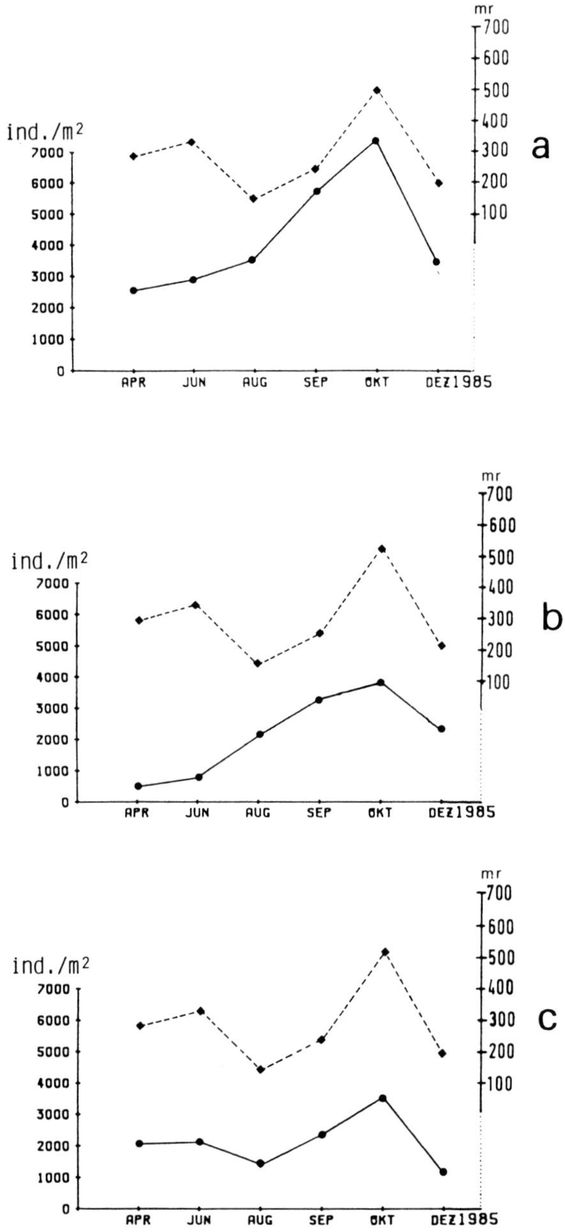

Figure 27.1: Frequency oscillations of finest roots with mycorrhiza (mr per 1000 cm^3 soil substrate, according to FEIL et al. 1986) – broken lined, and abundance of *Protura* (individuals per m^2, according to STUMPP 1987) in 1985/86 (spruce stand U1) – straight lined. A *Protura* total, b *Protura* imagines, c *Protura* larvae.

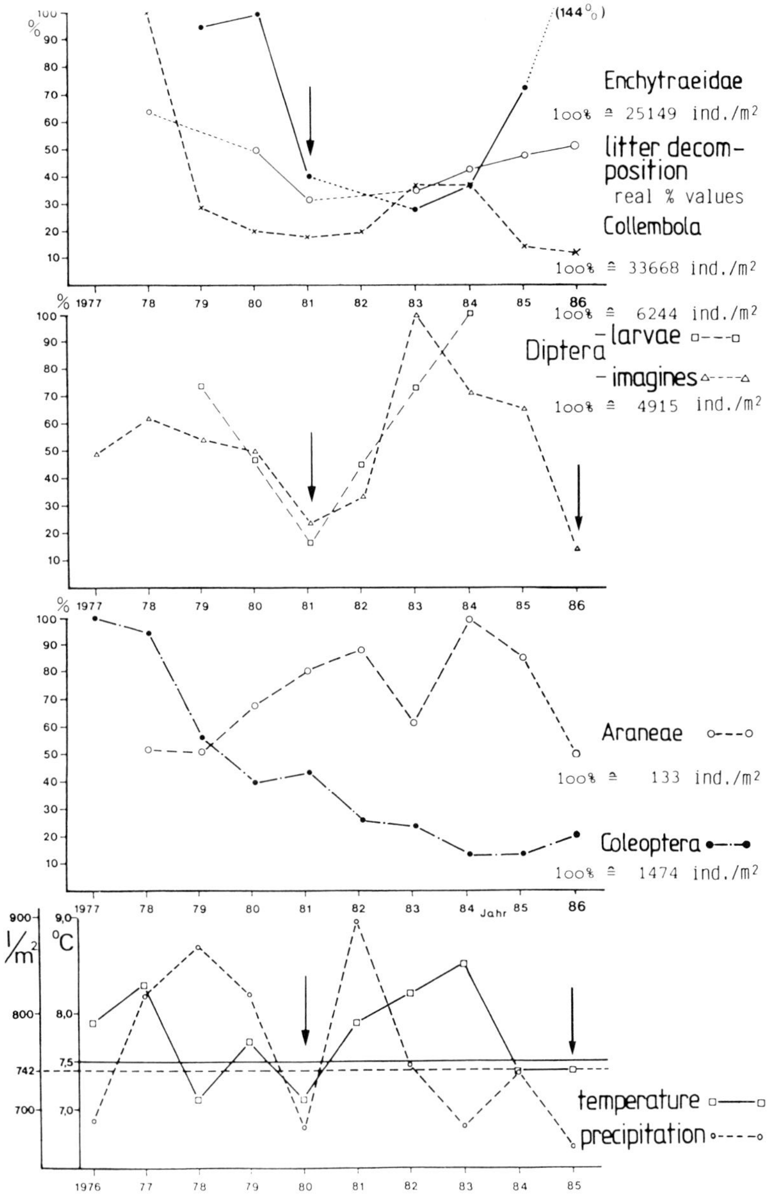

Figure 27.2: Fluctuations of litter decomposition and population density at U1NF (control area) in relation to weather conditions. Horizontal lines: mean values of annual average data (1976–1984) supplied by the meteorological service at Ulm, FRG. Arrows: max. decrease of abundance and litter decomposition, minima of temperature and precipitation.

Table 27.1: Relations between the stage of disease of various spruce stands in southern West Germany (and in the Harz Mountains) and coenosis data of *Protura* (according to Stumpp 1987). The symbols in the table mean:
o for description see FUNKE (1986)
oo according to FVA (Forstl. Versuchsanstalt) Freiburg, otherwise to STUMPP
* spring, ** summer, * * * autumn values, no symbol: annual means
Sites of investigation: U1 – near Ulm; Ux – near Ochsenhausen; Ks – near Freiburg; Fs – near Freudenstadt; Ed – near Stuttgart/Welzheim; Fg – near Bayreuth/Weissenstadt; Kk – near Bad Harzburg (see page 580);
A–F, X/X', Y/Y': sub-areas at different sites, for details see FUNKE 1986, STUMPP 1987.

++	$\geq$ 4 species	$\gg$ 2000 ind.·m^{-2}	$\geq$ 60% larvae	eval. very positive
+	$\geq$ 2 species	$\geq$ 2000 ind.·m^{-2}	$\geq$ 40% larvae	evaluation positive
–	0–2 species	$\leq$ 2000 ind.·m^{-2}	$\leq$ 40% larvae	evaluation negative
⊕		high shares of larvae		evaluation positive

site o	stage of disease	species number	*Protura* [indiv.·m^{-2}] ≥2000	<2000	larvae %	evaluation according to (3)–(5) 3	4	5	total
(1)	(2)	(3)	(4)		(5)	3	4	5	
U1	1–2 oo	3	×		47	+	++	+	++
Ux	1–2	6	×		49	++	++	+	++
OcA	1–2	3	×**		44	+	++	+	++
B	1–2	4		×**	43	+	–	+	+
C	1–2	3	×**		40	+	++	+	++
KsX/X'	3	1		×	39	–	–	–	–
Y/Y'	2–3	2		×	61	–	–	++	⊕
FsA	1–2 oo	2		×*	66	–	–	++	⊕
FsB	3 oo	3		×*	57	+	–	+	+
Ed	2	3		×	76	+	–	++	⊕
FgA/A'	4	1		×	20	–	–	–	–
FgB	3	4	×		31	+	+	–	+
FgC/C'	1–2	5	×		50	++	++	+	++
FgD/D'	2	3	×		49	+	–	+	+
FgE	4	2	×*		10	–	+	–	–
FgF	2	3		×*	11	+	–	–	–
KkA	3	–		–	–	–	–	–	–
B	2–3	1		×***	35	–	–	–	–

until 1983 (Fig. 27.3). Undoubtedly, the strong decrease of population density in 1984 can hardly be correlated with the weather, but rather with the over-population in 1983 and, subsequently, with density-dependent regulation-mechanisms (i.e. competition, predator–prey correlations). The unfavourable weather could not be without far-reaching effects on the vitality of trees. Their growth and power of resistance against plant and animal pests and toxic pollutions might have been decisively reduced. Analyses of weather data from recent years and furthermore recent decades have significantly demonstrated the following situation: The mean annual temperatures between 1977 and 1980 were far below the long-term mean value. The same holds true between 1980 and 1983 for precipitation (Fig. 27.4a). Furthermore, between 1961 and 1981 the number of days (per year) with temperatures above 25°C was drastically reduced (Fig. 27.4b). A similar observation was made with the duration of sunshine (Fig. 27.4c) and precipitation (FUNKE 1989).

27.3.3 Soil animals as indicators of alterations in forest soils, caused by liming, mineral-fertilization or by additional irrigation

Applications of lime and fertilizer (as well as additional irrigation) caused extreme alterations of zoocoenoses in U1 and Ux (see also ABRAHAMSEN 1983; BAATH et al. 1980; FUNKE and JANS 1989; HAGVAR 1984; SCHAUERMANN 1987 etc.). They gave rise to successions of unknown duration (FUNKE 1986). Earthworms especially showed positive effects after liming and fertilization (Fig. 27.5). Certainly, the population density and biomass (as well as species numbers) of *Lumbricidae* never reached the levels of the adjacent and less soil-acidic deciduous forests (FUNKE and JANS 1989). Since 1987 the differences between NF and DF have decreased. Long-term negative effects of liming and fertilization were observed above all in *Enchytraeidae* (BAATH et al. 1980; JANS and FUNKE 1989; SCHOCH-BÖSKEN and GREVEN 1987). From 1984 to 1988 one single species, *Cognettia sphagnetorum*, predominated the species-spectrum and number of individuals of U1NF. The other 8 species, which were found in U1NF had been of subordinate importance. In UxNF the individual spectrum was balanced. *C. sphagnetorum* and the other species together were always identical in quantity. After the application of lime the individual numbers of *C. sphagnetorum* decreased in UxDF. Since 1987 the decline of *C. sphagnetorum* has been compensated by two other species (*Enchytraeus buchholzi, Marionina cambrensis*, FUNKE and JANS 1989; FUNKE et al. 1989). Short-term tests showed that *Ciliata* reflect the alterations of the edaphon extremely quickly (LEHLE 1990; LEHLE and FUNKE 1989). Liming caused an increase, and fertilization a decrease of species and individuals. If liming is combined with fertilization, positive and negative effects seem to compensate each other. Therefore, in terrestrial ecosystems, especially in forests, *Ciliata* may give important information about the physico–chemical and biological properties of their environment (FOISSNER 1987), as it is known for limnic ecosystems. Surely, this information ought to be completed by investigations about *Nematoda*, *Enchytraeidae*, *Lumbricidae* and *Diptera*. With regard to trophic relationships, the

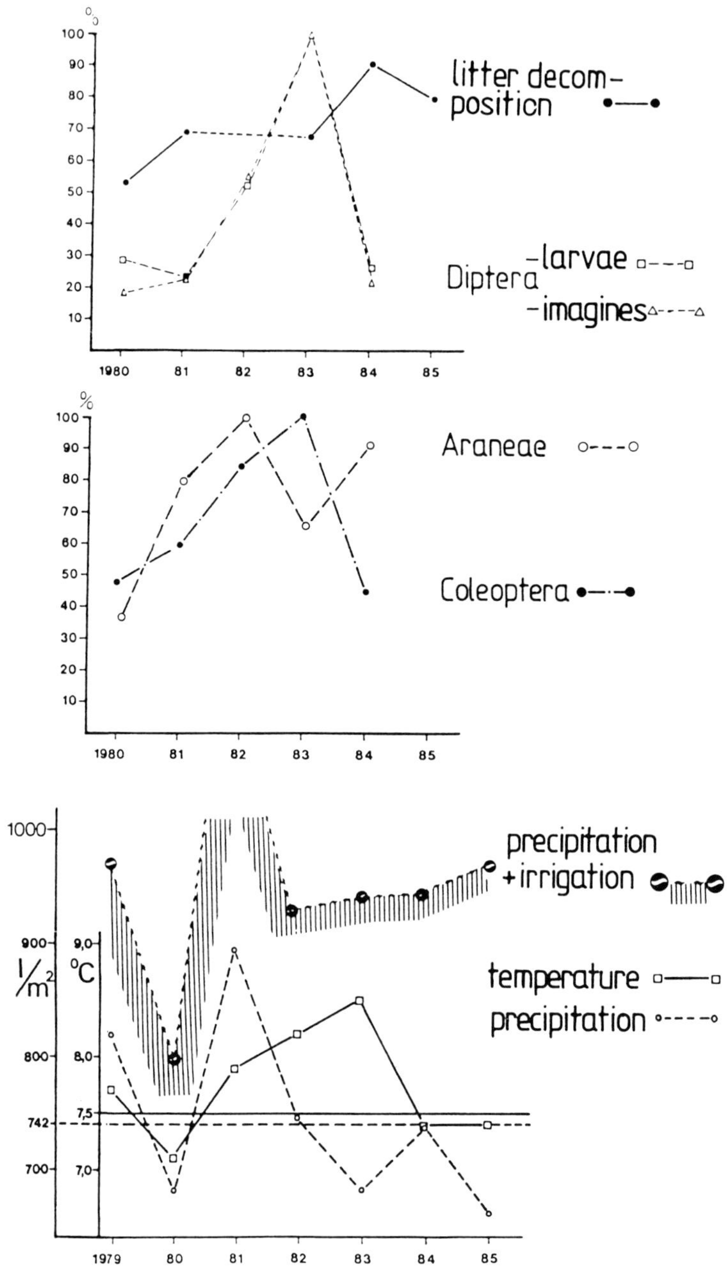

Figure 27.3: Fluctuations of litter decomposition and population densitiy at U1FF (irrigated area).

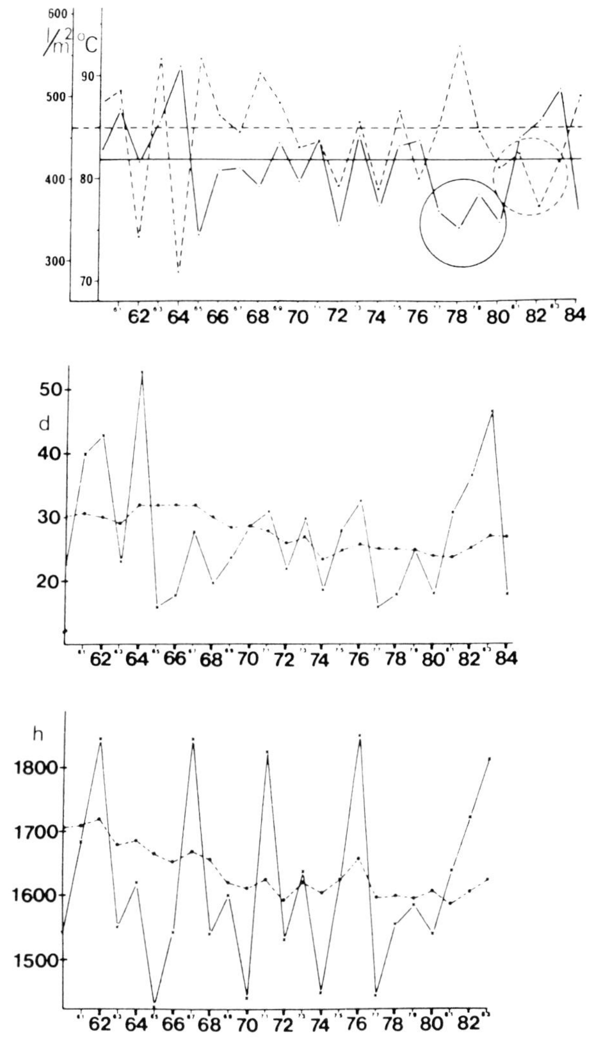

Figure 27.4: a) Sums of temperature (full lines) and precipitation (broken line) in summer between 1961 and 1981. Flat curves: sliding 10-year means (first value: mean of the years 1952–1961, second value: mean of the years 1953–1962 etc.).
b) Days with temperatures above 25°C.
c) Duration of sunshine (hours per year); broken lines: sliding 10-year means.
All data according to the meteorological service Stuttgart/Ulm, FRG.

Nematoda were of special interest. In U1NF and especially in U1DF and U1FF, bacteriovorous nematods predominated in the litter layer. Species feeding on roots

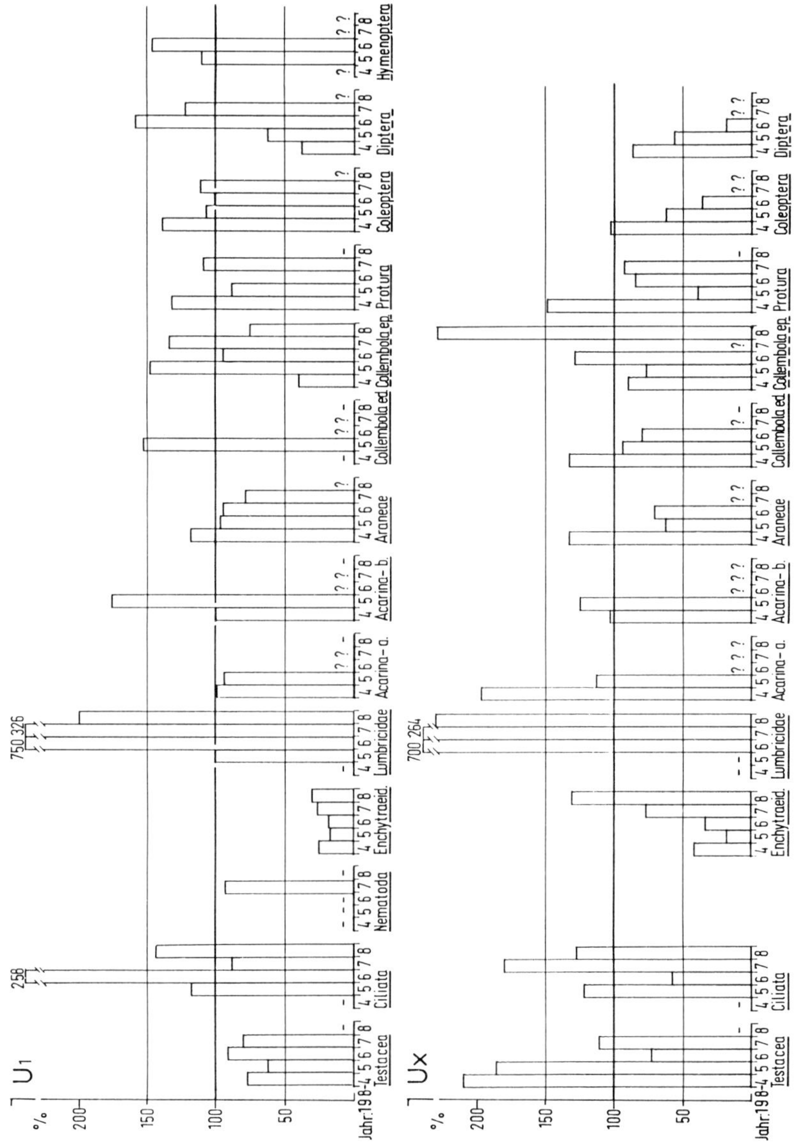

Figure 27.5: Effects of liming and fertilization on soil fauna at U1 and Ux. Values (individuals per m^2) of the control site NF are set to 100%. Deviations of more than 50% generally indicate significant differences between numbers of individuals at DF and NF.

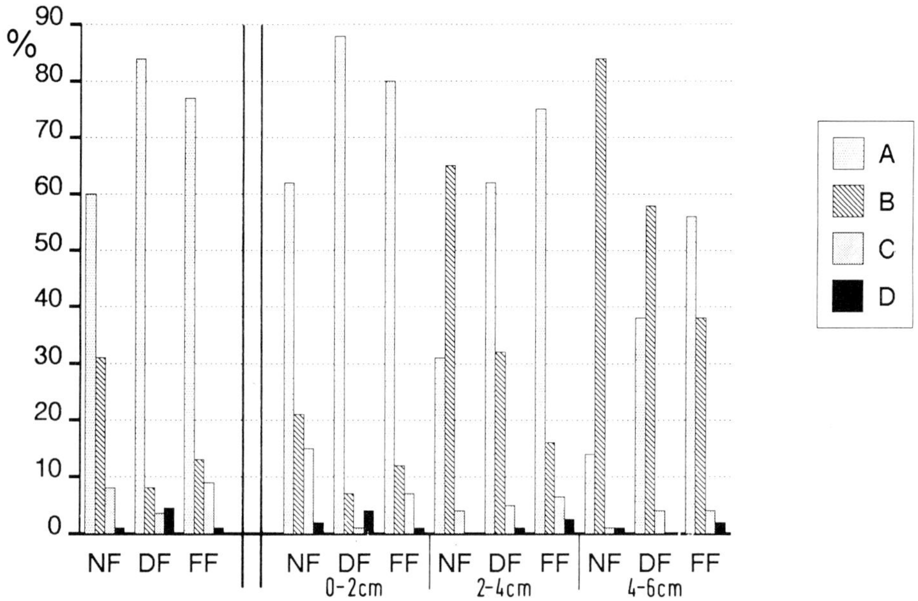

Figure 27.6: Amounts of nematodes in the different trophic groups at U1NF (control area), DF (limed and fertilized area) and FF (irrigated area).
On the left: 0–2cm depth between October 1986 and August 1987.
In the centre and on the right: in various soil depths in August 1987.
A – bacterivorous, B – root/fungivorous, C – omnivorous, D – predatory species (according to RATAJCZAK 1989).

or fungi were more numerous in NF than in DF and FF, above all in the humic and upper mineral layers of the soil (Fig. 27.6). The bacteriovorous species had been promoted by the application of lime and fertilizer (possibly by positive effects on basophil bacteria). But species feeding on roots and fungi had been injured – possibly by negative effects on acidophil fungi (FUNKE et al. 1989; RATAJCZAK 1989). The investigations showed that from the zoological point of view the application of lime and fertilizer in soil-acidic forests with regard to their effect upon ecosystems, should be interpreted very carefully (FUNKE and JANS 1989).

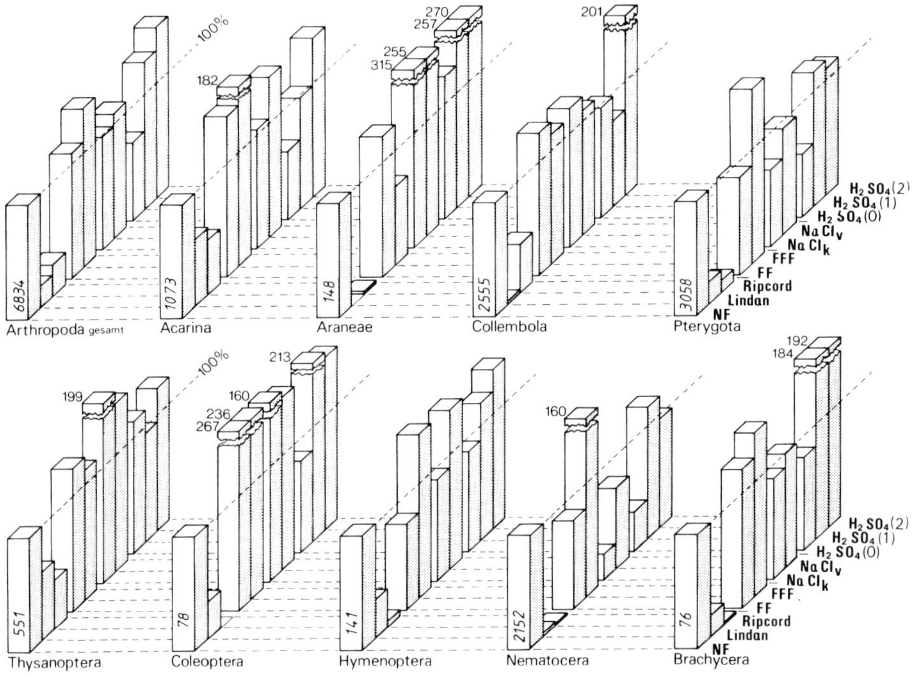

Figure 27.7: Abundance/activity density of arthropods (April–November 1985) at differently treated areas, covered by $0.2m^2$ photoeclectors; the numbers at each first column mean individuals per m^2 (set to 100%); NF – unpolluted control area. Lindan, Ripcord, $NaCl_v$, H_2SO_4 ; FFF – unique dose of 20 l aqua demin.; $NaCl_k$ – concentrated $NaCl_v$ – 20% solution; H_2SO_4 (O), (1), (2): acid solutions of pH 0 (ca. 10%), pH 1 (ca. 1%), pH 2 (ca. 0.1%) (according to SCHMITT 1989, see also FUNKE 1986).

27.3.4 Soil arthropods as sensitive indicators (test organisms) of the suitability of pesticides in nature

Conventional methods of testing the effects of pesticides are usually only applied to a small number of species. On the contrary, "photoeclector" and "soil core" methods exhibit the advantage of high efficiency with respect to a broad species-spectrum (FUNKE 1986, 1989; SCHMITT 1989; SCHMITT et al. 1989; VOGEL 1989; VOGEL and FUNKE 1989). The investigations demonstrated that Lindan and Cypermethrin (Ripcord 40) are extremely toxic to soil animals (Fig. 27.7). Even dilutions of a 0.5% solution of Ripcord with aqua demin in a ratio of 1:1000, as used

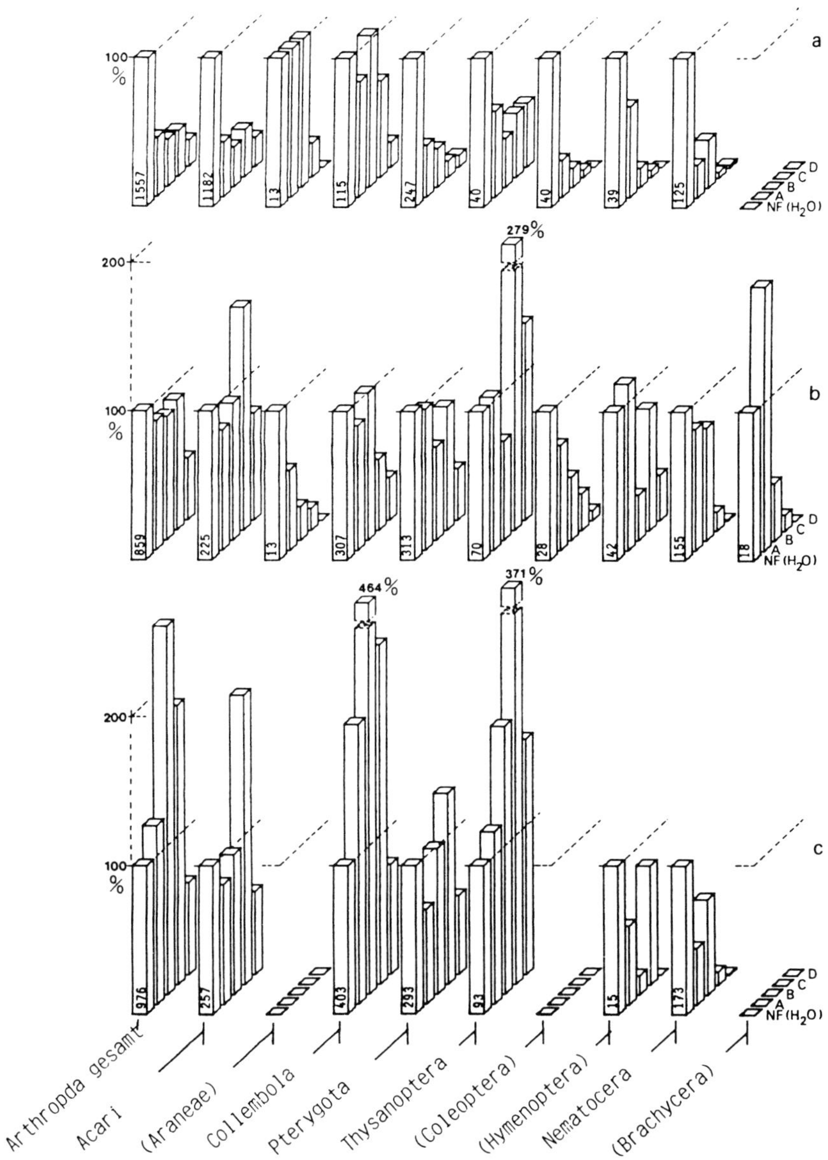

Figure 27.8: The influence of local contamination of different solutions of Ripcord 40 (Cypermethrin) on the abundance/activity density of arthropods (caught by $0.2 m^2$ photoeclectors) during three (a,b,c) 4-week intervals. A: 0.0005%, B: 0.005%, C: 0.05%, D: 0.5% solution of Ripcord (according to SCHMITT et al. 1989, see also FUNKE 1986).

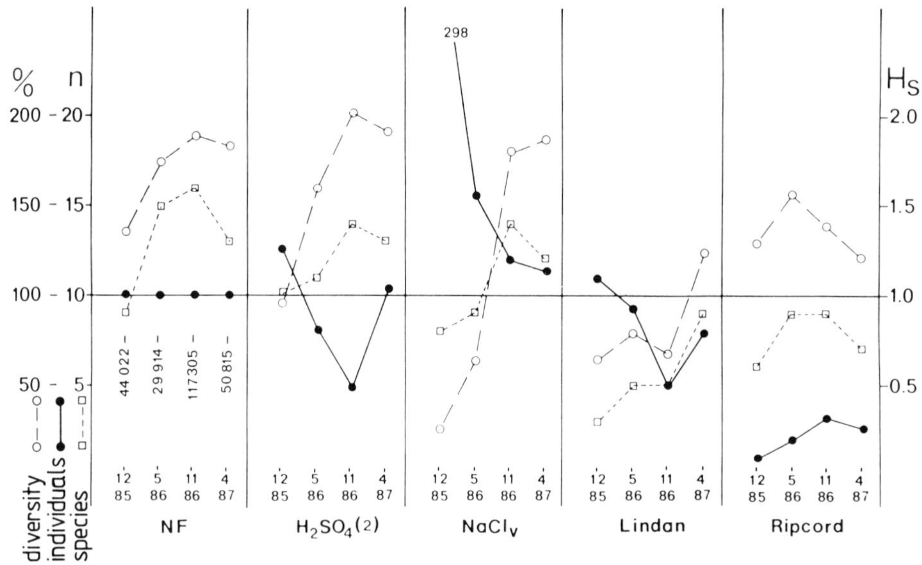

Figure 27.9: The influence of local contamination of different substances on eu-edaphic springtails between December 1985 and April 1987 after the sprinkling of 2 l per 0.2 m^2 in May 1985: NF – aqua demin., Lindan – 2.4 mg Nexit 80%, Ripcord 40 (Cypermethrin) – 0.5%, NaCl – 20%, H_2SO_4 – 0.1% (according to VOGEL and FUNKE 1989).

in the eclector-test, show extreme toxicity to arthropods (Fig. 27.8a). The reaction of the *Nematocera* was particularly sensitive. Because of their high abundance in forest soils, these animals may be very convenient test-organisms for the toxic effects of pesticides and other substances, perhaps even for acidic ingredients of natural precipitations. Populations of predatory arthropods (e.g. spiders, carabid and staphylinid beetles) are also highly affected, even after the application of highly diluted solutions of Ripcord. Their extinction brought about a model-like result. By the diminuation of the toxic effects of the diluted solution, small arthropods (obviously resistant developmental stages) appear in increasingly higher individual numbers than in the control areas (NF). This may be due to the exclusion of predators and/or to their longer periods of development in possibly deeper non-polluted soil layers (Fig. 27.8b, 27.8c). Investigations on edaphic springtails demonstrated that the decline of species numbers, individual numbers and diversity (especially after the application of Ripcord) extends over a long period of time (Fig. 27.9). Other

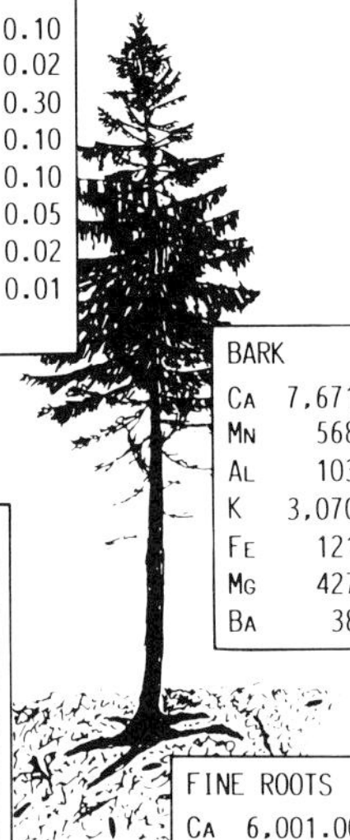

NEEDLES

Mn	2,200.00	Sb	0.10
Ba	24.00	Ag	0.02
Zn	30.00	Cd	0.30
Ni	2.50	Hg	0.10
Cu	3.00	Se	0.10
Cr	1.10	Cs	0.05
Rb	9.80	Sm	0.02
Co	1.50	Ta	0.01
La	0.50		

BARK

Ca	7,671.00	Pb	21.00
Mn	568.00	Zn	57.10
Al	103.50	Ni	19.60
K	3,070.00	Cu	4.30
Fe	121.00	Co	1.50
Mg	427.00	Cd	0.90
Ba	38.70		

LITTER

Ca	10,902.00	Ce	2.80
Mn	2,976.00	Co	1.80
Al	965.80	La	1.50
K	948.60	Mo	1.50
Fe	697.00	Sb	1.30
Mg	300.10	Ag	0.80
Ba	61.60	Cd	0.70
Pb	58.70	Hg	0.40
Zn	51.50	Se	0.40
Ni	9.50	Cs	0.30
Cu	9.40	Sm	0.20
V	4.50	Ta	0.10
Cr	3.50	Eu	0.06
Rb	3.40		

FINE ROOTS				FINEST ROOTS			
Ca	6,001.00	Ce	4.20	Ca	8,183.00	Ce	6.60
Mn	1,058.00	Co	0.90	Mn	668.00	Co	2.80
Al	318.50	La	1.90	Al	611.20	La	2.90
K	3,019.20	Sb	0.20	K	3,112.20	Sb	0.70
Fe	323.00	Ag	0.90	Fe	812.00	Ag	1.20
Mg	600.00	Cd	1.00	Mg	554.00	Cd	1.80
Ba	76.80	Hg	0.10	Ba	56.20	Hg	0.30
Pb	50.00	Se	0.10	Pb	106.40	Se	0.20
Zn	70.90	Cs	0.20	Zn	128.90	Cs	0.40
Ni	5.00	Sm	0.30	Ni	12.40	Sm	0.40
Cu	7.60	Ta	0.05	Cu	14.30	Ta	0.10
Cr	1.40	Eu	0.10	Cr	1.70	Eu	0.10
Rb	12.20			Rb	16.60		

Figure 27.10: Element content (ppm) of different compartments of *Picea abies* (spruce stand U1).

substances (e.g. NaCl, H_2SO_4) partly induce an increase in catch efficiencies of photoeclectors (Fig. 27.7). This is of great importance for biological production studies. Similar results are found by the soil-core test, especially after applications of diluted NaCl-solutions (Fig. 27.9).

27.3.5 Invertebrate animals as accumulative indicators of several elements

Arthropod food chains in spruce stands often begin with the standing crop of *Picea abies*. Thus, the element content of needles, bark, litter, fine and very fine roots form the basis of judgement, whether or not invertebrates are accumulative indicators of essential or toxic elements within the ecosystem (ROTH-HOLZAPFEL and FUNKE 1989). Within the tree compartments high element concentrations were

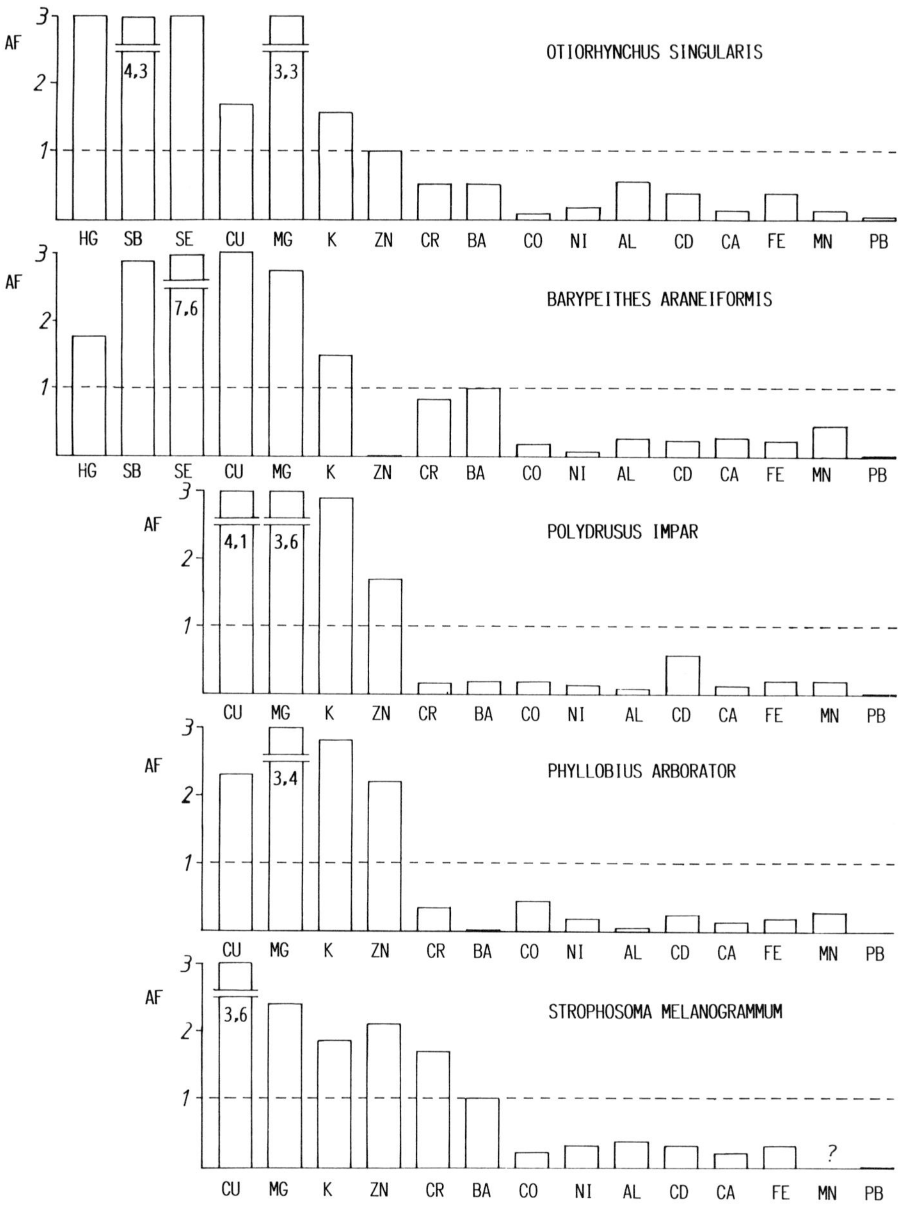

Figure 27.11: Accumulation factors (AF) of elements in weevils with rhizophagous larvae (spruce stand U11).

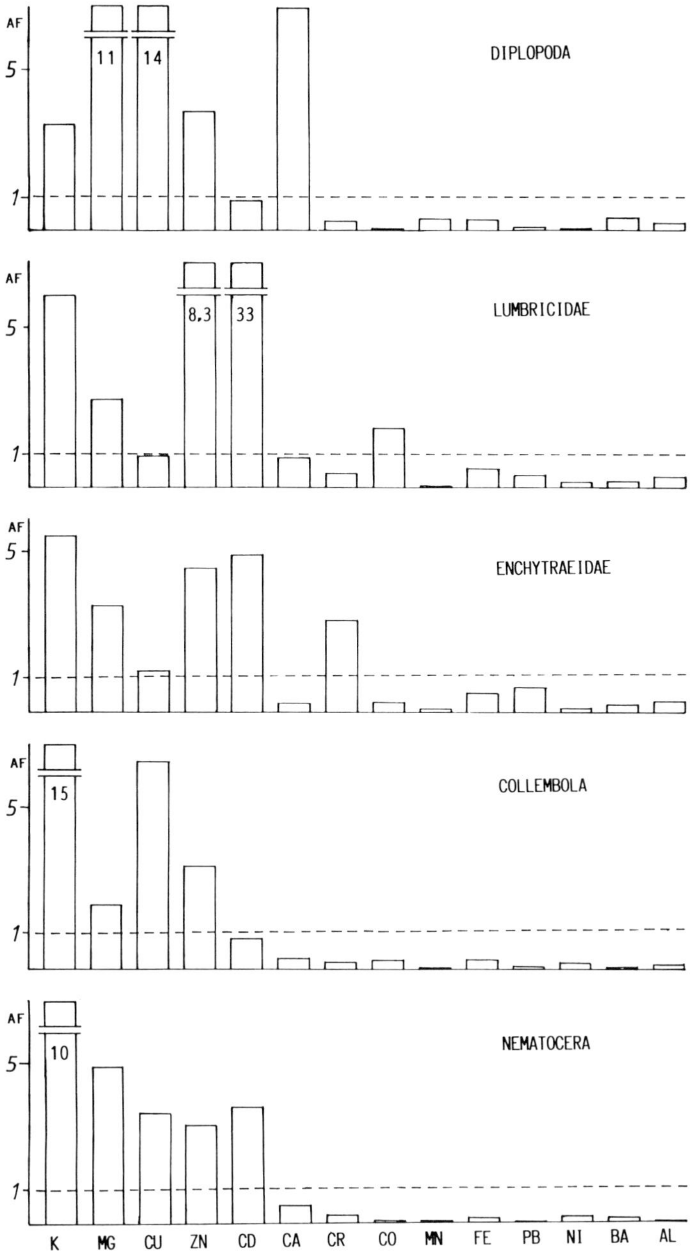

Figure 27.12: Accumulation factors (AF) of elements in saprophagous invertebrates (spruce stand U11).

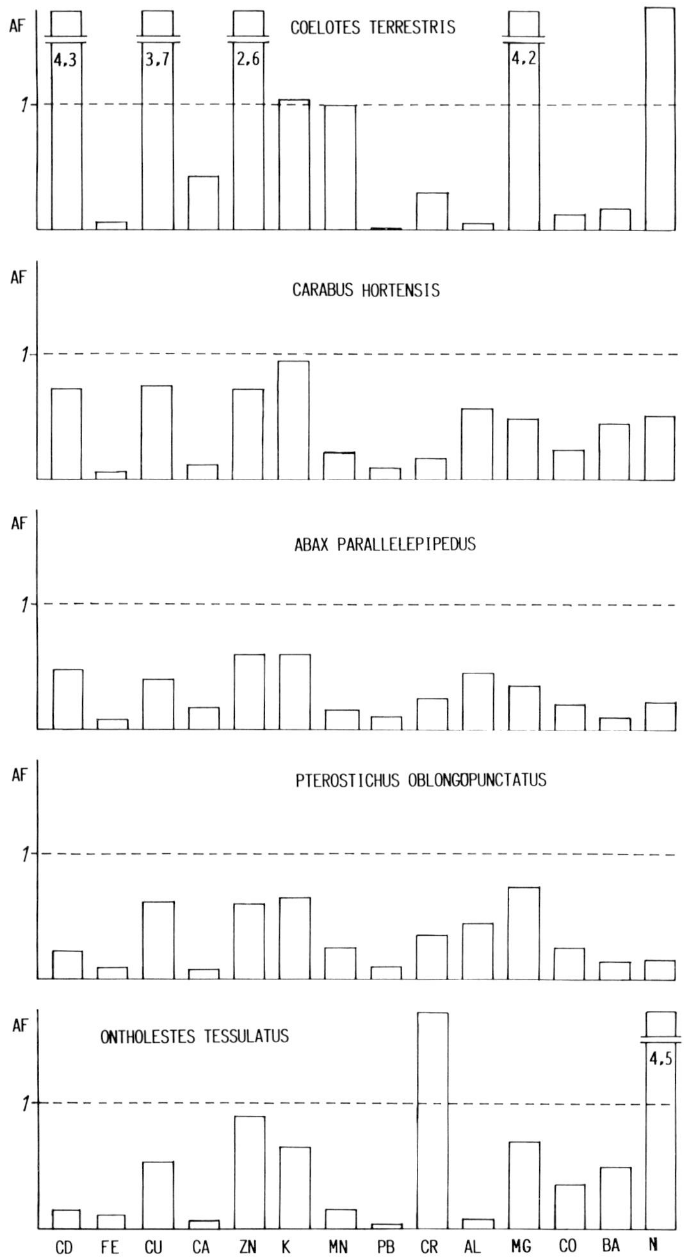

Figure 27.13: Accumulation factors (AF) of elements in zoophagous arthropods of soil surface (spruce stand U11).

found in litter and in very fine roots ($0 \leq 1$ mm). Low concentrations were found in needles and bark (Fig. 27.10). According to present investigations, invertebrates of the spruce stand accumulate the elements K, Mg, Ca, Cu, Zn, Cr, Ni, Se, Sb, Hg and Cd out of their food. The accumulation factors (AF), i.e. the ratio of element concentrations in animals to those in food are $\gg 1$. Most elements were accumulated by primary consumers (needle-eating larvae of *Geometridae*, bark-breeding *Scolytidae*, *Curculionidae* with rhizophagous larvae, saprophagous *Lumbricidae*, *Enchytraeidae*, *Nematocera*, *Collembola* and *Diplopoda*; ROTH-HOLZAPFEL and FUNKE 1989; ROTH-HOLZAPFEL 1990). High accumulation factors were found in weevils (Fig. 27.11) and – with respect to K, Mg, Cu, Zn and Cd – in saprophagous invertebrates (Fig. 27.12). Except with the spider species *Coelotes terrestris*, and the staphylinid beetle *Ontholestes tesselatus*, no accumulation of elements was found in animals of higher trophic level, such as predatory arthropods of the soil surface (Fig. 27.13). Moreover, the results demonstrate the obvious fact that a number of soil animals, due to their specific accumulation tendencies, may yield important information about the deposition of heavy metals. These data show that definite statements can only be made if investigations are based on species levels.

Bibliography

ABRAHAMSEN, G., 1983, Effects of lime and artificial acid rain on the enchytraeid (*Oligochaeta*) fauna in coniferous forest.
Holarct Ecol **6**, 247–254.

BAATH, E.; BERG, B.; LOHM, U.; LUNDGREN, B.; LUNDKVIST, H.; ROSSWALL, T.; SÖDERSTRÖM, B.; WIREN, A., 1980, Effects of experimental acidification and liming on soil organisms and decomposition in a Scots pine forest.
Pedobiologia **20**, 85–100.

BALOGH, J., 1958, Lebensgemeinschaften der Landtiere.
Akademie Verlag Berlin, p 560.

BORIN, A.; HERLITZIUS, H., 1987, Fluctuations of *Diptera* larvae in soil from a spruce forest in relation to climatic factors.
Biology and Fertility of Soils **5**, 41–48.

FEIL, W.; KOTTKE, J.; OBERWINKLER, F., 1986, Vergleichende Untersuchungen zur Wirkung von Trockenstreß auf Feinstwurzelsysteme.
KfK–PEF **4**, 324–330.

FOISSNER, W., 1987, Soil *Protozoa*: Fundamental Problems, Ecological Significance, Adaptions in *Ciliates* and *Testaceans*, Bioindicators and Guide to the Literature.
Progress in Protistology **2**, 69–212.

FUNKE, W., 1971, Food and energy turnover of leaf-eating insects and their influence on primary production.
Ecol Studies **2**, 81–93.

FUNKE, W., 1986, Tiergesellschaften im Ökosystem "Fichtenforst" (*Protozoa, Metazoa – Invertebrata*) – Indikatoren von Veränderungen in Waldökosystemen.
KfK–PEF **9**, p 150.

FUNKE, W., 1989, Tiere als Indikatoren von Veränderungen in unserer Umwelt. Untersuchungen in Waldökosystemen – Bioindikatoren – Symposium Graz **6**, p 30. Informationszentrale für Umweltschutz des Landeshygieneinstituts.

FUNKE, W.; BERNHARD, M.; HERLITZIUS, H.; HÖFER, H.; JANS, W.; LEHLE, E.; RATAJCZAK, L.; ROTH-HOLZAPFEL, M.; SCHMITT, G.; STUMPP, J.; VOGEL, J.; WANNER, M., 1987,
Protozoen und Metazoen als sensitive Indikatoren von Veränderungen im Ökosystem "Fichtenforst".
KfK–PEF **12**,1, 211–220.

FUNKE, W.; JANS, W., 1989, Kurz- und Langzeiteffekte von Kalk- und Mineraldüngergaben auf die Bodenfauna in Fichtenforsten.
IMA-Querschnittsseminar "Düngung geschädigter Waldbestände", Institut für Bodenkunde und Bodengeographie Bayreuth, (PEF), 61–74.

FUNKE, W.; JANS, W.; KÜHNER, M.; LEHLE, E.; RATAJCZAK, L.; SCHREIBER, R.; STUMPP, J.; VOGEL, J.; WANNER, M., 1989, Bodentiere als sensitive Indikatoren in Laub- und Nadelwäldern.
KfK–PEF **50,1**, 79–92.

HAGVAR, S., 1984, Effects of liming and artificial acid rain on *Collembola* and *Protura* in coniferous forest.
Pedobiologia **27**, 341–354.

HERLITZIUS, H., 1983, Biological decomposition efficiency in different wood land soils.
Oecologia **57**, 78–97.

HERLITZIUS, H.; BERNHARD, M.; BORIN, A.; FUNKE, W.; JANS, W., 1986, Streuabbau und Streuzersetzer in Waldökosystemen.
KfK–PEF **4,1**, 353–357.

JANS, W.; FUNKE, W., 1989, Die Enchyträen (*Oligochaeta*) von Laub- und Nadelwäldern Süddeutschlands und ihre Reaktion auf substantielle Einflüsse.
Verh. Ges. Ökol. **18**, 741–746.

LEHLE, E., 1990, Zur Ökologie von Bodenciliaten (*Protozoa*: *Ciliophora*) – Qualitativ–quantitative Untersuchungen in terrestrischen Ökosystemen.
Diss. Ulm, p 146.

LEHLE, E.; FUNKE, W., 1989, Zur Mikrofauna von Waldböden. I. *Ciliata* (*Protozoa*: *Ciliophora*).
Verh. Ges. Ökol. **17**, 385–390.

OBERWINKLER, F.; KOTTKE, J.; RITTER, TH.; FEIL, W., 1986, Vergleichende Untersuchungen der Feinwurzelsysteme und der Anatomie von Mykorrhizen nach Trockenstreß und Düngemaßnahmen.
KfK–PEF **4**, 315–330.

RATAJCZAK, L., 1989, Die Nematodenfauna eines Fichtenforstes.
Diplomarbeit Ulm.

ROTH, M., 1986a, Die Coleopteren im Ökosystem "Fichtenforst". I. Ökologische Untersuchungen.
Zoologische Beiträge NF **29**, 227–294.

ROTH, M., 1986b, Die Coleopteren im Ökosystem "Fichtenforst". II. Chemisch-analytische Untersuchungen.
Zoologische Beiträge NF **29**, 307–348.

ROTH-HOLZAPFEL, M., 1990, Multi-Element analysis of invertebrate animals in a forest ecosystem (*Picea abies* L.).
In: Lieth, H.; Markert, B. (eds.) Element Concentration Cadasters in Ecosystems. Methods of Assessment and Evaluation. Weinheim VCH, p 448.

ROTH-HOLZAPFEL, M.; FUNKE, W., 1989, Elementflüsse im Ökosystem "Fichtenforst" unter besonderer Berücksichtigung wirbelloser Tiere. KfK–PEF **50, 1**, 93–105.

SCHAUERMANN, J., 1987, Tiergesellschaften der Wälder im Solling unter dem Einfluß von Luftschadstoffen und künstlichem Säure- und Düngereintrag.
Verh. Ges. Ökol. **16**, 53–62.

SCHLECHTE, G., 1986, Zur Struktur der Basidomyceten-Flora in unterschiedlich immissionsbelasteten Forstökosystemen.
UBA, Querschnittseminar "Bioindikation", Berlin, 290–298.

SCHMITT, G., 1989, Über die Auswirkungen von Pflanzenschutzmitteln und anderen toxischen Substanzen auf Arthropoden eines Fichtenforstes.
Diss. Ulm, p 121.

SCHMITT, G.; FUNKE, W.; KRANIZ, V., 1989, Auswirkungen von Pflanzenschutzmitteln auf die Eklektorfauna eines Fichtenforstes.
Jber. naturw. Ver. Wuppertal **42**, 112–115.

SCHOCH-BÖSKEN, J.; GREVEN, H., 1987, Artenspektrum und Dominanzstruktur von Enchytraeiden einiger Buchen- und Fichtenwälder Nordrhein–Westfalens.
Verh. Ges. Ökol. (Gießen 1986) **16**, 79–84.

STUMPP, J., 1987, Zur Taxonomie und Ökologie einheimischer Proturen.
Diss. Ulm, p 133.

STURM, H., 1959, Die Nahrung der Proturen.
Die Naturwissenschaften **46**, 90–91.

VOGEL, J., 1989, Edaphische Collembolenzönosen deutscher Wälder. Reaktionen auf substantielle Einflüsse.
Diss. Ulm, p 113.

VOGEL, J.; FUNKE, W., 1989, Auswirkungen verschiedener Substanzen auf die euedaphische Collembolenzönose eines Fichtenforstes.
Jber. naturw. Ver. Wuppertal **42**, 116–118.

Part VI

The carbon cycle and climate change

Chapter 28

Carbon isotope fractionation during CO_2 fixation by plants

G. H. SCHLESER

28.1 Introduction

Carbon dioxide of the atmosphere contains about 1.1% of the heavy carbon isotope ^{13}C and 98.9% of the lighter isotope ^{12}C. This is very nearly the same isotopic composition as can be found in many carbonaceous materials. In 1939 it was observed for the first time that the heavy isotope ^{13}C is slightly depleted in plants when compared with carbonaceous materials (NIER and GULBRANSEN 1939). Subsequent studies revealed that the abundance ratio of the carbon isotopes, i.e. $^{13}C/^{12}C$ in terrestrial plants, is consistently lower than the corresponding ratio in marine limestone (WICKMAN 1952, CRAIG 1953). With the first detailed investigations on the abundance ratio of the carbon isotopes in the atmosphere it became clear that plants discriminate against ^{13}C during photosynthesis (CRAIG 1953, 1954, KEELING 1958).

It was soon found, however, that the $^{13}C/^{12}C$ ratios resulting from this discrimination are not constant. Plants exhibit a certain range in their isotopic carbon composition. In addition to this range it was found that some plants had anomalously high ^{13}C contents and that this was a characteristic of a few members of the *Gramineae* family, i.e. grasses, such as e.g. maize and sugar–cane (BENDER 1968, 1971, SMITH and EPSTEIN 1971). These observations could be explained by a new mechanism of primary carbon dioxide fixation which was only discovered in the mid–1960s, i.e. the C_4 pathway (KORTSCHAK et al. 1965, HATCH and SLACK 1966). C_4 plants fix carbon dioxide by the carboxylation of phosphoenolpyruvate as a first step. Refixation by ribulose bisphosphate carboxylase only takes place at a later stage.

The difference in isotopic composition between C_3 and C_4 plants is nowadays one of the standard methods by which C_4 plants can be distinguished from C_3 plants. The latter fix their carbon dioxide directly by the carboxylation of ribulose bisphosphate.

The carbon isotope discrimination in plants is closely related to the process of assimilation, and depends on a number of endogenous and exogenous factors. This fractionation of the carbon isotopes is therefore a very useful tool for investigating the conditions under which carbon dioxide is taken up. This is for example of special interest in relation to plant behaviour in climatically critical and/or anthropogenically disturbed environments.

The outline of a model is formulated here, which represents the use of a three step model that was developed many years ago. Some experimental work is added to elucidate its potentials and limitations of prediction.

28.2 Formulation of the model

Consider the pathway of carbon dioxide molecules from the free atmosphere into leaves as indicated in Fig. 28.1. Carbon dioxide molecules diffuse through the stomata of the leaf surface into the intercellular air spaces. The molecules are then absorbed at the interface air–mesophyll. By liquid diffusion, carbon dioxide molecules reach the chloroplasts. Diffusion thus comprises boundary layer transport and stomatal transport, as well as mesophyll transport. In the chloroplasts, carbon dioxide is directly used for the carboxylation of ribulose bisphosphate and therewith via phosphoglycerate for the synthesis of organic material. While this holds for C_3 plants, the carboxylation in C_4 plants is via phosphoenolpyruvate. Since the carboxylation step is irreversible, subsequent chemical reaction steps are not very important in the determination of the isotope fractionation. In connection with the carbon isotopes, the diffusive and kinetic steps of the molecules are of interest, since they control any isotope fractionation which occurs when proceeding from atmospheric to organic carbon. The liberation of carbon dioxide, either from mitochondria or from chloroplasts (dark and light respiration), constitutes an additional source for CO_2 fixation. Any changes in the internal carbon dioxide concentration C_i of a leaf, which is equivalent to the corresponding carbon concentration, will then be given by :

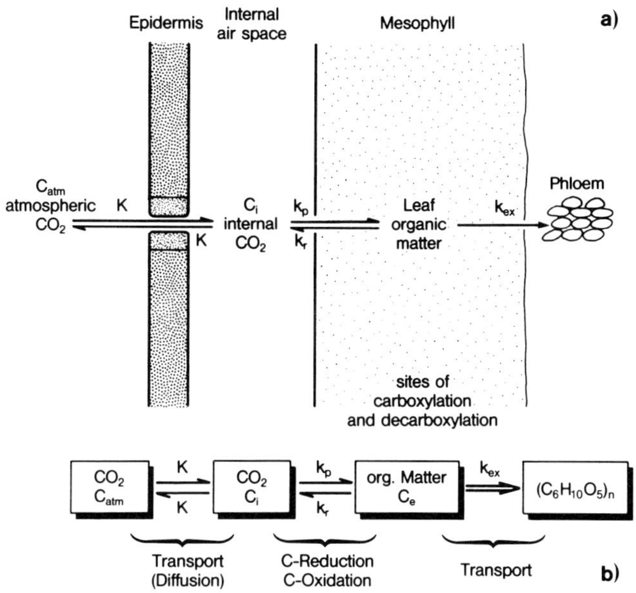

Figure 28.1: Model of CO_2 uptake into leaves for the verification of $\delta\,^{13}C$ values in plant organic matter. k_p, k_r and k_{ex} are the rate constants of carbon for assimilation, respiration and export respectively. K is the transfer coefficient for CO_2. For details cf. text.

$$\frac{dC_i}{dt} = K \cdot (C_{atm} - C_i) - k_p \cdot C_i + k_r \cdot C_e \tag{28.1}$$

In this equation K represents the transfer coefficient for carbon dioxide molecules along the path from the free atmosphere to the intercellular air spaces. C_{atm} and C_i are the carbon dioxide concentrations of the free atmosphere and the intercellular air spaces respectively. k_p and k_r represent the reaction rate constants for the processes indicated in Fig. 28.1 a. C_e summarizes the concentration of organic carbon, from which carbon dioxide can be released again. It will be noted that this formulation restricts diffusion to gaseous diffusion up to the interface air–liquid, i.e. air–mesophyll. All the subsequent steps such as transition at the air–liquid interface, equilibration of CO_2 with HCO_3^- and liquid diffusion are thus automatically represented by gaseous diffusion. This seems justified because isotope fractionations in conjunction with absorption and liquid phase diffusion should together have about the same value as is found for gaseous diffusion (VOGEL 1980).

Changes in the organic carbon content, C_e, will be given by:

$$\frac{dC_e}{dt} = k_p \cdot C_i - k_r \cdot C_e - k_{ex} \cdot C_e \tag{28.2}$$

Besides dissimilation, this equation also allows for export of organic material which is given by the last term.

It will be noted that the last term of (28.1) represents the rate of respiration R, whereas net photosynthesis, P_n, is given by the last term of (28.2). In this context the following relations hold:

$$\text{a) } R = k_r \cdot C_e \quad \text{b) } P_n = k_{ex} \cdot C_e \quad \text{c) } P_g = P_n + R \tag{28.3}$$

in which gross photosynthesis is given by P_g.

Apart from the initial phase of leaf development, i.e. the juvenile phase and the period of senescence, it is assumed that concentration changes of C_i and C_e do not occur. This implies steady state conditions for the major part of a leaf's life span, for which then holds:

$$\frac{dC_j}{dt} = 0; \quad j = i, e \tag{28.4}$$

28.3 Theoretical analysis

For steady state conditions, C_i and C_e are easily derived from (28.1) and (28.2). This leads to:

$$C_i = \frac{f_1 \cdot (1 + f_2)}{1 + f_1 \cdot (1 + f_2)} \cdot C_{atm} \tag{28.5}$$

$$C_e = \frac{k_p \cdot K}{[K \cdot (1 + f_2) + k_p] \cdot k_{ex}} \cdot C_{atm} \tag{28.6}$$

where $f_1 = K/k_p$ and $f_2 = k_r/k_{ex}$.

Equations (28.5) and (28.6) are valid for both the abundant light isotope ^{12}C and the heavier isotope ^{13}C. Writing similar equations for each isotopic species and dividing the corresponding isotope equations by each other gives:

$$\frac{C'_i}{C_i} = \frac{f'_1 \cdot (1 + f'_2) \cdot \{1 + f_1 \cdot (1 + f_2)\}}{f_1 \cdot (1 + f_2) \cdot \{1 + f'_1 \cdot (1 + f'_2)\}} \cdot \frac{C'_{atm}}{C_{atm}} \tag{28.7}$$

$$\frac{C'_e}{C_e} = \frac{k'_p \cdot K' \cdot k_{ex} \cdot \{K \cdot (1 + f_2) + k_p\}}{k_p \cdot K \cdot k'_{ex} \cdot \{K' \cdot (1 + f'_2) + k'_p\}} \cdot \frac{C'_{atm}}{C_{atm}} \tag{28.8}$$

where the dashed quantities are related to the heavier isotope.

Any changes of the isotope ratio are usually described by the isotope enrichment factor, or fractionation factor, α. This quantity is defined as the isotope ratio of the product, R_e, in this case represented by the corresponding value of the leaves, over the isotope composition of the substrate, i.e. that of the source which is given by the atmospheric value R_{atm}:

$$\alpha = \frac{R_e}{R_{atm}}$$

For a diffusion process, its value is the ratio of the respective diffusion constants, as can be derived from Fick's first law. In the case of chemical reactions, isotope fractionations are usually expressed in terms of "isotope effects". They are defined by the ratio of reaction rate constants for reactions of the corresponding isotopic substances, i.e. ${}^{13}k/{}^{12}k$. This ratio is identical to the definition of α as given above, if the source is a sufficiently large reservoir which remains largely unaffected by product formation.

For the present investigation, the main interest centers on the variation of the carbon isotopes in the organic matter, C_e, of leaves. According to the definition of α, its appropriate value follows directly from (28.8):

$$\alpha = \frac{R_e}{R_{atm}} = \frac{C'_e/C_e}{C'_{atm}/C_{atm}} = \frac{k'_p \cdot K' \cdot k_{ex} \cdot \{K \cdot (1 + f_2) + k_p\}}{k_p \cdot K \cdot k'_{ex} \cdot \{K' \cdot (1 + f'_2) + k'_p\}} \tag{28.9}$$

Rearranging this equation leads to:

$$\alpha = \frac{K' \cdot k_{ex} \cdot \{f_1 \cdot (1 + f_2) + 1\}}{K \cdot k'_{ex} \cdot \{f'_1 \cdot (1 + f'_2) + 1\}} \tag{28.10}$$

This value represents the overall enrichment factor which intrinsically comprises the α–values for every single step of transport and reaction that has been considered by (28.1, 28.2). If the appropriate fractionation factors for every single step are introduced and the dashed quantities f'_1 and f'_2 are expressed by the corresponding undashed quantities, (28.10) leads to:

$$\alpha = \frac{\alpha_D \cdot \{f_1 \cdot (1 + f_2) + 1\}}{\alpha_{ex} \cdot \{(\alpha_D/\alpha_p) \cdot f_1 \cdot [1 + (\alpha_r/\alpha_{ex}) \cdot f_2] + 1\}} \tag{28.11}$$

where $\alpha_D = K'/K, \alpha_p = k'_p/k_p, \alpha_r = k'_r/k_r$ and $\alpha_{ex} = k'_{ex}/k_{ex}$. The ratio of the transfer coefficients is equal to the ratio of the diffusion coefficients which is indicated by the appropriate subscript of the respective fractionation factor (α_D).

In general, α is only slightly different from unity. Therefore it is useful to introduce the fractionation ϵ, which specifies the deviation of α from unity:

$$\alpha = 1 + \epsilon$$

Bearing in mind that $\epsilon \ll 1$, α is given to a good approximation by:

$$\alpha \approx \frac{(1+\epsilon_D)}{(1+\epsilon_{ex})}\left(1-\frac{f_1 \cdot f_2 \cdot (\epsilon_r - \epsilon_{ex} + \epsilon_D - \epsilon_p) + f_1 \cdot (\epsilon_D - \epsilon_p)}{1+f_1(1+f_2)}\right) \qquad (28.12)$$

The denominator of the second part of (28.12), namely the expression $\{1+f_1 \cdot (1+f_2)\}$, is equivalent to that of (28.5) which contains the quantities C_i and C_{atm}. Introducing these quantities into (28.12) leads to:

$$\alpha \approx \frac{(1+\epsilon_D)}{(1+\epsilon_{ex})}\left(1-\frac{[f_2 \cdot (\epsilon_r - \epsilon_{ex} + \epsilon_D - \epsilon_p) + (\epsilon_D - \epsilon_p)] \cdot C_i}{(1+f_2) \cdot C_{atm}}\right) \qquad (28.13)$$

For the latter approximation it should be kept in mind that the undashed quantities of C_i and C_{atm} refer to the abundant isotope, while the quantities in (28.5) comprise both isotopic species according to their composition in the element of the particular reservoir. The corresponding difference has been disregarded because in general these ratios differ by less than 2%. By taking into account the relations (28.3 a, b, c), (28.13) can be simplified which leads to:

$$\alpha \approx \frac{(1+\epsilon_D)}{(1+\epsilon_{ex})}\left(1+(\epsilon_p - \epsilon_D) \cdot \frac{C_i}{C_{atm}} - (\epsilon_r - \epsilon_{ex}) \cdot \frac{R \cdot C_i}{P_g \cdot C_{atm}}\right) \qquad (28.14)$$

Introducing the overall fractionation ϵ into (28.14) and disregarding expressions of higher order which are small, results in:

$$\epsilon \approx \epsilon_D \cdot \left(1-\frac{C_i}{C_{atm}}\right) + \left(\epsilon_p - \epsilon_r \cdot \frac{R}{P_g}\right) \cdot \frac{C_i}{C_{atm}} - \epsilon_{ex} \cdot \left(1-\frac{R \cdot C_i}{P_g \cdot C_{atm}}\right) \qquad (28.15)$$

In all cases where the natural abundance of materials is considered, the isotope ratio is commonly expressed as a $\delta\,^{13}$C value which is defined by

$$\delta\,^{13}C = \left(\frac{R_{sample}}{R_{reference}} - 1\right) \cdot 1000$$

R represents the ^{13}C abundance ratio which is usually determined with a mass spectrometer by measuring carbon dioxide:

$$R = \frac{^{13}CO_2}{^{12}CO_2}$$

The reference is carbon dioxide obtained from a limestone, called PDB, which originates from the Pee Dee formation in South Carolina (CRAIG 1957). The unit of $\delta\,^{13}C$ is called "per mil" (‰). If R_e and R_{atm} are expressed with the help of $\delta\,^{13}C$, the following relations hold:

$$R_j = R_{PDB} \cdot \left(1 + \frac{\delta\,^{13}C_j}{1000}\right) \qquad j = e, atm$$

with $\delta\,^{13}C_j$ related to PDB. With these definitions it is convenient to express (28.14), and therewith the organic matter of leaves, in units of $\delta\,^{13}C$. Since α is the ratio of the carbon isotopes of organic matter divided by the corresponding ratio of atmospheric CO_2, as indicated by (28.9), (28.14) can be rewritten as:

$$\delta\,^{13}C_e \approx \delta\,^{13}C_{atm} + \epsilon_D \cdot \left(1 - \frac{C_i}{C_{atm}}\right) + \left(\epsilon_p - \epsilon_r \cdot \frac{R}{P_g}\right) \cdot \frac{C_i}{C_{atm}} - \epsilon_{ex} \cdot \left(1 - \frac{R \cdot C_i}{P_g \cdot C_{atm}}\right) \qquad (28.16)$$

Similar expressions have been derived by other workers, who have used different approaches (VOGEL 1980, O'LEARY 1981, FARQUHAR et al. 1982).

Equation (28.16) shows that the $\delta\,^{13}C$ value of leaf organic matter is given basically by two factors: a) source CO_2, namely the $\delta\,^{13}C$ value of the ambient atmosphere, and b) the physiological condition of leaves, expressed as the ratio of the CO_2 concentration, C_i/C_{atm}, and the ratio of R/P_g which will be termed the respiratory coefficient.

As mentioned earlier, the isotope ratio $^{13}C/^{12}C$ of PDB is consistently higher than that of organic matter. Therefore $\delta\,^{13}C$ values of plant materials, expressed in this nomenclature, are always negative.

Measurements of atmospheric $\delta\,^{13}C$ values revealed a range of about -6.4 to -10 ‰(CRAIG 1957, KEELING 1960, 1961, TANS 1978). Preindustrial $\delta\,^{13}C$ values were at the higher end of this range. Industrial activities which lead to increasing emissions of CO_2, due to the combustion of fossil fuels, caused the shift towards the more negative values. Since plant organic matter is basically derived from atmospheric CO_2, it is necessary for an assessment of $\delta\,^{13}C$ variations and its causes to determine the corresponding value of atmospheric CO_2.

Most important for the valuation of (28.16) is the knowledge of the fractionations ϵ. The fractionation occuring when CO_2 diffuses through air was first given as -4.4 ‰(CRAIG 1954). This value is basically derived from the binary diffusion coefficients for each isotopic species, namely $^{13}CO_2$ and $^{12}CO_2$ in air and assuming that the corresponding ratio of these diffusion coefficients only leaves the square root of the reduced mass terms. In this context the mean molecular weight of air is taken as 29. In principle, the fractionations ϵ_D for the transport of CO_2 from the atmosphere surrounding a leaf towards the fixation sites, should contain the value for gaseous diffusion, as well as the corresponding value which takes into account the slightly different solubility coefficients of the two molecular species in the cytoplasm and the fractionation for liquid diffusion. Estimates of the latter two fractionations produced together a value of about −4.2 ‰ (VOGEL 1980). For this estimate the value for liquid diffusion was also based on the square root of the ratio of reduced masses. This is, however, somewhat ambiguous because it is unclear whether or not this method is applicable to diffusion in the liquid phase. Probably, as has been pointed out by O'Leary (1984), another theoretical treatment has to be used, which considers also the self association of water molecules. Nevertheless it seems justified to assume at present that the fractionation of the molecular species of CO_2, due to the transport from the leaf surface to the fixation sites, is in the above stated order. But it should be clear that the representation of one mesophyll compartment, combined with the intercellular spaces, is a rather crude approximation.

Concerning the fractionation of ϵ_p which results from the enzymatic fixation of CO_2, different cases have to be considered: In C_3 plants, the isotope effect is associated with the enzyme catalysed carboxylation of ribulose bisphosphate. In vitro studies of the corresponding ratio of reaction rate constants $^{13}k/^{12}k$, resulted in ratios of about 0.96 to 0.98 (PARK and EPSTEIN 1960, WHELAN et al. 1973, ESTEP et al. 1978, WONG et al. 1979, WINKLER et al. 1982, ROESKE and O'LEARY 1984), thus indicating that this is by far the largest fractionation in C_3 plants. The most reliable value presently at hand is $\epsilon_p = -29$ ‰ (ROESKE and O'LEARY 1984). In this context, it should be mentioned that a number of investigators use the opposite sign convention. In C_4 plants, in which the initial carboxylation is catalysed by phosphoenolpyruvate carboxylase, only values in the range of $\epsilon_p = -2.5$ to -2.0 ‰ were found (WHELAN et al. 1973, REIBACH and BENEDICT 1977, SCHMIDT et al. 1978). With respect to the CO_2 fixation process, a third group of plants exists: the so called crassulacean acid metabolism (CAM) plants. They either fix CO_2 directly by using ribulose bisphosphate carboxylase or in a sequence by firstly carboxylating phosphoenolpyruvate to produce malate, which in a further step is decarboxylated. The resulting CO_2 is then taken up by ribulose bisphosphate carboxylase (KLUGE and TING 1978, OSMOND 1978). Correspondingly, ϵ_p of CAM plants can basically have values ranging anywhere between those of C_3 and C_4 plants.

The isotope composition of the CO_2 which is lost through respiratory processes, if known, should lead to ϵ_r, if the isotope composition of the proper plant material is known. However, numerous experimental problems have so far prevented a clear answer to the question of whether or not respiration leads to a fractionation of the carbon isotopes. It is, however, known that decarboxylation reactions discriminate against ^{13}C (O'LEARY 1976, 1977). The difficulty with respiration experiments stems from the fact that CO_2 originating from respiration cannot be separated from other contributions of CO_2. Photorespiration seems to affect the isotopic composition in C_3 plants. For *Atriplex petula*, for example, which is a C_3 species, a drop from 20% oxygen to 4% oxygen changed the $\delta^{13}C$ value of plant material by 2.1 to 4.3 ‰ in the direction of more positive values (BERRY et al. 1972). In other experiments using soybean leaves, a shift of similar magnitude in the opposite direction was observed (SMITH et al. 1976). In the C_4 species *Atriplex rosea*, a reduction of the partial pressure of oxygen had practically no effect on the isotope composition (BERRY et al. 1972). If respired carbon would be depleted in ^{13}C relative to the leaf ($\epsilon_r < 0$), an increase of the $\delta\,^{13}C$ value of leaf organic matter would follow, i.e. the leaf would become ^{13}C enriched, as is indicated by (28.16). At present ϵ_r is unknown. It should, however, be much smaller than ϵ_p and according to (28.16) is weighted with the respiratory coefficient, namely R/P_g. Since R/P_g is less than one (generally $R/P_g < 0.5$), the contribution of ϵ_r can therefore, according to present knowledge, only lead to a small correction of $\delta\,^{13}C$.

According to some investigations, an isotope fractionation step was proposed between carbon fixed in leaves and carbon laid down in wood cellulose (LEAVITT and LONG 1982). Cellulose of leaves was found to be isotopically lighter than that of the corresponding tree rings by an average of +2 ‰. In principle, the above model includes a fractionation step (ϵ_{ex}) at the point of material export from a leaf (see Fig. 28.1a), and its contribution is given by the last term of (28.16). However, there is as yet no evidence for a discrimination of the carbon isotopes at the interface leaf–stem (SCHLESER 1990). One interpretation of the observed difference between tree leaves and wood is that ^{12}C is preferentially laid down during the transport of molecules away from the leaves (SCHLESER 1990). The differences reported could, however, also be due to mixing effects of carbon with different $^{13}C/^{12}C$ ratios which originate from various assimilation points at different heights in a tree (SCHLESER 1989). Therefore, any fractionation ϵ_{ex} at the interface leaf–stem, i.e. phloem loading, if it exists at all, would have to be very small and can be disregarded in (28.16). The above considerations lead to the conclusion that the following equation is a good approximation of (28.16):

$$\delta\,^{13}C_e \approx \delta\,^{13}C_{atm} + \epsilon_D \cdot \left(1 - \frac{C_i}{C_{atm}}\right) + \epsilon_p \cdot \frac{C_i}{C_{atm}} \qquad (28.17)$$

This expression is equivalent to one already known in the literature (FARQUHAR et al. 1982). It shows that besides the big differences between C_3 and C_4 plants that have frequently been described (BENDER 1968, 1971, SMITH and EPSTEIN 1971, LERMAN and RAYNAL 1972, TROUGHTON et al. 1974, TROUGHTON and CARD 1975), and which are due to the known differences in their values of ϵ_p, an additional diversity exists. This diversity simply occurs because the values of ϵ_p, ϵ_D, and if at all operative also ϵ_r, are not fully effective but contribute with a certain fraction (C_i/C_{atm} or $1-C_i/C_{atm}$) to the $\delta\,^{13}C$ value of leaf organic matter, and this fraction changes with environmental conditions.

In general, it is assumed that the most important fractionation in C_3 plants, namely ϵ_p, is constant for the interpretation of $\delta\,^{13}C_e$ values. Investigations with ribulose bisphosphate carboxylase have shown, however, that ϵ_p varies with temperature, pH, metal ion concentration and other variables (WINKLER et al. 1982).

Therefore changes in the reaction conditions at the sites of CO_2 fixation could additionally affect the isotopic composition of organic matter.

Generally the interpretation of $\delta\,^{13}C$ variations in plant organic matter is not straight forward. Since variations of C_i/C_{atm} depend on many environmental factors which may partly lead to C_i/C_{atm} changes in the same direction, it is often difficult to trace and separate the various causes that determine $\delta\,^{13}C$. The succeeding details are therefore only intended to show some possible $\delta\,^{13}C$ variations in plants and possible interpretations with regard to (28.17) or (28.16).

28.4 Experimental analysis

Experiments have been performed, to analyze the isotope composition of leaf material from a forest tree. These investigations centered mainly on the question of how $\delta\,^{13}C_e$ varies during early leaf development and on the distribution of carbon isotopes in leaves.

28.4.1 Material and methods

A beech tree (*Fagus sylvatica* L.) from the forest of Hambach near Jülich, Fed. Rep. of Germany, served as research object. About 10 to 20 leaves from several heights in the tree, which grows on a pseudogley (soil with purched surface water), were pooled for a sample. In the case of leaf component experiments, leaves were always cut in half along the main rib. One half was used for intercostal tissue studies, while the other half remained for total tissue analysis. The latter half, excluding the main rib, is called leaf lamina. The tissue from the first half which remained after the removal of the main rib and ribs of first order was called intercostal tissue. Additionally, petiole, phyllopodium, and the initial part of the stem were separated.

Cellulose was also extracted from the various tissues, to compare the $\delta\,^{13}C_e$ data of one chemical compound from different leaf components, and to determine whether differences between total tissue and cellulose vary with different components. The method of cellulose extraction is described elsewhere (WIESBERG 1974).

For isotope analysis, dried material was ground to a powder, and cellulose was cut into small pieces. The samples were then combusted in an excess of oxygen. The results are reported against PDB (CRAIG 1957). In case of total organic material, no corrections for nitrogen oxides were necessary because these were separated from CO_2 by reduction with copper at 450 °C. $\delta\,^{13}C$ data are the mean of two to five repetitions. Sample replication, including all errors of sample preparation, was ± 0.12 ‰.

28.4.2 Results and discussion

Carbon isotope ratios of plant tissues in forests vary with the height from which the corresponding leaves originate. This finding is well known (VOGEL 1978, MEDINA and MINCHIN 1980, SCHLESER and JAYASEKERA 1985, EHLERINGER et al. 1986) and is also documented in Fig. 28.2. The gradient of such a curve can in part result from the presence of high concentrations of respired CO_2 at the forest floor; because degradation of soil organic matter produces CO_2 with carbon isotope ratios similar to that of the substrate (JACOBSON et al. 1970), i.e. with values in the range of about −24 ‰ to −28 ‰. While in some forests, particularly in temperate forests, soil respiration can be of minor importance (FRANCEY et al. 1985), it can contribute substantial amounts in tropical forests (DA SILVEIRA et al. 1989). In the presented case, however, measurements of air CO_2 did not yield substantially higher values within the canopy than in the open air.

The source value, namely $\delta\,^{13}C_{atm}$, of (28.17) should therefore not change drastically with height above the ground. This indicates that the more depleted values of leaf tissue in the lower leaf strata are to a large extent caused by responses of the tree to other environmental changes which obviously increase the ratio of C_i/C_{atm}. This is also seen from Fig. 28.3 in which $\delta\,^{13}C$ values at 7 m height are more depleted than at 6 m height. Observations have shown that leaves at 6 m height experienced sun intervals for restricted time periods, while leaves at 7 m height were permanently shaded. These differences in illumination may, according to (28.17), be responsible for slightly different ratios of C_i/C_{atm}, i.e. somewhat smaller ratios in leaves at 6m height, thus leading to this ^{13}C enriched organic matter. Investigations on the influence of light intensity partly corroborate this interpretation (SMITH et al. 1976).

The $\delta\,^{13}C$ values of cellulose from various tissues of a leaf (Fig. 28.3) indicate that variations of the carbon isotope ratio are, however, not only due to different environmental conditions. In this particular case the carbon isotope ratio of the various leaf tissues varies under the same micro–environmental conditions and carbon sources. Differences of up to 2 ‰ have been measured (Fig. 28.3). The reason for these variations is not known at present. Total organic matter gave values which are about 2 ‰ lower than the corresponding cellulose data but otherwise they exhibit almost the same pattern. Though these latter variations could be due to different compositions of chemical compounds, having different $\delta\,^{13}C$ values, the cellulose data do not support this interpretation.

It should be noted that $\delta\,^{13}C$ gradients as shown in Fig. 28.2 are not constant over the growing season, neither for total organic matter nor for cellulose. The levelling off of $\delta\,^{13}C$ values at the bottom and top of the tree was not found throughout the growing season. $\delta\,^{13}C$ values change with time, initially showing high values which gradually decrease till the end of the growing season (LOWDON and DYCK 1974). Fig. 28.4 documents the $\delta\,^{13}C$ results for the start of the growing season of a forest beech. While the first value directly after leaf emergence is low, the next one, only two days later, has risen considerably. Thereafter, a continuous decline of $\delta\,^{13}C$ values was measured.

In view of the model outlined, the reasons for changes should either be due to variations of the isotope composition of the source, or due to plant environmental

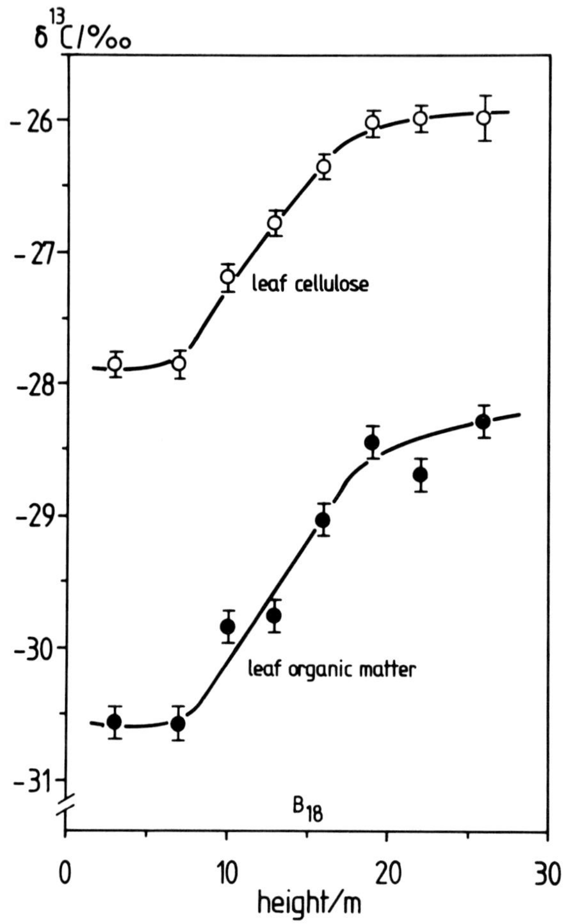

Figure 28.2: $\delta\,^{13}C$ values of leaf material from different heights in a forest beech. The corresponding values for extracted cellulose are also given.

changes which can alter the ratio of C_i/C_{atm}. Therefore, soil respired CO_2 was also measured and the flux of soil CO_2 is likewise plotted in Fig. 28.4. Its isotope composition was found to be between −24 ‰ and −26 ‰ (SCHLESER and JAYASEKERA, in preparation). Above the canopy, $\delta\,^{13}C$ data of air CO_2 showed values of −8.5 ‰ (SCHLESER and JAYASEKERA, in preparation). The flux of soil CO_2 increased during the whole period of the investigation, indicating a steady decrease of isotopically heavy atmospheric CO_2. Nevertheless, $\delta\,^{13}C$ values increased at first. There is no doubt that the very steep increase, even though it is only indicated by one data point on the curve, namely the second, is real. The $\delta\,^{13}C$ values

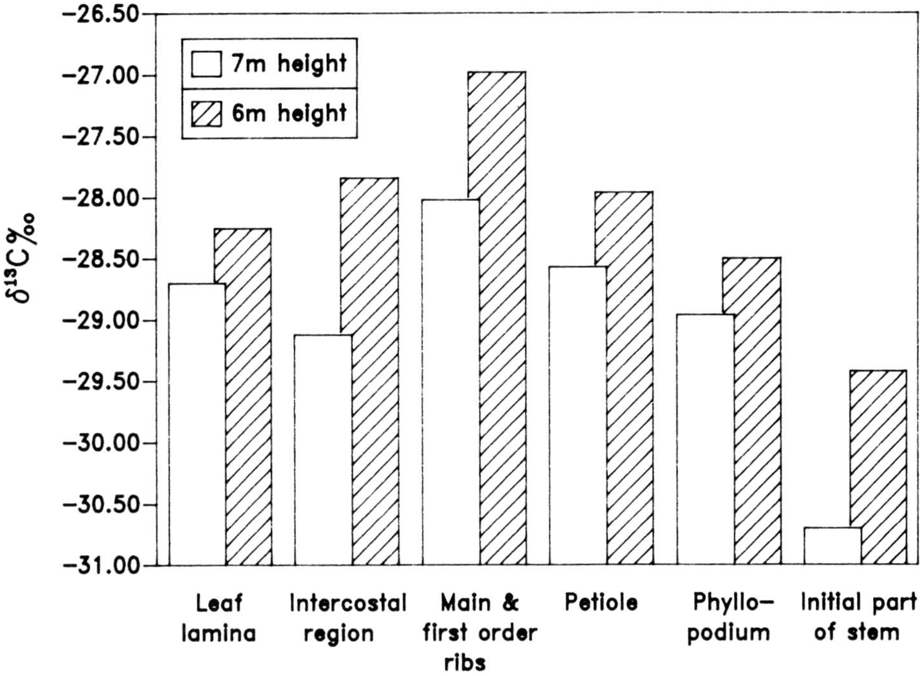

Figure 28.3: Carbon isotope composition of cellulose from various leaf parts of a forest beech. The leaves were collected at two different heights in the tree (6 m and 7 m).

of emerging leaves are comparable to those of stems (SCHLESER and JAYASEKERA, in preparation). The abrupt isotopic change that follows is therefore probably due to the very first incorporation of the isotopically much heavier carbon of air CO_2. Whether respiration which is extremely strong during the initial phase of leaf development (MÜLLER 1954) also contributes to this increase, see (28.16), cannot be answered at present.

It is difficult to conceive that the subsequent decrease of $\delta\,^{13}C$ values should, to a large extent, originate from the increasing flux of isotopically light CO_2 from the soil. Sampling was done at the top of the canopy, at a height of 26 m. At the time of first leaf development strong turbulent conditions prevailed in the vicinity of the tree. Therefore the air at the top of the canopy must have been well mixed. It is well known that in the juvenile phase of leaf development, pronounced biochemical, morphological and anatomical changes occur (GÄUMANN 1935). These changes can influence C_i/C_{atm} due to changing CO_2 resistances (FARQUHAR et al. 1982, SCHLESER et al. 1989) and therewith the $\delta\,^{13}C$ value according to (28.17).

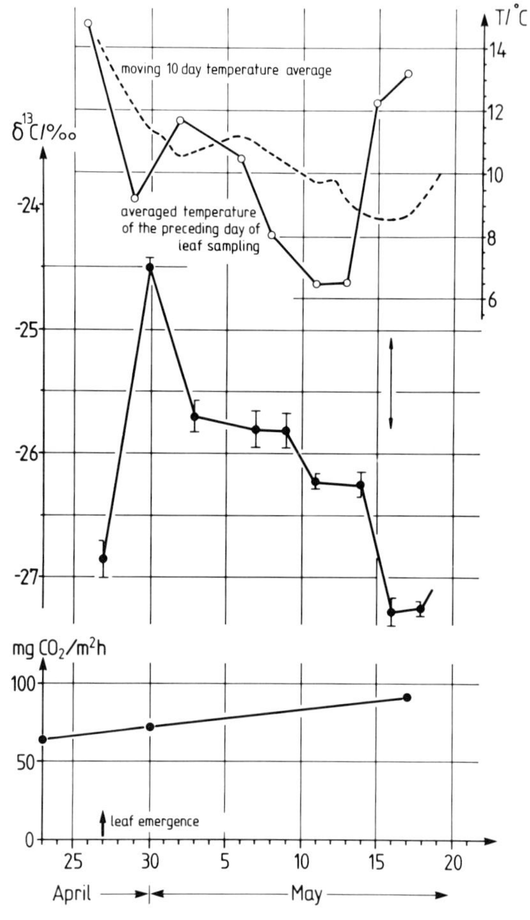

Figure 28.4: Carbon isotope composition in leaves during the juvenile phase of growth. The carbon dioxide flux out of the soil is also given, as well as the averaged daily temperature, including the 10 day moving average.

The leaf development is of course coupled to the changing climatical environment, especially the changing temperature conditions. Results of measurements of the temperature show a trend as indicated by Fig. 28.4. Daily temperature records are, however, not adequate for correlating $\delta\,^{13}C$ with temperature because $\delta\,^{13}C$ values always represent integrated data of the period in which plant organic matter has developed. Therefore a ten day moving average was calculated and plotted in Fig. 28.4. This curve shows a rather similar trend compared with the corresponding $\delta\,^{13}C$ curve, if the initial data point is discarded, indicating a positive temperature correlation. The corresponding correlation coefficient r amounts to $r = 0.886$, while the correlation coefficient based on averaged daily temperatures is $r = -0.485$. However, investigations on the influence of temperature on $\delta^{13}C$ values of plants have failed to find an effect as found above. According to some studies, $\delta^{13}C$ values become slightly more negative (by up to 2 ‰) with increasing temperature (SMITH et al. 1976, LEAVITT and LONG 1982, TROUGHTON and CARD 1975) but there were no pronounced trends in the $\delta^{13}C$ values as a function of temperature. A direct temperature effect on the isotope fractionation, if it is real, is probably very small. Changing humidity conditions which also affect C_i/C_{atm} have not been measured. Indirect effects, however, which involve plant physiological changes, such as are found during the juvenile phase of leaves, could have an additional influence. The positive temperature correlation which was found in this experiment probably reflects changes of quantities such as C_i/C_{atm} in (28.17).

It should, however, be kept in mind that the model is based on steady state conditions and is therefore not very well suited for interpretations as discussed above, thus indicating some of its limits. Especially during early stages of leaf development, total carbon is imported from sources which are not just atmospheric CO_2. Import and export processes are presently not well understood but they would have to be incorporated in more detail in models which are to explain the composition of the stable carbon isotopes in plants. Nevertheless, valuable applications based on this model have demonstrated that it provides some insight into the way in which the carbon isotope composition in plant organic matter is determined (FARQUHAR and RICHARDS 1984, HUBICK et al. 1986), and how it can be used for plant physiological studies.

Acknowledgments

I wish to thank Mr. Haustein for technical assistance, Dr. Jayasekera for the preparation of some of the samples and Dr. Bertram, Institute of Geology, University of Cologne, for help in the mass spectrometric analysis.

Bibliography

BENDER, M. M., 1968, Mass spectrometric studies of carbon 13 variations in corn and other grasses.
Radiocarbon **10**, 468–472.

BENDER, M. M., 1971, Variations in the $^{13}C/^{12}C$ ratios of plants in relation to the pathway of photosynthetic carbon dioxide fixation.
Phytochemistry **10**, 1239–1244.

BERRY, J.; TROUGHTON, J. H.; BJOERKMAN, O., 1972, Effect of oxygen concentration during growth on carbon isotope discrimination in C_3 and C_4 species of

Atriplex.
Carnegie Inst., Washington, Yearb. **71**, 158–161.

CRAIG, H., 1953, The geochemistry of the stable carbon isotopes.
Geochim. Cosmochim. Acta **3**, 53–92.

CRAIG, H., 1954, Carbon-13 in plants and the relationship between carbon-13 and carbon-14 variations in nature.
J. Geol. **62**, 115–149.

CRAIG, H., 1957, Isotopic standards for carbon and oxygen and correction factors for mass-spectrometric analysis of carbon dioxide.
Geochim. Cosmochim. Acta **12**, 133–149.

DA SILVEIRA, L.; STERNBERG, L.; MULKEY, S. S.; WRIGHT, S. J., 1989, Ecological interpretation of leaf carbon isotope ratios: influence of respired carbon dioxide.
Ecology **70**, 1317–1324.

EHLERINGER, J. R., FIELD, C. B.; ZHI-FANG LIN; CHUN-YEN KUO, 1986, Leaf carbon isotope and mineral composition in subtropical plants along an irradiance cline.
Oecologia (Berl.) **70**, 520–526.

ESTEP, M. F.; TABITA, F. R.; PARKER, P. L.; VAN BAALEN, CH., 1978, Carbon isotope fractionation by ribulose-1,5-bisphosphate carboxylase from various organisms.
Plant Physiol. **61**, 680–687.

FARQUHAR, G. D.; O'LEARY, M. H.; BERRY, J. H., 1982, On the relationship between carbon isotope discrimination and the intercellular carbon dioxide concentration in leaves.
Aust. J. Plant Physiol. **9**, 121–137.

FARQUHAR, G. D., RICHARDS, R. A., 1984, Isotopic composition of plant carbon correlates with water-use efficiency of wheat genotypes.
Aust. J. Plant Physiol. **11**, 539–552.

FRANCEY, R. J.; GIFFORD, R. M., SHARKEY, T. D.; WEIR, B., 1985, Physiological influences on carbon isotope discrimination in huon pine (*Lagarostrobos franklinii*).
Oecologia (Berl.) **66**, 211–218.

GÄUMANN, E., 1935, Der Stoffhaushalt der Buche (*Fagus silvatica* L.) im Laufe eines Jahres.
Ber. Schweizerische Bot. Ges. **44**, 157–334.

HATCH, M. D.; SLACK, C. R., 1966, Photosynthesis by sugarcane leaves. A new carboxylation reaction and the pathway of sugar formation.
Biochem. J. **101**, 103–111.

HUBICK, K. T.; FARQUHAR, G. D.; SHORTER, R., 1986, Correlation between water-use efficiency and carbon isotope discrimination in diverse peanut (*Arachis*) germplasm.
Aust. J. Plant Physiol. **13**, 803–816.

JACOBSON, B. S.; SMITH, B. N.; EPSTEIN, S.; LATIES, G. G., 1970, The prevalence of carbon-13 in respiratory carbon dioxide as an indicator of type of endogenous substrate.
J. Gen. Physiol. **55**, 1–17.

KEELING, C. D., 1958, The concentration and isotopic abundances of atmospheric carbon dioxide in rural areas.
Geochim. Cosmochim. Acta **13**, 322–334.

KEELING, C. D., 1961, The concentration and isotopic abundances of carbon dioxide in rural and marine air.
Geochim. Cosmochim. Acta **24**, 277–298.

KEELING, C. D., 1960, The concentration and isotopic abundances of carbon dioxide in the atmosphere.
Tellus **12**, 200–203.

KLUGE, M.; TING, I. P., 1978, Crassulacean Acid Metabolism, Springer, New York.

KORTSCHAK, H. P.; HART, C. E.; BURR, G. O., 1965, Carbon dioxide fixation in sugarcane leaves.
Plant Physiol. **40**, 209–213.

LEAVITT, S. W.; LONG, A., 1982, Evidence for $^{13}C/^{12}C$ fractionation between tree leaves and wood.
Nature **298**, 742–744.

LERMAN, J. C.; RAYNAL, J., 1972, Biologie végétale. – La teneur en isotopes stables du carbone chez les *Cyperacees*: sa valeur taxonomique.
C. R. Acad. Sc., Paris, Ser. D **275**, 1391–1394.

LOWDON, J. A.; DYCK, W., 1974, Seasonal variations in the isotope ratios of carbon in maple leaves and other plants.
Can. J. Earth Sci. **11**, 79–88.

MEDINA, E.; MINCHIN, P., 1980, Stratification of $\delta^{13}C$ values of leaves in amazonian rain forests.
Oecologia (Berl.) **45**, 377–378.

MÜLLER, D., 1954, Die Atmung der Buchenblätter.
Forstlige Forsoegsvaesen i Danmark, Kopenhagen **21**, 303–319.

NIER, A. O.; GULBRANSEN, E. A., 1939, Variations in the relative abundance of the carbon isotopes.
J. Am. Chem. Soc. **61**, 697–698.

O'LEARY, M. H., 1976, Carbon isotope effect on the enzymatic decarboxylation of pyruvic acid.
Biochem. Biophys. Res. Comm. **73**, 614–618.

O'LEARY, M. H., 1977.
In: VAN TAMELEN, E. E. (ed.), Bioorganic Chemistry. Vol. 3, Elsevier, Amsterdam, 259.

O'LEARY, M. H., 1981, Carbon isotope fractionation in plants.
Phytochemistry **20**, 553–567.

O'LEARY, M. H., 1984, Measurement of the isotope fractionation associated with diffusion of carbon dioxide in aqueous solution.
J. Phys. Chem. **88**, 823–825.

OSMOND, C. B., 1978, Crassulacean acid metabolism: a curiosity in context.
Ann. Rev. Plant Physiol. **29**, 379–414.

PARK, R.; EPSTEIN, S., 1960, Carbon isotope fractionation during photosynthesis.
Geochim. Cosmochim. Acta **21**, 110–126.

REIBACH, P. H.; BENEDICT, C. R., 1977, Fractionation of stable carbon isotopes by phosphoenolpyruvate carboxylase from C_4 plants.
Plant Physiol. **59**, 564–568.

ROESKE, C. A.; O'LEARY, M. H., 1984, Carbon isotope effects on the enzyme-catalyzed carboxylation of ribulose bisphosphate.
Biochemistry **23**, 6275–6284.

SCHLESER, G. H.; JAYASEKERA, R., 1985, δ^{13}C-variations of leaves in forests as an indication of reassimilated CO_2 from the soil.
Oecologia (Berl.) **65**, 536–542.

SCHLESER, G. H.; BERNHARDT, K.-G.; HURKA, H., 1989, Climatic adaptability of populations of *Diplotaxis erucoides* D.C. (brassicaceae) from Sicily, based on leaf morphology, leaf anatomy and δ^{13}C studies.
Int. J. Biometeorol. **33**, 109–118.

SCHLESER, G. H., 1989, δ^{13}C pattern in a forest tree as indicator of carbon fluxes and relative photosynthetic performance.
Ecology, submitted for publication.

SCHLESER, G. H., 1990, Investigations of the $\delta^{13}C$ pattern in leaves of *Fagus sylvatica* L.
J. Exp. Bot. **41**, 565–572.

SCHLESER, G. H.; JAYASEKERA, R., Seasonal variations of the stable carbon isotopes in a forest environment.
In preparation.

SCHMIDT, H.-L.; WINKLER, F. J.; LATZKO, E.; WIRTH, E., 1978, ^{13}C-kinetic isotope effects in photosynthetic carboxylation reactions and $\delta^{13}C$-values of plant material.
Isr. J. Chem. **17**, 223–224.

SMITH, B. N.; EPSTEIN, S., 1971, Two categories of $^{13}C/^{12}C$ ratios for higher plants.
Plant Physiol. **47**, 380–384.

SMITH, B. N.; OLIVER, J.; MCMILLAN, C., 1976, Influence of carbon source, oxygen concentration, light intensity, and temperature on $^{13}C/^{12}C$ ratios in plant tissues.
Bot. Gaz. **137(2)**, 99–104.

TANS, P. P., 1978, Carbon 13 and carbon 14 in trees and the atmospheric CO_2 increase.
Proefschrift, Univer. Groningen.

TROUGHTON, J. H.; CARD, K. A.; HENDY, C. H., 1974, Photosynthetic pathways and carbon isotope discrimination by plants.
Carnegie Inst. Washington, Yearb. **73**, 768–780.

TROUGHTON, J. H.; CARD, K. A., 1975, Temperature effects on the carbon-isotope ratio of C_3, C_4 and crassulacean-acid-metabolism (CAM) plants.
Planta (Berl.) **123**, 185–190.

VOGEL, J. C., 1978, Recycling of carbon in a forest environment.
Oecologia Plantarum **13**, 89–94.

VOGEL, J. C., 1980, Fractionation of the carbon isotopes during photosynthesis.
Sitzungsber. Heidelb. Akad. Wissensch., Springer, Berlin, Heidelberg, New York **3**, 111–135.

WHELAN, T.; SACKETT, W. M.; BENEDICT, C. R., 1973, Enzymatic fractionation of carbon isotopes by phosphoenolpyruvate carboxylase from C_4 plants.
Plant Physiol. **51**, 1051–1054.

WICKMAN, F. E., 1952, Variations in the relative abundance of the carbon isotopes in plants.
Geochim. Cosmochim. Acta **2**, 243–254.

WIESBERG, L., 1974, Die C-13 Abnahme in Holz von Baumjahresringen, eine Untersuchung zur anthropogenen Beeinflussung des CO_2-Haushalts der Atmosphäre. Thesis, Rheinisch-Westfälische Technische Hochschule, Aachen.

WINKLER, F. J.; KEXEL, H.; KRANZ, C.; SCHMIDT, H.-L., 1982, Parameters affecting the $^{13}CO_2/^{12}CO_2$ isotope discrimination of the ribulose-1,5-bisphosphate carboxylase reaction.
In: SCHMIDT, H.-L.; FÖRSTEL, H.; HEINZINGER, K. (eds.), Stable Isotopes. Proceedings 4th. International Conference, Jülich, F. R. Germany, 23–26 March 1981, Elsevier, Amsterdam, pp. 83–89.

WONG, W. W.; BENEDICT, C. R.; KOHEL, R. J., 1979, Enzymic fractionation of the stable carbon isotopes of carbon dioxide by ribulose-1,5-bisphosphate carboxylase.
Plant Physiol. **63**, 852–856.

Chapter 29

Carbon dioxide effects on vegetation

D. OVERDIECK and M. FORSTREUTER

29.1 Introduction

Since the beginning of continuous measurements of the tropospheric CO_2 concentration in 1958 at the Mauna Loa observatory in Hawaii (KEELING 1986), the upward trend of the CO_2 contents of the atmosphere has been evident. The annual mean of 344 ppm in 1984 (SOLOMON et al. 1985) increased to 351 ppm by 1988 (BODEN 1988, personal communication, Oak Ridge National Laboratory, TN, U.S.A.). This means a yearly average increment of 1.4 ppm for that period. Calculated using a separate data base, TRABALKA 1985 has estimated a mean global increase of 1.5 ppm. Analysis of air trapped in dated ice bubbles from Greenland and Antarctica ice sheets has revealed that the global preindustrial atmospheric CO_2 concentration might have been between 275 and 285 ppm (NEFTEL et al. 1985). From 275 ppm to today's Mauna Loa values this means a global increase of 76 ppm within about 100 years, equivalent to $\sim$ 28%, compared with present Mauna Loa values. Since the last century the continuous increment has been brought about by the intensified use of fossil fuels but also by a number of further activities, like the clearing of large parts of the tropical rain–forests, fire, soil erosion, conversion of arable land to pasture (and vice–versa), peat digging, drainage and flooding. The qualitative

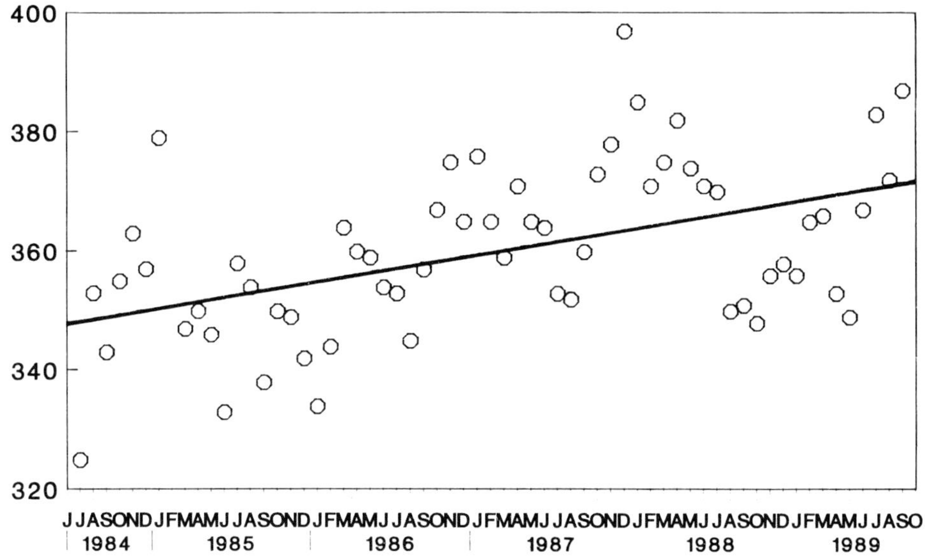

Figure 29.1: Arithmetical monthly mean of CO_2 concentration on the Westerberg (Osnabrück, Germany, 90 m above sea level, 52° 18'N, 8° 2'E) at 4 m above ground from July 1984 until October 1989; linear regression: $y = 348 + 0.38 \times x$. Data from OVERDIECK (1991).

and quantitative influence of these processes on the CO_2 exchange of terrestrial ecosystems is rated differently (AJTAY et al. 1979, HAMPICKE 1979, BROECKER et al. 1979, SEILER et al. 1980, SIEGENTHALER et al. 1980, ESSER 1987). Today the CO_2 concentrations are higher close to urban and industrial areas in northern latitudes than on Mauna Loa; and since 1984 they may have increased even more, which is suggested by monthly means, based on half-hourly means, from the "Westerberg" near Osnabrück (Fig. 29.1). There the yearly mean value amounted to approximately 348 ppm in the beginning of the measurement campaign and increased to 366 ppm by 1988 ($\sim$ +5%). This corresponds to an increment of 4.5 ppm in the yearly mean within that limited time span. It seems that the natural sinks for CO_2, i.e. oceans and marine and terrestrial ecosystems are not sufficient to decelerate effectively the continuous concentration increase. It is also likely that more CO_2 has been released into the atmosphere everywhere and that eventually the CO_2 concentration will be doubled by sometime in the next century, and that all plants and other organisms dependent on them, will be directly or indirectly affected.

The productivity of agriculturally used plants can be enhanced if more CO_2 is available (DAUNICHT 1971, DAUNICHT 1974). "CO_2-fertilization" of cultures in greenhouses has already been known since the last century (GODLEWSKI 1873, KREUSLER 1885). KIMBALL published literature reviews (1983a, 1983b) which are based on 430 and 770 single observations, mainly on agriculturally or horticulturally used plant species. According to these data, the yield will be increased by about 36% on the overall average if today's CO_2 concentration is doubled. Consequently, in many cases the actual CO_2 content of the troposphere seems to be suboptimal. Therefore, BOTKIN (1977) concluded that the total phytomass of the terrestrial vegetation would act as a sink for the anthropogenic CO_2 and that its mass would increase in the future. In spite of that knowledge and this founded hypothesis — from the ecological point of view — we are, however, still confronted with unsolved problems concerning direct CO_2 effects on plants.

Namely, during many investigations on crop species, CO_2 concentration levels were chosen which were partly much higher than the levels predicted for the next century. Additional studies within the range of 270 - 700 ppm are thus desirable.

Besides, studies on plant communities exposed for long periods, i. e. for months and years, to realistic CO_2 concentrations under field-like conditions are rather rare. Investigations on the terrestrial plant-community level are necessary because, from the results of the various CO_2 enrichment experiments with single individuals it is not possible to know if and when units of the upper soil layers and vegetation cover will be sources or sinks for additionally emitted CO_2. For these reasons we continuously exposed mixtures (1:1) of white clover and perennial ryegrass (*Trifolium repens / Lolium perenne*) and red clover and meadow fescue (*Trifolium pratense / Festuca pratensis*) for more than two years to elevated CO_2 concentrations under outdoor microclimatic conditions. Comparable investigations on the long-term effects of CO_2 on entire plant communities were conducted in the Alaskan Tundra and on a salt marsh of the eastern North America (HILBERT et al. 1987, CURTIS et al. 1989, DRAKE et al. 1989).

About 10% of the publications available to us deal with the effects of elevated CO_2 concentrations on woody plants which play the most important role in the terrestrial regional and global carbon cycle and could form a large sink for additional CO_2. The few recent long-term and autecological investigations on trees refer — just like the studies on herbs — to enhanced growth and increasing CO_2 net assimilation parallel to decreasing transpiration rates (ROGERS et al. 1983, DAHLMANN et al. 1985, HOLLINGER 1987, GAUDILLÈRE et al. 1989). Research about the effects on whole forest ecosystems have been based on experiments with single tree species only (WILLIAMS et al. 1986). Therefore, our team tried to broaden our understanding about the direct CO_2 effects on woody plants by means of long-term CO_2 enrichment experiments, including herbaceous plants of the understory.

Since the demand for minerals and water is probably linked to the potentially increasing CO_2 net assimilation, possibly essentially modifying the transformation

Figure 29.2: Field mini-greenhouses climatized according to outside conditions for the purpose of CO_2 enrichment studies (with open control plots in front).

of plant growth (WONG 1979), we also investigated about some aspects of mineral metabolism and water use.

29.2 Materials and methods

The experiments were conducted on the Westerberg near Osnabrück (52°18'N, 8°2'E). Between 1983 and 1989 we used 2–4 identical acrylic mini-greenhouses in the field which enclosed homogenized, natural, not fertilized top soil with intermediate mineral supply (Fig. 29.2). This soil was classified as "sandy clay" with a C/N ratio of 13:1. The closed soil volume amounted to 0.38 m^3 and the air volume to 0.51 m^3. The microclimate inside was immediately adjusted to the outside conditions by means of automatic electronic regulators (air temperature ±0.5°C, rel. air humidity: ±15%). The above-ground parts of the mini-greenhouses consisted of UV-permeable acrylic glass and, depending on the position of the sun, the interior light intensity was decreased by 2 – 18%. Until 1986 the air turbulence within the cabinet was constantly maintained at 0.5–1 m s^{-1}. Later the air velocity was automatically and proportionally adjusted to the wind speed outside up to a mean maximum of 2.5 m s^{-1} (compare FORSTREUTER 1990).

Mixtures (1:1, 40 seedlings dm^{-2}) of white clover (*Trifolium repens* L. var. Milka Paybjerg) and perennial ryegrass (*Lolium perenne* L. var. Printo) from the beginning and, since 1985, of red clover (*Trifolium pratense* var. Lero) and meadow fescue (*Festuca pratensis* Huds. var. Cosmos 2) in relation 1:1 (25 seedlings dm^{-2}) grew there until senescence in these greenhouses. The plants were irrigated corresponding to the precipitation rates outside. In addition, open plots with the same soil and plant mixtures were established in order to control possibly remaining glasshouse effects.

Since 1984 CO_2 concentrations were developed in these systems in accordance with the outside concentration and with elevations to 600 ppm CO_2. Since 1987 concentrations of ~ 450 and ~ 800 ppm were also maintained. In preliminary studies (1981/82) a similar experimental design had been used; but with only two CO_2 concentrations ~ 300 and ~ 600 ppm CO_2. The following text refers partly to the results from the first attempt. Some details about the construction and the working methodology of the experimental equipment are given by OVERDIECK et al. (1985) and OVERDIECK et al. (1987).

The experiments allowed us to work out results of two kinds:

1. CO_2 effects on development, growth, and performance under competition of the single plant species and

2. CO_2 effects on phytomass production of the entire vegetation stands and its long-term CO_2 and H_2O exchange rates under varying microclimatic conditions.

Microclimate and gas exchange data were obtained every 144 seconds; on this basis half-hour means were calculated and daily courses were plotted. Using daily courses for balances, monthly and yearly courses of the CO_2 and H_2O exchange values were composed.

The contents of the mineral elements N, P, K, Ca, Mg, and Zn and of the most important carbohydrates were determined in plant samples taken in chronologically different sequences from the mini-greenhouses. In addition, ribulose-1.5-bisphosphate (RuBP) carboxylase/oxygenase (Rubisco)-activity of leaf samples was measured.

During 1985–1988, seedlings and saplings (partly also cloned saplings) of *Acer pseudoplatanus* L. (mountain maple) and *Fagus sylvatica* L. (European beech) were also continuously exposed to the CO_2 concentration levels of ~ 390, ~ 520, and ~ 660 ppm and to different soil volumes in greenhouse compartments with an air volume of 100 m^3 each. The three compartments were connected with regulators to adjust the inside air temperature and relative air humidity instantly to the outside conditions. Thus the differences in temperature and rel. air humidity between the compartments could be maintained within the limits of the sensor sensitivities. The deviation from outside reached at most +1.8°C and +7% rel. air humidity. The photon flux density in the range of the photosynthetically active radiation was at the same level in all three compartments. On sunny, cloudless days the relative photon flux density inside was at 13–15% of outside values.

In April 1987, seedlings of the greater stitchwort (*Stellaria holostea* L.) and runner sections of the yellow archangel (*Lamium galeobdolon* (L.) NATHH.) were exposed to the three different CO_2 concentration levels. These plants grew for 4 months either in monoculture with 4 individuals or in 2:2 mixtures in 8 l-pots. All plants of this experimental design obtained the same natural soil of medium mineral supply, were not fertilized, but were watered regulary. In those greenhouse compartments, *Trifolium repens* also was cultivated, for anatomical studies (OVERDIECK et al. 1989).

29.3 Results and discussion

29.3.1 Carbohydrates

The usually great surplus of the enzyme Rubisco over its substrate CO_2 (HELD et al. 1986) seems to make the *in vivo* enhancement of the photosynthetic efficiency by elevated CO_2 concentrations quite understandable. It is difficult, however, to argue with experimental results which suggest that in many cases photosynthesis is already limited or optimized (OVERDIECK 1989) at CO_2 concentrations near the upper limit of the predicted tropospheric concentration increase. VON CAEMMERER et al. (1981) assume that the activation of RuBP is limited at higher CO_2 concentrations and therefore that the initial linear increase of CO_2 net fixation cannot be sustained. Possibly the capacity of the photosynthetic electron transport for the regeneration of the enzyme is restricted. We only found a slightly elevated, initial Rubisco activity in white-clover leaves under 600 ppm CO_2, in comparison with the control (340 ppm) at the very highest (> 1000 $\mu E \cdot m^{-2} \cdot s^{-1}$) photon flux densities of photosynthetically active radiation. Repetitions of the activity measurements together with determinations of the absolute amounts of Rubisco are still necessary to clarify the interrelations. Also, the influence of the O_2 partial pressure on this process warrants special research. Some experimentalists suppose due to increasing photosynthetic activity increasing amounts of assimilation starch accumulate in the chloroplasts leading to a negative feedback on photosynthesis (MADSEN 1968,CAVE et al. 1981,SASEK et al. 1985). In addition, some results from tissue analysis show that the carbohydrate contents of photosynthetically active organs increase under CO_2 enrichment: MAUNEY et al. (1979) found in cotton, soybean, sunflower and sorghum that the contents of starch but not those of the ethanol-soluble sugars increased after two months of exposure to 630 ppm CO_2. FINN et al. (1982) showed that the contents of the non-structural carbohydrates already increased after 36 h under 1000 ppm CO_2.

The results of our analysis of red clover corroborate these findings to some extent: namely that the mean contents of the ethanol–soluble monosaccharides glucose and fructose changed little. Compared to the control (350 ppm), however, higher saccharose– and starch–contents were obtained in the lamina which had been exposed to 450, 600 and 800 ppm CO_2 during their development. Also, in most cases, the cellulose content of the leaves harvested from our mini–greenhouses with elevated CO_2 levels exceeded that of the leaves from the control. Above all, the enhanced starch accumulation could restrict the capacity of photosynthesis.

29.3.2 Anatomical modifications

The enhancement of CO_2 uptake could be limited by a reduced number of stomata per unit leaf area if plants grow up at elevated CO_2 levels (WOODWARD 1986, WOODWARD 1987). This hypothesis did not remain undisputed after the investigation of 200 species by KÖRNER (1988). It is also conceivable that stomata become smaller or simply that they open less wide at elevated ambient CO_2 concentrations in general. We only studied the anatomy of white–clover leaves grown at 370 (control), 520 and 670 ppm CO_2 exactly enough and found that the stomatal density was not significantly decreased at elevated CO_2 concentrations (OVERDIECK et al. 1989). The mesophyll, however, was denser with more cells concentrated per unit space at the expense of intercellular holes. This corresponds with the results of THOMAS and HARVEY (1983) obtained for *Pinus taeda* and *Liquidambar styraciflua* (340 – 910 ppm) and can explain the increasing specific leaf weight (or decreasing specific leaf area) which was often found during CO_2 enrichment experiments (SIONIT 1983, OVERDIECK 1986). To all appearances, the anatomical changes in the mesophyll did not greatly reduce the positive effect at increasing CO_2 concentrations on growth and production of white clover.

29.3.3 Phytomass of herbaceous plants

Fig. 29.3 shows the time course of net primary production of white clover–perennial ryegrass mixtures (1:1) over 100 days. During this time the vegetation was cut four times. The upper part of the curves is based on the total above–ground dry mass and on samples taken at random from the below–ground phytomass; the lower parts of the curves rely totally on samples taken at random above and below ground.

Towards the end of the investigation period the difference between the two cultures (300 and 620 ppm CO_2) reached a constant value equivalent to a $\sim$ 68% higher weight at the elevated CO_2 level (OVERDIECK et al. 1986). Comparable amounts of additional CO_2 ($\sim$ +300 ppm) increased the phytomass of harvested systems by only 20–40%. In early phases of succession and shortly after cutting, which physiologically initializes to some respect a new beginning, grassland model ecosystems are a sink for the additional CO_2. The results of the studies on red

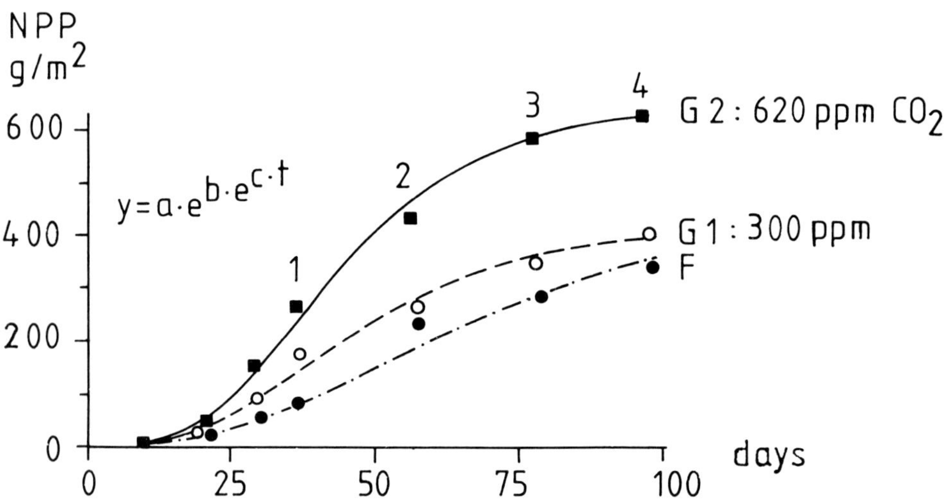

Figure 29.3: Total net primary production per m^2 ground area (NPP) from *Trifolium repens–Lolium perenne* mixtures (1:1) at 300 and 620 ppm CO_2 in mini-greenhouses (G1, G2) climatized according to the outside conditions; F: open field control; phytomass can be predicted by the equation shown in the figure; constants are as follows:
300 ppm: $a = 404$, $b = 7.5$, $c = -0.053$, $r = 0.992$;
620 ppm: $a = 648$; $b = -8.6$, $c = -0.058$, $r = 0.996$;
1–4: number of cut, data from OVERDIECK et al. 1986.

clover/meadow fescue mixtures (1:1) in which also the effects of 450 and 800 ppm CO_2 could be considered, showed a similar tendency: the elevated CO_2 concentration enhanced the dry matter accumulation the most in the beginning and also after complete cutting in the middle of the vegetation period. Dry weights did no longer differ considerably at the end of the vegetation period or at senescence of the plant stands. All weighings from 1984–88 were normalized to the control level which was set to 1.0, using the following equation:

$$F_{CO_2} = 1.65\left(1 - \frac{1}{\exp(0.0038(pCO_2 - 80))}\right) \tag{29.1}$$

CO_2 saturation occures at about 800 ppm (Fig. 29.4). The initial slope of the curve indicates that managed grassland accumulates more phytomass just at the tropospheric CO_2 concentration expected in a few decades (450 – 600 ppm).

Beyond the already mentioned physiological and anatomical limitations of the

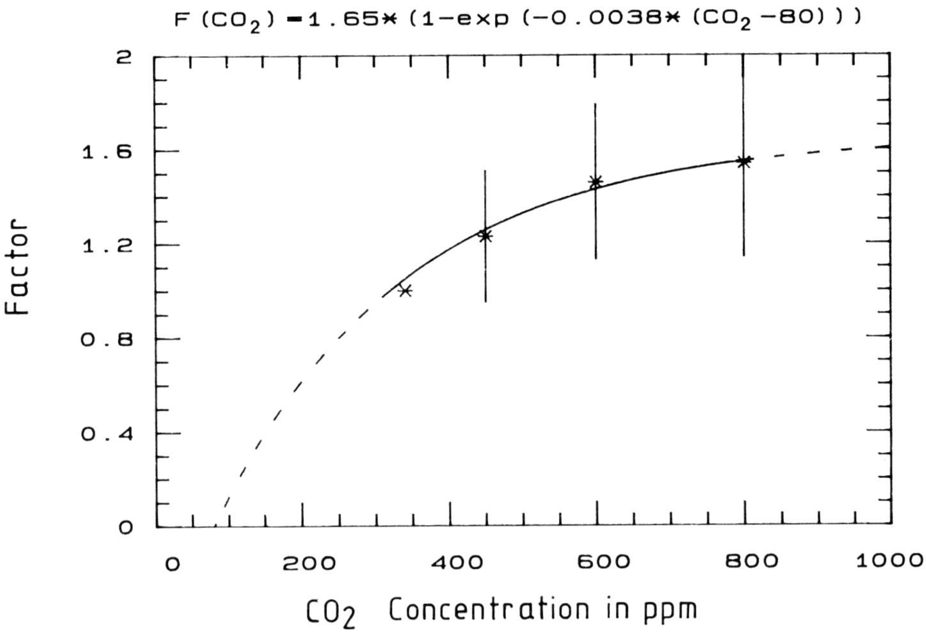

Figure 29.4: CO_2 fertilization factor (340 ppm = 1.0) as a function of the ambient CO_2 concentration (ppm) for the dry mass accumulation (above + below ground) of total plant stands formed by red clover and meadow fescue (1:1) from 1984–88. The data are from four different CO_2 concentrations in mini–greenhouses climatized according to the outside conditions (340 ppm: $n = 16$; 450 ppm: $n = 8$; 600 ppm: $n = 16$, 800 ppm: $n = 8$); after OVERDIECK (1991).

positive CO_2 effect around 800 ppm, one can also consider the more rapid increase in the "density" of the stand at elevated CO_2 concentrations leading to more shading of plant parts in the lower plant canopy, as well as different other forms of competition for limited environmental resources.

The mean dry weight of red clover seeds and caryopses of meadow fescue increased when the CO_2 concentration level was elevated from 340 (control) to 450 ppm (Tab. 29.1). Further CO_2 enrichment had no effect. A significant weight increase of 28% on the average could also be found with the seeds of white clover but not with the caryopses of perennial ryegrass grown at 600 ppm (OVERDIECK 1986). This means certain advantages in the long run for at least the two clover species studied if there is a close positive correlation between seed weight and initial growth of the seedlings (MYTTON 1973). The vegetative propagation, however, is

Table 29.1: Mean dry mass weight of seeds and caryopses after long–term exposure to different CO_2 concentration levels (n = 10 × 10 seeds or caryopses); data from OVERDIECK (1991).

Mean weight per thousand seeds or thousand–grain weight [g·1000^{-1}]

CO_2 concentration [ppm]	*Trifolium pratense*	*Festuca pratensis*
350	1.98 ± 0.10	1.05 ± 0.21
450	2.14 ± 0.13	1.62 ± 0.27
600	2.25 ± 0.09	1.40 ± 0.33
800	2.26 ± 0.10	1.63 ± 0.30

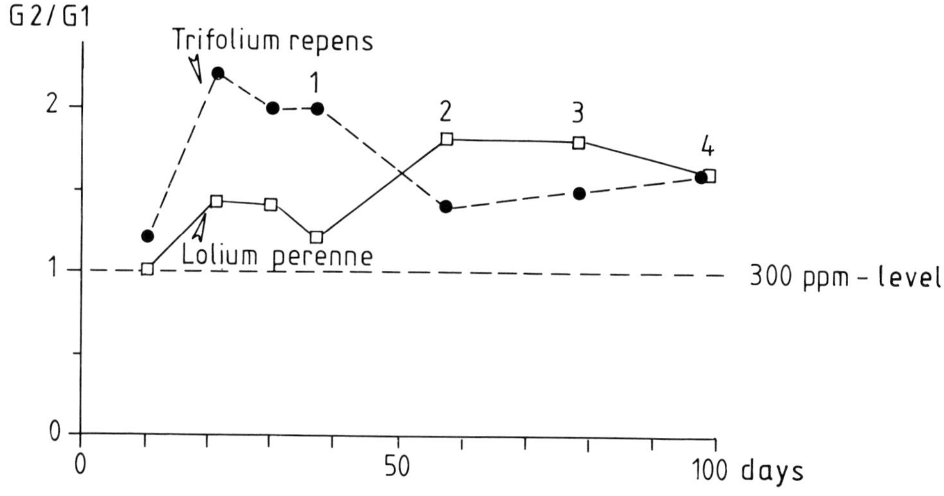

Figure 29.5: Enhancement of net primary production by ~ 600 ppm CO_2 (G2) relation to ~ 300 ppm CO_2 (G1) for *Lolium perenne* and *Trifolium repens* mixtures (1:1); 1–4: number of cuts, data from OVERDIECK et al. (1986).

more important than the generative in both species.

29.3.4 Competition

The species studied not only responded differently upon elevated CO_2 concentrations but also their performance changed as a result of regular cutting of the total above-ground phytomass (cf. Fig. 29.5). Without cutting (as in the beginning of the investigation) white clover was more enhanced by the additional CO_2 than the grass. After the first and the following complete cuttings (numbered 1–4) *Lolium* became the stronger competitor. Since perennial ryegrass normally grows faster than white clover without additional CO_2 (in the case of regular cuttings), the results had to be normalized to their growth at 300 ppm CO_2. Competition performance also had to be judged relative to the control situation, because the phenophases of the grass and the clover are dephased in the natural stand (Fig. 29.5). Thereby it could be shown that both species were enhanced by more CO_2 despite their mutual competition. Their dry mass per unit ground area, however, oscillated more than in the control (300 ppm CO_2). Though not proved after 100 days, it is possible that in the long run one of the two species will be suppressed to such a small population size that it will be unable to recover.

The growth of *Stellaria holostea* was evidently enhanced by CO_2 enrichment, whereas the CO_2 effect was negative with *Lamium galeobdolon* (Fig. 29.6). *L. galeobdolon* responds only weakly or not at all to additional CO_2 as we proved in monoculture, perhaps because this species is already adapted to higher CO_2 concentrations in the forest understory. Therefore it may lose in competition.

On the whole it can be concluded from these results that the increase of atmospheric CO_2 will lead to changes in species composition of grassland as well as of the herb layer of forest ecosystems.

29.3.5 Phytomass of woody plants

The growth of all measured plant organs of *Acer pseudoplatanus* and *Fagus sylvatica* was enhanced by greater CO_2 supply, whereby the increase in CO_2 from 390 (control) to 520 ppm enhanced growth and production of below and above ground parts, more, on the average than a further elevation to 660 ppm CO_2 (Fig. 29.7).

The greater the size of the rooting space of *A. pseudoplatanus* the more the phytomass increased under the influence of elevated CO_2 concentration (Fig. 29.8). In this case the CO_2 effect (390 – $\sim$ 660 ppm), in combination with increasing soil volumes (0.1–30 l per tree), could better be described by a non-linear three-dimensional model than by a three-dimensional linear model.

Since the absolute mean dry weight of *F. sylvatica* did not increase with elevated CO_2 as much as did *A. pseudoplatanus*, one can also suppose that the competitive performance of these two species will be influenced by the change in atmospheric CO_2.

With increasing soil volumes, however, the CO_2 effect became also more positive with *F. sylvatica* (Fig. 29.9). This contradicts the hypothesis that stress (in this case caused by decreased soil volumes for the roots) is clearly tempered with

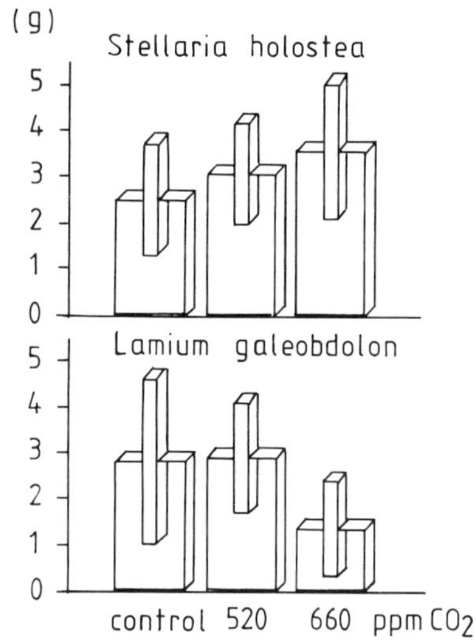

Figure 29.6: Mean dry weight of the above-ground parts at *Stellaria holostea* and *Lamium galeobdolon* in mixtures (1:1) at different CO_2 concentration levels.
Soil volume: 8 l; exposure 4 months (n/CO_2 level = 13 pots with 2 x 2 plants); data from OVERDIECK (1991).

additional CO_2 supply (SIONIT 1983). In principle it could be verified, however, that growth and production — of some important woody species too — are enhanced by elevated CO_2 concentrations (ROGERS et al. 1983, TOLLEY et al. 1984) as in most of the herbs that have been studied.

29.3.6 CO_2 gas exchange of "model ecosystems"

Fig. 29.10 is an example of the daily course of some microclimatic parameters in relation to the CO_2 gas exchange rates of the red clover/meadow fescue mixtures, including the upper soil in the control mini-greenhouse (350 ppm) and the greenhouse with a CO_2 concentration level of 600 ppm. The greatest positive deviation of CO_2 gas exchange from the control occurred in parallel with the maximum of photon flux density (PhAR). The nightly losses of CO_2 were always greater in the mini-greenhouse with the elevated CO_2 concentration.

On other days (not shown here), phases of CO_2 release to the ambient air coincided with periods of lower photon flux densities even during the day (Fig. 29.10). The maximum difference between the net CO_2 fixation rates was 1397 mg CO_2 m^{-2} h^{-1} at 14.30 h. This difference was equivalent to an increase of 29%. At

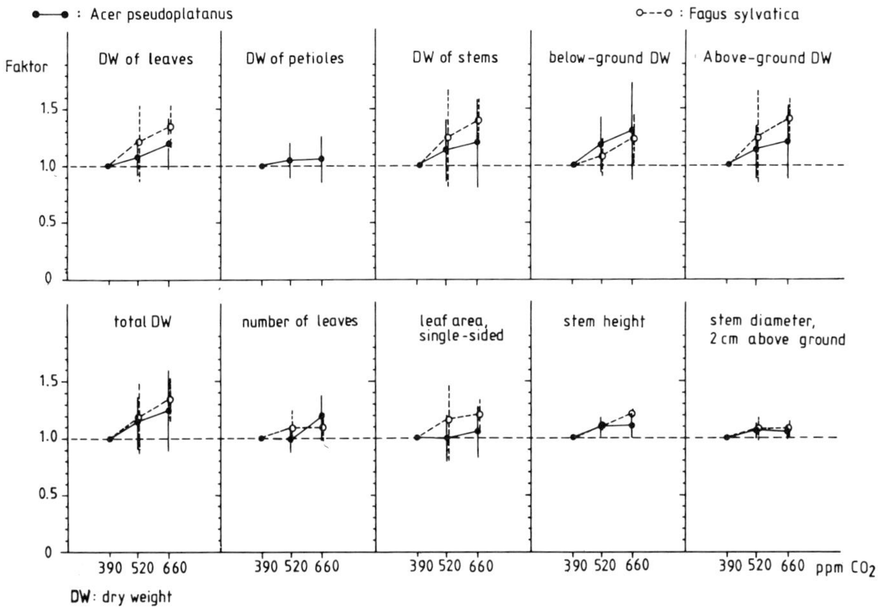

Figure 29.7: Mean CO_2 enrichment factors (CO_2 fertilization factors) for different parts of mountain maple (*Acer pseudoplatanus*) and beech saplings (*Fagus sylvatica*) for 2 years of CO_2 exposure in greenhouse compartments climatized to the outside conditions; mountain maple: 7.5 and 30 l soil volume/tree, 270–300 cm^2 ground area/tree, n = 10 and 30 trees/sampling/CO_2 level; beech: 12 and 32–36 l soil volume/tree, 260–660 cm^2 ground area/tree, n = 25–51 trees/sampling/CO_2 level; control (390 ppm) = 1.0; data from (OVERDIECK 1990).

the highest nighttime temperatures of this day (21.6 °C at 21.30 h) the difference between the respiration rates was the greatest; i.e. the respiration increased at the higher CO_2 concentration by 383 mg CO_2 m^{-2} h^{-1} equivalent to the enhancement of 62%.

In young cultures (and shortly after cutting) the differences between the set-ups exposed to high CO_2 and the control conditions were particularly large. Fig. 29.11 shows the complete daily courses of the CO_2 gas exchange at 300 and at 600 ppm CO_2 after the germination of white clover/ perennial ryegrass mixtures (1:1), registered and evaluated over three months in a preliminary experiment (1981). In this situation the mini-greenhouse with increased CO_2 supply very quickly reached essentially higher CO_2 net uptake rates during the day than the control. At the same time, the natural CO_2 loss rates also increased remarkably.

Around 12 h the half-hourly means of the CO_2 gas exchange of the entire plant stands at four CO_2 concentrations (Fig. 29.12) reached evidently higher levels with

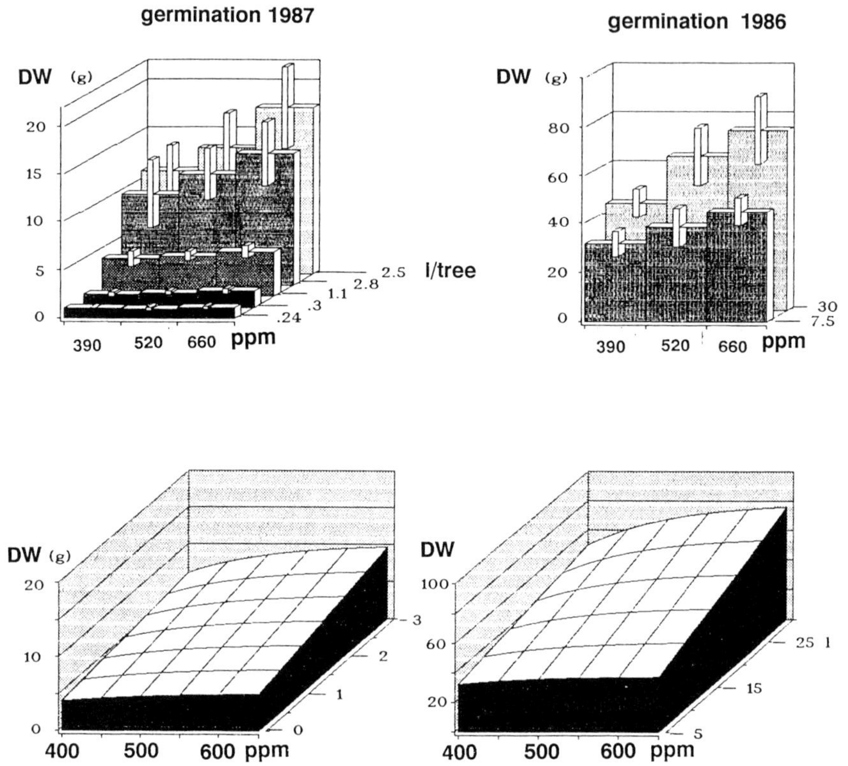

Figure 29.8: At the top: Mean dry weight (DW) of mountain maple (*Acer pseudoplatanus*) in greenhouse compartments, climatized according to outside conditions after 2 years of CO_2 exposure (harvest: Aug. 02 – Oct. 06) at different soil volumes; Model function calculated from the data at the bottom:
$y = c_0 + c_1 \tanh c_2 x_1 + c_3 x_2 + c_4 x_3 + c_5 x_2 \tanh c_2 x_1$
$R^2 = 0.91$, $S_{y,x} = 5.4$, $n = 198$ mean values for 10 trees;
y: dry weight DW [g], x_1: CO_2 [ppm], x_2: soil volume [l/tree], x_3: indicator variable, $x_3 = 1$ for the 1st year, $x_3 = 0$ for the 2nd year;
$c_0 = -23.9$, $c_1 = 29.0$, $c_2 = 0.005$, $c_3 = -24.1$, $c_4 = 24.6$, $c_5 = 25.5$. Data from REINING (1990).

increasing CO_2 concentrations up to 600 ppm (29–64 days after CO_2 exposure). Surprisingly, the CO_2 net fixation was not further enhanced by elevation of the CO_2 concentration level from 600 to 800 ppm, not even at the maximum of photon flux density. It may therefore be concluded that the maximum of CO_2 effect on net fixation of ecosystems would already be exceeded if today's atmospheric CO_2

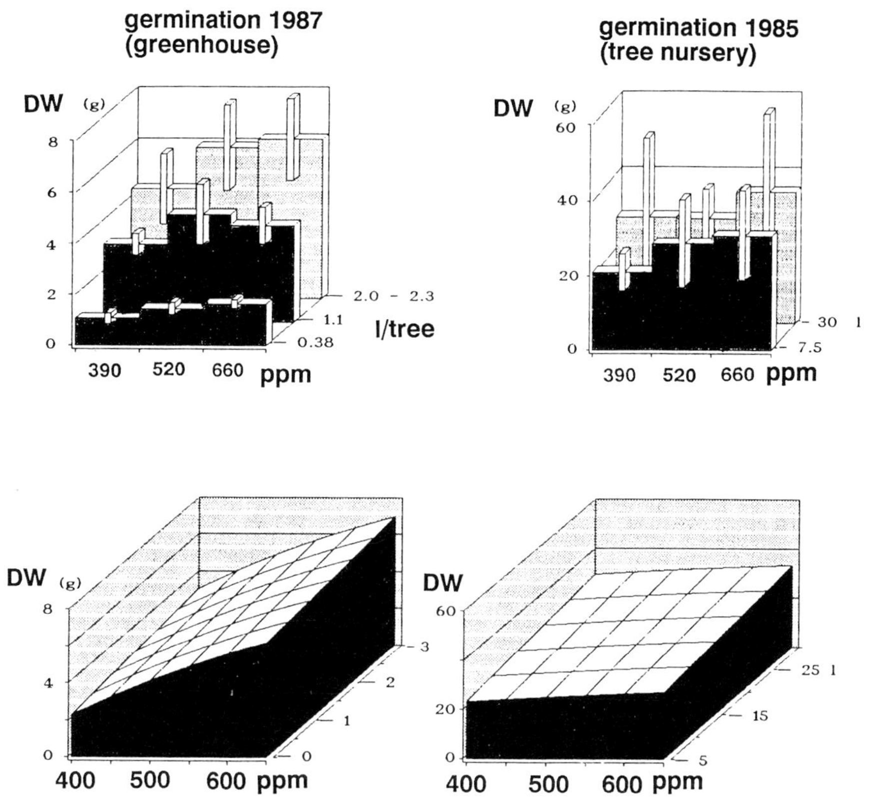

Figure 29.9: At the top: Mean dry weight (DW) of the beech (*Fagus sylvatica*) in greenhouse compartments climatized according to the outside conditions after two years of CO_2 exposure (harvest: Sept. 20 – Oct. 13) at different soil volumes; model function calculated from the data at the bottom:
$y = c_0 + c_1 \tanh c_2 x_1 + c_3 x_2 + c_4 x_3$
$R^2 = 0.7$, $S_{y,x} = 8.0$, $n = 106$ mean values for 10 trees;
y: dry weight: DW [g], x_1: CO_2 [ppm], x_2: soil volume [l/tree], x_3: indicator value: $x_3 = 0$ for the 2nd year, $x_3 = 1$ for the 1st year;
$c_0 = -8.9$, $c_1 = 18.7$, $c_2 = 0.0017$, $c_3 = 0.24$, $c_4 = 19.7$; data from REINING (1990).

concentration were doubled (cf. Fig. 29.13).

The mean maxima of net CO_2 fixation amounted to 4.4 (350 ppm), 5.6 (450 ppm), 6.3 (600 ppm) and 6.2 g CO_2 m^{-2} h^{-1} (800 ppm) and the difference between 600 and 800 ppm was not significant. The nocturnal respiration rates of the model ecosystems were significantly higher at 600 (+10%) and 800 ppm (+20%) as compared to 350 and 450 ppm CO_2.

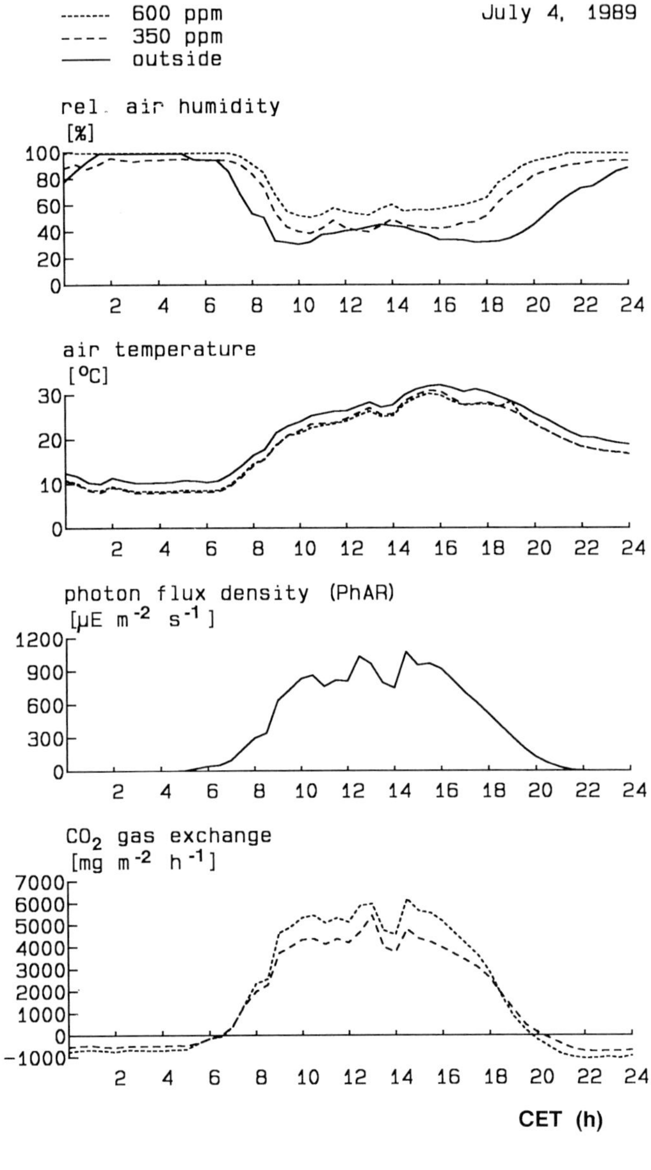

Figure 29.10: Day course (4 July 89) of the CO_2 gas exchange from half-hourly means (n = 12) of two model ecosystems composed of natural soil and the same mixtures (1:1) of *Trifolium pratense* and *Festuca pratensis* in mini-greenhouses climatized according to the outside conditions at 350 and 600 ppm CO_2. Above it, the corresponding microclimatic parameters; data from FORSTREUTER (1991).

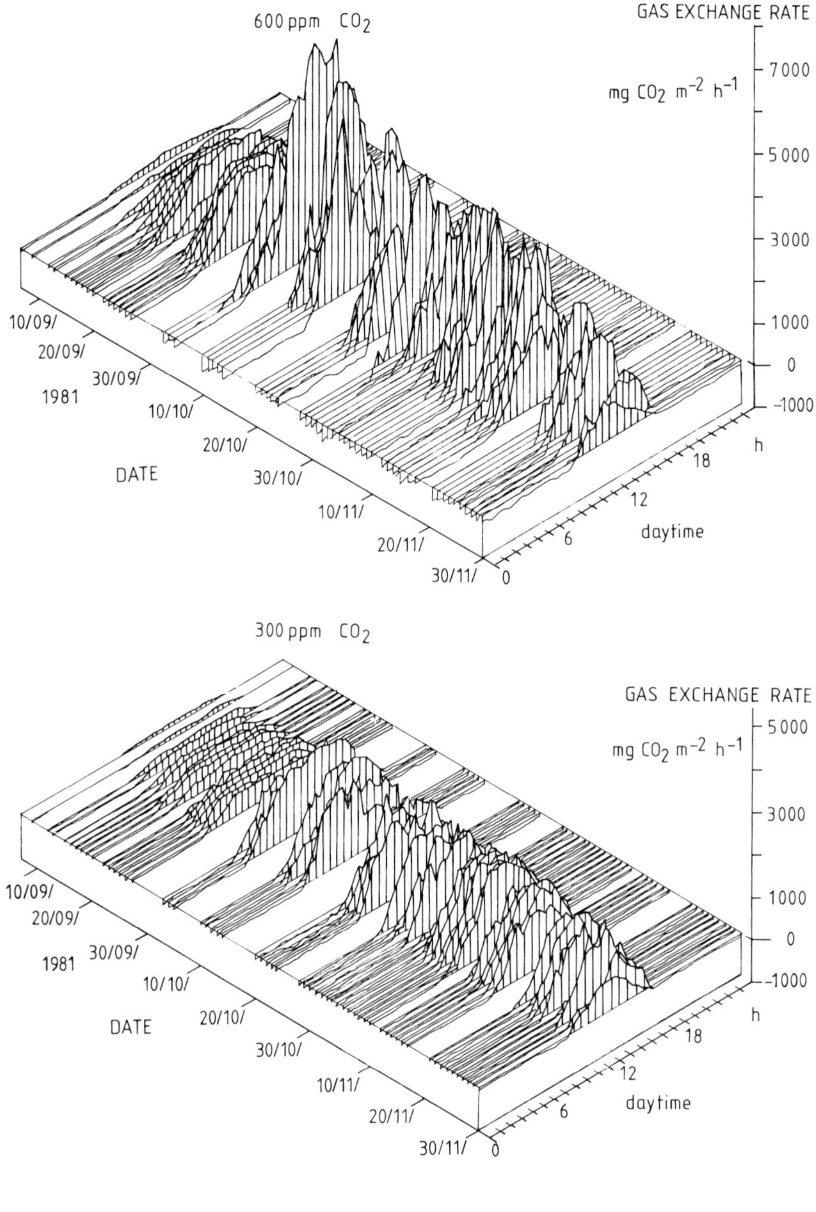

Figure 29.11: Daily courses (Sept.–Nov. 1981) of the CO_2 gas exchange per m^2 ground area of two model ecosystems composed of natural soil and the same mixtures (1:1) of *Trifolium repens* and *Lolium perenne*, grown at $\sim$ 300 and $\sim$ 600 ppm CO_2 in mini–greenhouses climatized according to the outside conditions; data from OVERDIECK 1991.

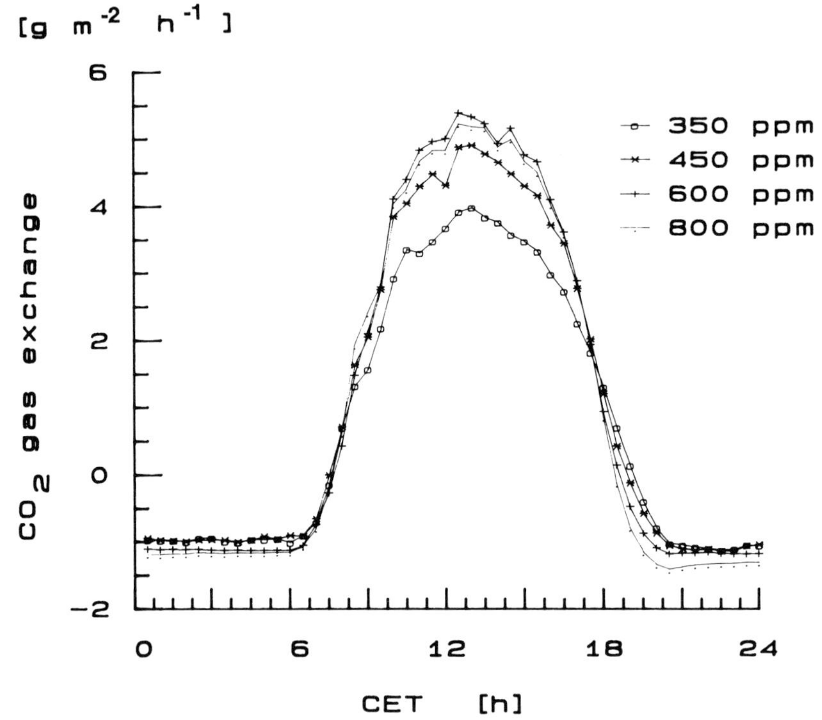

Figure 29.12: Mean daily courses of CO_2 gas exchange rates, on a ground area basis (m^2) of total stands of *Trifolium pratense* and *Festuca pratensis* (1:1) at four different CO_2 concentrations in mini–greenhouses climatized according to the outside conditions (means from Aug.–Sept. 1988, n = 35 days); data from FORSTREUTER (1991).

In order to answer the questions if and when such "model ecosystems" are a source or a sink for CO_2 at elevated concentrations the diurnal uptake rates were balanced against the nocturnal losses, and the daily differences (gain–loss), were summed over longer time periods. Fig. 29.14 shows the course of the summed differences from 1984 to 1986. The difference between the two compared mini–greenhouses (340 and 600 ppm CO_2) stayed positive until the end of the first year, i.e. the mixture of red clover and meadow fescue behaved like a sink for the additionally offered CO_2. Until the end of the second year 600 g more net CO_2 remained in the system with the elevated concentration. During the winter 1985/86, however, the additionally fixed CO_2 was completely returned to the atmosphere, and after the third vegetation period, in December only one–third of the level of the previous December was reached. First these results show that more CO_2 is taken

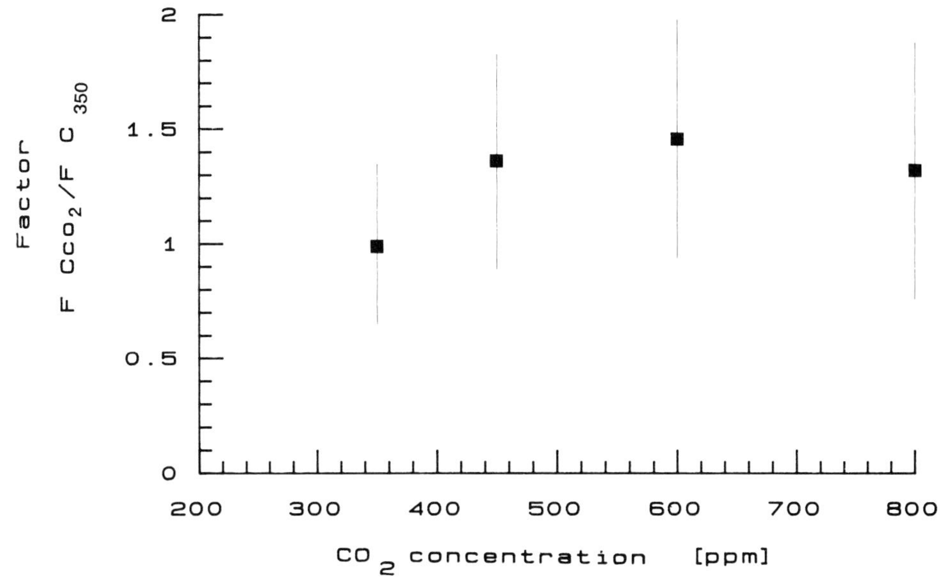

Figure 29.13: Mean factor for the CO_2 enrichment effect on CO_2 gas exchange relative to the control (350 ppm = 1.0) for total stands of *Trifolium pratense* and *Festuca pratensis* (1:1) at four different CO_2 concentrations in mini-greenhouses climatized according to the outside conditions (Aug.–Sept. 88: n = 35 days), reference basis: ground area [m^2]; data from FORSTREUTER (1991).

up at elevated CO_2 concentrations but that, in the long run, this additional uptake decreases. (During this long-term experiment the cultures were only cut once a year.)

This suggests that successionally young and herb-dominated vegetation can perform as sinks for additional CO_2 only to a rather limited extent. Although those systems form an important part of the biosphere in Central Europe, ecosystems dominated by woody plants are potentially more effective as sinks for the additionally emitted CO_2.

29.3.7 Mineral content

It could be shown for different herbaceous species that the Kjeldahl-nitrogen content (N) in percent decreases and the carbon/ nitrogen-ratio (Fig. 29.15) increases if the CO_2 concentration in the ambient air is elevated (OVERDIECK et al. 1986, OVERDIECK et al. 1988). If, for instance, the soybean is more productive under

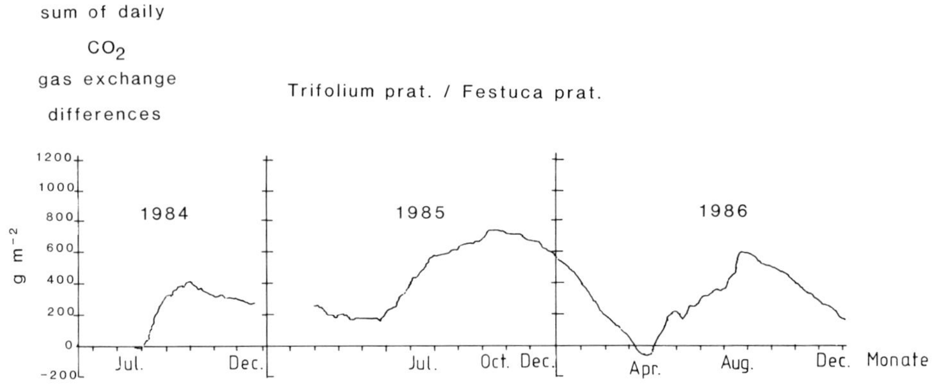

Figure 29.14: Differences in per–day balance of the CO_2 gas exchange rates summed for 3 vegetation periods of two "model ecosystems" with the same mixtures (1:1) of *Trifolium pratense* and *Festuca pratensis* at 340 and 600 ppm CO_2 grown in mini–greenhouses climatized according to the outside conditions; data from OVERDIECK (1991).

increased CO_2 in spite of a reduced N–pool in the substratum (SIONIT 1983), and if this is not due to N_2–fixing symbionts, it is only possible if the N–contents of their tissues can decrease within limits without important disturbances to metabolic processes. This seems to be not only true for N but also for other physiologically important minerals as potassium and others (OVERDIECK et al. 1990).

This phenomenon also occurs with woody plants (Fig. 29.16). In our experiments (520, 670 ppm CO_2) compared to the control (370 ppm CO_2), all mineral contents in percent and ppm, decreased in saplings of *Acer pseudoplatanus*. The same was found for *Fagus sylvatica*, with the exception of P and K. Only phosphorus seems to be an exception. In red clover, meadow fescue and mountain maple there was only a very slight decrease in P–concentration and in beech increased. Possibly plants are not very physiologically tolerant towards changes in P concentration.

On the other hand for elevated CO_2 concentrations whole plant communities appear to extract, always absolutely greater amounts of minerals from the soil per ground area unit because of increasing phytomass production, despite relatively slightly reduced mineral contents of their tissues in percent (for instance 10% less N and K at 600 ppm CO_2). Thus the statement of ZANGERL et al. (1984) is understandable, namely that the reaction of a whole plant community to CO_2

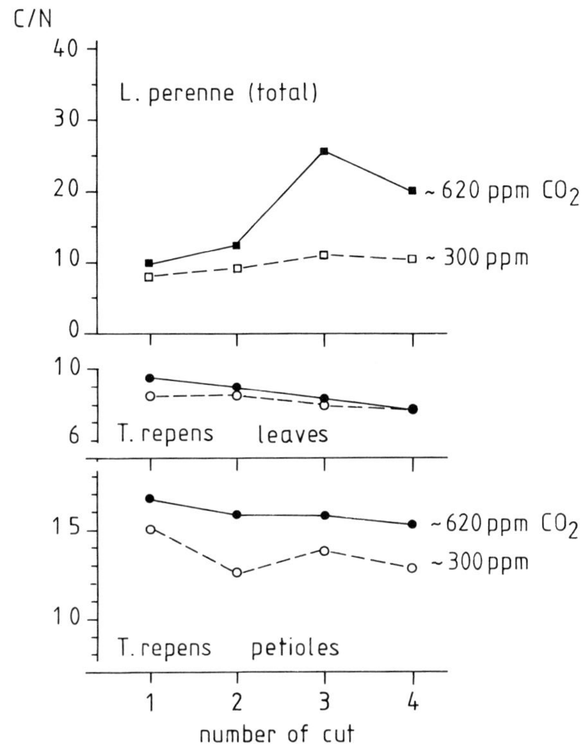

Figure 29.15: Carbon/nitrogen relationship of the dry mass (105 °C) of the above–ground parts of mixtures (1:1) of *Lolium perenne* and *Trifolium repens* at 300 (open symbols) and 620 ppm CO_2 (closed symbols) under field–like conditions (19–21 days between the cuts); data from OVERDIECK et al. (1986).

enrichment will not be affected — as much as expected — by the availability of minerals in the soil. In the case of the trees in our experiment, the basis for the calculation of average mineral uptake was the mean single sapling, not the whole stand (vegetation per m^2 ground area) as with the herbaceous plants. Both woody species accumulated absolutely more minerals per individual, however, on the average despite decreased concentrations.

From all the results it may be generally concluded that the enhancing effect on plant growth and production of increasing tropospheric CO_2 concentration will not be limited in soils of medium fertility because the mineral concentrations of the tissues can decrease at least to a certain degree without obvious negative results on growth. Two ecological consequences of these effects on mineral contents should be noted:

1. With decreasing mineral concentration the nutritive value of the food for

herbivores decreases.

2. For the same reason, microorganisms in the soil may decompose the litter more slowly and less effectively.

Since the absolute mineral amounts in the whole vegetation or in the single sapling are greater, it can be predicted that the flux rates of nutrients will increase in the biogeochemical cycles if the tropospheric CO_2 concentration continues to increase in the coming decades. This also means that more minerals will be taken up from the soil. Therefore nutrient–poor soils could become impoverished faster than before.

29.3.8 Water budget

Climatologists suppose that if the tropospheric CO_2 concentration continues to increase, the air layer close to the soil surface will warm up more and more. Thus global warming will cause higher actual evaporation and transpiration rates and drought stress will become more severe for natural vegetation and crops.

Higher evaporation and transpiration, on the other hand, would also increase the amounts of water involved in the atmospheric water cycle and the development of clouds would also be affected. More tropospheric clouds, however, would limit the temperature increase. So the water budget of plants and vegetation under increasing CO_2 concentrations warrants special scientific interest.

BIERHUIZEN et al. (1965) found, for cotton leaves which were exposed to up to 2000 ppm CO_2 in gas exchange cuvettes, that the relationship between the transpiration rate and the amount of carbohydrates produced decreases rapidly with increasing CO_2 concentration; i. e. under atmospheric CO_2 enrichment the water loss via transpiration is reduced whereas more phytomass is produced. ROSENBERG (1981) combined single results of mostly short–term CO_2 enrichment studies on herbaceous species (17 C_3, 8 C_4 species) in an overview which shows that the quotient of transpiration and CO_2 net assimilation rate decreases for C_3 as well as for C_4 plants under CO_2 enrichment (CO_2 concentration levels: 200–800 ppm).

Parallel measurements of the CO_2 gas exchange and transpiration of complete young individuals of the shrubby *Lyonia mariana* (Ericaceae) indicate (Fig. 29.17) that compared to the preindustrial CO_2 level of about 270 ppm three times as much CO_2 is fixed under 600 ppm CO_2 at the same water loss via transpiration (OVERDIECK 1989).

It was calculated from numerous publications (KIMBALL et al. 1983) that, on the total average, one has to expect a reduction of transpiration by 34% if today's ambient CO_2 concentration is doubled. Only about 8% of these studies — which they collected — were derived from woody plants, which play the most important role in the terrestrial water cycle.

The results so far available from autecological research can only be transferred to a very restricted extent to entire plant stands and plant communities in the field

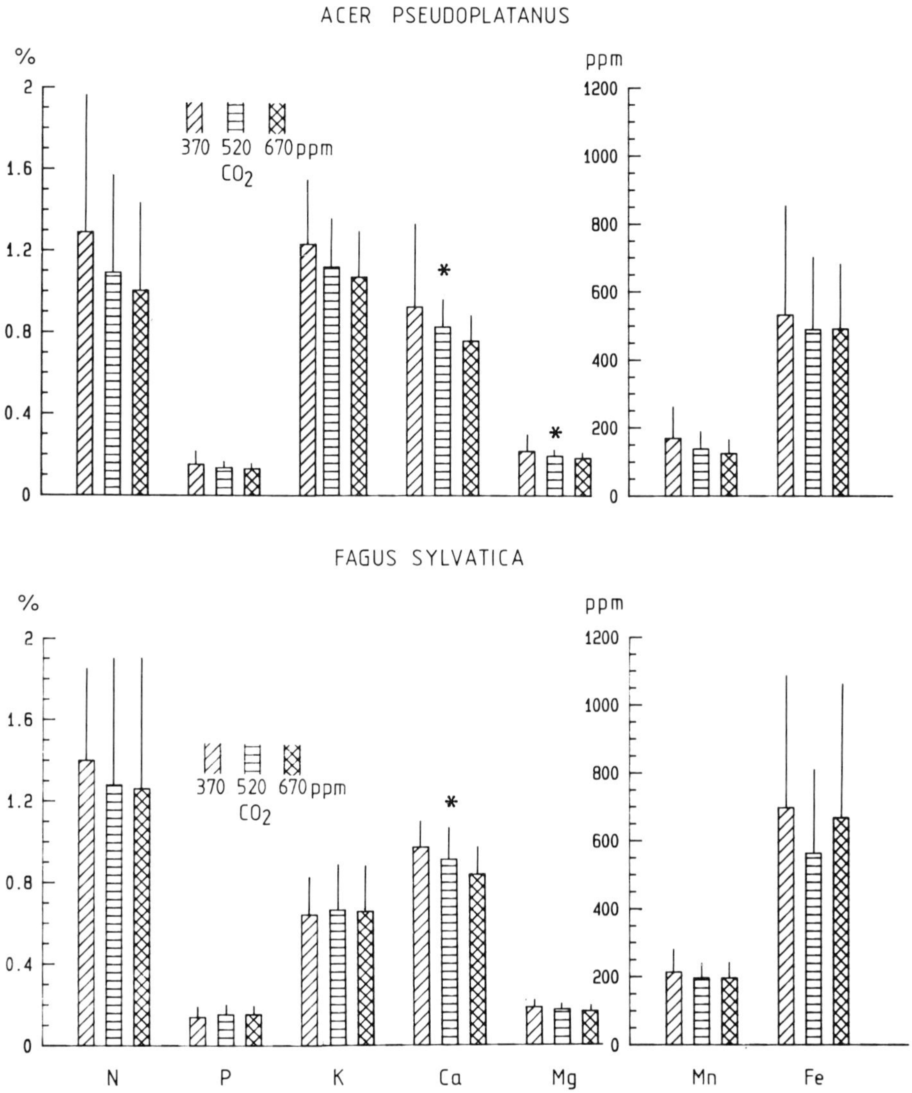

Figure 29.16: Mean total mineral concentrations (% and ppm) of the dry phytomass of mountain maple (*Acer pseudoplatanus*) and European beech (*Fagus sylvatica*) saplings after about two years of CO_2 enrichment under field-like conditions.

mountain maple: $n = 21$ harvests, total number of plants: 3000
soil volumes: 0.1–30 l/young tree; $* \ p < 0.05$;

European beech: $n = 12$ harvests, total number of plants: 1000
soil volumes: 0.1–30 l/young tree; $* \ p < 0.05$.

(Data from E. REINING, pers. comm., 1990).

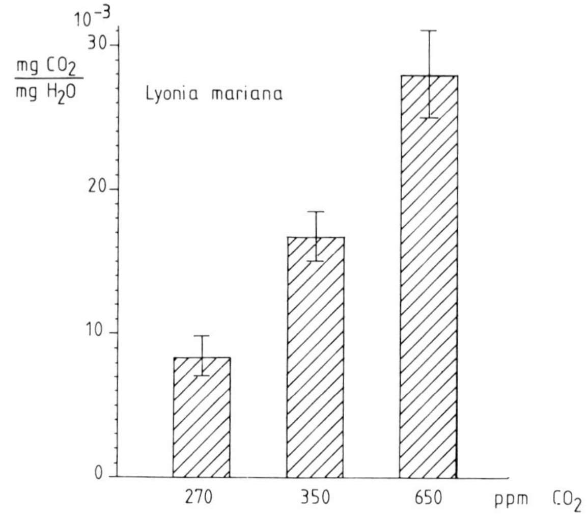

Figure 29.17: Mean water–use efficiency (mg CO_2/mg H_2O) at light saturation of the total above–ground parts of young *Lyonia mariana* shrubs after 2 months of exposure to different CO_2 levels at constant environmental conditions, data from OVERDIECK (1989).

because there the transpirating phytomass in the field increases under the influence of the atmospheric CO_2 enrichment and the free evaporation from the soil surface decreases. Also, greater amounts of litter accumulate which alter the water release from the soil particles.

In our experiment with "model ecosystems" composed of red clover, meadow fescue and natural, homogenized upper soil, the evapotranspiration at 450 ppm CO_2 was reduced by 10% as compared to the control (350 ppm). At higher CO_2 concentrations (600 and 800 ppm) the evapotranspiration rates reached the level of the control again. It can therefore be concluded that already at slight elevations of the CO_2 concentration plant stands will give less water to the atmosphere and that in this case the soil water supply would remain available for the vegetation over longer time spans, i.e. species could grow where it had been too dry previously (DOWNTON et al. 1980).

Because of the decreased evapotranspiration occurring together with increased CO_2 net fixation rates under CO_2 enrichment, the water use efficiency (WUE) of the total "model–ecosystem" increased maximally by 47% at 450 ppm, by 42% at 600 ppm and by 23% at 800 ppm (see Fig. 29.18). This corresponds approximately with the elevation of 24% found by JONES et al. (1985) for stands of soybeans at 800 ppm CO_2. At 600 ppm CO_2 monocultures of *Trifolium repens* and *Lolium perenne* showed a WUE increase of 35% or 25% (NIJS 1989).

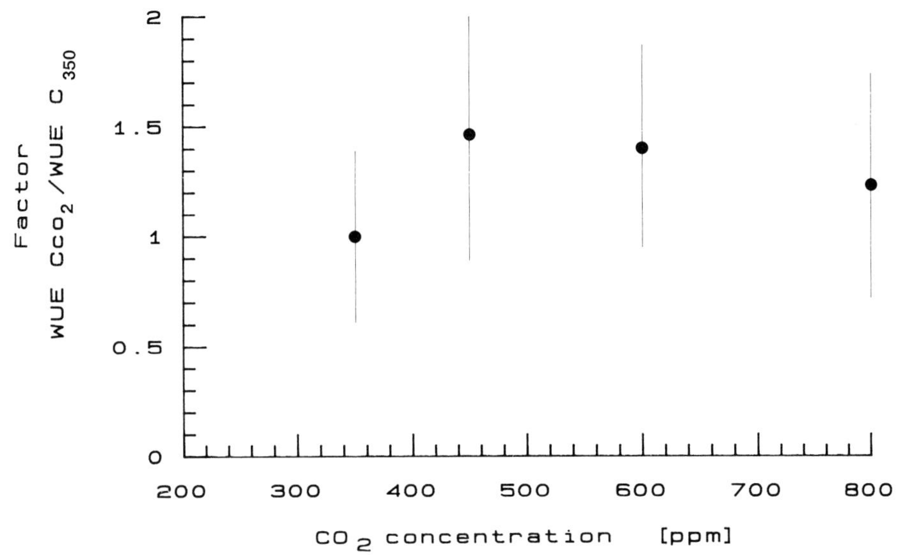

Figure 29.18: Mean factor for the CO_2 enrichment effect on water use efficiency (WUE) relative to the control (350 ppm: 1.0) for total stands of *Trifolium pratense* and *Festuca pratensis* (1:1) including the upper soil at four different CO_2 concentrations in mini–greenhouses climatized according to the outside conditions (Aug.–Sept. 1988, 35 days), reference basis: ground area [m^2]; data from FORSTREUTER (1991).

29.4 Modelling the CO_2 effect

In general, the CO_2 exchange rates of grassland vegetation in the field in response to changing microclimatic conditions and CO_2 concentrations can not be forecasted sufficiently enough. By reducing the CO_2 response in our small grassland "model ecosystems" in the experiments to simple two–dimensional functions of external parameters (saturation and maximum curves), the effects of microclimatic parameters could be partly quantified (OVERDIECK et al. 1985). However, that first modelling approach was not appropriate for the multifactorial field situation.

In a second approach we developed a new type of multi–dimensional model for grassland communities by modifying the modern mathematical technique of applying pseudo–cubic splines (BERLEKAMP et al. 1989) in order to have a method which is flexible enough to fit the real multi–factorial situation on the basis of the empirical data sets. Compared to the first method this attempt produced already realistic CO_2 response values with uncertainties at very high and low values of the external microclimatic parameters.

In a third attempt we chose the strategy of empirical modelling called "group method of data handling": GMDH after FARLOW (1984), in which regression functions are combined step-wise to polynomials of increasing complexity (IVAKHNENKO-polynomials). This third method is presented here (TAMURA et al. 1984).

The model was constructed on an IBM-compatible PC (AT) based on the 1986 data set from the experiment on *Trifolium pratense* and *Festuca pratensis* mixtures at 340 and 600 ppm CO_2.

In detail, the following parameters as daily means were monitored or calculated and used for building the model for the CO_2 exchange response of the "model ecosystems" [g CO_2 m^{-2} ground area d^{-1}]:

1. Photon flux density ($PhAR$) [$\mu E \cdot m^{-2} \cdot s^{-1}$]
2. Air temperature (T) [°C]
3. Relative air humidity ($r.\ H.$) [%]
4. Phytomass (m_{ph}) [$g \cdot m^{-2}$] dry matter
5. Leaf area index (LAI) —
6. Plant available water (PAW) [$kg \cdot m^{-2}$] H_2O
7. Phenological stage at a given date (based on SAUER 1978) (Phe) [1–15]

The actual number of available data was 2264. The data matrix was ordered in measured or calculated sets of empirical data in rows $z = 1 \ldots n$, and the dependent variable Y_z and the independent variable X_{zi} for $i = 1 \ldots m$ in the columns.

The following steps were repeated during the iterations:

1. The regression polynomials (29.2) were constructed (determination of the parameters) for each paired combination of the 1 ... m independent variables. The basic regression function is of the type:

$$Y(x_1, x_2) = ax_1^2 + bx_2^2 + cx_1 + dx_2 + ex_1x_2 + f \tag{29.2}$$

   with parameters a – f and variables x_1, x_2. Thus these

$$\binom{m}{2} = \frac{m\,(m-1)}{2} \tag{29.3}$$

   equations are estimations of the dependent variable.

2. The output values rY_{zr} of each of these regression surfaces were calculated from the independent variables $X_{zi} \ldots X_{zj}$ for $z = 1 \ldots n$ and different $i, j = 1 \ldots m$ and tested against the values of the dependent variable of the data set.

Therefore the total data set was used by means of the predicted residual sum of squares (PRESS) method (29.4).

$$PRESS = \sum_{z=1}^{n-1} (Y_z - rY_{zr})^2 \qquad (29.4)$$

for $r = 1 \ldots \binom{m}{2}$, with

Y_z value of the dependent variable of line z
rY_{zr} value of the regression r at the combination X_{zi} with X_{zj}

3. The output values of those equations which fitted the independent variable best, provided the new input variables (next generation). This variables replaced the variables of the previous generation. So the 1st generation variables were the observed variables. The variables of the following generations were calculated in the aboven given way.

Table 29.2: CO_2 response of red clover/meadow fescue "model ecosystems" at 340 and 600 ppm CO_2 predicted by multi-dimensional modelling (IVAKHNENKO–polynomialials) (after ZIMMERMANN, pers. comm., 1990).

Model Input

Parameter	PhAR	T	r. H.	m_{ph}	LAI	PAW	Phe
Unit	[$\mu E \cdot m^{-2} \cdot s^{-1}$]	[°C]	[%]	[$g \cdot m^{-2}$]	—	[$kg \cdot m^{-2}$]	—
Value	250	18	75	375	4	30	6

Predicted Response

	Value		Unit
CO_2–Conc.	340	600	[ppm]
Output	15.2	22.8	[$g \cdot m^{-2} \cdot d^{-1}$]

Utilizing each time the newly calculated variables the procedure 1...2 was repeated (3th, 4th generation and so on) as long as the value of the PRESS criterium (29.4) was decreased compared with the previous PRESS value. (If that value increases again the complexity of the model is getting to great).

Subsequently, regressions of the above given type (29.2) were connected to a complex polynomial of the general form:

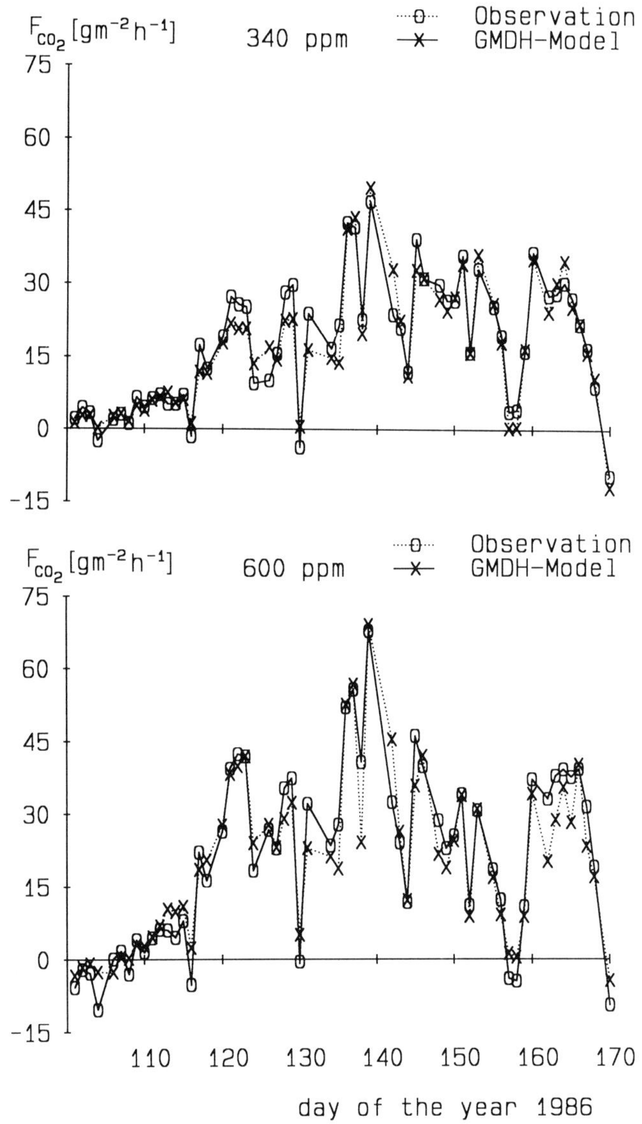

Figure 29.19: Original daily CO_2 gas exchange balances of red clover/meadow fescue "model ecosystems" at 340 and 600 ppm CO_2 circles and reconstructed values using the IVAKHNENKO modelling method (crosses). (Data from ZIMMERMANN, pers. comm., 1990).

$$Y_z = a + \sum_{i=1}^{m} b_i X_{zi} + \sum_{i=1}^{m}\sum_{j=1}^{m} c_{ij} X_{zi} X_{zj} + \sum_{i=1}^{m}\sum_{j=1}^{m}\sum_{k=1}^{m} d_{ijk} X_{zi} X_{zj} X_{zk} + \dots \quad (29.5)$$

(IVAKHNENKO–Polynomial) , with

line of the input set	z
model response value	Y_z
value of the input variables	$X_{zi},\ X_{zj},\ X_{zk} \dots$
stored parameters	$a,\ b_i,\ c_{ij},\ d_{ijk} \dots$

By means of the regression parameters which were stored during the run of the program (CO_2EXVEG., ZIMMERMANN 1990) the IVAKHNENKO polynomial could be determined after 7 iterations. Fig. 29.19 shows the course of the real CO_2 response data in comparison with the values derived from the modelling procedure for 340 and 660 ppm CO_2. In this case (1986) the mean deviation between the model and the empirical data amounted to 6–7%.

The two models permitted calculation of the CO_2 response of the "model ecosystems" as daily balances for every set of input data (7 parameters) within the span of measured or empirically determined values [Tab. 29.2]. The predicted difference in CO_2 response amounted to +50%.

Acknowledgment

We gratefully acknowledge A. Tschuschke and R. Stänner for much technical assistance and D. Rathmann and the electronic workshops at the University of Osnabrück for developing and constructing the electronic equipment of our experiments. The assistance of G. Postina with the data evaluation was appreciated. Some parts at the studies took place at the Duke University (North Carolina, USA) with the helpful support of Prof. Dr. B. R. Strain. We also are very grateful to Prof. Dr. E. Box from the University of Georgia (USA) and Prof. Dr. D. Wigham from the Smithsonian Institution (Maryland, USA) for spending so much time in improving the English essentially.

Our work was sponsored by the German "Umweltbundesamt", the "Bundesministerium für Forschung und Technologie", and the Commission of the European Communities.

Bibliography

AJTAY, G. L.; KETNER, P.; DUVIGNEAUD, P., 1979, Terrestrial primary production and phytomass.

In: Bolin B. (Hrsg.) The global carbon cycle. Wiley and Son, Chichester, New York, Brisbane, Toronto, 129–172.

BERLEKAMP, J.; OVERDIECK, D., 1989, Modelling the CO_2 gas exchange of grassland vegetation from experimental data.
Int. J. Biometeorol. **33**, 119–123.

BIERHUIZEN, J. F.; SLATYER, R. D., 1965, Effect of atmopheric concentration of water vapor and carbon dioxide in determining transpiration and photosynthesis relationship of cotton leaves.
Agricultural Meteorology **2**, 259–270.

BOTKIN, D. B., 1977, Forests, lakes and the anthropogenic production of carbon dioxide.
Bioscience **27**, 325–331.

BROECKER, W. S.; TAKAHASHI, T.; SIMPSON, H. J.; PENG, T.-H., 1979, Fate of fossil fuel carbon dioxide and the global carbon budget.
Science **206**, 409–418.

VON CAEMMERER, S.; FARQUHAR, G. D., 1981, Some relationships between the biochemistry of photosynthesis and the gas exchange of leaves.
Planta **153**, 376–387.

CAVE, G.; TOLLEY, L. C.; STRAIN, B. R., 1981, Effect of carbon dioxide enrichment and chlorophyll content, starch content and starch grain structure in *Trifolium subterraneum* leaves.
Physiologia Plantarum **51**, 171–174.

CURTIS, P. S.; DRAKE, B. G.; LEADLEY, P. W.; ARP, W. J.; WHIGHAM, D. F., 1989, Growth and senescence in plant communities exposed to elevated CO_2 concentrations on an estuarine marsh.
Oecologia (Berlin) **78**, 20–26.

DAHLMANN, R. C.; STRAIN, B. R.; ROGERS, H. H., 1985, Research on the response of vegetation to elevated atmospheric carbon dioxide.
Journal of Environmental Quality **14**, 1–8.

DAUNICHT, H. J., 1971, The CO_2 requirements at greenhouse crops.
Acta Horticulturae **17**, 62–71.

DAUNICHT, H. J., 1974, Control of plant growth and development by carbon dioxide concentration.
Acta Horticulturae **39**, 167–174.

DOWNTON, W. J.; BJORKMANN, O.; PIKE; C. S., 1980, Consequences of increased atmospheric concentrations of carbon dioxide for growth and photosynthesis of higher plants.

In: Pearman, G. I. (ed.) carbon dioxide and climate: Australian Research, Aust. Acad. Sci, Canberra, Australia, pp. 143–151.

DRAKE, B. G.; LEADLEY, P. W.; ARP, W. J.; NASSIRY, D.; CURTIS, P. S., 1989, An open top chamber for field studies of elevated atmospheric CO_2 concentration on salt marsh vegetation.
Functional Ecology **3**, 363–371.

ESSER, G., 1987, Sensitivity of global carbon pools and fluxes to human and potential climatic impacts.
Tellus **39 B**, 245–260.

FARLOW, ST. J., 1984, Self-organizing methods in modelling. GMDH type algorithmus.
Marcel Dekker, Inc. New York, Basel, pp. 1–23, 227–289.

FINN, G. A.; BRUN, W. A., 1982, Effect of atmospheric CO_2 enrichment on growth, nonstructural carbohydrate content and root nodule activity in soybean.
Plant Physiol. **69**, 327–331.

FORSTREUTER, M., 1991, Langzeitwirkungen der atmosphärischen CO_2–Anreicherung auf den Kohlenstoff– und Wasserhaushalt von Rotklee–Wiesenschwingelgemeinschaften.
Verh. Ges. Ökol. (Osnabrück 1989) **IXX** in print.

GAUDILLÈRE, J.–P.; MOUSSEAU, M., 1989, Short term effect of CO_2 enrichment on leaf development and gas exchange of young poplars (*Populus euramericana* cv I 214).
Acta Oecologica. Oecol. Plant. **10**, 95–105.

GODLEWSKI, E., 1873, Abhängigkeit der Sauerstoffausscheidung der Blätter von dem Kohlensäuregehalt der Luft.
Arbeiten des Botanischen Instituts in Würzburg, Leipzig, 343–370.

HAMPICKE, U., 1979, Man's impact on the earth's vegetation cover and its effects on carbon cycle and climate.
In: Bach, W. (ed.) Man's impact on climate. Elsevier, Amsterdam, 139–159.

HELDT, H. W.; FLÜGGE, U. I.; STITT, M., 1986, Kohlenhydratstoffwechsel der pflanzlichen Photosynthese.
Biologie in unserer Zeit **4**, 97–105.

HILBERT, D. W.; PRUDHOMME, T. I.; OECHEL, W. C., 1987, Response of tussock tundra to elevated carbon dioxide regimes, analysis of ecosystem CO_2 flux through nonlinear modeling.
Oecologia (Berlin) **72**, 466–472.

HOLLINGER, D. Y., 1987, Gas exchange and dry matter allocation responses to elevation of atmospheric CO_2 concentration in seedlings of three tree species.
Tree Physiology **3**, 193–202.

JONES, P.; ALLEN, L. H. JR.; JONES, J. W.; VALLE, R., 1985, Photosynthesis and transpiration responses of soybean canopies to short and long-term CO_2 treatment.
Agr. J. **77**, 119–126.

KEELING, C. D., 1986, Atmospheric CO_2 concentrations — Mauna Loa Observatory, Hawaii 1958–1986.
Prepared by T. A. Boden. U. S. Department of Energy (Oak Ridge National Laboratory. Environmental Science Division), Publication No. 2798.

KIMBALL, B. A., 1983a, Carbon dioxide and agricultural yield: An assemblage and analysis of 430 prior observations.
Agronomy Journal **75**, 779–788.

KIMBALL, B. A., 1983b, Carbon dioxide and agricultural yield: An assemblage and analysis of 770 prior observations.
WCL Report **14**, U. S. Dept. of Agric., Agric. Res. Serv., 71 ff.

KIMBALL, B. A.; IDSO, S. B., 1983, Increasing atmospheric CO_2: Effects on crop yield, water use and climate.
Agricultural Water Management **7**, 55–72.

KÖRNER, CH., 1988; Does global increase of CO_2 alter stomatal density?
Flora **181**, 253–257.

KREUSLER, U., 1885, Über eine Methode zur Beobachtung der Assimilation der Pflanzen und über einige diese Vorgänge beeinflussende Momente.
In: Thiel, H. (ed.) Landwirtschaftl. Jahrbücher **14**, 913–965.

MADSEN, E., 1968, Effect of CO_2 concentration on accumulation of starch and sugar in tomato leaves.
Physiologia Plantarum **21**, 168–175.

MAUNEY, J. R.; GUINN, G.; FRY, K. E.; HESKETH, J. D., 1979, Correlation of photosynthetic carbon dioxide uptake and carbohydrate accumulation in cotton, soybean, sunflower and sorghum.
Photosynthetica **13**, 260–266.

MYTTON, L. R., 1973, The effects of seed weight on the early growth and nodulation of white clover.
Ann. appl. Biol. **73**, 329–338.

NEFTEL, A.; MOOR, E.; OESCHGER, H.; STAUFFER, B., 1985, Evidence from polarice cores for the increase in atmospheric CO_2 in the past two centuries. Nature **315**, 45–47.

NIJS, I., 1989, De lange-termign impact van een verhoogh CO_2-gehalte in de atmosfeer planten.
Dissertation, Departement Biologie, Universiteit Antwerpen.

OVERDIECK, D., 1986, Long-term effects of an increased CO_2 concentration on terrestrial plants in model ecosystems. Morphology and reproduction of *Trifolium repens* L. and *Lolium perenne* L.
Int. J. Biometeor. **30**, 323–332.

OVERDIECK, D., 1989, The effects of preindustrial and predicted future atmospheric CO_2 concentration on *Lyonia mariana* L. D. DON.
Functional Ecology **3**, 569–576.

OVERDIECK, D., 1990, Direct effects of elevated CO_2 concentration levels on grass and clover in model ecosystems.
In: Beukema, J. (ed.) Expected effects of climatic change on marine coastal ecosystems, 41–47.

OVERDIECK, D., 1991, Direkte Wirkungen der atmosphärischen CO_2-Anreicherung auf die einheimische Vegetation (Pflanzengemeinschaften des Grünlandes und der Wälder).
Verh. Ges. Ökol. (Osnabrück 1989) **IXX**, in print.

OVERDIECK, D.; BOSSEMEYER, D., 1985, Langzeit-Effekte eines erhöhten CO_2-Angebotes auf den CO_2-Gaswechsel eines Modell-Ökosystems.
Angewandte Botanik **59**, 179–198.

OVERDIECK, D.; FORSTREUTER, M., 1987, Langzeit-Effekte eines erhöhten CO_2-Angebotes bei Rotklee-Wiesenschwingelgemeinschaften.
Verh. Ges. Ökol. (Gießen 1986) **XVI**, 197–206.

OVERDIECK, D.; IKELS, P., 1991, Mineralstoffe in Rotklee und Wiesenschwingel bei atmosphärischer CO_2-Anreicherung (*Trifolium pratense* L. und *Festuca pratensis* HUDS.).
Verh. Ges. Ökol. (Osnabrück 1989) **IXX**, in print

OVERDIECK, D.; REID, CH.; STRAIN, B. R., 1988, The effects of preindustrial and future CO_2 concentrations on growth, dry matter production and the C/N-relationship in plants at low nutrient supply: *Vigna unguiculata* (cowpea), *Abelmoschus esculentus* (okra) und *Raphanus sativus* (radish).
Angewandte Botanik **62**, 119–134.

OVERDIECK, D.; REINING, F., 1986, Effect of atmospheric CO_2 enrichment on perennial ryegrass (*Lolium perenne* L.) and white clover (*Trifolium repens* L.) competing in managed model ecosystems. I.– Phytomass production.
Acta Oecologica, Oecol. Plant. **7**, 357–366.

OVERDIECK, D., REINING, E., 1986; Effect of atmospheric CO_2 enrichment on perennial ryegrass (*Lolium perenne* L.) and white clover (*Trifolium repens* L.) competing in managed model-ecosystems. II.– Nutrient uptake.
Acta Oecologica, Oecol. Plant. **7**, 367–378.

OVERDIECK, D.; UNGEMACH, E., 1989, Wirkungen der atmosphärischen Kohlendioxid-Anreicherung auf die Blattanatomie des Weißklees (*Trifolium repens* L.)
Verh. Ges. Ökol. (Essen 1988) **XVIII**, 431–436.

REINING, F., 1990, Langzeiteffekte von erhöhtem CO_2-Angebot auf das Wachstum von *Acer pseudoplatanus* und *Fagus sylvatica*.
Dissertation, Osnabrück.

ROGERS, H. H.; BINGHAM, G. E.; CURE, J. D.; SMITH, J. M.; SURANO, K. A., 1983, Responses of selected plant species to elevated carbon dioxide in the field.
Journal of Environmental Quality **12**, 569–574.

ROSENBERG, N. J., 1981, The increasing CO_2 concentration in the atmosphere and its implication on agricultural productivity. I. Effects on photosynthesis, transpiration and water use efficiency.
Climatic Change **3**, 265–281.

SASEK, T. W., DE LUCIA, E. H.; STRAIN, B. R., 1985, Reversibility of photosynthetic inhibition in cotton after long-term exposure to elevated CO_2 concentrations.
Plant Physiology **78**, 619–622.

SAUER, R. H., 1978, A simulation model for grassland primary producer phenology and biomass dynamics.
In: Innis, G. S. (ed.), Grassland simulation model. Springer, New York, Berlin, Heidelberg, pp. 55–87.

SEILER, W.; CRUTZEN, P. J., 1980, Estimates of gross and net fluxes of carbon between the biosphere and the atmosphere from biomass burning.
Climate Change **2(3)**, 225–247.

SIEGENTHALER, U.; OESCHGER, H., 1980, Prediction of future CO_2 concentrations in the atmosphere.
Experientia **36**, 783–786.

SIONIT, N., 1983, Response of soybean to two levels of mineral nutrition in CO_2-enriched atmosphere.
Crop Science **23**, 329–333.

SOLOMON, A. M.; TRABALKA, J. R.; REICHLE, D. E.; VOORHEES, L. D., 1985, 1. The global cycle of carbon.
In: J. R. Trabalka (ed.) Atmospheric carbon dioxide and the global carbon cycle. United States Department of Energy, DOE/ER-0239, p. 5. Oak Ridge National Laboratory, Oak Ridge (USA).

TAMURA, H.; KONDO, T., 1984, On revised algorithmus of GMDH with applications.
In: Farlow, St. J. (ed.) Self-organizing methods in modeling. GMDH type algorithmus. Marcel Dekker Inc., New York, Basel, pp. 225–241.

THOMAS, J. F.; HARVEY, CH. N., 1983, Leaf anatomy of four species grown under continuous CO_2 enrichment.
Bot. Gaz. **144**, 303–309.

TOLLEY, L. C.; STRAIN, B. R., 1984, Effects of CO_2 enrichment on growth of *Liquidambar styraciflua* and *Pinus taeda* seedlings under different irradiance levels.
Can. J. For. Res. **14**, 343–350.

TRABALKA, J. R. (ed.), 1985, Atmospheric carbon dioxide and the global carbon cycle.
Oak Ridge National Laboratory, US DOE/ER-0239, 315 p.

WILLIAMS, W. E.; GARBUTT, K.; BAZZAZ; F. A.; VITOUSEK, P. M., 1986, The response of plants to elevated CO_2. IV. Two deciduous-forest tree communities.
Oecologia (Berlin) **69**, 454–459.

WONG, S. C., 1979, Elevated atmospheric partial pressure of CO_2 and plant growth.
Oecologia (Berl.) **44**, 68–74.

WOODWARD, F. I., 1986, Ecophysiological studies on the shrub *Vaccinium myrtillus* L. taken from a wide altitudinal range.
Oecologia **70**, 580–586.

WOODWARD, F. I., 1987, Stomatal numbers are sensitive to increase in CO_2 from preindustrial levels.
Nature **327**, 617–618.

ZANGERL, A. R.; BAZZAZ, F. A., 1984, The response of plants to elevated CO_2. II. Competitive interactions among annual plants under varying light and nutrients.
Oecologia **62**, 412–417.

Chapter 30

Tropical grasslands and their role in the global carbon cycle

D. O. HALL and J. M. O. SCURLOCK

30.1 The importance of tropical grasslands

More extensive practical work is required on the long–term primary productivity of natural grasslands, particularly those in the tropics, and on the role these ecosystems play in the global carbon cycle. Grasslands are important as a habitat to many people and wildlife, and their expansion by deforestation is a continuing phenomenon. We need to understand the dynamics of both temperate and tropical grasslands in relation to carbon cycling.

The atmosphere, biosphere and ocean surface layers contain approximately equal amounts of carbon, and the flux of carbon between these pools is dominated by the process of photosynthetic production, the majority of which is terrestrial. Global net primary production results in the fixation of 20 times as much carbon as is released annually by fossil fuel burning (HALL 1989), and the annual carbon flux through the biosphere is about one–eighth of total atmospheric carbon (Table 30.1).

Various authors estimate that tropical grasslands occupy about 15 million km^2

Table 30.1: Carbon fluxes and sinks — a synthesis of data (HALL 1989). 1 Pg = 10^{15} g = 1 Gt.

CO_2 in atmosphere (1986)	346 ppm (730 Pg carbon)
CO_2 in atmosphere (present annual increase)	1.5 ppm (3.2 Pg carbon)
CO_2 in atmosphere (preindustrial)	272 ppm (575 Pg carbon)
CO_2 in atmosphere (change from preindustrial to 1986)	74 ppm (27%; 155 Pg carbon remained in atmosphere)
CO_2 fossil fuel emissions (1986)	5.3 Pg carbon per year
CO_2 fossil fuel emissions (cumulative to date)	183 ± 15 Pg carbon
CO_2 biospheric emissions (present)	1.8 ± 0.9 (1.0–2.6) Pg carbon per year
CO_2 biospheric emissions (cumulative to date)	150 ± 50 Pg carbon
Fossil fuels carbon content (reserves)	770 Pg carbon
Fossil fuels carbon content (resources)	6500 Pg carbon
Biomass (terrestrial; 80% in trees)	560 Pg carbon
Biomass (aquatic)	3 Pg carbon
Gross primary production (terrestrial)	120 Pg carbon per year
Net primary production (terrestrial)	60 Pg carbon per year
Net primary production (aquatic)	46 Pg carbon per year
Net primary production (total)	106 Pg carbon per year
Soil organic carbon content	1515 Pg carbon
Ocean surface (75 m) carbon content	725 Pg carbon
Ocean intermediate and deep carbon content	38,000 Pg carbon
CO_2 uptake by ocean	1.9 Pg carbon per year
CO_2 uptake by freshwater	0.8 Pg carbon per year

and are nearly equal to tropical forests in terms of both land area and productivity, although the tropical forests contain 7–10 times as much standing biomass as tropical grasslands (Table 30.2). On the African continent in particular, tropical grasslands are much greater in extent than forests, and will play a most important role in regional reponses of the biosphere to climate change, yet they have received relatively little attention to date (HALL and ROSILLO-CALLE 1990). Our studies of net primary production in four contrasting tropical grassland communities show that when full account is taken of turnover, both above-ground and below-ground, these ecosystems are far more productive than previously appreciated (LONG et al. 1989). Net primary production at three of these grassland sites was up to five times higher than obtained using the standard IBP (International Biological Programme) procedure (Table 30.3). These findings have wide implications for the prediction of global carbon cycling and grassland ecosystem responses to climate changes.

Recent estimates of widespread burning of tropical grasslands suggest that gross emissions of carbon from this source are substantial (HAO et al. 1989). Frequent

Table 30.2: Selected world ecosystems; estimated annual net primary production, standing biomass and area. After WOODWELL (1986).

	Tropical forests	Tropical woodl.	Tropical grassl.	Total terrest.	Total marine	Total world
	Net primary production [Gt·a^{-1} dry matter]					
AJTAY et al. (1987)	30	13	26	133	44	177
OLSON et al. (1983)	21	19	16	134	–	–
	Standing biomass [Gt dry matter]					
AJTAY et al. (1987)	541	88	58	1244	–	–
OLSON et al. (1983)	364	177	50	1242	7	1249
	Area [10^6 km^2]					
AJTAY et al. (1987)	15	10	12	149	–	–
OLSON et al. (1983)	13	17	17	151	360	511

N. B. Many authorities still fail to agree upon the major categories of world ecosystems and their respective areas. For example, the area of "tropical forests" cited here is larger than recent estimates of about 8–10 × 10^6 km^2 for "tropical moist forests". Likewise the distinction between "tropical woodlands", "tropical savannas" and "tropical grasslands" is not always clear.

burning may result in a net loss of carbon from the soil, and volatilisation of soil nitrogen may have feedback effects through primary productivity. Grassland soils represent an extensive and very large carbon pool, which may behave as a net source or sink for atmospheric carbon.

The increase in CO_2 in the atmosphere (27% to 1987, with half this rise occurring since the mid–1950s) has been due to historic changes in land use, mostly deforestation, and the recent rapid rise in the use of fossil fuels. The net atmospheric gain is about 3 Gt carbon per annum (HOUGHTON and WOODWELL 1989). Combined with the effect of the other anthropogenic "greenhouse gases", this is now considered very likely to cause significant climate changes in the next 50 years. These changes will be manifest as a rise in global mean temperature, together with perturbation of rainfall patterns. Other modifications to the environment may follow in the form of nutrient availability and the rate of cycling of the major carbon and nitrogen pools in the biosphere, geosphere and atmosphere. It should be emphasised that the actual rate of change over the coming years will be crucial in

Table 30.3: Net primary production; a comparison of the estimates obtained by taking account of losses through death and below-ground production, with estimates from biomass change alone (LONG et al. 1989).

Net primary production [$g \cdot m^{-2} \cdot a^{-1}$]	Montecillos Mexico	Nairobi National Park Kenya	Klong Hoi Kong Thailand	Marchantaria Manaus Brazil
1) Accounting for mortality (including below-ground organs)	1741	1242	2220	9925
2) Accounting for mortality (above ground only)	1063 (39%)*	811 (35%)	1595 (28%)	9425 (5%)
3) IBP standard method (including below-ground organs)	740 (56%)	663 (47%)	570 (74%)	8680 (18%)
4) Maximum biomass (above-ground only)	430 (75%)	337 (75%)	442 (80%)	6300 (37%)

*Figures in parentheses indicate underestimation of productivity as a percentage of net primary production over the 12 months calculated as: 1) the sum of all changes in biomass and losses due to mortality (for both above and below ground); 2) as for 1), but for above-ground parts of vegetation only; 3) the sum of all positive changes in above- and below-ground biomass; 4) the peak biomass above-ground during the year.

determining whether ecosystems are able to adapt to climate changes.

However, the degree of uncertainty in estimating present pool sizes and rates of cycling of elements makes it hard to predict future changes for the carbon cycle alone (HALL and SCURLOCK 1990), let alone nitrogen or the other important elements of the biosphere. Estimates of the distribution of soil moisture or rainfall, long-term response of vegetation to elevated CO_2 concentration (CO_2 fertilisation), and even the present rate of carbon cycling through vegetation (primary production), are all subject to significant uncertainties. However, any attempt to stabilise the CO_2 levels in the atmosphere must recognise and attempt to quantify the CO_2 uptake, emissions and storage mechanisms in the vegetation, soils and oceans. Important decisions are already being made based on a poor understanding of these processes.

30.2 Accounting for carbon flows

We are confronted with the possibility that it may be necessary to limit the increase in atmospheric CO_2 to avoid adverse changes in the global climate. Information on carbon flows may be most important in the future for negotation of measures to limit national emissions of greenhouse gases. If "carbon taxes" are to be levied in the future, it is essential to know in advance whether countries or regions are net emitters or absorbers of CO_2.

In order to carry out a "carbon audit", whether at regional, national or local level, it is clear that we cannot concentrate on forests and fossil fuels alone. Careful analysis of carbon flows must include the role of grasslands and note any effect on the below–ground carbon pools, which are themselves influenced by grassland burning and management, and by changes in temperature and moisture content.

TANS et al. (1990) have suggested that the majority of the "missing sink" for carbon in the global CO_2 balance may be accounted for by terrestrial vegetation rather than by oceanic uptake. It is therefore important to include all major terrestrial ecosystems in attempts to model the global carbon cycle.

Our own rough estimates of carbon flows for the African continent (Tables 30.4, 30.5) show that the CO_2 emissions from deforestation (net), fossil fuel burning and burning of biomass fuels are roughly equivalent at about 0.2 Gt C per annum. These figures are of limited accuracy because estimates of woody biomass were determined by remote sensing for sub–Saharan Africa only, with limited "ground truth" work (ETC 1990). However, burning of African grasslands is estimated to result in gross CO_2 emissions of 1.1–1.4 Gt C per annum: the net emissions are unknown, depending upon the frequency and extent of overgrazing and erosion which produces loss of carbon from the soil. Whilst a true "carbon audit" on a continental scale would need to contain much more detail about local variations, we believe that these figures demonstrate the need for comprehensive accounting of all carbon flows.

These conclusions also have a bearing on the application of new information technologies such as satellite remote sensing and construction of environmental data bases. Although still in their infancy, these techniques have concentrated their studies of vegetation to date mainly on agriculture and forests. It is clear that monitoring of the dynamics of grasslands and soils is also required if a more comprehensive estimate of carbon fluxes is to be obtained.

30.3 Response to changes in atmospheric carbon dioxide

Most of the work to date on the response of plant productivity to increased CO_2 concentration has been conducted on agricultural species under laboratory, greenhouse or controlled field conditions (FORD and THORNE 1967; BAZZAZ and GAR-

Table 30.4: Biomass (live and dead vegetation) in African savannah grasslands and CO_2 emissions from burning (HALL and ROSILLO–CALLE 1990).

Nairobi National Park (UNEP Project Study Site):		
April 1985 (early long rains)	Maximum total dead	665 $g \cdot m^{-2}$
	live	180 $g \cdot m^{-2}$
	Total live + dead	845 $g \cdot m^{-2}$
Sept 1985 (end of dry season)	Maximum total dead	520 $g \cdot m^{-2}$
	live	338 $g \cdot m^{-2}$
	Total live + dead	858 $g \cdot m^{-2}$

Grassland Burning:

	UNEP Project	HAO et al. (1989)
Typical total above–ground vegetation dry matter	850 $g \cdot m^{-2}$	660 $g \cdot m^{-2}$
Area of African tropical grasslands (including savannah grasslands with trees)	5.91×10^6 km^2	5.91×10^6 km^2
Annual amount of African grassland dry matter burned (assuming 75% burned annually; 83% of matter is actually combusted)	3.13 Gt	2.43 Gt
Gross annual emissions of carbon from African grassland burning (assuming 45% of dry matter is carbon)	1.41 Gt C	1.09 Gt C

BUTT 1988; HENDREY et al. 1988). As well as direct stimulation of leaf photosynthetic rate, with reduced photorespiration and transpiration, there are secondary effects such as altered concentration, composition and allocation of photosynthates, modifed growth form and reproduction, changed water status and water use efficiency, and modified tolerance to gaseous atmospheric pollutants (STRAIN and

Table 30.5: CO_2 Flows in Africa: Energy use, forests and grasslands — a rough summary (HALL and ROSILLO–CALLE 1990).

		$g \cdot m^{-2} \cdot a^{-1}$	$Gt \cdot a^{-1}$ C
ENERGY			
Emissions from fossil fuels in 1988	(BP 1989)	–	0.18
Emissions from biomass fuels in 1988	(SCURLOCK and HALL 1990)	–	0.22
FORESTS			
Gross emissions	(MYERS 1989)		0.23
from burning	(HOUGHTON 1990)	–	0.21*
	(HAO et al. 1989)	-	0.18
Net primary production	(LIETH 1978; AJTAY et al. 1987)	1,600–2,200	2.1
Mean annual increment as wood	(ETC 1990)	–	1.4
GRASSLANDS			
Gross emissions	(HAO et al. 1989)	–	1.09
from burning	(LONG et al. 1989)	–	1.41
Net primary	(LIETH 1978)	700–900	1.86–2.39
production	(LONG et al. 1989)	1,200	3.19
Net exchange of carbon		–	?

*closed forests only

N. B.

area of African forests	2.67×10^6 km^2 (LANLY 1982, *sensu stricto*)
	1.37×10^6 km^2 (ETC 1990, Class 9)
standing biomass	12.2 Gt (equivalent to 8,800 $g \cdot m^{-2}$, ETC 1990)
area of African grasslands	5.91×10^6 km^2 (HAO et al. 1989)
	6.67×10^6 km^2 (ETC 1990, Classes 1+2)
standing biomass	4.5 Gt (6×10^6 km^2 with 750 $g \cdot m^{-2}$, based on LONG et al. 1989; HAO et al. 1989)

CURE 1985; KOHLMAIER et al. 1987). Plant community structure may also be affected, especially with respect to competition between C_3 and C_4 ecotypes (BAZZAZ and GARBUTT 1988).

It is difficult to extrapolate from such knowledge and experience to predict the response of unmanaged plant communities, although a limited amount of work has now been carried out on temperate grass communities in the USA (CURTIS et al. 1989, KANEMASU et al. 1989). These early findings suggest that C_3–type annual grasses experience enhanced growth with increased shoot densities and delayed senescence, leading to increased primary production. Root biomass of a C_3 grass showed a marked increase under elevated CO_2 (CURTIS et al., in press). Little or no response was detected in C_4 grasses, although the consequences of these results for C_3/C_4 competition must await more long-term study. The response of unmanaged ecosystems to CO_2 increase may have a significant effect on net carbon storage, depending upon whether increased productivity is sustained and what accompanying changes occur in decomposition rates (CURTIS et al. 1989).

Since many of the grassland species in the tropics and subtropics are perennials of the C_4 photosynthetic type, with seasonality of growth determined more by periods of rainfall than by marked annual changes in monthly temperature, there is an immediate need to determine their response to elevated CO_2, in terms of both growth rate and length of growing season. Techniques such as the use of open–top field chambers (DRAKE et al. 1989) may be put into practice in tropical grasslands, albeit with modifications to suit local climate, level of expertise and backup services.

GIFFORD (1980) estimated that CO_2 fertilisation may already be responsible for a net increase in carbon storage of 1.0 Gt per annum relative to 1958 CO_2 levels. In his model, tropical grasslands were considered to play a more important role than tropical forests. He also emphasised that the CO_2 response of leaf photosynthetic rate does not necessarily give an indication of the long term response of plant communities, where feedback through other environmental variables may have a greater effect.

The most important changes to the biosphere will involve shifts in the balance of gross production (total photosynthesis) and total respiration on a local, regional and global level (WOODWELL (in press)). This has also been emphasised by Gifford (GIFFORD (in press)), who notes that dark respiration may account for more than 50% of gross primary production, depending upon the type of ecosystem. Elevated CO_2 has been found to decrease respiration rates in some plants, which may cancel out the temperature effect discussed below. However, other studies suggest that respiration rates may actually increase; this discrepancy has not yet been resolved (GIFFORD (in press)).

The interaction between plant and soil respiration, CO_2 and temperature will have a significant impact upon net primary production. Since the flow of carbon through the biosphere is so large, it is likely that only a 1% global change in either photosynthesis or respiration will rapidly produce an effect on the atmospheric CO_2 concentration. Changes in land use are accompanied by changes in the amount of carbon held in both the vegetation and the soil, but these fluxes of carbon may be modified by feedback through the differential effect of climate change on respiration and photosynthesis. Although most work has concentrated on carbon fluxes between vegetation and the atmosphere, the effect of climate change on long-term storage of carbon in soils (which contain three times as much carbon as the vegetation) could potentially dwarf the fluxes of carbon considered so far (HOUGHTON (in press)).

30.4 Effect of changes in other environmental factors

Under future climate change, carbon storage and release in terrestrial ecosystems may be subject to major feedback phenomena via the stimulation of biospheric respiration by temperature. It is thought by Woodwell (WOODWELL (in press)) that a 1 °C rise in temperature may cause a 10–30% increase in respiration rate, which is unlikely to be matched by an equivalent increase in photosynthesis. However, in grasslands, where soil organic matter (SOM) is reasonably well-stabilised compared with forests, the effect of temperature and other physical factors on litter decomposition rate and SOM is considered relatively predictable compared with other ecosystems (ANDERSON (in press)). Nevertheless, changes in susceptibility to burning (more frequent fires in a hotter, drier climate) and the resultant effect on nitrogen availability may well provide important feedback mechanisms (see also page 669).

Water availability may also be significant, although there is still uncertainty over how patterns of rainfall will be affected by climate change. Both the total availability of soil water and the seasonality of water stress may affect grassland productivity, although reduced water availability may be partly mitigated by increased atmospheric CO_2 concentration (GIFFORD (in press)). Water stress is known to affect leaf growth before it limits photosynthetic rate, leading to preferential carbon allocation to roots, but its overall effect on shoot/root carbon allocation is poorly understood and can only be modelled empirically at present (LUDLOW (in press)).

The IUBS/UNESCO programme on Response of Savannas to Stress and Disturbance (RSSD) recognises soil moisture and soil nutrients as the key factors affecting primary productivity and plant quality in tropical savanna grasslands (FROST et al. 1986), whilst the IUBS/UNESCO programme on Tropical Soils Biology and Fertility (TSBF) has highlighted decomposition and root turnover as major factors governing soil fertility (SWIFT 1987). The efficiency of nutrient transfer between litter, soil and plants is considered to be much higher in natural ecosystems than in agriculture, with the decomposition of below-ground litter providing the most direct source of nutrients for new growth (SWIFT 1987). An increase in the rate of carbon input to the soil through root turnover and root exudation is likely to lead to an increase in activity of soil microorganisms, although a significant increase in soil organic matter content is less likely due to enhanced organic matter decomposition (VAN VEEN (in press)).

Despite these efforts to understand the response of natural grasslands to many aspects of climate change, and the existence of models which describe the dynamics of carbon and major nutrients in grassland soils (PARTON et al. 1988), it has been concluded that there is barely enough quantitative information to develop and validate models of the response of grassland productivity and carbon fluxes to climate change (GIFFORD (in press)). The next section of this review describes some of our own collaborative studies which attempt to address this problem.

30.5 The UNEP grassland productivity project

Estimates of grassland productivity worldwide come primarily from the extensive studies of the IBP (International Biological Programme) (COOPER 1975), although the results of the studies described here suggest that the IBP figures may be serious underestimates (LONG et al. 1989). Net primary production is the total photosynthetic gain by plant matter, less respiratory losses, expressed in terms of unit ground area (LIETH 1978; LINTHURST and REIMOLD 1978; ROBERTS et al. 1985). Over any one period, this must equal the change in live plant biomass plus any losses through death and decomposition, both above-ground and below-ground (ROBERTS et al. 1985).

The UNEP (UN Environment Programme) Project on Primary Productivity and Photosynthesis (1983–1988) concentrated on five contrasting grass ecosystems in different parts of the tropics and sub-tropics. Three terrestrial grasslands were studied; a dry savannah grassland within Nairobi National Park, Kenya; a semi-natural humid grassland in Southern Thailand; and a saline grassland near Mexico City. Additional studies were carried out on a stand of periodically-inundated emergent macrophyte vegetation in the River Amazon near Manaus, Brazil; and on a stand of sub-tropical tall bamboo in Zhejiang province, China.

In the terrestrial grasslands, monthly measurements were made of changes in live biomass and dead matter, above-ground and below-ground. Decomposition losses were estimated monthly using a litter bag technique. For the inundated site in Brazil and the tall bamboo in China, a revised methodology was used, but with the same intention of accounting for total turnover, above and below ground.

Results from the three terrestrial grassland sites suggest that their true productivity is up to five times higher than would be obtained using the standard IBP procedure, which measures positive increments in above-ground live biomass only (Table 30.3). However, the UNEP Project did not take into account carbon losses due to root exudation or to any pest attack between the sampling intervals; thus the NPP calculated for these tropical grasslands may be still further underestimated by about 25%. The emergent macrophyte grass in Brazil achieved a productivity of 99 t ha^{-1} $year^{-1}$, which approaches the maximum recorded for agricultural crops (LONG et al. 1989).

Thus the true input of organic matter to the other producer levels in grassland ecosystems will have been undervalued, including those within the soil, particularly in the case of the terrestrial grasslands. Furthermore, the assimilation of carbon into plant matter and the input to roots and rhizomes below-ground will also have been underestimated. In view of the extent of natural tropical grass ecosystems, this would have a significant effect on estimates of global carbon cycling and the biospheric feedback resulting from atmospheric CO_2 increases.

Previous studies suggested that productivity of terrestrial grasslands alone in the tropics is 13.5 Gt year^{-1} (LIETH 1978), about 9% of total global terrestrial production. The proportion attributable to grasslands would rise to more than 25% of total global terrestrial production if the minimum three-fold underestimate found for the UNEP Project's three terrestrial grassland sites is generally applicable. If the more modest underestimation of production in other grass ecosystems is also taken into account, the great significance of natural grass ecosystems to total terrestrial production becomes apparent.

An accurate estimate of productivity of grasses is essential for the establishment of a baseline against which the effects of the global rise of CO_2 levels may be assessed (WIGLEY et al. 1984), and to provide details of the amount of CO_2 which is cycled through vegetation. A simplified model of CO_2 fertilization has already suggested that tropical grasslands would be the largest sink for such additional CO_2, with 26% of the total extra carbon sequestered in such grasslands, compared to only 19% for tropical forests (GIFFORD 1980). Furthermore, it has been predicted that grasslands will increase in extent under increased atmospheric CO_2, whilst the extent of forests will decrease (Table 30.6).

These results also have implications for the remote sensing of vegetation by satellites and aircraft. Both techniques give estimates of standing biomass after calibration against "ground truth" measurements obtained by conventional methods. Productivity is then estimated by, for example, the maximum biomass method. However, on the basis of the findings of LONG et al. (1989), such methodology cannot be used alone to provide a reliable estimate of biological production in grasslands. Since it ignores death and below-ground production, this may lead to an unpredictable degree of underestimation. Development of techniques with an adequate degree of "ground truth" verification may provide a solution to this problem if turnover of plant matter in grasslands can be estimated, or by remote sensing of light interception together with estimation of light utilisation for photosynthetic production (TUCKER et al. 1986; WARRICK 1986).

30.6 Grassland burning and the global carbon cycle

Previous estimates of the area of tropical grasslands burned annually were about 600 million ha (SEILER and CRUTZEN 1980); this has been recently revised to 748 million ha, or about 50% of the world's total grassland area (HAO et al. 1989). Data extrapolated from the three UNEP Project terrestrial grassland sites suggests that the global flux of carbon into the atmosphere from the burning of tropical

Table 30.6: Predicted changes in extent of life–zones (land types) and climates resulting from a doubling of atmospheric CO_2. Note changes in precipitation were not taken into account when estimating changes in potential vegetation (BOLIN et al. 1986; UNEP 1987).

LAND TYPE	Present	Future
Forests	58%	53%
Grasslands	17%	22%
Deserts	19%	21%
Tundra, ice	6%	3%
Total	100%	99%*
CLIMATE	**Present**	**Future**
Tropical	26%	35%
Subtropical	16%	12%
Warm temperate	21%	21%
Cool temperate	15%	17%
Boreal	23%	16%
Total	101%*	101%*

*Rounding errors

grasslands falls in the range 2.4–4.2 Gt per annum, assuming that burning occurs at the time of maximum above–ground biomass and dead matter (near the start of the main dry season), or 1.5–2.9 Gt per annum based on maximum above–ground dead matter alone. These calculations consider that the carbon content of oven–dry biomass and dead matter is 45%, and that the fraction of vegetation burned is 0.83 (HAO et al. 1989).

HAO et al. (1989) independently estimate 1.7 Gt of carbon per annum from this source (3.7 Gt of dry matter). Compared with our UNEP estimates above, this difference may be accounted for by their low estimate of grassland dry matter at the time of burning (4.5 – 6.6 t ha^{-1}). Whichever estimate is used, the burning of tropical grasslands evidently plays a significant role in the global carbon cycle, compared with the net release of carbon from deforestation — estimated at 2.4 Gt per annum (MYERS 1989), or a likely annual range of 1.5 – 3.0 Gt (HOUGHTON (in press)) — and global fossil fuel emissions of 5.3 Gt carbon per annum.

HAO et al. (1989) consider that the gross flow of carbon to the atmosphere from the burning of tropical savanna grasslands is three to four times that from tropical deforestation, due to the large area of grasslands burned every year (100 times the burned area of forests) and the substantial fraction of plant matter burned (83% compared with 30–40% for forests). Half of this grassland CO_2 efflux is considered to come from African grasslands, where 75% of the humid grassland area is burned every year (MENAUT 1983).

It is important to distiguish between gross and net carbon emissions, but this raises a number of difficult questions. How much of the gross CO_2 emissions from grassland burning will remain as net emissions after regrowth? How much of this carbon would otherwise be returned to the atmosphere more slowly by decomposition instead of by burning, and what fraction would enter the pool of soil organic matter if burning did not take place?

Although the extent of net CO_2 emissions from grassland burning is hard to estimate, is it likely that too frequent burning, together with overgrazing, will result in degradation of the grasslands and a net loss of carbon from grassland soils. Both annually and over a number of years, net emissions depend upon the frequency of fire, overgrazing and drought, all of which factors determine the build-up or degradation of vegetation and soil. In order to determine the net flux of carbon, both vegetation and soil carbon balances need to be studied over many yearly cycles. This type of long-term data is only just becoming available.

Burning of tropical grasslands probably has a significant effect also on nitrogen cycling (J. C. MENAUT, pers. comm., P. J. CRUTZEN, pers. comm.). Nitrogen oxides and N_2 are released into the atmosphere by pyrodenitrification, and may result in soil depletion if not balanced by nitrogen fixation. As described above, the feedback effects through primary productivity may be complex, especially in those tropical soils which are already low in nitrogen. Trace emissions from burning, such as methane, CO and particulates, may also have environmental consequences through local and global air pollution.

30.7 Present and future work

A new phase of the UNEP grasslands project, entitled Environment Changes and the Productivity of Tropical Grasslands (1989–1992) is continuing to measure net primary production at the five study sites whilst providing data for the modelling of tropical grassland productivity and carbon fluxes under the guidance of the SCOPE (Scientific Committee on Problems of the Environment) Project on Effects of Climate Change on Production and Decomposition in Coniferous Forests and Grasslands. A version of the CENTURY model is being used for this purpose, which has the ability to simulate the effects of grassland burning (PARTON et al. 1988). Whilst subject to the weaknesses of an empirical model, the CENTURY model has the potential for adaptation both in the direction of site-specific physiological measurements and towards a correlative regional model. To date, the CENTURY model has been successfully applied to long-data production data on temperate grasslands in the USA, as well as to monthly changes in biomass and dead matter, above and below ground, for three contrasting tropical grassland sites under the UNEP Project in Kenya, Thailand and Mexico. Attention is now being paid to modelling of grassland burning, as well as application of the model to moist savanna sites in Ivory Coast, West Africa, dry savanna sites in South Africa, and a range of temperate grassland sites in the USSR.

Preparatory work has also commenced for determining tropical grassland responses to elevated CO_2 levels at the terrestrial grassland site in Mexico and at an estuarine grass site near Shanghai, China. Preliminary results from Shanghai suggest that stands of a C_3 estuarine reed show short-term increased productivity when the CO_2 concentration is doubled (Y. K. SHEN, pers. comm.).

30.8 Conclusions

Natural tropical grasslands are a close second to tropical forests in extent and may equal them in productivity. This ecosystem will play a most important role in the global biospheric response to climate changes, yet it has received relatively little attention to date. Regular burning of tropical grasslands is significant and an underestimated factor in the global carbon cycle, given the extensive pools of soil carbon which may be influenced by grassland management.

Modelling of the response of tropical grasslands to climate changes is of paramount importance, but there is barely enough data presently available to develop and validate such models. Studies of the response of natural ecosystems to increasing atmospheric CO_2 concentration have only recently commenced, but there are likely to be complex feedback effects via other environmental variables such as temperature, moisture and nutrients all of whose interactions are even less well understood. In tropical grasslands where C_4 species are common, competitive effects between C_3 and C_4 photosynthetic types will further complicate our efforts to understand them.

Deforestation in the tropics is likely to continue for some years yet, and the grasslands so formed will further augment this already widespread ecosystem. Recognition of the crucial role played by tropical grasslands in the global carbon cycle makes further study imperative.

Bibliography

AJTAY, G. L.; KETNER, P.; DUVIGNEAUD, P., 1987, Terrestrial primary production and phytomass,
In: BOLIN, B. et al. (eds.), The Global Carbon Cycle. SCOPE 13. John Wiley and Sons, Chichester, pp. 129–181.

ANDERSON, J.M., 1990, The effects of climate change on decomposition processes in grasslands, tundra and coniferous forests.
Presented at SCOPE Workshop on Ecosystem Response to Climate Change: The Effects of Climate Change on Production and Decomposition in Coniferous Forests and Grasslands.
Ecological Applications (in press).

BAZZAZ, F. A.; GARBUTT, K., 1988, The response of annuals in competitive neighbourhoods: effects of elevated CO_2.
Ecology **69**, 937–946.

BOLIN, B.; DOOS, B. R.; JAGER, J.; WARRICK, R. A., 1986, The Greenhouse Effect, Climate Change and Ecosystems.
SCOPE **29**. John Wiley and Sons, Chichester.

BP, 1989, British Petroleum Statistical Review of World Energy.
British Petroleum Company, London EC2Y 9BU.

COOPER, J. P. (ed.), 1975, Photosynthesis and Productivity in Different Environments.
IBP Vol. **3**. Cambridge University Press.

CURTIS, P. S.; DRAKE, B. G.; LEADLEY, P. W.; ARP, W. J.; WHIGHAM, D. F., 1989, Growth and senescence in plant communities exposed to elevated CO_2 concentrations on an estuarine marsh.
Oecologia **78**, 20–26.

CURTIS, P. S.; BALDUMAN, L. M.; DRAKE, B. G.; WHIGHAM, D. F., 1989, Elevated atmospheric CO_2 effects on below-ground processes in C_3 and C_4 estuarine marsh communities.
Ecology (in press).

DRAKE, B. G.; LEADLEY, P. W.; ARP, W. J.; NASSIRY, D.; CURTIS, P. S., 1989, An open top chamber for field studies of elevated CO_2 concentration on saltmarsh vegetation.
Functional Ecology **3**, 369–371.

ETC, 1990, Biomass assessment in Africa.
ETC Consultants, Newcastle NE30 1NQ, UK, for World Bank, Washington.

FORD, M. A.; THORNE, G. M., 1967, Effect of carbon dioxide concentration on growth of sugar beet, barley, kale and maize.
Annals of Botany **31**, 629–644.

FROST, P.; MEDINA, E.; MENAUT, J. C.; SOLBRIG, O.; SWIFT, M. J.; WALKER, B. (eds.), 1986, Responses of savannas to stress and disturbance: a proposal for a collaborative programme of research.
Biology International Special Issue No. **10**, 82 pp. International Union of Biological Sciences, Paris.

GIFFORD, R. M., 1980, Carbon storage by the biosphere.
In: PEARMAN, G. (ed.), Carbon Dioxide and Climate, 167–181. Australian Academy of Science, Canberra.

GIFFORD, R. M., 1990, The effects of CO_2 and climate change on photosynthesis, net primary productivity and ecosystem production.
Presented at SCOPE Workshop on Ecosystem Response to Climate Change: The Effects of Climate Change on Production and Decomposition in Coniferous Forests and Grasslands.
Ecological Applications (in press).

HALL, D. O., 1989, Carbon flows in the biosphere: present and future.
Journal of the Geological Society, London, **146**, 175–181.

HALL, D. O.; ROSILLO–CALLE, F., 1990, African forests and grasslands: sources or sinks of greenhouse gases?
In: JUMA, C. (ed.), Global Warming and Climate Change: African perspectives. African Centre for Technology Studies, Nairobi, Kenya.

HALL, D. O.; SCURLOCK, J. M. O., 1990, Climate change and productivity of natural grasslands.
Annals of Botany (in press).

HAO, W. M.; LIU, M. H.; CRUTZEN, P. J., 1989, Estimates of annual and regional releases of CO_2 and other trace gases to the atmosphere from fires in the tropics, based on FAO statistics for the period 1975–1980.
In: GOLDAMMER, J. G. (ed.), Proceedings of the Third International Synposium on Fire Ecology, Freiburg University, Germany; May 1989 Springer–Verlag, Berlin.

HENDREY, G. R.; LEWIN, K. F.; KOLBER, Z.; KOLBER, D.; LIPFERT, F. W.; DAUM, M., 1988, Field performance testing of a free–air controlled enrichment (FACE) system for the regulation of carbon dioxide concentration in a cotton field at Yazoo City, Mississippi.
Department of Applied Science, Brookhaven National Laboratory, Long Island, USA.

HOUGHTON, R. A.; WOODWELL, G. M., 1989, Global climatic change.
Scientific American **260** (4), 36–44.

HOUGHTON, R. A., 1990, Tropical deforestation and atmospheric carbon dioxide.
Climatic Change (in press).

KANEMASU, E. T.; KIRKHAM, M. B.; OWENSBY, C. E., 1989, Rangeland plant response to elevated CO_2.
In: Carbon Dioxide and Climate: Summaries of Research in FY 1989, 52–53. Report DOE/ER-0425, Carbon Dioxide Research Programme, US Dept. of Energy, Washington DC, USA.

KOHLMAIER, G. H.; BROHL, H.; SIRE, E. O.; PLOCHL, M.; REVELLE, R., 1987, Modelling stimulation of plants and ecosystem response to present levels of excess

atmospheric CO_2.
Tellus **39**B, 155–170.

Lanly, J. P., 1982, Tropical Forest Resources.
FAO, Rome. 102 pp.

Lieth, H. (ed.), 1978, Patterns of Primary Productivity in the Biosphere.
Hutchinson Ross, Stroudsberg, 342 pp.

Linthurst, R.; Reimold, R. J., 1978, An evaluation of methods for estimating the net primary production of estuarine angisperms.
Journal of Applied Ecology **15**, 919–931.

Long, S. P.; Garcia Moya, E.; Imbamba, S. K.; Kamnalrut, A.; Piedade, M. T. F.; Scurlock, J. M. O.; Shen, Y. K.; Hall, D. O., 1989, Primary productivity of natural grass ecosystems of the tropics: a reappraisal.
Plant and Soil **115**, 155–166.

Ludlow, M. M., 1990, Critique on the effects of water stress on carbon allocation in trees.
Presented at SCOPE Workshop on Ecosystem Response to Climate Change: The Effects of Climate Change on Production and Decomposition in Coniferous Forests and Grasslands. Ecological Applications (in press).

Menaut, J. C., 1983, The vegetation of African savannas.
In: Bourliere, F. (ed.), Tropical Savannas. Ecosystems of the World **13**, Elsevier Scientific, New York, pp. 109–149.

Myers, N., 1989, Deforestation rates in tropical forests and their climatic implications.
Friends of the Earth, London N1 7JQ.

Olson, J.; Watts, J. A.; Allison, L. J., 1983, Carbon in Live Vegetation of Major World Ecosystems.
Report TR 004, DOE/NRB–0037, US Dept. of Energy, Washington DC, USA, 152 pp.

Parton, W. J.; Stewart, J. W. B.; Cole, C. V., 1988, Dynamics of C, N, P and S in grassland soils: a model.
Biogeochemistry **5**, 109–131.

Roberts, M. J.; Long, S. P.; Tieszen, L. L.; Beadle, C. L., 1985, Measurement of plant biomass and net primary production.
In: Coombs, J. et al. (eds.), Techniques in Bioproductivity and Photosynthesis, pp. 1–19, 2nd edition, Pergamon Press, Oxford.

SCURLOCK, J. M. O.; HALL, D. O., 1990, The contribution of biomass to global energy use (1987).
Biomass **21**, 75–81.

SEILER, W.; CRUTZEN, P. J., 1980, Estimates of gross and net fluxes of carbon between the biosphere and the atmosphere from biomass burning.
Climatic Change **2**, 207–247.

STRAIN, B. R.; CURE, J. D. (eds.), 1985, Direct Effects of Increasing Carbon Dioxide on Vegetation.
Report DOE/ER–0238, Office of Basic Energy Sciences, US Dept. of Energy, Washington DC, 286 pp.

SWIFT, M. J. (ed.), 1987, Report of the third workshop of the Decade of the Tropics/TSBF Programme.
Biology International Special Issue No. **13**. 68 pp., International Union of Biological Sciences, Paris.

TANS, P. P.; FUNG, I. Y.; TAKAHASHI, T., 1990, Observational constraints on the global atmospheric CO_2 budget.
Science **247**, 1431–1438.

TUCKER, C. J.; FUNG, I. Y.; KEELING, C. D.; GAMMON, R. H., 1986, Relationship between atmospheric CO_2 variations and a satellite–derived vegetation index.
Nature **319**, 195–199.

UNEP, 1987, The Greenhouse Gases.
UNEP/GEMS Environment Library No. **1**, United Nations Environment Programme, Nairobi. 40 pp.

VAN VEEN, J. A., 1990, Carbon allocation to root exudates and mycorrhizae: controls and system–level consequences.
Presented at SCOPE Workshop on Ecosystem Response to Climate Change: The Effects of Climate Change on Production and Decomposition in Coniferous Forests and Grasslands.
Ecological Applications (in press).

WARRICK, R. A., 1986, Photosynthesis seen from above.
Nature **319**, 181.

WIGLEY, T. M. L.; BRIFFA, K. R.; JONES, P. D., 1984, Predicting plant productivity and water resources.
Nature **312**, 102–103.

WOODWELL, G. M., 1986, On the limits of nature.
In: REPETTO R. (ed.), The Global Possible, Yale University Press.

WOODWELL, G. M., 1990, Forests in a warming world. Climatic Change (in press).

Chapter 31

Osnabrück Biosphere Model: structure, construction, results

G. Esser

31.1 Introduction

Various trace gases are known to reduce the permeability of the atmosphere for long-wave infrared radiation, while the permeability for sunlight is essentially unchanged. Therefore, an elevated global mean temperature is necessary to restore the balanced global fluxes of short and long wave radiation. Although processes may exist, and some are already known, which enhance or reduce the primary effects of the climate relevant trace gases, those gases are suspected to cause changes of the global climate, if their concentration in the atmosphere continues to increase. Table 31.1 gives an overview over the most important anthropogenic "greenhouse gases" and their environmental behavior.

In addition to the gases in Table 31.1 water vapor is an important greenhouse gas, although its atmospheric concentration is not likely to be changed directly by human activities.

Carbon dioxide is presently the most important anthropogenic greenhouse gas, and in contrast to the other greenhouse gases, the human potential to rise its future

Table 31.1: Most important anthropogenic trace gases ("greenhouse gases", column 1), their effectiveness for the greenhouse effect relative to CO_2 (2), half-life time in the environment (3), estimated sources (4), present atmospheric concentration increase (5), and their contribution to the presently observed additional greenhouse effect (6). CO_2 carbon dioxide, CH_4 methane, N_2O dinitrogen-monoxide, FCKW fuoro-chloro-hydrocarbons, O_3 (tropos.) tropospheric ozone, UV ultraviolet radiation. Data sources: GUDERIAN (1985), RAMANATHAN (1987), WUEBBLES et al. (1988).

gas	rel. eff.	$t_{1/2}$ [a]	sources	trend [%·yr^{-1}]	contribution [%]
(1)	(2)	(3)	(4)	(5)	(6)
CO_2	1	$\simeq$ 140	80% fossil 20% deforest.	0.5	50
CH_4	32	5–7	ruminants rice fields garbage, mining natural gas	1.1	19
N_2O	150	150	deforestation agriculture burning	0.25	4
FCKW	15,000	75–110	aerosols plastic foams refrigerants	5	15
O_3 (tropos.)	3,000	0.1	waste gas +UV	0.5	8

atmospheric concentration seems to be very high since the fossil carbon resources (coal, oil, natural gas) probably exceed 6.5×10^{12} tons (6,500 Gt) of carbon.

The atmospheric CO_2 level has raised considerably since preindustrial times (Figure 31.1). In the same period the mean global temperature has increased by about 0.4–0.6 °C. The atmospheric CO_2 level and the temperature rise seem to be correlated. But correlations never establish cause/effect relationships. While we are unsure whether the observed CO_2 rise has caused the observed increase of the global mean temperature, it can be excluded with a high degree of reliability for the last few hundred years that the temperature rise has caused the CO_2 increase. We know the releases from fossil sources within a limit of uncertainty of about 10%. Those emissions are greater than, but parallel in time to, the buildup of atmospheric CO_2. Our knowledge further rests on the sophisticated models of the global carbon cycle and on data acquired by deconvolution of the atmospheric CO_2-trend and the ratio of carbon isotopes in the CO_2.

All known processes which influence the global energy budget may be put to-

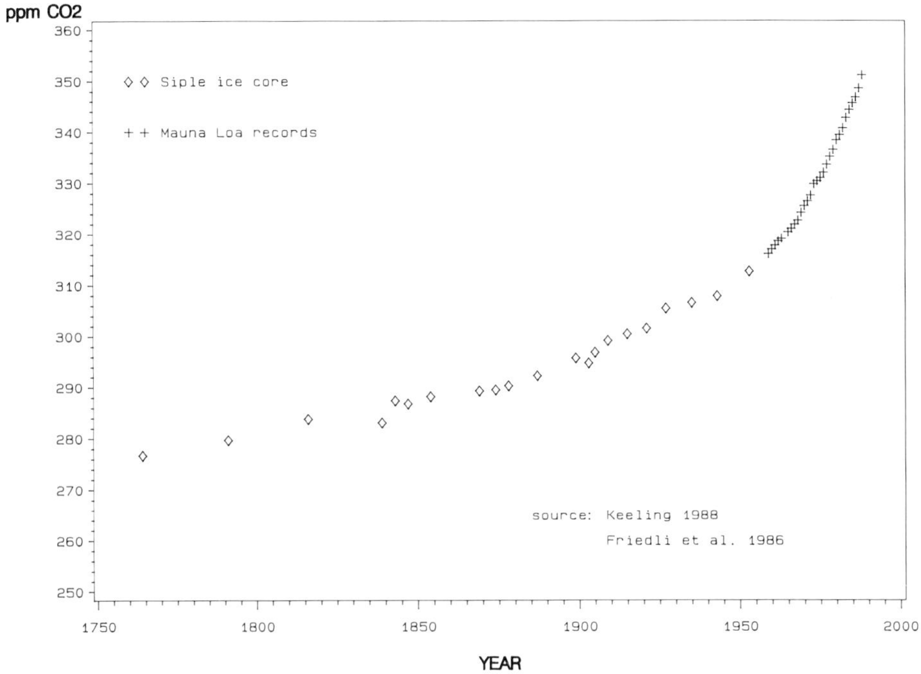

Figure 31.1: Observed atmospheric CO_2 concentration since preindustrial times. The rise prior to 1950 is probably mainly due to deforestation and its consequences. After 1950 anthropogenic emissions from fossil sources exceeded the emissions from deforestation considerably. Data prior to 1958 are from ice core analysis (Siple ice core, FRIEDLI et al. 1986), from 1958 on the records of the Mauna Loa observatory, Hawaii, have been considered (KEELING 1986).

gether quantitatively to create a comprehensive global climate model. Those models have been constructed by the scientific community in recent years and are said to be the most complex products of human intellect which have ever been developed. The temperature rise predicted by those models is in the same order of magnitude as the observed. For the next decades the predicted climate change is 10 to 100 times faster than the change at the end of the last glaciation.

Obviously, the interests of scientists and policy makers are concentrating on CO_2 since it has the strongest greenhouse potential. The total amount of contributing carbon is little less than 5×10^{13} tons, while the atmospheric pool presently amounts to only about 7.5×10^{11} tons carbon. Therefore small changes in the global carbon budget influence the atmosphere considerably and feedback effects are considered very important. Estimates for the carbon content of major carbon

Table 31.2: Major global pools of carbon and the respective annual fluxes which contribute to the global carbon budget. Values are given in Gt (10^9 t) respectively Gt·yr^{-1} of carbon. Sources: MAIER–REIMER & HASSELMANN (1987), ESSER (1990), based on model studies.

pools	ocean	39,000	Gt
	atmosphere	750	
	fossil fuel reserves	6,500	
	soil organic carbon	1,500	
	litter	100	
	live terrestrial phytomass	650	
fluxes	atmosphere → ocean	70–87	Gt·yr^{-1}
	ocean → atmosphere	70–87	
	net primary productivity (atmosphere → green plants)	~45	
	litter production (green plants → litter)	~45	
	soil organic carbon production (litter → SOC)	~10	
	litter depletion (litter → atmosphere)	~35	
	soil organic carbon depletion (SOC → atmosphere)	~10	
	fossil fuel burning (fuel → atmosphere)	~5.5	
	deforestation (green plants → atmosphere)	~1	

pools and the annual turnover of the fluxes between them may be found in Table 31.2. Small size variations of very large fluxes create sources or sinks for carbon in the related pools.

It has been the aim of the scientific community concerned with the climate problem since the beginning of the relevant discussion, to develop quantitative hypotheses for the dynamic functioning of the global carbon cycle. Those hypotheses, which resolve the relevant processes, are commonly called "carbon cycle models", and are used to demonstrate the consequences which follow from the assumptions made in the hypotheses. Those consequences could then be proved against independent data (i.e. data not used for the calibration of the model functions) to point out the shortcomings of the hypotheses.

One of those models, the Osnabrück Biosphere Model[1], is comprehensively

[1]The model was named after the city of Osnabrück, FR Germany, since it was presented the first time during a Symposium held in Osnabrück which was sponsored by the Commission of the European Communities in March 1983. Prof. Dr. Helmut Lieth is the initiator of the work. He contributed his MIAMI model, an empirical statement to calculate the flux net primary productivity of the potential natural vegetation from climate variables.

reported here. It has been developed to investigate the carbon budget of the terrestrial biosphere. It was developed at the University of Osnabrück, FR Germany, in the years 1980–1990 (ESSER 1984, 1986, 1987, 1990). The latest model version is No. 3. Version No. 4, which is presently programmed, will introduce considerable improvements, namely higher time resolution (1 month instead of 1 year), and higher spatial resolution (0.5 degrees latitude and longitude).

31.2 General model concept

The Osnabrück Biosphere Model (OBM) is geographically referenced. It rests on a global grid of 2.5 degrees longitude and latitude. For the land area except Antarctica 2433 grid elements were obtained. All the model functions were calculated to be valid in the entire span of the global array of driving forces which occurs in the grid.

The OBM does not follow the common practice of other models to assign certain flux values to vegetation units etc. for two reasons: (1) The number of vegetation units is much higher than the number of available data for the fluxes. Even for the key flux net primary productivity only about 100–150 values were measured globally, very few of them in remote regions. (2) The author is convinced that the correlation of ecological fluxes with vegetation units is much weaker than the correlation of such fluxes with the array of environmental influences (climate, soil, human impacts, and others). This assumption is strongly supported by data. Therefore the OBM uses those influences as driving functions for the fluxes. Then the extensive global climate data sets (based on more than 7,000 WMO standard stations) can be used to drive the model.

On the grid element level the usual systems analysis was carried out to determine the relevant carbon pools, fluxes, and control variables for the fluxes. The carbon pools are balanced by the fluxes. The system of differential equations is integrated using a standard forth–order Runge–Kutta method. It was attempted to replace the fluxes by functions of the pools. In this case, the obtained coefficients were expressed as depending on driving functions. If this was not possible, i.e. for the flux net primary productivity, then fluxes were derived from driving functions directly by means of empirical relationships.

The model is coupled to ocean models through a single–pool atmosphere or each grid element exchanges CO_2 with the respective atmospheric pool of a compartmented atmosphere, depending on the problem to be investigated. For complex experiments the OBM is coupled to the ocean and atmosphere models developed at the MPI for Meteorology in Hamburg.

Table 31.3: List of variables used in the model description in alphabetical order. Most of the variables are arrays which is indicated in the equations.

symbol	meaning	units
c	field crop (see Table 31.5)	
i	plant material (woody, herbaceous), $i = h, w$	
j	model year A.D., $j = 1860, \ldots, 2100$	
m	grid element, $m = 1, \ldots, 2433$	
o	soil unit, $o = 1, \ldots, O(m)$	
AA	agricultural area of m in a model year j	m^2
AG	total area of m	m^2
AGE	mean stand age of a vegetation unit	years
AS	area of soil type $o(m)$	m^2
AV	area covered by natural vegetation in m	m^2
c_1, c_2	parameters to determine slope and turning point of the clearing function (31.26)	–
CO_2	carbon dioxide concentration in the atmosphere	$\mu l \cdot l^{-1}$ (= ppmv)
DIS	annual discharge	$mm \cdot yr^{-1}$
DOC	dissolved organic C export	$g \cdot m^{-2} \cdot yr^{-1}$
f	soil factor of soil type o	–
F	weighted mean of all f of (m)	–
fa	conversion factor yield to NPP	–
FA	weighted mean of the $fa(c)$ of m	–
FCO_2	fertilizing factor of CO_2 on NPP	–
H	factor for sharing NPP into (i) compartments	–
j_{TR}	year of the turning point of the logistic clearing function (31.26)	–
kd	decay coefficient for litter	yr^{-1}
klp	litter production coefficient	yr^{-1}
$KORR$	factor to correlate $PART(j)$ $j = 1860, \ldots, 1980$ with $PART(1970)$; $0 \leq KORR \leq 1.5$ (see equation (31.25))	–
$ksocp$	soil organic carbon production coefficient	yr^{-1}
$ksoc$	soil organic carbon depletion coeficient	yr^{-1}
k_T	kd derived from T	yr^{-1}
k_{Pp}	kd derived from Pp	yr^{-1}

continued next page

Table 31.3 continued

symbol	meaning	units
L	litter pool	$g \cdot m^{-2}$
LD	litter depletion	$g \cdot m^{-2} \cdot yr^{-1}$
LP	litter production	$g \cdot m^{-2} \cdot yr^{-1}$
NPP	net primary productivity	$g \cdot m^{-2} \cdot yr^{-1}$
$NPPT$	NPP derived from T	$g \cdot m^{-2} \cdot yr^{-1}$
$NPPPp$	NPP derived from Pp	$g \cdot m^{-2} \cdot yr^{-1}$
$PART$	part of m under agricultural use	%
P	total above and below ground phytomass	$g \cdot m^{-2}$
$SOCP$	soil organic carbon production	$g \cdot m^{-2} \cdot yr^{-1}$
POC	particulate organic C export	$g \cdot m^{-2} \cdot yr^{-1}$
Pp	average annual precipitation	$mm \cdot yr^{-1}$
T	mean annual temperature	°C
WL	clearing probability derived from land-use in the period 1950–1980	–
WP	clearing probability derived from NPP	–
WS	clearing probability derived from soil fertility F	–
WU	clearing probability derived from land-use in surrounding grid elements	–
$yield$	yield of crops in m (weighted mean)	$g \cdot m^{-2} \cdot yr^{-1}$

31.3 Model construction

In this section it is discussed how the equations were developed which compute the model fluxes respectively the flux–coefficients on grid element level. The list of variables used in the description may be found in Table 31.3. The mass–balance of the model pools is carried out in the usual way, either by integrating the system of differential equations by using numeric routines, or by balancing the integrated fluxes explicitly at each time step, depending on the problem. The pool values necessary as initial conditions to start the model are computed using a fixed atmospheric CO_2 concentration. The atmosphere then acts as an "unlimited" carbon source to fill the pools. This "pre–run" procedure may need considerable computing time ($\sim 2,000$ model years) to get the large soil pools stable and to prevent model drift in the consecutive model run.

31.3.1 Net primary productivity

The total flux net primary productivity is calculated as an annually integrated two-dimensional array according to equation (31.1). The productivity share of the potential natural vegetation is basically supplied by the equations (31.2) and (31.3) which correspond to the original MIAMI model (LIETH 1975). The result is consecutively modified by influence of the soil (second line of equation (31.1)). The soil factor array $F(m)$ is the weighted mean (see (31.4)) of the individual soil factors of each soil type found in the respective grid element.

$$\begin{aligned} NPP(j,m) = & \Bigg(\min[NPPT(j,m), NPPPp(j,m)] \\ & \cdot F(m) \cdot \frac{AV(j,m)}{AG(m)} \\ & + yield(j,m) \cdot FA(m) \cdot \frac{AA(j,m)}{AG(m)} \Bigg) \\ & \cdot FCO_2(j) \end{aligned} \tag{31.1}$$

$$NPPT(j,m) = 3000/\{1 + \exp[1.315 - 0.119 \cdot T(j,m)]\} \tag{31.2}$$

$$NPPPp(j,m) = 3000 \cdot \{1 - \exp[-0.000664 \cdot Pp(j,m)]\} \tag{31.3}$$

$$F(m) = \frac{\sum_{o=1}^{O(m)} f(o) \cdot AS(m,o)}{\sum_{o=1}^{O(m)} AS(m,o)} \tag{31.4}$$

$$FCO_2(j,m) = A \cdot [1 - \exp\{-R \cdot (CO_2(j) - 80)\}] \tag{31.5}$$

$$A = 1 + \frac{F(m)}{4}$$

$$R = -\ln\left(\frac{A-1}{A}\right) \cdot \frac{1}{240}$$

The individual soil factors are empirical correction factors which relate the productivity measured on a given soil type to the MIAMI productivity. The values of $f(o)$ for the major FAO–Unesco soil units (FAO–Unesco 1974 ff.) are found in Table 31.4.

The third line of equation (31.1) supplies the productivity share of agriculturally used areas of the grid element. The term depends linearly on the yields of the main field crops in the grid element. The factor array $FA(m)$ relates the total productivity of the crops to their yields. The variables *yield* and FA are again weighted means of the crop yields and their factors $fa(c)$ in the grid element. For the factor array $fa(c)$ see Table 31.5.

Table 31.4: Soil factors $f(o)$ of the OBM which characterize the fertility of the main soil units of the world. Names of soil units according to FAO–Unesco (1974 ff.).

soil unit o	$f(o)$	soil unit o	$f(o)$
Gleyic Acrisol	0.87	Chromic Luvisol	1.04
Humic Acrisol	0.22	Ferric Luvisol	1.65
Orthic Acrisol	0.70	Gleyic Luvisol	2.78
other Acrisol	0.60	Orthic Luvisol	0.85
Dystric Cambisol	0.94	Dystric Histosol	1.39
Eutric Cambisol	1.69	Humic Podzol	0.56
Humic Cambisol	1.58	Orthic Podzol	0.61
Gelic Cambisol	0.76	other Podzols	0.55
Luvic Chernozem	0.99	Calcaric Regosol	1.61
Dystric Podzoluvisol	0.83	Eutric Regosol	1.14
Xanthic Ferralsol	0.55	Gelic Regosol	0.91
Humic Gleysol	0.47	other Regosol	1.20
Gelic Gleysol	0.57	Orthic Solonetz	0.59
other Gleysol	0.50	Vitric Andosol	1.65
Lithosol	0.52	Haplic Xerosol	0.42
Lithosol–Yermosol	1.14	Yermosol	0.30
Fluvisol	0.49	Haplic Yermosol	0.66
Eutric Fluvisol	0.61	Luvic Yermosol	0.23
other Fluvisol	0.55	Takyric Yermosol	0.09
Haplic Kastanozem	1.96	Orthic Solonchak	0.44
Luvic Kastanozem	1.61	Takyric Solonchak	0.03
other Kastanozem	1.80	other Solonchak	0.20
Albic Luvisol	0.34		

Finally, the term in line 4 of equation (31.1) is the CO_2 fertilization effect. In the OBM it is calculated from the actual atmospheric CO_2 concentration and the soil fertility according to equation (31.5). The implementation of the fertilization effect assumes that the productivity of the potential natural vegetation as well as of agricultural crops is enhanced similarly. The existence of such a fertilization effect in nature has been controversial since the beginning of carbon cycle research. In terms of plant physiology and ecophysiology, the processes are quite clear: the CO_2 concentration acts in three ways:

- Through a direct effect of the CO_2 partial pressure in the plant cell on the enzyme kinetics of the Ribulose–1,5–bisphosphate–carboxylase/oxygenase. The quantum–efficiency of photosynthesis is influenced directly by this effect.

Table 31.5: Factors $fa(c)$ for the calculation of NPP (dry weight) of some agricultural crops from the yields (fresh weight). From ASELMANN & LIETH (1983).

crop	$fa(c)$	crop	$fa(c)$
wheat	2.15[1]	cotton	5.00
barley	2.12[5]	sugar cane	0.44
oats	3.44[1]	mandioka (cassava)	0.64[4]
rye	2.65[5]	potato	0.54[2]
maize	2.46[1]	batate	0.55[3]
sorghum	3.44[1]	soybeans	2.46[1]
pearl millet	3.44[1]	Phaseolus beans	2.86[1]
paddy rice	2.86[1]	rape	2.81[5]
sugar beets	0.32[6]		

dry matter content in yield:
[1] 86%, [2] 32.5 %, [3] 30%, [4] 35%, [5] 87%, [6] 23%

- Through an indirect effect on the transpiration of a plant through the stomatal resistance. The elevated external CO_2 level raises the intracellular partial pressure of CO_2. Plants may counteract by raising the stomatal resistance to keep the internal CO_2 partial pressure constant. The reduced transpiration per unit leaf area allows the plant to increase the leaf area or to extend the vegetation period. Since most of the global land areas having mean annual temperatures above 5 °C are limited in water supply (ESSER 1987), this effect may be the most important on a global scale.

- The lack of minerals limits the fertilization effect. The flux net primary productivity must always be accompanied by adequate fluxes of minerals from the soil, since the ratios carbon/minerals within plant organs are constant within narrow limits (INGESTAD & LUND 1986; INGESTAD & ÅGREN 1988). Minor changes of the ratios probably occur in some plant species if grown at very high atmospheric CO_2 concentrations (OVERDIECK et al. this book, page 641).

Equation (31.5) was calibrated by use of physiological data. Physiological and ecophysiological investigations of the fertilization effect have been carried out at many institutions (see OVERDIECK et al., this book), but their value in determining the plant behavior in natural environments is limited. The function proposed here considers soil fertility, but the water interrelations are not considered explicitly. A plot of function (31.5) is found in Figure 31.2.

In the model the productivity is partitioned into two separate fluxes: herbaceous and woody productivity, by use of a simple linear sharing factor H:

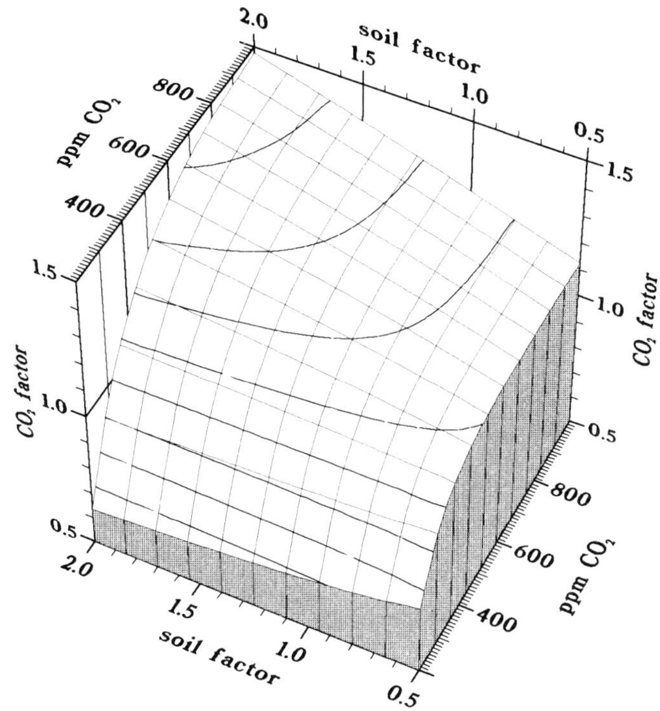

Figure 31.2: Plot of function (31.5), $FCO_2(j,m) = f[CO_2(j), F(m)]$, to calculate the CO_2 fertilization factor on net primary productivity. The CO_2 compensation concentration for entire plants is assumed to be 80 $\mu l \cdot l^{-1}$ ($FCO_2 = 0$ irrespective of F(m)). For 320 $\mu l \cdot l^{-1}$ CO_2 concentration $FCO_2 = 1.0$, since it was assumed that the data used to calibrate the functions (31.2) and (31.3) were measured mainly during the 1960s when CO_2 was about 320 $\mu l \cdot l^{-1}$. The function achieves saturation at 1.125 for $F(m) = 0.5$, at 1.250 for $F(m) = 1.0$, and at 1.375 for $F(m) = 1.5$. For the present atmospheric CO_2 of 355 $\mu l \cdot l^{-1}$ the function yields the factors 1.034 for $F(m) = 0.5$, 1.052 for $F(m) = 1.0$, and 1.065 for $F(m) = 1.5$.

$$NPP(i) = H(i) \cdot NPP \qquad (31.6)$$

The factors $H(i)$ are attributes of the vegetation units. In Table 31.6 the values for 31 vegetation formations used in the model are given.

Table 31.6: Mean stand ages $AGE(w)$ of woody material and factors $H(h)$ for separating herbaceous NPP for the 31 formations of the potential natural vegetation. $H(w)$ is simply calculated from $1 - H(h)$. Values were derived using our data base DATAVW, gaps were filled using the method of ranking. Ranked units carry an *. From ESSER (1984).

formation	mean stand age $AGE(w)$, years	herbaceous factor $H(h)$
tropical moist lowland forest	200	0.37
tropical dry lowland forest	*80	0.4*
tropical mountain forest	*80	0.37*
tropical savanna	5	0.98
tropical paramo woodland	*10	0.95*
tropical paramo grassland	*1	1.0*
Puna formation	*2	1.0*
subtropical evergreen forest	200	0.37
subtropical deciduous forest	150	0.44
subtropical savanna	5	0.90
subtropical halophytic formation	5	0.9
subtropical steppe and grassland	1	1.0
temperate steppe and meadow	1	1.0
subtropical semidesert	15	0.85
xeromorphic formation	20	0.4
desert (tropical, subtropical, cold)	*5	0.85*
Mediterranean sclerophyllous forest	*100	0.4*
Mediterranean shrub and woodland	15	0.47
temperate evergreen (coniferous) forest	130	0.29
temperate deciduous forest	150	0.38
temperate woodland	25	0.53
temperate shrub formation	*10	0.85
temperate bog and tundra	5	0.48
boreal evergreen coniferous forest	100	0.34
boreal deciduous forest	100	0.38
boreal woodland	*15	0.6*
boreal shrub formation	*10	0.85*
woody tundra	10	0.7
herbaceous tundra	2	1.0
azonal formation	*5	0.6*
mangrove	50	0.29

31.3.2 Litter production

The flux litter production is assumed to be proportional to the respective source pool phytomass. Separate fluxes for herbaceous and woody material are distinguished:

$$LP(i) = klp(i) \cdot P(i) \tag{31.7}$$

In order to derive the factors $klp(i)$ from the mean stand age of the plant material of the vegetation unit under consideration, an equation was used, which was originally developed to calculate phytomass from net primary productivity and stand age (ESSER 1984):

$$P = 0.59181 \cdot NPP \cdot AGE^{0.79216} \tag{31.8}$$

Reformulated for the biomass compartments and extended by the biomass share factors $H(i)$ this equation reads:

$$P(i) = 0.59181 \cdot \frac{NPP(i)}{H(i)} \cdot AGE(i)^{0.79216} \tag{31.9}$$

If we assume mature stands, the mass balance equation for phytomass may be set to zero:

$$\frac{dP(i)}{dt} = NPP(i,t) - LP(i,t) = 0 \tag{31.10}$$

LP is replaced by the term of equation (31.7):

$$NPP(i,t) - klp(i) \cdot P(i,t) = 0 \tag{31.11}$$

and with (31.9):

$$NPP(i,t) - klp(i) \cdot 0.59181 \cdot \frac{NPP(i,t)}{H(i)} \cdot AGE(i)^{0.79216} = 0 \tag{31.12}$$

Divided by $NPP(i,t)$ and solved for $klp(i)$:

$$klp(i) = \frac{H(i)}{0.59181 \cdot AGE(i)^{0.79216}} \tag{31.13}$$

At maturity, the results of equations (31.9) and after integration (31.10) are similar. The stand ages for woody material of the 31 vegetation formations are found in Table 31.6. For herbaceous material, a mean stand age of 1 year is assumed generally, except for evergreen rain forests (1.2 years) and for needle–leaved evergreen forests (2 years).

31.3.3 Litter depletion

It is assumed, that the decomposed amount of litter is proportional to the litter pool:

$$LD(t) = \sum_i -k(i) \cdot L(i,t) \qquad i = woody, herbaceous \tag{31.14}$$

The value of the coefficient $k(i)$ depends on the composition of the decomposed material (in the OBM woody and herbaceous materials are distinguished), and on the elements of the environmental vector (the OBM considers temperature and precipitation). $k(i)$ depends exponentially on the temperature while the relation to precipitation is a maximum function. The analysis of the data and derivation of the functions is described in detail by ESSER et al. (1982) and ESSER & LIETH (1988). The set of equations to predict the coefficients was calibrated by data listed in detail at ESSER (1986). The functions are (with $i = h, w$):

$$kd(h,j,m) = \min[k_T(h,j,m), k_{Pp}(h,j,m)] \tag{31.15}$$

$$k_T(h,j,m) = 0.1063 \cdot \exp[0.0926(T(j,m) + 6.41)] + 0.2365 \tag{31.16}$$

$$\begin{aligned} k_{Pp}(h,j,m) = & \left(\frac{0.4436}{0.0215 + \exp[4.2 - 0.0053 \cdot Pp(j,m)]} + 5.944\right) \\ & \cdot \left(\frac{0.094}{0.7 + \exp[0.0023 \cdot Pp(j,m) - 5.05]} + 0.076\right) \\ & \cdot (1 - \exp[-0.001 \cdot Pp(j,m)]) \end{aligned} \tag{31.17}$$

$$kd(w,j,m) = \min[k_T(w,j,m), k_{Pp}(w,j,m)] \tag{31.18}$$

$$k_T(w,j,m) = 0.037 \cdot \exp[0.0522(T(j,m) + 31.63)] - 0.0348 \tag{31.19}$$

$$\begin{aligned} k_{Pp}(w,j,m) = & \left(\frac{0.1927}{0.021 + \exp[8.53 - 0.0095 \cdot Pp(j,m)]} + 4.9352\right) \\ & \cdot \left(\frac{0.126}{1.51 + \exp[0.003 \cdot Pp(j,m) - 4.65]} + 0.05\right) \\ & \cdot (1 - \exp[-0.001 \cdot Pp(j,m)]) \end{aligned} \tag{31.20}$$

The precipitation–driven maximum functions (31.17) and (31.20) consist of three brackets: The term in the first bracket controls the ascend, the term in the second bracket the descend of the function. The third term forces the function to pass through the zero point.

31.3.4 Soil organic carbon production

It is assumed that the polyphenolic compounds in the litter (lignins) contribute to the soil organic carbon. Thus soil organic carbon production is part of the flux litter production:

$$SOCP(i,j,m) = ksocp(i) \cdot LP(i,j,m) \tag{31.21}$$

The sharing factor $ksocp(i)$, which represents the lignin content of the respective material, is set to 0.30 for woody material and 0.11 for herbaceous material. In the OBM it does not depend on the grid element or the model year.

31.3.5 Soil organic carbon depletion

It is assumed that lignin compounds have depletion coefficients $ksoc$ which amount to 1% of the coefficients of fresh herbaceous litter as given by the equations (31.15)–(31.17):

$$ksoc(j,m) = 0.01 \cdot kd(h,j,m) \tag{31.22}$$

31.3.6 Leaching of dissolved and particulate organic carbon (DOC and POC)

Dissolved and particulate organic carbon commonly occur as organic carbon freights in each body of fresh water. It could be shown by ESSER & KOHLMAIER (1990), who used data which were acquired by DEGENS et al. (1982, 1983, 1985), that the DOC and POC freights of a river mainly depend on its water discharge. In contrast, the correlations of transported DOC and POC with the extent of agricultural areas or their change in the watershed of the river were zero or even slightly negative. The authors established two equations to derive DOC and POC from the discharge of water from a m^2 of the watershed of a river:

$$DOC = 0.0064 \cdot DIS \tag{31.23}$$

$$POC = 0.0022 \cdot DIS \tag{31.24}$$

Since in many regions with high precipitation discharge correlates with precipitation, DIS may be replaced by mean annual precipitation.

31.3.7 Land–use changes and deforestation

The direct human influence on vegetation acts upon the variables $yield(j,m)$ and relative agricultural area $AA(j,m)/AG(m)$ in equation (31.1), and the stand age $AGE(i)$ in equations (31.13) and (31.9), respectively.

During the modelled period, the area $AA(j,m)$ of a grid element under agricultural use may increase by deforestation, or decrease by reforestation. The OBM uses historical data for the years up to 1980 and a scenario for the centuries after 1980 to obtain the relative agricultural area $PART(j,m) = AA(j,m)/AG(m)$ of each grid element in each model year. The land–use submodel is described in detail by ESSER (1989). Here only a short overview can be given.

Land–use changes 1860–1980. The basic land use information originates from the World atlas of Agriculture (Instituto Geographico de Agostini, 1969, 1971, 1973), which was digitized on the grid of the model. Five principal land–use classes where distinguished including arable land, permanent crops, meadows and pastures, forests, and rough grazing land. The land–use categories were digitized in terms of percentages of grid element area, which resulted in a finer areal resolution than suggested be the grid.

The areas digitized from the World Atlas of Agriculture were assumed to be valid for the year 1970. In order to model the changes within the period 1860–1980, the land–use areas given by the atlas are modified by a factor matrix $KORR(j,m)$:

$$PART(j,m) = PART(1970,m) \cdot KORR(j,m) \tag{31.25}$$

The factor matrix $KORR(j,m)$ was calculated by use of a data base published by RICHARDS et al. (1983). Since data were only available for the years 1860, 1920, 1930, 1950, 1961–1965, 1978, we interpolated the values for the years between in order to achieve at a complete sequence of $KORR$ factors. To test this approach, we began to evaluate Landsat images for land–use changes (ESSER & LIETH 1986). 934 scenes of South America were evaluated to find out the changes of the areas influenced by humans between the beginning 1970s and the beginning 1980s. In Table 31.7 The results are compared with the OBM modeled changes based on the RICHARDS data base.

Land–use development after 1980. The land–use scenario pertaining after the year 1980 must not be understood as an attempt to predict future developments, since scenarios are less predictions than logical assumptions about future developments. The basic concept is to derive $PART(j,m)$; $j = 1981,\ldots,2100$ from a logistic function for each grid element:

$$PART(j,m) = PART(1980,m) + \frac{PART_{max} - PART(1980,m)}{1 + \exp[c_1(m) - c_2(m)]} \tag{31.26}$$

The parameters $c_1(m)$, $c_2(m)$ determine slope and turning–point (i.e. year of maximum clearing rate) of the function. $PART_{max}$ is a user–defined upper limit of the relative agricultural area of the grid element. The year of the turning point of equation (31.26) is given by:

$$j_{TR}(m) = \frac{c_1(m)}{c_2(m)} \tag{31.27}$$

c_2 is a user–supplied value. If it is set to 0.2, the time necessary to clear the grid element in the limits $0.05 \leq PART(j,m) \leq 0.95$ is 30 years. In order to obtain c_1 the year j_{TR} must be determined by use of the global scenario functions. On global level, the two–step calculation includes:

Table 31.7: Areal changes of the natural vegetation of South American countries as derived from the evaluation of 934 Landsat scenes. Column (3) gives the percentage of the country area covered by the overlapped parts of scene pairs. Column (5) gives the respective estimate by RICHARDS et al. (1983) as mean of the period 1958–1978. From ESSER & LIETH (1986).

country	net area changed [$km^2 \cdot a^{-1}$]	evaluated area [%]	related phytomass changes (C) [$10^6 t \cdot a^{-1}$]	mean 1958–78 RICHARDS et al. 1983 [$km^2 \cdot a^{-1}$]
Brasil	6,689	76	80.	7,708
Argentina	1,516	85	12.	1,880
Paraguay	790	80	9.5	227
Venezuela	357	65	3.	960
Bolivia	76	40	1.	713
Chile	66	90	0.3	104
Columbia	3	30	0.02	280
Peru	0	34	–	649
Ecuador	0	6	–	187
Uruguay	0	95	–	–48
total	9,497	72	106.	12,660

- ranking the grid elements according to their individual "clearing probability".
- determination of the number $\Delta NT(j)$ of grid elements which have their turning point of function (31.26) in the model year j by use of the global logistic function.

The number of turning points occurring in the time span $j-(j-1)$ is:

$$NT(j) = \frac{2433 + NT_0}{1 + \left(\frac{2433+NT_0}{NT_0} - 1\right)\exp[-\beta(j-(j-1))]} \qquad (31.28)$$

This number of grid elements is taken successively from the ranked sequence of grid elements and their $c_1(m)$ are calculated after equation (31.27) with $j_{TR}(m) = j$.

Grid elements are ranked according to their "clearing probability", which is defined as the product of four individual probabilities:

$$WR(m) = WU(m) \cdot WP(m) \cdot WS(m) \cdot WL(m) \qquad (31.29)$$

The four individual probabilities are based on the following assumptions:

1. Clearing will occur earlier the more agricultural areas were present in the surrounding grid elements in 1970:

$$WU(m) = \frac{1}{N} \cdot \sum_{n=1}^{N} \left(\varrho(n) \cdot \frac{1}{K(n)} \cdot \sum_{k=1}^{K(n)} PART(1970, k) \right) \tag{31.30}$$

n	order of circles of grid elements surrounding element m $n = 1, \ldots, N$. The author used $N = 5$.
$k(n)$	current number of grid element in circle of nth order. $k(n) = 1, \ldots, K(n)$.
$\varrho(n)$	weighting factor; $\varrho(n) = \frac{1}{2^{n-1}}$

2. The higher the natural productivity $NPP(1970, m)$ the earlier clearing occurs:

$$WP(m) = \frac{NPP(1970, m) - NPP_{lim}}{NPP_{max} - NPP_{lim}} \tag{31.31}$$

NPP_{max}	the highest value of the matrix $NPP(1970, m)$
NPP_{lim}	The minimum natural productivity to allow economic use of the area, in the scenario set to either 0 or 500 [$g \cdot m^{-2} \cdot yr^{-1}$]

3. The higher the soil fertility expressed through the soil factor array $F(m)$ (see equation (31.4) and Table 31.4), the earlier clearing occurs:

$$WS(m) = \frac{F(m) - F_{min}}{F_{max} - F_{min}} \tag{31.32}$$

F_{max}	highest value of the array $F(m)$
F_{min}	lowest value of the array $F(m)$

4. The steeper the clearing increased in the respective grid element in the period 1950–1980, the earlier it is continued after 1980:

$$WL(m) = \frac{KORR(1980, m)}{KORR(1950, m)} \cdot \frac{1}{WL_{max}} \tag{31.33}$$

WL_{max}	the maximum value of the quotient–array of (31.33)
$KORR$	see equation (31.25).

31.4 Geographical data for running the model

To run the model all grid element related driving functions (driving variables which are arrays of (m)) must be set up as global patterns. Those variables include mean annual temperature, average annual precipitation, vegetation formation, soil class, agricultural crops. At present the information system rests on the following data sources:

Climate variables. Mean annual temperature and average annual precipitation were derived from the WMO standard net of climate stations (data set from NCAR, Boulder), the World Atlas of Climate Diagrams (WALTER & LIETH, 1960 ff.), and a data collection with climatic zones maps by MÜLLER (1982). The data were interpolated and corrected for the mean elevation of the grid element.

Soils. The Soil Map of the World (FAO–Unesco, 1974 ff.) was digitized on the grid. Areas covered by the soil units were expressed as percentage of grid element area. 106 soil units were considered. Associated soils were excluded.

Potential natural vegetation. The Vegetation Atlas which was published by SCHMITHÜSEN (1976) was digitized by use of the method described for soils. The 172 vegetation units which are included globally, were summarized to the 31 formations included in Table 31.6.

Land–use. World Atlas of Agriculture (Instituto Geographico de Agostini, 1969, 1971, 1973), FAO Production Yearbooks (FAO–Unesco, 1980 ff.), FAO Agro–Ecological Zones Project Results (FAO–Unesco, 1978 ff.). The country–related statistical data were distributed on the grid by use of the information supplied by the Agro–Ecological Zones Project.

Land–use changes. Factor array for 121 countries and each year of the period 1860–1980, based on data published by RICHARDS et al. (1983), and the evaluation of 934 Landsat images for the years 1972–1980 carried out by ESSER & LIETH (1986).

Crop yields. FAO Production Yearbooks (FAO–Unesco, 1980 ff.).

31.5 Fundamental model results

The important questions concerning the global carbon cycle of the terrestrial biosphere include:

- What is the importance of the CO_2 fertilization effect on a global scale and how is it geographically distributed.
- How contribute the tropical deforestation and other anthropogenic influences on vegetation to the carbon balance.
- How important is the climate—carbon cycle feedback.

Table 31.8: Comparison of the productivities of natural vegetation and agricultural crops for some tropical and extratropical countries. The productivities of the agricultural crops were calculated from the yields (FAO Production Yearbooks) by use of conversion factors given by ASELMANN & LIETH (1983). Values mean dry matter (for carbon multiply by 0.45). From ESSER (1990).

country	agricultural productivity [$g \cdot m^{-2} \cdot yr^{-1}$]	natural productivity [$g \cdot m^{-2} \cdot yr^{-1}$]	ratio agric./natural
Zaire	180	1960	0.10
Kenya	350	1300	0.13
Niger	150	890	0.17
Kampuchea	310	1800	0.17
Bolivia	280	1500	0.19
Brazil	310	1620	0.19
Spain	510	750	0.68
FR Germany	1130	1190	0.95
Belgium, Luxemburg	1290	1210	1.07

In the past years several results of the OBM which refer to those problems have been published (ESSER 1986, 1987, 1990). Here the author intends to draw some fundamental conclusions.

It is well documented that after deforestation the productivity of field crops is generally much lower than the productivity of the replaced natural vegetation. Table 31.8 shows that the productivity of agricultural crops is almost generally as low as 10–20% of the natural productivity. Very few industrialized countries which maximized the crop yields per unit area of their fields without considering expenses achieve 100% of the natural productivity. Thus deforestation does not only reduce the phytomass (through reduction of the stand–age). The net primary productivity is also reduced drastically. As a consequence, the input fluxes into the litter and soil organic carbon pools are reduced. Since the depletion coefficients are basically unchanged, the pools are reduced. On the other side, reforestation does not restore the production of litter and soil organic carbon immediately, since immature forests produce less litter. Thus reforestation of old fallow land, which has a productivity flux similar to that of the regular natural vegetation of the site, may even reduce the soil and litter carbon pools during the first decades.

The CO_2 fertilization effect increases the net primary productivity. As a consequence, the phytomass of mature stands grows if the stand age is unchanged. There are no experimental results available to support the assumption, that the stand age is not changed. If we assume, that the stand age of mature stands is not changed, a carbon sink is induced in the growing phytomass as long as the atmospheric CO_2 is rising. Subsequently, the productions of litter and soil organic carbon are also rising and thus the target pools. Deforestation reduces the fertilization–induced

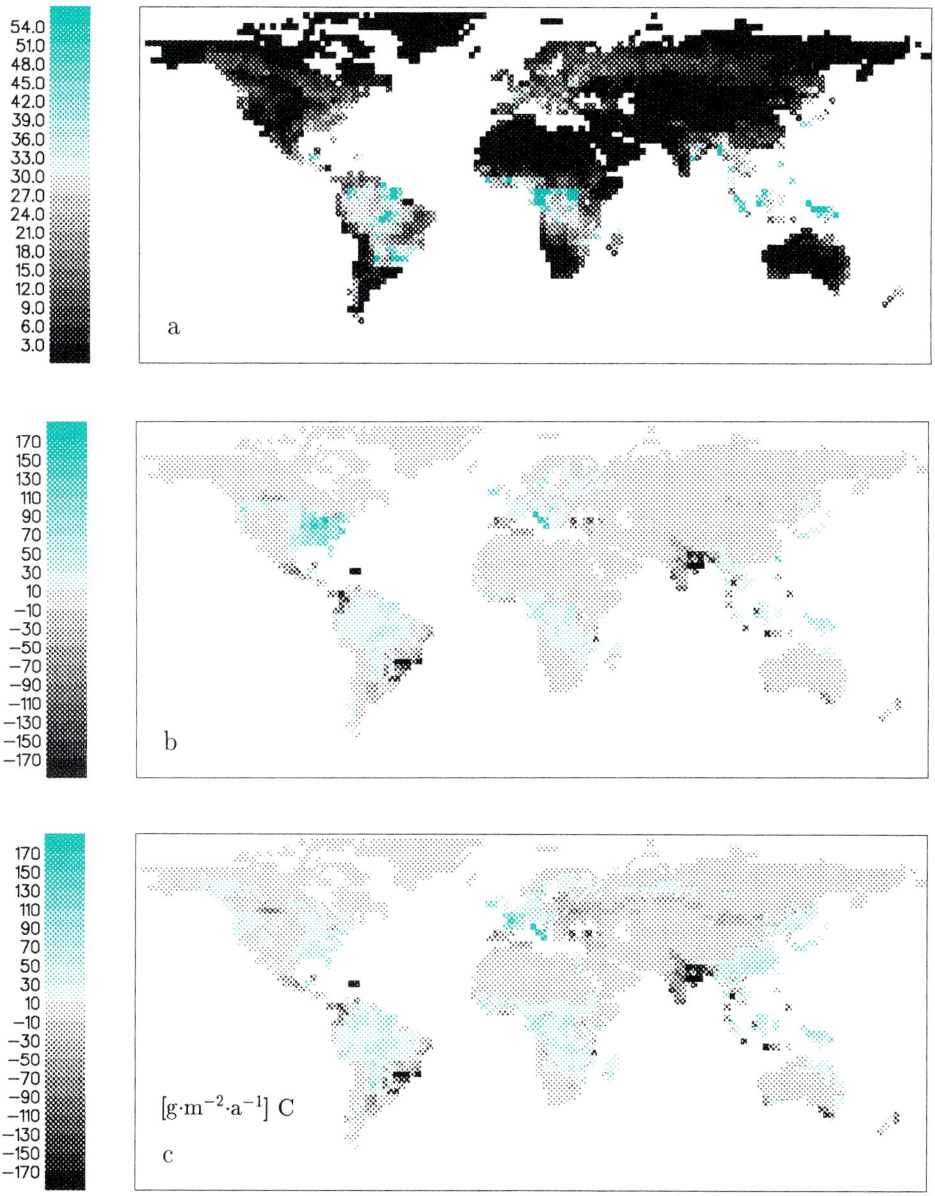

Figure 31.3: (a) CO_2 fertilization effect (phytomass increase), (b) phytomass change as balance of CO_2 fertilization and land–use changes (deforestation, reforestation), (c) total biospheric balance including phytomass, litter and soil. The reference year is 1980. The results are calculated with the Osnabrück Biosphere Model. Values mean carbon [$g \cdot m^{-2} \cdot yr^{-1}$]. Negative prefixes indicate losses to the atmosphere.

Table 31.9: Zonal integrals of the fertilization effect (a), the phytomass change due to fertilization effect and land–use changes together (b), and the total biospheric balance including phytomass, litter, and soil (c). The difference $(b - a)$ is the net effect of deforestation, reforestation, and land–use changes alone. The characters on top of the columns correspond to the world maps in Figure 31.3. Values are in 10^6 tons of carbon per year for zones of 10° latitude. Negative values are losses from the biosphere. The reference year is 1980.

hemisphere	latid. zone	fertilization	phytomass change	deforestation, reforestation	biosphere total
		(a)	(b)	$(b - a)$	(c)
north	70–80	1.1	1.0	−0.1	5.6
	60–70	40.4	34.9	−5.5	64.2
	50–60	80.7	62.6	−18.1	10.6
	40–50	102.5	171.1	68.6	143.9
	30–40	67.5	178.1	110.6	116.9
	20–30	80.5	−113.6	−194.1	−50.8
	10–20	82.3	−95.4	−177.7	−71.5
	0–10	233.9	79.5	−154.4	144.7
south	10–0	295.0	180.5	−114.5	215.0
	20–10	147.7	67.5	−80.2	115.4
	30–20	90.5	−102.3	−192.8	−57.3
	40–30	19.4	−20.9	−40.3	−31.0
	50–40	9.3	1.8	−7.5	1.6
	60–50	1.1	0.9	−0.2	1.9
total net biosph. sink		1251.9	445.7		608.2
total net biosph. source				−806.2	

sink in the phytomass pool (via the reduced stand age) as well as in the litter and soil organic carbon pools (via the reduced productivity).

Those processes are interacting. They may be deconvoluted by use of a model like the OBM. Results of such a deconvolution are shown in Figure 31.3: (a) demonstrates the regional importance of the fertilization effect. It is highest in the humid tropics where net primary productivity and phytomass are high. Moist subtropical or temperate forests fall behind the tropics. In the boreal coniferous forest regions (i.e. Siberia, Canada, Alaska, Northern Europe) the fertilization effect hardly exceeds 20 g carbon per year and m^2, frequently it is much lower. This is a consequence of the low productivity and phytomass in those regions. The zonal sums of the phytomass change caused by the fertilization effect are listed in Table 31.9 for the reference year 1980. In total, 1.25 Gt of carbon were additionally sequestered in the global phytomass, 0.76 Gt or 60% between 20° south and 20° north. In

Figure 31.3 (b) land use changes including deforestation are balanced against the fertilization effect. It is obvious that in regions with high deforestation rates net losses from the phytomass pool occur, although the fertilization effect alone may be high. An example for that is southeastern Brazil. In the southeastern United States carbon fixation by reforestation is more important than the fertilization effect. The fertilization effect in boreal zones is commonly below 10 $g \cdot m^{-2} \cdot yr^{-1}$ and therefore does not show up in Figure 31.3 (b). Globally, the 1.25 $Gt \cdot yr^{-1}$ of carbon sequestered by the fertilization effect are reduced to 0.45 $Gt \cdot yr^{-1}$, which means that 0.8 $Gt \cdot yr^{-1}$ were emitted from the living phytomass by land–use changes and deforestation globally. In the latitudinal belt 30° south to 30° north, the losses due to deforestation were 0.9 $Gt \cdot yr^{-1}$. In contrast, in the belt 30° to 50° north a net fixation of carbon of 0.2 Gt annually occurs probably due to reforestation (Fig. 31.9 $b - a$). Therefore, the total global losses due to deforestation (northern reforestation not taken in account) amount to 1.1 $Gt \cdot yr^{-1}$. In an recent paper HOUGHTON (1991) gives a range for the flux from changes in land use of 0.6 to 2.6 Gt C for 1980, although he stated in a personal communication that he subjectively considers 1.0 to 2.0 Gt C more likely. The OBM result is on the low side of this range.

If the secondary effects of CO_2 fertilization and land–use changes on soil organic carbon and litter are also considered (see Figure 31.3 (c) and Table 31.9 column (c)), then the global biospheric carbon balance is raised to 0.6 $Gt \cdot yr^{-1}$ fixation in 1980. It is interesting to compare the situations in the eastern United States and in western Europe. Obviously, considerable losses from soils and litter occur in the U.S., while the phytomass is a strong sink. This is probably because reforestation and low productive agriculture are present in the same grid elements. In western Europe, the phytomass is hardly increasing, while the soils and litter are sinks due to the increasing productivity of the agriculture in the EC. In east Asia and parts of the tropics, soils and litter are sinks. In general, the net effects on litter and soil are small (≈ 0.2 $Gt \cdot yr^{-1}$ in 1980).

Thus, we have the strange situation that although tropical deforestation was already a problem at the beginning of the 1980s, the fertilization effect together with reforestation over–compensated the losses. The terrestrial biosphere has been a sink for atmospheric CO_2.

TANS et al. (1990) compared observed atmospheric concentrations of CO_2 with boundary layer concentrations over the oceans, which were calculated with the transport fields generated by a general circulation model for specified source–sink distributions. In the model, the observed north–south atmospheric concentration gradient could be only maintained if sinks of CO_2 were greater in the extratropical northern than in the southern hemisphere. The authors concluded that there must be a terrestrial sink at temperate latitudes of the northern hemisphere to balance the carbon budget and to match the north–south gradient of atmospheric CO_2. We can compare their suggestion with the results of the OBM in this paper as listed in Table 31.9 column (c). The extratropical northern hemisphere (> 20° N)

Table 31.10: Development of the global sums of the major carbon pools of the terrestrial biosphere and the atmosphere in the period 1860–1980. Results of a model run with standard climate (no changes). The terrestrial pools are stated in Gt C, the atmosphere in $\mu l \cdot l^{-1}$. From Esser (1990).

year	pools					accum. fluxes	
	atmo-sphere $[\mu l \cdot l^{-1}]$	phytomass nat.	phytomass agri.	litter	organic C in soil [GtC]	deforest.	fertil. effect
1860	285.0	668	1.6	91	1536	−0	0
1870	286.7	663	1.7	92	1537	−7	2
1880	288.6	658	1.9	92	1538	−15	5
1890	290.7	654	2.1	92	1539	−22	8
1900	293.1	650	2.2	92	1539	−30	12
1910	296.1	646	2.4	92	1538	−37	15
1920	299.4	643	2.5	92	1538	−44	19
1930	303.2	639	2.6	92	1538	−54	25
1940	307.0	636	2.8	92	1537	−63	31
1950	311.6	633	2.9	92	1536	−73	38
1960	317.6	634	3.0	92	1535	−81	47
1970	326.5	637	3.1	92	1533	−89	58
1980	338.8	644	3.2	92	1531	−96	73

is a net sink of about 0.3 Gt C annually, while the similar southern latitudes are a source of about 0.08 Gt C. This northern hemispheric sink given by the OBM is considerably lower than the respective value estimated by TANS et al. (1990) of 2.0–2.7 $Gt \cdot yr^{-1}$. It can not be excluded at present, that the northern hemispheric sink might be larger than calculated by the OBM since this model does not yet consider the possible fertilization effect of anthropogenic emissions to the air besides CO_2, i.e. NO_x compounds.

Prior to TANS et al., KEELING et al. (1989) came to the conclusion, that the northern hemispheric sink necessary to explain the observed meridional atmospheric CO_2 gradient is only about 1 $Gt \cdot yr^{-1}$ C. They assumed the sink to be most likely mainly oceanic due to the observed $\delta\ ^{13}C$ profiles.

According to the OBM, the terrestrial biosphere has not always behaved as a sink for carbon. Time series of the pools calculated by the OBM for the period 1860–1980 (Table 31.10) show that until 1950 the phytomass decreased due to deforestation. The fertilization effect was too small to compensate for much of the carbon losses.

The litter and soil organic carbon show minor changes during the period 1860–1980. Therefore they probably contributed little to the carbon sequestering in the biosphere. Schlesinger (1990) comes to a similar conclusion in his recent paper in Nature. The results for the year 1980, which have been discussed in detail, suggest that the reduction of soil organic carbon which dominated since about 1900 has recently stopped.

In total, the terrestrial biosphere has been a small source for carbon ($\approx$ 26 Gt) between 1860 and 1980. Further back in time the fertilization effect was certainly negligible, while the deforestation rate already reached high values. Deconvolution of atmospheric CO_2 data including ice core data suggests a total net biospheric CO_2 source of about 93 Gt C in the period 1740–1980 (Keeling et al. 1989). This value is in good agreement with the OBM results as listed in Table 31.10, if we assume, that the net biospheric losses from deforestation prior to 1860 were in the same order of magnitude for each decade as in 1860–1870.

The results discussed so far were based on a standard climate data set with constant temperatures since 1860. It is probably more realistic to consider the global temperature rise since 1860. This would allow to investigate the temperature feedback effects. In this scenario the author used a temperature raise of 0.8 °C for the period 1860–1980, which is slightly higher than he observed (0.4–0.6 °C). The results are included in Table 31.11. The depletion fluxes of soil organic carbon and litter are enhanced while the losses from phytomass are 3 Gt less. This is due to the higher fertilization effect as consequence of the steeper CO_2 increase in the atmosphere. Although the effects are small, it nevertheless may be stated that there is a positive feedback of temperature changes.

The results in Table 31.11 also point out that it is probably unrealistic to ignore the CO_2 fertilization effect. If a fertilization effect is ignored it is necessary to assume an atmospheric CO_2 concentration of 261 $\mu l \cdot l^{-1}$ in 1860 to get the right value of 339 $\mu l \cdot l^{-1}$ for the year 1980. In contrast, the results of ice core analysis suggest 283 ± 3 ppm for 1860 (Friedli et al. 1986), which is in excellent agreement with model results which consider CO_2 fertilization and the temperature rise since 1860. For this scenario the CO_2 concentration of the atmosphere in the period of the Mauna Loa records (1958–1980) increased by 23.3 $\mu l \cdot l^{-1}$. The Mauna Loa records indicate an increase of 23.2 $\mu l \cdot l^{-1}$. This is not an absolute argument for the existance of a fertilization effect, since any other sink with a similar time behavior including an under–estimated oceanic uptake could have balanced the atmospheric CO_2. Further model studies are necessary which include seasonal cycles and the stable isotope fractionation.

31.6 Conclusions

The OBM model results suggest that 1980 taken as a reference year, the terrestrial biosphere has been a net sink for carbon (0.6 $Gt \cdot yr^{-1}$). Although the land–use changes (deforestation, reforestation, crop changes) led to an emission of 0.8 $Gt \cdot yr^{-1}$ (even 0.9 $Gt \cdot yr^{-1}$ between 30° south and 30° north and 1.1 $Gt \cdot yr^{-1}$ if northern reforestation is not considered) the losses were overcompensated by the

Table 31.11: Some model results based on three scenarios in the period 1860–1980. The atmospheric CO_2 concentration values given by the model are compared with measured data. Standard climate means the data set mentioned on page 697, temperature increase relates to a change of the mean annual temperature coupled to the atmospheric CO_2 to yield +0.8 °C in the period 1860–1980, no CO_2 fertilization assumes that NPP is uninfluenced by the atmospheric CO_2 level. Values are net changes in Gt carbon. Negative prefix indicates losses of the respective pool to the atmosphere. The CO_2 values are in $\mu l \cdot l^{-1}$. Mauna Loa means ΔCO_2 for the period 1958–1980 at the Mauna Loa observatory, Hawaii (KEELING, 1986). Ice core value after FRIEDLI et al. (1986). The ocean flux was calculated using a box diffusion model (OESCHGER et al. 1975).

	standard climate	temperature increase	no CO_2 fertil.
		$\mu l \cdot l^{-1}$ (ppmv)	
CO_2 1860	285	283	261
antarctic ice cores 1860		283 ± 3	
ΔCO_2 1958–1980	+22.6	+23.3	+37.4
Mauna Loa 1958–1980		+23.2	
changes 1860–1980		Gt C	
phytomass	−23	−20	−83
litter	+1	−0.5	−5
soil organic C	−5	−12	−26
ocean	+76	+78	+112
atmosphere	+114	+117	+164
fossil source		−163	

fertilization effect of the raised atmospheric CO_2, which amounts to 1.25 $Gt \cdot yr^{-1}$. Secondary effects in the litter and soil organic carbon bound another 0.2 $Gt \cdot yr^{-1}$.

Prior to 1950 the biosphere was a carbon source due to the low fertilization effect. In the entire period 1860–1980, the biosphere lost almost 30 Gt C as an overall net balance.

Concerning climate–biosphere feedback, the Osnabrück Biosphere Model calculates small additional losses of carbon, mainly from soils, for a temperature raise of 0.8 °C in the period 1860–1980.

It is not possible at present, to give any indication of the reliability of those model results in a statistical sense. If we compare the atmospheric CO_2 increase since 1860 calculated by the model with measurements, which were not used for the calibration of the model, i.e. the Mauna Loa records and ice core records of the Antarctic, then the most realistic model run is the one which considers a temperature raise of 0.8 °C in the period 1860–1980. In this model run, the CO_2 concentration in 1860 is equal to the measured value from ice cores (283 $\mu l \cdot l^{-1}$) and the raise 1958–1980 equals the corresponding value of the Mauna Loa records. If it is assumed, that no CO_2 fertilization effect exists, the atmospheric CO_2 calculated for 1860 is 261 $\mu l \cdot l^{-1}$ and the increase 1958–1980 is 37.4 $\mu l \cdot l^{-1}$. Either values are inconsistent with measurements. But it should be pointed out, that any other presently underestimated sink, i.e. the ocean, as well could compensate for the atmospheric surplus CO_2.

Although this might first be considered as a weak indication for the reliability of the model results, it is apprehended that model errors may compensate. It must be concluded that our understanding of the global carbon cycle is still very limited. Strong efforts should be made to (1) develop models which are more appropriate for validation (i.e. monthly rather than annual models) and (2) to develop proper validation techniques.

Acknowledgment

The author wishes to thank Charles D. Keeling, Scripps Institution of Oceanography, La Jolla (California), R. A. Houghton, The Woods Hole Research Center, Woods Hole (Massachusetts), William H. Schlesinger, Duke University, Durham (North Carolina), and Pieter Tans, NOAA, Boulder (Colorado), for careful reviews and helpful comments on the manuscript. The German Science Foundation (DFG), the German Federal Environmental Agency (UBA), the German Federal Ministry of Research and Technology (BMFT), and the Commission of the European Communities (CEC) strongly supported the work. I also wish to thank all colleagues in the German Climate Program and in the IGBP who promoted and encouraged the work by their assistance and critical discussion.

Bibliography

ASELMANN, I., LIETH, H., 1983, The implementation of agricultural productivity into existing global models of primary productivity.
In: Degens, Kempe, Soliman (eds.), Transport of carbon and minerals in major world rivers, Part 2, Mitt. Geolog.–Paläontolog. Inst. Univ. Hamburg, SCOPE/UNEP Sonderband, Heft **55**, 107–118.

DEGENS, E. T. (ed.) (1982). Transport of Carbon and Minerals in the Major World Rivers, Part 1. Mitt. Geol.–Paläont. Inst. Univ. Hamburg, SCOPE/UNEP Sonderband **52**.

DEGENS, E. T., KEMPE, S., AND SOLIMAN, H. (eds.) (1983). Transport of Carbon and Minerals in Major World Rivers, Part 2. Mitt. Geol.–Paläont. Inst. Univ. Hamburg, SCOPE/UNEP Sonderband **55**.

DEGENS, E. T., KEMPE, S., AND HERRERA, R. (eds.) (1985). Transport of Carbon and Minerals in Major World Rivers, Part 3. Mitt. Geol.–Paläont. Inst. Univ. Hamburg, SCOPE/UNEP Sonderband **58**.

ESSER, G., 1984, The significance of biospheric carbon pools and fluxes for the atmospheric CO_2: A proposed model structure.
Progress in Biometeorology **3**, 253–294.

ESSER, G., 1986, The carbon budget of the biosphere — structure and preliminary results of the Osnabrück Biosphere Model (in German with extended English summary).
Veröff. Naturf. Ges. zu Emden von 1814, New Series Vol. **7**, 160 pp. and 27 Figures.

ESSER, G., 1987, Sensitivity of global carbon pools and fluxes to human and potential climatic impacts.
Tellus **39**B, 245–260.

ESSER, G., 1989, Global land–use changes from 1860 to 1980 and future projections to 2500.
Eclogical Modelling **44**, 307–316.

ESSER, G., 1990, Modelling global terrestrial sources and sinks of CO_2 with special reference to soil organic matter.
In: Bouwman, A. F. (ed.), Soils and the Greenhouse Effect, Chapter 10., 247–261, John Wiley & Sons, Chichester · New York · Brisbane · Toronto · Singapore.

ESSER, G., ASELMANN, I., LIETH, H., 1982, Modelling the carbon reservoir in the system compartment "litter".
In: Degens, Kempe (eds.), Transport of carbon and minerals in major world rivers, Part 1, Mitt. Geolog.–Paläontolog. Inst. Univ. Hamburg, SCOPE/UNEP Sonderband, Heft **52**, 39–58.

ESSER, G.; KOHLMAIER, G., 1990, Modelling terrestrial Sources of Nitrogen, Phosphorus, Sulfur, and organic Carbon to Rivers.
In: DEGENS, KEMPE, RICHEY (eds.), Biogeochemistry of Maior World Rivers, SCOPE 42, Chapter 14, John Wiley & Sons, Chichester · New York · Brisbane · Toronto · Singapore.

ESSER, G., LIETH, H., 1986, Evaluation of climate relevant land surface characteristics from remote sensing.
Proc. ISLCP Conference, Rome, ESA Publ. SP–248 (May 1986), 205–211.

ESSER, G., LIETH, H., 1988, Decomposition in tropical rain forests compared with other parts of the world.
In: Lieth, Werger (eds.), Tropical rain forest ecosystems, Ecosystems of the World Vol. 14B, Elsevier Amsterdam.

FAO–Unesco, 1978 ff., Report on the Agro–Ecological Zones Project.
World Soil Resources Report **48/1** ff., Food and Agricultural Organization of the U. N., Rome.

FAO–Unesco, 1980 ff., Production yearbooks, Vol. 33 ff.
FAO Statistics Series No. **28** ff. Food and Agricultural Organization of the U. N., Rome.

FRIEDLI, H., LÖTSCHER, H., OESCHGER, H., SIEGENTHALER, U., STAUFFER, B., 1986, Ice core record of the $^{13}C/^{12}C$ ratio of atmospheric CO_2 in the past two centuries.
Nature **324**, 237–238.

GUDERIAN, R. (ed.), 1985, Air pollution by photochemical oxidants.
Ecological Studies **52**, Springer–Verlag Berlin · Heidelberg · New York · Tokyo.

HOUGHTON, R. A., 1991, Tropical deforestation and atmospheric carbon dioxide. Cliatic Change (in press).

INGESTAD, T., LUND, A.–B., 1986, Theory and techniques for steady–state mineral nutrition and growth of plants.
Scand. J. For. Res. **1**, 439–453.

INGESTAD, T., ÅGREN, G. I., 1988, Nutrient uptake and allocation at steady–state nutrition.
Physiol. Plant. **72**, 450–459.

Instituto Geographico de Agostini, 1969, 1971, 1973, World Atlas of Agriculture, Novara (Italy).

KEELING, C. D., 1986, Atmospheric CO_2 concentrations — Mauna Loa Observatory, Hawaii 1958–1986.
NDP–001/R1 Carbon Dioxide Information Centre, Oak Ridge, Tennessee (regularly updated).

KEELING, C. D.; BACASTOW, R. B.; CARTER, A. F.; PIPER, S. C.; WHORF, T. P.; HEIMANN, M.; MOOK, W. G.; ROELOFFZEN, H., 1989, A three–dimensional model of atmospheric CO_2 transport based on observed winds: 1. Analysis of observational data.
In: Peterson, D. H. (ed.), Aspects of climate variability in the Pacific and the western Americas. Geophysical Monograph **55**, 165–235, American Geophysical Union.

KEELING, C. D.; PIPER, S. C.; HEIMANN, M., 1989, A three–dimensional model of atmospheric CO_2 transport based on observed winds: 4. Mean annual gradients and interannual variations.
In: Peterson, D. H. (ed.), Aspects of climate variability in the Pacific and the

western Americas. Geophysical Monograph **55**, 305–363, American Geophysical Union.

Lieth, H., 1975, Modeling the primary productivity of the world.
In: Lieth, Whittaker (eds.), Primary productivity of the biosphere, Ecological Studies **14**, 237–283, Springer–Verlag, New York, Heidelberg, Berlin.

Maier–Reimer, E.; Hasselmann, K., 1987, Transport and storage of CO_2 in the ocean — an inorganic ocean-circulation carbon cycle model.
Climate Dynamics **2**, 63–90.

Müller, M. J., 1982, Selected climatic data for a global set of standard stations for vegetation science.
In: Lieth (ed.), Tasks for vegetation science, Vol. 5, Dr. W. Junk Publ., The Hague, Boston, London.

NCAR, National Center for Atmospheric Research, Data tape documentations TD–9645 and TD–9618.
Boulder, Co., U.S.A.

Oeschger, H.; Siegenthaler, U.; Schotterer, U.; Gugelmann, A., 1975, A box diffusion model to study the carbon dioxide exchange in nature.
Tellus **27**, 168–192.

Ramanathan, V., 1987, Climate–chemical interactions and effects of changing atmospheric trace gases.
J. Geophys. Res. **25**, 1441–1482.

Richards, J. F., Olson, J. S., Rotty, R. M., 1983, Development of a data base for carbon dioxide releases resulting from conversion of land to agricultural uses.
Institute for energy analysis, Oak Ridge Ass. Universities, ORAU/IEA–82–10(M); ORNL/TM–8801.

Schlesinger, W. H., 1990, Evidence from chronosequence studies for a low carbon–storage potential of soils.
Nature **348**, 232–234.

Schmithüsen, J., 1976, Atlas for Biogeography.
Meyers Grosser Physischer Weltatlas **3**, Bibl. Inst. Mannheim, Wien Zürich.

Tans, P. P.; Fung, I. Y.; Takahashi, T., 1990, Observational constraints on the global atmospheric CO_2 budget.
Science **247**, 1431–1438.

FAO–Unesco, 1974 ff., Soil Map of the World, 1:5,000,000.
Vol. I–X, Unesco Paris.

WALTER, H., LIETH, H., 1960 ff., World Atlas of Climate Diagrams. Gustav Fischer Verlag Jena, GDR.

WUEBBLES, D. J., EDMONDS, J., 1988, A Primer on Greenhouse Gases. Report for United States Dept. of Energy TR040, DOE/NBB-0083.

Chapter 32

Vegetation and climate: a tenuous link

J. GRACE

32.1 Introduction

According to current calculations, the continuing emissions of greenhouse gases will cause a mean global temperature rise of 0.3 °C per decade, with rates higher than this at northern latitudes (HOUGHTON, JENKINS and EPHRAUMS 1990). This represents a more rapid rate of change than has ever occured in the postglacial period (10,000 years ago to the present); and, coupled with direct effects of an elevation in CO_2, might be expected to bring about major shifts in vegetation patterns throughout Europe. This is illustrated particularly poignantly for the case of Great Britain. The current difference in July temperature between the far north (Stornaway, latitude 58.2 °N) and the south coast (Plymouth, 50.3 °N) is a mere three degrees Celcius. Thus, the thermal climate of northern Scotland may, within a hundred years, become similar to that of central or even southern England, and the English climate will become similar to that of southwestern France. Comparable transformations apply to other parts of northern Europe, as can be seen from a map of the July mean air temperature (Fig. 32.1). A shift in the boundaries of the major biomes will have wideranging implications for nature conservation, agriculture and forestry, as some species are gained and others lost.

Elsewhere, I used the well-known data set of LIETH (LIETH 1972, GRACE 1988) to estimate the effect of a change in temperature on the productivity of mid- and high latitude vegetation. There are difficulties in this approach, both in the extent of the data and its interpretation. The conclusion I came to from the scatter diagram in LIETH's paper, was that at the latitude of southern England, a 10 percent increase in productivity might be expected for every degree of warming. Since then, a number of other considerations have come to light; and I welcome the opportunity to re-examine the question.

32.2 Soils will change

As a result of enhanced microbial activity, organic matter in the soil will break down at a faster rate, releasing CO_2 and contributing further to the greenhouse effect (HENDRICKSON 1985, Fig. 32.2) Such positive feedbacks are not normally incorporated into GCMs, partly because of uncertainty about even the current rates of the process. The best estimates we have of the store of carbon in the soil suggest that it is almost twice that in the atmosphere (JARVIS 1989) so quite small increases in the rate of microbial activity might be expected to have a large impact on atmospheric CO_2. The soil is a source of other greenhouse gases, notably oxides of nitrogen and, in the case of anaerobic soils, methane.

The breakdown of organic matter in the soil will also release nitrogen, making it available for plant growth. Some idea of the rates of such processes is obtained from laboratory microcosm studies, in which soil or humus has been incubated at a range of temperatures. Both CO_2 and NH_4^+ are released at increasing rates over the temperature range of northern Europe (HENDRICKSON 1985). If warming is accompanied by drying, then the surface layers of many of the cold wet soils of the north and the uplands will become aerobic, leading to enhanced microbial activity. These soils, in agricultural or silvicultural terms, are currently 'difficult', but may be transformed in a warmer dryer climate (INESON et al. 1990).

Chemical weathering of mineral soils will increase with temperature. Overall, we expect an increase in the rate of supply of mineral elements to plants, which may be expected to increase plant growth rates proportionately. This aspect of climatic warming has undoubtedly been neglected, as workers have concentrated on the more easily studied, above ground, aspects of climate warming. The relevant experimental approach is that of Ingestad (INGESTAD 1982): in laboratory systems it can be shown that the relative growth rate of birch, willows and other species can be controlled by manipulating the relative addition rate of nitrogen. In the data of HENDRICKSON (1985) on forest humus, a three degree warming increased the release of ammonium by approximately 25% over the range 15–25 °C. If INGESTAD's hypothesis is applicable to the natural environment (this has scarcely been tested at all), we may expect commensurate increases in relative growth rate.

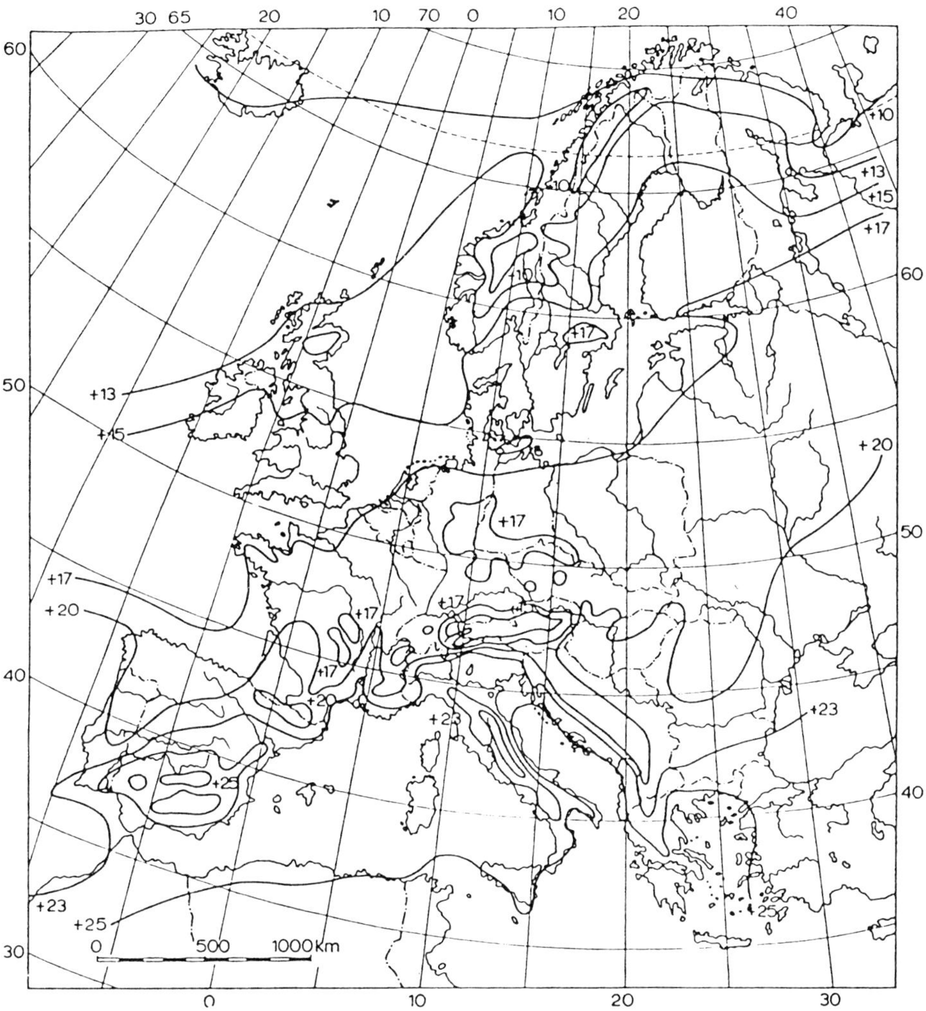

Figure 32.1: Mean temperature at station level, Europe, July. From LANDSBERG (1970).

32.3 Climate warming in the uplands

In the uplands, the growth of many species is limited by the temperatures of the plant tissues (GRACE et al. 1988). The familiar reduction in growth rate of trees

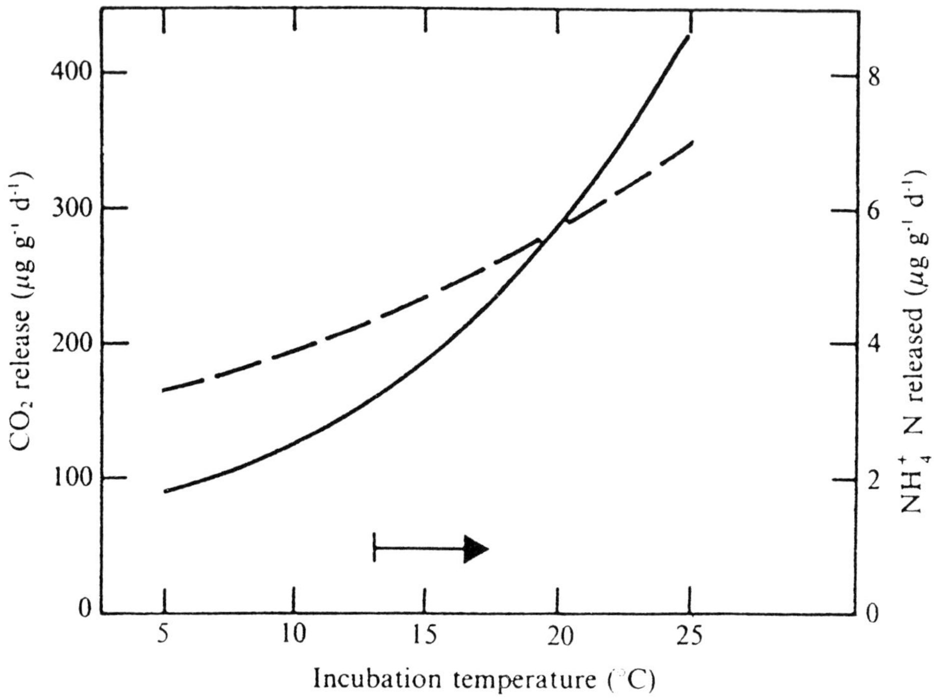

Figure 32.2: Breakdown of forest humus to yield carbon as CO_2 (solid line) and nitrogen as ammonium (broken line). From HENDRICKSON (1985).

with increase in altitude is caused by a combination of winter damage at high elevation (including frost–drought at the alpine tree line), the shorter season for growth, and the lower temperatures during the growing season. A demonstration of the primary importance of temperature, compared with other climatic variables, comes from a study of the growth rings of Scots pine near its climatic limit in the Cairngorm mountains of Scotland (GRACE et al. 1990, Fig. 32.3). Ring width in Scots pine was positively correlated with the temperatures of January and February (severe winters cause much damage to the trees) and July (warm July enables rapid extension growth of the main axis and the needles).

Another type of study provides evidence for the effect of temperature on Scots pine: Juntilla in Finland has obtained temperature response curves by growing pines in controlled environment rooms. He showed that an increase in temperature from 9 to 18 degrees caused a doubling in the number of needle fascicles produced

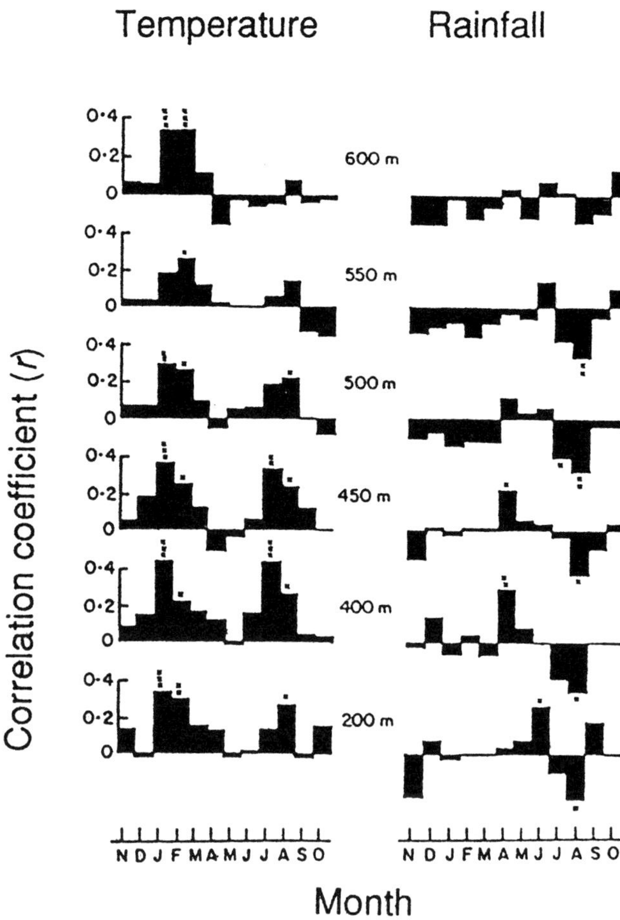

Figure 32.3: Plots of correlation coefficients between ring-widths of *Pinus sylvestris* and monthly climate. The data are from six altitudes at a native pine-wood in Scotland. From GRACE and NORTON (1990).

the following year (JUNTILLA 1986). Finally, a third type of evidence comes from studies of old tree stumps in the peat, mainly by Pears working in Scotland and Kullman in Scandanavia. They find that 6,000 years ago, when we think the temperature was a mere 1 °C warmer and perhaps dryer than it is today, the treeline was considerably higher in altitude and latitude than it is today (PEARS 1975, KULLMAN 1981). An important conclusion here is that the altitudinal and latitudinal limit for tree growth is likely to increase. If we take Koppen's rule that tree lines coincide with the July isotherm of 10 °C, and assume a temperature lapse rate of 7 °C km^{-1}, then we may expect the tree line to advance by 140 m for every

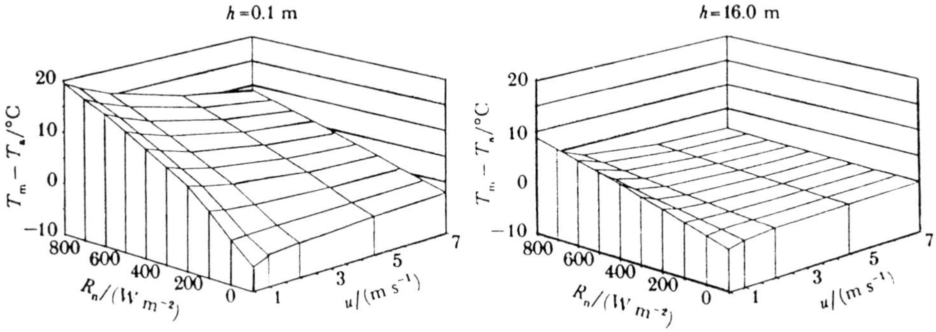

Figure 32.4: Influence of wind u and net radiation R_n on the difference in temperature between the terminal meristems and the air ($T_m - T_a$). The vegetational height h is set to 0.1 m and 16 m to represent dwarf shrubs and forest. From a model of heat balance, described in GRACE, ALLEN and WILSON (1989).

degree of warming.

Tissue temperatures of trees are closely coupled to air temperatures, as a result of the aerodynamic roughness of forests. In dwarf shrubs, and herbaceous vegetation, the tissue temperature is much more closely related to net radiation. This can be shown by observation and by application of the so–called 'big leaf' model in which the energy balance is solved for a structure with surface resistances corresponding to tall and short vegetation (KÖRNER et al. 1987; WILSON et al. 1987; GRACE et al. 1989). This has important consequencies for climate change, as it is the extent of cloudiness that has been particularly difficult to predict. If the cloudiness increases, the tissue temperatures are likely to fall despite a rise in the air temperature. Tissue temperature of dwarf shrubby vegetation is also a strong function of wind speed (Fig. 32.4). Thus any change in mean wind speed might also have a profound influence on tissue temperatures, possibly outweighing the effect of climatic warming.

32.4 Some negative effects of temperature on productivity

In lowland agriculture, the benefit of an increased temperature is less apparent. The most notable data set is the long record of wheat yields at the Broadbalk field, Rothamsted, England; where wheat was harvested in most years starting from 1853. The method of analysis was to correlate monthly values of rainfall, sunshine and temperature with final grain harvest of the crop, in an attempt to elucidate which is the most important variable, and in which month does it exert its strongest effect. The conclusion was that rainfall often accounted for a large part of the year–to–year variation, hours of sunshine sometimes did, but temperature never did (FISHER 1924; TIPPETT 1926; BUCK 1961). This result was considered by some people to be an anomoly, as Rothamsted is in one of the drier parts of the British Isles and perhaps not typical; until Monteith pointed out a negative correlation between yield of wheat crops in some other parts of southern England and the mean temperature of May, June and July (MONTEITH 1981). It was proposed that this may be the result of the determinate nature of development in such annual crops: with high temperatures, development proceeded at a faster rate. A consequence was that the duration of the crop was less in warm years and so the total photosynthesis and the yield of grain was also less.

Negative effects of temperature on yield are also well known for some fruit crops. Yields of the Cox's apple in Kent from 1950 to 1979 were reduced when the temperatures prior to blossoming were high (JACKSON et al. 1980, GOLDWIN 1982). The reduction in yield was about 10% for every 1 °C increase in temperature in the period February to April.

32.5 Northern and southern plants

Current geographical distribution patterns of plants furnish clues about what might happen to vegetation in the longer term. Such data are available as maps for individual countries and Europe as a whole (PERRING et al. 1962, JALAS et al. 1976). Some species have a markedly northern distribution, and so may fare less well in a warmer climate. The Bird cherry *Prunus padus* and the Rowan *Sorbus aucuparia* are distributed in the north of Britain, where they readily set seed. Very many more trees are southern, and do not reproduce in the north of Britain or are killed by cold winters. Southern trees include beech *Fagus sylvatica*, sweet chestnut *Castanea sativa*, apple *Malus sylvestris*, limes *Tilia cordata* and *T. platyphyllos*.

In some cases, detailed autecological investigations show cleary that the distribution is indeed related to summer temperature. In *Tilia cordata* for example, it is observed that at the northern limit, growth of the pollen tube is insufficient to effect fertilisation (PIGOTT et al. 1981). Not all species are climatically controlled in this way of course. There are others which cannot tolerate low temperatures in the winter; and probably many others whose distribution is much more restricted by soil factors than temperature (GRACE 1987). But given warmer temperatures over Europe as a whole, southern species now cultivated in southern Europe might become successful in northern Europe. It should be borne in mind that vineyards were successful all over England in the thirteenth century, when the temperature was perhaps one degree warmer on average than it is today. Given a three degree warming, vineyards could be as successful in sheltered parts of Scotland as they now are in Germany.

32.6 Beware of pests, diseases, droughts and winds

So far, the positive and primary influence of temperature on tree growth has been emphasised. But pests and diseases are likely to benefit too. Warm winters are likely to increase the number of insects surviving to the spring, and outbreaks of defoliating insects may become more common. One or two examples will show how important this may be. The green spruce aphid *Elatobium abietinum* overwinters in the adult/nymphal stage in maritime climates, and its survival is strongly affected by temperature (WATT et al. 1990). It is a severe defoliator of Sitka spruce, the most frequent species planted in British forestry, and any upsurge in its populations could devastate the forest estate. Similarly, the fecundity of the pine beauty moth, *Panolis flammea* is a strong function of temperature, and this species has already destroyed large areas of lodgepole pine in northern Scotland. Such outbreaks of insect pests are found not only in plantations, though it is here that they cause most concern.

If droughts become more frequent, as supposed in some scenarios, we may expect symptoms of forest decline. It is not understood how water stress affects tree performance over periods of several years, but there seems to be a carry over, perhaps the result of extensive embolism in the conducting tissues (PEÑA et al. 1986). The decline of red spruce in the Appalachians has occured over a century or more, and attributed mainly to warming, but might also have been the result of a decrease in the availability of water (HAMBURG et al. 1988). Some species are particularly intolerant of drought. Sitka spruce, for example, originates from the maritime zone of north America, and its growth is reduced when the atmosphere is dry. A combination of warming and drought will increase the likelihood of fires. Large areas of heather moorland and rough grazing as well as forest plantations are liable to destruction.

Even more worrying than drought, is the influence of gales. Gales may increase if atmospheric circulation patterns are appreciably altered. The north-west of Europe is currently one of the windiest regions of the world, and in the UK, 50 percent of the Forestry Commission's holdings are in windthrow hazard class 4–6, where trees cannot be grown more than 16–19 m tall. There is no safety margin: any increase in 'great gales' will affect the profitability of commercial forestry, as well as felling many amenity trees in gardens and parklands.

32.7 What should we do?

We may be entering a period of great climatic instability. The forest industry is particularly vulnerable, because of the longevity of the crop (CANNELL et al. 1989). The trees we are planting today will mature over the next 50 years, and so we should be attempting to exploit the increase in potential productivity, whilst reducing the possible adverse effects of pests, diseases, fire and wind damage. On the other hand, we must bear in mind that warming is not guaranteed. So we must not plant species that are too tender: we should hedge our bets.

Mixtures provide a way of doing this. Thus, new plantations might be composed of two or three genotypes in intimate mixture, one of them selected to exploit warming by producing either an increased yield, or a more valuable timber.

To avoid outbreaks of insect pests, native species should be chosen, on the grounds that natural predators for the insect pests are more likely to be present. Unfortunately, this is a very limiting restriction to apply to conifers in the UK, as there are only three natives, Juniper, Yew and Scots pine. Within Europe, there are many more species of conifer to chose from. However, the greatest benefit in the British Isles might come from an increased usage of broadleaved species, particularly oak, ash and beech in the north; and in the south of Britain the development of a rich diversity in community/amenity forests by the supplementation of the traditional diet with species widely used in southern Europe, such as *Quercus ilex, Prunus dulcis, Populus, Juglans regia, Ficus carica, Fagus orientalis* and *Castanea sativa*.

Agriculture is inherently more flexible than forestry because of the annual nature of the crop. It is easier to breed new genotypes, and southern crops may be soon taken on and grown by northern farmers. Over the world as a whole, the geography of food production may change dramatically. Northern regions will become more successful at producing all kinds of agricultural and horticultural crops, whilst the southern parts of Europe may be less productive and, if there is drought, more reliant on irrigation. This could mean a welcome economic improvement for some countries (Russia).

Nature conservationists will face the prospect of vegetational change. Nature reserves which have been selected as typical or representative examples of native vegetation will no longer be sustainable, and may be invaded by weeds and alien species. There will be excitement as rare species increase, but disappointment as other rarities decline. National surveillance sites are required to keep track of these changes, as nature reserves take on a new role as sensors of global change.

We face the challenge of adapting our patterns of land use to accommodate, and even benefit from, new species and varieties. This will be a risky and inherently unpredictable activity. Our best hope to succeed comes from the co-operation of many disciplines in the biological and physical sciences.

Bibliography

BUCK, S. F., 1961, The use of rainfall, temperature and actual transpiration in some crop-weather interactions.
Journal of Agricultural Science, Cambridge **57**, 355–365.

CANNELL, M. G. R.; GRACE, J.; BOOTH, A., 1989, Possible impacts of climatic warming on trees and forests in the United Kingdom: a review.
Forestry **62**, 337–264.

FISHER, R. A., 1924, The influence of rainfall on the yield of wheat at Rothamsted. Transactions of the Royal Society **213**, 89–142.

GOLDWIN, G. K., 1982, A technique for studying the association between components of the weather and horticultural parameters.
Scientia Hortensis **16**, 101–107.

GRACE, J., 1987, Climatic tolerance and the distribution of plants.
New Phytologist **106**, 113–130.

GRACE, J., 1988, Temperature as a determinant of plant productivity.
in: Long, S. P. and Woodward, F. I. (eds.), Plants and Temperature, The University Press, Cambridge, 91–107.

GRACE, J., 1989, Tree lines.
Philosophical Transactions of the Royal Society **B324**, 233–245.

GRACE, J.; ALLEN, S. J.; WILSON, C., 1989, Climate and the meristem temperatures of plant communities near the tree line.
Oecologia **79**, 198–204.

GRACE, J.; NORTON, D., 1990, Climate and growth of *Pinus sylvestris* at its upper altitudinal limit in Scotland.
Journal of Ecology **78**, 601–610.

GRACE, J.; UNSWORTH, M. H., 1988, Climate and microclimate of the uplands.
In: Thompson, D. and Usher, M. (eds.), The Changing Uplands, Blackwell Scientific Publications, Oxford. 137–150.

HAMBURG, S. P.; COGBILL, C. V., 1988, Historical decline of red spruce populations and climatic warming.
Nature **331**, 428–430.

HENDRICKSON, O. Q., 1985, Variation in the C:N ratio of substrate mineralised during forest humus decomposition.
Soil Biology and Biochemistry, **17**, 435–440.

HOUGHTON, J. T.; JENKINS, G. J.; EPHRAUMS, J. J., 1990, Climate Change — The IPCC Scientific Assessement.
The University Press, Cambridge.

INESON, P.; STEVENS, P. A., 1990, Soil Processes.
in: Cannell, M. G. R. and Cooper, M. D. (eds.), The Greenhouse effect and terrestrial ecosystems of the UK, HMSO, London, 14–17.

INGESTAD, T., 1982, Relative addition rate and external concentration; driving variables used in plant nutririon research.
Plant Cell & Environment **5**, 443–453.

JACKSON, J. E.; HAMER, P. J. C., 1980, The cause of the year–to–year variation in the average yield of Cox's Orange Pippin apple in England.
Journal of Horticultural Science **55**, 149–156.

JALAS, J.; SUOMINEN, J., 1976, Atlas Florae Europaeae: Distribution of Vascular Plants in Europe. Volume 3: Salicaceae to Balanophoraceae.
Societas Biologica Fennica Vanamo, Helsinki.

JARVIS, P. G., 1989, Atmospheric carbon dioxide and forests.
Transactions of the Royal Society of London **B324**, 369–392.

JUNTILLA, O., 1986, Effects of temperature on short growth in northern provenances of *Pinus sylvestris* L. .
Tree Physiology **1**, 185–192.

KÖRNER, CH.; DIEMER, M., 1987, *In situ* photosynthetic responses to light, temperature and carbon dioxide in herbaceous plants from low and high altitude.
Functional Ecology **1**, 179–194.

KULLMAN, L., 1981, Recent tree–limit dynamics of Scots Pine (*Pinus sylvestris* L.) in the southern Swedish Scandes.
Wahlenbergia **8**, 3–67.

LANDSBERG, H. E. (Ed.), 1970, World Survey of Climatology, Volume 5.
Elsevier, Amsterdam.

LIETH, H. 1972, Modelling the primary productivity of the world.
Tropical Ecology, **13**, 125–130.

MONTEITH, J. L., 1981, Climatic variation and the growth of crops.
Journal of the Royal Meteorological Society **107**, 749–774.

PEARS, N. V., 1975, Tree stumps in the Scottish hill peats.
Scottish Forestry **7**, 255–259.

PEÑA, J.; GRACE, J., 1986, Water relations and ultrasound emissions of *Pinus sylvestris* L., before, during and after a period of water stress.
New Phytologist **103**, 515–524.

PERRING, F. H.; WALTERS, S. M., 1962, Atlas of the British Flora.
Nelson, London.

PIGOTT, C. D.; HUNTLEY, J. P., 1981, Factors controlling the distribution of *Tilia cordata* at the northern limit of its geographical range III. Nature and causes of seed sterility.
New Phytologist **87**, 817–839.

TIPPETT, L. H. C., 1926, On the effect of sunshine on the wheat yield at Rothampsted.
Journal of Agricultural Science, Cambridge **16**, 159–165.

WATT, A. D.; WARD, L. K.; EVERSHAM, B. C., 1990,
In: Cannell, M. G. R.; Cooper, M. D. (eds.), The Greenhouse effect and terrestrial ecosystems of the UK, HMSO, London, 32–37.

WILSON, C.; GRACE, J.; ALLEN, J.; SLACK, F., 1987, Temperature and stature: a study of temperatures in montane vegetation.
Functional Ecology **1**, 405–413.

Part VII

Remote sensing and geographic modelling

Chapter 33

Global ecology: the role of remote sensing

D. E. WICKLAND

33.1 Introduction

Ecologists have striven for a global perspective on vegetation patterns and processes since the time of Alexander von Humboldt. Geographic and cartographic techniques were first used (HUMBOLDT and BONPLAND 1807), then general conceptual, organizing schemes were put forward (HOLDRIDGE 1947, DANSEREAU 1957), followed by modeling of global vegetation patterns based on simple climatic constraints (LIETH and BOX 1972, LIETH and WHITTAKER 1975, BOX 1976). It has only been since the advent of satellite-based remote sensing of the Earth in the 1960's and 1970's that ecologists have had a tool to synoptically observe global ecosystems (TUCKER et al. 1986, BOTKIN et al. 1984). One might have expected this new tool to have revolutionized vegetation science and to have led to new insights into the ecological processes which are manifest at large spatial scales. This has not happened – although one could argue it may be about to happen in the context of the International Geosphere–Biosphere Program (NRC 1986, IGBP 1988). This chapter will discuss the current capabilities of remote sensing for ecology and explore the reasons why remote sensing has not yet emerged as a principal tool employed by ecologists for developing understanding of global, or even local to regional, ecological processes. It is not intended that this chapter be a thorough review of the literature or of the state-of-the-art for all of remote sensing, but rather an essay assessing current capabilities of airborne and spaceborne remote sensing for ecology.

33.2 Ecological information from remote sensing

33.2.1 The interaction of electromagnetic radiation with the Earth's surface

Space-based remote sensing opens the possibility of acquiring ecological information over time scales and spatial scales never before possible. For the purpose of this discussion, remote sensing will be defined to include only measurements by airborne and spaceborne instruments of the reflected and emitted electromagnetic radiation from the Earth's surface. Sufficient solar radiation for remote sensing is transmitted through (down and back up to the sensor) the Earth's atmosphere in the following general wavelength regions: 0.4–2.5 μm, 3–5 μm, 8–14 μm, and 1–80 cm. Remote sensing instruments have been developed to measure radiation throughout these regions of the electromagnetic spectrum. Passive techniques, which rely exclusively on solar radiation or natural emissions, and active techniques, where the remote sensing instrument both provides the illumination source and measures the radiation returned (e.g., lidar and radar), have been employed.

Radiation reaching the Earth's surface can be either scattered or absorbed by the materials with which it interacts. For most surface materials, a portion of the incident radiation will be scattered and reflected back away from the Earth's surface, a portion will be absorbed, and another portion will be transmitted through the material. Remote sensing instruments detect that radiation which is reflected back through the Earth's atmosphere and reaches the sensor; this radiation is referred to as radiance. The energy from absorbed radiation can be emitted by surface materials at longer wavelengths. The radiance reaching a passive remote sensor is dominated by reflectance in the 0.4–2.5 μm wavelength region and by emittance in the 3 μm–80 cm wavelength region. Active sensors can examine reflectance throughout the available electromagnetic spectrum. The radiances measured using active microwave sensors, i.e., radar, are usually referred to as backscatter because they are dominated by scattering effects.

A great deal is known about the responses of vegetation, soils, and phytoplankton in the various wavelength regions measured by remote sensing instruments. As a general rule, scattering contains information about the structure of surface materials, and absorption contains information about their chemical composition. Electromagnetic radiation is scattered most strongly by structures that are of about the same size as or larger than the wavelength of the electromagnetic radiation. Thus, it is believed that molecules and biochemical compounds dominate scattering at short, visible wavelengths (0.4–0.7 μm), organelles and cells at near to shortwave infrared wavelengths (0.7–2.4 μm), and leaves and branches in the microwave region (1–80 cm). Absorption is believed to be dominated by pigments in the visible region and by water in the shortwave infrared (1.1–2.4 μm) through the microwave region. Similar biophysical understanding has been established for what is measured using differing viewing conditions and constraints, spatial and spectral resolutions, and temporal coverages. It should be noted that some of this understanding borders on dogma, not having been rigorously demonstrated, yet being generally believed to be true. This body of information will be examined in the sections which follow. Each section will describe first the applications of remote sensing that have been suggested in the literature and then will discuss the general validity of these assertions based on the available scientific evidence.

33.2.2 Basic observations across the electromagnetic spectrum

Surface cover pattern and identification

Unique identification of vegetation types and land cover classes has been promised from sensors working across the electromagnetic spectrum. In some cases, even species identification has been promised. Satellite data have been used to document phenological patterns on coarse and fine spatial scales and to compare interannual variability in these patterns.

For some parts of the world, it is possible to classify vegetation into broad community types with about 70–90 percent accuracy (WICKLAND 1989). In general, such broad vegetation types as grasslands, wetlands, deserts, shrublands and savannas, evergreen and needle-leaved forests, and deciduous and broad-leaved forests can be easily discriminated by a variety of remote sensing instruments. With high spatial and spectral resolution sensors, it is often possible to discriminate among different community types or species associations within forests, wetlands, grasslands, and savannas. The degree of specificity is dependent on the vegetation type, its phenological stage, and the sensor's sensitivity, wavelength range, and spatial resolution. For example, in the floristically rich Southern Appalachians of North America it is not normally possible to classify Landsat Thematic Mapper (TM) data into species association types, but Thematic Mapper Simulator data acquired at the time of peak fall color have been used successfully to do so (LANG et al. 1984).

It is possible to identify vegetation to the species level only under special conditions. Monocultures and single-species dominated communities can often be discriminated from other vegetation types, but for more mixed species associations, success is much less likely. High spectral resolution (20 nm resolution) data in the visible through the shortwave infrared region have been reported to be capable of species association discrimination in a complex mixed community (GOETZ et al. 1985), but this application has never been extended or generalized, and uses of the same sensor in other locations have not yielded similar success. Multi-date imagery documenting phenological differences among species or groups of species also shows promise for identifying species and species associations (TUCKER et al. 1985). Combining information from sensors spanning the visible through the infrared and microwave regions of the spectrum also may improve the accuracy of vegetation type identifications in ecosystems with mixed species composition. These approaches to species and species association identification remain to be tested.

Imagery acquired in the thermal infrared wavelengths (3–5 μm and 8–12 μm) shows patterns related to surface temperature and emissivity. In vegetated areas, where the emissivity is close to 1.0 and varies little, the main differences in thermal response are related to differences in surface temperatures. Clearings and bare areas are much warmer than forests during the daytime, and cooler at night. These patterns can be used to discriminate land cover types and, occasionally, to identify areas of stressed vegetation (with higher canopy temperatures related to closed stomata) (SADER 1986; MORRISSEY et al. 1986). For the most part, these applications are more qualitative than quantitative (although see "biochemical composition", page 732).

Microwave imagery contains useful information about land surface cover. The main information content in microwave wavelengths relates to surface water and surface structure or roughness. Microwave remote sensing has one great advantage in that it is generally insensitive to weather and can image the surface through clouds. This, of course, is of great advantage for observing land cover patterns in regions of the world that are frequently cloud-covered, such as the humid tropics. Radar discriminates major transitions in surface structural and textural properties and delineates edges well under certain viewing geometries. Microwave remote sensing, both active and passive, is particularly well-suited for discriminating wetlands from surrounding upland vegetation. Wetlands have strong emission and backscatter responses due to the high dielectric constant of water. Radar can be used to identify areas of flooding, even under fairly dense canopies. Widespread use of these techniques has been constrained by the limited availability of high spatial resolution airborne and spaceborne microwave data.

Fire detection

Satellite observations in the visible through the thermal infrared regions of the spectrum have been used to document the occurrence and timing of fires and their areal extent over large regions of the Earth's surface (MATSON and STEPHENS 1987, MALINGREAU et al. 1985). It also has been suggested that aerosol loadings and even trace gas emissions (e.g, methane, carbon dioxide, nitrous oxide) from fires can be inferred using remotely sensed data (KAUFMAN in press). Advanced very high resolution radiometer (AVHRR) data, because they are collected twice daily for every spot on the Earth's surface, can capture a significant amount of the dynamics of large fires. Both daytime and nighttime thermal data (in 3–5 μm and 8–12 μm wavelength regions) can be used to count fires and estimate their size and state. Imagery in the visible and near infrared channels can be used to infer certain properties of the fire (e.g., particulate emissions, burning efficiency) based on characteristics of the smoke plumes, and after the fire is over, to estimate the areal extent of burning. These techniques are extremely useful and are contributing greatly to understanding of biomass burning on a global scale. However, there are a few limitations that ought to be noted. Actual fire temperatures cannot be measured directly. Fires of limited extent and short duration are less likely to be observed because of the 12-hour time interval between satellite observations. Small, but intense, sub-pixel scale fires, which can saturate the thermal detectors, will appear to be larger than they actually are. If the effects of these types of fires cannot be quantified in some other fashion, the satellite generated estimates of the effects of biomass burning will be biased. Also, studies to estimate smoke plume aerosol properties are just beginning, and the inference of trace gas emissions is wholly untested. It is likely that several gross assumptions will have to be made in order to estimate trace gas emissions. The effort will be worthwhile only if global fluxes can be predicted with greater accuracy than before.

Change detection

It is possible to examine change over time on the Earth's surface using remotely sensed imagery acquired at multiple dates. Techniques to conduct these analyses are complicated because the calibration and/or intercalibration of the sensor(s) involved must be understood well and atmospheric effects must be removed from the data. When the image data have been adequately corrected, it is a relatively simple procedure to overlay digitally the images and quantitatively assess the areas of change. The types of changes detectable using such methods depend on the level of detail to which the original data could be classified.

The historical satellite record can be used to examine the changes in surface cover over time. The record available to most scientists consists of Landsat Multispectral Scanner System (MSS) data since 1972, Landsat TM data since 1982, System Probatoire d'Observation de la Terre (SPOT) data since 1986, Advanced Very High Resolution Radiometer (AVHRR) normalized difference vegetation index (NDVI) data since 1981, Scanning Multifrequency Microwave Radiometer (SMMR) data from 1978–1985, Special Sensor Microwave Imager (SSM/I) data since 1987, and Coastal Zone Color Scanner (CZCS) data from 1978–1986. The only available historical satellite radar data sets are those from the 3 month Seasat mission in 1979 and the Shuttle Imaging Radar Missions in 1981 and 1984.

It is now possible to measure quantitatively changes in vegetation and phytoplankton extent, distribution, and pattern over time scales of weeks, months, seasons, or years. These data allow ecologists to explore relationships between surface cover and such important environmental phenomena as regional drought, deforestation, catastrophic disturbances, and the El Niño – Southern Oscillation

(TUCKER et al. 1985, 1986, HALL et al. 1987, SADER and JOYCE 1988, MALINGREAU and TUCKER 1988). In time, it is likely these data will be used to produce improved maps of land vegetation, and seasonal phenological patterns and interannual variations will be used to delineate more accurately boundaries between types. TUCKER et al. (1985) has already demonstrated how phenological information can be used to produce an improved rendition of the vegetation types in Africa.

Terrestrial primary productivity and biomass

Simple vegetation indices derived from ratios of broad-band reflectance in red and near-infrared wavelengths have been reported as good estimators of leaf area index (LAI), vegetation greenness, absorbed photosynthetically active radiation (APAR), net primary productivity, canopy chlorophyll concentration, photosynthetic efficiency, and above ground biomass for land vegetation (WICKLAND 1989). At present, there is no clear community concensus about what is really being measured by these vegetation indices. It seems that, of the various vegetation indices reported in the literature, the majority of scientists prefer the normalized difference vegetation index (NDVI; which is the difference between the radiance at near infrared wavelengths and the radiance at red wavelengths divided by their sum), although most admit there is little difference among them. The best evidence to date suggests that the NDVI is a very good estimator of the percentage of photosynthetically active radiation that a canopy could absorb (i.e., its maximum potential for photosynthesis at an instant in time), and when integrated over a growing season, the NDVI is a very good estimator of net primary productivity (SELLERS 1985, GOWARD et al. 1985). There is a concern that in ecosystems with very high leaf area index values the NDVI may saturate and therefore be relatively insensitive to changes in productivity.

There are more serious concerns about the sensitivity of the NDVI when very little vegetation cover is present. It has been shown that differences in the soil or litter background can cause large differences in NDVI that do not correspond to real changes in vegetation cover or productivity (HUETE et al. 1985, HUETE 1989). HUETE (1988) has proposed a soil-adjusted vegetation index to address this problem with the NDVI in arid and semi-arid environments. CHOUDHURY and TUCKER (1987) have noted that a passive microwave (SMMR) derived 37 GHz (8.2 mm) polarization difference vegetation index is more sensitive to vegetation in arid and semi-arid regions and have suggested that it be used in conjunction with the NDVI to improve estimates of global terrestrial primary productivity. However, none of these refinements of the NDVI have been incorporated into global terrestrial productivity estimates yet.

In winter at high latitudes, spurious values of the NDVI result because of the extremely low signals received by the AVHRR. Surface temperature data have been used to correct these data by assigning an NDVI of 0.0 if the surface temperature is less than 0 °C; this assumes that essentially no photosynthesis occurs below this temperature (TUCKER et al. 1986). Similar problems with spurious NDVI's seem to occur in cold regions with evergreen vegetation.

Although empirical relationships have been demonstrated, it does not seem possible to determine biomass, or even above ground biomass, using vegetation indices derived from visible and infrared wavelength reflectances. However, it may be possible some day to estimate biomass using these indices in combination with remotely sensed estimates of vegetation height. Radar or lidar remote sensing observations of vegetation structure may be able to provide estimates, or even measures, of vegetation height. Airborne laser profilers have already provided extremely accurate measures of forest heights (NELSON et al. 1984). Radar altimeters may prove adequate for measuring canopy height along transects or at scattered points on the Earth's surface, but very little work has been done to demonstrate their utility for this purpose. At this time, there are no plans to develop satellite capabilities for synoptic measurement of canopy heights.

Models capable of accepting remotely sensed inputs have been developed to simulate ecosystem processes such as photosynthesis and evapotranspiration (RUNNING and COUGHLAN 1988). Remotely sensed vegetation indices, estimates of soil moisture, measures of surface temperature, and measures of other climatological or meteorological parameters can be used to drive such models. Integrated use of remotely sensed information on various aspects of ecosystems may ultimately yield the most robust and generally useful applications for estimating terrestrial primary productivity.

Oceanic primary productivity

For the ocean, there is general community concensus that one can measure ocean color (radiance) in visible wavelengths and then derive estimates of chlorophyll concentration. These estimates have been related to total phytoplankton biomass, productivity, and photosynthetic efficiency in the Earth's oceans.

Simple, empirical relationships developed through *in situ* work can be used to estimate chlorophyll-a concentrations from Coastal Zone Color Scanner (CZCS) measurements of backscattered radiation at 443, 520, and 550 nm. Ratios of the radiances of bands centered at 440 and 550 nm for low chlorophyll content waters and at 520 and 550 nm for high chlorophyll content waters are proportional to chlorophyll concentrations (PERRY 1986). These techniques yield quantitative estimates of phytoplankton pigment concentrations for the upper tens of meters of the open ocean and for somewhat lesser depths of coastal waters. Estimates of oceanic pigment concentrations derived from the CZCS have errors between 30 and 100 percent (LEWIS 1989). Although surface, *in situ* measures for individual stations are considerably better, the satellite estimates are the only ones that can cover large regions of the Earth's oceans. Thus, these estimates are of great value for understanding basin- to global-scale productivity patterns. However, the algorithms do not work well for coastal waters where high concentrations of suspended sediment, detritus, and pigments other than chlorophyll also affect the backscattered radiation. Current satellite data are not adequate to estimate pigment concentration in these situations.

Estimates of primary productivity are made by using the satellite derived estimate of near-surface chlorophyll concentration to provide an initial estimate of phytoplankton biomass as input to a mathematical model of primary productivity (LEWIS 1989). For many regions of the ocean, the results of this model agree well with *in situ* measurements of primary productivity. However, primary productivity extends deeper into the ocean than the CZCS is able to measure, and there are concerns that this productivity can never be accounted for in a model driven primarily by satellite observations. The potential errors have not been accurately estimated yet.

Laser-induced fluorescence from aircraft has also been used to estimate chlorophyll concentrations in near-surface waters, and sensors with high spectral resolution can discriminate among certain types of pigments (HOGE and SWIFT 1983). This technique is useful for oceanic transect observations with aircraft. *In situ* observations of laser-induced fluorescence can be used to estimate photosynthetic efficiency in phytoplankton (KOLBER et al. 1988). Similar observations have been made in the laboratory for terrestrial vegetation (CHAPPELLE and WILLIAMS 1986). Future airborne studies may explore this potentially very important application further, but it is doubtful that there will be a spaceborne capability in the foreseeable future.

Sea surface temperature can be measured with an accuracy of 0.5–1.5 °C using current thermal infrared and passive microwave satellite sensors (AVHRR and high resolution infrared sounder/microwave sounding unit (HIRS/MSU) on the NOAA polar orbiters and SMMR). Remotely sensed sea surface temperatures can be used to derive understanding of circulation and upwelling patterns which affect primary productivity in the ocean as well as the distribution of animal populations (NJOKU et al. 1985). Fisheries production studies have taken advantage of this data set. Integration of sea surface temperature and ocean color data for improved understanding of the processes affecting global marine primary production is also underway.

Biochemical composition

High spectral resolution reflectance data have been used to estimate vegetation canopy lignin, nitrogen, and cellulose content. The technique exploits absorption features in the visible through infrared region known to correspond to harmonics and overtones of electronic transitions or fundamental stretching frequencies of organic bonds. The extraction of this information is complicated because, in whole plants, these subtle absorption features are mixed and superimposed on each other, and the strong first-order absorption effects due to water and pigments dominate the reflectance spectrum (PETERSON et al. 1985). Stepwise multiple regression has been used successfully to identify and quantify lignin, nitrogen, and cellulose in vegetation canopies using high spectral resolution (20 nm resolution in the 0.4–2.45 μm region) airborne sensor data (PETERSON et al. 1985, WESSMAN et al. 1987, 1988, ELVIDGE 1988).

Because these estimates are based solely on regression equations, there is much skepticism concerning the validity and robustness of the relationship between reflectance and canopy biochemical composition. It is difficult to predict whether the technique will work for different vegetation types experiencing differing sets of environmental constraints. So far, attempts to extend the technique to new vegetation types and different ecological conditions have been generally successful. It is troubling that the same wavelengths are not consistently identified in the various studies which have been conducted. Recent studies to model the canopy water content, and then remove it from a canopy reflectance spectrum, have employed spectral matching techniques to identify in the residual spectrum canopy absorption features related to known absorption features of biochemical compounds (GOETZ pers. communication). If this line of evidence holds up, much of the skepticism concerning the relationship should be allayed.

An ability to assess quantitatively the biochemical composition of vegetation over large areas alone should be sufficient to integrate remote sensing into the ecologist's repertoire of tools. Knowledge of canopy lignin and nitrogen content is fundamentally important for estimating decomposition and nitrogen cycling rates. It also will prove useful in estimating the palatability of leaf tissues to various predators. Synoptic, remotely sensed information on canopy lignin and nitrogen content can be input to ecosystem and biogeochemical cycling models to scale decomposition and nitrogen cycling processes from the local, site-specific scale to larger, regional and global scales.

Soil composition and structure

Remote sensing has been used to assess the mineral composition, moisture content, and surface roughness of soils in agricultural and arid to semi-arid regions. Soil scientists long have used aerial photography (mainly black and white, but also color and color infrared) to aid in the mapping of soil units (IRONS et al. 1989). Soil color is a primary factor used in the classification of soils. Remote sensing can distinguish variations in soil color related to soil iron content. Differences related to iron oxides and the presence of ferric versus ferrous iron can be detected. Organic matter content can be quantified roughly; shortwave reflectance decreases as organic matter concentration increases. A number of minerals present in soils can be uniquely identified because of the presence of diagnostic absorption features in the shortwave infrared; these include calcite, carbonate, gypsum, and several clay minerals (e.g., montmorillonite and kaolinite) (GOETZ et al. 1985).

Multi-date remote sensing observations of soil moisture can be used to infer water holding capacity of soils (see "evapotranspiration", page 734). Integrated observations of surface slope, drainage patterns, and soil color could be used to develop understanding of soil erosion processes. Studies to exploit such observations have been conducted in agricultural regions, but very little has been done for natural ecosystems. Remote sensing of important soils properties is very much limited to areas where bare soils are exposed, i.e., agricultural areas with periodic harvesting and tilling, and arid regions. Although much indirect information about soils can be derived from remote sensing of the vegetation cover, it is not possible to derive direct information for most naturally vegetated regions of the Earth's surface.

Vegetation stress

Decreases in the absorption of visible light (blue and red wavelength regions) related to decreases in plant pigment concentrations have been observed in remotely sensed data of plant canopies. Changes in the near infrared reflectance related to changes in internal leaf structure and water content have also been reported. Canopies with increased temperatures due to closed stomata have been observed. All of these observations, as deviations from a normal condition in either space or time, can be used to detect stress in vegetation (ROCK et al. 1986, SADER 1986). It has been proposed that stresses caused by differing factors may have unique spectral signatures such that the cause of the stress can be diagnosed from the remote sensing observations alone (ROCK et al. 1986).

The location of the sharp rise in vegetation reflectance between 0.68 and 0.72 μm, the red edge, changes with vegetation condition. Its location shifts about 7–10 nm to shorter wavelengths, in the so-called blue shift of the red edge, in stressed vegetation (WICKLAND 1989). This effect was first reported as exclusive to heavy metal stress, but has since been observed in air pollution, acid deposition, and severely water stressed plants (ROCK et al. 1986, HORLER et al. 1980, WESTMAN and PRICE 1987). It seems likely that changes in pigment concentrations are responsible for this effect, but it has yet to be adequately understood and quantified. At present, it is not possible to identify uniquely the cause of a stress in vegetation without ground-based observations and studies.

Studies of stressed vegetation are usually directed toward identifying pre-visual symptoms of stress or developing a means of assessing the extent and severity of the phenomenon, both of which are extremely difficult to achieve on the ground. To date, more success has been had with assessing the extent and severity of the stress than in achieving a significant early warning capability.

Evapotranspiration

Vegetation acts as a uniform gray body in the thermal infrared wavelengths. Thus, thermal infrared emission measurements are used mainly to measure canopy (i.e., leaf surface) temperature. Such measures of canopy temperature from satellites have been used as inputs to models of evapotranspiration (JACKSON et al. 1983,

GURNEY and CAMILLO 1984, HOPE et al. 1986, LUVALL and HOLBO 1989). Surface temperature observations are used to calculate the sensible heat flux and then to determine the latent heat flux from the energy balance equation. Thermal bands on airborne sensors also have been used to estimate surface temperature (SADER 1986, LUVALL and HOLBO 1987, 1989). To be useful, the measurements of surface temperature must be accurate to within ± 0.5 °C. Only some of the airborne and satellite thermal sensors provide this accuracy. At least two observations per day are desired, and very few of the satellite sensors provide this coverage. Also, the modeling is challenging and requires additional inputs that are often not available from remote sensing. Thus, the application of thermal remote sensing to estimate evapotranspiration has not matured yet. The potential of remotely sensed thermal infrared data for estimating evapotranspiration will not be realized until appropriate spaceborne sensors are deployed and the required ancillary information can be provided with the same coverage and timing.

Another remote sensing approach to estimating evapotranspiration uses the NDVI and relies on the fact that the rates of photosynthesis and evapotranspiration are linked through the vital function of stomata in each process. The vegetation index can be used as an indicator of the potential for evapotranspiration as well as for carbon fixation (SELLERS 1985, HOPE et al. 1986). This estimate of potential evapotranspiration, presumably, could be used as an input to ecosystem models of the evapotranspiration process.

Soil and canopy water content

Leaf and canopy water content, moisture status, and water potential have been related to reflectance in the shortwave infrared (ROHDE and OLSON 1971, GOWARD 1985, ROCK et al. 1985, HUNT et al. 1987, HUNT and ROCK 1989). In the microwave region of the spectrum, vegetation emission (passive measurements) and backscatter (active measurements) are believed to be related to vegetation moisture status or water content and surface roughness. Active microwave sensors (radars) are widely touted as capable of providing information on surface moisture content, including soil moisture and vegetation water content, and surface roughness and texture. Passive microwave emissions are believed to provide primarily information on soil or vegetation (i.e., surface) moisture content.

Reflectance in the near and shortwave infrared is dominated by vegetation moisture content, as numerous studies of drying leaves have demonstrated (ROHDE and OLSON 1971, HUNT and ROCK 1989). Ratios of near-infrared and shortwave-infrared reflectance have been used to generate indices, believed to be indicative of moisture stress, which correlate well with leaf water potential measurements in stressed vegetation (ROCK et al. 1985, HUNT et al. 1987). However, an unequivocal connection to water stress or canopy water content has never been demonstrated. Other broad band and high spectral resolution remote sensing experiments to quantify leaf or canopy water content using these wavelength regions have yielded negative or inconclusive results (HUNT and ROCK 1989, GOWARD 1985, PIERCE et al. 1990). A likely explanation is that the range of water contents examined in the drying studies (5–95 percent) is far broader than the range of water contents likely to be experienced in living leaves (approximately 60–95 percent), even in severely stressed leaves. Thus, the only discriminable differences are likely to be those of little ecological significance. Atmospheric water vapor effects further complicate attempts to derive information about surface water content. At present, it seems that although gross broad-band near and shortwave infrared reflectance differences may be related to canopy water content, this relationship is not sufficiently sensitive to discriminate the range of water contents in living tissues. For high spectral resolution data, the same seems to be true, although new techniques using water absorption features and *in situ* measurements for calibration deserve further exploration (GOETZ pers. communication).

Radar backscatter measurements are believed to be influenced by two major characteristics of the surface: dielectric constant and roughness. Water has a very high dielectric constant (about 80) compared to surface materials devoid of water (about 3–4) and therefore absorbs a great deal of radar radiation, playing a major role in determining the overall radar response. Therefore, it is hoped that radar can be used to quantify the moisture content of surface soils and vegetation. It is important to note that because water absorbs so strongly, the radar signal will not penetrate soils very far; the most realistic expectations are that only the top few centimeters will be adequately characterized. Of course, in extremely arid regions, penetration can be much deeper (MC CAULEY et al. 1982). For certain conditions, strong relationships have been demonstrated between soil moisture in the upper 5 cm and particular wavelengths and polarizations of radar (BRADLEY and ULABY 1981). Much less work has been done to estimate canopy moisture using radar.

However, radar is also sensitive to surface roughness and the structure of material it penetrates (e.g., canopy structure). Leaves and branches are of the same length scales as radar wavelengths, on the order of centimeters to meters; therefore, they strongly scatter radar signals. In order to quantify definitively surface moisture content, the confounding response due to surface roughness and structure must be understood and removed. No one has been able to do this yet, but it is widely believed that various combinations of radar wavelengths, incidence angles, and polarizations will be the key to unmixing the two effects.

Passively measured electromagnetic energy emissions in the microwave region also relate strongly to surface soil moisture and surface roughness. The microwave emissions of soils vary with changes in the soils' emissivity, which is heavily influenced by water content. Microwave radiometers typically measure emissions in the 1–100 GHz range (30–0.3 cm); frequencies below 5 or 6 GHz (wavelengths longer than 6 cm) are most effective for soil moisture. Present capabilities are such that the moisture content of the surface 5 cm of soil can be measured with a relative error of 10–15 percent (SCHMUGGE 1990). These measurements can be made in all weather conditions and even when there is a moderate amount of vegetation cover.

Ancillary data from field measurements or even from radar sensors are being used to remove the effects of soil roughness and texture. This powerful application of remote sensing to the estimation of soil moisture is limited to aircraft sensors at present. Current and planned satellite passive microwave sensors do not operate at the appropriate wavelengths to measure surface soil moisture well, and their spatial resolution is quite coarse (on the order of 8–25 km).

Polarization differences at 37 GHz (8.2 mm) from the SMMR have been used to create a vegetation index which appears to be more sensitive to differences in vegetation cover in regions of low vegetation cover than is the NDVI (CHOUDHURY and TUCKER 1987, CHOUDHURY 1989, CHOUDHURY et al. 1987). The dominant attribute of the vegetation influencing this response is the canopy water content (SCHMUGGE 1990). Thus, passive microwave data provide an alternative vegetation index which is based on vegetation attributes different from those on which the NDVI is based. These results are quite new and efforts have not yet been made to apply them to providing a better estimate of vegetation cover or productivity. Presumably such applications will follow. The spatial resolutions of the SMMR and SSM/I are such that applications of these satellite observations will be limited to fairly coarse scales.

Surface and canopy structure

Some of the structural attributes of vegetation reported as derivable from radar observations are branch and leaf orientation (erectophile, planophile), branch and leaf angle distributions, vertical distribution of leaves and branches within the canopy, canopy density, canopy texture or roughness, and above ground biomass. It is widely believed that different wavelengths of radar penetrate vegetation canopies to differing levels; the longer the wavelength, the deeper the penetration. X-band (9.6 GHz or 3.1 cm) seems to interact primarily with the leaves at the surface of the canopy, C-band (5.3 GHz or 5.7 cm) with leaves and small branches, L-band (1.25 GHz or 25 cm) with branches and boles, and P-band (0.4 GHz or 68 cm) primarily with the boles. Of course, the overall density and layering of the canopy profoundly influences the actual depth of penetration. Field-based measurements that actually probe a canopy and measure the backscatter of various layers of a canopy provide evidence that such structural information would be extremely valuable (PARIS 1986). It is clear that radar responds to vegetation structure and many qualitative observations have been made to support this (FORD and WICKLAND 1985), but no one has yet been able to quantify the relationship or model the interactions for a real forest canopy. As noted above, the confounding effects of water absorption must be dealt with before useful quantification of canopy structure can be achieved.

Surface energy balance

Remote sensing provides a number of observations which can be used together to derive understanding of the Earth's energy fluxes and budget. Measurements of solar insolation and estimates of surface albedo from the weather satellites are probably the most straight-forward of these. Vegetation indices, surface cover pattern, surface soil moisture, and surface temperature are also useful for estimating land surface energy fluxes. Most of what has been done using satellite data for global energy balance has been done at very coarse spatial scales in the context of general circulation models to simulate and predict climate. A few recent studies have attempted to bring such observations and modeling down to spatial scales of interest for ecological processes. The HAPEX–MOBILHY (Hydrologic Atmospheric Pilot Experiment – Modelisation du Bilan Hydrique) and FIFE (First International Satellite Land Surface Climatology Project (ISLSCP) Field Experiment) field experiments to examine mass and energy fluxes over areas of 10,000 km^2 and 100 km^2, respectively, have provided the first integrated measurements and models of energy fluxes at meaningful scales for ecological processes (SCHMUGGE 1990, SELLERS et al. 1988). This type of integrated use of multisensor data, both measurements and derived estimates, from satellite and airborne platforms is exemplary of the highly integrated science that remote sensing can facilitate.

33.2.3 Specialized measurements and observation techniques

There are a few specialized measurements and methods for remotely sensing surface ecological and hydrological phenomena that have not been covered in the previous sections. They represent fairly new discoveries or techniques awaiting an airborne or spaceborne observation capability for testing.

Multi-angle observations

Almost all remotely sensed measurements at visible through thermal infrared wavelengths and using passive microwave sensors have been made looking straight down at the Earth's surface, i.e., viewing at nadir or a fixed range of angles centered on nadir. (Of course, imaging radar observations are made at an angle.) It is widely believed that instruments which could view a point on the Earth's surface at various angles would provide new information about the surface. Almost certainly, improved estimates of the albedo, important for understanding the Earth's energy budget, would result. Overlapping images taken at different angles can be used as stereo pairs to estimate surface topography and canopy height if the spatial resolution is sufficiently fine and the canopy is not too dense. Field and laboratory measurements taken at multiple angles suggest that additional information about canopy structure can be derived (SIMMER and GERSTL 1985, KIMES 1983, KIRCHNER et al. 1982). It has even been suggested that reflectance measured in the retro-reflection direction, the so-called hot spot, can be used to identify species

(GERSTL 1986). In layered canopies with low to intermediate leaf area indices, reflectance observations taken at low incidence angles will measure more radiation that has interacted with the leafy canopy than observations taken at nadir, which are more likely to have penetrated the canopy and have interacted with the soil/litter surface. Thus, these low incidence angle observations should be easier to analyze for information about canopy chemical composition and structure because confounding background effects will be minimized.

At present, there are no spaceborne sensors capable of viewing a portion of the Earth's surface from multiple angles almost simultaneously. The SPOT satellite can make observations at differing angles, but these observations can only be taken several days apart. SPOT's pointing capability is used mainly for targeting acquisitions and occasionally for acquiring stereo pairs. There are very few airborne sensors with pointing capability. Thus, the applications of multi-angle observations for looking at canopy structure, surface texture, and albedo over large regions have yet to be rigorously explored.

Polarization observations

Polarization is another aspect of the remotely sensed signal which is reported to provide useful information about terrestrial vegetation. In the visible to shortwave infrared, the polarized portion of the total reflectance represents that light which is reflected specularly from the canopy surface. For a leaf, this polarization depends on the angle of incident light on the leaf, the optical index of refraction of the leaf cuticle wax, and the surface roughness of the cuticle wax (VANDERBILT et al. 1985). Thus, it is possible that polarization data could be used to derive information about leaf surface properties and leaf geometric arrangement in the canopy. In addition, if the specular, polarized component of the reflectance signal could be removed, it might be possible to analyze separately the remaining diffuse component which contains information about leaf internal structure and chemical composition. At present, there are no spaceborne sensors, and very few airborne sensors, available to measure polarization in the visible and infrared regions of the electromagnetic spectrum.

Polarization differences in passive microwave observations have been used to investigate land surface moisture differences (CHOUDHURY and TUCKER 1987, CHOUDHURY 1989, also see sections "terrestrial primary productivity and biomass" on page 730 and "soil and canopy water content" on page 735). Fully polarimetric radar observations have been used to derive information about the type of scattering mechanism dominating the signal, and, hence, infer information about vegetation structure and moisture status (EVANS et al. 1988). Forested regions typically produce two types of polarization components, one is a two-bounce signal that indicates strong interaction between the tree trunk and the ground surface and the second is a diffuse component attributable to multiple scatter, probably within the leafy canopy (ZEBKER et al. 1987). The capability to acquire and analyze fully polarimetric data is so new, that many years of research will be required before the ultimate utility and suitability of polarization data for ecological applications will be known.

33.3 Discussion

It should be obvious from the ecological applications described in the previous sections that if remote sensing reaches its full potential, it must become an important data source for ecology. There are a number of useful applications which can be pursued now, with current understanding and technology, but there are many exciting applications that await the results of further research and testing or the deployment of appropriate sensors on aircraft or spacecraft. At present, it seems that the ecological community has been spending more time awaiting further developments than it has in exploiting proven remote sensing capabilities. In fact, the remote sensing community, composed primarily of physicists, engineers, and physical scientists, has played a greater role in applying remote sensing to ecological problems, or at least attempting to do so, than has the ecological community. It is not clear why this is the case, but some of the contributing factors will be discussed below.

Over the years, there have been many questions about how the various applications of remote sensing hold up after careful replication and testing. There have been questions about how generalizable remote sensing applications are across vegetation types and across the full range of environmental variability. In fact, there has been much concern that because the basic biophysics of many remote sensing applications often are not understood, the techniques may not be robust or extendable. It has been asserted that too much of the past work has been empirical, based on correlations and not on basic biophysical understanding of why the interaction of electromagnetic radiation with the surface would produce such an effect. All of these questions and concerns have served to generate confusion about the value and applicability of remote sensing techniques to ecology. The less familiar an ecologist is with remote sensing, its specialized journals, and its arcane vocabulary of acronyms and engineering terms, the greater his or her confusion and inability to discriminate the useful, well-proven techniques from the more exploratory, speculative ones.

Remote sensing has been oversold, and exceptional techniques and results have been generalized to conditions where they do not hold. Also, the promise of deriving useful ecological information based on simple theoretical models is sometimes touted as fact long before the observations and model inversions have been made to test the prediction. While this kind of false advertising is rarely found in the refereed literature, it has at times been employed to argue for new remote sensing instrumentation or to generate support for the remote sensing research community. As an example, the promise of being able to measure global primary productivity (with a little more research and the right instrumentation) is not quite the same as knowing it actually can be done within a known and acceptable level of error.

It is extraordinarily rare that a new airborne or spaceborne instrument is deployed without a strong and well-proven application in mind, but advertised additional uses of the sensor in discipline areas other than the primary one can be much less well-founded. To some extent, the technology of remote sensing has been driving the remote sensing scientists to find ecological applications that might not have otherwise been proposed or explored so vigorously, so soon. This is a major problem with respect to ecological applications, because there has never been a remote sensing instrument designed primarily for ecological research. To date, ecologists and remote sensing scientists have been doing their best to make progress using data from sensors not optimally designed for their applications.

Yet another factor impeding the incorporation of remote sensing into ecological research is the large investment of time and resources required to get involved initially. Ecologists rarely have educational backgrounds which include an introduction to remote sensing instrumentation, study of the biophysics of remote sensing, or experience with image analysis. Thus, they must invest time to learn about the science and techniques. They must also purchase, or acquire access to, sophisticated computer hardware and software. Remotely sensed data must be acquired and bought. High spatial resolution land remote sensing data (i.e., Landsat and SPOT data) are costly due to the fact that the data are provided by the commercial sector, and not the public sector. Securing access to experimental airborne sensors can be difficult. Thus, few ecologists can explore casually the potential of remote sensing for their research problems; they must be ready to make a serious commitment of time and resources.

The above factors may be a sufficient explanation for the slow incorporation of remote sensing techniques into ecology, but they may not be the only factors. A reluctance on the part of some ecologists to consider conducting research at spatial scales beyond a few tens of meters or to make the gross simplifications of complex ecological processes required to integrate information for coarse scale modeling could be a factor. The generally tight research budgets among the agencies which fund ecological research, and the consequent conservatism toward new ideas and approaches, could be another factor. Whatever the case, the situation seems to be changing slowly. The International Geosphere–Biosphere Programme (IGBP), or Global Change Program, has many people thinking about global ecological problems and is calling for new satellite observations as an integral component of the program to address them. Ecologists are involved in the advance planning for IGBP and their expertise is being incorporated into its plans (IGBP 1988).

33.4 Conclusions

The following ecological information can be obtained at reasonable accuracies with satellite remote sensing: 1) broad categories (e.g., grasslands, savanna, deciduous forest, coniferous forest, wetlands, desert, agriculture, urban) of land surface cover at several spatial scales of resolution (SPOT, Landsat TM & MSS, AVHRR with 10 m to 15 km pixels), 2) global phenological patterns at time steps of 2 weeks to 1 month (AVHRR and CZCS with 1 to 15 km pixels), 3) estimates of the frequency and extent of fires in selectable regions of the world from data collected twice daily (AVHRR with 1 km pixels), 4) estimates of land and ocean surface temperature (many sensors with differing spatial resolutions and sensitivities), 5) measurements of the extent and attributes of change over time for selectable regions of the Earth's land surface (SPOT, Landsat TM & MSS, AVHRR with 10 m to 15 km pixels), 6) vegetation indices (SPOT, Landsat TM & MSS, AVHRR with 20 m to 15 km pixels), 7) ocean color indices (CZCS with 1 km pixels), 8) evidence of stress in vegetation through symptoms which can be remotely sensed (e.g., increased canopy temperature, changes in color of the foliage, changes in canopy structure or water content), and 9) an index related to surface moisture content (SMMR with 25 km pixels). Airborne remote sensing adds to this list the following: 1) surface soil moisture for areas of low to moderate vegetation cover (microwave radiometers with 30–100 m pixels), 2) areal extent of inundation in wetlands and swamps (synthetic aperture radar with 5–50 m pixels), 3) measures of canopy height (with profiling lidar), and 4) broad categories of land surface cover in cloudy regions (synthetic aperture radar with 5–50 m pixels).

There is potential to obtain the following ecological information through remote sensing: 1) global, consistent estimates of net primary productivity both on the land and in the ocean, 2) detailed vegetation composition data at the species to species assemblage level (depending on the heterogeneity of the vegetation type), 3) identification of the type of stress vegetation is experiencing, 4) estimates of evapotranspiration, 5) estimates of canopy structural attributes (e.g., density, leaf and branch angle distributions and orientations, canopy surface roughness), 6) estimates of the emissions of aerosols and trace gases from biomass burning, 7) estimates of photosynthetic efficiency, 8) canopy biochemical contents (e.g., lignin, nitrogen, cellulose), and 9) vegetation water content. Much research and testing will be required before these applications of remote sensing can be achieved.

For most applications, both proven and potential, trade-offs must be made between accuracy at the local scale and consistent, synoptic, regional and global scale coverage. Point measurements taken on the surface are more accurate, but they cannot be directly extended to regional and global scales. Remotely sensed observations cover large areas of the surface consistently, but there is no direct means of judging their accuracy. It is not possible to assess whether global estimates from remotely sensed data are any more or less accurate than global estimates extrapolated from a set of surface measurements. The results of these two approaches will be compared and used as checks on each other. In time, scientists will gain confidence in how best to use remotely sensed observations and surface observations together.

The single most powerful current application of remote sensing to ecological problems is to document synoptically the cover of broad vegetation types on the

Earth's surface, and to be able do so at varying spatial scales that range from sampling resolutions of 10 m to 15 km. Never before has it been possible to make an unbiased, internally consistent measure (not estimate) of the spatial extent and pattern of vegetation and phytoplankton types for large, continuous regions of the Earth's surface and to do so over time intervals of days to months. This, in itself, achieves a new stage of advancement towards the global perspective on vegetation that ecologists and plant geographers have long sought. Sooner or later, the existence and utilization of these data sets will revolutionize vegetation ecology.

In the future, ecological research will require regular observations from multiple remote sensors operating at several spatial and temporal scales. The most powerful applications of remote sensing will come from the combined use of a variety of remotely sensed measurements taken across the electromagnetic spectrum to drive models of ecosystem function. Remote sensing makes it possible to operate mechanistic models at coarser spatial scales than ever before possible, without extrapolation or interpolation. No one parameter acquired consistently and synoptically is likely to be so powerful in the long run as a set of such parameters, each a measure of a different ecosystem attribute, which can be used to model or infer important ecological processes and which can be obtained repeatedly over time. In the upcoming decades of study of global change, these will be the applications of remote sensing that ecologists will pursue most vigorously.

Bibliography

BOTKIN, D. B.; ESTES, J. E.; MACDONALD, R. M.; WILSON, M. V., 1984, Studying the Earth's vegetation from space.
Bioscience **34**, 508–514.

BOX, E., 1976, Average energy fixation rate of natural vegetation ("Berlin Model" world map).
in: Lieth, H., Energy flow and efficiency differences in plants and plant communities, in: Lamprecht, I.; Schaarschmidt, B. (Eds.), Application of Calorimetry in Life Sciences, de Gruyter, Berlin–New York.

BRADLEY, G. A.; ULABY, F. T., 1981, Aircraft radar response to soil moisture.
Remote Sens. Environ. **11**, 419–438.

CHAPPELLE, E. W.; WILLIAMS, D. L., 1987, Laser-induced fluorescence (LIF) from plant foliage.
IEEE Trans. Geosci. Remote Sens. GE-25, 726–736.

CHOUDHURY, B. J.; TUCKER, C. J., 1987, Satellite observed seasonal and interannual variation of vegetation over the Kalahari, the Great Victoria Desert, and the Great Sandy Desert: 1979–1984.
Remote Sens. of Environ. **23**, 233–241.

CHOUDHURY, B. J.; TUCKER, C. J.; GOLUS, R. E.; NEWCOMB, W. W., 1987, Monitoring vegetation using Nimbus-7 scanning multichannel microwave radiometer's data.
Int. J. Remote Sens. **8**, 533–538.

CHOUDHURY, B. J., 1989, Monitoring global land surface using Nimbus-7 37 GHz data.
Int. J. Remote Sens. **10**, 1579–1605.

DANSEREAU, P., 1957, Biogeography: an ecological perspective.
Ronald Press, New York.

ELVIDGE, C. D., 1988, Vegetation reflectance features in AVIRIS data.
in: Proceedings of the Sixth Thematic Conference: "Remote Sensing for Exploration Geology", Houston, Texas, May 16–19, 1–14.

EVANS, D. L.; FARR, T. G.; ZYL, J. J. VAN; ZEBKER, H. A., 1988, Radar polarimetry: analysis tools and applications.
IEEE Trans. on Geosci. and Remote Sens. **26**, 774–789.

FORD, J. P.; WICKLAND, D. E., 1985, Forest discrimination with multipolarization imaging radar.
in: Digest, 1985 IEEE – International Geoscience and Remote Sensing Symposium, Amherst, Massachusetts, 462–469.

GERSTL, S. A. W., 1986, Off-nadir optical remote sensing from satellites for vegetation identification.
in: Proceedings of the International Geoscience and Remote Sensing Symposium, Zurich, Switzerland, 1457–1460.

GOETZ, A. F. H.; VANE, G.; SOLOMON, J. E.; ROCK, B. N., 1985, Imaging spectrometry for Earth remote sensing.
Science **228**, 1147–1153.

GOETZ, A. F. H., personal communication.

GOWARD, S. N., 1985, Shortwave infrared detection of vegetation.
Adv. Space Res. **5**, 41–50.

GOWARD, S. N.; TUCKER, C. J.; DYE, D. G., 1985, North American vegetation patterns observed with the NOAA-7 Advanced Very High Resolution Radiometer.
Vegetatio **64**, 3–14.

GURNEY, R. J.; CAMILLO, P. J., 1984, Modelling daily evapotranspiration using remotely sensed data.
J. Hydrol. **69**, 305–324.

HALL, F. G.; STREBEL, D..; GOETZ, S. J.; WOODS, K. D.; BOTKIN, D. B., 1987, Landscape pattern and successional dynamics in the boreal forest.
in: Proceedings International Geoscience and Remote Sensing Symposium, Ann Arbor, Michigan, 473–482.

HOGE, F. E.; SWIFT, R. N., 1983, Airborne dual laser excitation and mapping of phytoplankton photopigments in a Gulf Stream warm core ring.
Appl. Opt. **22**, 2272–2279.

HOLDRIDGE, L. R., 1947, Determination of world plant formations from simple climatic data.
Science **105**, 367–368.

HOPE, A. S.; PETZOLD, D. E.; GOWARD, S. N.; RAGAN, R. M., 1986, Simulated relationships between spectral reflectance, thermal emissions, and evapotranspiration of a soybean canopy.
Water Res. Bull. **22**, 1011–1019.

HORLER, D. N. H.; BARBER, J.; BARRINGER, A. R., 1980, Effects of heavy metals on the absorbance and reflectance spectra of plants.
Int. J. Remote Sens. **1**, 121–136.

HUETE, A. R., 1988, A soil-adjusted vegetation index (SAVI).
Remote Sens. Environ. **25**, 295–309.

HUETE, A. R., 1989, Soil influences in remotely sensed vegetation-canopy spectra.
in: Asrar, G. (Ed), Theory and Applications of Optical Remote Sensing, John Wiley & Sons, New York, 107–141.

HUETE, A. R.; JACKSON, R. D.; POST, D. F., 1985, Spectral response of a plant canopy with different soil backgrounds.
Remote Sens. Environ. **17**, 37–53.

HUMBOLDT, A. VON; BONPLAND, A., 1807, Essai sur la géographie des plantes.
Paris, Libraire Lébrault Schoell.

HUNT, E. R.; ROCK, B. N., 1989, Detection of changes in leaf water content using near- and middle-infrared reflectances.
Remote Sens. Environ. **30**, 43–54.

HUNT, E. R. JR.; ROCK, B. N.; NOBEL, P. S., 1987, Measurement of leaf relative water content by infrared reflectance.
Remote Sens. Environ. **22**, 429–435.

IGBP, 1988, Global Change Report No. 4, The International Geosphere–Biosphere Programme: A study of global change IGBP a plan for action.
IGBP, 1–200.

IRONS, J. R.; WEISMILLER, R. A.; PETERSON, . W., 1989, Soil reflectance.
in: Asrar, G. (Ed.), Theory and Applications of Optical Remote Sensing, John Wiley & Sons, New York, 66–106.

JACKSON, R. D.; HATFIELD, J. L.; REGINATO, R. J.; IDSO, S. B.; PINTER, P. J., 1983, Estimation of daily evapotranspiration from one time-of-day measurements.
Agric. Water Manag. **7**, 351–362.

KAUFMAN, Y. J., Satellite remote sensing of large scale air pollution – measurements of forest fire smoke.
J. Geophys. Res. (in press).

KIMES, D. S., 1983, Dynamics of directional reflectance factor distributions for vegetation canopies.
Appl. Opt. **22**, 1364–1372.

KIRCHNER, J. A.; KIMES, D. S.; MCMURTREY III, E., 1982, Variation of directional reflectance factors with structural changes of a developing alfalfa canopy.
Appl. Opt. **21**, 3766–3774.

KOLBER, Z.; ZEHR, J.; FALKOWSKI, P., 1988, Effects of growth irradiance and nitrogen limitation on photosynthetic energy conversion in photosystem II.
Plant Physiol. **88**, 923–929.

LANG, H. R.; CURTIS, J. B.; KOVACS, J. S., 1984, Lost River, West Virginia, petroleum test site report.
in: Abrams, M. J.; Conel, J. E.; Lang H. R. (Eds.), The Joint NASA/Geosat Test Case Project: Final Report, AAPG Bookstore, Tulsa, Oklahoma, 12–i – 12–96.

LEWIS, M. R., 1989, The variegated ocean: a view from space.
Newscientist, No. **1685**, 1989.

LIETH, H.; BOX, E., 1972, Evapotranspiration and primary productivity: C. W. Thornthwaite Memorial Model.
Publications in Climatology **25**, C. W. Thornthwaite Associates, Elmer, New Jersey, 37–46.

LIETH, H.; WHITTAKER, R. H., 1975, Primary productivity of the biosphere.
Ecological Studies **14**, Springer, Berlin–Heidelberg–New York.

LUVALL, J. C.; HOLBO, R., 1987, Using the TIMS to estimate evapotranspiration from a white pine forest.
ASB Bull. **34**, 69.

LUVALL, J. C.; HOLBO, H. R., 1989, Modeling surface temperature distributions in forest landscapes.
Remote Sens. Environ. **27**, 11–24.

MALINGREAU, J. P.; STEPHENS, G.; FELLOWS, L., 1985, Remote sensing of forest fires: Kalimantan and North Borneo in 1982–1983.
Ambio, 314–321.

MALINGREAU, J. P.; TUCKER, C. J., 1988, Large scale deforestation in the southeastern Amazon basin of Brazil.
Ambio **17**, 49–55.

MATSON, M.; STEPHENS, G., 1987, Fire detection using data from the NOAA-N satellites.
Int. J. Remote Sens. **8**, 961–970.

MCCAULEY, J. F.; SCHABER, G. G.; BREED, C. S.; GROLIER, M. J.; HAYNES, C. V.; ISSAWI, B.; ELACHI, C.; BLOM, R., 1982, Subsurface valleys and geoarcheology of the Eastern Sahara revealed by Shuttle radar.
Science **218**, 1004–1020.

MORRISSEY, L. A..; STRONG, L. L.; CARD, D. H., 1986, Mapping permafrost in the boreal forest with Thematic Mapper satellite data.
Photogram. Eng. and Remote Sens. **53**, 1513–1520

NELSON, R.; KRABILL, W.; MACLEAN, G., 1984, Determining forest canopy characteristics using airborne laser data.
Remote Sens. Environ. **15**, 201–212.

NJOKU, E. G.; BARNETT, T. P.; LAURS, R. M.; VASTANO, A. C., 1985, Advances in satellite sea surface temperature measurement and oceanographic applications.
J. Geophys. Res. **90**, 11,573–11,586.

NRC, 1986, Global change in the geosphere–biosphere: initial priorities for an IGBP.
National Academy Press, Washington, DC, 1–91.

PERRY, M. J., 1986, Assessing marine primary production from space.
Bioscience **36**, 461–467.

PARIS, J. F., 1986, Probing thick vegetation canopies with a field microwave scatterometer.
IEEE Trans. Geosci. and Remote Sens. **GE-24**, 886–893.

PETERSON, D. L.; MATSON, P. A.; LAWLESS, J. G.; ABER, J. D.; VITOUSEK, P. M.; RUNNING, S. W., 1985, Biogeochemical cycling in terrestrial ecosystems: modeling, measurement, and remote sensing.,
Proceedings of the 36th International Astronautical Federation Congress, Stockholm, Sweden.

Pierce, L. L.; Running, S. W.; Riggs, G. A., 1990, Remote detection of canopy water stress in coniferous forests using the NS001 Thematic Mapper Simulator and the Thermal Infrared Multispectral Scanner.
Photogram. Eng. and Remote Sens. **56**, 579–586.

Rock, B. N.; Williams, D. L.; Vogelmann, J. E., 1985, Field and airborne spectral characterization of suspected acid deposition damage in red spruce (*Picea rubens*) from Vermont.
in: Proceedings 1985 Machine Processing of Remotely Sensed Data Symposium, West Lafayette, Indiana, 71–81.

Rock, B. N.;, Vogelmann, J. E.; Williams, D. L.; Vogelmann, A. F.; Hoshizaki, T., 1986, Remote detection of forest damage.
Bioscience **36**, 439–445.

Rohde, W. G.; Olson, C. E., 1971, Estimating foliar moisture content from infrared reflectance data.
in: Third Biennial Workshop: Color Aerial Photography in the Plant Sciences and Related Fields, Amer. Soc. of Photogram., Falls Church, VA, 144–164.

Running, S. W.; Coughlan, J. C., 1988, A general model of forest ecosystem processes for regional applications I. Hydrologic balance, canopy exchange and primary production processes.
Ecological Modelling **42**, 125–154.

Sader, S. A., 1986, Analysis of effective radiant temperatures in a Pacific Northwest forest using thermal infrared multispectral scanner data.
Remote Sens. of Environ. **19**, 105–115.

Sader, S. A.; Joyce, A. T., 1988, Deforestation rates and trends in Costa Rica – 1940 to 1983.
Biotropica **20**, 11–19.

Schmugge, T., 1990, Measurements of surface soil moisture and temperature.
in: Hobbs, R. J.; Mooney, H. A. (Eds.), Remote Sensing of Biosphere Functioning, Springer-Verlag, New York, 31–63.

Sellers, P. J., 1985, Canopy reflectance, photosynthesis and transpiration.
Int. J. Remote Sens. **6**, 1335–1372.

Sellers, P. J.; Hall, F. G.; Asrar, G.; Strebel, D. E.; Murphy, R. E., 1988, The First ISLSCP Field Experiment (FIFE).
Bull. Amer. Meteorol. Soc. **69**, 22–27.

Simmer, C.; Gerstl, S. A. W., 1985, Remote sensing of angular characteristics of canopy reflectances.
IEEE Trans. Geosci. Remote Sens. **GE-23**, 648–658.

TUCKER, C. J.; VANPRAET, C. L.; SHARMAN, M. J.; ITTERSUM, G. VAN, 1985, Satellite remote sensing of total herbaceous biomass production in the Senegalese Sahel: 1980–1984.
Remote Sens. Environ. **17**, 233–249.

TUCKER, C. J.; TOWNSHEND, J. R. G.; GOFF, T. E., 1985, African land-cover classification using satellite data.
Science **227**, 369–375.

TUCKER, C. J.; FUNG, I. Y.; KEELING, C. D.; GAMMON, R. H., 1986, Relationship between atmospheric CO_2 variations and a satellite-derived vegetation index.
Nature **319**, 195–199.

TUCKER, C. J.; JUSTICE, C. O.; PRINCE, S. D., 1986, Monitoring the grasslands of the Sahel 1984–1985.
Int. J. Remote Sens. **7**, 1571–1581.

VANDERBILT, V. C.; GRANT, L.; DAUGHTRY, C. S. T., 1985, Polarization of light scattered by vegetation.
Proc. IEEE **73**, 1012–1024.

WESSMAN, C. A.; ABER, J. D.; PETERSON, D. L., 1987, Estimating key forest ecosystem parameters through remote sensing.
in: Proceedings International Geoscience and Remote Sensing Symposium, Ann Arbor, Michigan, 1189–1193.

WESSMAN, C. A.; ABER, J. D.; PETERSON, D. L.; MELILLO, J. M., 1988, Remote sensing of canopy chemistry and nitrogen cycling in temperate forest ecosystems.
Nature **335**, 154–156.

WESTMAN, W. E.; PRICE, C. V., 1987, Remote detection of air pollution stress to vegetation: laboratory-level studies.
in: Proceedings International Geoscience and Remote Sensing Symposium, Ann Arbor, Michigan, 451–456.

WICKLAND, D. E., 1989, Future directions for remote sensing in terrestrial ecological research.
in: Asrar, G. (Ed.), Theory and Applications of Optical Remote Sensing, John Wiley & Sons, New York, 691–724.

ZEBKER, H. A.; ZYL, J. J. VAN; HELD, D. N., 1987, Imaging radar polarimetry from wave synthesis.
J. Geophys. Res. **92**, 683–701.

Chapter 34

Search for geographic scale regularities in ecosystem processes

A. I. BREYMEYER

34.1 Introduction

The need for introducing results of ecological studies onto maps of various scales seems to be increasing in recent years; particularly the IGBP confronts poor possibilities of ecology in this matter with other earth sciences which have at their disposal results of studies set at a geographical perspective. Traditionally from among biological sciences the geographical scale is presented by zoogeography and phytogeography; also the ecology of populations, the smallest units studied by ecologists, may be partially identified with them. On the other hand, the largest units studied by ecologists, the landscapes, also constitute objects of study by geographers; the scope of activity of landscape ecology is not determined precisely.

In this paper attention is directed to the spatial patterns of ecosystems (understood in accordance with the below presented definitions). Two ecosystem processes, production and destruction of organic matter, have been chosen for an analysis in a geographical scale. The balance of those two processes belongs to the factors determining the layer of organic matter covering the surface of the earth.

It seems that ecology has a good position among other earth sciences and is competitive with them since it introduces to the maps of the earth the measurements of those two above mentioned processes.

34.2 Definitions and assumptions

In order to make further consideration more explicit, we should define (after BREYMEYER 1990) what is meant by an ecosystem. An ecosystem is an ecological system in which organic matter produced within this system (or partially coming from outside) is transformed at successsive trophic levels according to known and described patterns of matter and energy fluxes. This definition nearly fits the hierarchical approach presented by O'NEIL et al. (1986). Ecosystem is treated as a "medium number system", consisting of a number of smaller functional components. Such an explicitly functional definition of an ecosystem makes its delimitation difficult; generally speaking, so called "small cycles" of basic elements are included in this system as they are conditioned by biotic components of ecosystem. "Great cycles" of water and elements are dependent on atmosphere, bed–rock, and other geographic factors. They are external and can be considered only as an environment of ecosystem. The main inputs from this environment to the ecosystem are solar energy, O_2, CO_2, and water solutions of basic elements. Main outputs are the gases, the energy dispersed in metabolic processes, and the remains of organic matter falling into the soil or sediments. Interpreted this way the boundries of an ecosystem are designated in the diagram of carbon cycling (Figure 34.1). Considered ecological system operates within these limits, its environment is outside.

The creation and transformations of matter occuring in a set of functioning trophic levels are termed basic ecosystem processes. These are the production of organic matter, its consumption and incorporation into the bodies of successive consumers, and its breakdown by decomposers. The sequence of basic ecosystem processes included in the functional definition of an ecosystem is necessary for self–restoration of an ecosystem and for its theoretically unlimited duration. Organic matter production is not the only internal source of energy for ecosystem functioning. Also under conditions of increasing human impact, there is a diminution in the number of ecosystems functioning on the basis of their "own" organic matter produced within the system. Similarly, the process of decomposition that permits the release of nutrients necessary for biomass production and subsequent recycling is more or less supplemented by intended or unintended fertilizing. But in spite of the fact that basic ecosystem processes are intensified or limited from outside, the ecosystem function is still the same – energy flow and cycling of matter still follow the same pattern. According to our definition this is sufficient, and ecological systems characterized by the cycling of matter according to this pattern are called ecosystems.

By geographical scale we understand in this paper such a scale in which ecological processes are conditioned by the climate, i.e. in which their natural climatic

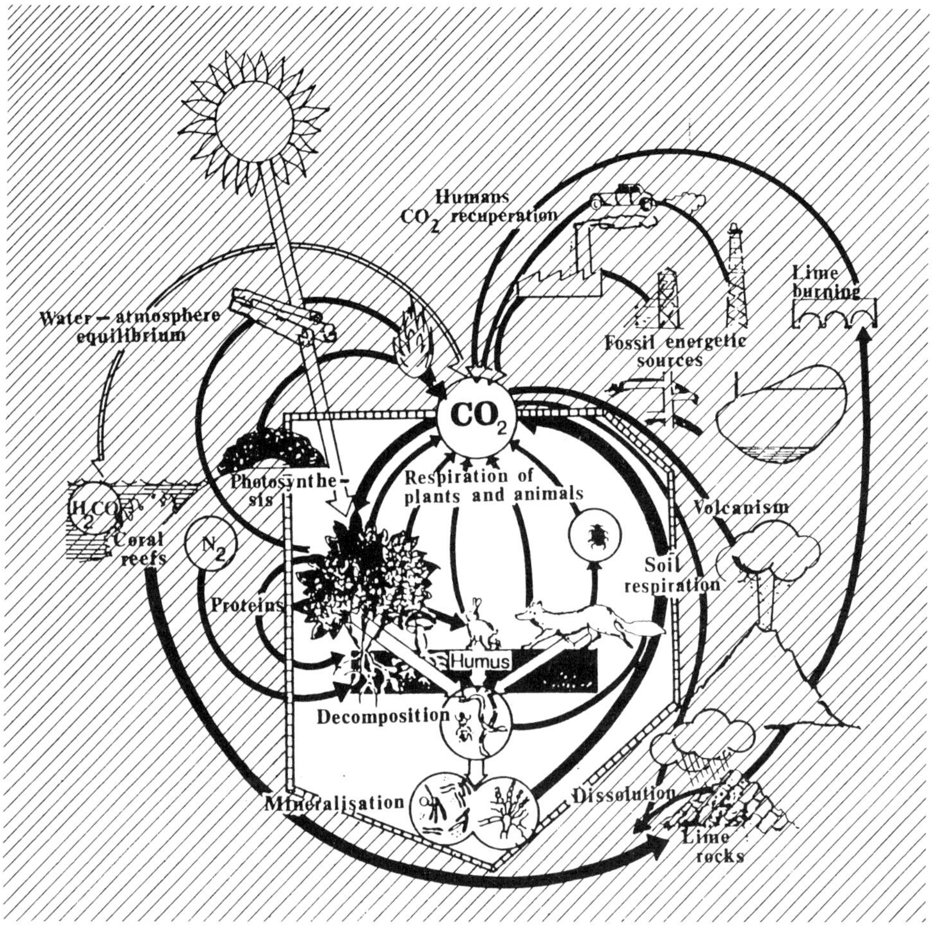

Figure 34.1: Boundries of an ecosystem marked on the scheme of carbon cycle. "Small" carbon cycle is framed by the ecosystem boundry; "large" carbon cycle connects the ecosystem with the outside environment. After DUVIGNEAUD and TANGHE (1962), adapted.

zonality occur. It ought to be stressed here that we are concerned above all with macroclimatic scales; the finer scales of topoclimate or microclimate (both terms are used by climatologists) are not taken into consideration in our study.

It was assumed that the analyzed processes, i.e. the production and decomposition of organic matter, are mainly conditioned by one element of the climate namely by the temperature. Examples of studies which will be analyzed, come from the

Northern Hemisphere, from the moderate cool climate; the main recurring climate rhythm is the winter–summer seasonality, and the length of the vegetation season constitutes the main condition of life activity. Supplies of water, even though naturally changeable during the year and in space, do not indicate such a distinct repeatable rhythm, and most probably a distinct lack of humidity limiting ecological processes seldom occurs on larger areas of this region.

This paper comprises two different parts: The first part is a discussion of geographical changeability of productivity and decomposition of organic matter on several selected examples from literature. Those are examples of creating maps of both processes in regional or global scales, based partly on direct measurements, and partly on models simulating transformations of ecosystem processes foreseen thanks to climatic changes.

The second part is an analysis of own authors materials collected recently on two transects differing through changes of thermal climate: the average annual temperatures of stands on those transects change regularly, and the differences between extremal stands amount 4 – 8 °C. This part of the paper contains unpublished materials, and consequently also short descriptions of methods and terrains.

34.3 Geographical changeability of production and decomposition of organic matter — review of selected maps

As the first cartographical presentations of organic matter distribution and changes on Earth we may consider a whole series of maps published in the seventies, connected with the name of LIETH and his collaborators (LIETH 1971, 72, 75; LIETH and BOX 1972), as well as of RODIN and BAZILEVICH (1968, 1970). To simplify, the concept on which the maps were based, were various efforts at stretching known relations between vegetation and climate onto increasing areas. It is well known today that such a generalization based on vegetation units brings forth numerous errors if interpreted as ecosystem characteristics. For example, it is worth looking through a fragment of the map of organic matter prepared by OLSON et al. (1983) closed by borders of plant formation – Fig. 34.2. The dark line delimits borders of the biome (formation) of boreal coniferous forests. On the other hand, in accordance with new data collected by OLSON transformed into a computer map of carbon in the live vegetation, the area of the coniferous forest formation is characterised by a considerable differentiation – the amount of C in biomass ranges here from 1 to 25 $kg{\cdot}ha^{-1}$. Thus the transferring of measures of ecological processes conducted in a "good", typical stand of coniferous forest of our climatic zone onto other, similarly typical stands of the same biome seems to be out of the question.

Even worse effects one can obtain if the map of potential vegetation of Eurasia is used.

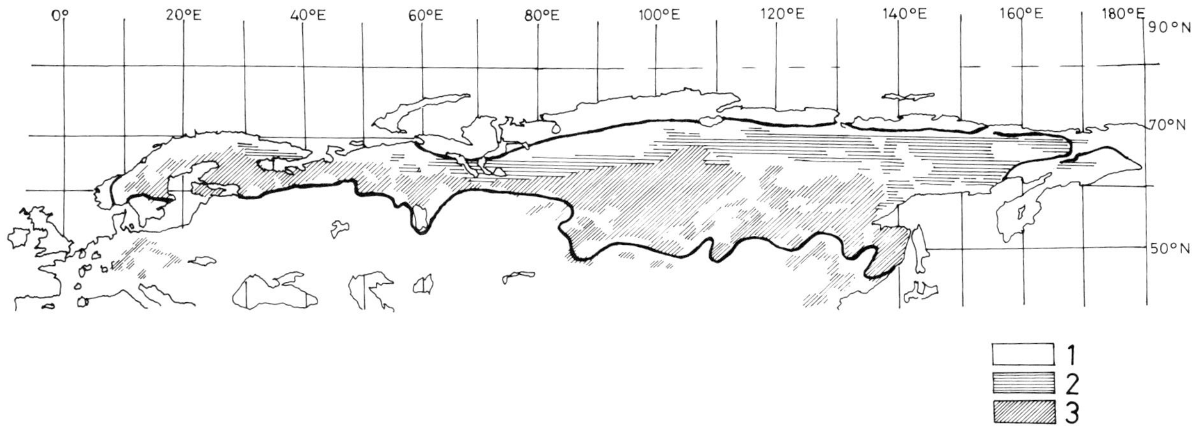

Figure 34.2: Coniferous forest biome and three groups of forest ecosystems classified by Olson et al. according to amount of carbon in live vegetation. 1: 0–1 kg·m^{-2}, 2: 3–6 kg·m^{-2}, 3: 4–25 kg·m^{-2} of carbon. After Olson et al. (1983), changed by introducing the boundries of biome acc. to UNESCO map of biomes and simplified.

Efforts of authors of ecological maps were directed at obtaining characteristics of an increasing number of vegetation units – in such a way the generalizations became gradually more close to reality. Also as the number of ecological measurements increased, it was possible to test and use more complex mathematical methods, as well as more complicated models. The models simulating development of previously registered relationships (processes) in changing conditions are probably the best tool in needed space and time scales.

Ågren et al. (in press) analysing the models of impact of climate on production–decomposition processes recognize several levels of resolution:

- Physiologically based models
- Population models
- Ecosystem models
- Regional or global models.

Physiologically–based models predict responses to environment of such processes as carbon uptake through photosynthesis, exchange of water vapour through transpiration or leaf energy balance. Ågren et al. choose the following models for forests: BACROS (de Wit et al. 1978; Goudriaan et al. 1984), FORGRO (Mohren et al. 1984, Mohren 1987), MAESTRO (Wang 1988), BIOMASS (McMurtie et al. 1988 a,b,c,), FOREST–BGC (Running and Coughlan 1988) and only one grassland model – BLUE GRAMA (Detling et al. 1979). From detailed analysis of Ågren et al. here we mention only some processes and their relationships of main environmental factors: process of photosynthesis is related to temperature as variation of photosynthetic parameters about some temperature optimum; the effects of water stress on photosynthesis are characterized by various approaches; dependence of stomatal conductance on variables like radiation, water status and humidity is described by appropriate equations and applied to the canopy by summation (a canopy is treated as single big leaf).

Population models are mainly tree growth models simulating effects of climate on single–species or multi–species populations. Regional stand tables are used as source of tree growth data and in this way the simulated effects of changing climate are registered over these regions.

The class of models called by Ågren et al. JABOVA–FORET seems interesting for our aims; this group of models begins with *a priori* assumptions on the influence of climate on tree growth and recruitment (Shugart 1984). This way of model building enables the incorporation of some ecological hypotheses at the beginning of modelling. However, as these are population models and they simulate the birth, growth and death of single trees they are not easily applicable to the analysis of ecosystem changeability with climate.

Ecosystem models (or ecosystem–tissue models) include the models based on carbon flows between the plant and soil systems. Compartment "plants" is considered as all green plant biomass; compartment "soil" connects the model to nutrient cycling model and only in this case the ecosystem model can simulate the impact of plant residue on nutrient availability and plant production. Ecosystem models based only on carbon flow (without linkage to nutrient flow) can simulate only short time ($<$ 10 years) response of ecosystem. Models predicting long term ecosystem dynamics use long (1–12 months) time steps and simulate long-term (10–100 years) ecosystem response to changing environment. According to ÅGREN et al. existing grassland models (ELM–INNIS 1978, SAGE–LAUENROTH and PRESTON 1984, SPUR–HANSON et al. 1988, AFRICA–COUGHENOUR et al. 1984, BLUE GRAMA–DETLING et al. 1979, ROOTS–PARTON et al. 1978) "are well structured to simulate the direct impact of different climatic conditions on short-term response of grasslands and are now being tested for their ability to simulate longer term (5–50 years) impacts".

The CENTURY model (PARTON et al. 1988) has recently been developed to simulate long-term (10–100 years) responses of ecosystems (grasslands and forests) to changing climatic conditions. This is low resolution simplified ecosystem model with long time step (monthly). CENTURY simulates regional patterns of plant production and organic matter levels in N. American Great Plains (PARTON et al. in press). The exercise with simulation of grassland response to increasing temperature (annual temperature growth 2 and 4 °C over a 50 year period) shows increased plant production and decreased soil organic matter.

Regional and global models predict the pattern of coincidence of vegetation and climate zones. The global vegetation model can be derived and parametrized by overlying vegetation maps with climate maps (if vegetation maps were prepared from real vegetation inventories, not from assumed relationship with climate). As an example the HOLDRIGE (1947, 1964) life zones classification system can be used; HOLDRIGE hypothesises that zonation of vegetation follows the gradient of average total annual precipitation, the ratio of potential evapotranspiration to precipitation, the mean annual biotemperature (average annual temperatures computed by setting all values below 0 °C to zero).

Also empirically derived regressions betweeen climate and ecological variables can be used; this was exercised for primary production (ROSENZWEIG 1968) and litterfall (MEENTEMEYER et al. 1982). This approach was used producing the global maps of primary production, litter fall, carbon storage and litter accumulation (LIETH 1972, 1975; BOX 1978, 1981; ESSER et al. 1982, MEENTEMEYER et al. 1982); the maps are based on global climate data sets, the regression equations predict the staedy states.

Let us analyse the results of some of above mentioned papers, let us analyse the maps.

Three consecutive papers of MEENTEMEYER (MEENTEMEYER et al. 1982, 1985; MEENTEMEYER 1984) introduce world maps of litter production (1982 – World Patterns and Amounts of Terrestrial Plant Litter Production), of decomposition rate (1984 – The Geography of Organic Decomposition Rates) and detrital carbon accumulation (1985 – World Patterns and Amounts of Detrital Soil Carbon). As

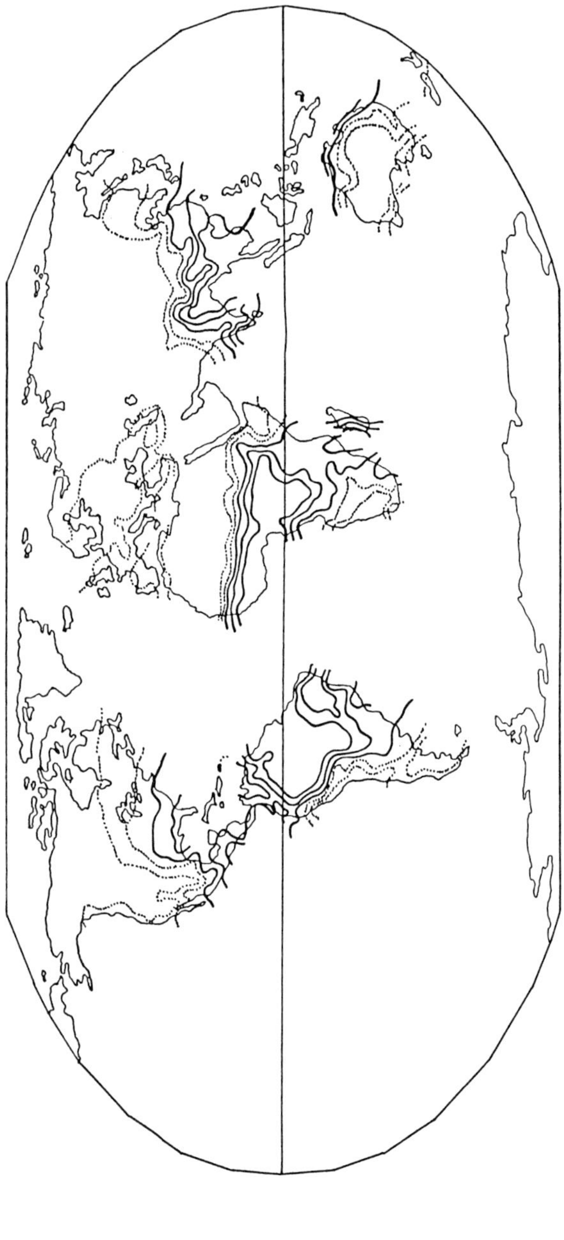

Figure 34.3: Computer generated map of total litter production. Data from 1225 climatic stations and regression equation *total litter production* = 1.0066*AET* − 183.92 were used. The solid isopleths represent the litter production above 500 $g{\cdot}m^{-2}{\cdot}yr^{-1}$, the dotted lines below 500 $g{\cdot}m^{-2}{\cdot}yr^{-1}$. After MEENTEMEYER et al. 1982, modified.

these are maps showing the geographical patterns of main ecosystem processes, it is worthwile to analyse some of them.

Using a world climatic data base (monthly temperatures and precipitation values for 1225 terrestrial stations), terrestrial plant litter production was estimated by climatic curve–fits and mapped by computer. Litter production was best related to AET (actual annual evapotranspiration) – the simplified map is shown with linear regression equation (Fig. 34.3). It is obvious that litter production pattern follows the pattern of primary productivity: increases in humid and warm climates. The European continent, where — according to my own knowledge and experience — a large amount of information exists, is poorly elaborated and only three isolines delimit rather low litter production values.

Accumulation of litter is considered in the papers of ESSER et al. (1982) and BAZILEVICH and RODIN (1971). The Northern Hemisphere parts of maps from these two papers are shown on Fig. 34.4. There are 8 signs of legend on the ESSER map and 10 signs on the Bazilevich map, which means that they use a similar number of vegetation units. ESSER's computations are based on the Hamburg model (LIETH and ESSER 1981 and LIETH 1971); the BAZILEVICH's map is produced on the basis of the author's own data, as well as literature data interpolated by cartographic methods. The obtained patterns of distribution of accumulated litter pool are different. BAZILEVICH's map shows the northward shift of pool of litter: this trend is not so distinct on ESSER's map. However, the boreal forest belt on the Eurasiatic continent shows similar litter accumulation on both maps. BAZILEVICH introduces larger ranges of litter pool mass (up to 10 kg·m^{-2}), while ESSER marks his highest unit as > 6 kg·m^{-2}. To conclude, the two pictures of litter pool distribution on the globe are different.

The interesting maps of production, decomposition rate and soil organic matter in US Great Plains grasslands are introduced by PARTON et al. 1987. The authors applied CENTURY model (with submodels SOM, plant, nitrogen) to simulate productivity, decomposition rate, and soil organic matter as final product resulting from changes of ecosystem processes over time. The maps of north–south directed isopleths were produced for the Great Plains. The excellent fit obtained for ABVG NPP simulation validation is a result of the facts that simulated plant production is highly correlated to simulated N inputs, which are direct functions of annual precipitation; the observed plant production was higly correlated with annual precipitation.

34.4 Production and decomposition of litter on two climatic transects

Studies were conducted of production and decomposition of organic matter on two transects:

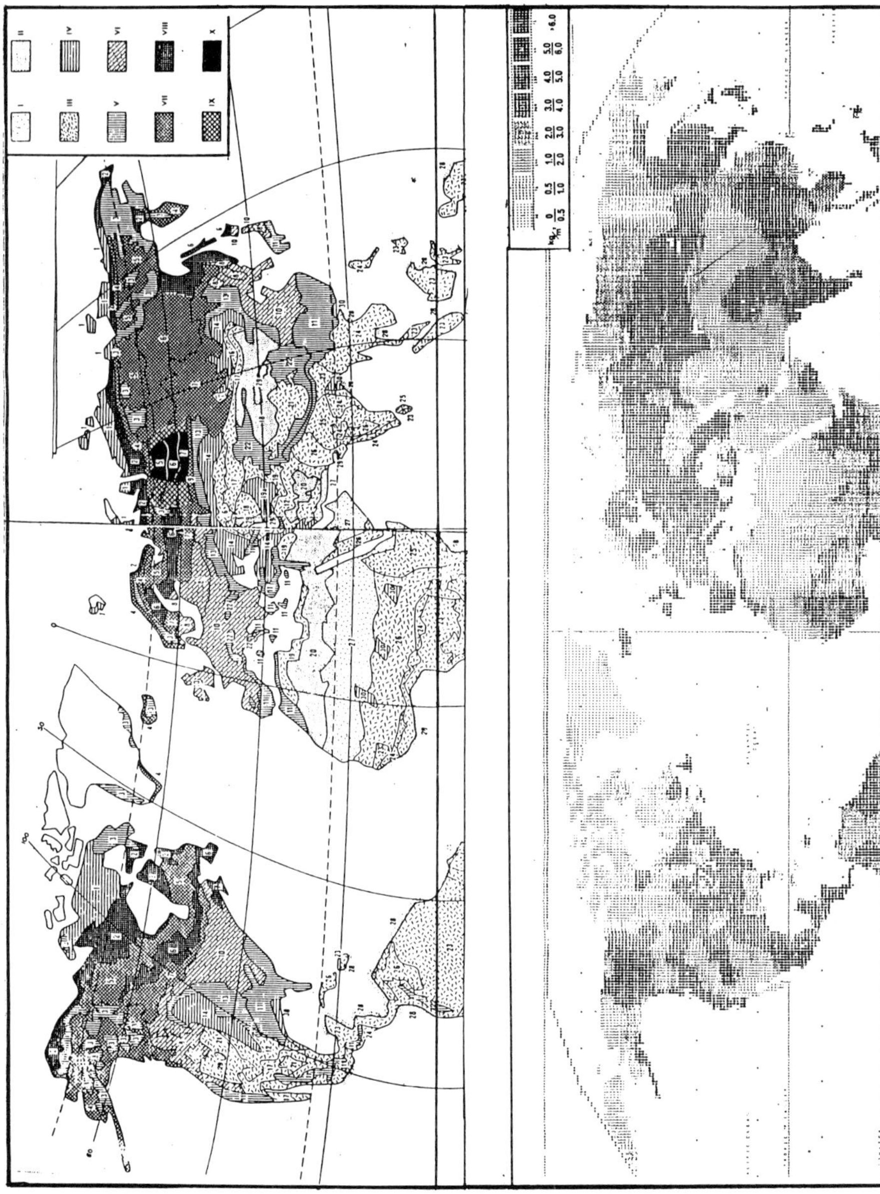

Figure 34.4: Accumulation of litter in the northern hemisphere terrestrial ecosystems acc. to BAZILEVICH and RODIN (1971) – the upper map, and ESSER et al. (1982) – the lower map. Legend for the upper map: I: < 0.01, II: 0.01 – 0.05, III: 0.06 – 0.25, IV: 0.26 – 0.75, V: 0.76 – 1.25, VI: 1.26 – 2.50, VII: 2.51 – 4.00, VIII: 4.01 – 6.00, IX: 6.01 – 10.00, X: > 10 [kg·m^{-2}] dry matter.

1. On the so-called North-South transect consisting of four stands of pine forests distributed from northern Sweden to central Poland

2. On the mountain transect, on which distributed were grassy stands in the vertical profile of the Tatra Mountains from 1200 to 1970 metres a.s.l.

The rate of decomposition was evaluated using the method of litter bags (forest transects) and the method of WIEGERT-EVANS, 1965 (grassy transect). The production od organic matter was also evaluated by the method of WIEGERT-EVANS on the mountain transect, while on the forest transect evaluated was the organic fall and it was treated as a production index (assuming that everything that falls into the ground in the forest must have previously been produced in that forest). Because results of studies from the Tatra Mountains are currently being printed (BREYMEYER and UBA a, b, in press) and those works contain methodical and field details, so in this paper we only present one summary table in order to characterise mountain transect (Table 34.4). Results of studies from the north-south transect have not previously been published, a short description of stands and methods is necessary.

34.4.1 North – South transect.

Stands. Studies were conducted in pine forests in Poland and Sweden. It was assumed that the main climatic factor influencing ecological processes is the regularly changing temperature. Efforts were undertaken to choose the stands in such a way which minimizes meridional deviations: in the central part of Europe the shifting of transect stands to the east or west would mean considerable change of continentalism.

Studies were carried out in the years 1986-88; the main climatic characteristics of all stands are presented in Table 34.1.

The most northerly stand, Torneträsk, is outside of the aknowledged ranges of pine forests; it is in the basin of the large Torneträsk lake, ca. 200 km north of the Polar Circle, at the elevation of about 300 metres a.s.l. Mean long-term temperatures on this area range from −1.0 to −0.5 °C (on the Torneträsk climatic post −0.7 °C), rainfall from 300 to 1000 mm annually (on the Torneträsk 470 mm). In the year 1986 in Torneträsk −1.1 °C and 476 mm were registered. The studied stand is situated several hundred metres from the edge of the lake, often fogs and ground-frosts occur on this area.

Soils were created here from poor, acid, glacial-fluvial sands originating from weathered material of siliceous rocks (SONESSON 1979). This is the border of forests range; the undulating terrain around the lake is covered by bogs and birch tundra (*Betula pubescens* EHRH. f. *tortuosa* LED.). Among the rather homogenous low birch tundra occur stands of *Pinus sylvestris* L. forest; they occur most frequently in the vicinity of the Torneträsk rail station and in the Abisko Valley. According to some opinions those are relict stands of pine, as years with sufficiently warm

Table 34.1: Main characteristics of forest sites on the North–South transect. The climate values are mean annual temperature and sum of annual precipitation.

stand	geographical position	altitude	climate
TORNETRÄSK	68°14'N 19°40'E	370 m a.s.l.	−0.7 °C 470 mm
JADRAAS	60°49'N 16°30'E	185 m a.s.l.	3.8 °C 607 mm
BORECKA	54°10'N 22°00'E	200 m a.s.l.	6.7 °C 579 mm
CZERLONKA	52°41'N 23°40'E	175 m a.s.l.	6.6 °C 626 mm
KAMPINOS	52°22'N 20°50'E	120 m a.s.l.	7.8 °C 552 mm
MIE.II LI.III[1]	52°20'N 23°00'E	165 m a.s.l.	7.2 °C 499 mm
PINCZOW	50°30'N 20°25'E	200 m a.s.l.	7.6 °C 595 mm

[1] MIERZWICE and LITEWNIKI PINE STANDS in EASTERN POLAND

temperatures for self–sowing of this species occur here very seldom. However our studies indicate that on stands of pine forest around Torneträsk there are numerous young and very young trees. SONESSON stated the same (1979): "In the last decades an increasing reproduction of the *Pinus sylvestris* L. pine tree was observed". Selected for the study was one of the largest stands of pine forest (about 1000 sq.m). The age of the oldest pines is estimated to 80 years. Understory plants were: *Betula nana, Empetrum hermaphroditum, Vaccinium victicoides* and *V. myrtyllus, Loiseleuria procumbens* and *Arctostaphylos alpinus*. It seems that this forest is in a good condition and ought to be functioning similarly as other pine forest ecosystems.

The next stand is the pine forest in Jadraas, in the central part of Sweden; geographic position, mean monthly temperature and atmospheric rainfall are presented in Table No. 34.1.

The five stands from the territory of Poland studied in the transect are situated in Borecka Forest (most northerly), in the laying on the bank of Bug river Mierzwice and Litewniki forests, in the Bialowieza Forest (Czerlonka stand) and in two pine forests in central Poland (Kampinos and Pinczow stands) (Tab. 34.1).

The transect extends between 68 to 50°N, and so it covers about 18°. The difference of temperatures between the most northern and the most southern stand amounts to 8.4 ° C.

All forests studied on this transect are situated on sandy soils, in the herb layer they have shrubberries and mosses. Admixtures of the other tree species are insignificant. Detailed descriptions of soils and vegetation are being prepared.

Methods. Organic fall was caught at each stand on 10–20 fall surfaces; they were constituted by smooth circular areas, cleared from vegetation and covered by sand, each 0.1 m^2. In such a way litter fall was gathered from the ground surface and

Table 34.2: Litter fall in coniferous forest. North–south transect.

Names of stands, numbers, years of sampling	Litter fall $g \cdot m^{-2} \cdot yr^{-1}$	% needles	% woody parts
TORNETRÄSK /1; 1986,87/	55.3; 61.5	47.0; 63.4	48.0; 37.0
JADRAAS /4; 1986,87/	112.6; 199.1	47.5; 46.3	52.5; 53.0
MIE.II LI.III /2; 1988/	224.0; 241.4	53.1; 60.0	25.7; 27.2
KAMPINOS[1] /2; 1975/	231.7; 244.6	63.9; 54.2	26.5; 29.5
INDEX I ($\frac{KAMP}{TORN}$):	$\frac{238.1}{58.4} = 4.08$	$\frac{58.0}{55.2} = 1.07$	$\frac{28.0}{42.5} = 0.66$
INDEX II[2] Rate of change per 1 °C:	$\frac{179.7}{8.5} = 21.1$	$\frac{3.8}{8.5} = 0.45$	$\frac{14.5}{8.5} = 1.71$

[1] After JÓZEFACIUKOWA (1975)
[2] Calculated as ratio of differences between extreme litter masses and temperatures $\frac{m_n - m_1}{t_n - t_1}$

additionally the production of the understory was observed, which is impossible with conventionally applied litter traps located at the hight of one metre. Once a year the dropped fragments of plants (seldom of animals) were collected and transported to the laboratory, fractioned, dried at a temperature of 90 °C, and then weighed.

The decomposition of organic matter was evaluated by the method of litter bags. On each stand, from litter collected during the year on litter fields, portions of 10 g were prepared, placed in nylon bags (net diameter 1 mm) and placed on the same stand for the period of one year. Samples of mixed litter, i.e. composition of various fragments as they fell onto the forest floor, were exposed in such a way; moreover, selected were various fractions of litter and then exposed similarly for decay.

Results. The comparison of litterfall/decomposition rates on the whole transect, from Torneträsk to Pinczow, was presented in Table 34.2 and 34.3.

Only 55–61 g of litter falls on each square metre of circumpolar forests in Tor-

Table 34.3: Decomposition in coniferous forests. North–south transect. INDEX I ist the relation $\frac{PINC}{TORN}$, INDEX II means rate of change per 1 °C. Mean values of the ranges were taken for the calculation.

names of stands, numbers, years of sampling	yearly decomposition rates % mixed litter	needles	branches	cones	birch leaves
TORNETRÄSK 1;1986,7	11.4–14.0	14.2–15.0	4.8	2.3–3.6	30.0
JADRAAS 4;1986,7	17.8–19.1	27.0–29.1	5.1–6.4	7.1–11.4	
BORECKA 2;1987	22.3–23.7	36.9	13.6	13.9	
CZERLONKA 1;1987,8	22.9–23.4	34.9	19.2	17.1	55.1
PINCZOW 1;1987,8	27.3–28.4	40.2	12.7	11.8	
INDEX I	2.15	2.75	2.64	4.0	1.83
INDEX II	1.80	3.08	0.95	1.06	3.02

neträsk; this litter decays slowly, 11–14% per annum. In Jadraas, in Central Sweden, the falling of litter is twice as large, the decay rate is also higher, although it does not change as rapidly as organic fall. It is to be seen clearly how interesting decomposition changes with transformations of the climate. Mixed litter (composition identical to that which falls on the bottom of the forest) has a hastened decomposition ranging from 11% in the north to 28% in the south. Most rapidly decay pine needles – from 14% in the north to 40% in the south. The biggest differences were observed in cones — if extremal values are to be considered — cones hasten their decomposition as much as 5–fold after shifting by 18 ° south of Torneträsk.

Thus after shifting by 18° to the south of Torneträsk the decomposition of mixed litter and birch leaves hastens twice, of needles and small twigs almost three times, and of cones – four times. The leaves of birch decay most rapidly: 30–50% per year; when the ratio of most southern to most northern stand is calculated (PINC/TORN INDEX) the birch leaves stay in the low position (INDEX I = 1.83). But when the INDEX II is calculated, the most rapidly decomposing birch leaves and pine needles shift on the first position: their decay rate changes about 3% per each 1 °C of temperature change.

Conclusions. Based on the evaluation of the data, the following conclusions may be drawn:

- Decay rate of woody litter fractions, particulary of cones and bark, is very low; for such slowly decaying parts as cones a period of about 30 years would be required for a complete decomposition.
- The decomposition of those woody fractions reacts most distinctly to warming or cooling of the climate.
- Remembering that it is acknowledged that the decomposition of litter encompasses at least two phases: fast initial decay of carbohydrates and slow decay of lignin and cellulose, it can be assumed that the same change of climate influences differently various decomposition phases (acceleration of lignin/cellulose phase is stronger).
- Litter fall of the northern stands comprises to a much greater part lignified, very slowly decaying plant fragments.
- As we move to the south, the composition of litter changes, it contains more remains of soft, green parts of plants; it is a litter richer in fast releasing nutrients.

34.4.2 Mountain transect

The second transect has been established in the Polish Tatra Mts.; stands have been distributed in a vertical profile from 1200 to 1970 metres a.s.l., differences of mean annual temperatures amounted to 5.7 °C (from −0.8 °C to 4.9 °C). It has been assumed that in the case of grassy ecosystems, plant mortality which in the WIEGERT–EVANS method (1964) is measured directly as a mass of dead plants may be considered as a counterpart of litter fall in forests.

Both processes — mortality and decomposition — indicate similar, rather regular tendencies for slowing down the rate with an increase of altitude (Table 34.4). On the basis of materials contained in the table, three indexes were calculated. The first one is the relation between values of measures taken from the lowest and highest elevations — decreasing of rate of both processes with the altitude increase is almost identical. The second index shows rate of changes of both processes on two parts of the studied transect: above and below the timber line. The forest border in the mountains is considered to be a distinct and significant line separating different environmental conditions in a vertical profile of the mountains. The processes of decomposition and mortality differ clearly above and below this line, but both change similarly. Also the third index, evaluation of rates of both processes per one degree of temperature change, is similar for mortality and decomposition (with the exception of one case — the rate of mortality below the timber line is distinctly higher). Despite that single deviation, both studied processes indicate

Table 34.4: Measurements of plant mass mortality and decomposition in mountain grasslands laying along altitudinal transect in Polish Tatras: annual values of temperature, mortality and decomposition (temperatures evaluated from long-term mean values and from a model, mortality measured in 1982–1984, decomposition, after one year of incubation measured in March 1982, April 1983, and May 1984).

altitude m	temperature mean °C	mortality $g \cdot m^{-2}$		decomposition $g \cdot m^{-2} \cdot yr^{-1}$
1970	−0.8	30.2		27.4
1820	0.3	90.6 80.9		100.5 80.2
1680	0.9	122.1		112.8
	timber line (TL)			
1520	2.4	223.2		266.1
		284.8		292.6
1200	3.3	346.4		319.2
INDEX I:	$\frac{1200m}{1970m}$	$\frac{364.4}{30.2} = 11.1$		$\frac{319.2}{27.4} = 11.6$
INDEX II:	$\frac{\text{belowTL}}{\text{aboveTL}}$	$\frac{248.8}{80.9} = 3.5$		$\frac{292.6}{80.2} = 3.6$
INDEX III:	Rate of	below TL	136.6	59.0
	change	above TL	54.0	50.2
	per 1 °C	transect	77.1	71.1

an astonishing similarity in the rate of changes with shifting along the transect. It ought to be added that measures of the rate of both processes have been conducted with different methods: litter decomposition was measured through evaluations of litter weight losses from nylon bags, while mortality was evaluated through a direct collection of dead plants on small experimental plots (the WIEGERT–EVANS method, 1964).

Summarizing results obtained on the transect, we can state that:

- At a elevation difference of 770 metres and temperature difference of 4.5 °C mortality and decay of plant mass change 11–fold;

- Changes of both processes with elevation are 3 times more rapid below the timber line;

- Rate of change of both processes per one degree of temperature amounts 70 $g \cdot m^{-2} \cdot yr^{-1}$.

Author will risk some reflections flowing from comparing broad scale picture of ecosystem processes with the same process changeability measured in the field:

In geography of decomposition usually the "first year decomposition rate" is used as a quantitative index of the rate of the process of organic matter decay. Most frequently the homogenous samples of leaves (needles) are exposed for decay as the substrate representing the litter. Our research shows that the woody fraction of litterfall can serve as a sensitive substrate reflecting the changing of climate.

What is more, if it is true that wood decomposition responds more sharply to climatic gradients than other, "soft" fractions of litter, some differing conclusions may be suggested. MEENTEMEYER (1984) in his model predicting the geography of decomposition rates of Canada and United States, assumes that lignin concentration in litter is an important predictor of the decomposition rate (many other authors use the same assumption in decomposition modelling). According to MEENTEMEYER's calculations, the slope of linear fit between percent of lignin and percent of annual weight loss by litter varies between −1.4 (the highest fit) and −0.5 (the lowest fit in the subarctic regions of North America, where the cold climate overwhelms all other regulators). From our results it seems to be obvious that litter fractions with the high lignin concentration (cones, bark) responds to a changing climate better than the other fractions. Does it mean that generalizations of MEENTEMEYER and our rough results are contradictory? Does it mean that for studying decomposition on a geographic scale one should use the late phases of decaying litter or simply decaying wood? Should we change our perspective and approach in this field? Surely we need more information, deeper understanding and more precise formulations of statements arising from our results, from models and maps.

Bibliography

ÅGREN, G. I.; MCMURTIE, R. E.; PARTON, W. J.; PASTOR, J., in press, Stata of the art of models of production–decomposition linkages in conifer and grassland ecosystems.
Proceedings of the SCOPE workshop "The effects of climate change on production and decomposition in coniferous forest and grasslands", Woods Hole, MA, 9–14 April 1989.

BAZILEVICH, N. I.; RODIN, L. YE., 1971, Geographical regularities in productivity and the circulation of chemical elements in the Earth's main vegetation types.
Soviet Geography: Review and Translation, published by American Geographical Society.

BOTKIN, D. B.; JANAK, J. F.; WALLLIS, J. R., 1972, Some ecological consequences of a computer model of forest growth.
Journal of Ecology **60**, 849–872.

Box, E., 1978, Geographical dimensions of terrestrial net and gross primary production.
Radiation and Environmental Biophysics **15**, 305–322.

Box, E. 1981, Macroclimate and plant forms: An introduction to predictive modeling in phytogeography.
D. W. Junk Publishers, The Hague, Netherlands.

Breymeyer, A. I., 1990, Measuring function and disfunction in ecosystems.
In: Grodzinski, W.; Cowling, E.; Breymeyer, A. (eds.),Ecological risks, concepts and impact studies. Perspectives from Poland and United States, US National Academy of Sciences, Washington.

Breymeyer, A.; Uba, L., in press, Response of mountain grassland ecosystems to climate change, the case of Tatra Mountains.
Proc MAB/UNU Conference "Transformation of mountain environments", TOME, Armenia 1989.

Breymeyer, A.; Uba, L., in press, Production and decomposition of organic matter in mountain transects– the comparative studies.
Ekologia Polska.

Coughenour, M. B.; McNaughton, S. J.; Wallace, L. L., 1984, Modelling primary production of perennial graminoids – uniting physiological processes and morphometric traits.
Ecological Modelling **23**, 101–134.

Detling, J. K.; Parton, E. W. J.; Hunt, H. W., 1979, A simulated model of *Bouteloua gracilis* biomass dynamics on the North American shortgrass prairie.
Oecologia **38**, 167–191.

de Wit, C. T.; Goudriaan, J.; van Laar, H. H.; Penning de Vries, F. W. T.; Rabbinge, R.; van Keulen, H., 1978, Simulation of assimilation, respiration and transpiration of crops.
PUDOC Wageningen, 141 pp.

Duvigneaud, P.; Tanghe, M., 1962, Ecosystèmes et biosphère.
Travaux du Centre d' Ecologie Générale Bruxelles.

Esser, G.; Aselmann, I.; Lieth, H., 1982, Modelling the Carbon Reservoir in the system compartiment "Litter".
In: Degens; Kempe (eds.), Transport of carbon and minerals in major world rivers, Part I, Mitt. Geolog–Pateontolog. Inst. Univ. Hamburg, SCOPE/UNEP, Sonderband Heft **52**, 39–58.

GOUDRIAAN, J.; VAN LAAR, H. H.; VAN KEULEN, H.; LOUWERSE, W., 1984, Simulation of the effect of increased atmospheric CO_2 on assimilation and transpiration of a closed crop canopy.
Wissenschaftliche Zeitschrift der Humboldt Universität Berlin, Math.-Nat. R. XXXII, 352–356.

HANSON, J. D.; SKILES, J. W.; PARTON, W. J., 1988, A multi-species model for rangeland plant communities.
Ecological Modelling **44**, 1–35.

HOLDRIGE, L. R., 1964, Life Zone Ecology.
Tropical Science Center, San Jose, Costa Rica.

INNIS, G. S., (ed.), 1978, Grassland Simulation model.
Springer, New York, 298 pp.

JOZEFACIUKOWA, W., 1975, Variation of the fall rate of plant debris from trees in the association Vaccinio myrtylli-Pinetum typicum (Kobenza 1930).
Br.-Bl. et Vlieger 1939 in the Kampinos National Park, Ekol. Pol. **23**, 93–101.

LAUENROTH, W. K.; PRESTON, E. M., (eds.), 1984, The effects of SO_2 on a grassland: A case study in the Northern great Plains of the United States.
Springer, New York, 207 pp.

LIETH, H., 1971, The net primary productivity of the earth with special emphasis on the land areas.
In: Whittaker, R. (ed.), Perspectives on primary productivity of the earth.

LIETH, H., 1972, Modeling the primary productivity of the world.
Nature and Resources **8**, 5–10.

LIETH, H., 1975, Primary productivity of the major vegetation units of the world.
In: Lieth, H.; Whittaker, R. H. (eds.), Primary productivity of the biosphere, Ecological Studies **14**, Springer Verlag, 203–217.

LIETH, H.; BOX, E., 1972, Evapotranspiration and primary productivity.
C. W. Thornthwaite Memorial Model Publications in Climatology **225** (2), 37–46.

LIETH, H.; ESSER, G., 1981, Zur Modellierung der Beziehung zwischen globaler Netto-Primärproduktivität und Umweltfaktoren.
III. Arbeitstagung Umweltbiophysik der DDR.

MCMURTRIE, R. E.; BENSON, M. L.; LINDER, S.; RUNNING, S. W.; TALSMA, T.; CRANE, W. J. B.; MYERS, B. J., in press, Water nutrient interactions affecting productivity of stands of *Pinus radiata* – Forest Ecology and Management.

MC MURTRIE, R. E.; LANDSBERG, J. J.; LINDER, S., 1988, Research priorities in field experiments on fast-growing tree plantations: implications of a mathematical model.
In: Pereira, J. S.; Landsberg, J. J. (eds.), Biomass production by fast growing trees, Martinus Nijhoff/Dr. W. Junk, The Netherlands.

MCMURTRIE, R. E.; ROOK, D. A.; KELLIHER, F. M., in press, Modelling the yield of *Pinus radiata* on a site limited by water and nitrogen.
Forest Ecology and Management.

MEENTEMEYER, V., 1984, The geography of organic decomposition rates.
Annals of the Assoc of American Geographers **74** (4), 551–560.

MEENTEMEYER, V.; BOX, E. O.; THOMPSON, R., 1982, World patterns and amounts of terrestrial plant litter production.
Bio Science **32**, 125–128.

MEENTEMEYER, V.; GARDNER, J.; BOX, E. O., 1985, World patterns and amounts of detrital soil carbon.
Earth surface processes and landforms **10**, 557–567.

MOHREN, G. M. J., 1987, Simulation of forest growth, applied to Douglas fir stands in the Netherlands.
Thesis Agricultural University Wageningen, 184 pp.

MOHREN, G. M. J.; VAN GERWEN, C. P.; SPITTERS, C. J. T., 1984, Simulation of primary production in even-aged stands of Douglas fir.
Forest Ecology and Management **9**, 27–49.

OLSON, J. S.; WATTS, J. A.; ALLISON, L. J., 1983, Carbon in live vegetation of major world ecosystems.
Oak Ridge National Laboratory.

O' NEILL, R. V.; DEANGELIS, D. L.; WAIDE, J. B.; ALLEN, T. F. H., 1986, Hierarchical concepts of ecosystems.
Monographs in Population Ecology **23**, Princeton Univ. press.

PARTON, W. J.; SINGH, J.; COLEMAN, D. C., 1978, A model of production and turnover of roots in shortgrass prairie.
Journal of Applied Ecology **45**, 515–542.

PARTON, W. J.; SCHIMEL, D. S.; COLE, C. V.; OJIMA, D. S., 1987, Analysis of factors controlling soil organic matter levels in Great Plains Grasslands.
Soil Sci. Soc. of America Journal **51** (5), 1173–1179.

PARTON, W. J.; STEWART, J. W. B.; COLE, C. V., 1988, Dynamics of CNP and S in grassland soil: a model.
Biogeochemistry **5**, 109–131.

PARTON, W. J.; COLE, C. V.; STEWART, J. W. B.; SCHIMEL, D. S., in press, Regional modelling of CN and P in temperate grassland soil.
Plant and Soil.

ROSENZWEIG, M. L., 1968, Net primary production of terrestrial communities: prediction from climatological data.
The American Naturalist **102**, 67–74.

RUNNING, S. W.; COUGLHAN, J. C., in press, A general model of forest ecosystem processes for regional applications. I. Hydrologic balance, canopy gas exchange and primary production processes.
Ecological Modelling.

SHUGART, H. H., 1984, A theory of forest dynamics.
Springer Verlag, New York, 278 pp.

SONESSON, M., 1979, Abisko Scientific Research Station: Environment and Research.
Holarctic Ecology **2**, 279–283.

WANG, Y. P., 1988, Crown structure, radiation absorption, photosynthesis and transpiration.
Ph. D. Thesis, University of Edinburgh, 188 pp.

WIEGERT, R. G.; EVANS, F. C., 1964, Primary production and the disappearance of dead vegetation in an old-field in south-eastern Michigan.
Ecology **45**, 49–63.

Chapter 35

Geographic modeling and modern ecology

E. O. BOX and V. MEENTEMEYER

35.1 Introduction

Modern ecology grew out of various traditions in natural history studies. One of the strongest of these traditions involved the European interest in plant and vegetation geography which resulted from the travels and writings of ALEXANDER VON HUMBOLDT (e. g. 1805, 1807). Throughout the Nineteenth Century, people interested in the ecology and geography of the natural world produced vegetation and other "ecological" maps, often at world scale (e. g. SCHOUW 1823; GRISEBACH 1866, 1872; ENGLER 1882). In addition they developed some of the early theoretical foundations for the science of ecology, often through comparative geographic approaches (e. g. DE CANDOLLE 1820, 1855; LIEBIG 1840; DARWIN 1859; WALLACE 1876; DRUDE 1890; WARMING 1985; SCHIMPER 1898). From the very beginning, "ecology" in Germany and neighboring parts of Europe thought in large terms, globally, and included a large geographic component. To this day, German ecology and botany have retained the strong geographic perspective which characterized their development (e. g. RÜBEL 1930; KÜCHLER 1949; KNAPP 1965, 1973; HUECK 1966; WALTER 1968, 1973, 1974; SCHMITHÜSEN 1959, 1976).

The first world map of primary productivity was produced by LIETH (1964), showing estimated patterns of annual net primary productivity (NPP) for both the land and ocean areas of the world. This map was digitized for computer operations (Lieth 1972), and a general computerized mapping system was developed. The mapping routines were then combined with climate–based models derived from at least quasi–global data–bases to produce predictive world maps based on simple formulas applied to large global climatic data–bases (Lieth 1973, Lieth and Box 1972). At this same time, Bazilevich et al. (1968, 1971) had produced world maps of NPP and related phenomena, Rosenzweig (1968) had produced a climate–based model of NPP, and Meentemeyer (1971) had produced a map of litter decomposition rates based on model results. It was Lieth, however, who saw the possibilities of combining these approaches, at global scale, to produce dramatic world maps of ecological phenomena which could not be mapped before because of lack of data. Relatively simple, deterministic, climate–based models could be used to extrapolate reasonably from the large climatic data–bases which were readily available to estimate global ecological patterns which were hitherto poorly understood. This contribution reinforced interest in large–area geographic relationships in ecology and established predictive geographic modeling as a fundamental feature of modern, holistic ecology.

We can now see that at least potential levels of many ecological phenomena seem to be controlled by climate, at least at more general scales (e. g. Troll 1960, Larcher 1976, Rosswall et al. 1988). This includes such basic processes as primary productivity, detrital decomposition, vegetation structure, soil physical and chemical properties, and biological diversity (see references in Table 35.1).

Continental or world maps have been produced for many of these phenomena, making it possible to see that many phenomena have similar geographic patterns. This paper focuses on geographic modeling and these basic geographic patterns of ecological phenomena, at global scale.

35.2 Predictive modeling

Predictive modeling is basically a methodology for using an information–rich database, in which many data are available, to provide information and insight for some less understood target phenomenon, for which few data are available. The target phenomenon may represent a structure, such as the amount or composition of standing vegetation, or an ongoing functional process, such as primary production or detrital decomposition and nutrient turnover. In a system–theoretic sense, "structure" may represent either configurations or pool sizes, while "function" generally represents processes and flows between compartments. Since structure often represents a balance among several individual processes, structures are generally more difficult to predict than are process rates. In any case, the geographic context is implicit, and it is often the geographic pattern of the target phenomenon which is the research objective, since this spatial pattern permits geographic inventories and inferences based on regional comparison.

Predictive geographic modeling may be useful for various reasons:

1. It can be used to generate new spatial patterns and site estimates for less understood phenomena.

2. The application of theory-based models to large global data-bases permits a form of hypothesis testing which is much more rigorous than normal statistical methods since it has a visible geographic context and results.

3. The new spatial patterns generated, once adequately evaluated, can be quantified to provide initial geographic inventories of phenomena previously unquantified.

4. Because of its ability to generate theory-based geographic patterns, predictive geographic modeling represents a natural complement to remote-sensing capabilities. Predictive modeling can suggest areas where more information is needed and can aid in pattern recognition and theoretical interpretation of satellite data. Remote sensing can provide independent site data and pattern information needed for both model validation and interpretation of predicted results.

Predictive geographic modeling is probably most useful over larger areas, where a greater degree of generalization is possible and where sufficient data cannot be gathered otherwise.

Initial, hypothesis-based maps ("research maps") may never be intended for publication but may be important in research. Hypotheses can be "tested" and models "validated" by comparing corresponding predicted and actual site values as well as spatial patterns. Such comparison, as well as comparison of different environments, may provide insight ("geographic inference") into causal relationships. Mathematically simple models seem to provide the best results, since they can express theory more clearly, are easier to understand (and develop), and permit more direct application.

It is important to note, however, that such modeling and model results are profoundly affected by the scale of the study (MEENTEMEYER 1989). KING (1986, pp. 193) and others have suggested that extrapolations from climate-based models of primary productivity may show "errors" of an order of magnitude at certain pixels within spatially heterogeneous landscapes. Global-scale climate-based models, however, can and are only intended to represent climatically constrained potential levels of their target phenomena. Such models cannot be expected to predict actual levels of NPP or other processes in each patch type of local landscapes, especially if these have been altered by human use or other disturbance. Even at the same global scale, however discrepancies in productivity estimates made by different methods, e.g. climatic versus satellite methods, can still reach a factor of 2–3 (BOX, in prep.). This is because climate can only estimate potential NPP levels by "natural" (i.e. climatic potential) vegetation, whereas satellite methods indicate features of the actual vegetation cover, whatever it may be. Potential values are useful as baseline data and often provide a more useful geographic context, especially at global scale, since actual local values may no longer show clear geographic patterns when viewed at more general scales.

35.3 Building geographic models

The most important criterion in developing predictive models for large–area geographic application is that the models be general. Many models predict accurately at one or a few sites, but such models are not useful to researchers working elsewhere. Models may be considered general if they include enough of the causal mechanisms and variation in input and output variables to be applicable to all of the environmental situations which may be present.

Model generality has two aspects: functional and geographic. Functionally general models should produce reasonable results in any situation. This can be assured by understanding the determinants, the scales at which they are important, and by including all of the most important determinants and mechanisms in the model, as best these can be judged. For terrestrial primary productivity, for example, water and energy appear to be the most important constraints at global scale. As a result, surprisingly general models can be constructed based on temperature and water–balance variables. At finer scales, however, more variables become important and must also be included in models in order to show local behavior (Meentemeyer and Box 1987).

Structure usually represents a balance among various processes. As a result, general structural models must include all of the important individual processes involved in the balance. For example, a model of leaf area index should at least include the cost of producing leaves (of the type required by the particular environment), the gain in photosynthesis by these leaves (including effects of stand architecture), and the potential water loss and desiccation resulting from a given leaf area, structure and deployment. Each of these functional aspects must be expressed via some common denominator, if such exists, or compared in some other way, such as by linear programming or simulational modeling. Models of structural phenomena are thus much more complex than models of individual process rates or totals.

Models which are functionally general still may not be geographically general, i. e. applicable over the full geographic range of situations. For example, a functionally general climate–based model of primary productivity will not accurately predict productivity in the tropics or polar regions if it was parameterized using only temperate–zone data. Lieth was thinking geographically, using global data–bases, and put an upper limit on NPP in the humid tropics, whereas Rosenzweig (1968) did not (see also the validations by Sharpe (1975) and O'Neill and DeAngelis (1981) of several published models).

The important criterion for geographically general models, assuming their functional logic is adequate, appears to be that parameterization must be based on a geographically representative data-base. As a first step the upper and lower end-points must be established correctly. For global models this requires including data from the wet tropics, arid deserts, and cold polar regions. More generally, a geographically representative data-base must:

1. cover the full range of variation in the input and output variables (i. e. include the extremes);
2. include all the important environmental situations or types which may occur (e. g. climate or biome types, major substrate types and anomalies, different management regimes); and
3. represent the different environmental situations in a geographically balanced way.

Item three means that data-bases should not be overloaded with data from the northern temperate zone, even though such data are often more numerous. If there are extra data from such areas or if data are otherwise spatially clumped, some of the "redundant" data should be held out of the model development process and saved for validating the model.

Over large areas, climate represents a constraint on all lower-level phenomena (cf. O'NEILL 1988) and is often the single most important environmental correlate of ecological phenomena (cf. MEENTEMEYER and BOX 1987). Climate also represented the only global data-base until NOAA and some other satellite data became available (cf. BOX et al. 1989). The most effective climatic variables may not be just temperature and precipitation, however, but rather integrative variables such as actual evapotranspiration (AET) or other water-balance variables. Where climate is the main control it also represents the best single basis for identifying different environmental situations for geographically representative data-bases. The commonly used quantitative climate classifications such as by KÖPPEN or THORNTHWAITE may not be as appropriate as more genetic classifications, which focus on climate-generating mechanisms and core areas rather than boundaries. An excellent genetic classification for this purpose is that of WALTER (1968, 1973, cf. WALTER and LIETH 1960-1967), which manages to cover all the major climatic types on earth with only ten basic climate units (plus altitudinal variants). The WALTER climate types correspond well to biomes or other large-area vegetation and landscape types.

35.4 Ecological phenomena and global patterns

There are a great many ecological phenomena, both functional and structural in nature, which can and to some extent have been related to environmental correlates in such a way that initial geographic models could be produced. Various important ecological phenomena, especially those for which maps are available, are listed in Table 35.1, along with the global geographic patterns which they seem to represent. These patterns are of course smoothed global-scale patterns, do not reflect local spatial heterogeneity or disturbance, and do not imply values at particular local sites or pixels (probably too small for world maps), which may be under local substrate, topographic or historical control. From the list in Table 35.1 one can recognize several basic patterns which recur frequently among both physical and ecological phenomena. The most fundamental of these appear to be the following four:

1. a thermal pattern, representing energy availability;
2. a moisture pattern, representing water availability;
3. a throughput pattern, representing simultaneous availability of warmth and moisture; and
4. an accumulation pattern, representing accumulations of positive balances over long periods of time, perhaps until equilibrium.

These hypothesized fundamental geographic patterns are discussed further below and are illustrated by world maps of particular representative phenomena. Since each phenomenon has its own seasonal variation, annual totals or averages show the geographic pattern most effectively. A similar climatic modeling basis for different phenomena often results in a similar predicted geographic pattern — but not always. Annual NPP and litter decomposition rate (k) are both well correlated with annual AET, for example, but only NPP is well correlated with satellite-observed surface greenness (Box et al. 1989). Table 35.1 includes references to maps of particular phenomena. Studies without maps may be useful, even famous, but are included only when maps are not available. The levels of the various thermal, moisture, throughput and accumulation patterns in different climate types (cf. biomes) are summarized in Table 35.2.

35.4.1 The thermal pattern

The thermal pattern represents energy availability, as reflected by phenomena such as temperature, insolation, and potential evapotranspiration, as well as energy-related ecological phenomena such as respiration rate. The global geographic pattern for energy-driven phenomena, of course, shows higher levels near the equator and decreases poleward. This thermal pattern is represented here by a world map of annual potential evapotranspiration (PET) (Figure 35.1). Not all phenomena of a given "type" will follow the geographic pattern perfectly. For example, in persistently cloudy tropical areas such as Liberia or parts of Indomalaysia, insolation may be much lower while temperatures remain high. This variation can be

Table 35.1: Important Physical and Ecological Phenomena and Their Apparent Global Geographic Patterns. References are to publications with world or at least continental-scale maps, unless indicated by cf. or parentheses. All processes represent annual totals or rates; structural entities represent amounts at time of maximum vegetation development, e. g. seasonal maximum leaf area index, equilibrium phytomass accumulation.

Phenomenon	Pattern	Reference
temperature levels	thermal	many, e. g. GEIGER 1965
insolation	thermal	GEIGER 1965
potential evapo-transpiration	thermal	THORNTHWAITE 1948, BAUMGARTNER & REICHEL 1975
respiration rate	thermal	(cf. LARCHER 1976, BOX 1978)
precipitation	moisture	many, e. g. GEIGER 1965
moisture balance (P/PET)	moisture	BOX 1981a
vegetative cover (%)	moisture	(?)
solar energy efficiency of NPP	moisture	BOX 1979a, UCHIJIMA and SEINO 1987
shoot-root biomass ratio	moisture (extra-polar only)	(cf. BJORKLAND 1984)
rooting depth	inverse moisture	(BOX, unpubl. map)
actual evapotranspiration	throughput	GEIGER 1965
soil texture	throughput	(BOX, herein)
soil field capacity	throughput	(BOX, unpubl. map)
time-integrated surface greenness (NDVI)	throughput	BOX et al. 1989, OHRING et al. 1989
net primary production	throughput	LIETH 1964, BAZILEVICH and RODIN 1971, LIETH and BOX 1972, ESSER et al. 1982, UCHIJIMA and SEINO 1987
gross primary production	throughput	LIETH and BOX 1977, BOX 1978
litter production	throughput	MEENTEMEYER et al. 1982, ESSER et al. 1982; cf. JORDAN 1971
decomposition rate (k) (litter, wood)	throughput	MEENTEMEYER 1979, 1984; cf. WHITFORD et al. 1981
decomposition amount	(throughput?)	(cf. BOX 1988)

continued next page

Table 35.1, continued.

Phenomenon	Pattern	Reference
soil CO_2 content	throughput	BROOK et al. 1983
soil CO_2 evolution	throughput	(cf. BOX et al. 1983)
soil pH	throughput	FOLKOFF et al. 1981
soil base saturation	throughput	(cf. WOJCIK 1983)
plant species richness	throughput(?)	CURRIE and PAQUIN 1987, cf. PIANKA 1966
standing biomass	accumulation	BAZILEVICH and RODIN 1971, OLSON et al. 1983
litter accumulation	accumulation	BAZILEVICH and RODIN 1971, MEENTEMEYER 1979
soil carbon	accumulation	MEENTEMEYER et al. 1985
net CO_2 flux	biseasonal throughput	BOX 1988
clay-mineral formation	biseasonal throughput	(cf. FOLKOFF and MEENTEMEYER 1987)
plant form richness (in canopy only)	non-forest(?)	BOX 1981a
litter turnover rate	??	BAZILEVICH and RODIN 1971, MEENTEMEYER 1979
nitrogen consumption for annual NPP	??	BAZILEVICH and RODIN 1971
phosphorus consumption for annual NPP	??	BAZILEVICH and RODIN 1971
mean ash content in biomass (annual NPP)	??	BAZILEVICH and RODIN 1971
leaf area index	???	(cf. TUGGLE 1981)

Figure 35.1: The "Thermal Pattern", as shown by Annual Potential Evapotranspiration. The thermal pattern represents energy availability. Where moisture is not limiting, many ecological processes may follow the thermal pattern, e. g. primary production and decompostion. Annual potential evapotranspiration is estimated as the higher of the THORNTHWAITE–MATHER (1957) and HOLDRIDGE (1959) estimates, at each of 1600 sites. This mechanism effectively selects the Thornthwaite estimate in higher latitudes, where it was developed (North American data–basis), and the Holdridge estimate in the tropics and subtropics, where it was developed (Neotropical data–basis). Each method gives relatively lower values, interpreted as underestimates, outside its own appropriate region (cf. DAMAGNEZ et al. 1963, STEPHENS and STEWART 1963, BOX 1979b, 1986).

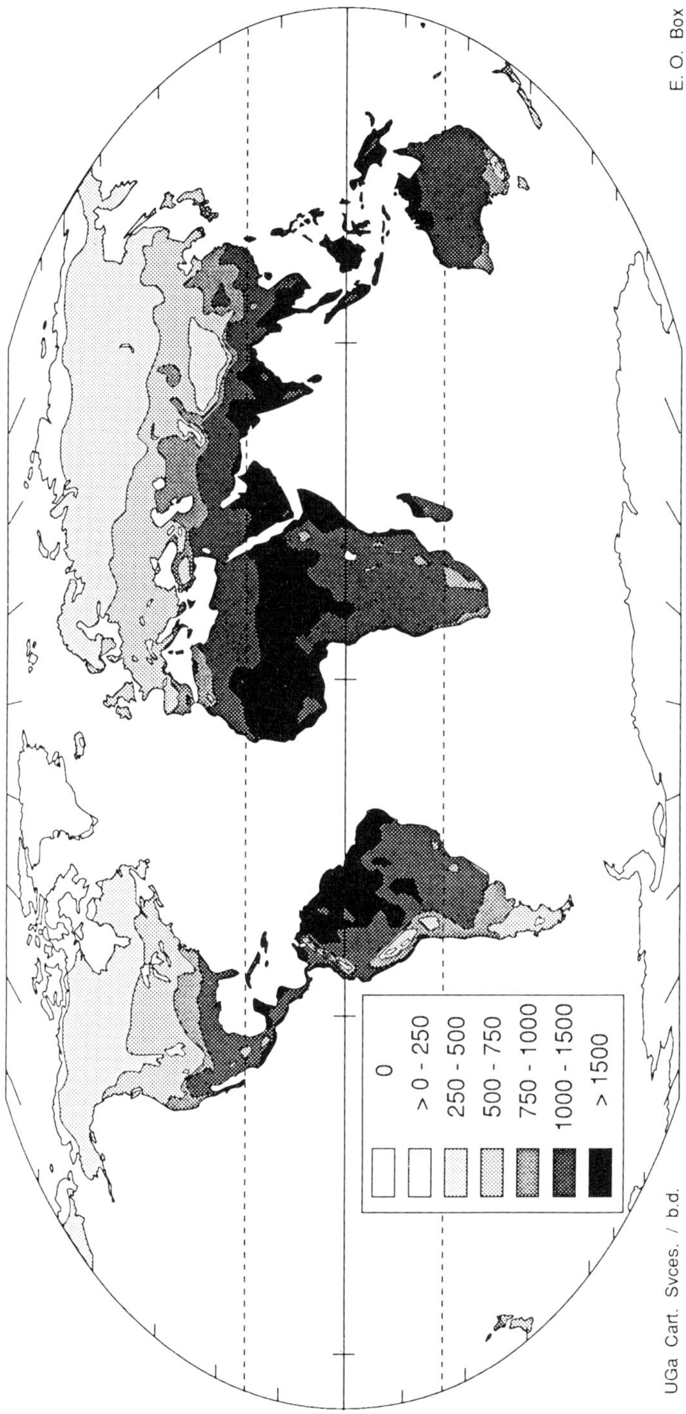

ANNUAL POTENTIAL EVAPOTRANSPIRATION (mm)

recognized but does not change the fundamental energy relationship or the basic type of geographic pattern which results. Autotrophic and heterotrophic respiration rates generally increase rapidly with increasing temperature (e. g. LARCHER 1976, cf. SCHLESER 1982). Where moisture is not limiting, many other ecological phenomena may be essentially energy-limited and thus follow the thermal geographic pattern, for example primary production and detrital decomposition in cooler climates.

35.4.2 The moisture pattern

The moisture pattern represents both moisture input (e. g. precipitation) and moisture availability (e. g. water balance). That the two produce similar geographic patterns, at least at global scale, can be verified by comparing a world map of average annual precipitation with the map of moisture balance or "moisture index" shown in Figure 35.2. Of course, 400 mm of precipitation may support a forest in Canada but only a dry savanna in tropical Africa, so water balance is more important than water input. Vegetative cover percentage in natural vegetation seems to follow the moisture balance up to a value of about one, beyond which all values show potentially forest vegetation with cover value essentially 100%. If photosynthetic efficiency is defined as NPP per unit of energy available (e. g. UCHIJIMA and SEINO 1987, cf. BOX 1979a), then this efficiency is largely moisture-dependent. Shoot-root biomass ratios are controlled mainly by water uptake versus loss and thus seem to follow the moisture-balance pattern (for vegetation stands, not always individual plants, cf. BJORKLAND 1984). Rooting depth by vegetation, on the other hand, seems to follow the inverse of the moisture pattern, being greater in drier climates and less where water is abundant (with various exceptions).

Figure 35.2: The "Moisture Pattern", as shown by Annual Moisture Index. The moisture pattern represents moisture availability, as illustrated here by the balance between precipitation and potential evapotranspiration (P/PET). Biological phenomena such as vegetation cover degree (percent) and the relative allocations to above-ground versus below-ground parts of plants tend to follow the moisture pattern. Forests, for example, occur naturally only at MI values above about 0.95 (as low as 0.8 in the tropics, with high dry-season PET deflating the MI fraction). The moisture index is truncated abruptly, without appreciable gradient, where permanently frozen conditions prevail. The moisture index P/PET is estimated here as annual precipitation divided by the THORNTHWAITE-HOLDRIDGE maximum PET (see Figure 35.1).

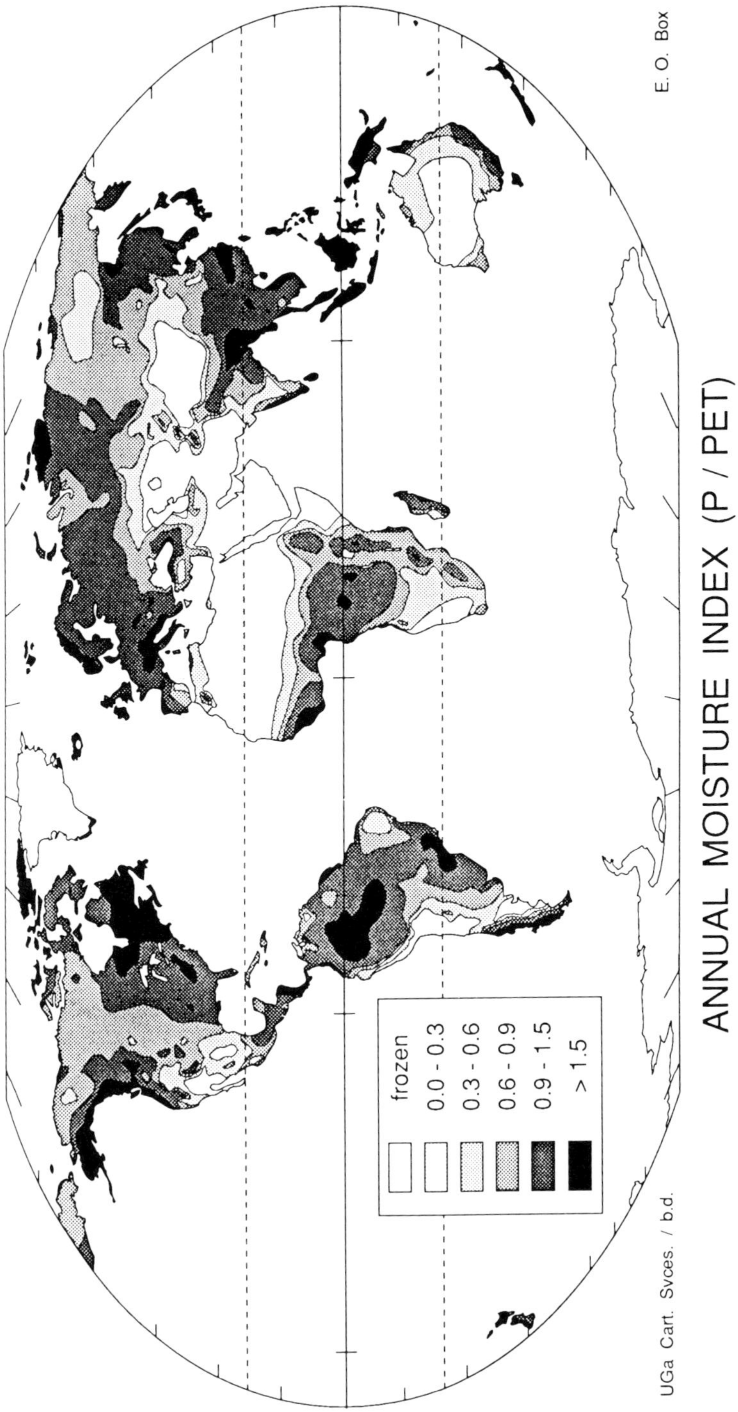
E. O. Box
ANNUAL MOISTURE INDEX (P / PET)
frozen
0.0 - 0.3
0.3 - 0.6
0.6 - 0.9
0.9 - 1.5
> 1.5
UGa Cart. Svces. / b.d.

35.4.3 The throughput pattern

The throughput pattern represents the simultaneous availability and action of energy and moisture, the two most important factors constraining many biological processes (e. g. primary productivity) as well as many physical processes (e. g. soil weathering). The most useful climatic variable for representing this synergism is actual evapotranspiration (AET), both seasonal and annual (cf. Major 1963). The geographic pattern is the result of more intense activity in the humid tropics and decreasing activity as one goes toward drier and/or colder environments. The global geographic pattern is represented here by a map of climatic potential annual net primary productivity (NPP), shown in Figure 35.3. Other world maps of NPP can be seen in Lieth (1964), Bazilevich and Rodin (1971), Esser et al. (1982), and Uchijima and Seino (1987), based on somewhat different data and environmental logic.

In addition to NPP and AET, various soil phenomena are controlled by the co-action of water and energy. Models of soil texture (Figure 35.4) and field capacity (Box, unpublished), based on temperature and precipitation variables alone, reproduce surprisingly well the global distributions of the main, traditional soil classes (e. g. FAO-UNESCO 1974). Other soil-related phenomena following the general throughput pattern include soil CO_2 content (Brook et al. 1983), CO_2 evolution from the soil (Box et al. 1983), soil pH (Folkoff et al. 1981), and soil base saturation content (Wojcik 1983).

Aspects of ecosystem carbon metabolism following the general throughput pattern include gross as well as net primary productivity (Lieth and Box 1977; Box 1978), litterfall (Esser et al. 1982, Meentemeyer et al. 1982), and detrital decom-

---

Figure 35.3: The "Throughput Pattern", as shown by Climatically Estimated Annual Net Primary Productivity. The throughput pattern represents the simultaneous availability of energy and moisture, the two most important factors for biological as well as many physical processes. Some "structural" phenomena also seem closely related to this throughput pattern, such as species richness among terrestrial plants. Ecological processes following the throughput pattern include primary productivity, detrital decompostion rate, soil CO_2 content and CO_2 evolution (except under strong nutrient limitations, cf. Box et al. 1983). Annual net primary productivity (NPP) is estimated here by averaging the "Miami Model" value (Lieth 1973; Box et al. 1971) and the NPP value given by an AET-based sigmoid model (Box, in prep.). This combination greatly reduces the overprediction by earlier models in cold and/or dry environments as well as the apparent underprediction in the humid tropics. Further improvements to the estimation of annual NPP can be made by including soil and other factors as secondary variables (e. g. Esser et al. 1982).

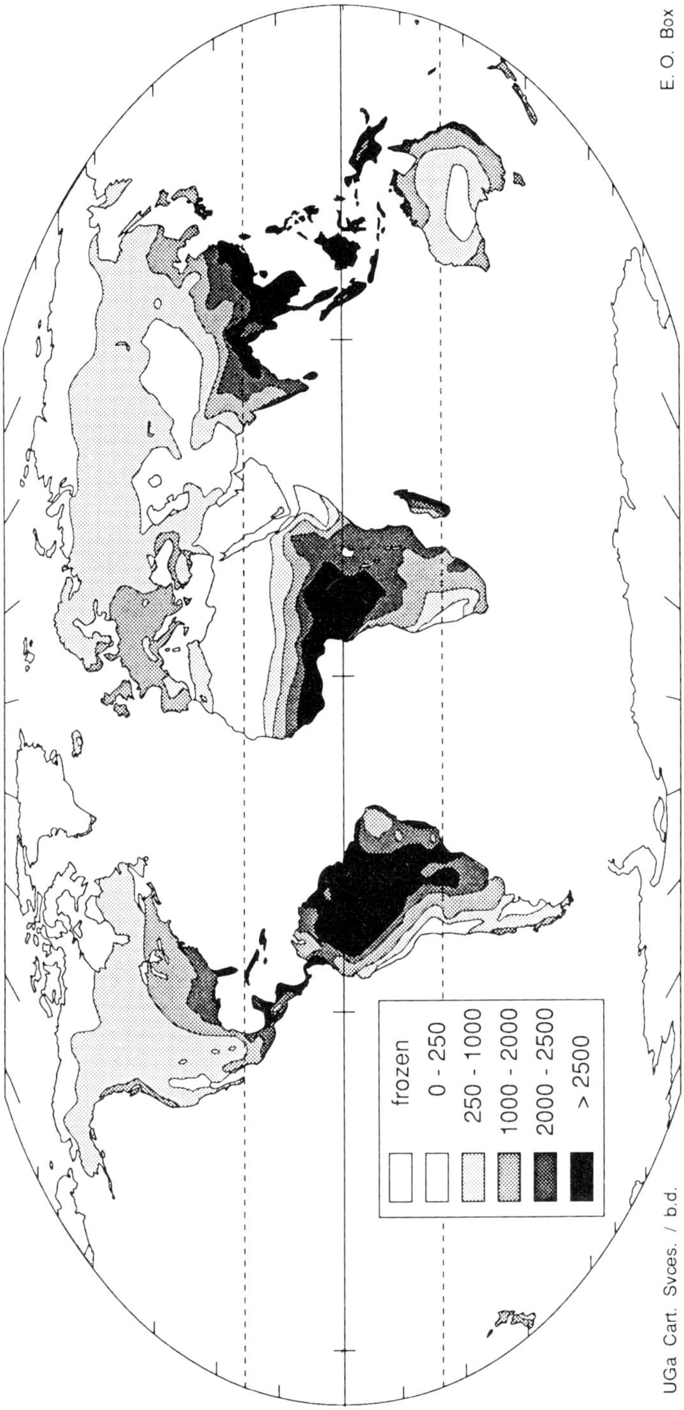

NET PRIMARY PRODUCTIVITY (g d.m. / m^2 / year)

Table 35.2: Levels of the Thermal, Moisture, Throughput, and Accumulation Patterns in Different Climatic Regions. The climatic regions are based on the climate types of Walter, with type V split into east–coast (warm–temperate) and west–coast (marine west–coast) sub–types.

Climate Type	Thermal	Moisture	Throughput	Accumulation
Equatorial (humid)	high	high	high	low–high
Tropical wet–dry	high	medium	high	low–med.
Subtropical arid	high	low	low	low
Mediterranean	medium	medium	medium	medium
Marine west–coast	medium	high	medium	high
Warm–temperate (east)	medium	high	med.–high	medium
Typical temperate	medium	high–med.	medium	med.–high
Temperate grassland	medium	low	med.–low	low–high
Temperate arid	medium	low	low	low
Boreal	low	high–low	low–med.	high
Polar (tundra)	low	high–low	low	high–low

position rates (Meentemeyer 1978, 1979, 1984, 1985). The geographic patterns of these phenomena are modified somewhat by factors such as nutrient limitations, especially in the case of CO_2 evolution (as seen from tropical field data) and detrital decomposition (cf. Whitford et al. 1981, Smith 1985).

Satellite–derived images of surface greenness (NDVI "vegetation index") and greenness duration (time–integrated NDVI) also follow the throughput geographic pattern (Ohring et al. 1989; cf. Box et al. 1989, Goward et al. 1985, Tucker et al. 1985). Total vegetation richness, both as species richness (cf. Currie and Paquin 1987) and as climatically predicted life–form richness (Box 1981a), generally follow the throughput pattern. This suggests that more favorable environments

Figure 35.4: The "Throughput Pattern", as shown by Climatically Estimated Mean Soil Texture of Mature Soils. Soil texture is represented by mean particle diameter of the solid mineral material. This mean particle size is estimated as the larger of two values, one a linear function of mean annual temperature and the other a saturation function of annual precipitation. Smaller particle sizes represent highly weathered soils such as clays and occur primarily in the humid tropics and subtropics. Coarser soils occur in climates which are either colder or drier (or both). This model is only applicable, of course, to mature soils which have been subject to weathering over long periods of time. Coastal and other young soils do not fit the model but, at global scale, cover relatively smaller areas.

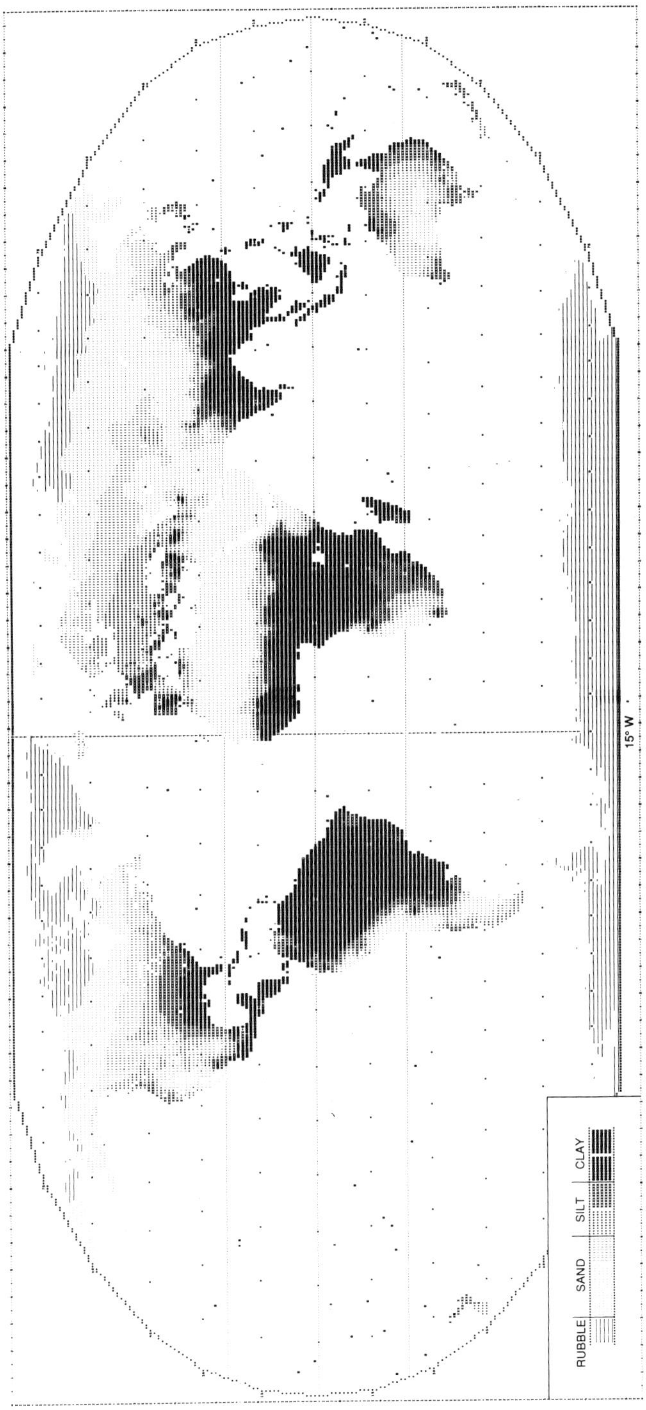
15° W
RUBBLE
SAND
SILT
CLAY

(higher throughput) permit greater differentiation by the vegetation, through a greater environmental wealth available to be partitioned into a greater number of exploitable "niches" (cf. Pianka 1966, Burger 1981, Stebbins 1981). The richness of plant communities may be significantly modifed, however, by competition between species (e. g. Keddy 1990), by successional status (e. g. Oosting 1942, Perino and Risser 1972), or other factors (cf. Hurlbert 1971, Barbour et al. 1980).

35.4.4 The accumulation pattern

The accumulation pattern represents the accumulation to equilibrium of positive annual increments which are themselves the balance of input and output processes. For example, annual increments to standing phytomass are the balance of annual NPP and annual litterfall (plus mortality of individuals, herbivory losses, etc.). Annual increments to soil carbon and to surface litter pools are the balance of detrital inputs and decompositional losses. The quantitative logic of detrital accumulations was worked out by Olson (1963).

The main phenomena which appear to follow the logic of the accumulation pattern are standing plant biomass, litter pools, and stored soil carbon, but these may give rather different geographic patterns. The maxima for stored soil carbon (Figure 35.5) occur in boreal and temperate grassland areas (subhumid prairies), while maxima for standing biomass occur in the temperate rainforests of marine west-coast climates and secondarily in tropical rainforests (Bazilevich and Rodin 1971, Cannell 1982, Olson et al. 1983). This provides a complication, but some similarities remain: minima are in deserts (dry and/or cold) and primary maxima are in cooler regions rather than in the regions of highest throughput (humid tropics).

Accumulations are not as quantitatively intuitive as individual processes or even

---

Figure 35.5: The "Accumulation Pattern", as shown by Climatically Estimated Steady-State Soil Carbon Accumulations in Undisturbed Soils. The accumulation pattern represents the accumulation to equilibrium of positive annual increments which are themselves the balance of input and output processes. For example, annual increments to standing phytomass are the balance of annual NPP, litterfall, mortality of individuals, etc. Annual increments to soil carbon are the balance of detrital inputs and decompositional losses. The main phenomena which appear to follow the accumulation pattern are standing plant biomass and stored soil carbon. Soil carbon is estimated here by the climatically based model of Meentemeyer et al. (1985), which predicts highest soil carbon accumulations at AET values near 400 mm/year, i. e. in both humid boreal and subhumid grassland (temperate) environments.

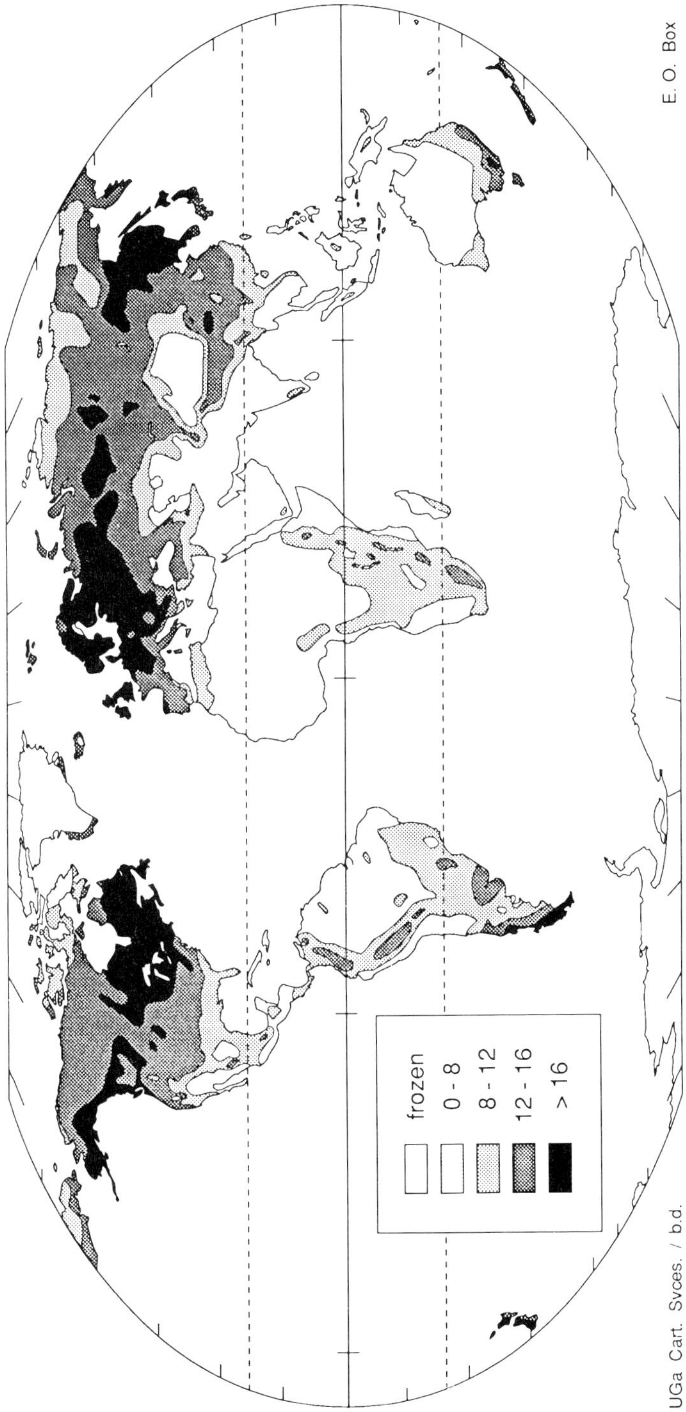

SOIL CARBON (undisturbed world) (kg / m^2)

process balances and may be quite sensitive to small input–output differences, since these are accumulated over long periods. Temperate rainforests, for example, do not have particularly high annual productivity but also have relatively low respiration, despite the huge biomass accumulations, because they occur in cool climates (cf. Gholz 1981, Grier 1976, 1977). A system model of vegetation biomass accumulation, driven by climate–based models of the individual metabolic processes, produced a global geographic pattern of equilibrium biomass accumulation (Box, unpublished; cf. 1981b and in press) similar to that of other biomass maps (cf. Olson et al. 1983).

35.4.5 Other phenomena

At the bottom of Table 35.1 are listed several ecological phenomena which do not fit clearly into any of the four main patterns. Seasonal activity may represent a variation of the throughput concept but may show quite different geographic patterns from that of annual totals (i. e. Figure 35.3). Maps of estimated seasonal net metabolic CO_2 flux by terrestrial vegetation were presented by Box (1988), illustrating a seasonal source–sink pattern. Many physical phenomena also show distinct seasonal activity patterns, such as soil moisture use and recharge (Thornthwaite and Mather 1955) and even the apparent seasonal formation and breakdown of certain clay minerals in soil (Folkoff and Meentemeyer 1987). Growth–form richness in the upper vegetation strata only ("physiognomic diversity") shows maxima in subhumid, more open environments where no form can completely dominate, suggesting a "non–forest" pattern (Box 1981a). Leaf area index, as the apparent balance among potential photosynthetic gains and transpirational losses (with complications by different leaf shapes and consistencies), represents an interesting case for which data are available but no clear geographic pattern emerges (Tuggle 1981, cf. Specht and Morgan 1981). Nutrient use and cycling patterns (cf. Bazilevich and Rodin 1971) represent a somewhat different dynamic but are not unrelated to throughput patterns.

35.5 Geographic inventories

An early by–product of the development of predictive computer mapping was the ability to quantify the raster–based maps by computerized planimetry (Box 1975, 1978), permitting the computer to estimate global totals and averages. Map planimetry could probably now be done in more sophisticated ways, but gray–scale isoline computer mapping, as needed for the natural sciences, has not kept pace with other aspects of computer graphics and mapping.

World maps of various carbon–balance components were quantified to estimate global totals, some of which are presented in Table 35.3. These estimates represent climatic potential amounts for "natural" ecosystems and may not agree with estimates made by methods based on actual land cover in disturbed environments.

Potential annual NPP appears to be only about 4% of the total undisturbed stock of soil carbon, and total annual litter production about 41% of NPP. The very large stock of soil carbon becomes especially significant when compared with total litter accumulation, which is only about 5% of undisturbed soil carbon.

The estimation of global totals of carbon flows and accumulations also permits estimation of global mean turnover rates. Since total litter production is about 33% of litter accumulation, the mean residence time of carbon in litter is about 3 years. Turnover rates have also been mapped (cf. BAZILEVICH and RODIN 1971) and compared with climatic factors directly (BJORKLAND 1984).

35.6 Validating geographic models

The problem of testing model performance, i. e. model validation, was perhaps first formally treated in ecology by GOODALL (1972), MANKIN et al. (1975), and CASWELL (1976). Various methodologies have been developed, but generally they all involve comparing model predictions against reality for a certain number of different situations, which for geographic models may be particular sites. If the validation sites represent independent data not used in model construction, then the test is more rigorous and one can speak of a "true validation." If independent data are not available, a model may be "verified" less rigorously by making the comparisons at a randomly selected subset of the sites used to develop the model.

Because of the wide range of values and situations possible in geographic modeling, it is necessary to be especially rigorous in the validation methodology. Functionally general models based on geographically representative data-bases may work well in some situations and still not in others, especially at world scale, since certain climate types (e. g. mediterranean, equatorial alpine, foggy coastal deserts) or edaphic situations (e. g. oligotrophic, disturbed, water-saturated) may be "unusual" and differ significantly from more normal environmental situations.

In order to validate geographic models with sufficient geographic rigor, the following criteria seem to be appropriate:

1. that the set of validation sites be at least as geographically representative and complete as was the data-base used for model development;

2. that the model be validated at the appropriate scale (i.e. not at sites much more local than can be shown on a map at the scale for which the model was developed); and

3. that the model be validated with data representing the same definition or aspect of the target phenomenon as in the original data-base (e.g. potential rather than actual vegetation or productivity if such was the original data-base).

Each different environmental situation should be validated individually, with at least one replicate. Walter or other fundamentally different genetic climate types,

Table 35.3: Global Carbon–Budget Totals for Land Areas Resulting from Climatic/Geographic Models.

ANNUAL NET PRIMARY PRODUCTION

59.4 to 62.3×10^9 mt C

Annual NPP is estimated to be about 4.1% of total soil C.

ANNUAL LITTER PRODUCTION

17.5×10^9 mt C leaf fraction

24.4×10^9 mt C total litter

Total litter production is about 41% of NPP and is less than 2% of soil C.

About 2.6% of the global biomass is lost annually to total litter production.

GLOBAL LITTER ACCUMULATION

75×10^9 mt C

Litter accumulation is about 5% of soil C.

Total litter production (annual) is equal to about 33% of litter accumulation. (Note: The litter accumulation estimate may be low because peatlands were underestimated, although WHITTAKER estimates 55×10^9 mt C in global litter.)

TOTAL DETRITAL SOIL CARBON

$1,457 \times 10^9$ mt C (Undisturbed world)

504×10^9 mt C (Fully disturbed world)

TOTAL CARBON IN SOIL RESPIRATION

62.7×10^9 mt C

Thus only about 39% of the C in soil respiration results from litter decomposition.

Total soil respiration is slightly higher than NPP.

Only about 4.3% of the detrital C is respired annually.

OTHER ESTIMATES

WHITTAKER and LIKENS' (1975) estimates indicate about 926×10^9 mt C in biomass.

Thus biomass C is only about 64% of soil C.

augmented by replicates for different substrate conditions, appear to provide the best basis for ensuring that all environmental situations have been included. Data for all situations may not always be available. In such cases, the generation of results at many sites over some large area, the whole globe if possible, with subsequent mapping of the predicted geographic pattern, represents a useful approach to geographic validation. This may be superior to purely statistical methods since, if there is a major problem with the model, it may become painfully visible on the predictive map. The spatial continuity in modeling provided by the mapping step thus may partly overcome data limitations and represents an important advantage of this geographic approach to biosphere modeling, as opposed to the discrete–biome approach.

35.7 Conclusion

Recent concern about environmental damage and global climate change is now fostering a global perspective in ecology, in which geographic considerations are fundamental (e. g. DELCOURT et al. 1983; PETERS and DARLING 1985; BOLIN et al. 1986; ROSSWALL et al. 1988, PETERS and LOVEJOY, in press; cf. FOSBERG 1976). LIETH (e. g. LIETH 1974, LIETH and WHITTAKER 1975) has contributed to this global perspective and perhaps put global ecological work in its proper, holistic context when he described unregulated growth of populations, GNP's, and profits as a "crime against our own children" (LIETH 1975b). At all scales, many of the tools newly available to ecologists are geographic in nature: satellite–based remote sensing, improved aerial photography, geographic information systems, etc. That ecologists recognize the need for geographic considerations even at local scales is seen in the recent emphases on landscape ecology (e. g. SHARPE et al. 1981, FORMAN and GODRON 1986, TURNER 1987) and on scale effects in ecological phenomena, data, and methods (e. g. ALLEN and STARR 1982, ROSSWALL et al. 1988).

It is not necessarily the purpose of this paper to encourage detailed taxonomies of global geographic patterns. Indeed there would probably be too many intermediate cases, with too many important secondary and tertiary factors, to make such classifications useful. Fundamental geographic patterns and the biophysical–ecological logic behind them, however, represent a useful global baseline framework for analyzing aberrations, the effects of other variables, and the robustness of the geographic pattern (i. e. of the ecological relationships) over different scales. This can be useful especially in interpreting satellite imagery and appears to fit some of the needs of global–change research.

Ecology operates at various scales, in environments which are heterogeneous and diverse. As a result, geography provides the variation necessary for testing the generality of ecological hypotheses. This may involve comparisons (i. e. testing) over the heterogeneity of local landscapes or "comparative ecology" involving similar environments on different continents.

All scientific work requires treating phenomena at the appropriate scale (MEENTEMEYER and BOX 1987, O'NEILL 1988). Meteorologists will eventually realize, for example, that it is not fruitful to attempt to infer the behavior of the entire biosphere from the hypothesized behavior of "stomatal resistances" (SELLERS et al. 1985, cf. Bolin 1988), if for no other reason than for lack of the enormous volumes of reliable stomate–scale data required to parameterize such models. Also, global models based on parameterization for a single region (e. g. Emanuel et al. 1985, based on Holdridge's (1959) Neotropical parameterization) may not work elsewhere (cf. Sargent 1988) and will eventually require more global bases.

A fundamental feature of modern ecology has and will continue to involve global models (e. g. Lieth 1985, Esser 1988) and maps based on models (e. g. Emanuel et al. 1985, Box 1988). It was Lieth who first demonstrated that such models are useful (e. g. Lieth 1975a) and that a process–based, spatially continuous approach to geographic modeling was fast, mathematically simple, and appears to be a useful method, within limits, for extrapolating across different scales, i. e. from sites to models to large regions of the globe. Lieth also used such global models to produce estimates of changes in NPP under different climatic conditions (Lieth 1976).

Acknowledgment

The authors would like to thank Brian Davis and (University of Georgia) Cartographic Services, as well as Jeff Torguson, for assistance in producing the final versions of the world maps.

Bibliography

Allen, T. F. H.; Starr, T. B., 1982, Hierarchy: Perspectives for Ecological Complexity.
University of Chicago Press. 310 pp.

Barbour, M. G.; Burk, J. H.; Pitts , W. D., 1980, Terrestrial Plant Ecology. Menlo Park (Calif.), Benjamin/Cummings. 604 pp.

Baumgartner, A.; Reichel, E., 1975, The World Water Balance.
Amsterdam, Elsevier. 179 pp., 3l maps.

Bazilevich, N. I.; Drozdov, A. V.; Rodin, L. Ye., 1968, (Productivity of the land vegetation covers, general distributional regularities and relations to climatic factors.)
Zh. Obshch. Biolog., 29(3), 262–271 (in Russian).

Bazilevich, N. I.; Rodin, L. Ye., 1971, Geographical regularities in productivity and the circulation of chemical elements in the earth's main vegetation types.
Sov. Geogr. Rev. Transl., 12, 24–53.

Bjorkland, R. A., 1984, Organic–matter turnover rates and their relationship with climate.
M. A. thesis, Univ. of Georgia. 233 pp.

BOLIN, B., 1988, Linking Terrestrial Ecosystem Process Models to Climate Models.
In: ROSSWALL, T. et al., (eds.), Scales and Global Change.
SCOPE 35, pp. 109-124. Chichester, Wiley and Sons.

BOLIN, B.; DÖÖS, B. R.; JÄGER, J.; WARRICK, R. A. (eds.), 1986, The Greenhouse Effect, Climatic Changes, and Ecosystems: A Synthesis of Present Knowledge.
SCOPE 29. Chichester, Wiley and Sons. 541 pp.

BOX, E. O., 1975, Quantitative Evaluation of Global Productivity Models Generated by Computers.
In: LIETH, H.; WHITTAKER, R. H. (eds.), Primary Productivity of the Biosphere. Ecological Studies, vol. **14**. New York, Springer–Verlag.

BOX, E. O., 1978, Geographic dimensions of terrestrial net and gross primary productivity.
Radiation and Envl. Biophysics **15**, 305–322.

BOX, E. O., 1979a, Estimated Average Photosynthetic Efficiency of Natural Terrestrial Vegetation (Map 4).
In: Use of Synagraphic Computer Mapping in Geoecology; Computer Mapping in Education, Research, and Medicine, pp. 11–27. Harvard Library of Computer Mapping, vol. **5**. Cambridge, (Mass.), Harvard University.

BOX, E. O., 1979b, Difference between Thornthwaite and Holdridge Estimates of Annual Potential Evapotranspiration (Map 7).
In: Use of Synagraphic Computer Mapping in Geoecology; Computer Mapping in Education, Research, and Medicine, pp. 11–27. Harvard Library of Computer Mapping, vol. **5**. Cambridge, (Mass.), Harvard University.

BOX, E. O., 1981a, Macroclimate and Plant Forms: An Introduction to Predictive Modeling in Phytogeography.
Tasks for Vegetation Science, vol. **1**. Amsterdam, Dr. W. Junk, Publishers.

BOX, E. O., 1981b, Climatic control of phytomass accumulation in terrestrial ecosystems.
Abstracts, 9th International Congress of Biometeorology (D. Overdieck et al., eds.), pp. 230–231. Swets and Zeitlinger.

BOX, E. O., 1986, Some climatic relations of the vegetation of Argentina.
In: ESKUCHE, U.; LANDOLT, E. (eds.), Contributions to Knowledge of the Flora and Vegetation of Northern Argentina. Veröff. Geobot. Inst. Rübel **91**, 181–216.

BOX, E. O., 1988, Estimating the seasonal carbon source–sink geography of a natural steady–state terrestrial biosphere.
J. Appl. Meteorol. **27**, 1109–1124.

Box, E. O.; Meentemeyer, V.; Brook, G., 1983, Climate and the world geography of soil carbon dioxide evolution.
Bull. Ecol. Soc. Amer. **64**(2), 123.

Box, E. O.; Holben, B. N.; Kalb, V., 1989, Accuracy of the AVHRR vegetation index as a predictor of biomass, primary productivity, and net CO_2 flux.
Vegetatio **80**, 71–89.

Box, E. O. (in press), Modeling Terrestrial Primary Productivity from Available Climatic Data.
In: Lieth, H. (ed.), Primary Productivity of the Biosphere, 2nd edition. Springer-Verlag.

Box, E. O. (in prep.), Mapping vegetation processes with climatic, satellite, and field data.
In: Falinski, J., Proceedings, 33rd Symposium of International Association for Vegetation Science.

Box, E. O.; Lieth, H.; Wolaver, T., 1971, Miami Model: Primary Productivity Predicted from Temperature and Precipitation Averages.
World map, reprinted by Lieth in Human Ecology **1**, 303–332.

Brook, G. A.; Folkoff, M. E.; Box, E. O., 1983, A world model of soil carbon dioxide.
Earth Surface Processes **8**, 79–88.

Burger, W. C., 1981, Why are there so many kinds of flowering plants?
BioScience **31**, 572–581.

de Candolle, Alphonse, 1855, Géographie botanique raisonée. Paris.

de Candolle, Augustin, 1820, Essai élémentaire de géographie botanique.
Dictionaire des Sciences Naturelles, vol. **18**. Paris.

Cannell, M. G. R., 1982, World Forest Biomass and Primary Production Data.
London, Academic Press. 391 pp.

Caswell, H., 1976, The Validation Problem.
In: Patten, B. C. (ed.), Systems Analysis and Simulation in Ecology, vol. **4** New York, Academic Press.

Currie, D.; Paquin, V., 1987, Large–scale biogeographical patterns of species richness of trees.
Nature **329**, 326–327.

Damagnez, J. et al., 1963, Estimation et measure de l'évapotranspiration poténtielle en Tunisie.
In: Assemblée Générale de Berkeley, Août 1963. Gentbrugge (Belgium): Assoc. Internat. d'Hydrologie Scientifique, publ. no. **62**.

DARWIN, C. R., 1859, On the Origin of Species by Means of Natural Selection, or the preservation of favoured races in the struggle for life.
London, John Murray.

DELCOURT, H. R.; DELCOURT, P. A.; WEBB, T. III, 1983. Dynamic plant ecology: the spectrum of vegetation change in space and time.
Quat. Sci. Rev. **1**, 153–175.

DRUDE, O., 1890, Handbuch der Pflanzengeographie. Stuttgart.

EMANUEL, W. R.; SHUGART, H. H.; STEVENSON, M. P., 1985, Climatic change and the broad-scale distribution of terrestrial ecosystem complexes.
Climate Change **7**, 29–43.

ENGLER, A., 1882, Versuch einer Entwicklungsgeschichte der Pflanzenwelt. Part I, 1879; part II, 1882 (with world map). Leipzig.

ESSER, G.; ASELMANN, I.; LIETH, H., 1982, Modelling the carbon reservoir in the system compartment "litter".
SCOPE/UNEP spec. vol. **52**, pp. 39–58. Univ. Hamburg, Geol.–Paläont. Inst.

ESSER, G., 1988, Sensitivity of global carbon pools and fluxes to human and potential climatic impacts.
Tellus **39**B, 245–260.

FAO–UNESCO 1974ff. Soil Map of the World. Vols. I-X. Paris, UNESCO.

FOLKOFF, M. E.; MEENTEMEYER, V.; BOX, E. O., 1981, Climatic control of soil acidity.
Phys. Geogr. **1**, 116–124.

FOLKOFF, M. E.; MEENTEMEYER, V., 1987, Climatic control of the geography of clay mineral genesis.
Annals Assn. Amer. Geogr. **77**, 635–650.

FORMAN, R. T. T.; GODRON, M., 1986, Landscape Ecology.
New York, Wiley and Sons, 619 pp.

FOSBERG, F. R., 1976, Geography, ecology, and biogeography.
Annals, Assn. Amer. Geogr. **66**, 117–128.

GEIGER, R., 1965, The Atmosphere of the Earth. Wall-map series, including temperature, precipitation, evapotranspiration, etc.
Darmstadt, Justus Perthes.

GHOLZ, H. L., 1982, Environmental limits on above-ground net primary production, leaf area, and biomass in vegetation zones of the Pacific Northwest.
Ecology **63**, 469–481.

Goodall, D. W., 1972, Building and testing ecosystem models.
In: Jeffers, J. N. R. (ed.), Mathematical Models in Ecology, Blackwell, Oxford, 173–194.

Grier, C. C., 1976, Biomass, productivity and nitrogen–phosphorus cycles in hemlock–spruce stands of the central Oregon coast.
In: Western Hemlock Management, pp. 71–81. Univ. Washington, Forest Resources, contrib. no. **34**.

Grier, C. C.; Logan, R. S., 1977, Old-growth *Pseudotsuga menziesii* communities of a western Oregon watershed: biomass distribution and production budgets.
Ecol. Monogr. **47**, 373–400.

Goward, S. N.; Tucker, C. J.; Dye, D. G., 1985, North American vegetation patterns observed with the NOAA Advanced Very High Resolution Radiometer.
Vegetatio **64**, 3–14.

Grisebach, A. R. H., 1866, Die Vegetations-Gebiete der Erde, übersichtlich zusammengestellt.
Petermann's Mitteilungen **12**, 45–53.

Grisebach, A. R. H., 1872, Die Vegetation der Erde nach ihrer klimatischen Anordnung. Leipzig.

Holdridge, L. R., 1959, A simple method for determining potential evapotranspiration from temperature data.
Science **130**, 572.

Hueck, K., 1966, Die Wälder Südamerikas: Ökologie, Zusammensetzung und wirtschaftliche Bedeutung.
Stuttgart, Fischer, 422 pp.

Humboldt, A. von, 1805, Essai sur la géographie des plantes. Paris.

Humboldt, A. von, 1807, Ideen zu einer Geographie der Pflanzen.
Tübingen: F. G. Cotta, reprinted 1963 by Wissenschaftl. Buchges. Darmstadt.

Hurlbert, S. H., 1971, The non-concept of species diversity: a critique and alternative parameters.
Ecology **52**, 577–586.

Jordan, C. F., 1971, A world pattern in plant energetics.
Amer. Scientist **59**, 426–433.

Keddy, P. A., 1990, Competitive Hierarchies and Centrifugal Organization in Plant Communities.
In: Grace, J. B.; Tilman, D. (eds.), Perspectives on Plant Competition, San Diego, Academic Press, 265–290.

KING, A. W., 1986, The seasonal exchange of carbon dioxide between the atmosphere and the terrestrial biosphere: extrapolation from site-specific models to regional models.
Ph. D. thesis, Univ. of Tennessee, Knoxville, 271 pp.

KNAPP, R., 1965, Die Vegetation von Nord- und Mittel-Amerika und der Hawaii-Inseln.
Stuttgart, Fischer, 373 pp.

KNAPP, R., 1973, Die Vegetation von Afrika unter Berücksichtigung von Umwelt, Entwicklung, Wirtschaft, Agrar- und Forstgeographie.
Stuttgart, Fischer, 626 pp.

KÜCHLER, A. W., 1949, "Natural Vegetation" (world map).
Goode's School Atlas, 8th edition. Chicago, Rand McNally.

LARCHER, W., 1976, Ökologie der Pflanzen.
2nd ed. Stuttgart, Ulmer.

LIEBIG, J., 1840, Organic Chemistry and its Applications to Agriculture and Physiology.
London, Taylor and Walton, 387 pp.

LIETH, H., 1964, Versuch einer kartographischen Darstellung der Produktivität der Pflanzendecke auf der Erde.
In: Geographisches Taschenbuch 1964/65, pp. 72–80. Wiesbaden, Steiner.

LIETH, H., 1972, Über die Primärproduktion der Pflanzendecke der Erde.
Z. Angew. Bot. **46**, 2–27.

LIETH, H., 1973, Primary production: terrestrial ecosystems.
Human Ecol. **1**, 303–332.

LIETH, H. (ed.), 1974. Phenology and Seasonality Modeling.
New York, Springer-Verlag, 444 pp.

LIETH, H., 1975a, Modeling the Primary Productivity of the World.
In: Lieth, H.; Whittaker, R. H. (eds.), Primary Productivity of the Biosphere, pp. 237–263. Springer, New York.

LIETH, H., 1975b, Preamble.
In: LIETH, H.; WHITTAKER, R. H. (eds.), Primary Productivity of the Biosphere, p. 2. New York, Springer-Verlag.

LIETH, H., 1976, Possible Effects of Climate Changes on Natural Vegetation.
In: KOPEC, R. J. (ed.), Atmospheric Quality and Climatic Change, University of N. Carolina, Studies in Geography, no. 9. Chapel Hill, Dept. of Geography, 150–159.

Lieth, H., 1985, A dynamic model of the global carbon flux through the biosphere and its relations to climatic and soil parameters.
In: Gloyne, R. W. et al. (eds.), Proceedings, 10th International Biometeorological Congress, Tokyo 1984, 17–31.

Lieth, H.; Box, E. O., 1972, Evapotranspiration and primary productivity; C. W. Thornthwaite Memorial Model.
Public. in Climatol. **25**(3), 37–46.

Lieth, H.; Box, E. O., 1977, The gross primary productivity pattern of the land vegetation: A first attempt.
Tropical Ecology **18**, 109–115.

Lieth, H.; Whittaker, R. H. (eds.), 1975, Primary Productivity of the Biosphere.
New York, Springer-Verlag. 339 pp.

Major, J., 1963, A climatic index to vascular plant activity. Ecology **44**, 485–498.

Mankin, J. B.; O'Neill, R. V.; Shugart, H. H.; Rust, B. W., 1975, New directions in the analysis of ecological systems, Part I.
In: Innis, G. S. (ed.), Simulation Councils Proceedings Series, vol. **5**, no. 1, Simulation Councils Inc., La Jolla (California).

Meentemeyer, V., 1971, The Prediction of Regional Differences in the Rate of Decomposition of Organic Debris.
Ph. D. dissertation, Southern Illinois University, Carbondale. 74 pp.

Meentemeyer, V., 1978, Macroclimate and lignin control of litter decomposition rates.
Ecology **59**, 465–472.

Meentemeyer, V., 1979, Global patterns in the production, turnover and accumulation of organic detritus.
Unpubl. extended abstract (with 3 world maps); short abstract in Program Abstracts, SE Div., Assn. Amer. Geogr. annual meeting, Orlando (Florida), p. 15.

Meentemeyer, V., 1984, The geography of organic decomposition rates.
Annals, Assn. Amer. Geogr. **74**, 551–560.

Meentemeyer, V., 1985, Climatic control of litter dynamics in tropical forest.
In: Misra, K. C. (ed.), Ecology and Resource Management in Tropics, Bhargava, Varanasi (India), 302–310.

Meentemeyer, V., 1989, Geographical perpectives of space, time, and scale.
Landsc. Ecol. **3**, 163–173.

MEENTEMEYER, V.; BOX, E. O.; THOMPSON, R., 1982, World patterns and amounts of terrestrial plant litter production.
BioScience **32**, 125–128.

MEENTEMEYER, V.; BOX, E. O., 1987, Scale Effects in Landscape Studies.
In: TURNER, M. G. (ed.), Landscape Heterogeneity and Disturbance, pp. 15–34.
New York, Springer.

MEENTEMEYER, V.; GARDNER, J.; BOX, E. O., 1985, World patterns and amounts of detrital soil carbon.
Earth Surf. Proc. Landforms **10**, 557–567.

OHRING, G.; GALLO, K.; GRUBER, A.; PLANET, W.; STOWE, L.; TARPLEY, J. D., 1989, Vegetation Index (world image).
Trans. Amer. Geophys. Union **70**(41), cover and pp. 889 ff.

OLSON, J. S., 1963, Energy storage and the balance of producers and decomposers in ecological systems.
Ecology **44**, 322–331.

OLSON, J. S.; WATTS, J. A.; ALLISON, L. J., 1983, Carbon in live vegetation of major world ecosystems.
Report DOE/NBB-0037. Washington, US Dept. of Energy, Carbon Dioxide Research Division.

O'NEILL, R. V., 1988, Hierarchy Theory and Global Change.
In: ROSSWALL, T., et al. (eds.), Scales and Global Change, pp. 29–45, SCOPE 35.
Wiley and Sons, Chichester.

O'NEILL, R. V.; DEANGELIS, D. L., 1981, Comparative Productivity and Biomass Relations of Forest Ecosystems.
In: REICHLE, D. E. (ed.), Dynamic Properties of Forest Ecosystems, IBP series, vol. **23**, Chapter 7. Cambridge University Press.

OOSTING, H. J., 1942, An ecological analysis of the plant communities of Piedmont, North Carolina.
Amer. Midl. Nat. **28**, 1–126.

PERINO, J. V.; RISSER, P. G., 1972, Some aspects of structure and function in Oklahoma old–field succession.
Bull. Torrey Bot. Club **99**, 233–239.

PETERS, R. L.; DARLING, J. D. S., 1985, The greenhouse effect and nature reserves.
BioScience **35**, 707–717.

Peters, R. L.; Lovejoy, T. (eds.) (in press), Consequences of the Greenhouse Effect for Biological Diversity.
Yale Univ. Press.

Pianka, E. R., 1966, Latitudinal gradients in species diversity: a review of concepts.
Amer. Naturalist **100**, 33–46.

Rosenzweig, M. L., 1968, Net primary productivity of terrestrial communities: prediction from climatological data.
Amer. Naturalist **102**, 67–74.

Rosswell, T.; Woodmansee, R, G.; Risser, P. G. (eds.), 1988, Scales and Global Change.
SCOPE 35. Chichester, Wiley. 355 pp.

Rübel, E., 1930, Pflanzengesellschaften der Erde. Bern, Huber.

Sargent, N. E., 1988, Redistribution of the Canadian boreal forest under a warmed climate.
Climatol. Bull. (Canada) **22**(3), 23–34.

Schimper, A. F. W., 1898, Pflanzengeographie auf physiologischer Grundlage. Jena.

Schleser, G. H., 1982. The response of CO_2 evolution from soils to global temperature changes.
Z. f. Naturforschung **A37**(2037), 1–5.

Schmithüsen, J., 1959, Allgemeine Vegetationsgeographie.
Berlin, Walter de Gruyter.

Schmithüsen, J. (ed.), 1976, Atlas zur Biogeographie.
Meyers Großer Physischer Weltatlas, Vol. **3**. Mannheim, Bibliographisches Institut.

Schouw, J. F., 1822. Grundträg til en almindelig Plantegeografie.
København.

Sellers, P. J.; Mintz, Y.; Sud, Y. C.; Dalcher, A., 1986, A simple biosphere model (SiB) for use within general circulation models.
J. Atmos. Sci. **43**, 505–531.

Sharpe, D. M., 1975, Methods of Assessing the Primary Production of Regions. In: Lieth, H.; Whittaker, R. H. (eds.) , Primary Productivity of the Biosphere, pp. 147–166. New York, Springer–Verlag.

SHARPE, D. M.; STEARNS, F. W.; BURGESS, R. L.; JOHNSON. W. C., 1981, Spatio–temporal Patterns of Forest Ecosystems in Man–Dominated Landscapes of the Eastern United States.
In: TJALLINGII, S. P.; DE VEER, A. A. (eds.), Perspectives in Landscape Ecology, pp. 109–116. Wageningen, Pudoc.

SMITH, D. D., 1985, The Effect of Climate and Substrate Quality on the Broad–Scale Geography of Wood Decomposition Rates.
M. A. Thesis, University of Georgia. 81 pp.

SPECHT, R. L.; MORGAN, D. G., 1981, The balance between the foliage projective covers of overstorey and understorey strata in Australian vegetation.
Austral. J. Ecol. **6**, 193–202.

STEBBINS, G. L., 1981, Why are there so many species of flowering plants?
BioScience **31**, 573–577.

STEPHENS, J. C.; STEWART, E. H., 1963, A comparison of procedures for computing evaporation and evapotranspiration.
In: Assemblée Générale de Berkeley, Août, 1963. Gentbrugge (Belgium), Assoc. Internat. d'Hydrologie Scientifique, publ. no. **62**.

THORNTHWAITE, C. W., 1948, An approach toward a rational classification of climate.
Geogr. Rev. **38**, 55–94.

THORNTHWAITE, C. W.; MATHER, J. R., 1955, The Water Balance.
Publications in Climatology **8** (1), 1–104. Elmer (New Jersey), C. W. Thornthwaite Associates, Laboratory for Climatology.

THORNTHWAITE, C. W.; MATHER, J. R., 1957, Instructions and Tables for Computing Potential Evapotranspiration and the Water Balance. Public. in Climatol. **5**(6), 181–311. (Climatology Laboratory, Elmer/New Jersey).

TROLL, C., 1960, Die Physiognomie der Gewächse als Ausdruck der ökologischen Lebensbedingungen.
In: Tagesber. Wiss. Abh. Dtsch. Geographentages 1959. Wiesbaden.

TUCKER, C. J.; TOWNSHEND, R. G.; GOFF, T. E., 1985, African land–cover classification using satellite data.
Science **227**, 369–375.

TUGGLE, S., 1981, An Examination of the Association of Selected Climatic Variables with Leaf Area Indices.
M. A. thesis, University of Georgia. 81pp.

TURNER, M. G. (ed.), 1987, Landscape Heterogeneity and Disturbance.
New York, Springer–Verlag. 239 pp.

Uchijima, Z.; Seino, H., 1987, Distribution Maps of Net Primary Productivity of Natural Vegetation and Related Climatic Elements on the Continents. National Inst. Agro. Envl. Scis., Kyushu Nat. Agr. Expt. Stn., Fukuoka (Japan). 102 pp.

Wallace, A. R., 1876, The Geographical Distribution of Animals. 2 vols. London, Macmillan.

Walter, H., 1968. Die Vegetation der Erde in ökophysiologischer Betrachtung. Vol **1**: Die gemäßigten und arktischen Zonen. Jena, Fischer, 1001 pp.

Walter, H., 1973, Die Vegetation der Erde in ökophysiologischer Betrachtung. Vol. **2**: Die tropischen und subtropischen Zonen. Jena, Fischer.

Walter, H., 1974, Die Vegetation Osteuropas, Nord– und Zentralasiens. Stuttgart, Fischer. 452 pp.

Walter, H.; Lieth, H., 1960–67. Klimadiagramm–Weltatlas. Stuttgart, Gustav–Fischer–Verlag. 8000 stations.

Warming, J. E. B., 1895, Plantesamfund, Grundtraek af den okologiske Plantgeografi. København, Philipsen.

Whitford, W. G.; Meentemeyer, V.; Seastedt, T. R.; Cromack, K.; Crossley, D. A.; Santos, P.; Todd, R. L.; Waide, J. B., 1981, Exceptions to the AET model: deserts and clear–cut forests. Ecology **62**, 275–277.

Whittaker, R. H.; Likens, G. E., 1975, The Biosphere and Man. In: Lieth, H.; Whittaker, R, H. (eds.), Primary Productivity of the Biosphere, pp. 305–328. Springer–Verlag, New York.

Wojcik, J., 1983, Climatic Control of the Geography of Base Saturation in Soils of the United States. M. A. thesis, University of Georgia. 116 pp.

Part VIII

Urban and landscape ecology

Chapter 36

Distribution patterns of flowering plants in the city of Zurich

E. LANDOLT

36.1 Introduction

The interest in studying the flora of cities is relatively new. Although at a much earlier time (NÄGELI and THELLUNG 1905), papers have been published on adventive flora in cities, their aim was purely floristic and not ecological. After World War II, vast areas in the cities of European belligerents were left covered with destroyed houses which could not be restored for many years. A spontaneous vegetation consisting of different native and adventitious elements developed on these ruins and debris-covered areas, encouraging closer study (e.g. ENGEL 1949, PFEIFFER 1957). Many investigations and inventories of urban flora and vegetation have since been published.

The urbanization of the landscape, the rationalization of modern agronomy and many other factors resulted in the destruction of biotopes and consequently in the disappearance of species, deprived of their living space. The possibility of finding compensatory ecological niches within the city was detected and studied further (e.g. JANSSEN and BRANDES 1984, REBELE and WERNER 1984, SUKOPP 1983).

Some years ago, the administrators of the City of Zurich initiated an inventory of the biologically most valuable areas, in order to protect and promote the city's biological diversity (BERNOWITZ and LEUTERT 1988). In the Geobotanical Institute we began to study diverse and rare vegetation types within the city (e.g. SAILER 1990, WISKEMANN 1990). The author of the present contribution has been mapping the vascular plants within the boundaries of the city. Although over a hundred species of the flora of 150 years ago have vanished, the biological diversity is still very rich and the total number of vascular plant species amounts to well over a thousand. This roughly corresponds to the species number of other central European cities (e.g. Brandes 1985). Distribution maps of plants in the city show interesting relations to certain environmental factors and give much information on the autecology of the different species, especially if we compare the behaviour of the species in different cities. Although the inventory of the species in Zurich is not yet finished, some patterns can already be recognized. They will be described in the following.

36.2 Methods of the inventory

A map of the city of Zurich was divided up into 120 sections of $1km^2$ each. Each segment was visited several times and all species observed were noted. Endangered and rare species were collected and graded according to their degree of rarity: 1) very rare (○ up to 20 individuals), 2) rare (◑ up to 200 individuals or very local) or 3) not rare (● more than 200 individuals and not local). At this time about half of the overview was complete. Until now all areas have been visited at least twice but with varying intensity. The nomenclature of the Latin names of plant species follows HESS et al. (1967–1972).

36.3 Some ecological characteristics of Zurich

Zurich (47°30' N latitude and 8°30' E longitude) is situated at the lower end of a long narrow lake ($88km^2$ surface area). The lowest point of the city lies in the northwest (395 m a.s.l.) and the highest point is the Uto Kulm (871 m a.s.l.), a mountain in the west of the city. Fig. 36.1 shows the segments reaching 600m a.s.l. and above and those with no areas below 500 m a.s.l. Some of the slopes, especially on the western mountain range are quite steep. Fig. 36.2 shows the segments where at least 5% of the WSW to SE area slopes with an inclination of over 40%. NW to NE slopes with an area of at least 10 ha and with an inclination of over 60% are also shown.

Most soils have developed on Würm moraine. Only the highest part of the city (Uto) consists of a rich base conglomerate. On steeper slopes there are tertiary sediments (mostly marl and sandstone). The prevalent soils are para–brownsoils

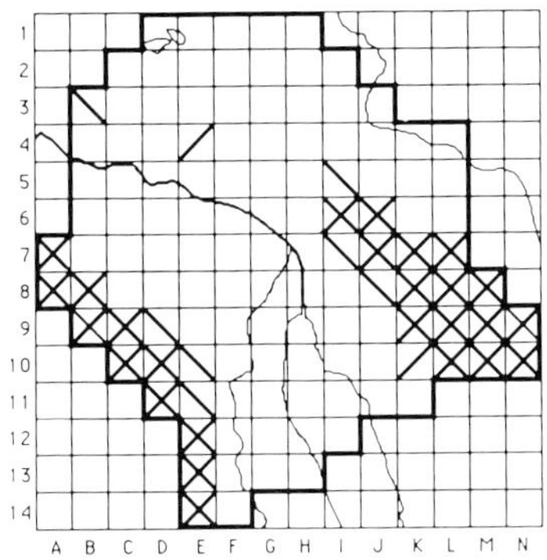

Figure 36.1: Segments with no elevations lower than 500 m (/), and segments with elevations over 600 m (\).

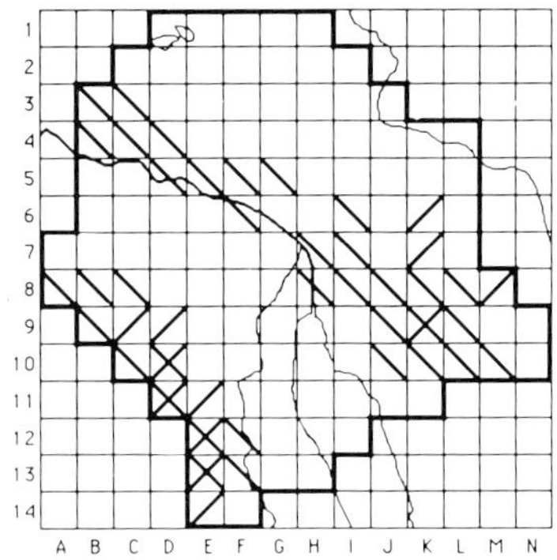

Figure 36.2: Segments where
>5% of the area are NW to NE slopes with an inclination of over 60% (/)
>5% of the area are WSW to SE slopes with an inclination of over 40% (\).

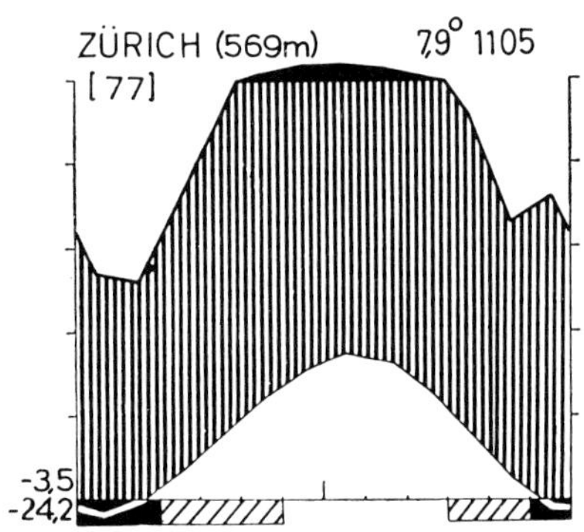

Figure 36.3: Climate diagram of Zurich (from Walter and Lieth 1960–1967).

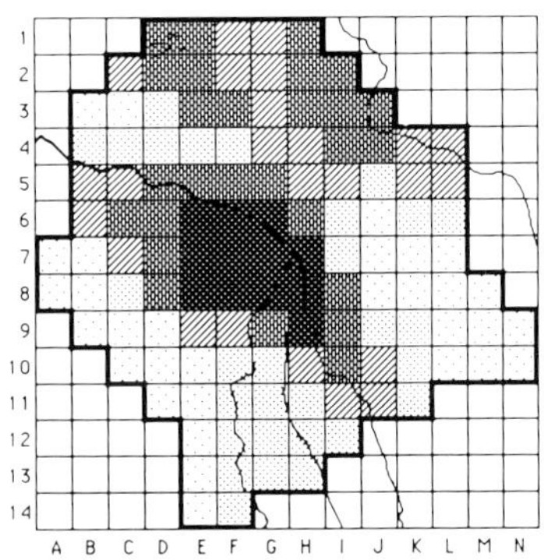

Figure 36.4: The temperature differences on a warm clear summer evening (according to Schlatter 1975). The warmest area lies in the centre of the city. The different segments contain places with

< 0.5 °C lower temperature
0.5 to 1 °C lower temperature
1 to 2 °C lower temperature
2 to 3 °C lower temperature
> 3 °C lower temperature

very often slightly hydromorphic and moderately acidic in the upper layers. In the inner–city the ground is very often comprised of deposits and debris.

The climate is of the Central European suboceanic type (Fig. 36.3). Wet and cool summers (mean July temperature about 18 °C), rather mild winters (mean January temperature about 0 °C) and dry periods normally no longer than 14 days during the vegetation period are characteristic. However, absolute minimal temperatures can fall down to −24 °C. Total precipitation reaches about 100 cm in the northeast and increases up to about 120 cm in the southwest. Since the sunshine is rather poor (around 50% of the maximum sunshine in July and August) and the foehn has little influence in the region, the lack of warm temperature is an important factor for the absence of many thermophilous species in the eastern part of the Swiss Plateau. Zurich's meteorological station is situated 570 m above sea level. The lower parts of the city have mean temperatures which are at least one degree higher than in the higher regions. As in other cities, the city climate differs from its surroundings by having milder winters and warmer summers (e.g. SUKOPP et al. 1974). Fig. 36.4 shows the temperature differences within the city on a warm clear summer evening (SCHLATTER 1975). The degree of soil sealing is demonstrated in Fig. 36.5. (UMWELTSCHUTZFACHSTELLE DER STADT ZURICH 1989).

25% of the city of Zurich is covered by forests. The percentage of forest coverage within the various segments is shown in Fig. 36.6. The zonal vegetation is the Fagion, but due to the relief and to extended ground water there are many other forest types present. Plantations of *Picea* and other trees are also frequent.

36.4 Some distribution patterns

The following lists contain typical examples. They are not complete.

36.4.1 Common species

This group comprises just over a hundred species which occur in all or nearly all of the segments. Important conditions for the occurrence of these species are:

- that the biotopes of the respective species occur relatively frequently and are well dispersed throughout the whole city.

- that the species are not restricted to an ecologically narrow and rare habitat, but grow under different conditions which occur frequently within the city. They have a tendency to grow and spread ruderally. Species which are well adapted to the local climate are likely to be included in this group. The division into subgroups is somewhat arbitrary since some of the species grow in more than one habitat and they are classified according to their main one. Neophytes are marked with (N).

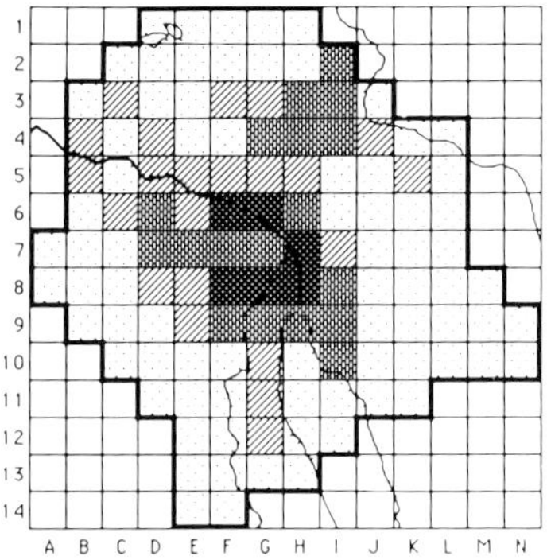

Figure 36.5: Percentage of soil sealing (according to Umweltschutzfachstelle der Stadt Zurich, 1989):
<35%
35–49%
50–65%
>65%

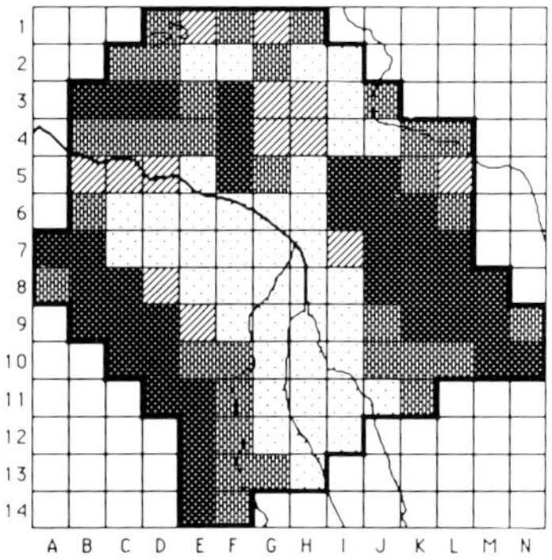

Figure 36.6: Percentage of forest coverage in the city of Zurich (estimation from the topographical map of Switzerland 1:25,000):
<5%
5–15%
15–45%
>45%

Forest species. The following species naturally occurring in forests or in undergrowth are to be found in nearly every segment:

Trees: *Acer platanoides, Acer pseudoplatanus, Betula pendula, Carpinus betulus, Fagus silvatica, Fraxinus excelsior, Hedera helix, Prunus avium, Quercus robur, Salix caprea, Ulmus scabra.*

Shrubs: *Corylus avellana, Cornus sanguinea, Crataegus monogyna, Lonicera xylosteum, Rubus caesius, Rubus fruticosus* s.l., *Sambucus nigra.*

Herbs and grasses: *Aruncus silvester, Anemone nemorosa, Brachypodium silvaticum, Carex silvatica, Circaea lutetiana, Dryopteris filix–mas, Geum urbanum, Mycelis muralis, Poa nemoralis, Potentilla sterilis, Ranunculus ficaria, Viola silvestris.*

Plants of nitrogen rich shady places. *Aegopodium podagraria, Alliaria officinalis, Cardamine flexuosa, Chelidonium majus, Convolvulus sepium, Epilobium montanum, Fragaria vesca, Geranium robertianum, Impatiens parviflora* (N), *Lampsana communis, Ranunculus repens, Stachys silvatica, Urtica dioeca, Veronica subulata.*

Plants of fat meadows and lawns. *Ajuga reptans, Arrhenatherum elatius, Bellis perennis, Bromus mollis, Cardamine pratensis, Cerastium caespitosum, Chaerophyllum silvestre, Crepis capillaris, Dactylis glomerata, Festuca pratensis, Festuca rubra, Galium album, Geranium pyrenaicum* (N), *Glechoma hederaceum, Her-*

acleum sphondyleum, Lolium perenne, Lotus corniculatus, Lysimachia nummularium, Medicago lupulina, Phleum pratense, Plantago lanceolata, Poa pratensis, Poa trivialis, Prunella vulgaris, Ranunculus friesianus, Rumex acetosa, Rumex obtusifolius, Taraxacum officinale, Trifolium repens, Trifolium pratense, Veronica filiformis (N), *Vicia sepium.*

Weeds and ruderals. *Capsella bursa-pastoris, Cardamine hirsuta, Chenopodium album, Cirsium arvense, Convolvulus arvensis, Equisetum arvense, Erigeron annuus* (N), *Erigeron canadensis* (N), *Euphorbia peplus, Galeopsis tetrahit, Plantago major, Poa annua, Polygonum aequale, Polygonum lapathifolium, Polygonum persicaria, Potentilla reptans, Sagina procumbens, Senecio vulgaris, Setaria viridis, Solidago canadensis* (N), *Solidago serotina* (N), *Sonchus asper, Sonchus oleraceus, Stellaria media, Tussilago farfara, Veronica hederifolia, Veronica persica* (N).

Most of these species have been frequent in the city of Zurich for a long time. But some species were not yet common 150 years ago. KÖLLIKER (1839) gives the frequency (cited in brackets) of the following species in Zurich in 1839.

Cardamine hirsuta (rather frequent), *Cardamine flexuosa* (rare), *Erigeron annuus* (–), *Geranium pyrenaicum* (not so frequent), *Impatiens parviflora* (–), *Circaea lutetiana* (rather frequent), *Veronica persica* (–), *Veronica filiformis* (–), *Solidago canadensis* (–), *Solidago serotina* (–). (–: not listed in KÖLLIKER (1839)).

Veronica filiformis, today one of the most common and successful plant species in the whole area, was not observed in Switzerland until 1925. *Erigeron annuus* and *Veronica persica* had already reached today's distribution in 1900.

36.4.2 Frequent but not common species

These species occur in between 50% and 80% of the segments. They are mostly bound to special ecological conditions which cannot be found throughout the whole city area.

Forest species. The following forest species are either moderately urbanophob or have special ecological demands which are not realized in each segment, e.g. high soil humidity, rich base soils. *Acer campestre, Allium ursinum, Arum maculatum, Athyrium filix-femina, Carex digitata, Carex flacca, Carex pendula, Carex remota, Clematis vitalba, Convallaria majalis, Crataegus oxyacantha, Deschampsia caespitosa, Equisetum maximum, Evonymus europaeus, Festuca gigantea, Galium odoratum, Lamium montanum, Ligustrum vulgare, Mercurialis perennis, Moehringia trinerva, Oxalis acetosella, Paris quadrifolia, Phyteuma spicatum, Polygonatum multiflorum, Primula elatior, Rosa arvensis, Scrophularia nodosa, Taxus baccata, Viburnum lantana, Viburnum opulus, Vinca minor, Viola riviniana.* A few of these species (e.g. *Convallaria* and *Vinca*) which are also found outside the forests are often escaped garden plants. *Leucojum* can be found growing naturally in the forests of the Uto mountains. However, in most localities it is a garden escapee. In many cases it is difficult to distinguish between a natural and an introduced occurrence of a species.

As an example of the category of forest species which occur frequently, *Primula elatior* is shown in Fig. 36.7: Until about 30 years ago, *P. elatior* also grew outside the forests in not too sunny, fat meadows and in older parks. But the intensification of agriculture, including the early mowing of fat meadows, suppressed the species from these vegetations. It also disappeared from parks, mainly due to change of management and to the advance of *Primula vulgaris*, which forms hybrids with *P. elatior*. Today the distribution correlates positively with forest areas (Fig. 36.6).

Species of meadows and lawns. The species of this group are restricted to certain types of meadows.

Achillea millefolium, Alopecurus pratensis, Bromus erectus, Chrysanthemum leucanthemum, Crepis biennis, Daucus carota, Festuca arundinacea, Holcus lanatus, Hypochoeris radicata, Primula vulgaris, Sanguisorba minor, Trifolium dubium, Veronica chamaedrys, Veronica serpyllifolia, Viola odorata. Some species in this group were more common 40 years ago (e.g. *Bromus erectus, Chrysanthemum leucanthemum, Crepis biennis, Sanguisorba minor, Veronica chamaedrys*). Others (*Alopecurus pratensis, Primula vulgaris, Trifolium dubium*) are more frequent now. The distribution of *Primula vulgaris* is shown in Fig. 36.8. It is very often planted in lawns and completely naturalized throughout the city, even in meadows where it was never planted. To compete with other plants the lawn needs to be mowed several times during the vegetation period, but not too early in spring. Unlike its closest native places in the Foehn valleys of the northern Alps, it does not penetrate the forests in Zurich. It is negatively correlated with forest areas (Fig. 36.6). This might be explained by the too cool temperatures in the local climate of forests outside of Foehn influence and by the competition of *Primula elatior* in the forest. KÖLLIKER (1839) does not mention *P. vulgaris* in Zurich at all, and according to NÄGELI and THELLUNG (1905) it escaped from culture only on one place, in a cemetery. Another garden plant behaving very similar to *Primula vulgaris*, although not as frequent, is *Galanthus nivalis*.

Plants of swamps and wet meadows. Only very few plant species of this category are distributed throughout more than half of the segments. These are mostly species of habitats rich in nutrients: *Agrostis stolonifera, Angelica silvestris, Carex acutiformis, Cirsium oleraceum, Epilobium hirsutum, Filipendula ulmaria, Juncus effusus, Juncus inflexus, Lythrum salicaria, Veronica beccabunga.* Species of this group usually avoid areas with high soil sealing (cf. Fig. 36.5). However, the Sihl and the Limmat rivers provide the opportunity to reach the centre of the city. The distribution of *Cirsium oleraceum* is shown in Fig. 36.9. It resembles the distribution of *Primula elatior* (Fig. 36.7).

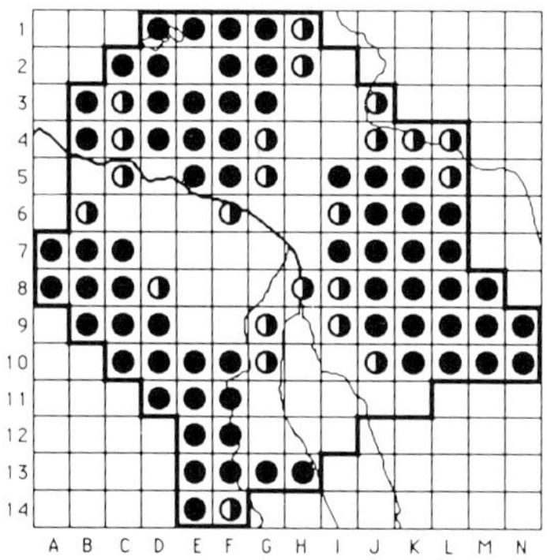

Figure 36.7: *Primula elatior.*

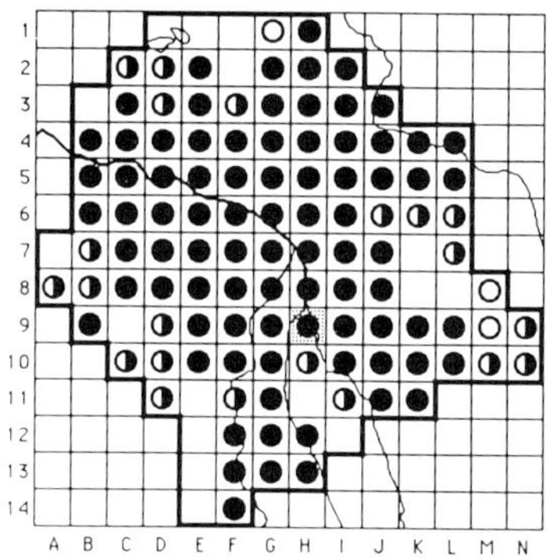

Figure 36.8: *Primula vulgaris.*

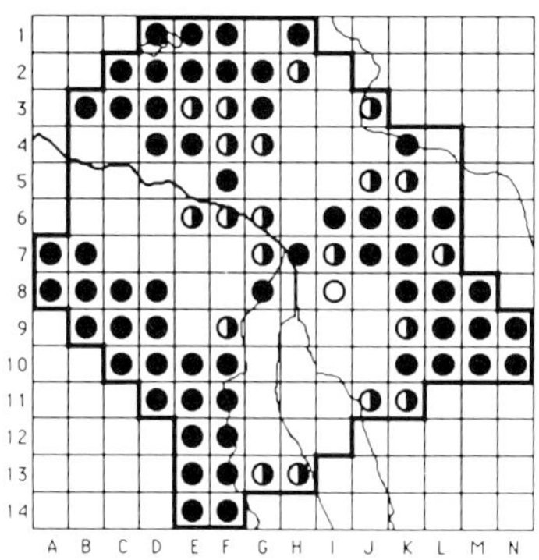

Figure 36.9: *Cirsium oleraceum.*

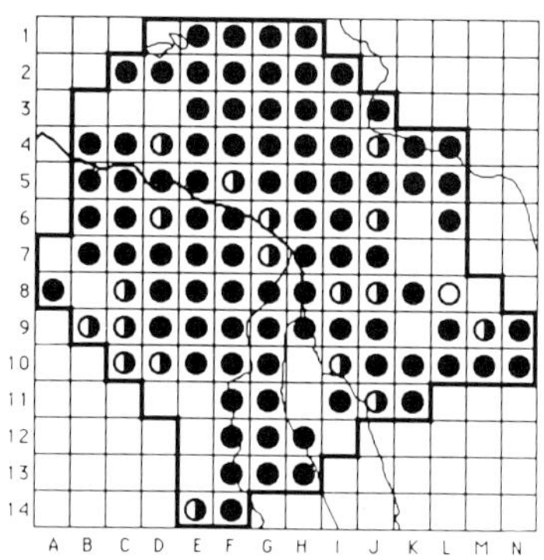

Figure 36.10: *Sinapis arvensis.*

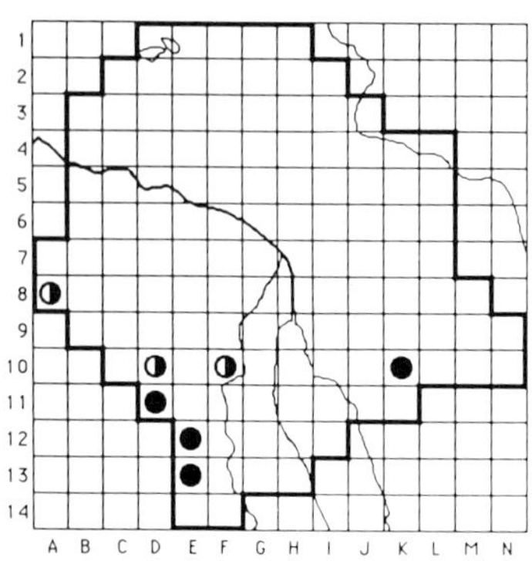

Figure 36.11: *Centaurea scabiosa.*

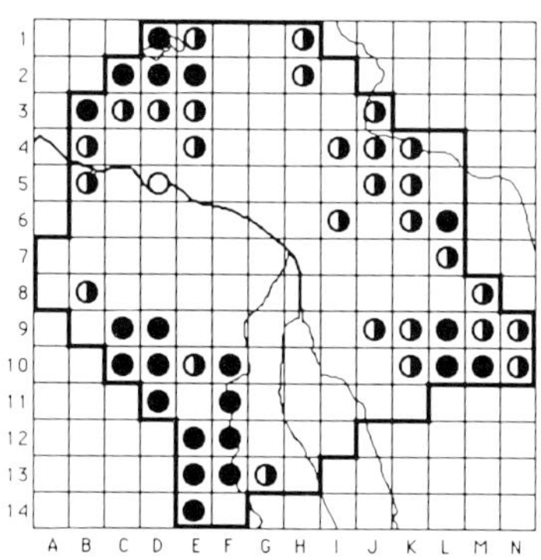

Figure 36.12: *Colchicum autumnale.*

Weeds and ruderals. In this category, species are included which are widespread but which have narrower ecological demands. These cannot be fulfilled in all segments, e.g. at sites subject to high radiation or on sandy soils. The species particularly avoid forested areas: *Agropyrum repens, Amaranthus retroflexus, Arabidopsis thaliana, Arenaria serpyllifolia, Cerastium glomeratum, Chenopodium polyspermum, Cirsium vulgare, Echinochloa crus-galli, Epilobium roseum, Galinsoga ciliata* (N), *Lamium purpureum, Linaria minor, Panicum capillare* (N), *Plantago intermedia, Polygonum monspeliense, Sinapis arvensis, Solanum nigrum, Sonchus arvensis, Tripleurospermum inodorum* (N), *Veronica arvensis.* The distribution of *Sinapis arvensis* is shown in Fig. 36.10. It is negatively correlated with forest areas (Fig. 36.6).

36.4.3 Plants common or rather common 150 years ago which have become infrequent

The changing of agricultural management resulted in the disappearance of many habitats. Some rather isolated biotopes still exist as nature reserves. A few individuals have been able to survive along the edges of forests and on embankments of roads and rivers. But the populations are so small that the species will eventually die out in most cases. Today the species occur only in less than 40% of the segments. In KÖLLIKER's time (1839) the species were much more frequent (frequency in brackets).

Species of extensively utilized, rather dry meadows and pastures. *Agrimonia eupatoria* (frequent), *Anthyllis vulgaris* (frequent), *Brachypodium pinnatum* (frequent), *Briza media* (common), *Campanula glomerata* (frequent), *Campanula rotundifolia* (common), *Centaurea scabiosa* (common) (Fig. 36.11), *Galium pumilum* (frequent), *Koeleria pyramidata* (frequent), *Ononis repens* (frequent), *Ononis spinosa* (frequent), *Pimpinella saxifraga* (everywhere), *Plantago media* (common), *Primula veris* (frequent), *Salvia pratensis* (everywhere), *Scabiosa columbaria* (common), *Thymus pulegioides* (everywhere).

Species of rather fat meadows

Campanula patula (infrequent; today nearly extinct), *Centaurea jacea* (common), *Cruciata laevipes* (common), *Knautia arvensis* (everywhere), *Helictotrichon pubescens* (frequent), *Tragopogon orientalis* (common), *Silene dioeca* (no frequency indication), *Alchemilla sp.* (frequent), *Colchicum autumnale* (frequent) (Fig. 36.12). It is negatively correlated with soil sealing (Fig. 36.5).

Species of straw meadows. *Gentiana verna* (frequent) (Fig. 36.13). *Primula farinosa* (not so frequent; today only in one locality), *Pinguicula vulgaris* (not so frequent; today very rare), *Bellidiastrum michelii* (not rare; today only along the Uto mountain range), *Parnassia palustris* (frequent).

Many species are not shade tolerant and disappear if the meadows become too densely populated in spring. *Gentiana verna* (Fig. 36.13) is an example of this group; it has lost almost its whole distribution area within the city. The only locality left is a relatively small area within a nature reserve which has always been managed as a straw meadow (never manured, cut in autumn).

36.4.4 Typical species of the developed areas

Species of this group are rather widespread but confined to developed areas. They inhabit pavements, roadsides, railway areas, walls, debris, lawns and are very rarely found in natural habitats of the region or in open fields, agriculturally managed meadows or forests. The occurrence coincides with the segments of highest soil sealing (Fig. 36.5). Since the degree of soil sealing correlates with elevated local temperatures (Fig. 36.4) it is difficult to decide whether the presence of a species is due to the warmer climate or to the special habitats in sealed areas. Very many rare species which can be found only in the warmest areas, such as railroads, ought to be classified in this category.

Ailanthus glandulosa (N), *Arenaria serpyllifolia, Cerastium glomeratum, Cerastium semidecandrum, Corydalis lutea* (N), *Diplotaxis muralis, Diplotaxis tenuifolia, Epilobium dodonaei, Eragrostis minor, Erophila praecox, Erophila verna, Galeopsis angustifolia, Lepidium virginicum* (N), *Malva neglecta, Oxalis corniculata* (N) (Fig. 36.16, *Paulownia tomentosa* (N), *Plantago intermedia, Poa bulbosa, Poa compressa, Saxifraga tridactylites.* (N = neophyte.) Many of these species were introduced incidentally or as garden plants (N) and originate from rocks or screes or from Mediterranean areas. *Oxalis corniculata* (Fig. 36.14) is an example of a garden plant, which correlates with soil sealing (Fig. 36.5). It was not recorded in KÖLLIKER (1839) (contrary to *O. europaea*). NÄGELI and THELLUNG (1905) do not cite any locality within the city of Zurich either. More restricted in its distribution and probably more dependent on higher temperatures is *Ailanthus glandulosa* (Fig. 36.15, cf. Fig. 36.4). Originally an ornamental tree, it spreads along roads, on debris and in gardens. *Eragrostis minor* (Fig. 36.16) is typically found along pavements and railroads. With increasing urbanization it could spread. An example of a species which is restricted to railroads is *Galeopsis angustifolia* (Fig. 36.17). Originally growing on calcareous scree of the lower belts in Central European mountains it expanded as a weed to dry fields. In this habitat it became almost extinct within the last 50 years. *Corydalis lutea* (Fig. 36.18) has been naturalized on walls of the inner-city for more than 150 years. It is a scree plant of the Southern Alps that was introduced as a garden plant.

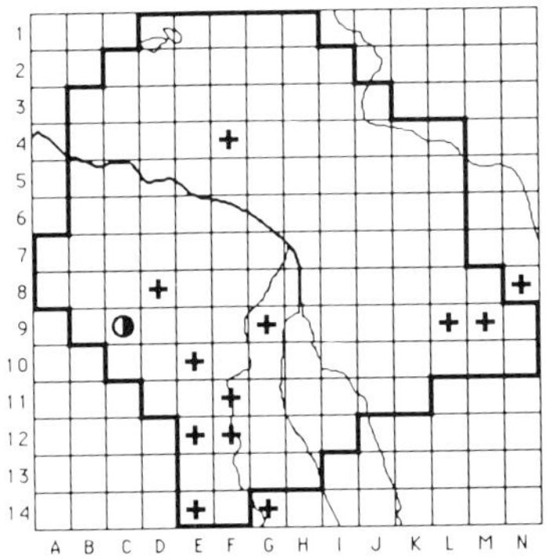

Figure 36.13: *Gentiana verna* + extinct.

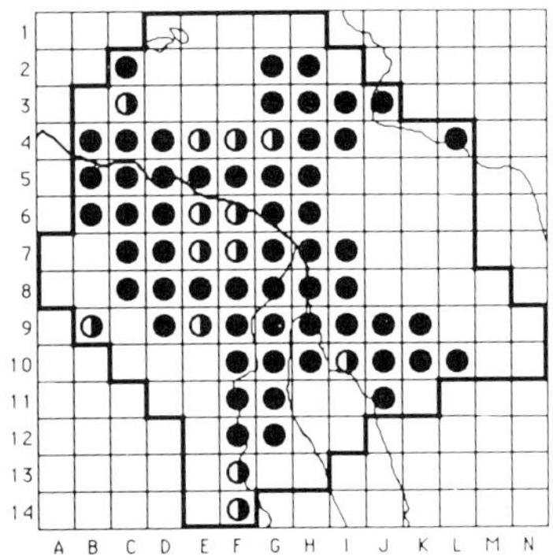

Figure 36.14: *Oxalis corniculata*.

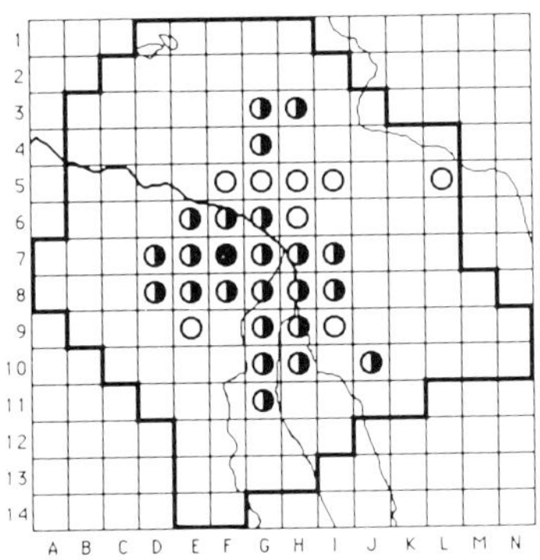

Figure 36.15: *Ailanthus glandulosa*.

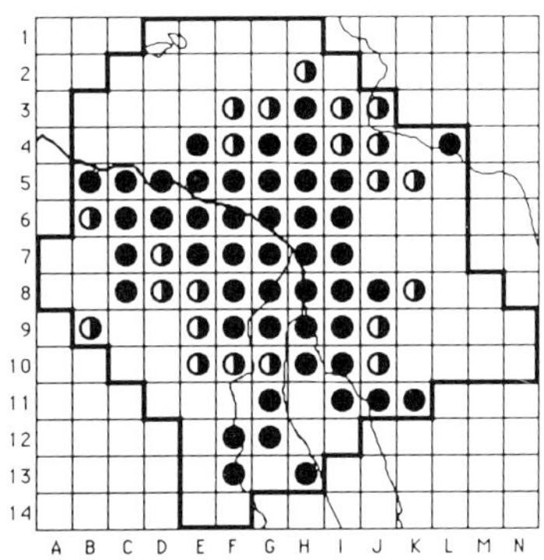

Figure 36.16: *Eragrostis minor*.

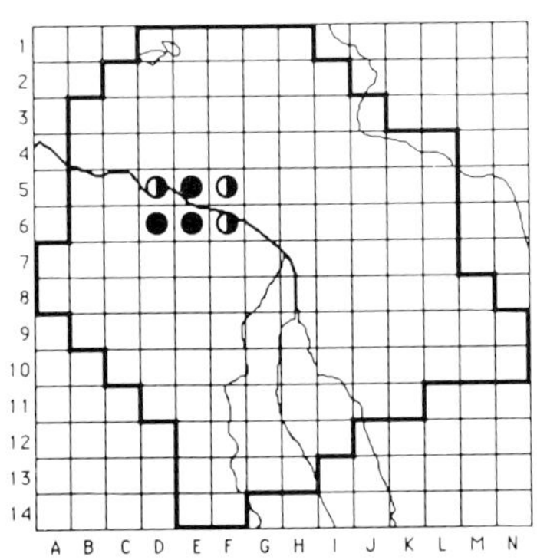

Figure 36.17: *Galeopsis angustifolia*.

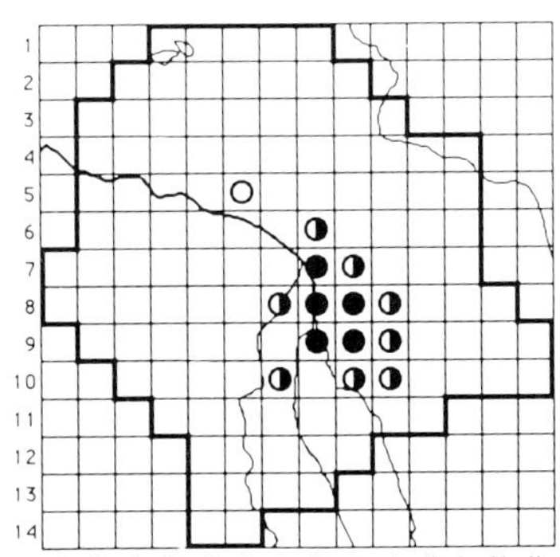

Figure 36.18: *Corydalis lutea*.

36.4.5 Species of southern slopes

A typical representative of this group is *Viola alba* (Fig. 36.19). It is restricted to the edges of forests, undergrowth and certain types of southern exposed lawns (cf. Fig. 36.2).

36.4.6 Species of northern slopes and higher altitudes

Species of this group probably avoid warm regions. *Aconitum vulparia, Actaea spicata, Lonicera alpigena, Sambucus racemosa, Vicia silvatica, Vicia dumetorum, Chaerophyllum cicutarium.* Most extreme is *Lonicera alpigena* (Fig. 36.20), which can only be found on steep northern exposed slopes, where it never receives sunshine (cf. Fig. 36.2). *Sambucus racemosa* (Fig. 36.21) is more generally distributed in the cooler regions, in higher altitudes (cf. Fig. 36.1) or in northern exposed regions (Fig. 36.2).

36.4.7 Species with a southern limit

The amount of precipitation, and probably connected with this factor, the degree of cloudiness, increases from northeast to southwest (towards the Alps). A few species have a southern boundary within the city limits of Zurich, which is probably due to the above-mentioned factor. *Ranunculus auricomus* s.l., a plant of moist forests and shady meadows, is an example of this group (Fig. 36.22).

36.4.8 Mountain species

The species of this group need rocks or steep slopes which are not densely forested. They are restricted to the Uto mountain range in the west of the city: *Petasites paradoxus, Campanula cochleariifolia, Leontodon hyoserioides, Saxifraga mutata, Saxifraga aizoides, Thesium alpinum, Sesleria coerulea, Thymus polytrichus, Pinus montana, Polygala chamaebuxus, Laserpitium latifolium, Gentiana ciliata, Vincetoxicum officinale, Chrysanthemum adustum, Carlina stricta, Carduus defloratus, Centaurea montana, Buphthalmum salicifolium. Sesleria coerulea* is typical of this group (Fig. 36.23).

36.4.9 Shore species

Some species grow only along the shores of Lake Zurich and the Limmat (some are also rarely found along the shores of Katzensee). This is a very small relic group which had been much larger 150 years ago, but was diminished by the destruction of the original shores changing the water regime and sealing the banks of the lake and river: *Butomus umbellatus, Lycopus europaeus, Rorippa amphibia, Scutellaria galericulata.* The distribution of *Rorippa amphibia* (Fig. 36.24) shows the position

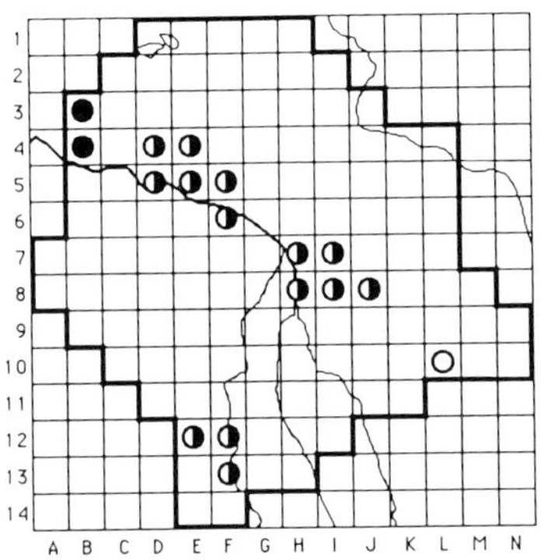

Figure 36.19: *Viola alba.*

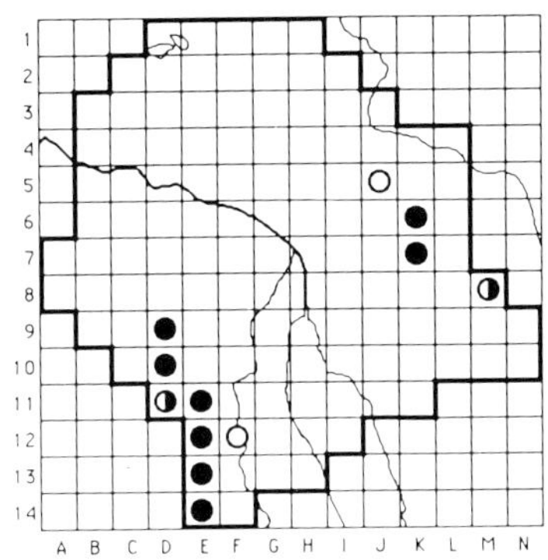

Figure 36.20: *Lonicera alpigena.*

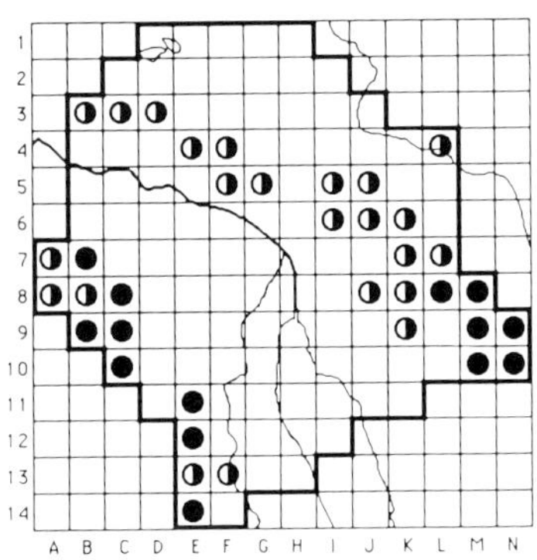

Figure 36.21: *Sambucus racemosa.*

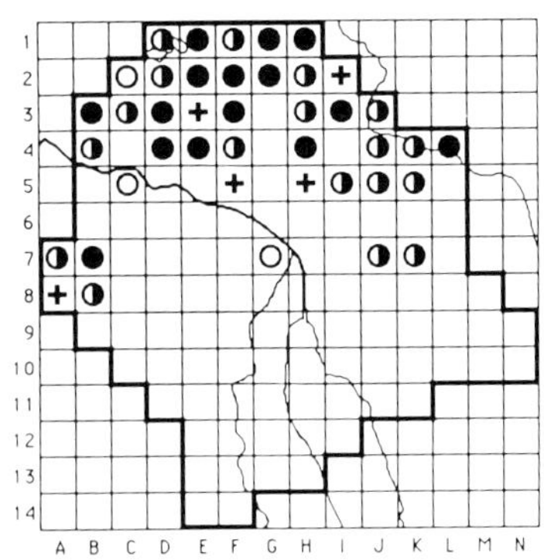

Figure 36.22: *Ranunculus auricomus* s.l.

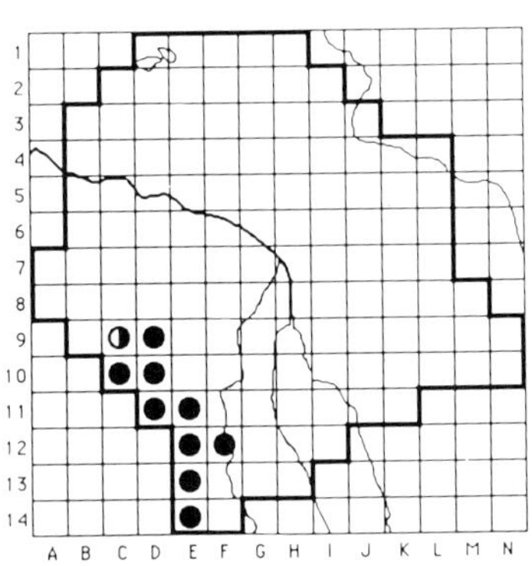

Figure 36.23: *Sesleria coerulea.*

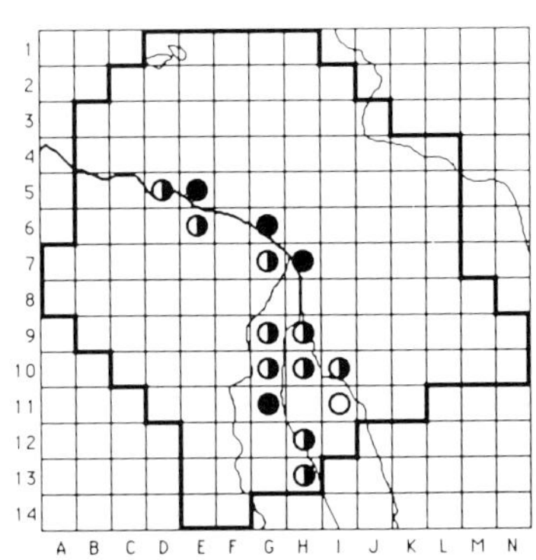

Figure 36.24: *Rorippa amphibia.*

of the shores of Lake Zurich and the Limmat. In the centre of the city, where the riverbanks are sealed, the species is absent.

36.4.10 Species introduced by the Sihl river

A hundred years ago the Sihl (on the map: the river left of Lake Zurich) was an important source of alpine plant propagules within the city of Zurich. Today it is dammed in the upper part, and only a few seeds reach Zurich. Only a few species can survive for a long time along the river banks, e.g. *Senecio alpinus, Carduus personata, Ranunculus aconitifolius, Petasites hybridus, Mentha longifolia, Chaerophyllum cicutaria.*

36.5 Neophytes

Apart from the very frequent species introduced during the 19th century or earlier (*Geranium pyrenaicum, Veronica persica, Solidago serotina, Solidago canadensis, Erigeron canadensis, Erigeron annuus, Amaranthus retroflexus, Oxalis europaea, Impatiens parviflora, Matricaria suaveolens, Tripleurospermum inodorum*) the following species have spread out during this century and are to be found in more than half of the segments: *Buddleja davidii, Cotoneaster horizontalis, Fragaria indica, Galinsoga ciliata* (*G. parviflora* is very rare in Zurich), *Heracleum mantegazzianum, Lepidium virginicum, Oxalis corniculata* (Fig. 36.14), *Polygonum cuspidatum, Veronica filiformis.*

The following Neophytes with limited distribution show a tendency to spread but are still limited in their distribution. Some prefer the warmest places (cf. Fig. 36.4) and are also cited as typical species of developed areas (see page 816). Others are restricted to places with more oceanic climates (northern shaded slopes, shady walls) (e.g. *Hypericum*):

Veronica peregrina, Ailanthus glandulosa (Fig. 36.15), *Artemisia verlotorum, Hypericum androsaemum, Mahonia aquifolia, Cotoneaster salicifolia, Cornus stolonifera, Prunus laurocerasus* (during recent years this species has begun to spread out in forests and shady gardens but does not yet reproduce autonomously), *Impatiens glandulifera.*

36.6 Discussion

This preliminary survey of the flora of Zurich shows that there is a stock of more than 100 species. This is about 10% of the total number of species occurring within the city limits of Zurich, or if we consider the different segments, one third of the total number of species. These species which grow frequently and in nearly every segment must have a climatic optimum in the region of Zurich. In addition, they are not very limited in their ecological tolerances. The group includes forest plants as well as weeds or meadow plants. The other species are either ecologically restricted to habitats which are not evenly distributed within the city or are on the limits of their climatic occurrence. To find out correlations between the distribution area and some climatic or edaphic factors is one of the most fascinating aspects of the study of urban vegetation. In comparing the results with those of other cities it will be possible to learn much more about the biological characteristics of many plant species. But this will be the occasion of later interpretations. Compared with other Central European cities (e.g. Berlin), the number of forest plants penetrating the inner-city is much larger. On the other hand, continental species which are widespread in Berlin are absent or restricted to the warmest places in Zurich. Cologne (KUNICK 1984) seems to be quite similar to Zurich, although some forest plants (e.g. *Brachypodium silvaticum, Geranium robertianum, Mycelis muralis, Carex silvatica, Circaea lutetiana*) can be found in nearly every segment in Zurich, and only in less than half of those in Cologne. Some other species like *Ballota nigra, Berteroa incana, Euphorbia esula*, which are not rare in Cologne but absent in Zurich, show that the climate is generally warmer and drier in Cologne.

Further discussion will be possible when the mapping is completed. More species with a general distribution will probably turn up, since a systematic search for missing species in all segments has not yet taken place.

Acknowledgments

I am grateful to M. Bichsel for the development of a programme to delineate the maps.

Bibliography

BERNOWITZ, K.; LEUTERT, F., 1988, Karte der erhaltens- bzw. schutzwürdigen Biotoptypen der Stadt Zürich.
Gartenbauamt der Stadt Zürich.

BRANDES, D., 1985, Pflanzen in der Stadt.
Staatl. Naturhist. Museum Braunschweig. 64 p.

ENGEL, H., 1949, Trümmerpflanzen von Münster.
Natur und Heimat **9** (2), 1–12.

HESS, H. E.; LANDOLT, E.; HIRZEL, R., 1967–1972, Flora der Schweiz und angrenzender Gebiete.
Birkhäuser, Basel. 3 volumes.

JANSSEN, C.; BRANDES, D., 1984, Struktur und Artenvielfalt von Randzonen der Großstädte — Dargestellt am Beispiel Münster.
Braunschw. Naturk. Schr. **2**, 57–97.

Kölliker, A., 1839, Verzeichnis der phanerogamischen Gewächse des Cantons Zürich.
Orell Füssli, Zürich. 154 p.

Kunick, W., 1984: Verbreitungskarten von Wildpflanzen als Bestandteil der Stadtbiotopkartierung, dargestellt am Beispiel Köln.
Verh. Ges. Ökologie **12**, 269–275.

Nägeli, O.; Thellung, A., 1905, Die Flora des Kantons Zürich. I. Teil: Die Ruderal- und Adventivflora des Kantons Zürich.
Raustein, Zürich. 82 p.

Pfeiffer, H., 1957, Pflanzliche Gesellschaftsbildung auf dem Trümmerschutt ausgebombter Städte.
Vegetatio **7**, 301–320.

Rebele, F.; Werner, P., 1984, Untersuchungen zur Ökologischen Bedeutung industrieller Brach- und Restflächen in Berlin.
Förderungsprogr. Freie Univ. Berlin für junge Wissenschaftler **3**, 169 p.

Sailer, U., 1990, Vegetationentwicklung auf Brachflächen der Stadt Zürich.
Ber. Geobot. Inst. ETH, Stiftung Rübel, Zürich **56**, 78–117.

Schlatter, B., 1975, Zum Stadtklima von Zürich.
Diplomarbeit Geographisches Institut Univ. Zürich. 113 p. (Polycopy)

Sukopp, H., 1983, Ökologische Charakterisierung von Grossstädten. Grundriss der Stadtplanung.
Akad. für Raumforschung u. Landesplanung, Hannover. 51–82.

Sukopp, H.; Blume, D.; Chinnow, D.; Kunick, W.; Runge, M.; Zacharias, F., 1974, Ökologische Charakterisierung von Grossstädten, besonders antropogene Veränderungen von Klima, Boden und Vegetation.
Z. TU Berlin **6**, 469–488.

Umweltschutzfachstelle der Stadt Zürich, 1989, Umweltbericht 1988.
Stadt Zürich. 74 S.

Walter, H.; Lieth, H., 1961–1967, Klimadiagramm-Weltatlas.
VEB Fischer, Jena.

Wiskemann, Ch., 1990, Vegetation auf verdichteten Böden in der Stadt Zürich.
Ber. Geobot. Inst. ETH, Stiftung Rübel, Zürich **56**, 118–141.

Chapter 37

Urban ecosystems and coastal management

M. NUMATA

37.1 Introduction

The role of man in an ecosystem. Our research team on the structure, function and dynamics of urban ecosystems in the Tokyo Metropolis was organized in 1971 in conjunction with the MAB Project No. 11. It was a developmental phase of IBP with the object of making fundamental analyses and estimations of biological productivity on earth. In the MAB Projects, both "Man" and the "Biosphere" have been the main objects of study, which differs from the approach excluding man, such as in the IBP. Even in biology textbooks, the role of man in the ecosystem has been too small and should be properly evaluated.

The urban ecosystem. An urban ecosystem is an analogy of a natural ecosystem. An ecosystem is usually understood to be composed of producers, consumers, decomposers and abiotic factors. In addition to this, an ecosystem is a relatively closed system, such as a lake with a lakeside. Terrestrial ecosystems do not, in general, have a clear border such as the land–water interface, and yet we call them forest ecosystems, grassland ecosystems, desert ecosystems, and so on. The boundary of an urban ecosystem is even more ambiguous. However, a climatological dome with an inversion layer is sometimes seen as a limit to the range of an urban ecosystem. Of course, in a broad sense, it is composed of producers, consumers, decomposers and abiotic factors, and functions as an anthropocentric ecosystem (NUMATA 1983).

The land–water interface surrounding the coast line. Since 1978 we have been studying the ecological and multi–disciplinary factors of the coastal cities that face Tokyo Bay. In Japan there are many large seaside cities, such as Sendai, Tokyo, Yokohama, Nagoya, Osaka, Kobe, Fukuoka and Niigata amongst others. A few are also inland, such as Kyoto and Sapporo. Many seaside cities are bay–shore cities. The definition of the coast should be examined in relationship to this. The terrestrial coast or shoreland is a sandy or rocky shore beyond the coastline. On the sandy shore, there are vegetation belts such as unstable or frontal belts, semistable or intermediate belts and stable or rear belts. The stable belt is composed of grassland, pine forest, etc.

The sea coast is sometimes the terrestrial coast (or shoreland), while on the other hand it may mean the part of the coast under sea or coastal water. In this case, the sea coast includes a backshore and foreshore. The backshore is between the coastline (the highest high–tide line) and the mean high–tide line, and the foreshore is between the mean high–tide line and the lowest low–tide line. The front of the foreshore is the offshore. Recently, coastal waters have tended to be legally defined as 200 nautical miles in width. In a word, the coast is classified into terrestrial portion and marine sections (NUMATA 1977).

37.2 The history of Tokyo Bay and its coastal zone

The area of Tokyo Bay is ca. 1,500 km^2 with an average depth of 15 m and a maximum depth of 70 m. It consists of an inner and an outer bay. Freshwater is supplied by rivers to a total amount of 101 hundred million $km^3 \cdot a^{-1}$ (300 $m^3 \cdot sec^{-1}$). The distance across the southern part of the inner bay (north of the line between Kannonzaki and Futtsu) is only 6 km. At its broadest, the bay is 30 km wide.

Old Tokyo Bay existed from the Miocene age of the Tertiary Period till the diluvial epoch. Old Tokyo Bay was an estuary open to the east. After that, through transgression and regression by the effect of glaciation and deglaciation, the embryonal form of the present Tokyo Bay was formed some 10,000 years ago.

In the middle of the Jomon Period (from 7500 to 300 BC), the seacoast retrogression began, and wetlands formed at the edge of the terrestrial area. In the Yayoi Period (from 300 BC to 300 AD), the Tokyo Bay reached its present state. Several rivers flowed into the Bay, but the Tone River was changed in the 17th century to flow in its present direction into the Pacific Ocean. In 1590, Shogun

Ieyasu Tokugawa came to Edo Castle (the Imperial Place at the present time). Edo city, however, was frequently flooded (KONDO 1979).

The inner part of Tokyo Bay had a wide muddy flat along the seacoast where sea weed (Nori) was cultivated, and small fish and shells were harvested in abundance. However, landfilling started during the 17th century and 125 ha had been reclaimed by the first half of the 18th century. After that, 578 ha were filled from 1913 to 1955 in order to build the Keihin (Tokyo–Yokohama) Industrial Zone. Landfills before World War II amounted to a total area of 3700 ha, and a total of 20,000 ha were reclaimed during the 30 years after the war. The Keiyo (Tokyo–Chiba) Industrial Zone was built on landfills after 1950. These landfills were supplied to enterprises, particularly for steel manufacturing, oil refining, electricity, etc. Industrial water was also guaranteed (OTSUKA 1979).

Due to landfilling along the coast of Tokyo Bay, the natural seacoast came to be covered by a high concrete sea wall, and the mud flats, where shells, crabs, lugworms, small fish, shrimps and many other creatures were living, were lost. People used to be able to swim, collect shells and fish there. Due to the loss of access to the seacoast, the "Coastal Access Rights Movement" was formed in 1973 as an environmental rights movement (SHAPIRO 1980). Over 40% of the nation's 120 million inhabitants live and work in Japan's coastal cities, with an average population density of over 30 times that of in the United States. The average length of coastline per capita in Japan has been estimated to be 30 cm (SHAPIRO 1984). The per capita length of natural coastline is well under half of that.

According to the State of Land Utilization Survey of the Shoreland (Environmental Agency 1975), cities and industrial areas make up 59.6% of the Chiba Prefecture, and 80% of the Kanagawa Prefecture (95.4% of the Osaka Prefecture).

South of the Futtsu Cape, bordering the inner and the outer bays, a branch current of the Kuroshio (warm current) flows anti–clockwise, whilst a branch current in the northern inner bay flows clockwise. The westward growth of the Futtsu sand bar was promoted by both of these currents. This fact is verified by past topographical maps, and when I studied the sandy vegetation there after the war (NUMATA 1949), the sand bar stretched westward directly to an artificial island (the so–called "First Marine–Fort"). After that, the current of the inner bay weakened, and the sand bar curved to the north away from the island.

Dokan Ohta established Edo Castle in 1457 when the coastline was further inland than the present one (MASAI 1980). After the shogunate of Edo was established in 1603, there were many warehouses along the waterways.

37.3 The Keiyo Industrial Zone and bay–coast cities

It was anticipated that much water would be used in the Keiyo Industrial Zone that was built after World War II. However, there are no big rivers within a radius

Figure 37.1: The bayshore cities of the Keiyo Industrial Zone

of 50 km from Chiba City, except the lower reaches of the Edo and Tone Rivers. The Tama, Sagami, Ara, Edo and Tone Rivers are all within a 50 km radius of Tokyo. The introduction of water from the River Tone and Lake Imba and the utilization of water from medium and small rivers were considered during the final stage of construction of the Keiyo Industrial Zone.

The watersheds are situated 10–15 km and 15–20 km inland from the coastline of Tokyo Bay in the Chiba–Ichihara and Ichihara–Kisarazu areas respectively (Fig. 37.1). Here the alternate beds of clay and sand gradually incline to the northwest and contain a large amount of groundwater. YAMAMOTO (1980) calculated that the amount of groundwater in the Keiyo Industrial Area was 137×10^3 $m^3 \cdot d^{-1}$, which is the amount being pumped up at maximum utilization in Chiba City. The amount of groundwater in the loamy plateau of the Kanto District is 1–2 $mm \cdot d^{-1}$ and the amount of groundwater per km^2 is 1×10^3–2×10^3 $m^3 \cdot d^{-1}$. To obtain 137×10^3 $m^3 \cdot d^{-1}$ of groundwater, an area of 137–69 km^2 is necessary. The width of the outcrop of the sandy layer is 5 km at its widest point. Therefore the outcrop should be 14 to 28 km long. However, there is not such a large outcrop for Chiba City, and so a plan to carry industrial water was promoted.

Since 1951, various factories have been built on coastal landfills. These included steel works, oil refineries, power plants, etc. The Boso Seaside Industrial Water Supply System was begun by the Chiba Prefectural Government in 1970 with the aim of supplying 740×10^3 $m^3 \cdot d^{-1}$ by 1979, mainly from the Tone River as well as from the Obitsu and Yoro Rivers. A survey on the demand for the water needed by Chiba bayshore factories in 1968 showed the need for a large amount of water, 1400×10^3 $m^3 \cdot d^{-1}$. However, the Prefectural Government decided to supply 420×10^3 $m^3 \cdot d^{-1}$ for 44 enterprises in the Keiyo Industrial Area in 1974, just after the oil embargo of 1973. After that decision, Bosuikyo, an organization of 44 enterprises, requested that the Prefectural Government decrease industrial water to 110×10^3 $m^3 \cdot d^{-1}$ which is less than tenth of the original 1968 requirement.

There were many self–welling pipes in the Keiyo Industrial Area. The utilization of groundwater for industry was sustained at a level of 60×10^3 $m^3 \cdot d^{-1}$ until 1963, when the amount increased greatly, particularly during 1969 and 1974 to an average of 100×10^3 $m^3 \cdot d^{-1}$. Also, the water table has risen due to the decrease of water requirement, the recycling of water, the regulation of groundwater utilization by law and an increase in self–welling water since 1982.

The Sumida River, which was the most polluted of the rivers flowing into Tokyo Bay is a medium sized river of 50 km in length with a drainage basin of 600 km^2. Upstream is a bed town of the national capital region supporting more than 2 million people. Between 1970 and 1980 the population there increased rapidly, by 36%. The spreading of the sewerage is therefore slow, and untreated domestic sewage flows into this river. On the other hand, 3 million people are concentrated in the lower reaches of the drainage basin, where there is a good distribution of sewers, and most of the domestic sewage flows into the river after having been treated. The amount of water flowing into the river from the drainage basin are: rain water: 5–6 $\times 10^8$ $m^3 \cdot a^{-1}$; untreated sewage: 16×10^8 $m^3 \cdot a^{-1}$; treated sewage: 7×10^8 $m^3 \cdot a^{-1}$ and water for purification from the Ara River: 2.5×10^8 $m^3 \cdot a^{-1}$. About a half of the water supplied to the river originated from the sewage, and the pollutant load is 32,000 $t \cdot a^{-1}$ (90 $t \cdot d^{-1}$) of BOD. The branches of the upper stream section are among the most polluted streams, with 40–100 ppm of BOD. The main river stream has 10–20 ppm of BOD except in the tidal area and fish can live in the water at the river mouth as it is diluted with sea water (FUKUSHIMA and HANYA 1984).

As a result of polluted water flowing into the sea, the decrease in dissolved oxygen on the sea bed is observed, particularly in the summer (FUROTA 1985). The decrease in the amount of oxygen was remarkable on the lower part of a slope and a connected flat, where in August the amount of oxygen was 0.3–0.9 $mg \cdot l^{-1}$. On the substratum consisting of particles finer than sand and with little physical disturbance as such waves, five categories of indicator benthic animal were recognized: 0: no organism; I: strongly polluted area; II: weakly polluted area; III: strongly hypertrophic area; and IV: weakly hypertrophic area. Corresponding to the anaerobic conditions on the sea bed, attached animal communities, such as *Mytilus edulis galloprovicialis, Balarus eburneus, Hydra* sp., etc. degenerated and disappeared.

There are only a limited number of animals, such as lugworms, tubificids and midges, etc. living in the lower reaches of the rivers that flow into Tokyo Bay, such as the Yoro and Obitsu, etc. The scantiness of benthic animal communities in rivers was caused by pollution caused by the development of the coastal cities and watershed areas. Stenithermal aquatic insects with branchial respiration and high oxygen demand in the lotic environment are limited in the polluted rivers. On the other hand, eurythermal aquatic insects with low oxygen demand living in the lentic environment are relatively resistant to pollution (HISAI 1981).

37.4 Concluding remarks

The changes and consequences of the land–water system in relationship to urbanization and industrialization in the coastal areas of the Tokyo Bay were discussed in this paper. It was stated earlier that the countermeasures against the lowering of the water table, land subsidence, the shortage of industrial and domestic water, etc. were legal regulations against the utilization of groundwater and the transfer of surface water from neighbouring areas. Due to these measures, the water table has returned to its former position. Furthermore, saving water and particularly the recycling of water is necessary. Industrial water is, in many cases, not input, but "throughput" for negative production from low entropy clean water to high entropy polluted water (TAMANOI 1981). Therefore, water as a throughput should be reused as recycled water.

The impact of industrial development and urbanization on river and coastal waters can be seen by the use of bio-indicators (NUMATA 1984). Measures to reduce the impacts mentioned above include the maintenance of the natural mud flats, the construction of man-made flats along the landfills, the prevention of rivers being covered by concrete and planting vegetation on man-made beaches, etc. In Chiba City, the planting of a pine forest on a man-made beach by the citizens has been successful.

The impact of industrial development and urbanization on the land–water system of coastal cities can be seen in the atmosphere, soils, plants, animals and citizens (particularly in their environmental perception). In this paper, however, consideration and discussion have been limited to a water-oriented approach.

Bibliography

FUKUSHIMA, K.; HANYA, T., 1984, Chemical impact of human activities in urban areas on the environment through the utilization of water.
In: Numata, M. (ed.), Water-Oriented Urban Ecosystem Studies, II (Jap. Version), 57–61, (in Japanese).

FUROTA, T., 1985, Living conditions of benthic periphytons in the Inner Bay of Chiba Prefecture of the degeneration as an indicator of anaerobic water zone.
In: Numata, M. (ed.), Impact of Coastal Development on Plants and Animals in Chiba Prefecture, **12**, 94–132, (in Japanese).

HISAI, N., 1981, Changes of benthic animal communities of rivers caused by urbanization and their characteristics.
In: Numata, M. (ed.), Integrated Ecological Studies in Bay–Coast Cities, III (Jap. Version), 39–40, (in Japanese).

KONDO, K., 1979, The present and past of the Tokyo Bay.
In: Nihon–Kagakusha–Kaigi (ed.), The Tokyo Bay, 1–7, (in Japanese).

MASAI, Y., 1980, Coastal change of Edo: Facts and effects.
In: Numata, M. (ed.), Integrated Ecological Studies in Bay–Coast Cities, II, 69–74.

NUMATA, M., 1949, Studies on the structure of plant communities, I. Especially on the sand vegetation by the seaside in the vicinity of Futtsu Cape.
Physiol. and Ecol. Japan **3** (1/2), 47–65 (in Japanese with English summary).

NUMATA, M., 1977, Ideas and facts on coastal parks and reserves.
Bull. Marine Park Res. Sta. **1**, 80–82.

NUMATA, M., 1983, Urban ecosystems and environmental science.
A paper for the International Symposium on Urban Ecosystem and Environmental Science, Yokohama, November 7–9, 1983.

NUMATA, M., 1984, Bio–indicators of human environment, particularly of urbanization.
Radovi (Sarajevo) **76**, 139–145.

OHTSUKA, H.,1979, History of landfills in the Tokyo Bay.
In: Nihon–Kagakusha–Kaigi (ed.), The Tokyo Bay, 84–104, (in Japanese).

SHAPIRO, H. A., 1980, The coastal access rights movement in Japan.
Coastal Zone Mgt. J. **8**, 1-43.

SHAPIRO, H. A., 1984, Coastal area management in Japan. An overview.
Coastal Zone Mgt. J. **12**, 19–56.

TAMANOI, Y., 1981, Negative production neglected by economics.
Kagaku Asahi Jul. 1981, 57–61, (in Japanese).

YAMAMOTO, S., 1980, Water problems in the Keiyo bayshore area.
In: Numata, M. (ed.), Integrated Ecological Studies in Bay–Coast Cities, II (Jap. Version), 181–185, (in Japanese).

Chapter 38

Landscape ecology — fundamentals, aims and perspectives

H. LESER and H. RODD

38.1 What is "landscape ecology"?

In their book "Landscape Ecology — Theory and Application" NAVEH and LIEBERMAN (1984,21) stated: "Landscape ecology is presently viewed in Europe as the scientific basis for land and landscape planning, management, conservation, development and reclamation. As such, it has overstepped the purely natural realm of classical bioecological sciences and has entered the realm of human-centered fields of knowledge — the socio-psychological, economic, geographic and cultural sciences connected with modern land uses."

The founder members of landscape ecology - Carl TROLL (1939, 1950), Josef SCHMITHÜSEN (1942) and Ernst NEEF (1956, 1967), NEEF, SCHMIDT and LAUCKNER (1961) - put forward an allround definition of landscape ecology and its subject matters. This definition, however, was harshly criticized in the course of steadily growing scientific specialization. The view that our environment is a holistic ecological system predominantly influenced by man reconsiders these ideas again (LESER 1985). For research and application this, however, means that landscape ecology

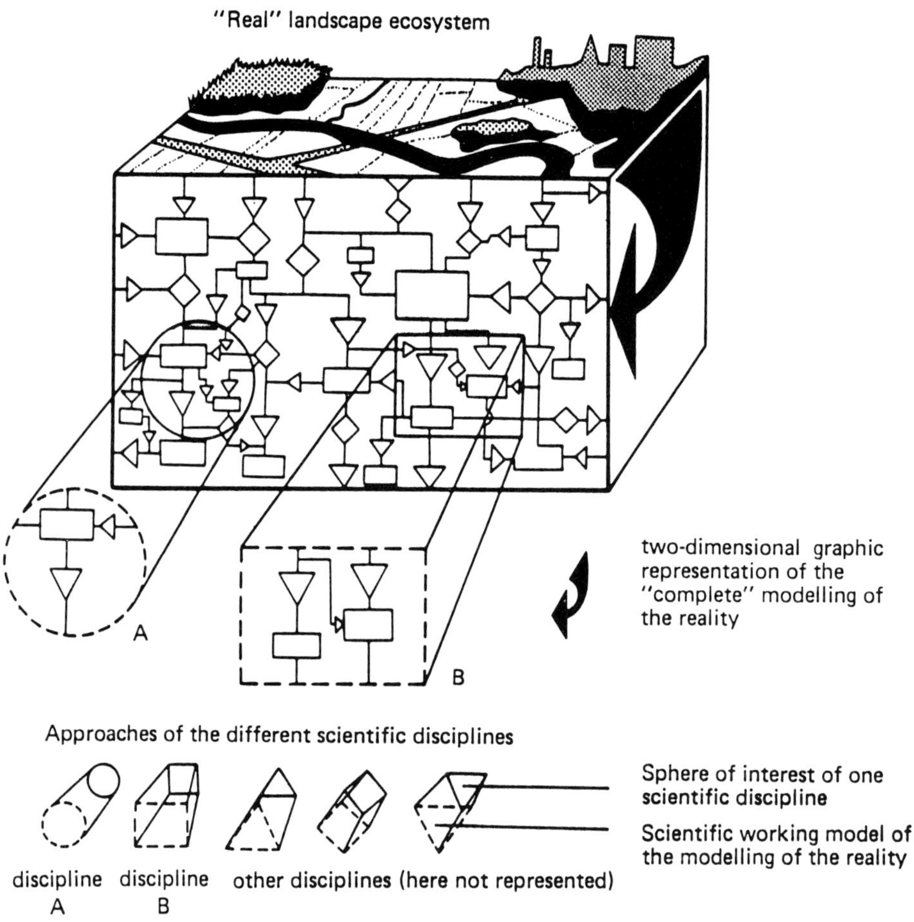

Figure 38.1: The reality of the landscape, the model of the landscape ecosystem and the approaches of the different ecological disciplines.

cannot possibly be the concern of one scientific discipline alone.

Just as ecology is being dealt with in various disciplines apart from biology, the landscape eco-system is being dealt with from the points of view of different fields of research as well as of ... application. This corresponds with the models of metascientific theory, system analysis and cybernetics (Fig.38.1).

38.1.1 Definition of the subject matter of landscape ecology

This definition (LESER 1984) is based on the analysis of extensive literature (LESER 1978 2nd ed.; LESER 1991; NAVEH and LIEBERMAN 1984 and FINKE 1986). The "environment", the "biotope" and the "geobiosphere" are looked upon as the subject matters of landscape ecology. This is the reason why we speak of "environmental systems" or "landscape ecosystems". Within the theory of landscape ecology there are highly complex systems. The landscape ecosystem can be defined as follows:

- The landscape ecosystem is a highly complex pattern of physiogenetic, biotic and anthropogenic factors directly or indirectly correlated with one another, thus forming a paramount functional correlation, represented by "landscape". Yet for reasons of methodology and the specific interests of the various disciplines, the landscape ecosystem can be defined more deliberately. This means that only a section of reality is examined, evaluated and planned.

Fig. 38.2 shows this clearly. In simple terms this means: the landscape ecosystem is made up of the following three major subsystems: geosystem, biosystem and anthroposystem. However, it is also possible to stress partial aspects of such models. NAVEH and LIEBERMAN (1984), for example, define the landscape ecosystem as the "Total Human Ecosystem". Here, only the anthropo–ecological component is stressed. The holistic–ecological approach remains the same as in the above–mentioned definition of the landscape ecosystem.

38.1.2 Landscape ecology as a scientific branch

The branch of landscape ecology is to be dealt with in this article in detail. The structures, perspectives, possibilities and limits of this branch will be taken into consideration. The idea that it is impossible to get off the "spacecraft earth", although it is charged to an extent so far unknown in the history of mankind, forms the background to these considerations. However, before this discussion is led, a definition of the branch of landscape ecology shall be attempted. It, too, draws on the interpretation of the literature mentioned on pages 833ff.

- Landscape ecology deals with the interdependence of the factors marking the landscape ecosystem. They present themselves functionally and visually in what we call "landscape", for example in a complex territorial structure. The variety of the aspects of consideration and examination is dealt with in different disciplines. As they are interested in different things and therefore apply different methods, they can — for methodical reasons — only examine more or less complex parts of the landscape ecosystem. This is done for scientific, as well as for practical purposes, when planning and making use of the landscape.

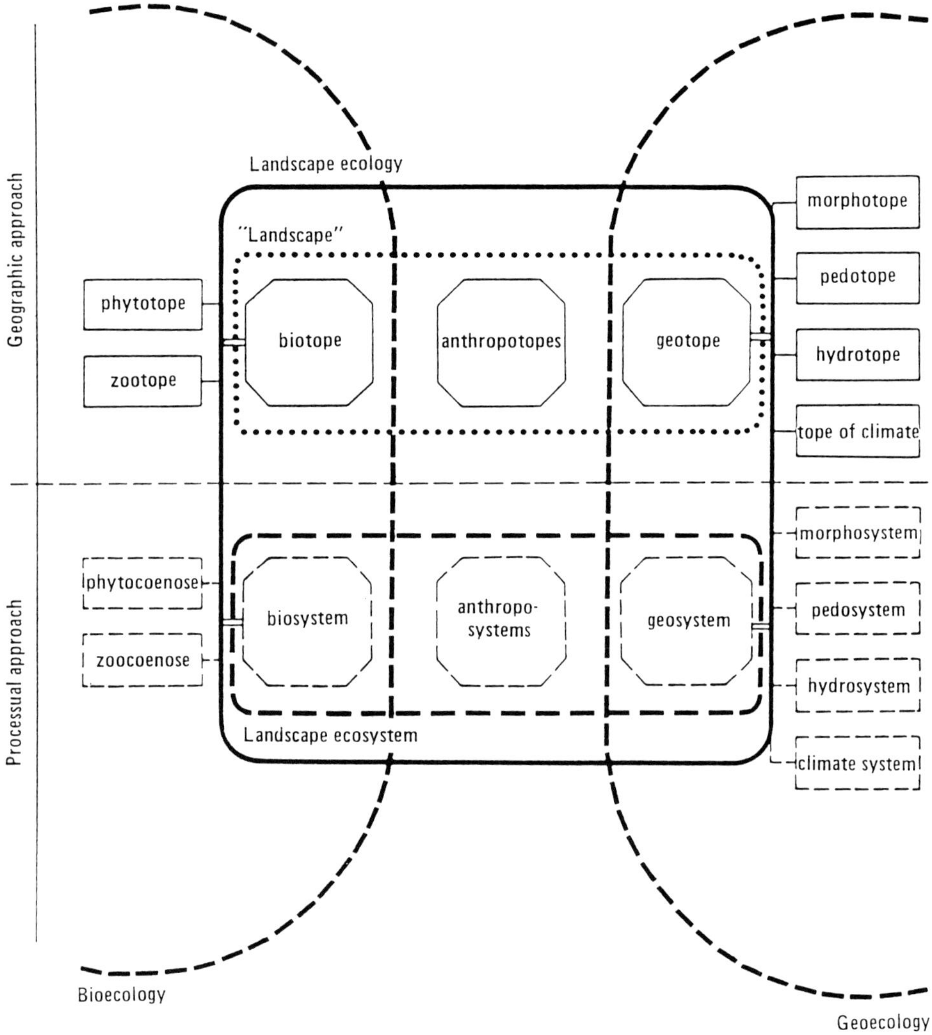

Figure 38.2: Landscape ecology, bioecology and geoecology: Approaches to systems, subsystems and sub–subsystems.

It becomes obvious that the branch of landscape ecology overlaps many other disciplines. This is the reason for the manifold approaches in landscape ecology towards research work and application.

38.2 Methodical and methodological perspectives of landscape ecology

This section deals with the question of the "operators" of landscape ecology and with the problems of research work in this field. The methodical consequences of a landscape ecology that is both easy to handle and relevant to society will be also discussed.

38.2.1 Who practices landscape ecology?

This question cannot be answered easily, as, when it comes to practice in the field of research and application, the model of the landscape ecosystem is often talked about but not treated accordingly. The problem is additionally aggravated by the fact that a sharp contrast between "practice/application" on the one hand and "science/basic/research" on the other, is artificially created. This is particularly valid for continental Europe, whereas in the Anglo-American world a higher degree of pragmatism is favoured ("Landscape ecology is what landscape ecologists do".)

In Continental Europe as well as in the Anglo-American world the appliers and users of landscape ecological basics turn out to be a very heterogeneous circle.

The decisive factor about this is that it is the appliers in particular who coin the image of landscape ecology in the public eye and thus make possible the feedback to science and landscape-ecological basic research. In so far it is important for the "scientific" landscape ecologist to realize the development with respect to landscape ecological practice and to keep close contact to it. Based on the lack of such contact certain gaps unfortunately exist between the practice and the science of landscape ecology.

Different branches of the practice operate in charge of landscape ecology. They do it in a rigorous methodical and methodological way. For that there are quite a number of good, as well as not so good reasons. In spite of this, the scientific landscape ecologist neither can, nor will ignore the ideas of the practician, as he stands for a large potential of practical experience, with the help of which the development of methodology and theory have to be accelerated.

Landscape ecology is in practice carried out on the following levels:

- Direct landscape ecology in the form of research in various disciplines, such as geography and biology. Here further different integrating natural sciences, as well as interdisciplinary institutions of human sciences, have to be mentioned. As a whole, these fields cannot be precisely summed up in the term "ecologies". They pursue different aims and follow totally different ecological approaches, which in turn leads to the fact that different definitions of the ecosystem exist, even among the various sciences. The aim, however, is more or less the same: these "ecologies" look upon the environmental systems as relatively complex and thus try to define them as integrated and complex matters. The allocation of such fields to "ecology in a wider sense" is mostly justified.

- Landscape ecology is also carried out indirectly, for example in different, partly classical special sciences, including hydrology, climatology, pedology and biology. Every now and again they succeed in producing an integrated approach in the sense of the aim "ecologies in a wider sense". Normally, however, they examine complex single factors, for example the ground–water household, ground–water regimes, functions of the climate etc. These examinations are based on an approach that is of a system–analytic nature, and for that reason it is not justified to call these examinations "ecological" with regard to the definition of the "subject matter of landscape ecology" (see pages 833ff.).

These special sciences, however, supply landscape ecology with basic results. What is even more important is that the methodic experiences gained there can be transferred to the model and methodology of landscape ecology on an integrated basis. If you take the scales of such examinations into consideration, you must admit that the results often lack the necessary relevance as to territory. Referring to NEEF (1967), the results often rank below the topic dimensions of the "Theory of geographic dimensions".

- The third field dealing with landscape ecology is the field of practice "per se", that is to say of the applier and user of landscape ecological methodology and their results (Fig. 38.3). They live on basic research work and put their results into practice when solving existing problems in man's environment. This field of "practice" includes administrators, consultants, politicians, authorities, and private offices. They make use of approaches, methods and results of the two other groups of landscape ecological users. They are far away from the theoretic postulates of landscape ecology. Often they do not even know the theory of landscape ecology. – But the importance of this third group lies in the fact that it is the one that transfers landscape ecological approaches, methodologies and results in the proper–scientific field to other fields, so that they can be used for planning, development aid, the protection of resources, the exploitation of the soil, the protection of nature and of our environment, as well as for recultivation, to name but a few.

38.2.2 Problems of landscape ecological methodology seen from the practical point of view

Considerations as to the methodology of landscape ecology must be based on the landscape ecosystem per definitionem. It is defined as a complex functional pattern. Consequently the methodology used must be complex, too, as it comprises and quantifies the various single factors of the system: storage regulator and processes.

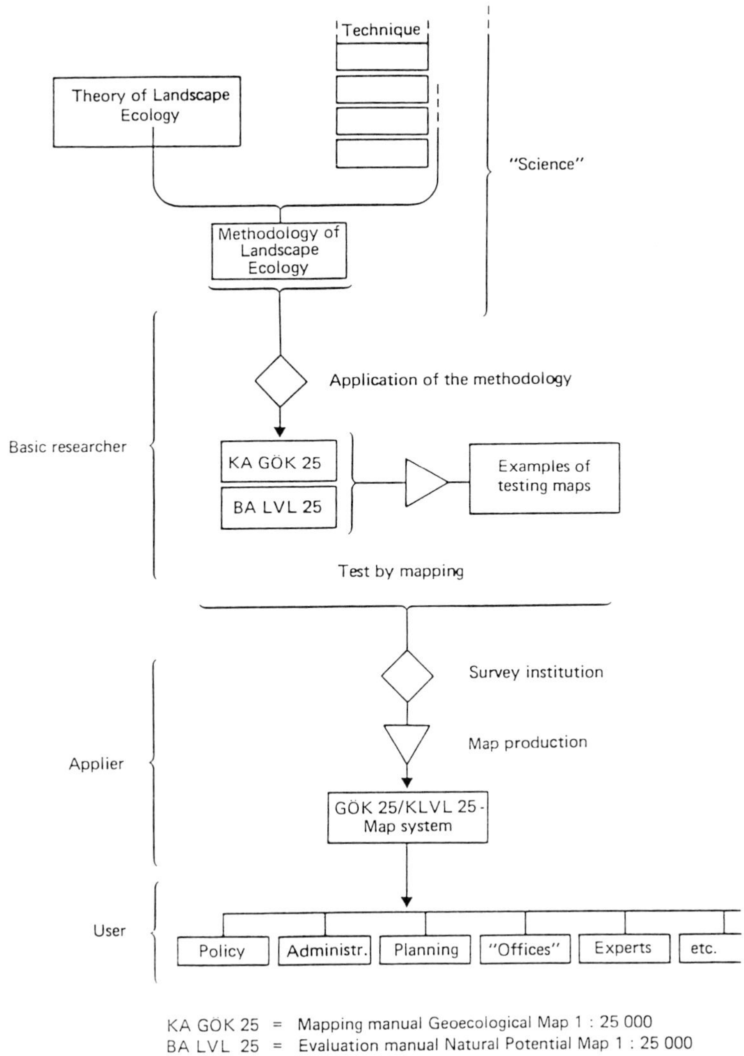

Figure 38.3: Methodic levels of scientific and practical landscape ecology: Researcher, applier and user.

As the sections of the environment differ in their complexity and size, there cannot possibly be "the one" landscape ecological methodology. But there are some basic rules summed up from a methodical point of view by MOSIMANN (1978, 1984 a,b). The different working methods are aggregated into disciplinary landscape ecological methods on a problem — orientated, as well as on an aim–orientated way. Irrespective of the mission in question, these landscape ecological methodologies are characterized by defining factors relevant to for both the applier and the user in the practical field of landscape ecology:

- Possibility of quantitative statements concerning storage regulator and processes in the landscape ecosystem.

- Large and larger–scale statements.

- Results valid for large territories.

- The results do not only refer to static facts of environmental systems, but also to the processes and functions of the submodels of the geo–, bio– and anthroposystem.

These methodical possibilities are only rarely made use of by the appliers and users. It is in the field of basic landscape ecological research at best, that work is methodically more comprehensive. Thus the highbrow integrating approach and the definition of the landscape ecosystem are done justice to. There are various reasons why the methodical possibilities and thus qualities of the landscape ecological approach and methodology are not fully utilized. They are

- methodical or methodological reasons: when only particular approaches are followed, specifically appropriate to a problem;

- pragmactic–practical reasons: for the applier or user deliberately neglects particular parts of the instrument panel, either because they promise no basic insights or because the problem can be solved in a less extravagant way;

- economic reasons: the applier or user has, for instance, not much time to tackle the problem, so all ballast, no matter whether existing or simply presumed, is thrown overboard — a fact lowering the seriousness of the work and its results;

- political reasons: that is to say that the findings of landscape ecological research may in practice result in disagreeable conclusions. Therefore this or that item is neglected as to the methodology, so that more "flexible" results for plans or political decisions are available.

These considerations make it plain to the reader that it is not only the one who does basic research work in landscape ecology but also the applier and user who is highly responsible when choosing an approach or a methodology with respect to his aims. Thus a few ethic perspectives can be derived. They will be discussed on pages 839–841.

38.3 Landscape ecology – "ecological policy" – methodology: Is there some interdependence?

The landscape ecology dealing with the interdependence of nature, technology and society (NEEF 1967) in the "Total Human Ecosystem" (NAVEH and LIEBERMAN (1984) influences the moulding and the preservation of man's environment. This at the same time implies a high degree of responsibility for the further existence of our "spaceship earth".

The descriptions of environmental systems delivered again and again by landscape ecology make it plain that the charges are increasing, whereas the minimum stress is decreasing. In spite of threatening or already existing environmental crises the mentality of advancing one's claims and real demands on the space of environment is spreading. This is documented, for example, by the fact that the landscape ecosystem and its compartments are no longer considered to be of actual value.

The landscape ecosystems are relevant because of their basic functions as man's habitat and economic area. It is a "utility in a wider sense". This usefulness, however, is regarded as an economic one by the public, by planners, and especially by politicians. Hardly ever does the idea arise that it is a matter of preserving man's habitat and its quality. This does not only concern the phenomenon of its offered goods. The quality of life also includes clean water, clean air, a good bioclimate and natural biotopes with animals and plants.

38.3.1 "Ecological planning" and "ecological policy"

Meanwhile, these ideas have found a broad forum amongst the public. The results are demands directed to the planners and politicians as it is they who take the decisions as to environmental systems and their functions. These decisions are far away from the sphere of the scientist, who does basic research work in landscape ecology. They have become independent and have led to slogans such as "ecological planning" or "landscape policy". It is the task of landscape ecology to hint at these terms and to make it plain that they are not only scientific terms but that:

- they imply a specific meaning with consequences as to methodology and decisions and that, therefore,
- "ecological planning" and "ecological policy" have to meet the demands of science and human ecology.

To get away from these slogans, two definitions are offered, that are linked with the foundations of landscape ecology, as well as with the ethical values and standards that planners and politicians take into account, at least verbally:

- Ecological planning: This takes into consideration the single parts of animate and inanimate life, in order to measure future territorial and functional realities, as to their usefulness for man and as to their harmony with nature.

Ecological planning has to be integrated planning, that is — all planning has to be carried out in the form of integrated, all–comprising environmental planning. All administrations and decisions have to adapt to it.

Only on condition that the above–defined planning is practised and serves as the basis of politics can we righteously speak of a true ecological policy. From the point of view of landscape ecology and of commonly valid ethical norms, it is defined as follows:

- Ecological policy is a policy whose decisions must be based on the result of ecological planning. The landscape ecological approach seeing the environmental systems as man's habitat has to be acknowledged. The thus developed ecological basis of decision has also got to be accepted. Only a decision of that kind stands for a real contribution to the further existence and preservation of our "spaceship earth".

A thus defined policy is still unacceptable for today‘s politicians, as every landscape ecologist and planner knows. The landscape ecologist has to realize that politicians, scientists and modern planners obviously all see and evaluate "their" landscape ecosystems differently. This means that any further introduction of landscape ecological findings into practice and then into the reality of environmental systems of our habitat is prevented. The existing demand in landscape ecological findings is politically run down or simply negated.

38.3.2 Possibilities and limits of landscape ecological methodology

There is a direct interdependence of landscape ecology as a science — its methodology and planning on the one hand and politics on the other. Perspectives of landscape ecology and their findings have to take into account the fact of the growing world population – which is a political problem.

The growth of the world population consequently means further claims on mineral resources and much more additional stress on the landscape ecosystems of the earth – and this in all climates around the globe. Dubious technologies meanwhile result in the stress reaching to the highest hights of the atmosphere as well as to the deepest depths of the oceans. Consequently man should use the earths' landscape ecosystems more wisely, i.e. "in a better way". This, however, means the fulfilment of a catalogue of demands based on the findings, the possibilities, and the limits of landscape ecological methodology. At the same time, it stands for particular fronts as to research in landscape ecology, as well as for important aspects concerning the preservation of quality in "spaceship earth". This would correspond to REMMERT's "And now? Ecosystem Research!" (1984):

- Research work in landscape ecology must:

- quantitatively carry out the "descriptions of the status quo" of landscape ecosystems on all levels
- It has to ensure that the quality of the landscape ecosystem will be safeguarded, because it firstly often cannot be depicted quantitatively, because it secondly often serves as a function for a function and thirdly also for the experience of the landscape ecosystems.
- It has to ensure that the multi–disciplinary approach of landscape ecology becomes common in all those practice fields of research work and practice directly or indirectly dealing with man's environment and man's position in the ecosystem.

• Landscape ecological research must refine and rationalize its methodologics, that is to say:

- a world–wide net of areas under permanent observation has to be installed in order to be able to compare the ecological situations;
- facts have to be gathered on a world–wide basis, especially with the help of remote sensing methods linked with geographical information systems, which then supply the material for model simulations — from topological to geospheric dimensions;
- networks of geographical information systems and of landscape ecological databanks, that allow supra–regional ecological data exchange as well as supra–regional model simulations, have to be installed.

• Landscape ecological research must develop and introduce methodologies in the form of "language with practitioners", f. ex. by:

- methods of translation for landscape ecological basic research work or for practice–orientated methodologies;
- methods of measuring and evaluating the results of landscape ecological work with respect to practical problems.

The above–mentioned developments have already set in. Nevertheless, they exist as unsolved problems, in so far as these developments are either uncoordinated or are seen under too strong a regional aspect.

Thus, the idea of the installation of regional and functional networks of environmental systems is neglected. Only a coordinated, scale–related and methodically clean cooperation of all landscape ecologists promises a coherent and quantitative description of landscape ecosystems. This also means that they have to be dealt with in the right way, as far as planning, management and policy are concerned.

38.4 Landscape ecology — Where does it lead to?

More than any other discipline or science, it is landscape ecology that is closely linked with man's environment. Thus it has its share in those fields that are concerned with the planning and moulding of landscapes and areas — no matter whether they are towns, mining areas, industrial regions, holiday resorts or rural areas. New perspectives for landscape ecology, as well as for the further development of already visible approaches of environmental research and planning, will only be possible if there is a change in the interplay science—practice—public.

This means that the branch of scientific landscape ecology together with the landscape ecologist of the various branches will have to propagate the respective ideas. They will also have to spread them amongst those who are either still ignorant or who want to ignore them. Consequently, the following basic perspectives for the multidisciplinary field of landscape ecology can be derived:

- The keeping up of and the further development of this multi–disciplinary aspect;
- a consolidation and enlargement of the interdependence of basic research, practice, and policy, aiming at an "ecological policy" that is only possible via ecological planning based on landscape ecological foundations.
- Also, more ethical–moral values and ideas have to find their way into landscape ecology, due to the methodological postulate of the interdependence of the systems of nature, technology and society. It is true that these ethical–moral aspects are already at the basis of landscape ecological theories but they have to be taken more seriously — in science as well as in practice. At the same time, they would have to be given a philosophical basis by the foundation of a "landscape ecological ethic".

Up until the present day, landscape ecology has been lacking a philosophical basis, although the basic idea is already to be found in the fields of natural sciences and human ecology, and also partly in politics. The necessity of a philosophical basis is consequently derived from the subject matter "landscape ecosystem" that stands for man's habitat, thus touching upon his existence.

Such a philosophical foundation would have to take into consideration that the scientific plans of the landscape ecosystem have to be the methodical–methodological principles for actions and decisions in science, practice and politics. The multi–disciplinary character of landscape ecology favours these ideas, thus securing their broad effect.

Landscape ecology is not only the subject matter of those ecological sciences, but also the subject matter of those fields that are concerned with its appliers, ranging from regional planning to politics. The special sciences, too, have to be integrated more strictly into landscape ecology and have to be made aware of the basic rules of ecological thinking and action. It is common truth that they do contribute to landscape ecology, but the methodologies of these fields (of research, appliers and users) are in most cases orientated towards a very particular aim or problem. This stands for a very limited view — as to the contents and quantity of the matter. Reconsideration of the integrative approach of landscape ecology and other ecologies, and the integration of specific scientific results into the model of the landscape ecosystem do help the sciences to become "more ecological".

Everywhere on earth the landscape ecosystems are damaged and put under more and more stress by the steadily growing world population. This implies danger to the ecosystem earth as a whole. This danger can only be met by political decisions based on ecological thinking. Unfortunately, the political moral has not yet developed so far as to guarantee the further existence of our "spaceship earth", even to guarantee conditions worthy of man. This would mean taking political decisions founded on the concerns of landscape ecology and to guarantee them to mankind.

Landscape ecology is moving towards that ethical-moral basis. Such a change, however, cannot yet be seen in politics. This change can only become reality if the politicians take landscape-ecological findings into account and if they measure their decisions with the help of the new ethical norms. As pointed out above, this change cannot be detected amongst the politicians. This, however, should be no reason for the landscape ecologist not to continue his work even more intensively.

Bibliography

FINKE, L., 1986, Landschaftsökologie. Das Geographische Seminar.
Höller & Zwick, Braunschweig.

LESER, H., 1978, Landschaftsökologie
2nd edn, UTB **521**. Ulmer, Stuttgart.

LESER, H., 1984, Zum Ökologie-, Ökosystem- und Ökotopbegriff.
Natur und Landschaft **59**, 351–357.

LESER, H., 1985, Ernst Neef und die landschaftsökologische Forschung. Gedanken zum wissenschaftlichen Wirken eines großen deutschen Landschaftsforschers.
Die Erde **111**, 1–6.

LESER, H., 1991, Landschaftsökologie. Ansatz, Methodik, Modelle, Anwendung
3rd edn., UTB **521**. Ulmer, Stuttgart.

MOSIMANN, TH., 1978, Der Standort im landschaftlichen Ökosystem. Ein Regelkreis für den Strahlungs-, Wasser- und Nährstoffhaushalt als Forschungsansatz für die komplexe Standortanalyse in der topologischen Dimension.
Catena, 351–364.

MOSIMANN, TH., 1984a, Landschaftsökologische Komplexanalyse.
Franz Steiner Verlag, Wiesbaden.

MOSIMANN, TH., 1984b, Methodische Grundprinzipien für die Untersuchung von Geoökosystemen in der topologischen Dimension.
Geomethodica, Veröffentlichungen des 9. Basler Geomethodischen Colloquiums **9**, 31–65.

NAVEH, Z.; LIEBERMAN, A. S., 1984, Landscape Ecology — Theory and Application.
Springer Series on Environmental Management. Springer, New York, Berlin, Heidelberg.

NEEF, E., 1956, Einige Grundfragen der Landschaftsforschung.
Wiss. Zeitschrift der Universität Leipzig, Math.–naturwissenschaftl. Reihe **5**, 531–541.

NEEF, E., 1967, Die theoretischen Grundlagen der Landschaftslehre.
VEB Hermann Haack Geographisch–Kartographische Anstalt, Gotha Leipzig.

NEEF, E.; SCHMIDT, G.; LAUCKNER, M., 1961, Landschaftsökologische Untersuchungen an verschiedenen Physiotopen in Nordwestsachsen.
Abhandlungen d. Sächsischen Akad. d. Wiss. zu Leipzig, Math.–nat. Klasse, Bd **47/1**.

REMMERT, H., 1984, And now? Ecosystem Research!
In: Cooley; J. H.; Golley, F. B. (eds) Trends in Ecological Research for the 1980s. New York, London.

SCHMITHÜSEN, J., 1942, Vegetationsforschung und ökologische Standortslehre in ihrer Bedeutung für die Geographie der Kulturlandschaft.
Zeitschrift der Gesellschaft für Erdkunde zu Berlin (no Vol. number), 113–157.

TROLL, C., 1939, Luftbildplan und ökologische Bodenforschung.
Zeitschrift der Gesellschaft für Erdkunde zu Berlin (no Vol. number), 241–298.

TROLL, C., 1950, Die geographische Landschaft und ihre Erforschung.
Studium generale III, 163–181.